THE EVOLUTIONARY BIOLOGY OF FLIES

The Evolutionary Biology of Flies

EDITED BY

David K. Yeates and
Brian M. Wiegmann

Columbia University Press
NEW YORK

Columbia University Press
Publishers since 1893
New York   Chichester, West Sussex

Library of Congress Cataloging-in-Publication Data

The evolutionary biology of flies / edited by David K. Yeates and Brian M. Wiegmann.
p. cm.
Includes bibliographical references and index.
ISBN 978-0-231-12700-4 (cloth : alk. paper)
1. Diptera. 2. Diptera—Evolution. I. Yeates, David K. II. Wiegmann, Brian M., 1963–
QL531.E94 2005
595.77′138—dc22   2005041379

Columbia University Press books are printed on permanent and durable acid-free paper.

Printed in the United States of America

10 9 8 7 6 5

CONTENTS

PREFACE

The true flies (Diptera) are immediately familiar because they are ubiquitous and cosmopolitan, and they have had tremendous impacts on human civilization. Mosquitoes and tsetse flies transmit important diseases, such as malaria and sleeping sickness, to humans and animals. Scientists focusing on the dipteran model organism *Drosophila* have provided the breakthroughs and insights that have driven genetics and developmental biology for the past century. Many other flies perform ecological roles, such as nutrient recycling and pollination, that are essential for the sustainability of managed and wild ecosystems.

Probing deeper into evolutionary time, flies represent one of the largest radiations of eukaryotic organisms, making up about 15% of animal species. They have been buzzing around terrestrial environments since at least the Permian geological period, 250 Mya, and have evolved into hundreds of thousands of species today. During this prolonged evolutionary radiation, flies have survived two mass extinction events, seen the angiosperms arise, and watched the dinosaurs evolve, rule, and disappear. In fact, fly maggots probably fed on the carcass of the last dinosaur. During this extensive evolutionary history, flies have been transformed into the exuberant array of different shapes and sizes that we see today, classified into 150 families and 150,000 described species. Flies are among the most abundant arthropods found in biodiversity surveys and have a wide variety of feeding strategies, including predation; detritivory; plant and animal parasitism; or feeding on plant nectar, pollen, and other biological exudates. Magnifying this ecological diversity, Diptera have a complex holometabolous life cycle, and their larvae (maggots) and adults have entirely different anatomy and behavior, separate ecological requirements, and occupy different niches.

This book is the result of our growing belief that the evolutionary biology of flies is entering a renaissance fueled by two important scientific innovations: the explosion of genetic information arising from dipteran genomics and developmental biology, and improved phylogeny estimation that relies on large amounts of new molecular data and quantitative, statistical analytic methods. Even more than before, dipteran model systems represent some of the most compelling and tractable models for macro- and microevolutionary research, and provide the language for communication and integration between these domains. Our knowledge is such that the genetics and development of such complex phenotypic traits as behavior will probably be first unraveled in multicellular organisms through a dipteran model system.

We have divided the chapters of the book into three sections: Phylogeny (three chapters), Genomics and Developmental Biology (five chapters), and Evolutionary Ecology and

Biogeography (six chapters). In the Phylogeny section, Mike Whiting reviews the phylogenetic relationships of the order, focusing on the recent controversial hypothesis that Strepsiptera and Diptera form a clade, Halteria. We rely on recently developed supertree methods to generate a new synthesis of dipteran phylogeny, and review other new phylogenetic hypotheses for flies. Rudolf Meier examines the role of dipterist Willi Hennig, the father of phylogenetic systematics, in the development of phylogenetic theory, and traces the development of Hennig's key phylogenetic concepts through his books and papers.

In the Genomics and Developmental Biology section, Michael Ashburner reviews dipteran genomics, concentrating on the recently completed genomes of *Drosophila* and *Anopheles*. Rob DeSalle's essay examines eukaryote developmental biology from a comparative and evolutionary perspective, using the *Drosophila* model system. Margaret Kidwell examines the effect that transposable elements, such as long interspersed nuclear elements and short interspersed nuclear elements, have on the evolution of the dipteran genome, again focusing attention on the well-studied *Drosophila* systems. David Merritt's chapter focuses on the evolution and development of the dipteran nervous system and poses the question: How representative is *Drosophila?* There are a wide variety of sex determination mechanisms in flies, and it appears to be a labile evolutionary feature, even varying within families, such as Tephritidae. Neil Davies and George Roderick's chapter focuses on the evolution of sex determination systems and the utility of the transformations found in Diptera for understanding sex determination.

The Evolutionary Ecology and Biogeography section begins with Conrad Labandeira's review of the fossil history and evolutionary ecology of flies and their associations with plants. His chapter traces their impact on freshwater ecosystems, nectar and nectarlike fluid feeding, and the variety of modes in which fly larvae have fed on the internal tissues of plants. Peter Cranston assesses the contribution of Diptera to ecological and historical biogeography, including the works on Chironomidae by Lars Brundin and sciaroids by Loic Matile, and reviews the most common continental-scale biogeographic patterns found in Diptera.

Gerald Wilkinson and Philip Johns review the evolution of sexual selection and mating systems in flies, examining the degree to which ecological factors have influenced pre- and postcopulatory activities. Fly model systems have contributed greatly to the development of mating system theory through organisms such as the dung fly *Scathophaga*. The genetic basis of host use is the focus of the chapter by Ken Filchak, Bill Etges, Nora Besansky, and James Feder, with emphasis on host specificity in mosquitoes, *Drosophila, Rhagoletis,* and the Hessian fly, *Mayetiola*. Sonja Scheffer's chapter shows that molecular genetic markers are extremely important tools for studying cryptic and invasive species, and reviews recent empirical findings in a number of phytophagous species, such as Mediterranean fruit flies and leafmining flies, as well as mosquitoes. The rainforest is the laboratory of Roger Kitching, Daniel Bickel, and Sarah Boulter, and in their chapter they examine the community ecology of flies, assessing the degree to which guild analyses can help us understand fly biodiversity.

We have deliberately chosen contributions that provide a comparative and evolutionary perspective on fly biology. As phylogenetic information on flies improves and becomes more widespread in the order, this kind of perspective will provide even greater instruction and insight. Our hope is that these chapters demonstrate that there is much of fundamental importance to be gained if the deep insights from model systems are compared in an evolutionary framework. We hope that this book provides the biological community with compelling examples of the utility of dipteran model systems in evolutionary biology. By studying the humble fly, we can see the first evidence of a grand synthesis of genotype, development, and phenotype in a phylogenetic and evolutionary framework.

We used a number of reviewers during the process of editing the book. In addition to the authors of various chapters, we thank the following for reviewing chapters on behalf of all authors: D. Amorim, Universidad de São Paulo; E. Ball, Australian National University; A. Borkent, Salmon Arm, British Columbia; M. F. Freidrich, Wayne State University; R. Gagné, U.S. Department of Agriculture, Systematic Entomology Laboratory; J. W. Mahaffey, North Carolina State University; H. Robertson, University of Illinois; F. C. Thompson, U.S. Department of Agriculture, Systematic Entomology Laboratory; and J. Tu, Virginia Tech.

DAVID K. YEATES AND BRIAN M. WIEGMANN
Raleigh, North Carolina

PART I

Phylogeny

CHAPTER ONE

Phylogenetic Position of Diptera: Review of the Evidence

Michael F. Whiting

Deciphering phylogenetic relationships among insect orders has challenged entomologists for well over two centuries. The monophyly of the majority of the 30-plus insect orders is well established via morphological characters (Hennig 1981; Kristensen 1991) and more recently by DNA sequence information (Whiting et al. 1997; Wheeler et al. 2001), indicating that these are natural groupings. Insect orders which have been established as being nonmonophyletic include Psocodea (book lice), which is paraphyletic with respect to Phthiraptera (true lice); Mecoptera (scorpionflies), which is paraphyletic with respect to Siphonaptera (fleas); Blattodea (roaches), which is paraphyletic with respect to termites; and probably Zygentoma, which is paraphyletic with respect to the placement of the odd-ball *Tricholepidion,* perhaps the most primitive insect. However, our modest understanding of phylogenetic relationships among the insect orders continues to be refined as additional data are gathered and analyzed in new ways. Molecular data have certainly provided some important —and sometimes surprising—insights into our understanding of these relationships. But we must acknowledge that from a genomic perspective, insect interordinal phylogenetics is still in its infancy, and we are still constrained by the challenges of adequately sampling the enormous number of species representing extant insect diversity, by our reliance on only a few genetic markers to make these inferences, and by limitations on our computational abilities. These obstacles become less problematic with each passing year, and entomology is finally reaching a point where a robust view of the basic pattern of insect diversification is emerging.

The difficulty of inferring a phylogeny for the insect orders has long been recognized, and it is in fact the quest to understand insect phylogeny that gave rise to the most important innovations in phylogenetic theory over the past century. Willi Hennig, the renowned German dipterist, is appropriately credited as the first to clearly annunciate the necessity of phylogeny as the most natural way of organizing the diversity of life (see Meier, Chapter 3). His insistence that all taxonomic groups must be monophyletic to be natural, and that this monophyly can only be indicated via the presence of derived characters (synapomorphy), revolutionized systematic theory and formed the underpinnings for the modern practice of phylogenetic systematics (Hennig 1979). Hennig worked specifically on insect ordinal phylogeny, and the work he produced still provides an outstanding discussion of some of the morphological evidence supporting these relationships. This blossoming of theory was all a result of Hennig's passion to solve the conundrum of insect phylogeny, a problem whose solution, in many ways, still eludes us today.

So why is the problem of insect ordinal phylogeny so hard to solve? Every insect order has a very specialized set of morphological features that are easily distinguished from other insect orders—beetles have elytra, fleas are laterally flattened, butterflies and moths have scales on their wings—but these morphologies are so specialized that they retain few clues of their phylogenetic history. Although any child can tell the difference between a beetle, a fly, and a butterfly, even the erudite entomologist struggles to decipher which two insect groups are more closely related. There has been a wealth of research on the morphology of particular insect orders: this book attests to the amount of work that has been performed on Diptera. But comparative morphology among the orders has been largely neglected, and the position of many groups is still obscure. Molecular data have provided insights, but in many cases, all available sources of data simply do not provide satisfactory answers.

Controversy has surrounded the phylogenetic placement of Diptera; historically, it has been difficult to settle on a single hypothesis. Part of the problem is that finding a robust position for Diptera not only requires establishing its immediate sister taxon, but also requires finding a robust placement for the other orders that are closely related to Diptera. Diptera is simply one piece of the much larger problem of holometabolan phylogeny, and understanding where Diptera fits requires some knowledge of the placement of other holometabolous insects. For example, the order Siphonaptera (fleas) has historically been closely associated with Diptera, either as the immediate sister group to Diptera (Boudreaux 1979; Wood and Borkent 1989) or nested directly within flies rendering Diptera paraphyletic (Byers 1996). Mecoptera (scorpionflies) has been closely associated with Diptera, either with the entire order as the immediate sister group to Diptera (Hennig 1981; Kristensen 1991) or with one of the mecopteran families considered as sister groups to Diptera (Wood and Borkent 1989; Blagoderov et al. 2002). Most recently, evidence presented from molecular analyses supporting Strepsiptera as a sister group to Diptera (Whiting et al. 1997) not only rekindled the controversy over dipteran affinities, but was used as a case in point for competing methods of phylogenetic inference (Huelsenbeck 1997; Whiting 1998a). One cannot divorce a discussion of the phylogenetic placement of Diptera from the larger context of holometabolan phylogeny.

Diptera Among the Holometabola

Holometabola is composed of 11 orders, representing 80% of insect diversity at the species level and accounting for more than 50% of all animal species (Wilson 1988; Kristensen 1999). Each constituent order has been established as monophyletic, with the notable exception of Mecoptera, which includes Siphonaptera as a sublineage, sister group to Boreidae (Whiting 2002a). The four major orders—Coleoptera, Hymenoptera, Lepidoptera, and Diptera—include the vast majority of holometabolan species, and as further monographic work is done on these groups, these numbers are certain to increase. The monophyly of Holometabola is well established; this group is supported by a series of unique morphological characters (Kristensen 1999) and is recovered in every major molecular analysis performed on these insects (Whiting et al. 1997; Wheeler et al. 2001; Whiting 2002c). There is no reason to doubt that Holometabola is monophyletic and that Diptera is a member of this well-established clade.

There are approximately 34.5 million ways that the 11 holometabolous orders can be arranged on a bifurcating, rooted topology. Fortunately, relationships among the holometabolous insect orders are already partially resolved, helping to narrow down the number of unique topologies that include Diptera. The monophyly of Neuropterida (Neuroptera, Megaloptera, and Raphidioptera) appears well supported via morphological (Kristensen 1999; Aspöck 2002) and molecular data (Whiting et al. 1997; Wheeler et al. 2001; Whiting 2002c). Coleoptera has traditionally been placed as a sister group to Neuropterida based on specializations of the ovipositor (Achtelig 1975) and most recently, by characters associated with the base of the wings (Hörnschemeyer 2002). Molecular data have never independently supported this relationship, but the morphology is perhaps sufficiently well established to accept Neuropterida + Coleoptera as monophyletic. The monophyly of Trichoptera + Lepidoptera forming the group Amphiesmenoptera is the best-supported sister group relationship among all insect orders, with a wealth of morphological (Boudreaux 1979; Hennig 1981; Kristensen 1999) and molecular (Wheeler et al. 2001; Whiting 2002a; Wiegmann et al. 2002) data bolstering this conclusion. Recently, a close affinity between Siphonaptera and Mecoptera has been convincingly demonstrated via morphology (Bilinski et al. 1998) and molecular data (Whiting 2002a), rendering Mecoptera paraphyletic, but making the clade including Mecoptera and Siphonaptera monophyletic. It is safe to say that there is a general consensus among entomologists that the relationships described above are relatively well established. Assuming that these clades are monophyletic, the 11-taxon statement is reduced to a six-taxon statement, resulting in 105 possible placements for Diptera; a vast improvement over the previous tally. The real questions, and the areas of greatest controversy, surround the relationships of the clades listed above with one another and the orders Hymenoptera, Strepsiptera, and of course, Diptera.

Diptera is widely considered a member of the superordinal group Mecopterida, which also includes the orders Trichoptera, Lepidoptera, Siphonaptera, Mecoptera, and perhaps Strepsiptera. From a morphological standpoint, the monophyly of this group is supported by the insertion of a pleural muscle on the first axillary sclerite and characters associated with a reduction in larval mouthpart musculature (Kristensen 1999). The former character is a relatively prominent feature present in these orders except for Siphonaptera (which lacks wings) and Strepsiptera (as described below). The monophyly of this group, however, has never been independently supported by molecular data (Whiting 2002c). Mecopterida is traditionally divided into two major clades: Amphiesmenoptera (Lepidoptera + Trichoptera) and Antliophora (Diptera, Mecoptera, Siphonaptera, and perhaps Strepsiptera). The monophyly of the former group is well established as described above, but the monophyly of the latter is more nebulous and based on general reductions of morphology, such as larval mouthparts without lateral labral retractor, hypopharyngeal retractor, and ventral salivarium dilator muscles; imaginal mandibles (lost in the Siphonaptera) slender, anterior articulations weakly developed or lost; and prelabium without endite lobes and associated muscles (Kristensen 1991). An additional apparent synapomorphy is a pleural ridge/scutum muscle insertion on the posterior notal wing process (Kristensen 1999), although this is, of course, absent in fleas. The monophyly of Antliophora has never been independently confirmed with molecular data (Whiting 2002c). Diptera has traditionally been associated with the orders that comprise Antliophora, and most hypotheses take varying views of how Diptera is associated with flea and mecopteran taxa. The relative support for each of these hypotheses is discussed here in greater detail.

Alternate Hypotheses of Sister-Group Relationship with Diptera

FLEA-FLY HYPOTHESIS

The first dominant theory in dipteran phylogeny suggests that there is a close association between Siphonaptera and Diptera. Proponents of this position argue that either Diptera is the sister group to Siphonaptera, as advocated most strongly by Boudreaux (1979) but also promoted by Wood and Borkent (1989); or Siphonaptera is nested somewhere within Diptera, as advocated by Byers (1996). Both hypotheses rely on similar morphological characters to establish the flea-fly relationship and will be treated together.

Boudreaux (1979) suggested that Diptera and Siphonaptera formed a sister group, which he termed "Haustellodea," in reference to insects with an "obvious sucking beak" (Fig. 1.1A). He rejected any close association between Diptera and Mecoptera by criticizing the siphonapteran/mecopteran synapomorphies as either "primitive insect characters" or convergences. Byers (1996) followed Boudreaux in suggesting that Siphonaptera has features more in common with nematocerous Diptera than with Mecoptera, and argued for the placement of Siphonaptera within Diptera, somewhere near the Mycetophilidae, rendering flies paraphyletic. Wood and Borkent (1989) followed Boudreaux, and placed fleas as sister group to Diptera, although they did split Nannochoristidae out of the Mecoptera and placed this family as sister to the Diptera + Siphonaptera clade (Fig. 1.1B). Advocates of the flea-fly hypothesis take the position that the weight of morphological evidence does not support the placement of fleas as a sister group to Mecoptera, and that characters shared by Mecoptera and Siphonaptera are likely to be convergences. Hence fleas are a sister group to Diptera, or placed within Diptera almost by default. The arguments put forth by these authors are not particularly convincing, in large part because they are not framed in terms of specific characters supporting particular relationships evaluated in a cladistic context, but are rather based on untested hypothetical scenarios of character evolution.

Boudreaux (1979) argued that the ground plan condition in Siphonaptera and Diptera is the presence of piercing-sucking mouthparts consisting of at least a pair of maxillary lacinial stylets in adults, and the presence of apodous larvae. These characters have been adequately discussed and dismissed by Kristensen (1991). Byers (1996) observed that in fleas and some dipteran lineages, the mandibles are lost, but they are always retained in mecopteran lineages, although he noted their reduction within the mecopteran family Nannochoristidae. The shared loss of mandibles, by itself, is not a particularly convincing synapomorphy to support this relationship.

Boudreaux (1979) and Byers (1996) find it significant that hind wings are reduced in Diptera and wings are entirely absent in fleas. Presumably the character these authors offer is the shared propensity toward wing loss in both groups. However, dipterans have not truly lost their wings, as they retain a full-sized mesothoracic wing and a highly derived metathoracic wing forming the haltere. I would suggest this form of wing modification is qualitatively different from the complete absence of wings in fleas. Moreover, wing loss has occurred thousands of times independently in nearly every insect order for multiple reasons (Wagner and Liebherr 1992; Roff 1994), and wing reduction and loss have occurred in many mecopteran

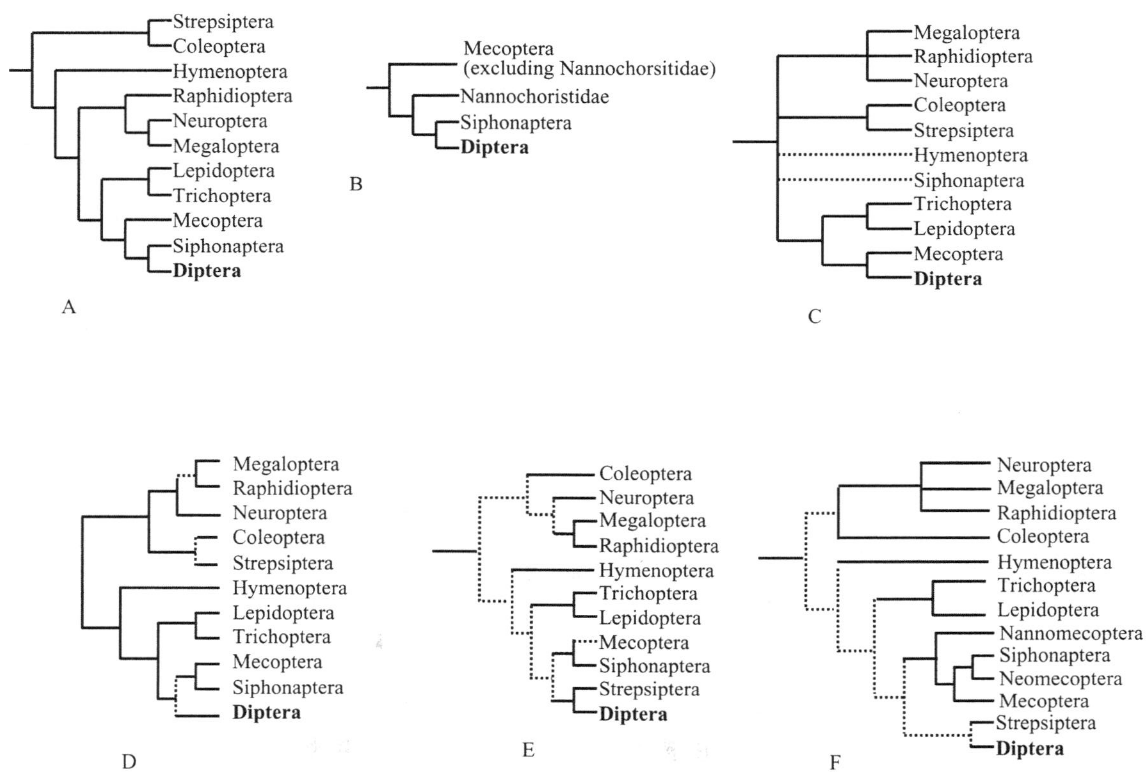

FIGURE 1.1. Previous phylogenetic hypotheses for the placement of Diptera among the holometabolous insect orders. (A) Boudreaux (1979), based on morphology; (B) Wood and Borkent (1989), based on morphology; (C) Hennig (1981), based on morphology; (D) Kristensen (1991), based on morphology; (E) Whiting et al. (1997), based on morphology and DNA; (F) summary tree, based on current molecular and morphological data. Dashed lines represent relationships that are considered poorly supported.

lineages including within Boreidae, Apteropanorpidae, Panorpodidae, and at least three times independently in Bittacidae, suggesting an even greater propensity toward winglessness in Mecoptera than in flies, if such a character were to have any merit.

Boudreaux (1979) argued that the adult antennae in fleas and flies are short and never as long as the wings, in contrast to the long antennae of scorpionflies. Clearly this character is difficult to assess in fleas, which lack wings. Byers (1996) expanded this observation by noting that the antennae of fleas are short, wide, and compact, and that similar-shaped antennae with short, broad flagellomeres occur in some Mycetophilidae, but he suggests that these are most likely instances of convergences. This is also a problematic character, as the antennae of ectoparasitic insects are almost always reduced (e.g., Phthiraptera, Nycterobiidae, Hippoboscidae, Polyctenidae, Hemimeridae), and in fleas, they serve the peculiar function of grasping during copulation (Traub and Starcke 1980), raising doubts about the phylogenetic utility of this character for grouping flies.

Byers (1996) suggested that because fleas are ectoparasites of birds and mammals, one might expect the ancestors of fleas to be nest dwellers. There are no Mecoptera that are nest dwellers, but scatopsid flies do occur in nests, which, according to Byers, bolsters the argument that fleas are closely related to a subgroup of flies. However, nest dwelling is not a ground plan condition in Diptera, and other insects are also nest dwellers (lineages within Coleoptera, Hemiptera, and Phthiraptera). In a similar vein, Byers (1996) argued that fleas and some flies are blood feeders, but no Mecoptera are blood feeders, suggesting that this mode of feeding supports the flea-fly hypothesis. Although blood feeding is a ground plan condition for Siphonaptera, it certainly is not the ground plan condition in Diptera and can only be secondarily derived.

It should be recognized that neither Boudreaux nor Byers performed any sort of phylogenetic analysis to test the utility of the characters they embraced or rejected, but rather, they presented arguments based on presumed scenarios of character evolution. Byers explicitly states that he does not follow the principles of cladistics, and although he expresses admiration for Hennig as a dipterist, he admits that he is nonetheless "untroubled by the idea of paraphyly" (1996: 276). But more importantly, these arguments are now somewhat dated, as they were formed prior to the wealth of molecular and morphological data that have emerged for a placement of fleas within Mecoptera as the sister group to Boreidae (described below), rendering the flea association with Diptera as unrealistic. In light of these new data, it seems the flea-fly hypothesis can now be rejected.

FLY-SCORPIONFLY HYPOTHESIS

The second hypothesis centers on the close affiliation that Diptera may have with Mecoptera, although the actual placement of the order relative to Mecoptera differs among workers. Hennig (1981) preferred a placement of Diptera directly as the sister group to Mecoptera (Fig. 1.1C). Kristensen (1991) suggested a placement as sister to Mecoptera + Siphonaptera (Fig. 1.1D). Later (1999), he revised this by recognizing that fleas are nested within Mecoptera, but he still favored the placement of Diptera as sister to this group. As discussed above, Wood and Borkent (1989) placed their Diptera + Siphonaptera clade as sister group to the unique mecopteran family Nannochoristidae (Fig. 1.1B). Finally, Diptera has been placed as a sister group to the mecopteran family Bittacidae (Blagoderov et al. 2002). The characters used to place Diptera as sister to Mecoptera are simply those used to support Antliophora, with arguments centering around the particular arrangement of these taxa.

The exact position of Diptera among these mecopteran taxa centers directly on the particular phylogeny that one advocates for Mecoptera. As mentioned above, the preponderance of evidence suggests that fleas are nested within Mecoptera as a sister group to Boreidae, rendering Mecoptera paraphyletic. From a molecular standpoint, this is supported by four genetic loci (18S rDNA, 28S rDNA, cytochrome oxidase II, and elongation factor 1-alpha) (Whiting 2002a). A sister group relationship between Boreidae and Siphonaptera is also supported by morphological evidence. The process of resilin secretion in the flea (pleural arch) and *Boreus* (wing base) is similar, and differs from that of the locust and dragonfly (Rothschild 1975; Schlein 1980). The unusual proventricular spines in fleas and boreids are morphologically similar (Richards and Richards 1969). Both groups have multiple sex chromosomes (Bayreuther and Brauning 1971) and also have eyes in a "skeletal socket" (Schlein 1980). The most convincing morphological evidence comes from recent research on ovarioles, which demonstrates that boreid ovarioles are fundamentally different from those in other Mecoptera, but similar to those found in fleas. Mecoptera possess polytrophic-meroistic ovarioles, whereas the ovarioles in *Boreus* are devoid of nurse cells and therefore are panoistic (Bilinski et al. 1998). Fleas and boreids share the following ovariole characteristics: (1) secondary loss of nurse cells; (2) completion of initial stages of oogenesis during postembryonic development; (3) occurrence of rDNA amplification and resulting appearance of multiple nucleoli; (4) differentiation of the late previtellogenic ooplasm into two clearly recognizable regions; and (5) presence of accumulations of membrane-free, clathrinlike cages (Bilinski et al. 1998). This combination of morphological and molecular evidence provides a compelling argument for a sister group relationship between Boreidae and Siphonaptera (Fig. 1.1E).

The other critical taxon in this discussion is the placement of Nannochoristidae among other Mecoptera. Nannochoristids are a small group of southern hemisphere insects that exhibit a unique combination of morphological and life history characteristics. Current morphological (Willmann 1987) and molecular data (Whiting 2002a) support a sister relationship of Nannochoristids relative to the remainder of mecopteran and flea taxa, but it is not entirely clear whether nannochoristids are basal within other flea and mecopteran taxa or are a sister group to Boreidae + Siphonaptera. Recent work on nannochoristid ovarioles suggests that they are panoistic and similar to those found in fleas and boreids (Simiczyjew 2002) but quite different from ovarioles in other mecopteran taxa. These data suggest a basal placement for nannochoristids, but do not resolve whether they are a sister group to Boreidae + Siphonaptera, sister group to the remainder of Mecoptera, or sister to all mecopteran and flea taxa. This is because the panoistic ovariole may be the plesiomorphic state in Mecoptera, and as such, would not necessarily support a sister group relationship between Nannochoristidae and Boreidae + Siphonaptera. Given that there is a desire among many in the entomological community to retain fleas as a legitimate insect order, this rearrangement of taxa requires the designation of two additional holometabolous orders: Neomecoptera (=Boreidae) and Nannomecoptera (=Nannochoristidae; Fig. 1.1F).

A robust phylogeny for Mecoptera allows evaluation of the different versions of the fly-scorpionfly hypothesis. The suggestion that Diptera was derived from the extinct mecopteran families Robinjohniidae or Permochoristidae is based on presumed similarities in wing venation and leg elongation in these bittacid-like mecopterans and nematoceran Diptera (Blagoderov et al. 2002). In essence, the argument is that if you pluck off the hindwings of a bittacid, you get a tipulid. This, however, implies that the clade Mecoptera + Neomecoptera + Nannomecoptera is paraphyletic, suggesting that the morphological and molecular characters supporting the monophyly of this group are homoplasious. Moreover, this scenario would imply that bittacids have a more basal placement in mecopteran phylogeny than appears to be the case, given current morphological and molecular evidence. The placement of Diptera as sister to Nannochoristidae (but not to fleas, as in Wood and Borkent 1989) may have more merit, but it again requires that Mecoptera + Neomecoptera + Nannomecoptera is a paraphyletic group and the preponderance of evidence does not support this conclusion.

DIPTERA AS A SISTER GROUP TO STREPSIPTERA

The hypothesis that has received the greatest attention in the past few years has been that the enigmatic insect order Strepsiptera is a sister group to Diptera, forming the group Halteria. This result created a stir within the entomological community because it challenged some traditional views of strepsipteran affinities; it created controversy within the phylogenetic community by raising important issues regarding competing methods of phylogenetic inference and generated excitement within the developmental biology community because it implied that a major homeotic shift in haltere formation might be responsible for the diversification of an entire order of insects.

The conclusion was first proposed by Whiting and Wheeler (1994) and was elaborated in a more extensive analysis (Whiting et al. 1997; Fig. 1.1E). This result has been attributed to an artifact of parsimony analysis, and for a while was the poster child for long-branch attraction (Felsenstein 1978; Carmean and Crespi 1995; Huelsenbeck 1997). I have argued elsewhere that this relationship is most congruent with morphological data and that it should not be surprising to find sister taxa with elevated substitution rates (Whiting 1998a,b; Sidall and Whiting 1999). Indeed, despite earlier claims that this is the classic case of long-branch attraction, highlighting the failings of parsimony (Huelsenbeck 1997), reanalysis of the more extensive Whiting et al. (1997) dataset by Huelsenbeck (Huelsenbeck 1998), with likelihood methods that account for rate heterogeneity, supported Halteria, although not significantly. I have recently reanalyzed these data using standard maximum likelihood methods on a supercomputer with much more computational power than was available 5 years ago. Out of 100 replicates, Strepsiptera was placed as the sister group to Diptera 92 times, and in the other eight cases in which Strepsiptera nested elsewhere, the topology was always suboptimal. Further analyses with more extensive taxon sampling (Whiting 2002b,c) produce similar results, suggesting that regardless of mode of analysis, these data support Halteria, although one could argue whether this support is weak or strong depending on one's analytical preferences.

Hwang et al. (1998) approached the "Strepsiptera problem" by generating sequence data for a portion of 28S and 5.8S for a small sample of holometabolous taxa (11 exemplars). They found that these data supported Halteria when analyzed via parsimony, but that they did not support Halteria when analyzed via maximum likelihood, and again attribute this result to long-branch attraction. However, because their analyses in fact supported no interordinal holometabolous relationships (as indicated by their fully unresolved consensus cladogram for holometabolan phylogeny), they were unable to retrieve even those that are groups well supported in other molecular and morphological analyses, suggesting that their study provides very little insight into deciphering the phylogenetic position of Strepsiptera.

I have argued elsewhere that a monophyletic Halteria is congruent with the morphological characters supporting the placement of Strepsiptera within Mecopterida and Antliophora (Whiting 1998b). Kristensen (1999) has discussed these characters and suggests that they are inconclusive regarding the placement of Strepsiptera within these groups of orders. A series of wing venation characters used to support the placement of Strepsiptera as sister group to Coleoptera (Kukalova-Peck and Lawrence 1993) was analyzed and rejected elsewhere (Whiting and Kathirithamby 1995). Despite Kukalova-Peck's rebuttal that there are additional wing venation characters supporting placement of Strepsiptera with Coleoptera (Kukalov-Peck 1997), I agree with Kristensen (1999) that these venation characters are unpersuasive.

Perhaps the most intriguing morphological feature is the similarity in the form and function of the mesothoracic haltere in Strepsiptera and the metathoracic haltere in Diptera. The strepsipteran haltere has been demonstrated to function as a gyroscopic balancing organ, as it does in Diptera (Pix et al. 1993). I have observed in many male strepsipterans in flight that the rotation and vibration of the haltere is very different from the simple elevation of the elytra in beetles (see inbio.byu.edu/faculty/mfw2/whitinglab for some posted movies). Moreover, it is clear that from a morphological standpoint, the strepsipteran haltere is not simply a reduced elytron, as proposed by Crowson (1960). Further work is needed to establish the morphological and functional similarities between the halteres in Diptera and Strepsiptera, but the hypothesis that the strepsipteran forewing is a modified elytron can be confidently discarded.

Perhaps the real question is not whether the current molecular data support Halteria—under any mode of analysis, one arrives at the same result—but rather, whether the current data are sufficient to robustly support this relationship. It is clear that additional data are needed to test further the monophyly of Halteria, both in terms of additional genetic loci and more careful morphological analysis, before the issue can be put to rest.

Conclusions

Diptera is a major order of insects, and to better understand the rise and diversification of these remarkable creatures, this group needs to be placed phylogenetically within the context of other insect orders. It is very clear that Diptera belong among the holometabolous insect orders, and is most probably a member of Mecopterida and Antliophora. Recent data on mecopteran and flea phylogeny narrow down the possibilities by establishing that fleas are a sister group to Boreidae, thus allowing rejection of the flea-fly hypotheses. In addition, these data establish the monophyly of the Mecoptera + Neomecoptera + Nannomecoptera complex, making it appear unlikely that Diptera is subordinate anywhere within these orders. The placement of Strepsiptera as a sister group to Diptera remains controversial, but even if Halteria is monophyletic, this group would be a sister group to the entire mecopteran complex, as described above. A summary tree representing the current state of holometabolan phylogeny is given in Fig. 1.1F. Whatever the true sister group to Diptera may be, it is clear that the phenomenal success of the megadiverse Diptera had its origins from relatively humble beginnings.

REFERENCES

Achtelig, M. 1975. Die Abdomenbasis der Neuropteroidea (Insecta, Holometabola). *Zoomorphologie* 82: 201–242.

Aspöck, U. 2002. Phylogeny of the Neuropterida (Insecta: Holometabola). *Zool. Scripta* 31:51–56.

Bayreuther, K., and S. Brauning. 1971. Die Cytogenetik der Flöhe (Aphaniptera). *Chromosoma* (Berl.) 33:19–29.

Bilinski, S., J. Bünnig, and B. Simiczyjew. 1998. The ovaries of Mecoptera: Basic similarities and one exception to the rule. *Folia Histochem. Cytobiol.* 36:189–195.

Blagoderov, V. A., E. D. Lukashevich, and M. B. Mostovski. 2002. Order Diptera Linne, 1758: The true flies. In A. P. Rasnitsyn and D. L. J. Quicke, eds., *History of Insects*, pp. 227–239. Dordrecht: Kluwer Academic Publishers.

Boudreaux, H. B. 1979. *Arthropod Phylogeny with Special Reference to Insects.* New York: John Wiley.

Byers, G. W. 1996. More on the origin of Siphonaptera. *J. Kans. Entomol. Soc.* 69:274–277.

Carmean, D., and B. J. Crespi. 1995. Do long branches attract flies? *Nature* 373:666.

Crowson, R. A. 1960. The phylogeny of Coleoptera. *Annu. Rev. Entomol.* 5:111–134.

Felsenstein, J. 1978. Cases in which parsimony or compatibility methods will be positively misleading. *Syst. Zool.* 27:401–410.

Hennig, W. 1979. *Phylogenetic Systematics.* Urbana: University of Illinois Press.

Hennig, W. 1981. *Insect Phylogeny.* New York: Academic Press.

Hörnschemeyer, T. 2002. Phylogenetic signficance of the wing-base of the Holometabola (Insecta). *Zool. Scripta* 31:17–30.

Huelsenbeck, J. P. 1997. Is the Felsenstein zone a fly trap? *Syst. Biol.* 46:69–74.

Huelsenbeck, J. P. 1998. Systematic bias in phylogenetic analysis: Is the Strepsiptera problem solved? *Syst. Biol.* 47:519–537.

Hwang, U. W., W. Kim , D. Tautz, and M. Friedrich. 1998. Molecular phylogenetics at the Felsenstein zone: Approaching the Strepsiptera problem using 5.8S and 28S rDNA sequences. *Mol. Phylo. Evol.* 9: 470–480.

Kristensen, N. P. 1991. Phylogeny of extant hexapods. In CSIRO, *The Insects of Australia*, 2nd ed., pp. 125–140. Carlton: Melbourne University Press.

Kristensen, N. P. 1999. Phylogeny of endopterygote insects, the most successful lineage of living organisms. *Euro. J. Entomol.* 96:237–253.

Kukalov-Peck, J. 1997. Arthropod phylogeny and "basal" morphological structures. In R. A. Fortey and R. H. Thomas, eds., *Arthropod Relationships*, pp. 269–279. London: Chapman and Hall.

Kukalova-Peck, J., and J. F. Lawrence. 1993. Evolution of the hind wing in Coleoptera. *Can. Entomol.* 125:181–258.

Pix, W., G. Nalbach, and J. Zeil. 1993. Strepsipteran forewings are haltere-like organs of equilibrium. *Naturwissenschaften* 80:371–374.

Richards, P. A., and A. G. Richards. 1969. Acanthae: A new type of cuticular process in the proventriculus of Mecoptera and Siphonaptera. *Zool. Jb. Anat. Bd.* 86:158–176.

Roff, D. A. 1994. The evolution of flightlessness: Is history important? *Evol. Ecol.* 8:639–657.

Rothschild, M. 1975. Recent advances in our knowledge of the order Siphonaptera. *Annu. Rev. Entomol.* 20:241–259.

Schlein, Y. 1980. Morphological similarities between the skeletal structures of Siphonaptera and Mecoptera. In *Proceedings of the International Conference on Fleas*, pp. 359–367. Rotterdam: A. A. Balkema.

Sidall, M. E., and M. F. Whiting. 1999. Long-branch abstractions. *Cladistics* 15:9–24.

Simiczyjew, B. 2002. Structure of the ovary in *Nannochorista neotropica* Navas (Insecta: Mecoptera: Nannochorsittdae) with remarks on mecopteran phylogeny. *Acta Zool.* 83:61–66.

Traub, R., and H. Starcke. 1980. *Fleas. Proceedings of the International Conference on Fleas.* Rotterdam: A. A. Balkema.

Wagner, D. L., and J. K. Liebherr. 1992. Flightlessness in insects. *Trends Ecol. Evol.* 7:216–220.

Wheeler, W. C., M. F. Whiting, Q. D. Wheeler, and J. M. Carpenter. 2001. The phylogeny of the extant hexapod orders. *Cladistics* 17:113–169.

Whiting, M. F. 1998a. Long-branch distraction and the Strepsiptera. *Syst. Biol.* 47:134–138.

Whiting, M. F. 1998b. Phylogenetic position of the Strepsiptera: Review of molecular and morphological evidence. *Int. J. Morphol. Embryol.* 27:53–60.

Whiting, M. F. 2002a. Mecoptera is paraphyletic: Multiple genes and phylogeny of Mecoptera and Siphonaptera. *Zool. Scripta* 31:93–104.

Whiting, M. F. 2002b. Phylogeny of the holometabolous insect orders based on 18S ribosomal data: When bad things happen to good data. In R. DeSalle, W. C. Wheeler, and G. Giribet, eds., *Molecular Systematics and Evolution: Theory and Practice,* pp. 69–84. Basel: Birkhauser.

Whiting, M. F. 2002c. Phylogeny of the holometabolous insect orders: Molecular evidence. *Zool. Scripta* 31:3–16.

Whiting, M. F., and J. Kathirithamby. 1995. Strepsiptera do not share hind-wing venational synapomorphies with the Coleoptera. *J. N.Y. Entomol. Soc.* 103:1–14.

Whiting, M. F., and W. C. Wheeler. 1994. Insect homeotic transformation. *Nature* 368:696.

Whiting, M. F., J. C. Carpenter, Q. D. Wheeler, and W. C. Wheeler. 1997. The Strepsiptera problem: Phylogeny of the Holometabolous insect orders inferred from 18S and 28S ribosomal DNA sequences and morphology. *Syst. Biol.* 46:1–68.

Wiegmann, B. W., J. C. Regier, and C. Mitter. 2002. Combined molecular and morphological evidence on the phylogeny of the earliest lepidopteran lineages. *Zool. Scripta* 31:67–81.

Willmann, R. 1987. The phylogenetic system of the Mecoptera. *Syst. Entomol.* 12:519–524.

Wilson, E. O. 1988. The current state of biological diversity. In E. O. Wilson, ed., *Biodiversity,* pp. 3–18. Washington, D.C.: National Academy Press.

Wood, D. M., and A. Borkent. 1989. Phylogeny and classification of the Nematocera. In J. F. McAlpine and D. M. Wood, eds., *Manual of Nearctic Diptera,* pp. 1333–1370. Ottawa: Agriculture Canada Research Branch.

CHAPTER TWO

Phylogeny and Evolution of Diptera: Recent Insights and New Perspectives

David K. Yeates and Brian M. Wiegmann

The insect order Diptera (the true flies) is one of the most species-rich, anatomically varied, and ecologically innovative groups of organisms, making up 10–15% of known animal species. An estimated 150,000 species of Diptera have been described (Groombridge 1992; Thompson 2004); however, the actual total number of extant fly species is many times that number. The living dipteran species have been classified into about 10,000 genera, 150 families, 22–32 superfamilies, 8–10 infraorders and 2 suborders (McAlpine and Wood 1989; Yeates and Wiegmann 1999; Thompson 2004; Fig. 2.1), and around 3,100 fossil species have been described (Evenhuis 1994). The monophyly of Diptera is well established, with a number of complex morphological modifications recognized as synapomorphies, including the transformation of the hindwings into halters and the development of the mouthpart elements for sponging liquids (Hennig 1973; Wood and Borkent 1989; Kristensen 1991; Kukalova-Peck 1991; Wood 1991; Griffiths 1996).

The German entomologist Willi Hennig (1913–1976) was the preeminent systematist of the twentieth century. His methodological advances (1950, 1966) fueled the phylogenetic renaissance in systematics over the past three decades. Hennig spent much of his life applying his new system of phylogenetics to the Diptera (see Meier, Chapter 3) and placed Diptera classification on a firm phylogenetic footing for subsequent generations of dipterists. His phylogenetic analysis is one of the main sources of data that we use to assess the relationships of the Diptera today. Since Hennig's work, major advances in dipteran systematics have been made through a relatively small number of extensive phylogenetic treatments using morphological data. In a recent review of the systematics of the order (Yeates and Wiegmann 1999: Fig. 1), we developed a qualitative phylogenetic hypothesis that summarized much of this information for the entire order. Here we develop this summary quantitatively, with a supertree analysis of fly relationships using matrix representation with parsimony (MRP) coding (Baum 1992; Ragan 1992; Sanderson et al. 1998), encompassing a series of major phylogenetic analyses as input trees that includes the phylogenetic arrangement of Hennig (1973). We use the resulting supertree as a reference point in our review of the current status of dipteran higher-level phylogenetics, identifying relationships that are well established, and also identifying relationships that are proving difficult to resolve.

Recent research into the higher phylogeny of Diptera has been characterized by more sophisticated methods of analyzing traditional morphological characters (e.g., Oosterbroek and Courtney 1995; Yeates 2002), the inclusion of ever larger volumes of molecular sequence data (e.g., Collins and Wiegmann 2002a,b), and the introduction of a surprising number of new and extremely well-preserved fossils, dating back to the Cretaceous (e.g., Grimaldi and Cumming 1999; Borkent 2000). Most studies focus on relationships below the family level, and few studies attempt to reconstruct relationships at higher taxonomic levels. There has also been some exploration of novel morphological character systems, especially soft tissue anatomy, and also novel gene sequences, particularly single copy nuclear genes (e.g., Moulton 2000). The most rigorous dipteran systematics at present reflects that of the discipline, synthesizing all available data from multiple molecular and/or morphological partitions and analyzing them quantitatively. Taxon sampling strategies are becoming more sophisticated and intensive, and sensitivity analyses determine the effect of critical parameters on the topology and support for phylogenetic relationships (e.g., Meier and Baker 2002) and the strength of the phylogenetic signal contributed from different data partitions. Some studies are also beginning to use phylogenetic trees to elucidate the evolution of behavior and other traits and to examine the temporal context of the deep divergence events in the Diptera (Wiegmann et al. 2003) using molecular data. With the increasing number of dipteran genomes becoming available (see Ashburner, Chapter 4), new insight into the relationships of flies will increasingly come through comparative genomics (see DeSalle, Chapter 5).

Current State of Knowledge: A Dipteran Supertree

The greatest advances in dipteran phylogenetics over the past four decades have been made by a relatively small number of authors attempting syntheses of the entire order, or large components of it, using Hennigian methods. Supertree methods have emerged in recent years as a rigorous approach to summarizing phylogenetic information to produce more inclusive phylogenies (Bininda-Emonds et al. 2002). We have used MRP (Baum 1992; Ragan 1992; Sanderson et al. 1998) to produce a supertree of the Diptera (Fig. 2.1) and provide a quantitative summary of recent dipteran phylogenetics using morphological data. This supertree was produced using a matrix compiled from the nine primary relationship sources (Griffiths 1972; Hennig 1973; McAlpine 1989; Wood and Borkent 1989; Woodley 1989; Sinclair et al. 1994; Cumming et al. 1995; Oosterbroek and Courtney 1995; Yeates 2002) produced over the past three decades. The trees that were coded to be included in this supertree analysis were all based on morphological evidence and include "qualitative" trees produced by hand without an explicit data matrix, as well as those produced from a more explicit quantitative analysis of morphological evidence. Hennig's analysis, and those found in volume 3 of the *Manual of Nearctic Diptera* (McAlpine and Wood 1989) are qualitative analyses that cover the entire order. Griffiths (1972) provided a radical viewpoint on the relationships of the Cyclorrhapha based on a novel interpretation of male genitalic homologies. Oosterbroek and Courtney (1995) and Yeates (1992) provide modern quantitative analyses of the Lower Diptera and Lower Brachycera, respectively. The trees in Sinclair et al. (1994) and Cumming et al. (1995) range from the Lower Brachycera to the Lower Cyclorrhapha and are largely qualitative, although Cumming et al. (1995) back their analysis of the Lower Cyclorrhapha with a quantitatively analyzed data matrix. The supertree matrix was coded at the family level (151 families were included); Fig. 2.1 is presented at superfamily or infraorder level. Names for higher categories follow our recent review of this subject (Yeates and Wiegmann 1999) and the full family level supertree can be viewed at the website www.inhs.uiuc.edu/cee/FLYTREE.

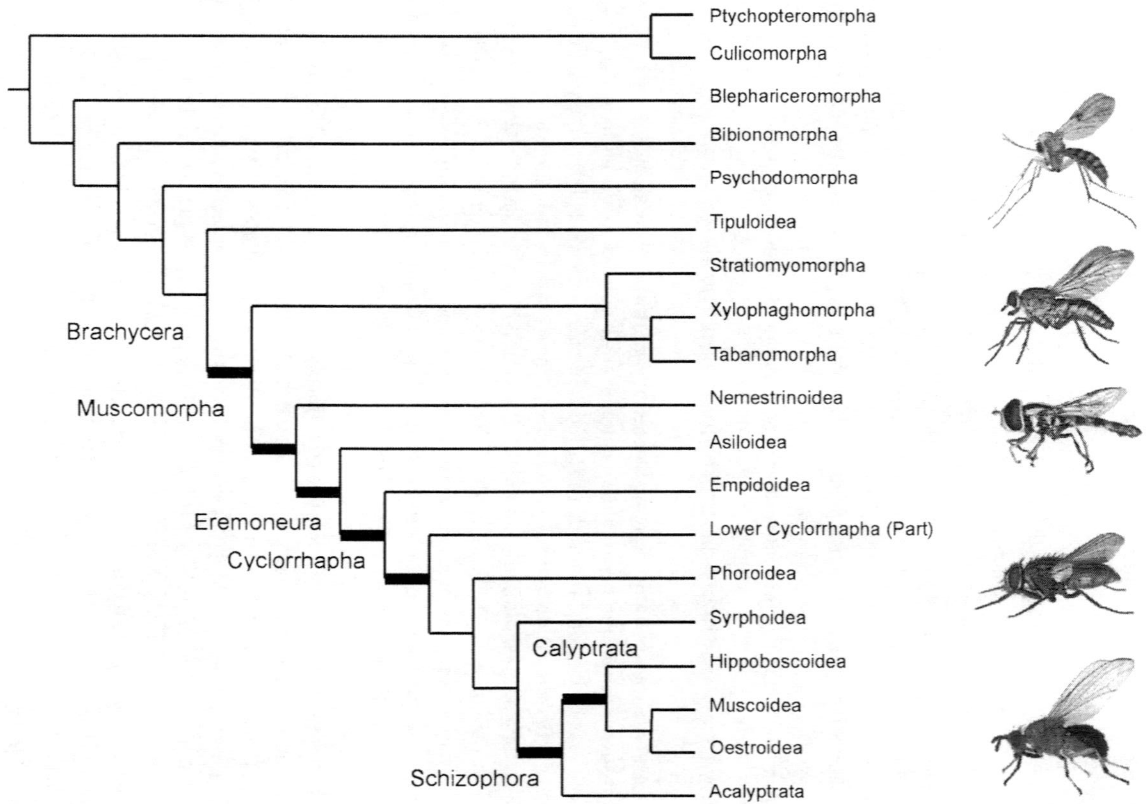
Ptychopteromorpha
Culicomorpha
Blephariceromorpha
Bibionomorpha
Psychodomorpha
Tipuloidea
Stratiomyomorpha
Xylophaghomorpha
Tabanomorpha
Nemestrinoidea
Asiloidea
Empidoidea
Lower Cyclorrhapha (Part)
Phoroidea
Syrphoidea
Hippoboscoidea
Muscoidea
Oestroidea
Acalyptrata
Brachycera
Muscomorpha
Eremoneura
Cyclorrhapha
Calyptrata
Schizophora

Results of our supertree analysis show that major dipteran higher categories (e.g., Culicomorpha, Bibionomorpha, Brachycera, Eremoneura, Muscomorpha, Cyclorrhapha, Schizophora, Acalyptrata, Calyptrata) are monophyletic, and Psychodomorpha, Tipulomorpha, Nematocera, Orthorrhapha, Aschiza are paraphyletic, as presaged in a recent literature review (Yeates and Wiegmann 1999). More detailed results are discussed in the relevant sections below.

Fossil Evidence and Divergence of Major Lineages

Dipteran stem group fossils with four wings belonging to the family Permotipulidae are known from the Late Permian (250 Mya) (Hennig 1981; Willman 1989; Wootton and Ennos 1989; Krzeminski 1992a,b), and a large proportion of fossil Diptera are known from the Mesozoic (Hennig 1981; Evenhuis 1994; Labandeira 1994; see Labandeira, Chapter 9). The main Lower dipteran lineages are known to have evolved by the Late Triassic, perhaps only 25–40 My after the existence of the stem lineage (Woodley 1989; Krezminiski 1992a,b; Fraser et al. 1996; Kremininski and Kremininska 1996; Friedrich and Tautz 1997a,b). A dipteran proboscis designed for lapping evolved 100 My before the appearance of the angiosperms (Labandeira 1997), and extrafloral sources of nectar, such as nonangiospermous anthophytes or hemipteran honeydew, may have been the original carbohydrate source for adult flies (Downes and Dahlem 1987; Labandeira 1998; see Labandeira, Chapter 9). The first brachyceran fossils are known from the Early Jurassic, and the group probably arose in the Triassic (208–245 Mya) (Kovalev 1979; Woodley 1989). Well-preserved tabanids, nemestrinids, bombyliids, and mydids have been recovered from the Late Jurassic of China (Ren 1998). The Asiloidea may not have diversified until the Early Cretaceous (Grimaldi and Cumming 1999), at the same time as the major Angiosperm radiation (Grimaldi 1999). The origin and diversification of Eremoneuran lineages is thought to have begun in the Early Cretaceous (100–140 Mya), with major lineages of Empidoidea and Lower Cyclorrhapha in the Early to Middle Cretaceous. There is abundant fossil evidence for extant families, such as Drosophilidae and Muscidae, not appearing until the Eocene (Beverly and Wilson 1984; Grimaldi and Cumming 1999).

Nucleotide data from the 28S rDNA when analyzed using a Bayesian divergence time estimation procedure, which does not require a molecular clock assumption, largely support the dates reported above for major brachyceran clades and provide a quantitative upper and lower bound for gene-based date estimates (Wiegmann et al. 2003). These data suggest that nearly all of the major brachyceran lineages above the family level (Stratiomyomorpha,

---

FIGURE 2.1. Supertree for Diptera based on MRP coding of 313 nodes found in 12 primary trees listed in the text. The MRP supertree matrix (available from the senior author) was analyzed with PAUP* 4.0B10 (Swofford 2002) using Goloboff's weighting function, 10 random addition sequences, and NNI branch swapping. The tree is stable to values of *k* ranging from 1 to 8. The figure represents a semistrict consensus of 1879 trees (each cost, –295.34). Goloboff's weighting scheme downweights characters with homoplasy during tree search, and *k* describes the shape of the weighting function, or the severity with which homoplasious characters are downweighted. Lower values of *k* discriminate most strongly against homoplasy, but the tree is insensitive to a range of different weighting functions. In terms of a MRP matrix, homoplasy can be interpreted as input tree nodes that are incongruent with other input tree nodes. This weighting scheme tends to prefer congruent nodes over incongruent ones.

Xylophagomorpha, Tabanomorpha, Muscomorpha, Nemestrinoidea, Heterodactyla, Eremoneura; Fig. 2.1) originated before the earliest age estimates for the appearance of flowering plants (Wikström et al. 2001; Wiegmann et al. 2003).

Lower Diptera

The paraphyly of this assemblage ("Nematocera") was suspected for three decades (Hennig 1968, 1973, 1981; Wood and Borkent 1989) and demonstrated in recent cladistic analyses (Sinclair 1992; Oosterbroek and Courtney 1995). Although there have been a few modern phylogenetic analyses of the relationships between the Lower dipteran families, using both morphological (Wood and Borkent 1989; Oosterbroek and Courtney 1995) and molecular (Friedrich and Tautz 1997b) data, there is little consensus on relationships (Yeates and Wiegmann 1999). However, the supertree is well resolved in the Lower Diptera, largely reflecting Oosterbroek and Courtney's (1995) tree.

Some of the traditionally recognized Lower dipteran infraorders near the origin of the Brachycera are not monophyletic in the supertree—Psychodomorpha and Tipulomorpha form a paraphyletic grouping, and the superfamily Tipuloidea is placed as sister group to the Brachycera. This arrangement of Tipulomorpha and Psychodomorpha reflects the incongruence between the trees of Wood and Borkent (1989) and Oosterbroek and Courtney (1995). The Culicomorpha and Ptychopteromorpha form a monophyletic group that is the sister lineage to all other Diptera, and the Culicomorpha contains two sister superfamilies, Culicoidea and Chironomoidea.

Culicomorpha is a well-supported clade containing most bloodsucking Lower dipterans. This group includes the families Culicidae (mosquitoes), Dixidae, Corethrellidae, Chaoboridae, together comprising Culicoidea; and families Thaumaleidae, Simuliidae (black flies), Ceratopogonidae (biting midges), and Chironomidae (midges), together comprising Chironomoidea (Hennig 1981). Most recent phylogenetic studies in Lower Diptera have focused molecular sequence data on issues within the Culicomorpha, especially the Culicidae. A number of studies have examined the relationship between Culicomorpha using sequence data from ribosomal genes (Miller et al. 1996; Pawlowski et al. 1996). The latter results generally did not support the morphology-based tree of Oosterbroek and Courtney (1995). Saether (2000a,b) reexamined culicomorph relationships using 81 morphological characters, including a number of new characters not considered by previous authors. Results varied, depending on specific weights and transformation models applied to characters, suggesting that support for critical nodes may be weak for this dataset. In Saether's tree, Thuamaleidae or (Thaumaleidae + Nymphomyiidae) was the sister to all other culicomorph families. Chironomidae and Simuliidae formed a sister clade to the remaining families in the infraorder, and this clade sometimes included the Ceratopogonidae. The Chironomoidea was paraphyletic with respect to the Culicoidea. Beckenbach and Borkent (2003) use mtDNA to resolve the phylogeny of Ceratopogonidae and in so doing, address the position on the family within the infraorder (Fig. 2.2). Their results are congruent with earlier morphological analyses of the family and infraorder in suggesting that the Ceratopogonidae is sister to the Chironomidae, and that Simuliidae is sister to this combination. It appears that the mtDNA evolves at a higher rate in Ceratopogonidae and Chironomidae than in the other families sequenced.

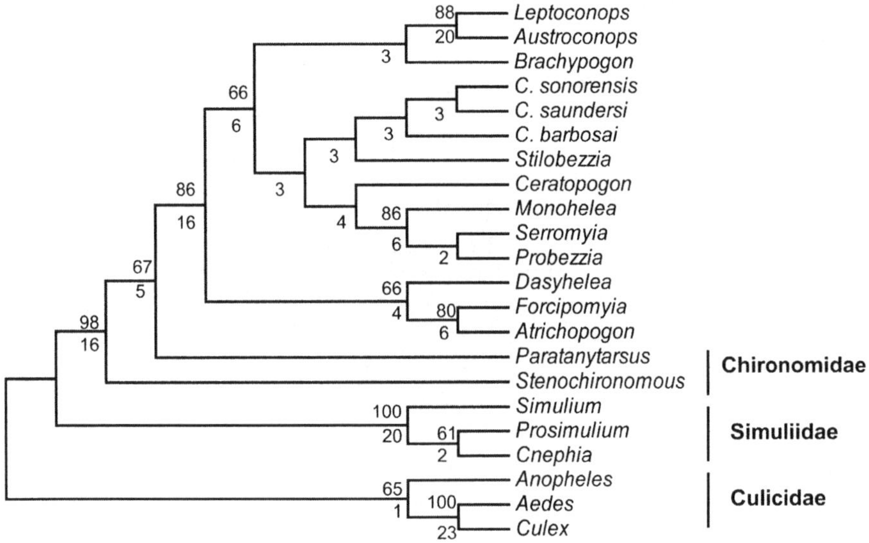

FIGURE 2.2. Molecular phylogeny of the Culicomorpha using 690 bp of COII mtDNA, from Beckenback and Borkent (2003). Parsimony analysis with a 2:4:1 weighting of codon positions, including only nonsynonymous first positions, all second position variation, plus third position transversions.

More recent studies have examined relationships of and within the Culicidae, using both mitochondrial (Beebe et al. 2000; Krzywinski et al. 2001a; Mitchell et al. 2002; Sallum et al. 2000) and single copy nuclear genes (Besansky and Fahey 1997; Krzywinski et al. 2001b) and using morphological data (Harbach and Kitching 1998; Anthony et al. 1999; Sallum et al. 2000). Subfamily relationships of Chironomidae were articulated using a matrix of 89 morphological characters (Saether 2000a,b) compiled from previous studies. Results of this quantitative study were broadly comparable to previous, nonquantitative approaches. Simultaneous phylogenetic analysis of 28S, elongation factor-1α, phosphoenolpyruvate carboxykinase (PEPCK), and dopa decarboxylase (DDC) sequences yielded concordant results with morphological studies for the oldest divergences in the Simuliidae (Moulton 2000).

Blephariceromorpha is made up of three families—Blephariceridae, Deuterophlebiidae, and Nymphomyiidae—united by a number of morphological characteristics generally associated with their larval habitat preference for swift-flowing streams (Wood and Borkent 1989; Courtney 1990a,b, 1991; Arens 1995; Oosterbroek and Courtney 1995). The infraorder is also monophyletic in the current supertree analysis (Fig. 2.1).

Bibionomorpha includes Bibionidae, Pachyneuridae, Mycetophilidae, Sciaridae, and Cecidomyiidae (Wood and Borkent 1989; Blaschke-Berthold 1994), and Axymyidae was added recently (Oosterbroek and Courtney 1995). Evidence from 28S rDNA sequence (Friedrich and Tautz 1997b) supported an expanded concept (Hennig 1981) of Bibionomorpha that also contains the families Anisopodidae and Scatopsidae from Psychodomorpha. Chandler (2002) examined the relationships of Sciaridae, Mycetophilidae sensu stricto, and their relatives, discussing the distributions of 21 adult morphological characters. A number of extant genera normally placed in the Sciaridae or Diadocidiidae showed greater affinities with extinct Mesozoic families of Sciaroidea. Malaise trapping in New Zealand temperate forests revealed a new family-level lineage of sciaroids, the Rangomaramidae, or long-winged fungus gnats (Jaschhof and Didham 2002). The small, little-known family, Axymyidae, is placed as sister group to the remaining Bibionomorpha on the full supertree.

Psychodomorpha includes the families Psychodidae, Perissommatidae, Anisopodidae, Scatopsidae, and Synneuridae and was considered monophyletic, based on synapomorphies of the larval mouthparts (Krivosheina 1988; Wood and Borkent 1989). These synapomorphies have been criticized because of their widespread distribution in other infraorders (Griffiths 1990). More recent morphological studies have found that Psychodomorpha is paraphyletic with respect to Tipulomorpha and Brachycera (Oosterbroek and Courtney 1995). Molecular analyses have suggested that Psychodomorpha is polyphyletic, and Anisopodidae and Scatopsidae have closest relatives in Bibionomorpha (Friedrich and Tautz 1997b). The relationships of the Synneuridae were examined using 59 adult morphological features (Amorim 2000). The Canthyloscelidae is now recognized as a family-level taxon and is placed as sister group to the Scatopsidae (Hutson 1977; Amorim 2000). A phylogeny of the genera of Mycetophilidae sensu stricto genera Soli (1997) did not support the three commonly recognized subfamilies.

The composition and phylogenetic position of Tipulomorpha have come under detailed scrutiny in the past decade (Oosterbroek and Theowald 1991; Oosterbroek and Courtney 1995). A number of synapomorphies were proposed for Tipulomorpha, containing Tipulidae and Trichoceridae (Dahl 1980; Hennig 1981; Griffiths 1990; Oosterbroek and Courtney 1995). Tipulomorpha is paraphyletic in an analysis of 28S rDNA sequence data (Friedrich and Tautz 1997b) and also in the supertree analysis (Fig. 2.1). The Trichoceridae nests within the Psychodomorpha in the supertree.

The search for the sister group of Brachycera among subgroups of Lower Diptera began relatively recently. The root of Brachycera has been localized within the Psychodomorpha in most studies (Wood and Borkent 1989; Woodley 1989; Sinclair 1992; Michelsen 1996), or is shared with the Psychodomorpha and Tipulomorpha together (Oosterbroek and Courtney 1995). Some studies favor Anisopodidae over other families of Psychodomorpha as the sister group of Brachycera (Krivosheina 1988; Woodley 1989; Oosterbroek and Courtney 1995). Synapomorphies proposed to link Anisopodidae and Brachycera include the loss of mandibular prostheca in the larva, larval head with membranous ventral region, larval anal papillae absent (Oosterbroek and Courtney 1995), veins R_4, M_3, and discal cell present in the adult wing, and three spermathecae present in the adult female (Woodley 1989). The supertree analysis places the Tipuloidea as sister to the Brachycera (Fig. 2.1).

Brachycera

The basalmost lineage of Brachycera in the supertree analysis contains three infraorders, Stratiomyomorpha plus (Xylophagomorpha + Tabanopmorpha), reflecting the results of Yeates et al. (2002) and Yeates (2002). The Nemestrinoidea, Asiloidea, and Empidoidea are monophyletic, arising sequentially from the main stem of the Brachycera. The Cyclorrhapha is sister to the Empidoidea. A lower cyclorrhaphan grade comprises three separate lineages, with the Syrphoidea placed as the closest relatives of the major cyclorrhaphan group, Schizophora. This grouping of Syrphoidea and Cyclorrhapha has been called "Eumuscomorpha." The supertree reflects the emerging consensus of various datasets supporting this clade (Wada 1991; Skevington and Yeates 2000; Collins and Wiegmann 2002b). The Calyptrata is made up of the monophyletic superfamilies Hippoboscoidea plus (Muscoidea + Oestroidea). The acalyptrate groups Nerioidea, Diopsoidea, Conopoidea, Tephritoidea, Lauxanioidea, Sciomyzoidea, Opomyzoidea, Carnoidea, Sphaeroceroidea, and Ephydroidea are also monophyletic. The arrangement of acalyptrate superfamilies reflects the views of McAlpine (1989), with conopoids + tephritoids forming a clade sister to the nerioids + diopoids. The lauxanioids + sciomyzoids together are placed as sister to the sphaeroceroids + ephydroids and carnoids + opomyzoids.

LOWER BRACHYCERA

The Brachycera is certainly a monophyletic group, with a large number of undisputed synapomorphies (Hennig 1973; Woodley 1989; Sinclair 1992; Sinclair et al. 1994; Griffiths 1996).

The phylogeny of the Lower Brachycera ("Orthorrhapha") has been scrutinized intensively over the past 15 years. A quantitative reanalysis of 101 morphological characters used to define relationships between the lower brachyceran families attempted to summarize and synthesize this research (Yeates 2002; Fig. 2.3). This study revealed weak evidence for the monophyly of a clade containing Xylophagomorpha, Stratiomyomorpha, and Tabanomorpha, and weak evidence for a monophyletic Asiloidea. These findings are reflected in the supertree. Stuckenberg (1999) reassessed the evolution of the antenna in Brachycera, suggesting that the antenna evolved through progressive fusion of segments and specialized sensory functions; he proposed that the antenna is divided into a postpedicel and stylus in the Brachycera.

XYLOPHAGOMORPHA

Most authors prefer to arrange constituent species into a single family Xylophagidae (Yeates and Wiegmann 1999), with synapomorphies including some extremely distinctive features of the predatory larvae. Discovery of larvae of *Exeretonevra* clearly showed that the genus belonged to the Xylophagidae (Palmer and Yeates 2000). Xylophagomorpha and Tabanomorpha have been united based on synapomorphies of the male genitalia: a membranous outer wall of aedeagus and the development of an endophallic guide inside the sperm pump (Griffiths 1994). These two infraorders have been united with Stratiomyomorpha based on results of a study of the ventral nerve cord (Yeates et al. 2002; see Merritt, Chapter 7).

STRATIOMYOMORPHA

There are numerous synapomorphies for Stratiomyidae and Xylomyidae (Woodley 1989; Sinclair 1992; Sinclair et al. 1994), but fewer for the infraorder once Pantophthalmidae are added (Griffiths 1990; Sinclair 1992; Nagatomi and Liu 1995; Yeates and Wiegmann 1999). The phylogeny of the subfamilies of Stratiomyidae elucidated by quantitative analysis of 20 morphological characters showed that Parhadrestiinae is sister to the remaining subfamilies (Woodley 2001).

TABANOMORPHA

Tabanidae, Pelecorhynchidae, Rhagionidae, Athericidae, and Vermileonidae are united by a compelling suite of morphological characters (Hennig 1973; Woodley 1989; Sinclair 1992;

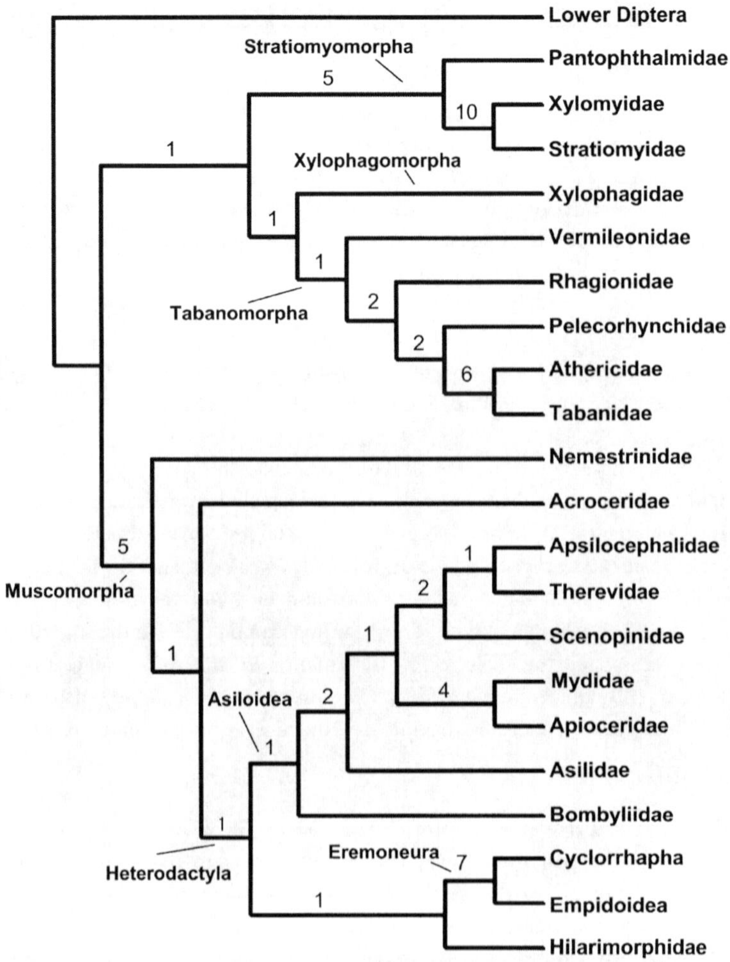

FIGURE 2.3. Most parsimonious tree from PAUP analysis of data from 101 morphological characters for Brachycera, from Yeates (2002), with Bremer support mapped onto nodes.

Sinclair et al. 1994). Tabanomorph relationships revealed by 28S ribosomal DNA sequence data (Wiegmann et al. 2000) were similar to those generated using morphological data (Woodley 1989; Yeates 2002), among them monophyly of the infraorder and its families, including the Vermileonidae. Critical studies of the morphology of flies in the Rhagionidae were used to support the division of Rhagionidae into three families (Spaniidae, Austroleptidae, and Rhagionidae) and the reassignment of the Pelecorhynchidae as a subfamily of Rhagionidae (Stuckenberg 2001), contradicting the results of Wiegmann et al. (2000). Recent reevaluations of the Tabanomorpha based on molecular and morphological data support an alternative arrangement of these families incorporating a new perspective on the lineages comprising the Rhagionidae sensu lato (Hibbs 2002). These studies support the family-level ranking of Austroleptidae, Spaniidae, and Pelecorhynchidae (Kerr-Hibbs, pers. comm.).

MUSCOMORPHA

The infraorder Muscomorpha (Fig. 2.1) contains all brachyceran families except those belonging to Stratiomyomorpha, Xylophagomorpha, and Tabanomorpha (Woodley 1989), and is a well-supported clade found on the supertree. Nemestrinidae and Acroceridae have been united into the superfamily Nemestrinoidea, based on their shared parasitic larval lifestyle (Hennig 1973; Woodley 1989), but authors have also found the superfamily paraphyletic (Yeates 1994) or suggested the group may be better placed in Tabanomorpha (Nagatomi 1992; Griffiths 1994). Hennig (1973) placed Bombyliidae in a group with Nemestrinoidea because of their parasitic larva, but recent treatments have placed Bombyliidae in Heterodactyla (Woodley 1989; Nagatomi 1992, 1996; Yeates 1994). Nemestrinoidea is monophyletic in the supertree analysis. Muscomorpha excluding Nemestrinoidea is united in a clade called Heterodactyla (Woodley 1989), also present in the supertree analysis.

The families Asilidae, Apioceridae, Mydidae, Scenopinidae, Therevidae, and Bombyliidae have been united in Asiloidea on the basis of the apomorphic position of the larval posterior spiracles in the penultimate abdominal segment (Woodley 1989; Yeates 1994). Bombyliidae alone (Woodley 1989), or with Hilarimorphidae (Yeates 1994) has been considered the sister group to the remaining Asiloidea. A number of asiloid families have received critical phylogenetic scrutiny in recent years, partly because of their proximity to Eremoneura.

The monophyly of Therevidae is also not well supported (Yeates 1994), raising the possibility that Scenopinidae may have arisen from them (Woodley 1989). The genus *Apsilocephala* was excluded from Therevidae (Irwin and Lyneborg 1981), and the genus and its relatives were given family status (Nagatomi et al. 1991). The affinities of this group remain obscure, with some authors placing them inside or near Therevidae (Sinclair et al. 1994; Yeates 1994). The paraphyly of Apioceridae was suspected based on the male genitalia (Sinclair et al. 1994), and subsequently, the subfamily Megascelinae was transferred to Mydidae (Yeates and Irwin 1996). Irwin and Wiegmann (2001) used morphology and 28S RNA to confirm the placement of the Rhaphiomidinae and Megascelinae in the Mydidae rather than in the Apioceridae, in agreement with the findings of Yeates and Irwin (1996). Relationships of the Asilidae were analyzed using a combination of four genes: 16S rDNA, 18S rDNA, 28S rDNA, and cytochrome oxidase II (Bybee et al. 2004). Results confirmed that Leptogastrinae was the sister to the remaining asilids, but suggested that the current subfamily classification of asilidae only partially reflected its phylogeny.

The relationships and systematics of the therevoid group of families (Therevidae, Apsilocephalidae, Scenopinidae, and Ocoidae) have been intensively scrutinized in recent years. The molecular phylogeny of Yang et al. (2000) of the therevoid families using 28S RNA and elongation factor-1α resulted in a sister group relationship between the Scenopinidae and Therevidae, as predicted by morphology (Yeates 2002), and a sister group relationship between the clade containing these two families and the Apsilocephalidae. The recently discovered, enigmatic Chilean therevoid family Ocoidae (Yeates et al. 2003) is an independent lineage of asiloids closely related to Apsilocephalidae and Scenopinidae, judging by morphological and 28S rDNA nucleotide data. The therevid subfamilies Phycinae and Therevinae are monophyletic on the tree by Yang et al. (2000); however, Winterton et al. (2001), using a combination of adult morphology and elongation factor-1α, separated some Australian genera traditionally placed in the Therevinae to the Agapophytinae. Numerous other phylogenetic studies have been conducted in the Therevidae (e.g., Gaimari and Irwin 2000; Winterton et al. 2000). There has been increasing use of data from multiple genes and morphology (both separately and in combination) in the Lower Brachycera, and an examination of the relative strength of signal from each data source, including partitioned Bremer support (e.g., Lambkin and Yeates 2003).

Bombyliidae is the sister to the remaining asiloid families in the supertree (Woodley 1989; Yeates 2002), and recent work combining 16S mtDNA molecular and morphological data (Lambkin and Yeates 2003) has confirmed the relationships of the tribes of Anthracinae found by Yeates (1994).

EREMONEURA

"Eremoneura" is the name given to the muscomorphan lineage containing Empidoidea + Cyclorrhapha (Fig. 2.1). This is the best-supported higher-level brachyceran clade with many synapomorphies (Chvála 1983; Griffiths 1984; Meinertzhagen 1989; Woodley 1989; Marois and Meinertzhagen 1990; Sinclair 1992; Wiegmann et al. 1993; Griffiths 1994; Cumming et al. 1995). Recent morphological work has emphasized male genitalic characters for phylogenetic reconstruction in Eremoneura (Griffiths 1972; Chvála 1983; Wiegmann et al. 1993; Shatalkin 1994; Cumming et al. 1995; Griffiths 1996; Zatwarnicki 1996); however, some analyses of molecular data are beginning to appear. Empidoidea is now widely accepted as monophyletic and the sister to Cyclorrhapha (Chvála 1983; Griffiths 1984, 1996; Sinclair 1992; Wiegmann et al. 1993; Cumming et al. 1995; Daugeron 1997; Collins and Wiegmann 2002a; Moulton and Wiegmann 2004). There seems to be strong evidence for the monophyly of Empidoidea, and Atelestidae, Hybotidae, Empididae, and Microphoridae + Dolichopodidae from both morphological (Chvála 1983; Wiegmann et al. 1993; Cumming et al. 1995) and molecular (Collins and Wiegmann 2002a; Moulton and Wiegmann 2004) data (Fig. 2.4).

Cyclorrhapha

Cyclorrhaphan monophyly is well supported (Griffiths 1972; Stoffolano et al. 1988; McAlpine 1989; Cumming et al. 1995; Melzer et al. 1995). Reduction of the larval head capsule and larval feeding structures, as well as pupation within the puparium, are the most recognizable features of this landmark in dipteran evolution. Over the past 40 years, only three workers have attempted to synthesize phylogenetic evidence on cyclorrhaphan relationships in a comprehensive fashion (Hennig 1958, 1971, 1973, 1976; Griffiths 1972; McAlpine 1989). The results of these landmark studies are synthesized in the supertree. Exploration of new character systems—for example, from egg and larval morphology (Meier 1995b, 1996; Meier and Hilger 2000), female genitalia, and internal morphology (e.g., King 1989, 1991; Kotrba

---

FIGURE 2.4. Trees for the Eremoneura from Moulton and Wiegmann (2004), comparing phylogenetic signals from two genes. (A) The most parsimonious tree inferred from 3,880 CAD nucleotides using 5:25:1 weighting of codon positions. (B) The most parsimonious tree inferred from regions D–Z of 28S rDNA. Numbers above and below the nodes are bootstrap and Bremer support values, respectively.

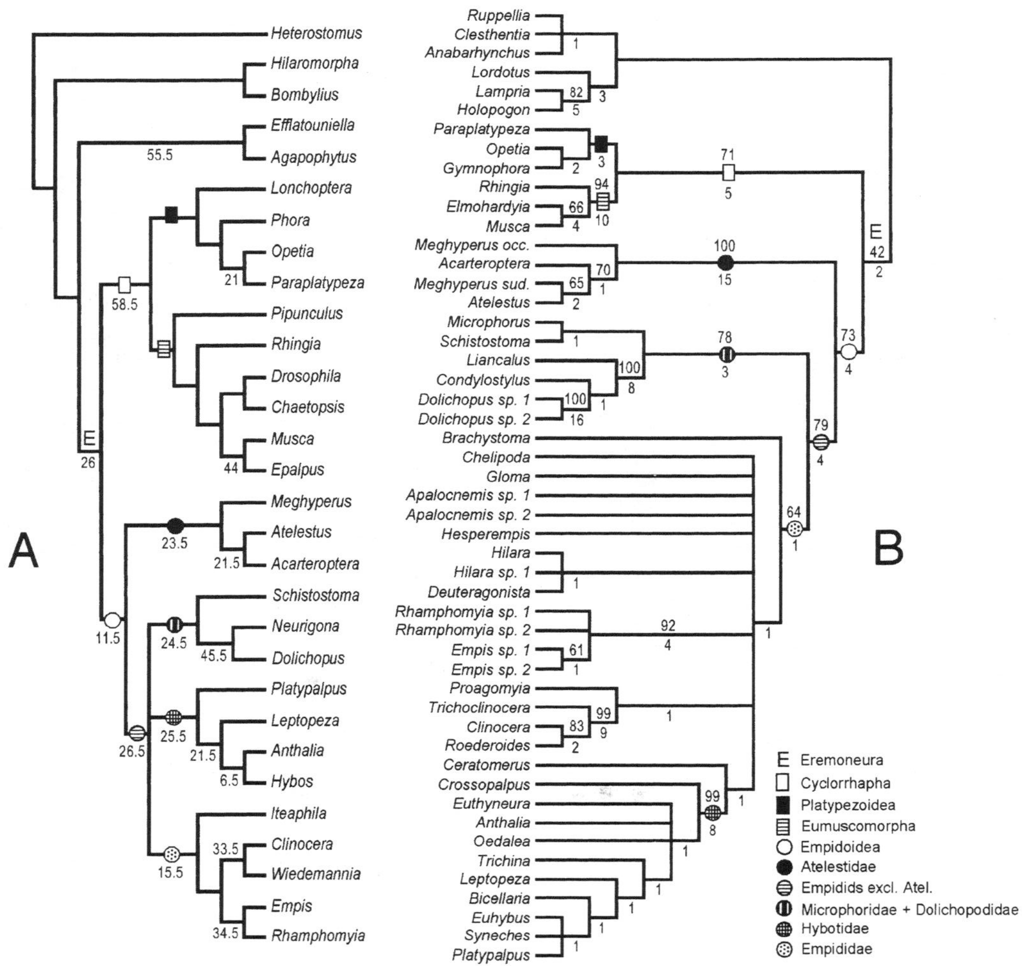
A
Heterostomus
Hilaromorpha
Bombylius
Efflatouniella
Agapophytus
Lonchoptera
Phora
Opetia
Paraplatypeza
Pipunculus
Rhingia
Drosophila
Chaetopsis
Musca
Epalpus
Meghyperus
Atelestus
Acarteroptera
Schistostoma
Neurigona
Dolichopus
Platypalpus
Leptopeza
Anthalia
Hybos
Iteaphila
Clinocera
Wiedemannia
Empis
Rhamphomyia
B
Ruppellia
Clesthentia
Anabarhynchus
Lordotus
Lampria
Holopogon
Paraplatypeza
Opetia
Gymnophora
Rhingia
Elmohardyia
Musca
Meghyperus occ.
Acarteroptera
Meghyperus sud.
Atelestus
Microphorus
Schistostoma
Liancalus
Condylostylus
Dolichopus sp. 1
Dolichopus sp. 2
Brachystoma
Chelipoda
Gloma
Apalocnemis sp. 1
Apalocnemis sp. 2
Hesperempis
Hilara
Hilara sp. 1
Deuteragonista
Rhamphomyia sp. 1
Rhamphomyia sp. 2
Empis sp. 1
Empis sp. 2
Proagomyia
Trichoclinocera
Clinocera
Roederoides
Ceratomerus
Crossopalpus
Euthyneura
Anthalia
Oedalea
Trichina
Leptopeza
Bicellaria
Euhybus
Syneches
Platypalpus
E Eremoneura
Cyclorrhapha
Platypezoidea
Eumuscomorpha
Empidoidea
Atelestidae
Empidids excl. Atel.
Microphoridae + Dolichopodidae
Hybotidae
Empididae

1995), and nucleotide sequences (e.g., Smith et al. 1996; Han and McPheron 1997; Collins and Wiegmann 2002b; Meier and Wiegmann 2002)—should provide important new evidence on relationships when applied broadly across cyclorrhaphan groups.

LOWER CYCLORRHAPHA

Cyclorrhapha has traditionally been divided into two groups, "Aschiza" and Schizophora, based on the absence or presence, respectively, of a ptilinal fissure (McAlpine 1989). The nonschizophoran families are Opetiidae (Chandler 1981), Platypezidae, Lonchopteridae, Sciadoceridae, Ironomyiidae, Phoridae, Pipunculidae, and Syrphidae. Most recent studies instead have concluded that "Aschiza" is probably paraphyletic with respect to Schizophora (Griffiths 1972, 1990; Wada 1991; Cumming et al. 1995; Zatwarnicki 1996), and this holds true in the supertree analysis. The supertree analysis divides the Lower Cyclorrhapha into three separate lineages, the Opetiidae and Platypezidae, the Phoroidea (including the Lonchopteridae), and the clade most closely related to the Schizophora, the Syrphoidea (Fig. 2.1).

Basal Cyclorrhapha has received increased phylogenetic scrutiny in recent molecular systematic studies. Datasets based on nuclear 28S rDNA (Collins and Wiegmann 2002a) and the nuclear protein encoding locus CAD (Moulton and Wiegmann 2004) support a monophyletic grouping of the Phoridae, Sciadoceridae, Platypezidae, Opetiidae, and Lonchopteridae, but break up the Syrphoidea, with the Syrphidae and Pipunculidae forming separate lineages basal to the origin of Schizophora. Results from a phylogenetic study using mitiochondrial 12S and 16S ribosomal DNA showed a monophyletic Syrphidae and Pipunculidae (Skevington and Yeates 2000). The genes provided excellent information to resolve relationships within Pipunculidae, but not within Syrphidae. Results of combining molecular and morphological data produced a paraphyletic Nephrocerinae, with Chalarinae the sister to the remaining Pipunculidae. Mitochondrial DNA has been used to construct phylogenies within Syrphidae (Ståhls and Nyblom 2000), and Katzourakis et al. (2001) use a near-complete morphological phylogeny of Syrphidae to test ideas for morphological, life-history, and ecological correlates of diversity. An extensive study combining nuclear and mitochondrial DNA data with information on larval and adult morphology for 51 syrphid species resulted in a basal Microdontini, monophyletic Eristalinae, and Pipizini as the sister to the Syrphinae (Ståhls et al. 2003).

SCHIZOPHORA

Schizophora is classified into at least 80 families and make up just over half the family-level diversity in Diptera (McAlpine 1989; Colless and McAlpine 1991; Yeates and Wiegmann 1999). Schizophoran flies emerge from the puparium by inflation of a membranous head sac, the ptilinum. The major synapomorphies for Schizophora are features associated with this method of emergence (McAlpine 1989). Griffiths (1972) and Cumming et al. (1995) also list as evidence the unique 360° rotation of the male genital capsule within the puparium.

Traditional views of schizophoran classification depend on the size of the lower calypter; hence, the names for the two divisions of the group. It has long been recognized that this character is too variable in both groups to be a reliable synapomorphic or diagnostic feature. The works that provide major evidential reviews of Schizophora (Griffiths 1972; Hennig 1973; McAlpine 1989) are included in the supertree analysis. The synthetic revisions of schizophoran classification by Griffiths (1972) and McAlpine (1989) both endeavor to build on Hennig's (1958, 1971, 1973) framework. Griffiths (1972) added detailed reinterpretation and scorings of male genital characters along with other morphological features, and McAlpine's (1989) fully resolved phylogenetic arrangements draw on most morphological character systems, as well as aspects of fly biology. McAlpine divided the group into 13 superfamilies; these are found in the supertree: Nerioidea, Diopsoidea, Conopoidea, Tephritoidea, Lauxanioidea, Sciomyzoidea, Opomyzoidea, Sphaeroceroidea, Carnoidea, Ephydroidea, Hippoboscoidea, Muscoidea, and Oestroidea. Generally, McAlpine's classification maintains Hennig's groupings, whereas Griffiths's is a more radical restructuring of the higher-level schizophoran framework.

ACALYPTRATAE

The search for convincing synapomorphies uniting acalyptrate families has been difficult. Apparent homoplasy in most character systems makes assignment of synapomorphies within Acalyptratae in qualitative analyses contradictory. There are still no comprehensive quantitative phylogenetic analyses of Schizophora. McAlpine (1989) and, less strongly, Hennig (1971, 1973), favored a monophyletic Acalyptratae as sister group to Calyptratae. Griffiths (1972) argued against acalyptrate monophyly, but the monophyletic viewpoint is preserved in the supertree. More than 50% of acalypterate species diversity is contained in just six large families: Tephritidae, Lauxaniidae, Agromyzidae, Chloropidae, Drosophilidae, and Ephydridae.

The group of long-bodied flies called "Nerioidea" is recognized in all three competing schizophoran arrangements, and synapomorphies include the elongate male and female genitalia of member species. These families have been further divided; for example, Calobatidae, Taeniapteridae from Micropezidae (Hennig 1973), and Pseudopomyzidae from Cypelosomatidae (Shatalkin 1994; McAlpine 1996). Griffiths (1972) included the families of Nerioidea in his muscoid prefamily Micropezoinea. McAlpine (1996) recently reviewed nerioid relationships, listed synapomorphies of Pseudopomyzidae and considered Heliomyzoidea (Sphaeroceroidea of McAlpine) their sister group, and provided a phylogenetic analysis of the Micropezidae (McAlpine 1998).

McAlpine's (1989) concept of Diopsoidea followed Hennig's (1958, 1973) grouping Nothyboidea. Synapomorphies in McAlpine's system include the reduction to two or three orbital bristles and fusion of male sternites 7 and 8. Griffiths (1972) included these families in three different higher-level groupings, prefamilies Tanypezoinea (Tanypezidae, Heteromyzidae, Strongylophthalmidae) and Diopsoinea (Diopsidae, Syringogastridae), and the superfamily Nothyboidea (for Nothybidae, Psilidae, Somatiidae, Periscelididae, Teratomyzidae, and Megamerinidae).

Conopoidea have been a critical taxon for higher-level classifications of Schizophora. The theca, a specialized modification of female abdominal sternites 5 and 6, is a striking and unique synapomorphy of the group. In earlier arrangements, conopids were often considered the sister group to all other Schizophora (Hennig 1958) or were grouped with Syrphoidea (e.g., Stone et al. 1965). Hennig (1952, 1958, 1973) erected a superfamily for the family Conopidae and suggested affinities with Tephritoidea. Griffiths (1972) presented evidence from the male genitalia, including the hingelike swinging mechanism used in extension of the aedeagus, to support this placement. McAlpine (1989) tentatively placed conopids inside or sister to tephritoids, and they appear in this position in the supertree analysis.

The cluster of tephritoid families closely related to Tephritidae—including Tachiniscidae, Pyrgotidae, Platystomatidae, and Otitidae—is a monophyletic component of each of the three major classifications. Griffiths (1972) placed all these families in his Tephritidae, and one synapomorphy of the group is the absence of spiracles on male abdominal segments 6 and 7. This group (including Richardiidae) was called "Otitoidea" by Hennig (1958, 1973). At variance in the competing systems however, is McAlpine's inclusion of Lonchaeidae as the basal tephritoid lineage, and Piophilidae + Pallopteridae as sister group to the tephritoid family Richardiidae. The latter three families form Hennig's (1973) Pallopteroidea. Griffiths (1972) also included Piophilidae and Pallopteridae in the Tephritidae family group, but placed Lonchaeidae together with Cryptochaetidae in his superfamily Lonchaeoidea. Korneyev (2000) analyzed 43 morphological characters in a quantitative cladistic analysis of tephritoid families. He found five synapomorphies from the male genitalia and a feature of the female ovipositor in support of tephritoid monophyly. Korneyev (2000) breaks the superfamily into two sections, a monophyletic "higher Tephritoidea" comprising the Ulidiidae (=Otitidae =Pterocallidae), Platystomatidae, Pyrgoitidae, and Tephritidae, and a paraphyletic "lower Tephritoidea" comprising the Lonchaeidae, Piophilidae, Pallopteridae, and Richardiidae. Morphological support is limited for the exact resolution of the relationships among the lower tephritoid families. In Griffiths's (1972) system, the Tephritidae family group is one of eight clades making up the Tephritoinea section of his superfamily Muscoidea.

Baker et al. (2001) used nucleotide sequence data from three mitochondrial genes (COII, 16S, and 12S) and three nuclear genes (ef-1α, wingless and white) to construct a phylogeny of the Diopsidae. They found that by virtually any measure, the nuclear genes provided greater utility than did the mitochondrial genes. These data were analyzed in combination with the morphological data of Meier and Hilger (2000) to produce a combined molecular and morphological phylogeny (Meier and Baker 2002). In the combined analysis, morphological and molecular data agree on all major clades, and the average morphological character produces twice the node support of an average molecular character (Fig. 2.5). The Syringogastridae are likely to be the sister of the Diopsidae (Meier and Hilger 2000), although McAlpine favored the Megamerinidae + Syringogastridae. Both molecular and morphological data have been applied extensively to resolve relationships within the Tephritidae (cf. Aluja and Norrbom 2000 and references therein). Relationships in the dacine tephritids were studied using mitochondrial DNA sequence data (Smith et al. 2002; Smith et al. 2003), and this phylogeny in turn was used to study the evolution of the male lure response.

Lauxanioidea is a well-supported superfamily present in all three major classifications (Hennig 1958, 1973; Griffiths 1972; McAlpine 1989), containing the Chamaemyiidae, Lauxaniidae, Eurychoromyiidae, and Celyphidae. Hennig (1973) placed the monotypic Bolivian Eurychoromyiidae in Sciomyzoidea. Synapomorphies for the group include convergent

---

FIGURE 2.5. Phylogenetic hypothesis for Diopsidae, based on 54 morphological characters and 1,195 base pairs of mtDNA and 966 base pairs of nuclear DNA sequence data, from Meier and Baker (2002). Bootstrap/Jacknife/Bremer support values are listed to the right of the respective node numbers, followed by the analysis conditions *not* supporting the node. Abbreviations: 1&2, first and second position costs in protein-coding genes; tv, transversion cost; 3rd tv+1&2 = transversion cost in third positions of protein coding genes and change cost for remaining DNA characters; the numbers following the abbreviations specify the costs.

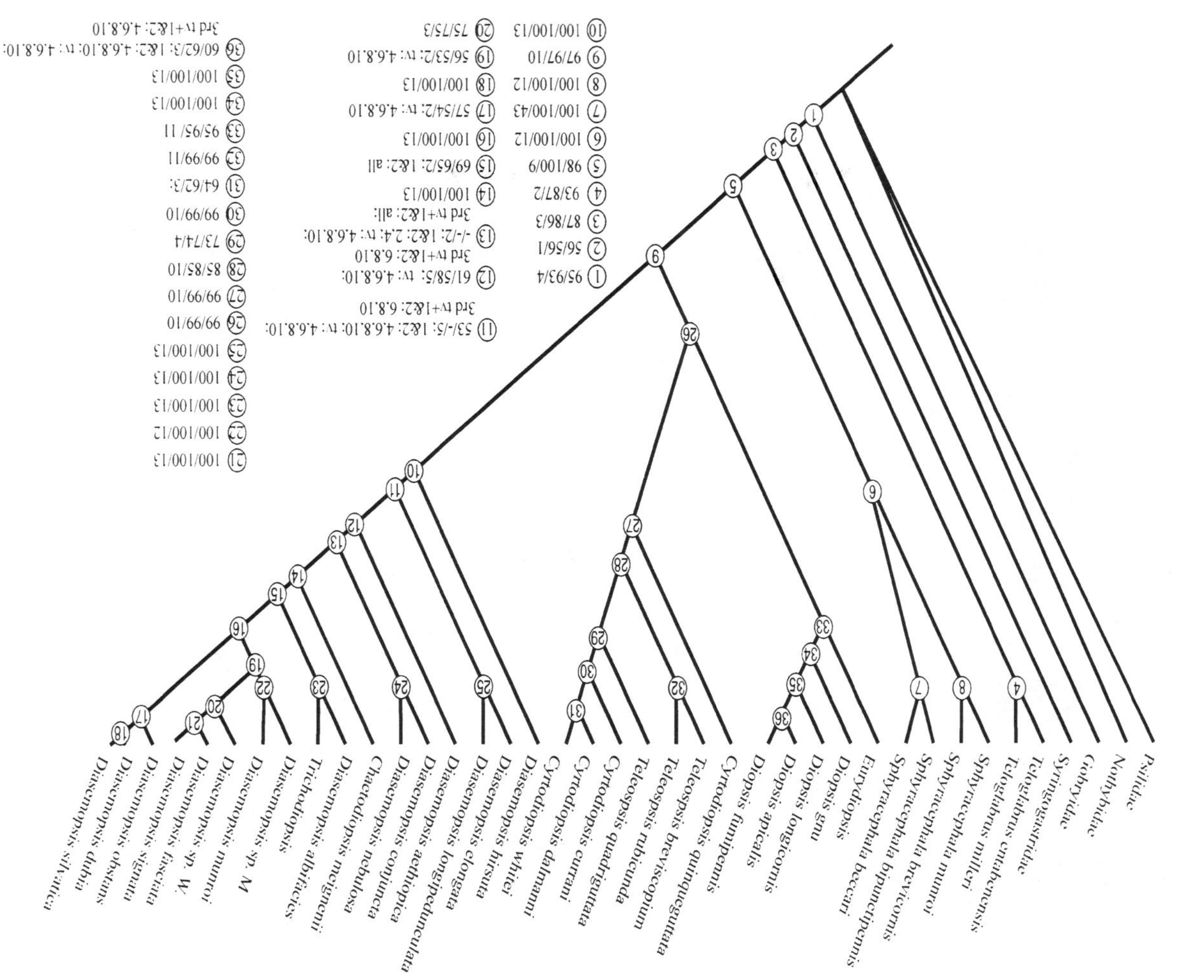

Diasemopsis silvatica
Diasemopsis dubia
Diasemopsis obstans
Diasemopsis signata
Diasemopsis fasciata
Diasemopsis sp. W.
Diasemopsis munroi
Diasemopsis sp. M
Diasemopsis
Trichodiopsis
Diasemopsis albifacies
Diasemopsis meigenii
Chaetodiopsis nebulosa
Diasemopsis conjuncta
Diasemopsis aethiopica
Diasemopsis longipedunculata
Diasemopsis elongata
Diasemopsis hirsuta
Diasemopsis whitei
Cyrtodiopsis dalmanni
Cyrtodiopsis currani
Cyrtodiopsis quadriguttata
Teleopsis rubicunda
Teleopsis breviscopium
Teleopsis quinqueguttata
Cyrtodiopsis fumipennis
Diopsis apicalis
Diopsis longicornis
Diopsis gnu
Diopsis
Eurydiopsis
Sphyracephala beccari
Sphyracephala bipunctipennis
Sphyracephala brevicornis
Sphyracephala munroi
Sphyracephala milleri
Teloglabrus entabenensis
Teloglabrus
Syringogastridae
Gobryidae
Nothybidae
Psilidae
1 95/93/4
2 56/56/1
3 87/86/3
4 93/87/2
5 98/100/9
6 100/100/12
7 100/100/43
8 100/100/12
9 97/97/10
10 100/100/13
11 53/-/5: 1&2: 4.6.8.10: tv: 4.6.8.10:
3rd tv+1&2: 6.8.10
12 61/58/5: tv: 4.6.8.10:
3rd tv+1&2: 6.8.10
13 -/-/2: 1&2: 2.4: tv: 4.6.8.10:
3rd tv+1&2: all:
14 100/100/13
15 69/65/2: 1&2: all
16 100/100/13
17 57/54/2: tv: 4.6.8.10
18 100/100/13
19 56/53/2: tv: 4.6.8.10
20 75/75/3
21 100/100/13
22 100/100/12
23 100/100/13
24 100/100/13
25 100/100/13
26 99/99/10
27 99/99/10
28 85/85/10
29 73/74/4
30 99/99/10
31 64/62/3:
32 99/99/11
33 95/95/ 11
34 100/100/13
35 100/100/13
36 60/62/3: 1&2: 4.6.8.10: tv: 4.6.8.10:
3rd tv+1&2: 4.6.8.10

postocellar bristles, an abbreviated anal vein, and fusion of male abdominal tergites 7 and 8. Celyphidae have an enlarged scutellum that may cover the entire abdomen; Lauxaniidae are morphologically diverse, species-rich, and most species are saprophagous (Shewell 1987).

Sciomyzoidea also forms a component of all three acalyptrate arrangements, containing nine families, including Sciomyzidae, Sepsidae, and Coelopidae. All three authors point out, however, that explicit evidence for the group is at best weak (Hennig 1958; Griffiths 1972; McAlpine 1989). Proposed synapomorphies include the face desclerotized along the vertical midline and the frontal vitta densely and strongly setose. In Griffiths's (1972) arrangement, this same cluster of families comprises the prefamily Sciomyzoinea of Muscoidea. Coelopid relationships were studied using a combination of morphological characters and mitochondrial and nuclear genes (Meier and Wiegmann 2002). Relationships of the Sepsidae have been studied in detail by Meier (1995a,b, 1996), producing a phylogenetic analysis of the family based on morphological data with a special emphasis on the egg and larval stages. Knutson and Vala (2002) studied the evolution of behavior in Sciomyzidae using the morphological phylogeny of Marinoni and Mathis (2000).

McAlpine (1989) grouped 13 families into the Opomyzoidea, including Agromyzidae, Anthomyzidae, and Asteiidae. Synapomorphies for this diverse assemblage include wing contrastingly patterned, anepisternum with raised ridge along upper margin and male abdominal segment 7 reduced and fused with sternite 8 (McAlpine 1989). Family-level composition of the four suprafamilies is: Clusionea for Clusiidae + Acartophthalmidae; Agromyzoinea for Odiniidae, Agromyzidae, and Fergusoninidae; Opomyzoinea for Opomyzidae + Anthomyzidae; and Asteoinea for Aulacigastridae, Periscelididae, Neurochaetidae, Teratomyzidae, Xenasteidae, and Asteidae. The families included in McAlpine's Clusionea and Opomyzoinea were placed in Hennig's (1973) family group Anthomyzidea of the superfamily Anthmoyzoidea. McAlpine's phylogenetic hypothesis for opomyzoids, however, does not support a sister group relationship for Clusionea + Opomyzoinea. Additional differences between Hennig's and McAlpine's interpretation of this group are Hennig's placement of Fergusoninidae in Drosophiloidea and his placement of the asteioine families in his family group Periscelididea of Anthomyzoidea. The sister group relationship between Anthomyzidae and Opomyzidae was confirmed by cladistic analysis (Rohacek 1998). Larvae of most opomyzoid families are associated with living plants, such as agromyzid leafminers (Spencer 1990) and fergusoninid larvae associated with gall-forming nematodes on *Eucalyptus* (Colless and McAlpine 1991).

Griffiths's (1972) assessment of the phylogenetic relationships of the opomyzoid families differs markedly from that of McAlpine (1989). Based on differences in the structure of the aedeagus, Griffiths listed seven synapomorphies uniting Clusiidae with Agromyzidae and doubted the grouping of Odiniidae + Agromyzidae. He placed Odiniidae in Tephritoinea, Acartophthalmidae in the Chloropidae family group (see below), and placed Fergusoninidae incertae sedis. Griffiths included Opomyzidae, Anthomyzidae, Asteiidae, Chyromyidae, and Aulacigastridae in his muscoid prefamily Anthomyzoinea along with Sphaeroceridae, Helcomyzidae, Rhinotoridae, Borboropsidae, and Trixoscelididae.

McAlpine's (1989) thorough review of the fluid history of alternative classificatory positions of the families he included within the superfamily Carnoidea is illustrative of the difficulties faced in all higher-level phylogenetic studies of acalyptrates. Synapomorphies for this grouping are mostly found in the details of head and thorax setation. Chloropidae and

Milichiidae are well-supported sister groups (Brake 2000), but the position of these flies and identification of their closest relatives has been elusive. Hennig (1973) grouped Chloropidae and Milichiidae together with Carnidae as Chloropoidea, and placed the remaining carnioid families in Drosophiloidea. Griffiths's (1972) Chloropidae family group is one of eight clades of Tephritoinea and included Chloropidae, Milichiidae, Risidae, Carnidae, and Acartophthalmidae. Griffiths placed tethinids in Tephritoinea, the Braulidae and Canacidae were placed as incertae sedis, Cryptochaetidae was placed as a sister group to Lonchaeidae, and Australimyzidae as a separate prefamily of Muscoidea, the Australimyzoinea. Chloropidae and Milichiidae are the most species-rich families in the superfamily; Milichiidae is very diverse in morphology and Chloropidae is one of the most easily recognized acalypterate groups.

McAlpine (1989) grouped the diverse families Heleomyzidae and Sphaeroceridae along with Chryomyidae and monotypic Mormotomyidae in Sphaeroceroidea, but evidence for this clade is weak. Sphaeroceridae and Heleomyzidae have a history of alternative placements and have been divided into a variety of subgroups. For example, Griffiths (1972) divided Heleomyzidae into eight families and divided these into three different superfamilies. This division has been challenged in a later family-level revision of Heleomyzidae (McAlpine 1985). Griffiths (1972) included Heleomyzidae sensu stricto, Sphaeroceroidea, Chyromyidae, as part of the muscoid group Anthomyzoinea. Hennig (1973) grouped heleomyzids, sphaerocerids, and chyromyids within his broader concept of Drosophiloidea. Mormotomyidae, a monotypic and little-known family of African, bat-associated flies, occupies three entirely different phylogenetic positions in the three major classifications. Hennig considered these flies a basal lineage of Calyptratae, based on the presence of a cleft on the pedicel. Griffiths and McAlpine challenged this placement and provided morphological evidence that the group is more correctly placed among acalyptrates, either in Sphaeroceroidea (McAlpine 1989), or among the Tephritoinea grouped with several former Heleomyzidae (Griffiths 1972).

Ephydroidea (=Drosophiloidea of Griffiths 1972) are among the better-supported acalyptrate clades. Morphological evidence for the superfamily is strong, with nine synapomorphies recognized by McAlpine (1989) and seven by Griffiths (1972), including a single proclinate fronto orbital bristle, the presence of precoxal bridges, and female sternite 8 reduced or absent. A recent reevaluation of ephydroid relationships (Grimaldi 1990) preferred an arrangement of two clades, (Curtonotidae + Drosophilidae) and another clade including Ephydridae, Diastatidae, and Camillidae. Hennig (1973) included these groups together with nine additional families in the larger group Drosophiloidea.

The relationships of the Drosophilidae have been studied by numerous authors over the past 15 years. Grimaldi's (1990) morphological analysis of the family has been complemented by a number of analyses of subgroups of the family using molecular data, with the intention of discovering the relationships of the Hawaiian *Drosophila* radiation (Remsen and DeSalle 1998; Remsen and O'Grady 2002) and also the relationships of the species related to the model organism *Drosophila melanogaster* (Schawaroch 2002). The most recent synthetic analyses, using both multiple molecular markers and morphological characters, found that the Hawaiian *Drosophila* is monophyletic, but the genus *Drosophila* is not monophyletic (Remsen and O'Grady 2002), and that taxonomic revision will be necessary to provide a useful and phylogenetically sound nomenclature for this group. Judging by the distance between the type species of the genus *Drosophila* (*funebris*) and *Drosophila melanogaster* on Remsen and O'Grady's summary phylogeny, it is likely that *melanogaster* will no longer belong to the genus *Drosophila* after the group is divided into monophyletic genera. In a detailed and impressive study, Schawaroch (2002) studied the relationships of 49 species in the *Drosophila melanogaster* species group, using one mitochondrial and two nuclear gene loci. The evolution of three morphological structures (sex comb, epandrium, and midtibia) was mapped onto the tree. Tamura et al. (2004) used a molecular clock–based method and multiple genes to estimate divergence times for various *Drosophila* species, species groups, and closely related genera.

RELATIONSHIPS BETWEEN ACALYPTRATE GROUPS

Griffiths (1972) divided the acalyptrates among five superfamilies but did not resolve the relationships among them. He placed most of the acalyptrates (including the prefamilies Tanypezoinea, Micropezoinea, Australimyzoinea, Diopsionea, Sciomyzoinea, Anthomyzoinea, Agromyzoinea, and Tephritoinea) along with the Calyptratae in his superfamily Muscoidea. Synapomorphies for Muscoidea included characters of the male abdominal segment 7 and sternite 6. McAlpine's (1989) fully resolved phylogenetic arrangement yielded two main acalyptrate assemblages, and these are preserved in the supertree analysis; a clade including Nerioidea, Diopsoidea, Conopoidea, and Tephritoidea; and a clade comprising Lauxanioidea, Sciomyzoidea, Opomyzoidea, Carnoidea, Sphaeroceroidea, and Ephydroidea. The former clade was supported by a specialized development of female abdominal segment 7 to form a bulbous oviscape, and the male aedeagus elongate, flexible, and tending to be looped or coiled. The latter clade was supported by a reduced male sternite 6. Within the first subgroup, conopoids and tephritoids are united by a piercing ovipositor, and Nerioidea and Diopsoidea are united by three synapomorphies of wing venation. In the second subgroup, lauxanioids and sciomyzoids are united by characters of setation of the head and legs. The four superfamilies Opomyzoidea, Carnoidea, Sphaeroceroidea, and Ephydroidea share well-developed vibrissae and subcostal break of the wing. Sphaeroceroids and ephydroids share convergent postocellar bristles, preapical dorsal tibial bristles, wing vein R_1 bare, and a reduced male tergite 6.

CALYPTRATAE

Calyptratae have long been recognized as a major lineage of higher Diptera, and the morphological support for this clade is stronger than for any other schizophoran group (Hennig 1971; Griffiths 1972; McAlpine 1989). Calyptratae include some of the more diverse and successful fly families, including Calliphoridae, Sarcophagidae, Tachinidae, Anthomyiidae, and Muscidae, as well as the more specialized Streblidae, Nycteribiidae (bat ectoparasites), Hippoboscidae (bird parasites), Glossinidae (tsetse flies), and Oestridae (bot flies). McAlpine (1989) recognized three constituent superfamilies of Calyptratae: (1) Hippoboscoidea, including Glossinidae, Hippoboscidae, Streblidae, and Nycteribiidae; (2) Muscoidea, including Scathophagidae, Anthomyiidae, Fanniidae, and Muscidae; and (3) Oestroidea, including Calliphoridae, Mystacinobiidae, Sarcophagidae, Rhinophoridae, Tachinidae, and Oestridae. Muscoidea and Oestroidea were united as sister groups; McAlpine's views have been preserved in the supertree analysis. Griffiths (1972) included all the families of Oestroidea in Tachinidae and all the Hippoboscoidea in his Hippoboscidae family group. Griffiths (1972, 1982) later revised the concept of Tachinidae to constitute a Tachinidae family group. The calyptrates were monophyletic, but the Muscoidea arose from within the Oestroidea, making the latter paraphyletic, in an analysis of 18S and 16S rDNA (Nirmala et al. 2001), and Bernasconi et al. (2000) found a similar result using COI and COII mtDNA sequences.

HIPPOBOSCOIDEA

The clade Hippoboscoidea (Glossinidae, Hippoboscidae, Streblidae, and Nycteribiidae) was formerly called "Pupipara" and is supported most notably by the development of the larva within the body cavity of the female by adenotrophic viviparity. McAlpine (1989) lists loss of the salivary pump, palp modified to sheath the proboscis, and other features associated with obligate blood feeding as evidence of their monophyly. The Streblidae and Nycteribiidae are bat ectoparasites and live continuously on their hosts. The Glossinidae are the only free-living members of the superfamily; they transmit sleeping sickness to humans.

McAlpine (1989) and Hennig (1973) unite Scathophagidae, Anthomyiidae, Muscidae, and Fanniidae in the superfamily Muscoidea. Synapomorphies include the anus of the male situated above the cerci, male sternite 10 forming bacilliform sclerites, and female abdominal spiracle 7 located on tergite 6. Griffiths (1972) did not regard this grouping of families as monophyletic, and they were included in his prefamily Calyptratae. Relationships within Muscidae are increasingly scrutinized, and the millipede parasitic tribe Eginiini are now considered part of the Muscidae rather than a separate family (Skidmore 1985; Carvalho 1989). Sequence data from mtDNA COI gene was insufficient to resolve relationships between 22 genera of Scathophagidae; however, the monophyly of many genera was confirmed except for *Scathophaga* (Bernasconi et al. 2000).

Review and analysis of morphological variation have generated competing hypotheses of relationships for Oestroidea (Griffiths 1972; McAlpine 1989; Pape 1992; Colless 1994). Pape (1992) presented a quantitative cladistic analysis of morphological evidence on oestroid relationships, results that differ considerably from the conclusions of McAlpine (1989). Pape's results support two major clades, Tachinidae + Sarcophagidae, and Rhinophoridae + (Oestridae + Calliphoridae). Alternatively, McAlpine's (1989) arrangement unites Sarcophagidae + Calliphoridae, and includes Oestridae as sister group to Rhinophoridae + Tachinidae. Synapomorphies for Oestroidea include a vertical row of bristles present on the meron and wing vein M_1 deflected forward to join C before the wing apex (McAlpine 1989; Pape 1992). Rognes (1986, 1991, 1997) has argued that Rhinophoridae may actually belong within Calliphoridae, and seems to favor Sarcophagidae + Tachinidae + Calliphoridae as a monophyletic lineage with Oestridae their sister group. A key issue in the debate over oestroid relationships is the monophyly of Calliphoridae. Griffiths (1982) concluded that the New Zealand bat fly family Mystacinobiidae should be considered a synonym of Calliphoridae, not a close relative of Drosophilidae, and this conclusion was confirmed using 16S mtDNA (Gleeson et al. 2000). Rognes (1986) listed morphological features supporting the monophyly of the major oestroid families.

Using 118 morphological characters from all developmental stages analyzed in a quantitative analysis, Pape (2001) produced a phylogeny of Oestridae giving the four major clades of the analysis (Cuterebrinae (Gasterophilinae (Hypodermatinae + Oestrinae))) subfamilial rank. In the first use of molecular data to reconstruct evolutionary relationships in the very large family Tachinidae, Stireman (2002) was able to recover monophyly of the family, and subfamily Exoristinae, and the Winthemiini, Exoristini, and Blondeliini using ef-1α and 28S rDNA.

Recent Insights and New Perspectives

The increasing number of new phylogenetic hypotheses based on rigorous quantitative analyses at all levels of dipteran phylogeny provide a context for many evolutionary studies. These range from studies of key ecological and morphological innovations (e.g., King 1991) to new assessments of the paleontological history of flies (Grimaldi and Cumming 1999; Blagoderov et al. 2002), and comparative genomics and gene evolution (Bolshakov et al. 2002; Zdobnov et al. 2002; Severson et al. 2004). The number of genes being used for dipteran systematics, and the taxonomic range in their application, is also steadily increasing. These genetic comparisons across phylogenetic histories allow the application of evolutionary rates to estimate divergence times. Gene-based divergence time estimates provide a quantitative range with which to gauge fossil-based hypotheses for the origin and radiation of many dipteran groups, especially those for which a fossil record is lacking or severely limited. Methods for estimating dipteran divergence times have employed both traditional molecular clock assumptions (Beverly and Wilson 1984; Powell 1997; Gaunt and Miles 2002; Tamura et al. 2004) and Bayesian methods that allow evolutionary rates to vary across phylogeny (Thorne et al. 1998; Wiegmann et al. 2003). Table 2.1 lists some recent hypotheses of divergence times for key dipteran clades, based on genetic analyses of evolutionary rates, and a comparison with the age of key fly fossils.

A rich source of phylogenetic data is available from developmental biology, and dipteran phylogeny can be used as a framework for developmental gene evolution (Schmidt-Ott 2002). Stauber et al. (2002) found that the *Hox3* gene has been duplicated in the Cyclorrhapha, producing *zerknüllt* and *bicoid* copies. This gene duplication can be considered both as a putative synapomorphy of the Cyclorrhapha and as the reduction of the embryonic amnion and serosa into an amnioserosa (Schmidt-Ott 2000). Detailed comparative studies of external morphology, such as mid-leg articulations (Frantsevich and Gladun 2002), and internal anatomy, such as ovary development and oogenesis (Kubrakiewicz et al. 1998; Meier et al. 1999), and musculature (Michelsen 1996; Ovtshinnikova and Yeates 1998; Palmer et al. 2000) are being explored for phylogenetic data. However, results to date have been mixed, with only low levels of congruence with data from other morphological character systems, suggesting widespread convergence and reversal. Notions of homology and phylogenetic coding need to be explored and reevaluated for these new characters (e.g., Rozowski 2002), because they have not been subject to the decades of scrutiny applied to more traditional character systems. The evolution of the ventral nerve cord (VNC) and neural organization of the visual system in Brachycera shows a greater level of concordance with phylogenetic relationships than do other internal anatomical systems. A number of authors have now studied the architecture of the VNC of adult (Yeates et al. 2002) and larval (Melzer et al. 1995) flies in a phylogenetic context. Adult VNC architecture shows a high degree of conservatism, with a general trend toward fusion of segmental ganglia through dipteran evolution. The fusion of all ganglia into a synganglion, as is found in *Drosophila,* has evolved independently three or four times. Larval and adult VNC architecture rarely match and it is likely that they are decoupled in ontogeny. A parsimony analysis of 32 neuroanatomical characters of the fly visual system from representatives of 23 families produced results that were in general agreement with the accepted phylogenetic framework (Buschbeck 2000).

TABLE 2.1. Divergence Time Estimates for Fly Lineage Based on Macromolecular Sequences and the Fossil Record

Taxon	Estimated Divergence Time from Molecular Data (Mya)	Earliest Known Fossil (Mya)
Basal Diptera	247.7–282.8[a]	233[b]
Culicomorpha	—	165[c]
Brachycera	216 (194, 241)[d]	187 (208)[c]
Muscomorpha	216 (194, 241)[d]	144[c]
Stratiomyomorpha	204 (176, 232)[d]	187[c]
Xylophagomorpha	192 (160, 224)[d]	187[c]
Tabanomorpha	192 (164, 227)[d]	187[c]
Heterodactyla	202 (179, 226)[d]	144[c]
Eremoneura	166 (143, 192)[d]	150[e]
Empidoidea	143 (122, 172)[d]	130[f]
Cyclorrhapha	143 (122, 172)[d]	130[f]
Schizophora	86 (71, 113)[d]	80[f]
Dros/Musca	48 (29, 76)[d]	70[f]
Drosophilidae	99[g]	30[c]
Drosophila	40–60[h]	30[c]
D. subgenera	62.9[j]	30[c]
D. melanogaster	5.4[j]	—

[a] Gaunt and Miles 2002.
[b] Krzeminski and Evenhuis 2002.
[c] Evenhis 1994.
[d] Posterior means for Bayesian divergence time estimates followed in parentheses by 95% credibility intervals from Wiegmann et al. 2003.
[e] Nagatomi and Yang 1998.
[f] Grimaldi 1999; Grimaldi and Cumming 1999.
[g] Beverley and Wilson 1984.
[h] Powell and DeSalle 1995.
[j] Tamura et al. 2004.

Tipulidae were relatively derived in the Lower Diptera, in accordance with some recent results obtained from external morphology (Fig. 2.1), and Brachycera was monophyletic on the neuroanatomical tree; however, some Syrphidae were placed well within the Calyptratae.

Analytical trends in the use of sequence data for dipteran phylogenetics include employing a greater range and variety of single-copy nuclear genes (e.g., Moulton 2000 on Simuliidae; Moulton and Wiegmann 2004 on Eremoneura), and in some cases, these multiple-gene regions are analyzed simultaneously with morphological data (e.g., Meier and Baker 2002 on Diopsidae). The field is also moving toward a much more structured and detailed taxon sampling strategy (compare Vossbrinck and Friedman 1989 with Schawaroch 2002). Dipterists can also benefit from the completion of the *Drosophila melanogaster* and *Anopheles gambiae* genomes (see Ashburner, Chapter 4) with the identification of both mitochondrial and especially nuclear genes and primers useful for phylogenetics (Bonacum et al. 2001), or to survey widely within the genome by shotgun sequencing using universal priming sites (Zilversmit et al. 2001). Methods of analysis are also becoming more sophisticated, with most studies now being conducted in a quantitative framework based on an explicit data matrix and analysis methods, and support for nodes on trees being estimated using contemporary repeatable measures, such as bootstrap values and Bremer support. Diptera systematists are much better able to decide the degree to which a dataset can distinguish between different hypotheses of relationship in a statistical framework (e.g., Collins and Wiegmann 2002a; Meier and Baker 2002; Moulton and Wiegmann 2004; Laamanen et al., 2005).

Conclusions

We have summarized recent phylogenetic results in the Diptera using a supertree approach (Fig. 2.1). This method synthesizes information in different trees, and there is no direct connection to the original data; hence, we are unable to estimate levels of support in the primary data for nodes in the supertree. However, we do know that levels of support for different nodes vary widely; for example, the Brachycera is an extremely well-established node supported in all relevant input trees, but many of the acalyptrate groupings are only supported by a few morphological characters in any input tree. We predict that the arrangement and composition of superfamilies will be most fluid in two of the three paraphyletic grade groups identified earlier: the psychodomorphs + tipulomorphs and the lower cyclorrhaphans. The third paraphyletic grade group, the Lower Brachycera, have been well studied in recent years, and the composition of the superfamilies is reasonably stable now, with the exception of the Asiloidea. The relationships among the Lower Brachyceran infraorders Xylophagomorpha, Tabanomorpha, and Stratiomyomorpha are likely to remain unstable until appropriate character systems can be found and analyzed.

ACKNOWLEDGMENTS

We are pleased to acknowledge Anne Hastings and Jacquie Recsei (CSIRO Entomology) for assistance with the assembly of Fig. 2.1. Andy Beckenbach, Art Borkent, Rudolf Meier, and Kevin Moulton all kindly provided copies of their figures for reproduction here. The work for this chapter was supported by grants from the National Science Foundation (DEB-9521925, DEB-9977958, DEB-0089745, and EF-0334948).

REFERENCES

Aluja, M., and A. L. Norrbom, eds. 2000. *Fruit Flies (Tephritidae): Phylogeny and Evolution of Behavior.* Boca Raton, Fla.: CRC Press.

Amorim, D. 2000. A new phylogeny and phylogenetic classification for the Canthyloscelidae (Diptera: Psychodomorpha). *Can. J. Zool.* 78:1067–1077.

Anthony, T. G., R. E. Harbach, and I. J. Kitching. 1999. Phylogeny of the pyretophorus series of *Anopheles* subgenus *Cellia* (Diptera : Culicidae). *Syst. Entomol.* 24:193–205.

Arens, W. 1995. Structure and evolution of spiracular gills in pupae of net-winged midges (Nematocera: Blephariceridae): Part I. Paulianina and Edwardsina (subfamily Edwardsininae). *Can. J. Zool.* 73: 2318–2342.

Baker, R., G. S. Wilkinson, and R. DeSalle. 2001. Phylogenetic utility of different types of molecular data used to infer evolutionary relationships among stalk-eyed flies (Diopsidae). *Syst. Biol.* 50:87–105.

Baum, B. R. 1992. Combining trees as a way of combining data sets for phylogenetic inference, and the desirability of combining gene trees. *Taxon* 41:3–10.

Beckenbach, A. T., and A. Borkent. 2003. Molecular analysis of the biting midges (Diptera: Ceratopogonidae), based on mitochondrial cytochrome oxidase subunit 2. *Mol. Phylogen. Evol.* 27:21–35.

Beebe, N., R. Cooper, D. Morrison, and J. Ellis. 2000. A phylogenetic study of the *Anopheles punctulatus* group of malaria vectors comparing rDNA sequence alignments derived from the mitochondrial and nuclear small ribosomal subunits. *Mol. Phylogen. Evol.* 17:430–436.

Bernasconi, M. V., J. Pawlowski, C. Valsangiacomo, J. Piffaretti, and P. Ward. 2000. Phylogeny of the Scathophagidae (Diptera, Calyptratae) based on mitochondrial DNA sequences. *Mol. Phylogen. Evol.* 16:308–315.

Besansky, N. J., and G. T. Fahey. 1997. Utility of the white gene in estimating phylogenetic relationships among mosquitoes (Diptera: Culicidae). *Mol. Biol. Evol.* 14:442–454.

Beverly, S. M., and A. C. Wilson. 1984. Molecular evolution in *Drosophila* and the higher Diptera II. A time scale for fly evolution. *J. Mol. Evol.* 21:1–13.

Bininda-Edmonds, O. R. P., J. L. Gittleman, and M. A. Steel. 2002. The (super) tree of life: Procedures, problems, and prospects. *Annu. Rev. Ecol. Syst.* 33:265–289.

Blagoderov, V. A., E. D. Lukashevich, and M. B. Mostovski. 2002. Order Diptera Linne, 1758. The true flies (=Muscida Laicharting, 1781). In A. P. Rasnitsyn and D. Quick, eds., *History of Insects,* pp. 227–240. Dordrecht: Kluwer.

Blaschke-Berthold, U. 1994. Anatomie und Phylogenie der Bibionomorpha (Insecta, Diptera). *Bonn. Zool. Monogr.* 34:1–206.

Bolshakov, V. N., P. Topalis, C. Blass, E. Kokoza, A. della Torre, F. C. Kafatos, and C. Louis. 2002. A comparative genomic analysis of two distant diptera, the fruit fly, *Drosophila melanogaster,* and the malaria mosquito, *Anopheles gambiae. Genome Res.* 12:57–66.

Bonacum, J., R. DeSalle, R. T. O'Grady, D. Olivera, E. H. Wintermute, and M. Zilversmit. 2001. New nuclear and mitochondrial primers for systematics and comparative genomics in Drosophilidae. Drosophila *Inf. Serv.* 84:201–204.

Borkent, A. 2000. Biting midges (Ceratopogonidae: Diptera) from the Lower Cretaceous Lebanese amber with a discussion of the diversity and patterns found in other ambers. In D. Grimaldi, ed., *Studies in Fossils in Amber, with Particular Reference to the Cretaceous of New Jersey,* pp. 355–451. Leiden: Backhuys.

Brake, I., 2000. Phylogenetic systematics of the Milichiidae (Diptera, Schizophora). *Insect Sys. Evol.* 57: 3–120.

Buschbeck, E. 2000. Neurobiological constraints and fly systematics: How different types of neural characters can contribute to a higher level dipteran phylogeny. *Evolution* 54:888–898.

Bybee, S. M., S. D. Taylor, C. R. Nelson, and M. F. Whiting. 2004. A phylogeny of robber flies (Diptera: Asilidae) at the subfamilial level: Molecular evidence. *Mol. Phylogenet. Evol.* 30:789–797.

Carvalho, C. 1989. Classificação de Muscidae (Diptera): uma proposta através da análise cladística. *Rev. Bras. Zool.* 6:627–648.

Chandler, P. 1981. *Nemedina alamirabilis* sp. n., a new genus and species of Diptera Eremoneura, of uncertain affinities, from Hungary. *Acta Zool. Acad. Sci. Hung.* 27:103–113.

Chandler, P. 2002. Heterotricha Loew and allied genera (Diptera: Sciaroidea): Offshoots of the stem group of Mycetophilidae and/or Sciaridae. *Ann. Soc. Entomol. France* 38:101–144.

Chvála, M. 1983. The Empidoidea (Diptera) of Fennoscandia and Denmark. II. General part. The families Hybotidae, Atelestidae and Microphoridae. *Fauna Entomol. Scand.* 12:1–297.

Colless, D. 1994. A new family of muscoid Diptera from Australasia, with sixteen new species in four new genera (Diptera: Axiniidae). *Invertebr. Taxon.* 8:471–534.

Colless, D., and D. McAlpine. 1991. Diptera (Flies). In CSIRO, *The Insects of Australia,* 2nd ed., pp. 717–786. Carlton: Melbourne University Press.

Collins, K. P., and B. M. Wiegmann. 2002a. Phylogenetic relationships and placement of the Empidoidea (Diptera: Barachycera) based on 28s rDNA and EF-1α sequences. *Insect Syst. Evol.* 33:421–444.

Collins, K. P., and B. M. Wiegmann. 2002b. Phylogenetic relationships of the Lower Cyclorrhapha (Diptera: Brachycera) based on 28S rDNA sequences. *Insect Syst. Evol.* 33:445–456.

Couri, M., and A. Pont. 2000. Cladistic analysis of Coenosiini (Diptera : Muscidae :Coenosiinae). *Sys. Entomol.* 25:373–392.

Courtney, G. 1990a. Cuticular morphology of larval mountain midges (Diptera: Deuterophlebiidae): Implications for the phylogenetic relationships of Nematocera. *Can. J. Zool.* 68:556–578.

Courtney, G. 1990b. Revision of Nearctic mountain midges (Diptera: Deuterophlebiidae). *J. Nat. Hist.* 24:81–118.

Courtney, G. 1991. Phylogenetic analysis of the Blephariceromorpha, with special reference to mountain midges (Diptera: Deuterophlebiidae). *Syst. Entomol.* 16:137–172.

Cumming, J., B. Sinclair, and D. Wood. 1995. Homology and phylogenetic implications of male genitalia in Diptera-Eremoneura. *Entomol. Scand.* 26:120–151.

Dahl, C. 1980. Comparison of postembryonic organization of the genital segments in Trichoceridae, Tipulidae, and Anisopodidae (Diptera, Nematocera). *Zool. Scr.* 9:165–185.

Daugeron, C. 1997. Evolution of feeding and mating behaviors in the Empidoidea (Diptera: Eremoneura). *Mem. Mus. Natl. Hist. Nat.* 173:163–182.

Downes, W., Jr., and G. Dahlem. 1987. Keys to the evolution of Diptera: Role of Homoptera. *Environ. Entomol.* 16:847–854.

Evenhuis, N. 1994. *Catalogue of the Fossil Flies of the World (Insecta: Diptera).* Leiden: Backhuys; updates at hbs.bishopmuseum.org/fossilcat.

Frantsevich, L., and D. Gladun. 2002. Evolution of the middle leg basal articulations in flies (Diptera). *Acta Zool.* 83:125–147.

Fraser, N., D. Grimaldi, P. Olsen, and B. Axsmith. 1996. A Triassic Lagerstätte from eastern North America. *Nature* 380:615–619.

Friedrich, M., and D. Tautz. 1997a. An episodic change of rDNA nucleotide substitution rate has occurred during the emergence of the insect order Diptera. *Mol. Biol. Evol.* 14:644–653.

Friedrich, M., and D. Tautz. 1997b. Evolution and phylogeny of the Diptera: A molecular phylogenetic analysis using 28S rDNA. *Syst. Biol.* 46:674–698.

Gaimari, S. D., and M. E. Irwin. 2000. Phylogeny, classification and biogeography of the Cycloteline Therevinae (Insecta: Diptera: Therevidae). *Zool. J. Linn. Soc.* 129:129–240.

Gaunt, M. W., and M. A. Miles. 2002. An insect molecular clock dates the origin of insects and accords with paleontological and biogeographic landmarks. *Mol. Biol. Evol.* 19:748–761.

Gleeson, D., R. Howitt, and R. Newcomb. 2000. The phylogenetic position of the New Zealand batfly, *Mystacinobia zelandica* (Mystacinobiidae; Oestroidea) inferred from mitochondrial 16S ribosomal DNA sequence data. *J. R. Soc. New Zealand* 30:155–168.

Griffiths, G. 1972. *The Phylogenetic Classification of Diptera Cyclorrhapha, with Special Reference to the Structure of the Male Postabdomen.* The Hague: Junk.

Griffiths, G. 1982. On the systematic position of Mystacinobia (Diptera: Calliphoridae). *Mem. Entomol. Soc. Wash.* 10:70–77.

Griffiths, G. 1984. Discussion on "Male genitalia in the classification of Chloropidae," XVII International Congress of Entomology, Hamburg.

Griffiths, G. 1990. Book review: *Manual of Nearctic Diptera,* vol. 3. *Quaest. Entomol.* 26:117–130.

Griffiths, G. 1994. Relationships among the major subgroups of Brachycera (Diptera): A critical review. *Can. Entomol.* 126:861–880.

Griffiths, G. 1996. Review of papers on the male genitalia of Diptera by D. M. Wood and associates. *Studia Dipterol.* 3:107–123.

Grimaldi, D. 1990. A phylogenetic, revised classification of genera in the Drosophilidae (Diptera). *Bull. Amer. Mus. Nat. Hist.* 197:1–139.

Grimaldi, D. 1999. The co-radiations of pollinating insects and angiosperms in the Cretaceous. *Ann. Missouri Bot. Gard.* 86:373–406.

Grimaldi, D., and J. Cumming. 1999. Brachyceran Diptera in Cretaceous ambers and Mesozoic diversification of the Eremoneura. *Bull. Am. Mus. Nat. Hist.* 239:1–121.

Groombridge, B. 1992. *Global Biodiversity: Status of the Earth's Living Resources.* London: Chapman and Hall.

Han, H.-Y., and B. McPheron. 1997. Molecular phylogenetic study of Tephritidae (Insecta: Diptera) using partial sequences of the mitochondrial 16S ribosomal DNA. *Mol. Phylogenet. Evol.* 7:17–32.

Harbach, R., and I. Kitching. 1998. Phylogeny and classification of the Culicidae (Diptera). *Syst. Entomol.* 23:327–370.

Hennig, W. 1948–1952. *Die Larvenformen der Dipteren: eine Übersicht über die bisher bekannten Jungendstadien der zweiflügeligen Insekten.* Berlin: Akademie-Verlag.

Hennig, W. 1950. *Grundzüge einer Theorie der phylogenetischen Systematik.* Berlin: Deutscher Zentralverlag.

Hennig, W. 1958. Die Familien der Diptera Schizophora und ihre phylogenetischen Verwandtschaftsbeziehungen. *Beitr. Entomol.* 8:505–688.

Hennig, W. 1966. *Phylogenetic Systematics.* Urbana: University of Illinois Press.

Hennig, W. 1968. Kritische Bemerkungen über den Bau der *Flügelwurzel* bei den Dipteren und die Frage nach der Monophylie der Nematocera. *Stuttg. Beitr. Naturkd.* 195:1–23.

Hennig, W. 1971. Insektfossilien aus der unteren Kreide. III. Empidiformia ("Microphorinae") aus der unteren Kreide und aus dem Baltischen Bernstein; ein Vertreter der Cyclorrapha aus der unteren Kreide. *Stuttg. Beitr. Naturkd.* 232:1–28.

Hennig, W. 1973. Diptera (Zweiflügler). *Handb. Zool. (Berlin)* 4:1–200.

Hennig, W. 1976. Das Hypopygium von Lonchoptera lutea Panzer und die phylogenetischen Verwandtschaftsbeziehungen der Cyclorrhapha (Diptera). *Stuttg. Beitr. Naturkd.* 283:1–63.

Hennig, W. 1981. *Insect Phylogeny.* New York: Wiley.

Hibbs, P. 2002. Phylogeny of the Rhagionidae sensu lato inferred from molecular and morphological evidence. In *Abstracts of the 5th International Congress of Dipterology,* p. 101, Brisbane, September 2002.

Hutson, A. M. 1977. A revision of the families Synneuridae and Canthyloscelidae. *Bull. Br. Mus. (Nat. Hist.) Entomol.* 35:67–100.

Irwin, M., and L. Lyneborg. 1981. The genera of Nearctic Therevidae. *Bull. Ill. Nat. Hist. Sur.* 32:193–277.

Irwin, M. E., and B. M. Wiegmann. 2001. A review of the southern African genus *Tongamya* (Diptera: Asiloidea: Mydidae: Megascelinae), with a molecular assessment of the phylogenetic placement of *Tongamya* and the Megascelinae. *Afr. Invert.* 42:225–253.

Jaschhof, M., and R. K. Didham. 2002. New Zealand and implications for the phylogeny of the Sciaroidea (Diptera: Bibionomorpha). *Studia Dipt. Supp.* 11:1–60.

Katzourakis, A., A. Purvis, S. Azmeh, G. Rotheray, and F. Gilbert. 2001. Macroevolution of hoverflies (Diptera: Syrphidae): The effect of using higher-level taxa in studies of biodiversity, and correlates of species richness. *J. Evol. Biol.* 14:219–227.

King, D. 1989. Phylogenetic diversity of cellular organization in the cardia of muscoid flies (Diptera: Schizophora). *J. Morphol.* 202:435–455.

King, D. 1991. The origin of an organ: Phylogenetic analysis of evolutionary innovation in the digestive tract of flies (Insecta: Diptera). *Evolution* 45:568–588.

Knutson, L., and J. Vala. 2002. An evolutionary scenario of Sciomyzidae and Phaeomyiidae (Diptera). *Ann. Soc. Entomol. France* 38:145–162.

Korneyev, V. A. 2000. Phylogenetic relationships among higher groups of Tephritidae. In M. Aluja and A. Norrbom, eds., *Fruit Fies (Tephritidae): Phylogeny and Evolution of Behavior,* pp. 73–112. Boca Raton, Fla.: CRC Press.

Kotrba, M. 1995. The internal female genital organs of Chaetodiopsis and Diasemopsis (Diptera: Diposidae) and their systematic relevance. *Ann. Natal Mus.* 36:147–159.

Kovalev, V. 1979. The main aspects in the evolution of Diptera Brachycera in the Mesozoic Era. In O. A. Skarlato, K. V. Skufjin, E. P. Narchuk, O. P. Negrobov, and M. N. Kandybina, eds., *Ecological and Morphological Principles of Dipteran Systematics,* pp. 35–37. Leningrad: Akademia Nauk SSSR, Zoologicheskii Institut.

Kristensen, N. 1991. Phylogeny of extant hexapods. In CSIRO, *The Insects of Australia,* 2nd ed., pp. 125–140. Carlton: Melbourne University Press.

Krivosheina, N. 1988. Approaches to solutions of questions of classification of the Diptera. *Entomol. Obozr.* 67:378–390.

Krzeminski, W. 1992a. Triassic and Lower Jurassic stage of Diptera evolution. *Mitt. Schwiez. Entomol. Ges.* 65:39–59.

Krzeminski, W. 1992b. The oldest Polyneura (Diptera) and their importance to the phylogeny of the group. *Acta Zool. Cracov.* 35:45–52.

Krzeminski, W., and N. L. Evenhuis. 2000. Review of Diptera paleontological records. In L. Papp and B. Darvas, eds., *Contributions to a Manual of Palaearctic Diptera,* vol. 1, pp. 535–564. Budapest: Science Herald.

Krzeminski, W., and E. Krzeminska. 1996. The origin of flies. *Kosmos (Warsaw)* 45:701–714.

Krzywinski, J., R. Wilkerson, and N. J. Besansky. 2001a. Evolution of mitochondrial and ribosomal gene sequences in Anophelinae (Diptera: Culicidae): Implications for phylogeny reconstruction. *Mol. Phylogen. Evol.* 18:479–487.

Krzywinski, J., R. Wilkerson, and N. J. Besansky. 2001b. Towards understanding Anophelinae (Diptera, Culicidae) phylogeny: Insights from nuclear single-copy genes and the weight of evidence. *Syst. Biol.* 50:540–556.

Kubrakiewicz, J., S. M. Bilinski, and M. Mazurkiewicz. 1998. Diptera—Ovary structure and oogenesis in midges and flies. *Folia Histochem. Cytobiol.* 36:197–203.

Kukalova-Peck, J. 1991. Fossil history and the evolution of hexapod structures. In CSIRO, *The Insects of Australia,* pp. 141–179. Carlton: Melbourne University Press.

Laamanen, T. R., R. Meier, M. A. Miller, A. Hille, and B. M. Wiegmann. 2005. Phylogenetic analysis of *Themira* (Sepsidae: Diptera): Sensitivity analysis, alignment, and indel treatment in a multigene study. *Cladistics,* in press.

Labandeira, C. 1994. A compendium of fossil insect families. *Milwaukee Public Mus. Contrib. Biol. Geol.* 88:71.

Labandeira, C. 1997. Insect mouthparts: Ascertaining the paleobiology of insect feeding strategies. *Annu. Rev. Ecol. Syst.* 28:153–193.

Labandeira, C. 1998. How old is the fly and the flower? *Science* 280:57–59.

Lambkin, C. L., and D. K. Yeates. 2003. Combined morphological and molecular analysis of the Australian Anthracinae (Diptera: Bombyliidae). *Invert. Syst.* 17:735–891.

Marinoni, L., and W. N. Mathis. 2000. A cladistic analysis of Sciomyzidae Fallen (Diptera). *Proc. Biol. Soc. Wash.* 113:162–209.

Marois, R., and I. Meinertzhagen. 1990. The visual system of flies: Analysis of the number, specificity, plasticity, and phylogeny of identified synapses. *NATO Adv. Sci. Inst. Ser.,* Ser. A (Life Sciences) 192:127–139.

McAlpine, D. K. 1985. The Australian genera of Heleomyzidae (Diptera: Schizophora) and a reclassification of the family into tribes. *Rec. Aust. Mus.* 36:203–251.

McAlpine, D. K. 1996. Relationships and classification of the Pseudopomyzidae (Diptera: Nerioidea). *Proc. Linn. Soc. N.S.W.* 116:223–232.

McAlpine, D. K. 1998. Review of the Australian stilt flies (Diptera : Micropezidae) with a phylogenetic analysis of the family. *Invert. Taxon.* 12:55–134.

McAlpine, J. 1989. Phylogeny and classification of the Muscomorpha. In J. McAlpine and D. Wood, eds., *Manual of Nearctic Diptera,* vol. 3, pp. 1397–1518. Ottawa: Agriculture Canada Research Branch.

McAlpine, J., and D. Wood, eds. 1989. *Manual of Nearctic Diptera,* vol. 3. Ottawa: Agriculture Canada Research Branch.

Meier, R. 1995a. Cladistic analysis of the Sepsidae (Cyclorrhapha: Diptera) based on a comparative scanning electron microscopic study of the larvae. *Syst. Entomol.* 20:99–128.

Meier, R. 1995b. A comparative SEM study of the eggs of the Sepsidae (Diptera) with a cladistic analysis based on egg, larval and adult characters. *Entomol. Scand.* 26:425–438.

Meier, R. 1996. Larval morphology of the Sepsidae (Diptera: Sciomyzoidea), with a cladistic analysis using adult and larval characters. *Bull. Am. Mus. Nat. Hist.* 228:1–147.

Meier, R., and R. Baker. 2002. A cladistic analysis of Diopsidae (Diptera) based on morphological and DNA sequence data. *Insect Syst. Evol.* 33:325–336.

Meier, R., and S. Hilger. 2000. On the egg morphology and phylogenetic relationships of Diopsidae (Diptera : Schizophora). *J. Zool. Syst. Evol. Res.* 83:1–36.

Meier, R., and B. M. Wiegmann. 2002. A phylogenetic analysis of Coelpidae (Diptera) based on morphological and DNA sequence data. *Mol. Phylogenet. Evol.* 25:393–407.

Meier, R., M. Kotrba, and P. Ferrar. 1999. Ovoviviparity and viviparity in the Diptera. *Biol. Rev. Cam. Phil. Soc.* 74:199–258.

Meinertzhagen, I. 1989. Fly photoreceptor synapses: Their development, evolution, and plasticity. *J. Neurobiol.* 20:276–294.

Melzer, R., A. Panzinger, F. Reckel, and U. Smola. 1995. Central nervous system of brachyceran larvae (Insecta, Diptera). *Zool. Anz.* 234:113–123.

Michelsen, V. 1996. Neodiptera: New insights into the adult morphology and higher level phylogeny of Diptera (Insecta). *Zool. J. Linn. Soc.* 117:71–102.

Miller, B., M. Crabtree, and H. Savage. 1996. Phylogeny of fourteen *Culex* mosquito species, including the *Culex Pipiens* complex, inferred from the internal transcribed spacers of ribosomal DNA. *Insect Mol. Biol.* 5:93–107.

Mitchell, A., F. Sperling, A. H. Hickey, and D. A. Hickey. 2002. Higher-level phylogeny of mosqitoes. *Insect Syst. Evol.* 33:163–174.

Moulton, J. 2000. Molecular sequence data resolves basal divergences within Simuliidae (Diptera). *Syst. Entomol.* 25:95–113.

Moulton, J. K., and B. M. Wiegmann. 2004. Evolution and phylogenetic utility of CAD (rudimentary) among Mesozoic-aged Eremoneuran Diptera (Insecta). *Mol. Phylogenet. Evol.* 31:363–378.

Nagatomi, A. 1992. Notes on the phylogeny of various taxa of the orthorrhaphous Brachycera (Insecta: Diptera). *Zool. Sci.* 9:843–857.

Nagatomi, A. 1996. An essay on phylogeny of the orthorraphous Brachycera (Diptera). *Entomol. Mon. Mag.* 132:95–148.

Nagatomi, A., and N. Liu. 1995. Spermatheca and female terminalia of Pantophthalmidae and Xylophagidae s. lat. (Diptera). *Ann. Entomol. Soc. Am.* 88:603–626.

Nagatomi, A., and D. Yang. 1998. A review of extinct Mesozoic genera and families of Brachycera (Insecta, Diptera, Orthorrhapha). *Entomol. Mon. Mag.* 134:95–192.

Nagatomi, A., T. Saigusa, H. Nagatomi, and L. Lyneborg. 1991. Apsilocephalidae, a new family of the orthorrhaphous Brachycera (Insecta, Diptera). *Zool. Sci.* 8:579–591.

Nirmala, X., V. Hypsa, and M. Zurovec. 2001. Molecular phylogeny of Calyptratae (Diptera: Brachycera): The evolution of 18S and 16S ribosomal rDNAs in higher dipterans and their use in phylogenetic inference. *Insect Mol. Biol.* 10:475–485.

Oosterbroek, P., and G. Courtney. 1995. Phylogeny of the Nematocerous families of Diptera (Insecta). *Zool. J. Linn. Soc.* 115:267–311.

Oosterbroek, P., and B. Theowald. 1991. Phylogeny of the Tipuloidea based on characters of larvae and pupae (Diptera, Nematocera). *Tijdschr. Entomol.* 134:211–267.

Ovtshinnikova, O. G., and D. K. Yeates. 1998. Male genital musculature of Therevidae and Scenopinidae (Diptera : Asiloidea): Structure, homology and phylogenetic implications. *Aust. J. Entomol.* 37:22–33.

Palmer, C. M., and D. K Yeates. 2000. Phylogenetic importance of immature stages: Solving the riddle of Exerotonevra Macquart (Diptera: Xylophagidae). *Ann. Entomol. Soc. Am.* 93:15–27.

Palmer, C. M., O. G. Ovtshinnikova, and D. K. Yeates. 2002. Male genitalia of Exeretonevra Macquart (Diptera: Xylophagidae): Structure, function and phylogenetics implications. *Aust. J. Entomol.* 39: 270–274.

Pape, T. 1992. Phylogeny of the Tachinidae family-group (Diptera: Calyptratae). *Tijdschr. Entomol.* 135: 43–86.

Pape, T. 2001. Phylogeny of Oestridae (Insecta: Diptera). *Syst. Entomol.* 26:133–171.

Pawlowski, J., R. Szadziewski, D. Kmieciak, J. Fahrni, and G. Bittar. 1996. Phylogeny of the infraorder Culicomorpha (Diptera: Nematocera) based on 28S RNA gene sequences. *Syst. Entomol.* 21:167–178.

Powell, J. R. 1997. *Progress and Prospects in Evolutionary Biology: The Drosophila Model.* New York: Oxford University Press.

Ragan, M. A. 1992. Phylogenetic inference based on matrix representation of trees. *Mol. Phylogenet. Evol.* 1:53–58.

Remsen, J., and R. DeSalle. 1998. Character congruence of multiple data partitions and the origin of the Hawaiian Drosophilidae. *Mol. Phylogenet. Evol.* 9:225–235.

Remsen, J., and P. O'Grady. 2002. Phylogeny of Drosophilinae (Diptera: Drosphilidae) with comments on combined analysis and character support. *Mol. Phylogenet. Evol.* 24:264.

Ren, D. 1998. Flower associated brachyceran flies as fossil evidence for Jurassic angiosperm origins. *Science* 280:85–88.

Rognes, K. 1986. The systematic position of the genus Helicobosca Bezzi with a discussion of the monophyly of the calypterate families Calliphoridae, Rhinophoridae, Sarcophagidae and Tachinidae (Diptera). *Entomol. Scand.* 17:75–92.

Rognes, K. 1991. *Blowflies (Diptera, Calliphoridae) of Fennoscandia and Denmark.* Leiden: E. J. Brill and Scandinavian Science Press.

Rognes, K. 1997. The Calliphoridae (blowflies) (Diptera: Oestroidea) are not a monophyletic group. *Clad. Int. J. Willi Hennig Soc.* 13:27–66.

Rohacek, J. 1998. Taxonomic limits, phylogeny and higher classification of Anthomyzidae (Diptera), with special regard to fossil record. *Eur. J. Entomol.* 95:141–177.

Rozowski, M. 2002. Establishing character correspondence for sensory organ traits in flies: Sensory organ development provides insight for reconstructing character evolution. *Mol. Phylogenet. Evol.* 24: 400–411.

Saether, O. 2000a. Phylogeny of Culicomorpha (Diptera). *Syst. Entomol.* 25:223–234.

Saether, O. 2000b. Phylogeny of the subfamilies of Chironomidae (Diptera). *Syst. Entomol.* 25:393–403.

Sallum, M., T. Schultz, and R. Wilkerson. 2000. Phylogeny of Anophelinae (Diptera: Culicidae) based on morphological characters. *Ann. Entomol. Soc. Am.* 93:745–775.

Sanderson, M. J., A. Purvis, and C. Henze. 1998. Phylogenetic supertrees: Assembling the trees of life. *Trends Ecol. Evol.* 13:105–109.

Schawaroch, V. 2002. Phylogeny of a paradigm lineage: The *Drosophila melanogaster* species group (Diptera: Drosophilidae). *Biol. J. Linn. Soc.* 76:21–37.

Schmidt-Ott, U. 2000. The amnioserosa is an apomorphic character of Cyclorrhaphan flies. *Dev. Genes Evol.* 210:373–376.

Schmidt-Ott, U. 2002. Evolution of *Hox3* in flies. *Am. Zool.* 41:1579.

Severson, D. W., B. deBruyn, D. D. Lovin, S. E. Brown, D. L. Knudson, and I. Morlais. 2004. Comparative genome analysis of the yellow fever mosquito *Aedes aegypti* with *Drosophila melanogaster* and the malaria vector mosquito *Anopheles gambiae. J. Heredity* 95:103–113.

Shatalkin, A. 1994. Palaearctic species of Pseudopomyzidae (Diptera). *Russian Entomol. J.* 3:129–145.

Shewell, G. E. 1987. Lauxaniidae. In J. McAlpine, B. Peterson, G. Shewell, et al., eds., *Manual of Nearctic Diptera,* vol. 2, pp. 951–964. Ottawa: Agriculture Canada Research Branch.

Sinclair, B. 1992. A phylogenetic interpretation of the Brachycera (Diptera) based on the larval mandible and associated mouthpart structures. *Syst. Entomol.* 17:233–252.

Sinclair, B., J. Cumming, and D. Wood. 1994. Homology and phylogenetic implications of male genitalia in Diptera–Lower Brachycera. *Entomol. Scand.* 24:407–432.

Skevington, J., and D. K. Yeates. 2000. Phylogeny of the Syrphoidea (Diptera) inferred from mtDNA sequences and morphology with particular reference to classification of the Pipunculidae (Diptera). *Mol. Phylogen. Evol.* 16:212–224.

Skidmore, P. 1985. *The Biology of the Muscidae of the World.* Dordrecht: Junk.

Smith, P. T., B. A. McPheron, and S. Kambhampati. 2002. Phylogenetic analysis of mitochondrial DNA supports the monophyly of Dacini fruit flies (Diptera: Tephritidae). *Ann. Entomol. Soc. America* 95: 658–664.

Smith, P. T., S. Kambhampati, and K. A. Armstrong. 2003. Phylogenetic relationships among *Bactrocera* species (Diptera: Tephritidae) inferred from mitochondrial DNA sequences. *Mol. Phylogenet. Evol.* 26:8–17.

Smith, S., P. Fuerst, and K. Mechklenburg. 1996. Mitochondrial DNA sequence of cytochrome oxidase II from *Calliphora erythrocephala:* Evolution of blowflies (Diptera: Calliphoridae). *Ann. Entomol. Soc. Am.* 89:28–36.

Soli, G. 1997. The adult morphology of Mycetophilidae (S. Str.), with a tentative phylogeny of the family (Diptera, Sciaroidea). *Entomol. Scand.* suppl. 50:5–55.

Spencer, K. 1990. *Host Specialisation in World Agromyzidae (Diptera).* Dordrecht: Kluwer Academic.

Ståhls, G., and K. Nyblom. 2000. hylogenetic analysis of the genus *Cheilosia* (Diptera, Syrphidae) using mitochondrial COI sequence data. *Mol. Phylogenet. Evol.* 15:235–241.

Ståhls, G., H. Hippa, G. Rotheray, J. Muona, and F. Gilbert. 2003. Phylogeny of Syrphidae (Diptera) inferred from combined analysis of molecular and morphological characters. *Syst. Entomol.* 28:433–450.

Stauber M., A. Prell, and U. Schmidt-Ott. 2002. A single *Hox3* gene with composite bicoid and zerknüllt expression characteristics in non-Cyclorrhaphan flies. *Proc. Natl. Acad. Sci. USA* 99:274–279.

Stireman, J. O., III. 2002. Phylogenetic relationships of tachinid flies in subfamily Exoristinae (Tachinidae: Diptera) based on 28S rDNA and elongation factor-1 alpha. *Syst. Entomol.* 27:409–435.

Stoffolano, J., N. Woodley, A. Borkent, and L. Yin. 1988. Ultrastructural studies of the abdominal plaques of some Diptera. *Ann. Entomol. Soc. Am.* 81:503–510.

Stone, A., C. Sabrosky, W. Wirth, R. Foote, and J. Coulson. 1965. *A Catalog of the Diptera of America North of Mexico.* Washington, D.C.: U.S. Department of Agriculture.

Stuckenberg, B. R. 1999. Antennal evolution in the Brachycera (Diptera), with reassessment of terminology relating to the flagellum. *Studia Dipt.* 6:33–48.

Stuckenberg, B. R. 2001. Pruning the tree: A critical review of classifications of the Homeodactyla (Diptera, Brachycera), with new perspectives and an alternative classification. *Studia Dipt.* 8:3–42.

Swofford, D. L. 2002. *PAUP*: Phylogenetic analysis using parsimony (*and other methods),* ver. 4.10B10. Sunderland, Mass.: Sinauer.

Tamura, K., S. Subramanian, and S. Kumar. 2004. Temporal patterns of fruit fly (*Drosophila*) evolution revealed by mutation clocks. *Mol. Biol. Evol.* 21:36–44.

Thompson, F. C. 2004. *Biosystematic Database of World Diptera.* www.diptera.org.

Thorne, J. L., H. Kishino, and I. S. Painter. 1998. Estimating the rate of evolution of the rate of molecular evolution. *Mol. Biol. Evol.* 15:1647–1657.

Vossbrinck, C., and S. Friedman. 1989. A 28S ribosomal RNA phylogeny of certain cyclorrhaphous Diptera based upon a hypervariable region. *Syst. Entomol.* 14:417–431.

Wada, S. 1991. Morphologische Indizien für das unmittelbare Schwestergruppenverhältnis der Schizophora mit den Syrphoidea ("Aschiza") in der phylogenetischen Systematik der Cyclorrhapha (Diptera: Brachycera). *J. Nat. Hist.* 25:1531–1570.

Wiegmann, B. M., C. Mitter, and F. Thompson. 1993. Evolutionary origin of the Cyclorrhapha (Diptera): Tests of alternative morphological hypotheses. *Cladistics* 9:41–81.

Wiegmann, B. M., S. C. Tsaur, D. W. Webb, and D. K. Yeates. 2000. Monophyly and the relationships of the Tabanomorpha (Diptera: Brachycera) based on 28s ribosomal gene sequences. *Ann. Entomol. Soc. Am.* 93:1031–1038.

Wiegmann, B. M., D. K. Yeates, J. L. Thorne, and H. Kishino. 2003. Time flies, a new molecular time-scale for brachyceran fly evolution without a clock. *Syst. Biol.* 52:745–756.

Wikström, N., V. Savolainen, and M. W. Chase. 2001. Evolution of the angiosperms: Calibrating the family tree. *Proc. R. Soc. Lond.* B 268:2211–2220.

Willmann R. 1989. Rediscovered: *Permotipula patricia,* the oldest known fly. *Naturwissenschaften* 76: 375–377.

Winterton, S. L., J. H. Skevington, M. E. Irwin, and D. K. Yeates. 2000. Phylogenetic Revision of *Bojeania* Irwin and Lyneborg (Diptera: Therevidae). *Syst. Entomol.* 25:1–30.

Winterton, S. L., L. L. Yang, B. M. Wiegmann, and D. K. Yeates. 2001. Phylogenetic revision of Agapophytinae subf. n. (Diptera: Therevidae) based on molecular and morphological evidence. *Syst. Entomol.* 26:173–211.

Wood, D. 1991. Homology and phylogenetic implications of male genitalia in Diptera. The ground plan. In I. Weismann, L. Országh, and A. Pont, eds., *Proceedings of the Second International Congress of Dipterology, Bratislava, Czechoslovakia,* pp. 255–284. The Hague: SPB Academic.

Wood, D., and A. Borkent. 1989. Phylogeny and classification of the Nematocera. In J. McAlpine and D. Wood, eds., *Manual of Nearctic Diptera,* vol. 3, pp. 1333–1370. Ottawa: Agriculture Canada Research Branch.

Woodley, N. E. 1989. Phylogeny and classification of the "Orthorraphous" Brachycera. In J. McAlpine and D. Wood, eds., *Manual of Nearctic Diptera,* vol. 3, pp. 1371–1395. Ottawa: Agriculture Canada Research Branch.

Woodley, N. E. 2001. A world catalog of the Stratiomyidae. *Myia* 11:473.

Wootton, R., and A. Ennos. 1989. The implications of function on the origin and homologies of the dipterous wing. *Syst. Entomol.* 14:507–520.

Yang, L., B. Wiegmann, D. K. Yeates, and M. Irwin. 2000. TI Higher-level phylogeny of the Therevidae (Diptera: Insecta) based on 28S ribosomal and elongation factor-1 alpha gene sequences: *Mol. Phylogenet. Evol.* 15:440–451.

Yeates, D. K. 1992. Towards a monophyletic Bombyliidae (Diptera): The removal of the Proratinae (Diptera: Scenopinidae). *Am. Mus. Novit.* 3051:1–30.

Yeates, D. K. 1994. Cladistics and classification of the Bombyliidae (Diptera: Asiloidea). *Bull. Am. Mus. Nat. Hist.* 219:1–191.

Yeates, D. K. 2002. Relationships of the Lower Brachycera (Diptera): A quantitative synthesis of morphological characters. *Zool. Scr.* 31:105–121.

Yeates, D. K., and M. E. Irwin. 1996. Apioceridae (Insecta: Diptera): Cladistic reappraisal and biogeography. *Zool. J. Linn. Soc.* 116:247–301.

Yeates, D. K., and B. M. Wiegmann. 1999. Congruence and controversy: Toward a higher-level classification of Diptera. *Annu. Rev. Entomol.* 44:397–428.

Yeates, D. K., D. J. Merritt, and C. H. Baker. 2002. The adult ventral nerve cord as a phylogenetic character in brachyceran Diptera. *Org. Divers. Evol.* 2:89–96.

Yeates, D. K., M. E. Irwin, and B. M. Wiegmann. 2003. Ocoidae, a new family of asiloid flies (Diptera: Barachycera: Asiloidea), based on *Ocoa Chilensis* gen. and sp. nov. from Chile, South America. *Syst. Entomol.* 28:417–431.

Zatwarnicki, T. 1996. A new reconstruction of the origin of the eremoneuran hypopygium and its implications for classification (Insecta: Diptera). *Genus* 3:103–175.

Zdobnov, E. M., C. von Mering, I. Letunic, D. Torrents, M. Suyama, R. R. Copley, G. K. Christophides, D. Thomasova, R. A. Holt, G. Mani Subramanian, H.-M. Mueller, G. Dimopoulos, J. H. Law, M. A. Wells, E. Birney, R. Charlab, A. L. Halpern, E. Kokoza, C. L. Kraft, Z. Lai, S. Lewis, C. Louis, C. Barillas-Mury, D. Nusskern, G. M. Rubin, S. L. Salzberg, G. G. Sutton, P. Topalis, R. Wides, P. Wincker, M. Yandell, F. H. Collins, J. Ribeiro, W. M. Gelbart, F. C. Kafatos, and P. Bork. 2002. Comparative genome and proteome analysis of *Anopheles gambiae* and *Drosophila melanogaster. Science* 298:149–159.

Zilversmit, M. P., R. T. O'Grady, M. Russello, and R. DeSalle. 2001. High throughput sequencing protocols for a survey of genomic characters in the family Drosophilidae. Drosophila *Info. Serv.* 84:199–201.

CHAPTER THREE

Role of Dipterology in Phylogenetic Systematics: The Insight of Willi Hennig

Rudolf Meier

The movie industry has profited handsomely from the idea of time travel, and we probably all have wished at times that we owned a time machine ourselves. In science, the next best solution to time travel is the more mundane task of reading the old scientific literature. It reflects major changes in research agenda and methods in the different fields. If we study the literature in systematic biology and evolutionary biology of the twentieth century, we notice a dramatic change in content and technique after 1966 (Hull 1989). Many of these changes were initiated from within the field of dipterology by Willi Hennig, one of the premier systematists of the twentieth century. In this chapter, we follow his career, learn about his main research interests, and realize how his contributions opened the gate to many new disciplines that are today flourishing within systematics and evolutionary biology.

Career Outline

When researching the life of important scientists, we almost always find that their success was ultimately the result of ingenuity, dedication, hard work, and living at the right time in the right place. We also notice that many discovered their life's mission early, often during their high school years. Hennig is no exception. He was an extremely dedicated and hard-working scientist and an excellent, logical thinker. His theoretical work was done at a time when evolutionary theory was finally accepted within biology, but its consequences for tree reconstruction and classification had not yet been explored. He discovered his passion for systematics and insects early in his life. His childhood years are now fairly well known due to research by Vogel and Xylander (1999) and Schmitt (2001). Hennig was born into a working class family in a small town close to Dresden (Germany). He was an excellent student in primary school and high school (except for sports) and even skipped one grade. His academic talents eventually let him to attend a boarding school for academically gifted pupils in Dresden-Klotzsche, and it was here that he met his first important mentor in biology, M. Rost, who made several important introductions. He kindled Hennig's interest in systematic entomology by encouraging him to collect insects and introduced him to W. Meise from the Museum für Tierkunde in Dresden, where Hennig would volunteer as a high school student. By the time Hennig was in the ninth grade, he already considered himself well accustomed to museum entomology (Schmitt 2001).

Meise was primarily interested in flying vertebrates, and his influence explains Hennig's early publications on reptiles (Hennig and Meise 1932; Hennig 1936c,d). One of these publications is already typical for Hennig's entire oeuvre (1936d), in that it combines actual taxonomic revisionary work with treating theoretical topics in systematics (Schmitt 2001). This interest is also evident from a high school essay on theoretical systematics that Hennig wrote at the age of 18 (Schlee 1978), which in many ways reads like a research agenda for the rest of his life (Hennig 1978): Hennig was determined to rebuild the reputation of systematics as a science by creating a solid theoretical foundation and removing all obviously subjective elements. This urge is apparent from the introductory sections of many of his publications.

Given that Hennig initially worked on reptiles, nondipterists might ask why he changed to Diptera. It appears that his early introduction to entomology by Rost and his contact with the entomologist at the Museum für Tierkunde, F. I. von Emden, created a lasting affinity for insects. Hennig's first major research project in Diptera was a taxonomic revision and morphological study of what is today known as the Micropezidae (see Anonymous 1978 for references). He went on to become a student at the University of Leipzig, where he earned a doctoral degree at the mature age of 23 (Schlee 1978), with a dissertation on the genitalia and morphology of the postabdomen of cyclorrhaphan flies (Hennig 1936e). The morphology of these structures continued to interest him throughout his entire career: 11 of his more than 160 publications are primarily dedicated to this subject. A subsequent postdoctoral fellowship led him to Berlin, the center of pre-war German science. Here he worked at the Deutsches Entomologisches Institut, where he quickly moved from a postdoctoral position to full-time employment. It was during his time in Berlin that the Second World War broke out and he was drafted into the army. After being wounded in Russia, he eventually resumed duty as an army scientist working on malaria control. It was in this function that he became a prisoner of war in 1945 while working in Italy (Schlee 1978).

Hennig only held two long-term positions during his entire life. The first was a scientist position at the Deutsches Entomologisches Institut, to which he returned after the war upon completion of a short employment as a temporary lecturer at the University of Leipzig. The institute itself had come back to Berlin after having been evacuated to Northern Germany during the war years (Schmitt 2001). Its pre-war location had been in West Berlin, but unfortunately, the American allied command had not released the original building for use by the institute (Schlee 1978). As a consequence, it moved to a location in East Berlin (Friedrichshagen), so that Hennig, who lived in West Berlin, became a daily commuter between East and West. However, his private and political convictions were incompatible with the political system of East Germany, and although he was appointed professor at the University of Potsdam (at the Brandenburgische Landeshochschule Potsdam in East Germany) and nominated to become the next director of the Deutsches Entomologisches Institut, he fled East Berlin soon after the erection of the Berlin Wall in 1961 (Schlee 1978). After two years of temporary employment, he became head of a newly created section for phylogenetic research at the Museum für Naturkunde in Stuttgart. Here he died of a heart attack at the age of 63 (Schlee 1978).

Development of "Phylogenetic Systematics"

It was during the war years in Italy that Hennig bought a 170-page notebook (Schmitt 1997). On its first pages he outlined the table of contents for a book that has since been perused countless times and is known as one of the milestones in systematics: *Grundzüge einer The-*

orie der phylogenetischen Systematik (Hennig 1950a). The handwritten manuscript has survived. It was written between spring 1945 and October 1945; that is, much of it during Hennig's time as a prisoner of war, and the pages of the notebook reveal that the text was written with few changes. Although edited in 1948 (under candle-light conditions: Schlee 1978), the book was not printed until 1950, due to a paper shortage. This "1950 book" is not only one of the key publications in systematic biology, it is probably also one of Hennig's works that has often been incorrectly cited. It is often assumed that it contains roughly the same content as its 1966 "translation." However, this is incorrect, because the translation was prepared based on an extensively revised version of the 1950 book manuscript (publication in German: Hennig 1982).

Why is the 1950 book so frequently incorrectly cited? If we were to ask systematists for a list of essential concepts in cladistics, the list would probably be headed by: (1) "phylogenetic relationship," (2) "monophyly," (3) "paraphyly," (4) "apomorphy and plesiomorphy," (5) "cladism or cladistics," (6) "parsimony," and (7) "outgroup comparison or outgroup rooting." Upon actually reading the 1950 book, we realize that only three of these concepts are mentioned (1, 2, 4), and only two are "properly" used. It is instructive to study which they are.

Hennig was the first systematist to precisely and narrowly define both of the terms "monophyly" (term 2 above) and "phylogenetic relationship" (term 1) (Nelson 1972). By providing strict definitions, he was also the first systematist to clearly spell out the goals of a strictly phylogenetic systematics; that is, the discovery of monophyletic groups and their sister group relationships. In the 1950 book, Hennig explicitly removes "paraphyletic groups" from the cover of the traditionally only loosely defined concept of "monophyly" and denotes them with the almost derogatory term "rest bodies" ("Restkörper": Hennig 1936a, 1948, 1950a). This redefinition caused considerable irritation with many systematists of the time (Mayr 1974; see Hull 1989; Farris 1990). The term "paraphyletic group" (term 3) for such "rest bodies" was only introduced well after the 1950 book was published (Hennig 1962).

But what about the remaining important concepts of cladistics? "Apomorphy and plesiomorphy" (4) were first published by Hennig in 1949 (1949a). The terms are inconsistently used in this publication and the 1950 book. In a few instances, the concepts are applied to character states. However, more often they are used for taxa. The latter is a mortal sin in phylogenetics (Hennig 1984), because no taxon is entirely apomorphic or plesiomorphic. Hennig was well aware of this phenomenon ("Spezialisationskreuzungen" in 1949a), but apparently in 1950, he did not yet realize that the only logically sound application for the terms "apomorphy" and "plesiomorphy" would be for character states. Obviously, the transition from the traditional taxon-level thinking to character-level thinking had not yet fully occurred. This is also apparent from other publications of this time, in which character discussions do not distinguish between apomorphic and plesiomorphic states (e.g., Hennig 1949b).

However, as early as 1936, Hennig (1936a) was already well on this way of finding new tree reconstruction methods. He had broken with the traditional view that overall similarity can be used for determining phylogenetic relationship (e.g., Hennig 1936a, 1943; see Dupuis 1984). He realized that similarities have to be judged based on quality and not on quantity (e.g., Hennig 1943, 1948, 1950a,b). However, in his early publications, it remains unclear what the quality criterion should be (Richter and Meier 1994; Schmitt 2001). It is only today that we know it to be apomorphy and plesiomorphy. Overall, we therefore have to conclude that in the 1950 book, Hennig only defined the goals of a strictly phylogenetic systematics, but left the reader uncertain about how these goals could be attained.

A precise use of "apomorphy" and "plesiomorphy" only evolved in the early 1950s. Particularly important is one of Hennig's most significant early contributions to dipterology, the three-volume *Die Larvenformen der Dipteren* (*The Larval Forms of Diptera;* Hennig 1948, 1950b, 1952). It is in the third volume that Hennig states categorically that only apomorphic character states can support monophyly (Hennig 1952: 103), whereas the terms "apomorphy" and "plesiomorphy" were still explicitly introduced for taxa in the second volume of the same work (Hennig 1950b: VI–VII). After 1952, the concepts are properly used throughout all of his publications.

But what about the remaining concepts? "Cladistics" (5) was initially a derogatory term that Mayr (1974) applied to Hennig's brand of "Phylogenetic Systematics" (for a more complete etymology, see Dupuis 1984). As is often the case during paradigm shifts in science, such derogatory terms can easily turn into badges of honor proudly worn by a young generation of revolutionaries in a field that is undergoing major changes. Such happened here. A strict concept of "parsimony" (6) is also lacking in the 1950 book, but we find important precursors in Hennig's insistence that systematists should always assume homology for two similar structures unless proven otherwise (e.g., Hennig 1953, 1965a; see also Farris and Kluge 1997). Combined with his concept of reciprocal illumination ("wechselseitige Erhellung"), they provided a foundation for the parsimony concept of Farris (1983).

What about outgroups and character polarization (term 7)? Problematic in all of Hennig's writings are his techniques for determining character polarities. Four criteria are variously mentioned (e.g., Hennig 1966). One is the ontogenetic criterion whose value remains unclear and disputed (Meier 1997). Another, the paleontological criterion, is based on the sequence of occurrence of different character states in the fossil record. The criterion is theoretically sound, but only applicable for taxa with good fossil records. A third is the criterion of character correlation. It is questionable at best (Schmitt 2001), because it proposes that if the polarity of one character is known, all other characters with the same character state distribution will have the same polarity. The fourth criterion is based on biogeographic distributions and has been entirely abandoned. The most surprising omission in Hennig's theoretical publications is thus "outgroup comparison" (Schmitt 2001). Of course, if polarity decisions are at all explicitly discussed in his empirical studies, they are often implicitly based on outgroup comparison, but the theoretical justification for the method is strangely missing in Hennig's writings. The closest approximation is found in Hennig and Schlee (1978), which recommends a comparison of character conditions within and outside the ingroup for reconstructing the direction of character evolution. A formal description of outgroup comparison only appears late in the history of phylogenetics (Watrous and Wheeler 1981) and has in the meantime been replaced by outgroup rooting (Nixon and Carpenter 1993).

Immediate Impact of Hennig's Theoretical Publications

The main theoretical elements of Hennig's phylogenetic systematics were in place two years after the publication of the 1950 book (Hennig 1952). Yet his proposals did not have a major impact on systematics until more than 10 years later (Dupuis 1984, 1990; Andersen 2001). This delay requires an explanation. I would argue that new ideas and methods in science spread most effectively when they are promoted by vocal researchers with high public visibility who are employed by internationally renowned academic institutions, and who popularize their findings through a large number of graduate students and publish easily accessible accounts in prominent scientific journals that publish in the leading science language of the day. In all points, Hennig's phylogenetic systematics was facing a severe uphill battle. Clearly, Hennig was not a skillful promoter of his own ideas.

Hennig avoided lectures in front of large audiences (Schmitt 2002) and on many occasions he declined, often for health reasons, invitations to conferences that would have given him the opportunity to spread his ideas. He also undertook only two major research-related oversea journeys (Schmitt 2001), thus depriving himself of the opportunity to personally convince influential scientists abroad. It was not that he was unwilling or incapable of defending his ideas in word (Schlee 1978; but see Hull 1989) or print. Indeed, throughout his career, he challenged well-known senior scientists and fiercely defended his own ideas against an attack by Mayr (e.g., Hennig 1943 vs. Thienemann and Krüger 1937; Hennig 1936b vs. Enderlein 1913, 1936; Hennig 1957a vs. Paclt 1954; Mayr 1974 vs. Hennig 1974, 1975). But arguments in private or print are unfortunately not as effective in communicating ideas as are public lectures.

To understand the slow spread of his ideas, we should probably also consider that in the 1950s, he was operating from the "wrong" side of the iron curtain and that the reputation of German academic institutions had seriously suffered during the war. Furthermore, Hennig was only able to advise students late in his career (after 1971) and even then, he apparently spent little time with them (Schmitt 2001, 2002). None of these students pursued a very successful university career, so that Hennig did not start his own "school of thought" at any of the German institutions. Moreover, his publication strategy was less than optimal as judged from today's point of view. In retrospect, the main problems were choice of language, journals, and style. In the 1950s and 1960s, Hennig summarized his main theoretical concepts in several excellent publications (Hennig 1953, 1955, 1957a, 1962). However, many failed to reach a wide readership, because they were in German (see Hull 1989), published in journals of a decidedly low visibility, and/or addressed an exclusively entomological audience. Furthermore, despite early criticism for this habit (Hennig 1943), Hennig continued to intertwine empirical and theoretical work in the same publications (Schmitt 2001), often without properly highlighting the theoretical aspects in their titles. For example, the first volume of *Die Larvenformen der Dipteren* starts with an abridged version of the 1950 book (Hennig 1948), but this introduction was lost on all but the most enthusiastic aficionados of maggots.

In the 1960s, Hennig published two major publications in English. One is an excellent but short description of his theory and methods in the *Annual Review of Entomology* (Hennig 1965a) and the second is the well-known 1966 book (*Phylogenetic Systematics*). This book attracted a wide readership, despite its notoriety for being difficult to read and oddly structured (Hull 1989). Many of these problems are the result of its contorted history. Hennig had been asked for permission to translate the 1950 book into English. At this point, he had already extensively rewritten whole sections (Hull 1989) to reflect his great methodological advances since 1945. It was this revised manuscript of 1961 (Schlee 1978; published in Hennig 1982) that became the basis for the translation, which was not published until 1966, because one of the translators died while working on the project (Hull 1989). In retrospect, these revisions to the text of the 1950 book were probably not extensive enough. A more lucid and consistent account would have surely ensured a wider readership. Furthermore, the 1966 book was published 15 years after Hennig had largely completed his theoretical work, so that one would argue in hindsight that too little was done too late.

The slow spread of Hennig's ideas in the 1950s and 1960s is thus in part the result of bad salesmanship (cf. Ernst Mayr's promotion of the "Biological Species Concept"). However, one cannot avoid thinking that it might have also been intentional on Hennig's side. He was mainly dedicated to Diptera systematics (see also Schmitt 2001) and probably wanted to maximize his time dedicated to this activity. During his relatively short life, he published more than 160 articles and books comprising more than 9,000 pages, which included in excess of 3,000 illustrations (Schlee 1978), and only a small fraction of these publications have a predominantly theoretical focus. Hennig was a driven practicing systematist who even continued his work during his days of active duty in the army and under the harsh postwar conditions in Berlin (Schlee 1978). Theory was for Hennig not a goal in itself. It was carried out to obtain the concepts and methods for conducting scientifically defensible systematic work on Diptera.

The spread of Hennig's ideas was thus to an unusual extent the work of third parties who relentlessly promoted and defended his ideas (see Hull 1983, 1989 for details). This dissemination was very uneven across different taxa (see Schuh and Wygodzinsky 1977; Schmitt 2001). For example, early converts were particularly common in entomology (Kiriakoff 1959; Brundin 1965, 1966) and later in ichthyology (Nelson 1972; Kühne 1978), after Nelson read and discussed Hennig's concepts with Brundin (Hull 1989). Especially important for the success of cladistics was the early adoption of Hennig's phylogenetic systematics by Colin Patterson at the Natural History Museum in London (Hull 1989), by a number of scientists at the American Museum of Natural History (e.g., Eldredge, Cracraft, Gaffney, Wiley, Rozen, Nelson, and later Platnick, and Schuh), some of whom later became influential editors of *Systematic Zoology* (Hull 1983). Equally important were the contributions by Farris and Kluge, who vigorously promoted Hennig's ideas in the 1970s (Hull 1989). In Germany, Klaus Günther (1956, 1962) and later Peter Ax (e.g., 1987) were instrumental in carrying Hennig's ideas into the German universities. Notoriously late in accepting Hennig's ideas were the majority of scientists working on the pet groups of the staunchest opponents of Hennig's Phylogenetic Systematics; that is, plants, fossils, and birds (e.g., Simpson 1961; Mayr 1974; Cronquist 1987).

Paradigm Shifts and the Great Leap Forward

The number of scientists practicing Hennigian phylogenetics has dwindled to a mere handful. Modern phylogenetic systematics or cladistics owes much of its success to later additions to the theoretical foundation that Hennig had laid out in the 1950s (Dupuis 1984). These additions will here be only briefly discussed because they are not the main subject of this chapter. We all know that modern cladistic analyses are carried out using numerical techniques with parsimony as the criterion for choosing among cladograms (quantitative cladistics). These numerical techniques and a strict concept of parsimony were largely developed by J. S. Farris (Kluge and Farris 1969; Farris 1970, 1983). His earliest publications describe important precursors to the same numerical techniques that are so important for modern numerical phylogenetic analyses (Kluge and Farris 1969; Farris 1970). In these papers, Hennig is not cited, and apparently the fusion between Hennigian philosophy and Farris's concept of parsimony and his numerical techniques only occurred in the early 1970s (Farris et al. 1970; see discussion in Hull 1988, 1989). Once combined, a powerful arsenal of theoretical concepts and methods had been assembled that quickly swept through systematics and revolutionized the field (Andersen 2001).

But did this sweep constitute a paradigm shift sensu Kuhn (1970)? A fully satisfying answer to this question will require more research and goes well beyond the scope of this chapter. In brief, I believe we have to distinguish between classification and tree reconstruction on the one hand and the relative importance of the two on the other. Prior to Hennig, classifications were the hallmark of systematics, and it was considered sound if they contained a mixed bag of paraphyletic and monophyletic groups (Mayr 1974). Phylogenetic trees were rarely presented, and the relationship between the published trees and the supporting evidence was at best opaque. After Hennig, only monophyletic groups have a right to existence in classifications, and phylogenetic trees are based on explicit character information. Initially, this sounds like a major paradigm shift, but the real shift is not in requirements for classifications and the techniques used for tree reconstruction—it is in the relative importance of the two endeavors. Since Hennig, the importance of phylogenetic reconstruction has dramatically increased at the expense of classifying; that is, there are ever more biologists building trees and ever fewer who are producing classifications. In retrospect, Hennig's real importance is thus in redefining the research agenda within systematics.

Hennig's work falls within a period of great turmoil in systematics and evolutionary biology. This becomes evident when we consult leading systematics textbooks of the so-called "New Systematics" school. The books are either rather technical in nature or they tend to treat extensively nomenclatural and alpha-level taxonomy. If they have a more theoretical or evolutionary orientation, they concentrate on species-level problems (e.g., Huxley 1940; Mayr et al. 1953; but see Simpson 1961). The latter is not surprising, as one of the main achievements of evolutionary biology in the 1940s was the development of species concepts based on reproductive isolation. In the same textbooks, higher-level classification and tree reconstruction occupies little space (but see Simpson 1961) and is considered an art and not a science.

It is with this backdrop that we have to judge Hennig's work, which initiated a major change in the research agenda within systematics (Kühne 1978). It leads away from species-level studies and toward phylogeny reconstruction (Richter and Meier 1994). We should not forget that Hennig also extensively treated species-level phenomena in his empirical and his theoretical papers (1950a, 1966). But this work mainly sets the stage for discussing why constructing phylogenetic trees and classifications are science and not art (Schuh and Wygodzinsky 1977). He explicitly explains why taxa above the species level are real and have individuality. He thus demonstrates that a "scientific" systematics does not have to restrict itself to phenomena below the species level (e.g., Hennig 1947: 279). Help for this cause of establishing an objective basis for systematics came from an unlikely source. Parallel to Hennig, pheneticists pursued similar goals in the anglophone world (Hull 1988, 1989) and the issues were thus already on the table when Hennig entered the stage. For a while, three competing schools of thought within systematics ("phenetics," "phylogenetics," and "evolutionary systematics") battled viciously for supremacy (Hull 1988, 1989), before phylogenetics emerged as the victor; however, not without having absorbed some ideas from its much maligned competitors (e.g., numerical approaches to tree reconstruction; Hull 1989). Hennig's work reoriented systematics toward phylogeny reconstruction. But this major change also sent ripples throughout evolutionary biology, and in retrospect, his work had an even larger impact here than in the comparatively small world of systematics. Today no research in macroevolution can afford to ignore phylogenetic trees, and techniques for reconstructing cladograms belong to the standard repertoire of evolutionary biologists and ecologists. This development was anticipated by Hennig (e.g., Hennig 1969 on "Evolutions ökologie").

Abandoned/Forgotten Concepts and Recent Changes

The work of important scientists often also contains elements that have fallen into disregard. In the case of Hennig, several concepts come to mind. One is the distinction between "tokogenetic" and "phylogenetic" relationships (but see Davis and Nixon 1992). Tokogenetic relationships are netlike and exist among individuals and populations of biparental species, whereas phylogenetic relationships are hierarchical and only begin above the species level (e.g., see Hennig 1957a, 1966). Numerous phylogenetic analyses are today carried out on taxa within biparental species. The resulting "cladograms" are potentially spurious, as they cannot reflect netlike relationships (Hennig 1957a). We shall see how long it will take until the distinction between tokogenetic and phylogenetic is again taken more seriously.

Other examples of almost abandoned/forgotten concepts are Hennig's "internodal" species concept (but see "Hennigian Species Concept" in Wheeler and Meier 2000) and the dichotomy principle. Hennig's species concept uses consecutive speciation events as borders of species in time, using reproductive isolation as the species criterion for populations living at the same time horizon. His species concept attracted much criticism (e.g., Mayr 1974), because it proposes that a stem species dissolves during speciation; that is, that it ceases to exist and thus all descendent species must be considered new species even if one was genetically and morphologically indistinguishable from its ancestor. In retrospect, it appears strange that this element of Hennig's phylogenetic systematics attracted so much attention, because it is not directly related to tree reconstruction. However, we need to remember that the criticism predominantly came from the very systematists who considered species-level work to be at the heart of systematics. The same reason probably spawned the criticism of Hennig's dichotomy principle; it proposed that stem species give rise to exactly two daughter species (instead of more than two). From today's point of view, both concepts are not essential for understanding and appreciating Hennig's phylogenetic systematics.

Also abandoned is Hennig's theoretically sound proposal for assigning ranks/categories based on age (e.g., Hennig 1960, 1965c, 1969). Hennig discussed this idea in numerous publications, which implies that he considered it a very important building block for a theoretically sound phylogenetic systematics. However, his proposal has all but disappeared from the literature. I believe three factors can explain this development. First, the poor fossil record for many taxa makes the proposal unrealistic. Second, the number of available ranks/categories is insufficient to reflect accurately the branching order and ages of all taxa on a tree of life with at least 10 million surviving leaves. Third, the issue is no longer discussed in much detail because it has simply lost its urgency with the reduced interest in classifications.

Ranks/categories are today either omitted (Ax 1987) or their subjectivity is tolerated. Ranks/categories are thus one area in which Hennig failed to remove art from systematics.

Vibrant sciences like systematics are constantly changing. These changes also affect Hennig's legacy. One major change in post-Hennigian cladistics concerns the relationship between systematics and evolutionary biology. Hennig's discussions do not strictly distinguish between evolutionary processes and the patterns that are produced by these processes. For example, Hennig frequently uses process-based arguments to reconstruct character polarities. Such arguments would be unacceptable to a large number of today's cladists (pattern cladists), who insist on keeping process-based arguments separate from tree reconstruction methods (e.g., Platnick 1979). However, not all systematists would agree, and a growing number even insist that explicit evolutionary models should be used when reconstructing trees based on DNA sequence data (e.g., Swofford et al. 1996).

Hennig's reliance for tree reconstruction on apomorphy for supporting monophyletic groups is also currently under threat. This threat is coming from two sources. The availability of abundant DNA data makes it impractical to map characters onto trees, and in most modern cladistic analyses, clade support is no longer expressed by apomorphy, but instead by more abstract values, such as bootstrap, jackknife, and Bremer supports. The second and more fundamental threat originates from model-based tree reconstruction methods, such as maximum likelihood (e.g., Swofford et al. 1996) and Bayesian likelihood (e.g., Huelsenbeck et al. 2001). Both techniques are also theoretically divorced from any apomorphy-plesiomorphy scheme of argumentation and abandon yet another key element of Hennigian phylogenetics. We thus have to ask whether there would still be a legacy of Hennig, if model-based reconstruction methods were to indeed replace traditional cladistics. I would argue, yes. One legacy is the demonstration that the reconstruction of phylogenetic trees and phylogenetic classifications are science (Schuh and Wygodzinsky 1977). This demonstration started the complete reorientation of the research agenda in systematics. The second legacy is found in the fundamental concepts of "monophyly," and "phylogenetic relationship"; that is, in the very concepts that were also the key innovations of Hennig's earliest major contribution to systematic theory in his 1950 book *Grundzüge einer Theorie der phylogenetischen Systematik.*

Would Hennig be horrified by these recent developments? The answer is necessarily speculative. I think Hennig might have been concerned about the lack of classifications in many publications, but I believe that his early manuscripts clearly indicate that he was mostly interested in reconstructing phylogenetic trees and not so much in the reconstruction methods themselves, as long as they were theoretically sound. So, if other methods were shown to be better estimators of phylogeny than his own, I would surmise that Hennig would have embraced them. I am almost certain that he would have welcomed DNA sequence data. Hennig repeatedly emphasized that all features of a species should be considered when reconstructing its phylogenetic relationships (see his "holomorphology" concept; e.g., Hennig 1966).

I believe that another shift in the research agenda of systematics would have been more worrying for Hennig. The success of phylogenetics coincides with a further decline of taxonomy. Indeed, most of the positions for systematists and much of the funding in systematics is now devoted to phylogenetics. Is this concomitant shift toward phylogenetics thus partially responsible for the decline in taxonomy? To me, it appears doubtful that without this shift, taxonomy would have fared better. But this is just a guess.

Hennig's Impact on Animal Classification

I have argued previously that living at the right time in the right place can give scientists an enormous competitive advantage over colleagues from the same field. A good example is Darwin, who, during the 30 years between becoming convinced of evolution and natural selection and the publication of *The Origin of Species,* had an extraordinary amount of time to reevaluate the biological data gathered over the course of several hundred years. Hennig had a similar head start over colleagues when it came to reevaluating the available morphological evidence from a phylogenetic point of view. This head start was considerable. His theoretical concepts of monophyly and phylogenetic relationship were complemented with appropriate methods by 1952, and they did not become popular until the late 1960s. In quick succession, Hennig proposed phylogenetic classifications for Diptera, Insecta, and eventually all of Animalia.

INSECTA

Insecta was the target of his first phylogenetic reevaluation. First trees and classifications were published in 1953, and the seminal papers of 1953, 1955, and 1962 culminated in a book on insect phylogeny (Hennig 1969, 1981). It has become the basis for all future work on insect phylogeny (e.g., Kristensen 1975, 1981, 1991, 1995, 1999; Boudreaux 1979). Indeed, many characters have, almost unmodified, found their way into the most recent character matrices for the reconstruction of the phylogenetic relationships of insects based on morphological and molecular data (Wheeler et al. 2001). Interestingly, many of Hennig's hypotheses have since been confirmed with molecular data (e.g., paraphyly of Apterygota, Thysanura; doubtful monophyly of Paurometabola), and many of the open relationship questions in Hennig's publications remain unresolved to this day (e.g., monophyly of Palaeoptera, relationship within the hemimetabolous "orders").

DIPTERA

The next group targeted for phylogenetic reevaluation by Hennig were the Diptera. The earliest publication covering the entire order is based on a comprehensive reevaluation of dipteran wing morphology (Hennig 1954). This investigation presents numerous original observations and a discussion of Rohdendorf's interpretations of compression fossils. Further high-level work within the order includes the seminal papers on the monophyly of Nematocera (Hennig 1968) and the relationships within the Schizophora (Hennig 1958, 1965b, 1971). In these works, Hennig reviewed the literature data pertaining to Cyclorrhapha and added numerous new characters that he had generated during his extensive work on the morphology of the male and female postabdomen. New relationship hypotheses were proposed, old ones reviewed, and a first phylogenetic framework for this particularly poorly understood section of the order emerged. Further refinement and revisions were made after Hennig started to study amber fossils. Toward the end of his career, he also included data from musculature and spent more time on Muscidae and Anthomyiidae (e.g., Hennig 1964; 1965b, 1976a).

However, we cannot understand Hennig's contributions to Diptera phylogenetics without considering the numerous short publications targeting smaller taxa and/or issues. It is here that most of his phylogenetic hypotheses are found. They are found in the descriptive publications with a morphological theme like the postabdomen in Cyclorrhapha, in the large number of publications containing descriptions of new species and discussions of the species' phylogenetic placement (Schmitt 2001), and they are in yet another class of publications; that is, those on the placement of morphologically aberrant Diptera (e.g., Hennig 1934, 1937a, 1938a,b, 1941b, 1976b). Of course, we should not forget the monographs for 14 families in the series *Die Fliegen der Paläarktischen Region.* All contain phylogenetic hypotheses and some cover very speciose and taxonomically difficult groups (e.g., Muscidae: Hennig 1965b, 1974; Anthomyiidae: Hennig 1976a). Overall, Hennig was undoubtedly also one of the most important alpha-level taxonomists of the twentieth century (Schmitt 2001). Although he does not rank among the top ten descriptive taxonomists in Diptera, he excelled by describing species in many different families (extant and extinct) and carrying out many complex taxonomic revisions (e.g., Hennig 1964 and 1976a are each approximately 1,000 pages in length). All this taxonomic and phylogenetic work culminates in the classification of Diptera presented in a monograph for the series *Handbuch der Zoologie* (Hennig 1973). Here, a phylogenetic classification and a family-level identification key for the Diptera is presented. Moreover, the ecology, economic significance, and morphology of all life history stages are reviewed.

It is thus not surprising that even today, a study on the phylogenetic relationships of a dipteran taxon often starts with a review of what Hennig has said about the group. Hennig was the first to discuss the monophyly of major groupings, such as the Nematocera, Orthorrhapha, and Aschiza. For some groupings, he was able to propose phylogenetic alternatives. For example, he supported the sister group relationship between Syrphidae + Pipunculidae and the Schizophora and provided character support for what is today known as the Eremoneura. In other cases, his proposals were at best controversial (e.g., groupings within Nematocera), but they certainly encouraged further research. The Diptera thus quickly became the phylogenetically best-studied megadiverse order of insects (e.g., Griffiths 1972; McAlpine 1989; see review in Yeates and Wiegmann 1999). Even the controversies surrounding the homologies of male genitalia across the order (see references in Yeates and Wiegmann 1999) can be traced back to Hennig.

ANIMALIA

Being a dipterist and thus an entomologist, it is not surprising that Hennig was familiar with the morphological data for Diptera and the insects. But why was Hennig also more or less simultaneously capable of producing revised phylogenetic classifications for all Metazoa (Hennig 1957b, 1959)? It is important to remember that Hennig held temporary positions as a lecturer at the Universities of Leipzig and Potsdam. Part of his duties were lectures in systematic zoology, invertebrate systematics, vertebrate systematics, and systematic entomology (Schmitt 2001). During his time in Stuttgart, he also taught taxon-specific seminars at the University of Tübingen (Schlee 1978). He was thus intimately familiar with the literature on a wide variety of animal groups.

His phylogenetic publications on Animalia start with several volumes on invertebrates (Hennig 1957b, 1959; updated in Hennig 1979, 1986) and culminate in a posthumously published book on the phylogeny of Chordata (Hennig 1983). These books have been widely used in teaching at German universities, but they had little effect on the phylogenetic work of specialists abroad. Indeed, from today's point of view, neither the style of the books, usually being mostly composed of long lists of characters, nor their content is entirely convincing. The latter point leads us to the observation that many of Hennig's initial phylogenetic hypotheses from the 1950s have not survived scrutiny. This applies to some degree to his work on Animalia and Diptera and, to a lesser degree, to his work on insect relationships. We might ask why so many of Hennig's hypotheses have not survived scrutiny. In Hennig's publications, there is an overly optimistic reliance on a few characters for supporting relationships. The research of the past 25 years has shown that, regardless of how much care is given to establishing homology, homoplasy is a very common phenomenon at all levels (DNA to morphology), and more than a few characters are usually required to confidently support taxa. Not unlike his contemporaries, Hennig simply underestimated the abundance of homoplasy.

Biogeography

Biologists who are troubled by the plethora of methods for tree reconstruction have not taken a closer look at the most chaotic area of systematics: biogeography (Humphries 2000). Yet no discussion of Hennig's contributions to Dipterology and Systematics would be complete without touching on this subject. Hennig's fascination with biogeography goes back to his earliest publications (e.g., Hennig 1936a,d), and he is here already looking for common geographic patterns in different groups of acalyptrate flies. This theme is repeated in many of his Diptera publications. Occasionally, Hennig even uses biogeographic arguments to support the monophyly of taxa (Neriidae: Hennig 1937b). He also considered biogeographic evidence particularly important for establishing the age of taxa (e.g., Hennig 1960). The latter was important for his goal of establishing objective ranks and/or categories.

Undoubtedly, his most important biogeographic contribution is the treatise on New Zealand Diptera (Hennig 1960). In this work, he clearly outlines the goals of a scientific biogeographic analysis. He argues that it should start with identifying the endemic taxa, it should establish the geographic distribution of their sister groups, and eventually it should identify and explain the predominant patterns. Ultimately the minimum ages of the different faunal layers should be estimated. Accordingly, his publication on New Zealand Diptera begins with a list of the endemic taxa. He then discusses for many endemics their putative sister groups, as well as their distributions. Different hypotheses for the distributional patterns found in the New Zealand fauna are discussed and minimum ages for the taxa proposed. In this publication, many goals of a cladistic vicariance biogeography (Nelson and Platnick 1981) are anticipated, and Hennig explained what data would be needed to resolve the various unresolved matters. Hennig intended to carry out similar analyses for Taiwan and the Sunda Islands, but the outbreak of the Second World War prevented the completion of the project, so that only descriptive manuscripts were published (Hennig 1941c,d).

The biogeographic analysis of the New Zealand Diptera is an early highlight of cladistic biogeography. But unfortunately, the image of Hennig as a biogeographer is tainted by the treatments of the subject in the 1950 and 1966 books. Both largely reflect Hennig's thinking of the 1940s and they fail to impress because of an overreliance on dispersal ideas.

Larval Morphology

Larvae and adults of the same species have the same phylogenetic history. Yet if we use similarity to classify the larvae and the adults of the same taxa, we frequently come to very different classifications. Hennig argued as early as 1936 that obviously similarity is an imperfect estimator of evolutionary relationship (Hennig 1936a, 1943). This observation was of vital importance for Hennig's challenge to the generally held view that overall similarity in homologous characters can be used to reconstruct phylogenetic trees. It was this inexplicable inconsistency between larval and adult classifications for the same groups that ultimately motivated Hennig's search for alternative tree reconstruction techniques. As early as 1938 (Hennig 1938a), he insists that incongruence between larval and adult classifications is due to error; by 1948, he suspects that paraphyletic groups ("Restkörper") might be to blame, and in 1957, he explicitly attributes incongruence between larval and adult classifications to plesiomorphic similarities (Hennig 1957a).

Hennig's interest in larvae started early (Hennig 1943) and continued throughout his life. At the time of his death, he was working on a new edition of his most ambitious Diptera project *Die Larvenformen der Dipteren* (Schlee 1978). The first edition had been published in three volumes (1948–1952) and was expressly launched to test for congruence between larval and adult classifications and to critique the current state of systematic theory (Hennig 1948). This also explains the extensive discussion of theoretical concepts found throughout all three volumes of the work. A test of phylogenetic ideas based on larvae was also a main motivator for Hennig's publication on Diptera wing venation (Hennig 1954). Unfortunately, the lack of data on the larvae for many Diptera groups made it difficult for Hennig to extract the desired information and the situation has hardly improved since (Ferrar 1987).

Amber Fossils

Laypersons are often surprised when they realize that the study of fossils resides in geology departments, whereas living species are studied in biology departments. Hennig (1936a) recognized early in his life that both recent and fossil material can provide important evidence in the quest for reconstructing phylogenetic relationships and determining the age of taxa. He thus started to work on both the living and extinct representatives of the same group, thus challenging this traditional division of labor. Indeed, in the twentieth century, Hennig was one of the first and foremost entomologists who extensively integrated work on fossil and extant species (Andersen 2001). Initially, he considered fossils particularly important for testing relationship hypotheses based on recent species (e.g., Hennig 1936a: 166) and for clarifying biogeographic questions (Hennig 1936a, 1938a). Later, he emphasized the value of fossils for determining character polarities and the age of taxa (e.g., Hennig 1954, 1965a, 1966, 1969, 1983). Toward the end of his career, he emphasized that the same methods should be used for fossil and recent species (Hennig and Schlee 1978), thus anticipating the kind of integrated study that is favored today (e.g., Donoghue et al. 1989). Hennig's own research on fossils was largely restricted to amber, but in one of his main publications on Diptera (Hennig 1954) and in his book on insect phylogeny (1969), he also extensively discussed compression fossils.

It is not generally known that his work on amber fossils started early (Hennig 1939, 1940, 1941a), because his studies only started in earnest after moving to the Museum für Naturkunde in Stuttgart. He was now cut off from the extensive Diptera collections and the excellent entomological library of the Deutsches Entomologisches Institut and was looking for a new research subject (Schlee 1978). He seized on the opportunity to study amber when a colleague from the University Tübingen, A. Seilacher, told him that the famous Königsberg amber collection had survived the Second World War and was deposited at the University of Göttingen (Schlee 1978; Schmitt 2001). Hennig extensively studied this and other collections and in quick succession, he published a large number of extremely important articles that set a high standard for work on fossil Diptera (e.g., Hennig 1965c). New fossil species were not merely described. Instead, the descriptions and illustrations are accompanied by a discussion of putative phylogenetic relationships.

Conclusions

So what was the role of dipterology in the development of phylogenetic systematics? What were the insights of Willi Hennig? Ever since high school, Hennig was a systematist at heart, and he dedicated his life to building a solid scientific foundation for systematics and for classifying animals. Luckily for Dipterology, he chose flies as his favorite group, and his desire to propose a scientifically defensible classification for the Diptera also motivated much of his innovative theoretical research. The impact of his research on systematics and evolutionary biology cannot be overestimated. His theoretical breakthroughs led to a redefinition of the research agenda in systematics, with important ramifications for all of evolutionary biology. His biogeographic work on the fauna of New Zealand foreshadows vicariance biogeography; his simultaneous treatment of extant and extinct species bridged traditional borders between palaeontology and biology. But let us not forget that he was also instrumental in establishing the main clades within the Diptera, which comprise roughly 10% of all known animal species, and that he himself described numerous new species. He also revised a wide variety of taxonomically challenging taxa, and his dipterology works alone would have occupied the lifetime of a productive dipterist (Schuh and Wygodzinsky 1977). Any one of these achievements would have secured his place in the history of biology, but being able to achieve all these goals leaves our and future generations of dipterologists and systematists bewildered. How could a single scientist possibly be so productive within a single, rather short lifetime?

REFERENCES

Andersen, N. M. 2001. The impact of W. Hennig's "phylogenetic systematics" on contemporary entomology. *Eur. J. Entomol.* 98:133–150.

Anonymous (Wolfgang Hennig). 1978. In memoriam: Willi Hennig (20.4.1913–5.11.1976). *Beitr. Entomol.* 28:169–177.

Ax, P. 1987. *The Phylogenetic System. The Systematization of Organisms on the Basis of Their Phylogenesis.* Chichester: Wiley.

Boudreaux, H. B. 1979. *Arthropod Phylogeny with Special Reference to Insects.* Chichester: Wiley.

Brundin, L. 1965. On the real nature of transantarctic relationships. *Evolution* 19:496–505.

Brundin, L. 1966. Transantarctic relationships and their significance, as evidenced by chironomid midges. *K. Svenska Vetensk. Akad. Handl.* 11:1–472.

Cronquist, A. 1987. A botanical critique of cladism. *Bot. Rev.* 53:1–52.

Davis, J. A., and K. C. Nixon. 1992. Populations, genetic variation, and the delimitation of phylogenetic species. *Syst. Biol.* 41:421–435.

Donoghue, M., J. Doyle, J. Gauthier, A. Kluge, and T. Rowe. 1989. The importance of fossils in phylogeny reconstruction. *Annu. Rev. Ecol. Syst.* 20:431–460.

Dupuis, C. 1984. Willi Hennig's impact on taxonomic thought. *Annu. Rev. Ecol. Syst.* 15:1–24.

Dupuis, C. 1990. Hennig, Emil Hans Willi. In F. L. Holmes, ed., *Dictionary of Scientific Biography,* vol. 17, suppl. 2, pp. 407–410. New York: Charles Scribner's Sons.

Enderlein, G. 1913. Dipterologische Studien II. Zur Kenntnis der Tanypezinen. *Zool. Anz.* 42:224–229.

Enderlein, G. 1936. Klassifikation der Tanypezidae. *Deutsch. Entomol. Zeitschr.* 1936:39–47.

Farris, J. S. 1970. Methods for computing Wagner trees. *Syst. Zool.* 19:83–92.

Farris, J. S. 1983. The logical basis of phylogenetic inference. In N. I. Platnick and V. A. Funk, eds., *Advances in Cladistics,* vol. 2: *Proceedings of the 2nd Meeting of the Willi Hennig Society,* pp. 7–36. New York: Columbia University Press.

Farris, J. S. 1990. Haeckel, history, and Hull. *Syst. Zool.* 39:81–88.

Farris, J. S., and A. G. Kluge. 1997. Parsimony and history. *Syst. Zool.* 46:215–218.

Farris, J. S., A. G. Kluge, and M. J. Eckhart. 1970. On predictivity and efficiency. *Syst. Zool.* 19:363–372.

Ferrar, P. 1987. A guide to the breeding habits and immature stages of Diptera Cyclorrhapha. *Entomonograph* 8:1–907.

Griffiths, G. C. D. 1972. The phylogenetic classification of Diptera Cyclorrhapha, with special reference to the structure of the male postabdomen. *Ser. Entomol.* 8:1–340.

Günther, K. 1956. Systematik und Stammesgeschichte der Tiere 1939–1953. *Fortschr. Zool.,* N. F. 10:33–278.

Günther, K. 1962. Systematik und Stammesgeschichte der Tiere 1954–1959. *Fortschr. Zool.,* N. F. 14:268–547.

Hennig, W. 1934. Über Bau und Verwandtschaft der Kerguelenfliege *Calycopteryx moseleyi* Eat. *Zool. Anz.* 108:196–201.

Hennig, W. 1936a. Beziehungen zwischen geographischer Verbreitung und systematischer Gliederung bei einigen Dipterenfamilien: Ein Beitrag zum Problem der Gliederung systematischer Kategorien höherer Ordnung. *Zool. Anz.* 116:161–165.

Hennig, W. 1936b. Beiträge zur Kenntnis des Kopulationsapparates und der Systematik der Tanypezidae (Dipt., Acalypt.). *Deutsch. Entomol. Zeitschr.* 1936:27–38.

Hennig, W. 1936c. Revision der Gattung *Draco* (Agamidae). *Temminckia* (Leiden) 1:153–220.

Hennig, W. 1936d. Über einige Gesetzmäβigkeiten der geographischen Variation in der Reptiliengattung *Draco* L.: "parallele" und "konvergenete" Rassenbildung. *Biol. Zentralblatt* 56:549–559.

Hennig, W. 1936e. Beiträge zur Kenntnis des Kopulationsapparates der cyclorrhaphen Dipteren. *Zeitschr. Morphol. Ökol. Tiere* 31:328–370.

Hennig, W. 1937a. Die morphologische Deutung des männlichen Kopulationsapparates der Gattung *Glossina. Zeitschr. Parasitenkunde* 9:345–350.

Hennig, W. 1937b. Übersicht über die Arten der Neriiden und über die Zoogeographie dieser Acalyptratengruppe (Dipera). *Stettiner Entomol. Zeitg.* 98:46–50.

Hennig, W. 1938a. Die Gattung *Rhachicerus* Walker und ihre Verwandten im Baltischen Bernstein (Dipt.). *Zool. Anz.* 123:33–41.

Hennig, W. 1938b. Zur Frage der verwandtschaftlichen Stellung von *Braula coeca* Nitzsch (Diptera). *Arbeiten Morph. Taxon. Entomol.* (Berlin) 5:164–174.

Hennig, W. 1939. Über einen Floh aus der Bernsteinsammlung des Herrn Scheele (Aphaniptera: Ctenopsyllidae). *Arbeiten Morph. Taxon. Entomol.* (Berlin) 6:330–332.

Hennig, W. 1940. Über die systematische Stellung von *Palaeotimia hoesti* Meun. *Arbeiten Morph. Taxon. Entomol.* (Berlin) 7:89–91.

Hennig, W. 1941a. Beiträge zur Kenntnis des Kopulationsapparates und der Systematik der Acalyptraten, III. Pallopteridae, Theyreophoridae, Diopsidae, *Pseudopomzya, Pseudodinia* (Diptera). *Arbeiten Morph. Taxon. Entomol.* (Berlin) 8:54–64.

Hennig, W. 1941b. Die Verwandtschaftsbeziehungen der Pupiparen und die Morphologie der Sternalregion des Thorax der Dipteren. *Arbeiten Morph. Taxon. Entomol.* (Berlin) 8:231–249.

Hennig, W. 1941c. Verzeichnis der Dipteren von Formosa. *Entomol. Beihefte* 8:1–239.

Hennig, W. 1941d. Dipteren von den kleinen Sunda-Inseln. Aus der Ausbeute der Sunda-Expedition Rensch, I. Einleitung und Bearbeitung der Familien Sciomyzidae, Tylidae, Lonchaeidae, Pyrgotidae und Platystomidae. *Arbeiten Morph. Taxon. Entomol.* (Berlin) 8:16–23.

Hennig, W. 1943. Ein Beitrag zum Problem der "Beziehungen zwischen Larven- und Imaginalsystematik." *Arbeiten Morph. Taxon. Entomol.* (Berlin) 10:138–145.

Hennig, W. 1947. Probleme der biologischen Systematik. *Forsch. Fortschr.* 21/23:276–279.

Hennig, W. 1948. *Die Larvenformen der Dipteren,* vol. 1. Berlin: Akademie Verlag.

Hennig, W. 1949a. Zur Klärung einiger Begriffe der phylogenetischen Systematik. *Forsch. Fortschr.* 25: 136–138.

Hennig, W. 1949b. Sepsidae. In E. Lindner, ed., *Die Fliegen der Paläarktischen Region,* Teil. 39a (Lieferungen 157, 159). Stuttgart: Schweizerbartsche Verlagsbuchhandlung.

Hennig, W. 1950a. *Grundzüge einer Theorie der phylogenetischen Systematik.* Berlin: Deutscher Zentralverlag.

Hennig, W. 1950b. *Die Larvenformen der Dipteren,* vol. 2. Berlin: Akademie Verlag.

Hennig, W. 1952. *Die Larvenformen der Dipteren,* vol. 3. Berlin: Akademie Verlag.

Hennig, W. 1953. Kritische Bemerkungen zum phylogenetischen System der Insekten. *Beitr. Entomol.* 3:1–85.

Hennig, W. 1954. Flügelgeäder und System der Dipteren unter Berücksichtigung der aus dem Mesozoikum beschriebenen Fossilien. *Beitr. Entomol.* 4:245–388.

Hennig, W. 1955. Meinungsverschiedenheiten über das System der niederen Insekten. *Zool. Anz.* 155: 21–30.

Hennig, W. 1957a. Systematik und Phylogenese. *Ber. Hundertjahrfeier Dtsch. Entomol. Ges.* (Berlin) 50–71.

Hennig, W. 1957b. *Taschenbuch der Zoologie,* vol. 2: *Wirbellose I,* 1st ed. [3d ed., 1976] Leipzig: Thieme Verlag.

Hennig, W. 1958. Die Familien der Dipterea Schizophora und ihre phylogenetischen Verwandtschafts-beziehungen. *Beitr. Entomol.* 8:505–688.

Hennig, W. 1959. Taschenbuch der Zoologie, vol. 3: *Wirbellose II,* 1st ed. [3d ed., 1968] Leipzig: Thieme Verlag.

Hennig, W. 1960. Die Dipteren-Fauna von Neuseeland als systematisches und tiergeographisches Problem. *Beitr. Entomol.* 10:221–329.

Hennig, W. 1962. Veränderungen am phylogenetischen System der Insekten seit 1953. *Deutsche Akad. Landwirschaftswiss. Berlin Tagungsber.* 45:29–42.

Hennig, W. 1964. Muscidae. In E. Lindner, ed., *Die Fliegen der Paläarktischen Region,* Teil. 63b (Lieferungen 1955–1964). Stuttgart: Schweizerbart'sche Verlagsbuchhandlung.

Hennig, W. 1965a. Phylogenetic Systematics. *Annu. Rev. Entomol.* 10:97–116.

Hennig, W. 1965b. Vorarbeiten zu einem phylogenetsichen System der Muscidae (Diptera: Cyclorrhapha). *Stuttgarter Beitr. Naturkunde,* Ser. A (Biol.) 141:1–100.

Hennig, W. 1965c. Die Acalyptratae des Baltischen Bernsteins und ihre Bedeutung für die Erforschung der phylogenetischen Entwicklung dieser Dipteren-Gruppe. *Stuttgarter Beitr. Naturkunde,* Ser. A (Biol.) 145:1–215.

Hennig, W. 1966. *Phylogenetic Systematics.* Urbana: University of Illinois Press.

Hennig, W. 1968. Kritische Bermerkungen über den Bau der Flügelwurzel bei den Dipteren und die Frage nach der Monphylie der Nematocera. *Stuttgarter Beitr. Naturkunde,* Ser. A (Biol.) 193:1–23.

Hennig, W. 1969. *Die Stammesgeschichte der Insekten.* Frankfurt: Waldemar Kramer & Co.

Hennig, W. 1971. Neue Unterschungen über die Familien der Diptera Schizophora (Diptera: Cyclorrhapha). *Stuttgarter Beitr. Naturkunde,* Ser. A (Biol.) 226:1–76.

Hennig, W. 1973. Diptera (Zweiflügler). In J.-G. Helmcke, D. Starck, and H. Wermuth, eds., *Handbuch der Zoologie,* pp. 1–337. Berlin: Walter de Gruyter.

Hennig, W. 1974. Kritische Bemerkungen zur Frage "cladistic analysis or cladistic classification." *Z. Zool. Syst. Evolutionsforsch.* 12:279–284.

Hennig, W. 1975. Cladistic analysis or cladistic classification—a reply to Ernst Mayr. *Syst. Zool.* 24:244–256.

Hennig, W. 1976a. Anthomyiidae. In E. Lindner, ed., *Die Fliegen der Paläarktischen Region,* Teil. 63a (Lieferungen 1966–1976). Stuttgart: Schweizerbart'sche Verlagsbuchhandlung.

Hennig, W. 1976b. Das Hypopygium von *Lonchoptera lutea* Panzer und die phylogenetischen Verwandtschaftsbeziehungen der Cyclorrhapha (Diptera). *Stuttgarter Beitr. Naturkunde,* Ser. A (Biol.) 283:1–63.

Hennig, W. 1978. Die Stellung der Systematik in der Zoologie. *Entomol. Germ.* 4:193–199.
Hennig, W. 1979. *Taschenbuch der Speziellen Zoologie,* Teil 1: *Wirbellose I* (Wolfgang Hennig, ed.), 4th ed. Jena: Fischer Verlag.
Hennig, W. 1981. *Insect Phylogeny.* Chichester: Wiley.
Hennig, W. 1982. *Phylogenetische Systematik* (Wolfgang Hennig, ed.). Berlin and Hamburg: Paul Parey Verlag.
Hennig, W. 1983. Stammesgeschichte der Chordaten (Wolfgang Hennig, ed.). *Fortschr. Zool. Syst. Evolutionsforsch.* 2:1–108.
Hennig, W. 1984. *Aufgaben und Probleme stammesgeschichtlicher Forschung* (Wolfgang Hennig, ed.). Berlin and Hamburg: Paul Parey Verlag.
Hennig, W. 1986. *Taschenbuch der Speziellen Zoologie,* Teil 2: *Wirbellose II* (Wolfgang Hennig, ed.), 4th ed. Jena: Fischer Verlag.
Hennig, W., and W. Meise. 1932. Die Schlangengattung *Dendrophis. Zool. Anz.* 99:273–297.
Hennig, W., and D. Schlee. 1978. Abriß der phylogenetischen Systematik. *Stuttg. Beitr. Naturk.,* Serie A (Biol.) 319:1–11.
Huelsenbeck, J. P., F. Ronquist, R. Nielsen, and J. P. Bollneck. 2001. Bayesian inference of phylogeny and its impact on evolutionary biology. *Science* 294:2310–2314.
Hull, D. L. 1983. Thirty-one years of Systematic Zoology. *Syst. Zool.* 32:315–342.
Hull, D. L. 1988. *Science as a Process.* Chicago: University of Chicago Press.
Hull, D. L. 1989. The evolution of phylogenetic systematics. In B. Fernholm, K. Bremer, and H. Jörnwall, eds., *The Hierarchy of Life,* pp. 3–15. Amsterdam: Elsevier.
Humphries, C. J. 2000. Form, space and time; which comes first? *J. Biogeogr.* 27:11–15.
Huxley, J. S. 1940. Introductory: Towards the new systematics. In J. S. Huxley, ed., *The New Systematics,* pp. 1–46. Oxford: Oxford University Press.
Kiriakoff, S. G. 1959. Phylogenetic systematics versus typology. *Syst. Zool.* 8:117–118.
Kluge, A. G., and J. S. Farris. 1969. Quantitative phyletics and the evolution of anurans. *Syst. Zool.* 18:1–32.
Kristensen, N. P. 1975. The phylogeny of hexapod "orders." A critical review of recent accounts. *Z. Zool. Syst. Evolut. Forsch.* 13:1–44.
Kristensen, N. P. 1981. Phylogeny of insect orders. *Annu. Rev. Entomol.* 26:135–157.
Kristensen, N. P. 1991. Phylogeny of extant hexapods. In CSIRO, *The Insects of Australia,* 2nd ed., pp. 125–140. Carlton: Melbourne University Press.
Kristensen, N. P. 1995. Forty years' insect phylogenetic systematics: Hennig's "Kritische Bemerkungen . . ." and subsequent developments. *Zool. Beitr.,* N. F. 36:83–124.
Kristensen, N. P. 1999. Phylogeny of endopterygote insects, the most successful lineage of living organisms. *Eur. J. Entomol.* 96:237–253.
Kuhn, T. S. 1970. *The Structure of Scientific Revolutions,* 2nd ed. Chicago: University of Chicago Press.
Kühne, W. G. 1978. Willi Hennig 1913–1976: Die Schaffung einer Wissenschaftstheorie. *Entomol. Germ.* 4:374–376.
Mayr, E. 1974. Cladistic analysis or cladistic classification. *Z. Zool. Syst. Evolutionsforsch.* 12:94–128.
Mayr, E., E. G. Linsley, and R. L. Usinger. 1953. *Methods and Principles of Systematic Zoology.* New York: McGraw-Hill.
McAlpine, J. F. 1989. Phylogeny and classification of the Muscomorpha. In J. F. McAlpine, ed., *Manual of Nearctic Diptera,* pp. 1397–1518. Ottawa: Agriculture Canada Research Branch.
Meier, R. 1997. A test and review of the empirical performance of the ontogenetic criterion. *Syst. Biol.* 46:699–721.
Nelson, G. J. 1972. Comments on Hennig's "Phylogenetic Systematics" and its influence on ichthyology. *Syst. Zool.* 21:364–374.
Nelson, G. J., and N. I. Platnick. 1981. *Systematics and Biogeography: Cladistics and Vicariance.* New York: Columbia University Press.
Nixon, K. C., and J. M. Carpenter. 1993. On outgroups. *Cladistics* 9:413–426.
Paclt, J. 1954. Zum phylogenetischen System der niederen Insekten. *Zool. Anz.* 153:275–281.
Platnick, N. I. 1979. Philosophy and the transformation of cladistics. *Syst. Zool.* 28:537–546.
Richter, S., and R. Meier. 1994. The development of phylogenetic concepts in Hennig's early theoretical publications (1947–1966). *Syst. Biol.* 43:212–221.
Schlee, D. 1978. In memoriam Willi Hennig 1913–1976. Eine biographische Skizze. *Entomol. Germ.* 4: 377–391.

Schmitt, M. 1996. Klaus Günthers Bedeutung für die Phylogenetische Systematik. *Sber. Ges. Naturf. Freunde Berlin,* N. F. 35:13–25.

Schmitt, M. 1997. Willi Hennig handschriftliche Urfassung der "Grundzüge einer Theorie der Phylogenetischen Systematik." In C. von Campenhausen, ed., *Andenken zoologischer Vergangenheit,* pp. 42–43. Mainz: Katalog zur Ausstellung im Naturhistorischen Museum.

Schmitt, M. 2001. Willi Hennig (1913–1976). In *Darwin & Co. Eine Geschichte der Biologie in Portraits,* pp. 316–343. München: Beck Verlag.

Schmitt, M. 2002. Willi Hennig (1913–1976) als akademischer Lehrer. In J. Schulz, ed., *Fokus Biologiegeschichte—Zum 80. Geburtstag der Biologiehistorikerin Ilse Jahn,* pp. 53–64. Berlin: Akadras Verlag.

Schuh, R. T., and P. Wygodzinsky. 1977. Willi Hennig. *Syst. Zool.* 26:104–105.

Simpson, G. G. 1961. *Principles of Animal Taxonomy.* New York: Columbia University Press.

Swofford, D. L., G. J. Olsen, P. J. Waddell, and D. M. Hillis. 1996. Phylogenetic inference. In D. M Hillis, C. Moritz, and B. K. Mable, eds., *Molecular Systematics,* pp. 407–430. Sunderland: Sinauer.

Thienemann, A., and F. Krüger. 1937. "*Orthocladius*" *abiskoensis* Edwards und *rubicundus* (Mg.), zwei "Puppen-Spezies" der Chironomiden (Chironomiden aus Lappland II). *Zool. Anz.* 117:257–267.

Vogel, J., and W. R. Xylander. 1999. Willi Hennig—Ein Oberlausitzer Naturforscher mit Weltgeltung. Recherchen zu seiner Familiengeschichte sowie Kinder- und Jugendzeit. *Ber. Naturf. Ges. Oberlausitz* 7/8:131–141.

Watrous, L. E., and Q. D. Wheeler. 1981. The out-group comparison method of character analysis. *Syst. Zool.* 30:1–11.

Wheeler, Q. D., and R. Meier, eds. 2000. *Species Concepts and Phylogenetic Theory: A Debate.* New York: Columbia University Press.

Wheeler, W. C., M. Whiting, Q. D. Wheeler, and J. M. Carpenter. 2001. The phylogeny of the extant hexapod orders. *Cladistics* 17:113–169.

Yeates, D. K., and B. M. Wiegmann. 1999. Congruence and controversy: Toward a higher-level phylogeny of Diptera. *Annu. Rev. Entomol.* 44:397–428.

PART II

Genomics and Developmental Biology

CHAPTER FOUR

The Genomes of Diptera

Michael Ashburner

Our knowledge of dipteran genomes has been revolutionized by the completion of the sequences of the genomes of three flies, those of *Drosophila melanogaster*, *D. pseudoobscura*, and *Anopheles gambiae*. The impetus for these major projects has been the extraordinary utility of the first species as a model organism for both basic and biomedical research, the utility of the second species for comparative purposes, and the importance of the last as a disease vector. These achievements must not, however, blind us to the fact that these are but three of an estimated 250,000 extant species (admittedly from two very different lineages) of Diptera, and that more "normal" research on aspects of the genome biology of flies has been, and remains, an active research field. More detailed information concerning drosophilid genomes can be found in Ashburner et al. (2005).

Genome Size

The size of the haploid genome in the Diptera reportedly extends over the range from 0.09 pg (90 Mb) in the cecidomyiid *Mayetiola destructor* (see Ma et al. 1992) to 1.90 pg (1.9 Gb) in *Aedes zoosophus* (Rao and Rai 1987). A summary of the available data can be found in the database maintained by Gregory (2001), and for two species, at least, are remarkably accurate. For *D. melanogaster*, careful experimental estimates were of a genome size of 0.18 pg (180 Mb); data from the sequence project estimate a genome of 175 Mb (Celniker et al. 2002; Bennett et al. 2003). For *An. gambiae*, estimates from biochemical analysis were of a genome size of 0.27 pg (278 Mb) (Besansky and Powell 1992); this is now estimated to be 260 Mb (Holt et al. 2002).

Within groups of related species, small variations in genome size will, by and large, be the result of variations in copy number of transposable elements (see below and Vieira et al. 2000, 2002). One of the most telling cases is that of *Chironomus thummi thummi* and *C. t. piger*, subspecies that produce fertile hybrids; the former subspecies has 27% more DNA than the latter (Keyl 1965). This increase is due, in part at least, to an increase in transposable element copy number. For example, there are some 10,000 copies (all centromeric) of the *C1a* element in *C. t. piger*, whereas there are 70,000 (dispersed) copies in *C. t. thummi* (Ross et al. 1997). There are suggestions that, within a species, increases in transposable element copy number may occur when a species adapts to a new environment (see Vieira et al. 2002). If published reports are any guide, then intraspecific variation in genome size can be quite marked. Populations of *Aedes albopictus*, a vector of dengue, are said to vary in genome size by almost threefold, from 0.62 pg to 1.66 pg, due to variations in the content of repetitive DNA sequences (Rao and Rai 1987; Black and Rai 1988; Kumar and Rai 1990).

In the genus *Drosophila,* haploid genome sizes vary between 158 Mb in *D. simulans* (J. Spencer Johnston, pers. comm.; see Ashburner et al. 2005) and 779 Mb in *D. nasutoides,* where as much as 60% of the DNA from diploid cells is satellite, all located on a single heterochromatic chromosome (Cordeiro et al. 1975; Zacharias 1986, 1990). With the recently released genomic sequence of *D. pseudoobscura,* an opportunity to study in detail the reason for wide intraspecific variations in genome size now exists, as R. Dawley and colleagues (unpublished, quoted by Powell 1997: 303–306) have found about 10% variation in haploid DNA content among strains of this species.

Genomic Sequences

It became obvious in the mid-1990s that the complete genomic sequence of *D. melanogaster* had become a research priority. Efforts were initiated by both the Berkeley *Drosophila* Genome Project (BDGP), in the United States, and the European *Drosophila* Genome Project (EDGP), in Europe, to work toward this end, by the then-conventional method of sequencing a minimum tiling path of clones (in P1, cosmid, and, latter, BAC vectors) that had been assembled into contiguous physical maps of the genome.

Early physical contigs of cloned DNA had been assembled by chromosome walking. Now clones can be assembled into contigs either by fingerprinting or by sequence-tagged site content mapping. These methods, coupled with the development of P1 (in particular, BAC) vectors, have made physical mapping rapid, especially in genomes that are relatively poor in repeated sequences. In many Diptera, there is the additional advantage that these maps can be verified, albeit at relatively low resolution, by in situ hybridization of the clones to polytene chromosomes. For *D. melanogaster,* physical maps were constructed with both cosmid (e.g., Madueño et al. 1995) and P1 (Smoller et al. 1991) clones. These have, by and large, been superseded by the sequenced BAC clones (Hoskins et al. 2000; Peter et al. 2002). Physical maps have also been constructed for *D. virilis* (Viera et al. 1997), *D. repleta* (Ranz et al. 2001; González et al. 2002), and *D. buzzatii* (A. Ruiz, pers. comm.), and are in progress for *An. gambiae.* For maximum utility, these maps must be coupled with a mechanism to distribute the clones to researchers.

Although the conventional clone-by-clone sequencing of the genome of *D. melanogaster* had considerable success (Ashburner et al. 1999; Benos et al. 2000, 2001; Celniker et al. 2002), the efforts of both the BDGP and EDGP were rendered ineffective by the announcement, in May 1998, by C. Venter (Venter et al. 1998) that he would, through his new company Celera, complete the sequence of this species by whole-genome shotgun (WGS) sequencing. At the time, there was considerable skepticism that this method could be applied to a large and complex genome (Green 1997). By September 1999, the first assembly of the WGS of *D. melanogaster* was available (Myers et al. 2000). Collaborating with both the BDGP and EDGP, and incorporating much of their clone sequence data and BAC-end sequences, Release 1 of the sequence of this species was annotated and released in March 2000 (Adams et al. 2000; Rubin et al. 2000a). Even though the sequence coverage of the genome was high (about 13-fold; Myers et al. 2000), this assembly was very much a preliminary, albeit very useful and interesting, effort: it covered only the euchromatin, the transposable elements were represented by metasequences (although correctly located), there were a large number of sequence and physical gaps and many repeated sequences had been collapsed into one by the assembler software. A somewhat improved assembly (WGS2) was released in October 2002, but the major improvement only came with the efforts of the BDGP, collaborating with the Human Genome Sequencing Center at the Baylor College of Medicine, to close the gaps and improve sequence quality by directed sequencing of BAC clones. This release (Release 3) was annotated by FlyBase and published in December 2002 (Celniker et al. 2002; Misra et al. 2002). Further improvements are still required. Seven physical and 44 sequence gaps remained within the "euchromatic" sequence at the time the Release 3.1 data were frozen for analysis (November 2001). Moreover, although this release extends some 2 Mb or so, on each of the major chromosome arms, into the "heterochromatin," the bulk of the heterochromatin (including the Y chromosome and telomeres) remains as 20.7 Mb of unmapped sequence fragments (Hoskins et al. 2002; Carvalho et al. 2003). About half of these gaps have been closed in the 118.4-Mb Release 4 (April 2004) and efforts are under way to rectify the outstanding problems, as well as to sequence the complex heterochromatin (www.dhgp.org/).

A WGS of the genome of a second drosophilid is now completed: that of *D. pseudoobscura* (hgsc.bcm.tmc.edu/projects/drosophila/; S. Richards, pers. comm.). This has assembled 139 Mb into contigs, and 10,516 orthologs of *D. melanogaster* genes have been identified. There are not, to my knowledge, any plans to complete this sequence, although particular regions will undoubtedly be finished. Shotgun sequences of six other drosophilids (*D. ananassae, D. erecta, D. mojavensis, D. simulans, D. yakuba,* and *D. virilis*) are now available from the Trace Repositories (www.ncbi.nlm.nih.gov/Traces/trace.cgi?) and those of several further species (*D. willistoni, D. grimshawi, D. persimilis, D. sechellia,* and *D. willistoni*) are expected by 2005 (see rana.lbl.gov/drosophila/multipleflies.html, a site that also gives access to the assemblies of the genomes of drosophilids).

The efforts of an international consortium were rewarded in 2001 by the assembly of a tenfold WGS sequence of the genome of *An. gambiae* (Holt et al. 2002). The strain sequenced, PEST, would appear to include regions from more than one population of *An. gambiae,* as some genomic regions were found to be hypervariable and to assemble into distinct haplotypes. These are a very valuable source of single nucleotide polymorphisms (SNPs) for mosquito researchers. An 8X WGS of the genome of *Aedes aegypti* is now in progress (see www.nd.edu/%7Edseverso/genome.html and Severson et al. 2004 for a review of the genome of this species), and plans are being prepared or have been announced for WGS sequencing of the genomes of *Sciara coprophila, Glossina morsitans,* and *Musca domestica.*

Genomic sequencing remains an expensive exercise. It is estimated that, with today's technology, a sixfold coverage WGS costs on the order of U.S. $3.5 million for a 100-Mb genome. Although sequencing costs continue to decrease, the interested research community must clearly make a strong case for a whole genome sequence of their favorite species, as has been done for *Ae. aegypti* (Knudson et al. 2002). The *Drosophila* community has made this case for ten species (see above) (Begun and Langley 2002). In addition, some 0.2–0.3 Mb of sequence, from targeted fosmid clones, are available for four species of *Drosophila: D. erecta, D. willistoni, D. pseudoobscura,* and *D. virilis* (Bergman et al. 2002). There were several justifications for sequencing other selected species of *Drosophila* in particular and other Diptera in general. Sequence comparison is enormously informative with respect to the prediction of protein coding genes (see below). In the absence of experimental data, comparative sequencing is also the most robust method we now have for predicting noncoding regions, including regulatory regions and genes encoding nontranslated RNA species (other than tRNAs that are easy to predict computationally). Above all, however, comparative sequence analysis will be essential if we are to begin to understand the evolution of genomes (as opposed to the evolution of particular genes or gene families) and the significance of sequence change and phenotypic divergence.

Other Genomic Resources

Two general classes of genomic resource are of extraordinary utility in the absence of extensive or complete genomic sequence information. These are high-quality gridded clone libraries and complete or partial cDNA sequences. Robust methods are now available for the construction of high (e.g., tenfold)-coverage BAC libraries with an average insert size of 150 kb or so; the cost is on the order of $15,000 for a 100-Mb genome. These can be replicated and gridded at high density on nylon filters and made available to the research community (see www.chori.org/bacpac/). BAC clones of other species of *Drosophila* are available from the Tucson *Drosophila* Genomics Consortium (tdgc.arl.arizona.edu/). Apart from *Drosophila,* BAC libraries are now available for *An. gambiae* (www.malaria.mr4.org/mr4pages/index.html) and *Ae. aegypti* (www.tigr.org/tdb/e2k1/aabe/intro.shtml).

Careful choice of taxa is clearly a prerequisite for any serious application to construct genomic resources or sequence the genomes of further Diptera. But this choice is not sufficient. If large-scale shotgun sequence assembly is to be attempted, then the choice of genotype is critical. Many, if not most, natural populations of Diptera will be polymorphic for chromosomal inversions; because of the suppression of exchange caused by inversions, sequence divergence can occur between haplotypes to the extent that sequence assembly may become difficult, if not impossible. However, the absence of inversions in a strain is by no means an indication of an absence of sequence divergence, as the experience with the PEST strain of *An. gambiae* has shown (Holt et al. 2002). In this case, the divergence is probably the result of introgression between different populations. If possible, highly inbred lines initiated from a single or very small number of females should be used. Even in this case, the strain must be checked for balanced inversions.

Large-scale partial cDNA sequencing (EST) is now acknowledged to be both a rapid route to gene discovery and an essential resource for modeling gene structures. More than 260,000 EST sequences, and more than 5,000 different full-length cDNAs, are now available for *D. melanogaster* (Rubin et al. 2000b; Stapleton et al. 2002a,b). Smaller collections of ESTs for those species of *Drosophila* being sequenced are also available or will be so within a year or so. In addition, smaller-scale EST projects for *An. gambiae* (Dimopoulos et al. 2002), *Ae. aegypti* (www.tigr.org/tdb/e2k1/aabe/intro.shtml), *Glossina morsitans* (Lehane et al. 2003), *Rhagoletis pomonella* (Roethele et al. 2001), and *Culicoides sonorensis* are under way or have been completed (data from dbEST; see Boguski et al. 1993; www.ncbi.nlm.nih.gov/dbEST/dbEST_summary.html). DNA microarrays, either on glass slides or Affymetrix chips, have been constructed for *D. melanogaster,* and on glass slides for *An. gambiae.*

In addition to these physical resources, the software developed for the annotation of the genome of *D. melanogaster,* in particular APOLLO (Lewis et al. 2002), will be of immeasurable benefit to other genome sequencing projects, as will both the design and some of the components of the computational pipeline for genome analysis (Mungall et al. 2002). Many of the tools required for the computational analysis and display of genomes are now available from the Generic Model Organism Database project (gmod.sourceforge.net/).

Databases

Comprehensive databases of biological data are now essential for research. These are, broadly, of two classes: the "horizontal" databases that are narrow in depth but broad in width (e.g., the international nucleic acid sequence and protein databases), and the "vertical" databases that are broad in depth, but narrow in width (e.g., single-taxon databases). For the Diptera, comprehensive databases of genetic and genomic information are available for the family Drosophilidae (FlyBase 2003) and for mosquitoes: AnoBase (skonops.imbb.forth.gr/AnoBase/) and AgaDB (mosquito.colostate.edu/acedb/AgaDB-acedb.html) for *An. gambiae* and the Mosquito WWW Server (klab.agsci.colostate.edu/) for other species, including *Ae. aegypti* (AeDB). A new database, VectorBase (vectorbase.nd.edu/) is now being established for genomic data of invertebrate vectors of human pathogens (including mosquitoes and tsetse flies) (see also www.genedb.org/genedb/glossina/index.jsp).

Chromosome Number and Synteny

Haploid chromosome number varies little in the Diptera: from $n = 2$ to $n = 6$. Presumably, the primitive diploid karyotype was three pairs of metacentric chromosomes. The major exceptions to the stability of chromosome number in the Diptera are found in the Cecidomyiidae, Sciaridae, and the parthenogenetic orthocladine Chironomidae. In these species, the somatic chromosome number is typical of other Diptera, but in the germ-line, there are between two and 52 extra chromosomes eliminated from the soma during the embryonic cleavage divisions (for reviews, see White 1973: Chapter 14; Matuszewski 1982; Gerbi 1986). These "E-chromosomes" are presumably homologous with the somatic chromosomes (see Staiber 1991), but the two classes of chromosome have diverged by, among other mechanisms, the accumulation of transposable elements (Staiber et al. 1997; Staiber and Schiffkowski 2000; Staiber 2002). True polyploidy is rare in the Diptera; the few triploid species are all parthenogenetic—for example, *Psychoda parthenogenetica* (Troiano 1978), *Phytomyza crassiseta* (Block 1969), and in the genus *Prosimulium* (Chubareva et al. 1974, 2001). B-chromosomes have been found in several families of Diptera; the available data are summarized in Zhimulev (1998: Table 24).

Comparisons of genetic linkage maps gave the first indication that there has been considerable conservation of synteny in the Diptera. In the genus *Drosophila,* the primitive haploid karyotype is six acrocentric elements. These combine in various combinations to form metacentric chromosomes and, hence, reduce the chromosome number to five, four, or three. Muller (1940) emphasized the homologies of the chromosome arms in all species (see also Sturtevant and Novitski 1941 and Matthews and Munstermann 1994 for mosquitoes). These analyses suffer from problems in "homologizing" genes from their mutant phenotypes, yet have been extended to demonstrate the conservation of synteny from the drosophilids to both the tephritids (Malacrida et al. 1985, 1986) and the calyptrate Diptera (Foster et al. 1981). In the absence of a genome project, EST sequences are very useful for the determination of syntenic relationships between species. This utility is for the obvious reason that sequence homology is most readily detected in coding regions (e.g., Roethele et al. 2001). However, even genomic clones can be useful for mapping homologies, at least within the Drosophilidae (e.g., Segarra and Aguadé 1992; Lozovskaya et al. 1993; Segarra et al. 1995, 1996; Ranz et al. 1997; González et al. 2002). Such comparisons within the Drosophilidae not only confirm paracentric inversions to be the major event remodeling chromosomes on an evolutionary time scale, but also show that their rate is on the order of one inversion/My/chromosome element (Ranz et al. 1997, 2001; González et al. 2002). This rate, which assumes an evolutionary distance between subgenera of drosophilines of some 62 My, is equivalent to 0.05–0.07 breaks/Mb/My. It is probable, however, that there is considerable heterogeneity in this rate. Rates of inversion fixation are higher on the X chromosome than on the autosomes. Comparisons of *D. melanogaster* and *D. repleta* (or its close relative, *D. buzzatii*) suggest a greater than threefold difference in breakpoint density, from 2.57 to 8.50/Mb, in different chromosome regions (González et al. 2000). The transposition of genes between chromosome elements is a very rare event in the evolution of *Drosophila* (Ranz et al. 2003). This conclusion is confirmed by the analysis of the draft *D. pseudoobscura* genome. The number of transposition events detected is small, and most of these seem to have involved retrotransposition, most often from the X chromosome (see also Betrán et al. 2003; S. Richards, pers. comm.).

A comparison of genomic sequences between *D. melanogaster* and *An. gambiae*, first in a limited chromosome region (Bolshakov et al. 2002), and then across the entire genomes of these species (Zdobnov et al. 2002), has revealed very extensive microsynteny between these species; that is, conservation of gene neighbors. At the chromosomal level, there is also very extensive conservation of chromosome arm synteny, although blocks are clearly transposed between chromosome arms. This is not surprising, given the evolutionary distance (~250 Mya; Yeates and Wiegmann 1999) between these species; it may well be a consequence of the combined effects of paracentric inversions and Robertsonian translocations.

In addition to being fixed between species, paracentric inversions are very commonly polymorphic in natural populations of Diptera (see Krimbas and Powell 1992). There has been evidence for some time that these inversions result from recombination between a pair of transposable elements arranged in inverted sequence with respect to one another (e.g., Lyttle and Haymer 1992). The most convincing data come from the study of polymorphic inversions in the cactophilic species *D. buzzatii* (Cáceres et al. 1999; Casals et al. 2003). In the cases of both *In(2)j* and *In(2)q*^{7}, polymorphic inversions in this species, the inversion breakpoints are associated with a complex array of transposable elements, including the *Galileo* element, a member of the *Foldback* class. Homologous exchange between two transposable elements should lead to an inversion in which the target site duplications of the original elements are now exchanged (i.e., if the elements of the progenitor chromosome have target site duplications *aa* and *bb*, then the inverted chromosomes should have the elements bounded by the target sites *ab′* and *a′b*, where *a′* and *b′* are inverted complements of *a* and *b*, respectively; Cáceres et al. 1999). This is exactly what is found in the case of *In(2)j*, whereas for *In(2)q*^{7}, one of the target sites seems to have been destroyed by the subsequent insertion of other transposons. Transposable elements also appear to be involved in the generation of the intraspecific polymorphic inversions of *D. pseudoobscura* (S. Richards, pers. comm.).

Composition of the Genome

The sequence complexity of genomes was first demonstrated by the study of the reassociation kinetics of denatured genomic DNA. At a first approximation, this study allows four classes of DNA sequence to be distinguished, because the rate of reassociation is a function of sequence complexity. The first is a fraction that reassociates with zero-order kinetics; that is to say, the sheared single-stranded DNA molecules reassociate by an intramolecular reaction —their sequences are now known as "foldback DNA." A second fraction reassociates very rapidly, and often corresponds to a DNA fraction that has a distinct density on a neutral CsCl gradient—satellite DNA. A third fraction reassociates with kinetics suggesting that it has a moderate sequence complexity; we now know that this includes some, but not all, of the sequences of transposable elements. Finally, there is a fraction that reassociates with kinetics, suggesting that it is composed of sequences present in very few copies per haploid genome. The relative proportion of genomic DNA contained in each fraction varies greatly among species. In *D. melanogaster,* some 21% of the DNA is satellite (Lohe and Brutlag 1987), about 6% foldback (Schmid et al. 1975), about 12% middle repetitive (Laird and McCarthy 1969), and most of the rest is "unique." Similar proportions of "unique" DNA (60%) have been reported for *Ae. aegypti* (with 20% satellite DNA; Warren and Crampton 1991), but this seems low (Brown et al. 2001 suggest that the euchromatin of this species is minimally 45% of the genome); in *Ae. albopictus* and *Culex pipiens,* the values are 33% and 22%, respectively (Black and Rai 1988).

The relationship between repetitive sequences and heterochromatin is well known. However, heterochromatin is not homogeneous; nor are there sharp boundaries between heterochromatin and euchromatin. The structure of the centric heterochromatin of *D. melanogaster* has been studied using the 1.3-Mb *Dp(1;f)1187* "minichromosome" by Karpen and colleagues (Le et al. 1995; see below).

FOLDBACK DNA

DNA sequences that are tandemly repeated in inverted order (e.g., *abc . . . cba*) will, as single-stranded DNA molecules, hybridize with themselves, forming a hairpin structure with any nonrepeated sequence forming a single-stranded loop. The kinetics of reassociation of foldback DNA allows its isolation. Despite indications from older studies that such sequences are abundant in the genome of *D. melanogaster* (e.g., Schmid et al. 1975), there has been little global analysis of foldback sequences in the completed genomes of this species, or that of *An. gambiae.* One foldback sequence has, however, been reasonably well characterized from *D. melanogaster*—that of the *foldback* (*FB*) element (Potter et al. 1980; Truet et al. 1981; Potter et al. 1982). Foldback elements have also been characterized in other drosophilids, in particular in *D. buzzatii* (see below). In various species of mosquito, the proportion of the genome that is composed of foldback sequences has been estimated to lie between 0.04% and 0.11% (summary of data in Knudson et al. 1996: Table 13.1).

SATELLITE DNA

Satellite DNAs were first recognized by their distinct densities in neutral density gradients of sheared genomic DNA. In general, they are characterized by their abundance and relative simplicity of sequence. Typical satellite DNAs have a very short (5–12-bp) sequence motif, although there is also a class of satellites with more complex repeats, often about 150 or 300 bp in length. Satellite DNAs are usually found in large (megabase) blocks in the pericentromeric regions of chromosomes. The proportion of the genome that is satellite varies enormously. In *D. melanogaster,* it is about 20%; in *D. nasutoides,* it is 56%, but in *D. ezoana,* no satellite has been detected, although its relatives in the *D. virilis* group have as much as 46% of their DNA as satellite (Holmquist 1975; Cohen and Bowman 1979; see also Biessmann et al. 2000). Satellite DNAs may be quite specific in chromosome location. For example, the so-called "1.688 satellite" of *D. melanogaster* (a complex sequence with a 359-bp repeat) is found only distal to the X chromosome centromere, and the *dodeca* satellite only on chromosome *3* (Lohe et al. 1993). A further important feature of satellite DNAs is that they are generally very different in their abundances among closely related species. For example, the $[AAGAG]_n$ satellite represents 5.6% of the genome of *D. melanogaster,* but only 0.71% of that of *D. simulans* (Lohe and Roberts 1988).

The pericentromeric location of the major blocks of satellite DNA suggests a function that may be related to that of the centromeres themselves. Indeed, Csink and Henikoff (1998) suggest they function to insulate the centromere from sequences that might disrupt its function—for example, by altering its normal time of replication, which is very late in the cell cycle. It is intriguing that, during mitosis, different ubiquitously expressed DNA binding proteins (often proteins known to have specific regulatory functions at other times) bind the pericentromeric satellite sequences. These binding proteins differ in different species. For example, the PROD protein binds the pericentric regions of mitotic chromosomes, due to an affinity with the $[AATAACATAG]_n$ satellite in *D. melanogaster,* but not in *D. simulans* (Platero et al. 1998; Torok et al. 2000). Because these satellite sequences must condense at mitotic metaphase, it seems that they "borrow" whatever proteins are available to them. This obviously puts constraints on satellite DNA evolution.

TRANSPOSABLE ELEMENTS

The genomes of all sexual organisms harbor transposable elements (for reviews, see Craig et al. 2002; see also Kidwell, Chapter 6). Eukaryotic transposable elements are divided between those that transpose via an RNA intermediate (class I), the retrotransposons, and those that transpose by DNA excision and repair (class II), the nonretrotransposons. Within the retrotransposons, the major division is between those that possess long terminal repeat (LTR) elements and those that do not (non-LTR): LINE-like elements and SINE elements (Deininger 1989). Among the nonretrotransposons, the majority transpose via a DNA intermediate; encode their own transposase; and are flanked by relatively short, terminally inverted repeat structures (TIR elements). "Foldback" elements are characterized by their property of reannealing after denaturation with zero-order kinetics (see above).

Analysis of Release 3 of the genomic sequence of *D. melanogaster* identified 1,572 different transposable elements belonging to 96 different families (Kaminker et al. 2002). Although this sequence extends into the pericentromeric heterochromatin, the bulk of the heterochromatin, which we know to have very high numbers of transposable elements, was not included in this analysis (see Hoskins et al. 2002). The majority (about two-thirds) of the transposable elements are incomplete, due to truncation, deletion, or by having suffered the insertion of other transposable elements, leading to nests of elements. Transposable element density and the proportion of nested elements are both reasonably uniform on the chromosome arms until the basal (centromeric) 2 Mb or so, where it rises several-fold; the overall density on the assembled chromosome arms is between 10 and 15 elements/Mb, except on the small chromosome *4*, where it is 82/Mb. Interestingly, the high density of elements on chromosome *4* is due only to an increase in number of non-LTR and TIR elements, not of LTR-retrotransposons. This estimate of transposable element copy number, and diversity, is almost certainly too low. Not only has the heterochromatin yet to be analyzed in any detail, but also, as Quesneville et al. (2003) demonstrated, the use of novel algorithms uncovers many more elements, especially those that are incomplete. In *An. gambiae,* 41 families of transposable elements have been identified, representing a higher proportion of the euchromatic genome than in *D. melanogaster*—16% vs. 3.9% in the latter species (Holt et al. 2002; Kaminker et al. 2002). As with *D. melanogaster,* novel strategies for transposable element discovery uncover many more elements than those that were found at the stage of genome annotation: indeed, Quesneville et al. (2003) have evidence for more than 400 families of transposable elements in *An. gambiae.* General overviews of the transposable elements of mosquitoes are given by Severson et al. (2001) and Tu and Coates (2004; see also Kapitonov and Jurka 2003; Kapitonov and Jurka 2004 describe some 115 families of transposable elements in *An. gambiae*).

In *D. melanogaster,* the size of families of transposable elements varies between zero and 146 in the euchromatin. The most abundant element, *roo,* shows signs of having recently invaded the genome of *D. melanogaster,* as the variation in sequence between its copies is very low (see also Bowen and McDonald 2001). Some elements of *D. melanogaster* are absent from the strain sequenced. These include the *P-element,* long known to be absent from old established laboratory strains (although now ubiquitous in natural populations). Other elements (e.g., *R2, ZAM*) may be entirely heterochromatic in location and therefore not included in the Release 3 assembly. In *An. gambiae,* some elements have very low copy numbers (1–4), but others are very high; for example, 1184 *gypsy*-like elements and 886 *Pao*-like elements. The SINE200 element has a copy number of more than 20,000 (Holt et al. 2002).

In addition to the *P-element,* several other elements in *D. melanogaster* are known to show very high variation in copy numbers among strains. These include two other elements known to be responsible for hybrid dysgenesis, the *I-element* and *hobo,* the LTR-retrotransposons *gypsy* and *ZAM,* and the enigmatic *NOF* element. In the case of *gypsy,* it is known that strains with a high copy number carry a variant version of the element, known to correlate with a high transposition frequency (Lyubomirskaya et al. 2001).

Very little is known about many aspects of transposable element behavior—in particular, how they invade foreign genomes (see below) and how their copy number is regulated, as assuredly it is. Many elements in *D. melanogaster,* as in other species, show sequence specificity of insertion site. This pattern may be rather general; for example, for A + T-rich sequences, very specific, as in the case of the *R1* and *R2* elements, which insert at specific sites within 28S genes (Jakubczak et al. 1991), or a specificity that seems to be related not to the sequence but to the physical properties of the insertion site. For example, the *P-element* is well known to insert much more frequently near or in the 5′-ends of genes than elsewhere (Spradling et al. 1995; Bellen et al. 2004). The specificity of *P-element* insertion is probably due both to structural features of the DNA (Liao et al. 2000) and to variation in chromatin accessibility. Similarly, the *pogo* element inserts preferentially in regions of higher, and the *roo* element in regions of lower, than average denaturation temperatures (Kaminker et al. 2002).

Retroviral-Like Elements. The two transposable elements first characterized in *Drosophila, copia* and *412,* belong to a very widespread class of elements that are related to true retroviruses in structure. Their termini are long direct repeats (i.e., LTRs) and their genomes encode several proteins, including a reverse transcriptase. There are three major clades of retroviral-like elements (Eickbush and Malik 2002): in the *gypsy* and *BEL* clades, the reverse transcriptase, RNAse H and integrase domains are in this order; in the *Ty1/copia* clade, the integrase occurs before the reverse transcriptase and RNAse H domains. Members of all three clades may, in some species at least, also carry an *envelope* gene (see below). LTR elements are related not only to the true retroviruses of vertebrates but also to hepadnaviruses and caulimoviruses.

The annotated genome of *D. melanogaster* includes 682 retroviral-like elements in 49 families (Kaminker et al. 2002); in *An. gambiae,* there are 3,9301 euchromatic LTR-elements in eight families. In the yeast *S. cerevisiae,* LTR elements cluster near tRNA genes; moreover, the great majority of elements are solo LTR sequences, presumably arising by recombination between the two LTR sequences of an element; neither is the case in *Drosophila.* LTR elements have been identified in all dipteran genomes that have been studied.

Retroviruses. The majority of retroviral-like elements in the Diptera do not encode an envelope protein, nor is there any strong evidence that their genomes can be packaged into viral particles. A few are, however, exceptional: they are true retroviruses and are now classified as Errantiviridae. These differ from the retroviral-like elements in that their genomes encode a functional *envelope* gene; although different viruses may acquire this gene independently, Malik et al. (2000) argue that these *envelope* genes come from other viruses (e.g., baculoviruses). The best-characterized insect retrovirus is the *gypsy* element of *D. melanogaster,* which is present in most strains as a very few integrated genomic copies but which can, in a permissive genotype, produce infectious particles in the ovary (Pelisson et al. 1994; Prud'homme et al. 1995; Lecher et al. 1997; Pelisson et al. 2002). There are probably other retroviruses in drosophilids. For example, the *ZAM* element is also packed into particles in the ovary, although there is no evidence yet for infectivity (Leblanc et al. 1997, 2000), and the *Oswaldo* element of *D. buzzatii* displays features of a retrovirus (Pantazidis et al. 1999).

Non-LTR Elements. First identified in mammals as long interspersed repeat elements (LINEs), non-LTR elements are almost ubiquitous in the metazoa (the bdelloid rotifers are devoid of any retroelement, but this group seems to have given up sex in favor of parthenogenesis some 40 Mya; Arkhipova and Meselson 2000). Like retroviral-like elements, they transpose via an RNA intermediate, catalyzed by an element-encoded reverse transcriptase. They lack LTR sequences, however, and usually terminate in a poly(A) sequence (from which their reverse transcription was primed). Because of their mode of transposition, non-LTR elements suffer extensive 3′-truncations; in *D. melanogaster,* only 12 of 69 *jockey* elements (the most abundant non-LTR element) are full length (Kaminker et al. 2002). Five major evolutionary clades of non-LTR elements are recognized on the basis of their reverse transcriptase protein domains (Eickbush and Malik 2002); members of all occur in the Diptera, with 27 families in *D. melanogaster* (Kaminker et al. 2002) and 12 in *An. gambiae* (Holt et al.

2002). The non-LTR element *NLRCth1* is widely distributed in the Chironomidae (Papusheva et al. 2004). The endonuclease encoded by some non-LTR elements displays very high target sequence specificity. This is best seen in *R1* and *R2,* which insert at different sites within the 28S rDNA genes of all insects. An unusual non-LTR element, *CM-gag,* has been described from *C. pipiens;* like the telomeric *HeT-A* element of *Drosophila,* it encodes a *gag* protein but not a polymerase (Bensaadi-Merchermek et al. 1997).

SINE Elements. SINES are short-interspersed nucleotide elements. They are nonautonomous retroelements, the best known being the human *Alu* element. The majority of SINES are derived from RNA polymerase III transcripts, and it is thought that they are retrotranscribed using a reverse transcriptase from a non-LTR element, at least in mammals. SINE elements are not well characterized in the drosophilids (see below), but have been described from culicids and chironomids. In *Ae. aegypti,* the *Feilai* element is clearly tRNA-related in sequence; its copy number is estimated at 59,000—some 2% of the genome (Tu 1999). The *Maque* SINE of *An. gambiae* has a more reasonable copy number, about 200/genome, but that of the *SINE2000* element is more than 20,000 (Holt et al. 2002); like *Feilai* elements in *Ae. aegypti,* its insertion is biased to high A + T sequences (Tu 2001a). Kapitonov and Jurka (2003) describe a SINE element from *An. gambiae* that has a copy number of about 10,000 and, unusually, has no sequence similarity to any known RNA polymerase III transcript. SINES are also known in chironomids, where they show site-specific insertion in two classes of centromeric tandem repeat (He et al. 1995; Liao et al. 1998; see also Papusheva et al. 2004). The *Twin* SINE of *C. pipiens* is unusual in containing two tRNA-related regions; its copy number is about 500 (Feschotte et al. 2001). Elements that are SINE-like, but do not posses an RNA polymerase III promoter, have been described as *mini-me, NAREP1,* or *INE-1* elements in *Drosophila,* muscids, and calliphorids (Locke et al. 1999; Wilder and Hollocher 2001). Their copy number can be high, estimated to be more than 3,000 in *D. melanogaster.* It has been suggested by Kapitonov and Jurka (2003) that these are remnants of a *Penelope*-like element (see below).

Intron-Containing Retroelements. A novel class of retroelement, one that contains an intron, has recently been described (Arkhipova et al. 2003). The best-known representative of this class is the *Penelope* element of *D. virilis* and other drosophilids (Evgen'ev et al. 1997). Known for some time to be an odd, non-LTR element, its endonuclease having similarities to the homing endonucleases of group I introns, it is now known to have relatives in very different taxa. How it can retain an intron and yet be retrotransposed (if that is indeed its method of transposition) is not known. Some populations of *D. virilis* lack the *Penelope* element, suggestive of a recent horizontal transfer into this species (Evgen'ev et al. 2000). A *Penelope* element has also been found in *Ae. aegypti* (Tu and Coates 2004).

Helitrons. Helitrons are elements that transpose by a rolling-circle replication; autonomous elements encode a DNA helicase and a DNA nuclease/ligase. These elements lack terminal inverted repeats but have conserved 5′- and 3′-termini of TC and CTRR, respectively (Kapitonov and Jurka 2001). First discovered in prokaryotes, several families of helitron have now been characterized in the *An. gambiae* genome (Kapitonov and Jurka 2001, 2003, 2004). In the genome of *D. melanogaster,* a sequence that might be a remnant helitron has also been identified by Kapitonov and Jurka (2003).

TIR Elements. TIR elements are DNA transposons; their transposition does not involve an RNA intermediate but occurs through a transposon-encoded transposase (for a review, see Robertson 2002). Generally, these elements encode only their transposase and their ends are exact inverted repeats, which may be quite short (e.g., 12 bp) or a few hundred bp. Several different clades have been identified by phylogenetic analyses. Nineteen different TIR-element families are known in *D. melanogaster,* with 372 elements; in *An. gambiae,* there are 634 euchromatic elements in 12 families (Holt et al. 2002). One of the largest clades of TIR element is the *Tc1/mariner* superfamily, which includes many families of the ubiquitous *mariner* element, as well as other elements in *Drosophila* and other Diptera (e.g., *Topi1* in *An. gambiae*). A second very large superfamily is *hAT,* which includes a dipteran specific family, *hobo,* with members in *Drosophila* (*hobo*), *Lucilia* (*hermit*), tephritids, and *Musca* (*hermes*) (Warren et al. 1994; O'Brochta et al. 1996).

A third group of TIR elements, known from Lepidoptera, is the *piggyBac* family; these show insertion specificity at TTAA sites and possess a single open reading frame encoding a transposase (Fraser 2000). This family of transposons is widespread; in the Diptera, related sequences are known from *An. gambiae* (Holt et al. 2002) (one of ten copies may be active; Sarkar et al. 2003a), *D. melanogaster* (the *looper* element; Kapitonov and Jurka 2002) and *Batrocera dorsalis* (Handler and McCombs 1999) (see Sarkar et al. 2003a). The *piggyBac* elements are useful transformation vectors (Handler and Harrell 1999; Handler 2002) and have been used to transform Diptera as different as *Ae. aegypti* (Lobo et al. 1999, 2002), *D. melanogaster* (e.g., Horn et al. 2003; Bonin and Mann 2004), and *Lucilia cuprina* (Heinrich et al. 2001).

The *P-elements* constitute a fourth family of TIR elements. These are specific to the Diptera, best known in the drosophilids; the only related sequences are known from the Muscomorpha (Perkins and Howells 1992; Lee et al. 1999; C.-L. Chen, cited in Robertson 2002) and anophelene mosquitoes (de Carvalho et al. 2003; Sarkar et al. 2003b). The *P-element* of *D. melanogaster* occupies a special place in our affections, as it has proved to be both of extraordinary evolutionary interest and exceptionally useful as a tool for germ-line transformation (for reviews of *P-element* biology, see Engels 1989 and Rio 2002). These elements must have invaded the genome of this species very recently (see below).

MITEs. MITEs (Miniature inverted-repeat elements) were identified as small elements (<500 bp) with short inverted repeat ends (Feschotte et al. 2002). Whether they warrant their own class is debatable; they would appear to be grossly deleted TIR elements, as evidenced by the existence of what are clearly active (i.e., full-length) forms (e.g., Jiang et al. 2003). However, one characteristic of some families of MITEs is that they can accumulate to very high copy numbers; others are that they are relatively homogeneous in size within families, have a high A + T content, and that the sequences of many can (computationally) form stable secondary structures (Wessler et al. 1995). A number of different families of MITEs, or related elements, has been identified in mosquitoes (Tu 1997, 2000, 2001a,b; Tu and Orphanidis 2001), and seven different families are recognized in the whole-genome analysis of *An. gambiae* (Holt et al. 2002). As examples of accumulation, the *TA-I* element of *An. gambiae* has an estimated copy number of 1,747 (Tu 2001b; Holt et al. 2002), the *Momo* element of *C. pipiens* a copy number of about 1,000 (Feschotte and Mouchès 2000), and the

Pony element of *Ae. aegypti* may reach a copy number of 10,000—more than 1% of the genome (Tu 1997, 2000). The extent to which the very high numbers of MITEs in the genome of *Ae. aegypti* contribute to its short-period interspersion pattern (see Davidson et al. 1975) remains a speculation (see Tu 1997). MITEs have also been identified in drosophilids; in particular, the *SGM* family in European members of the *D. obscura* species group (Miller et al. 2000) and those of the *hAT* family in *D. willistoni* (Holyoake and Kidwell 2003). These are active transposons in *D. subobscura* and *D. madeiriensis,* but not in *D. guanche,* where they have accumulated to be as much as 10% of the genome. In the latter species, these elements form a centromere-associated satellite DNA sequence.

FB *Elements.* The major features of *FB* elements were summarized above. In the Release 3 genome of *D. melanogaster,* there are 13 *FB* elements. Their termini are not only long (up to a few Kb) inverted repeats, but the sequence within each repeat shows a complex pattern of internal repeats (Potter et al. 1982). The inverted repeats may be juxtaposed or there may be other sequences between the arms; in particular, an enigmatic element known as *NOF* (Goldberg et al. 1982) and the TIR element *HB* (Truet et al. 1981). In some strains of *D. melanogaster,* the composite *FB-NOF* element is very active and may increase in copy number from the usual 1–2/genome to 30/genome (Harden and Ashburner 1990). It may then catalyze both the transposition of large segments of DNA and the formation of chromosome aberrations (see Lovering et al. 1991). Elements similar to *FB* have been characterized in *D. repleta* group species and other drosophilids (Silber et al. 1989; Marin and Fontdevila 1995; Cáceres et al. 2001). *FB* elements have also been characterized in *D. buzzatii,* where there is evidence that they are implicated in the generation of chromosomal inversions (Cáceres et al. 2001; Casals et al. 2003), and in chironomids (Hankeln and Schmidt 1990). How *FB* elements transpose is not known.

Horizontal Transfer of Transposable Elements. There is abundant phylogenetic evidence that transposable elements may be transferred horizontally, in Diptera and other taxa. In *D. melanogaster,* the best data are for the *P-element.* There are several different sequence subfamilies of drosophilid *P-element* and different types of genomic organization of these elements in the genomes of different species (e.g., dispersed vs. blocks of *P-element* sequences). The canonical *P-element* family is restricted to two neotropical species groups of *Sophophora,* the *saltans* and *willistoni* species groups, with the notable exception of *D. melanogaster* (see Silva and Kidwell 2000). Kidwell has made a strong case for several independent horizontal transfer events of this element within the drosophilids (see Clark and Kidwell 1997; Pinsker et al. 2001). Perhaps the best evidence is that this element has an identical sequence in *D. melanogaster* and *D. willistoni* (Daniels et al. 1990). The latter species is northern neotropical in its distribution, and these species overlap (e.g., in the southeastern United States), although the former species has only been present in this region in the past few hundred years (following its introduction from West Africa; see Lachaise et al. 1988).

The retrotransposon *copia* is identical in sequence in *D. melanogaster* and *D. willistoni,* although a phylogenetic analysis would indicate a transfer from *D. melanogaster* (Jordan et al. 1999). The eggs of both of these drosophilids can be attacked by the macrochelid mite *Proctolelaps regulis;* this has been suggested as a potential agent for the horizontal transfer of transposable elements between these species (Houck et al. 1991). In general, parasitoid wasps are also a potential agent for horizontal transfer of genes; these are but ubiquitous hypodermic needles on wings. That a *mariner* element has been found to have 97.6% sequence identity in the moth *Adoxophyes hommai* and its parasitoid *Ascogaster reticularis* (Yoshiyama et al. 2001) suggests that this speculation is not a fantasy.

Organization of the Genome

STRUCTURAL FEATURES

Telomeres. Most eukaryotes have telomeres that consist of arrays of a simple sequence whose length and integrity are maintained by telomerase, an enzyme that reverse transcribes its internal RNA and adds repeats to telomeric DNA. This is apparently so in most insect orders, but not in the Diptera (Okazaki et al. 1993). In *D. melanogaster,* the telomere consists of an ordered array of two non-LTR transposable elements, *HeT-A* and *TART* (for reviews, see Biessmann and Mason 1997; Pardue and DeBaryshe 2003). Both are somewhat unusual non-LTR elements, in having long untranslated 3′-regions; moreover, *HeT-A,* unlike most non-LTR elements, does not encode its own reverse transcriptase. The telomeres are maintained by the transposition of these elements into the telomeres. In addition, these elements can capture a "naked" chromosome end and repair it with a new telomere (Biessmann et al. 1990, 1992; Levis et al. 1993; Sheen and Levis 1994). Chromosomes lacking telomeres resect at the rate of about 75 bp/generation (Biessmann and Mason 1988). A remarkable feature of the *HeT-A* element, at least, is that it evolves very rapidly; that in *D. yakuba* has only 55% nucleotide similarity with its *D. melanogaster* homolog (Casacuberta and Pardue 2002). Proximal to the arrays of *HeT-A* and *TART* elements at *Drosophila* telomeres there are blocks of a complex satellite-like sequence, known as the "telomere-associated satellite" (TAS) (Karpen and Spradling 1992; Thompson-Stewart et al. 1994). The origin of the TAS repeat is unknown, but R. Levis (pers. comm.) has found sequence similarity with the *invader4* transposable element.

Telomere lengths in *D. melanogaster* vary between 26 Kb and 147 Kb, and are variable between strains (see Melnikova and Georgiev 2002; Siriaco et al. 2002). Abad et al. (2004a,b) have cloned and characterized six of the telomeres of the sequenced strain of this species. The TAS sequences vary between 16 Kb and 25 Kb in length, but are absent from two telomeres. Remarkably, Abad et al. (2004a,b) have discovered four copies of a new element, *TAHRE,* which carries a reverse transcriptase similar to that of the *TART* element and a gaglike protein (and UTR sequences) similar to those of *HeT-A*. One of the four *TAHRE* elements appears to be functional and is suggested to provide the reverse transcriptase for *HeT-A* transposition.

In chironomids, telomeres are composed of blocks of tandem repeats on the order of 300-bp long (see Martinez et al. 2001). How these blocks are maintained is not known, perhaps by both gene conversion and unequal crossing-over, although there is a tantalizing hint that reverse transcription may be involved, as reverse transcriptase has been detected immunologically at the telomeres of chironomid chromosomes (López et al. 1996, 1999; see Rosén et al. 2002). The nature of the telomeres of mosquitoes is unclear; Roth et al. (1997; also Biessmann et al. 1998) provide evidence that neither simple sequence repeats nor transposable elements occur at one telomere of *An. gambiae.* They suggest this telomere may be maintained by unequal recombination.

Centromeres. Attempts to define centromeres at the sequence level in the higher eukaryotes have been fraught with difficulties. Some progress is, however, being made in *D. melanogaster,* and perhaps some of the reasons for the previous difficulties are becoming clearer. Karpen and colleagues have defined the sequences necessary for centromere function on the small chromosome *Dp(1;f)1187* to within a 420-Kb region (Murphy and Karpen 1995) and have recently extended their analysis to sequencing some of the satellite and most of the complex DNA of this centromere (Sun et al. 2003). They conclude that there will be no "centromere-specific" sequences in *Drosophila,* or other metazoa; rather, that the centromere is defined by a combination of higher-order structural features and epigenetic mechanisms (Karpen and Allshire 1997; Sullivan et al. 2001). In an alternative view, Henikoff and colleagues (Csink and Henikoff 1998; Henikoff et al. 2001) point to the rapid coevolution of not only the pericentromeric satellite DNAs and their binding proteins (see above) but also of the remarkably rapidly evolving centromeric Histone H3 variant, CID (Malik and Henikoff 2001; Malik et al. 2002). They stress that homologous centromeres are competing for inclusion in a functional egg nucleus, in animals and plants, and that any factor that increases centromere "strength" will drive through a population. However, homologous centromeres cannot be allowed to diverge, as this would lead to nondisjunction, and the coevolution of the sequence of the *N*-terminal tail of the CID protein may be the compensatory mechanism that maintains equality of homologous centromeres. A nice feature of this model is that it gives a basis for the sequence variation (e.g., in satellite DNA content) of centromeres.

In polytene nuclei of drosophilids, the embedding of the centromeres in heterochromatin means that these regions do not polytenize; indeed, all of the chromosome arms are held together at a rather amorphous chromocenter. This is not the case in all Diptera. In simuliids (Chubareva et al. 2003) and chironomids, for example, chromocenters are absent, and the centromere may be represented by typical bands, as in *Camptochironomus pallidivitatus.* In this species, all centromeres include large (~50-Kb) blocks of a 155-bp tandem repeat (Rovira et al. 1993); in addition, on chromosome *3* alone, there are some 100 copies of a 375-bp tandem repeat (He et al. 1998). Chromosome specific repeats mapping close to centromeres have also been characterized in *D. melanogaster;* for example, a 359-bp repeat (aka "1.688 satellite") on the X chromosome (Gall et al. 1971), the *Rsp1* and *Porto* repeats on chromosome *2* (Wu et al. 1988; Coelho et al. 1996), and the *dodeca* satellite repeat on chromosome *3* (Abad et al. 1992).

Organization of the Genes

NONPROTEIN CODING GENES

Ribosomal RNA Genes. Clusters of genes encoding the major—18S and 28S—ribosomal RNAs are found at specialized nuclear structures, the nucleoli. In *D. melanogaster,* there are two nucleolar organizers, on the X and Y chromosomes, each with 200–250 copies of the 28S + 18S gene, forming arrays greater than 2 Mb in length. Two non-LTR transposable elements, *R1* and *R2,* interrupt a proportion of the 28S genes, each inserting at a specific site (Eickbush 2002). These interrupted genes are not functional. Because of their value for phylogenetic analyses, the rDNA sequences of more than 700 Diptera, from over 60 families, have been sequenced (GenBank, 1 March 2003). However, in the majority of cases, these sequences are not complete. The 5S rRNA is also encoded by a tandemly repeated gene with, in *D. melanogaster*, about 165 copies clustered on an autosomal arm *2R*. In some species, the 5S rRNA genes are split among two or more sites (e.g., Kress et al. 2001).

One expected function of the rDNA arrays, at least in *Drosophila*, is that they are essential for the normal disjunction of the X and Y chromosomes during male meiosis. A deletion of the entire X chromosome rDNA array results in a high frequency of sex-chromosome nondisjunctions; a single rDNA array, ectopically located on the X chromosome, can correct this defect. The rDNA sequences responsible are within the spacer region (IGS) between rDNA genes (McKee and Karpen 1990; McKee 1996; Ren et al. 1997).

Transfer RNA Genes. The tRNA encoding genes of *D. melanogaster* are probably fully known. There are 290 predicted in the Release 3 sequence (Misra et al. 2002). This is a relatively low number, similar to that in *S. cerevisiae* (275), and far fewer than in *C. elegans* (588) (GtRDB 2003). A characteristic of tRNA genes in *Drosophila* is that they often occur in clusters that may include both similar and different tRNAs (with respect to their specificities). In *D. melanogaster*, all ten tyrosyl-tRNA genes have introns (20-113-bp), as do two of the 11 isoleucyl-tRNAs and four of the 23 leucyl-tRNAs (see Trotta and Abelson 1999).

"Small RNA" Genes. The biological role of noncoding RNAs (other than the well-known tRNAs and rRNAs) is of increasing interest (Pasquinelli 2002). Small (21–22-nucleotide) RNAs were first identified in *C. elegans*, and some are known to have important roles as regulators of developmental timing. More than 110 miRNA genes are estimated for *D. melanogaster* (Lagos-Quintana et al. 2001; Bae et al. 2002; Aravin et al. 2003; Enright et al. 2003; Lai et al. 2003). These RNAs are cleaved from longer precursors by the RNA interference pathway using the *Dicer* RNA-processing enzyme. Two genes encoding this ribonuclease III enzyme are known in *D. melanogaster*, as are many of the components of the RISC complex, required for the interactions between miRNAs and their target mRNAs (Caudy et al. 2002; Kennerdell et al. 2002). These miRNAs may function as regulators of mRNA translation (see, e.g., Brennecke et al. 2003; Xu et al. 2003), and several attempts have been made to computationally predict miRNA targets (Enright et al. 2003; Stark et al. 2003; Rajewsky and Socci 2004).

Small RNAs are also associated with transposable elements; these have been called "repeat-associated small interfering RNAs" (rasiRNAs) and are between 24 and 26 nt in length (Djikeng et al. 2001). These rasiRNAs match both the sense and antisense strands of repetitive elements and may be associated with either the silencing of element activity or the establishment of chromatin structure (see Aravin et al. 2001; Reinhart and Bartel 2002; Kogan et al. 2003). Aravin et al. (2003) identified 178 rasiRNAs, from 40% of the known transposable elements and other repeated sequences, in *D. melanogaster*. These RNAs are most abundant in early embryos and testes RNA preparations. There is increasing evidence that these transcripts are involved in the regulation of transposable elements. It is interesting that antisense transcripts of the telomeric *TART* element are far more abundant in nuclei than are sense transcripts (Danilevskaya et al. 1999). Moreover, mutations in *spn-E*, which encodes for an RNA helicase implicated in RNAi-mediated translational suppression, increases the steady-state level of the transcripts of several transposable elements (Aravin et al. 2001; Stapleton et al. 2001; Kogan et al. 2003). The organization, and sequences, of miRNA genes seem to evolve rapidly, at least according to preliminary comparisons of these in *Drosophila* and *Anopheles* (Lai et al. 2003).

Other major classes of nontranslated RNA include those that are components of the spliceosome complex and those required for the processing of the primary ribosomal RNA transcripts. The genome of *D. melanogaster* encodes at least 28 of the former, the snRNAs (Misra et al. 2002). Eukaryotic genomes have two classes of intron: a major class spliced by the U2-spliceosome and a rare class spliced by the U12-spliceosome. Both are known in *Drosophila* (see below). One "snRNA," U7, is not involved in splicing but in the processing of the 3′-end of the cell-cycle specific histone mRNAs (Dominski et al. 2003). Most snRNA genes are transcribed by RNA polymerase II, but the U6 genes use RNA polymerase III. The promoters of both classes are characterized by PSE elements (Jensen et al. 1998; McNamara-Schroeder et al. 2001).

The snoRNAs function as "guides" for the modification and folding of ribosomal and snRNAs. Yuan et al. (2003) have identified 35 snoRNA genes, and a further 16 were identified by Tycowski and Steitz (2001). Most snoRNAs are encoded by genes within introns of other genes, either conventional coding genes (in which case, they usually encode proteins involved in aspects of protein synthesis or protein or RNA metabolism), or noncoding "host" genes, such as *Uhg1* and *Uhg2* (Tycowski and Steitz 2001).

Components *roX1* and *roX2* are untranslated RNA components of a ribonucleoprotein complex responsible for the hyperactivation of the male X chromosome, associated with dosage compensation in *D. melanogaster* (for a review, see Meller and Kuroda 2002). These RNAs are male specific, at least as stable transcripts. Their precise roles are unclear; certainly *roX1* is not required for dosage compensation (Meller et al. 1997), although in its absence, the *roX2* RNA may compensate (Franke and Baker 1999). Other nontranslated RNAs appear to be associated with another ribonucleoprotein complex, the pole granules. These include the *pgc* RNA (Nakamura et al. 1996; Martinho et al. 2004) and the large mitochondrial rRNA (Iida and Kobayashi 1998). Another nontranslated RNA is that transcribed at a major heat-shock puff, *Hsro,* in drosophilids (Prasanth et al. 2000) and in chironomids (Morcillo et al. 1993).

PROTEIN CODING GENES

Gene Number. Gene number in *Drosophila* was first estimated by genetic methods based on the rates of mutation of individual genes and the average rate of mutation (under the same conditions) of all genes on the X chromosome that yielded lethal alleles. Such experiments gave an estimate of about 5,000 genes, which (unhappily) was the same number as that of polytene chromosome bands. Although it was realized perfectly well that not all genes were vital, there was not, at that time, any good estimate of just what fraction of the genes was essential for viability. In fact, this number is relatively low: 24% (Ashburner et al. 1999). The estimate of gene number from genome sequencing relies, of course, on the accuracy of methods of gene prediction (see Ashburner 2000a; Lewis et al. 2000). From the Release 3.2.0 sequence and from the analysis of the heterochromatic sequences so far available, the predicted number of protein coding genes in *D. melanogaster* is 13,651 (data from FlyBase). In addition, there are more than 400 nontranslated RNA genes (this number is an underestimate). The estimate of the number of protein coding genes in the genome of *An. gambiae* is very similar: 13,683 (Holt et al. 2002).

There has been some controversy about the numbers of protein coding genes estimated from sequenced genomes. This is an inevitable consequence of the imperfection of gene prediction programs and a paucity of experimental data. The FlyBase annotation of the genome of *D. melanogaster* was, by design, conservative. Hild et al. (2003; see also Gopal et al. 2001) have combined other gene prediction programs with experimental data and estimate that this genome may include 16,000–17,000 genes. This may be too liberal an interpretation: a reasonable ballpark estimate for this genome would, today, be 15,500. It is interesting that a detailed comparison of the genomes of *C. elegans* and *C. briggsae* increased the number of genes in the former species by about 1,800 (Stein et al. 2003). Similar comparative data from drosophilids may do likewise; indeed, there are indications from the analysis of the sequence of *D. pseudoobscura* that this is so (S. Richards, pers. comm.).

Gene Families. Many genes are organized as families by the criterion of the similarities of their products; often these products are also related in function. In general, gene families are either dispersed or clustered. The seven actin coding genes of *Drosophila* are a good example of a small dispersed family; other distributed families (e.g., the trypsin-like genes and cytochrome P450s—90 genes and >165 genes, respectively) may be much larger and include small clusters. It is expected that these evolutionary expansions of particular gene families will be adaptive, and therefore, will differ in different species of fly. Indeed, this can be seen in the hemoglobin-containing chironomids, where there are about 40 different globin genes arranged in a few clusters (see below).

Clustered families presumably evolve by gene duplication. The best-known clustered families are of the genes encoding the ribosomal RNAs. The maintenance of homogeneity within large arrays may be due to their concerted evolution (Smith 1974); that is, the combined effects of gene conversion and array expansion/contraction due to unequal crossing-over. At the major rRNA locus in *Drosophila,* unequal crossing-over is, indeed, an important process (for a review, see Ashburner et al. 2005). There is also evidence for the role of gene conversion in homogenizing the much smaller arrays of *Hsp70* genes in *Drosophila* (e.g., Bettencourt and Feder 2002).

Large arrays may include more than one gene; the best-known example of these is perhaps the histone cluster in *Drosophila:* in *D. melanogaster,* the array consists of about 100–200 tandemly repeated copies of 4.8-Kb and 5-Kb modules that carry all five major histone genes (Lifton et al. 1978). This cluster is not continuous; subclusters are flanked by nonhistone gene DNA (Saigo et al. 1981; Colby and Williams 1993). Interestingly, the organization of the cell-cycle histone genes is quite dynamic in drosophilids (for a recent discussion, see Ranz et al. 2003). Whether concerted evolution is responsible for maintaining homogeneity of the coding regions of the genes of the histone cluster is not yet clear. Piontkivska et al. (2002) argue that the evolution of the histone H4 family is best described by the "birth-and-death" model (Nei and Hughes 1992). In this model, in contrast to concerted evolution, each member of the array evolves independently, and homogeneity is maintained by purifying selection. These models can be tested when more of the sequence of the histone array is available (its repeated nature poses major problems for the assembly of the array, and at present, this region is a sequence gap in the *D. melanogaster* genome). Concerted evolution would predict homogeneity even at the level of synonymous sites; the birth-and-death model would predict that the number of synonymous differences will be high (unless the duplications are very young; see Nei et al. 2000).

A surprise of the first large-scale annotation of genomic DNA in *D. melanogaster* was how common small, tandemly arrayed clusters of genes are (Ashburner et al. 1999). This was confirmed at the whole-genome level (Rubin et al. 2000a). In most of these tandem arrays, all members are transcribed from the same strand, consistent with an origin by unequal exchange. These arrays can be quite large; for example, one of genes of unknown function with 17 genes and two clusters of ten glutathione S-transferase genes. Nevertheless it has been argued (Gu et al. 2002) that, in comparison with *Saccharomyces cerevisiae* and *C. elegans*, the rate of gene duplication in *D. melanogaster* is very low. This conclusion is supported by the analyses of Lynch and Conery (2000), who deduce that the rate of duplication in *D. melanogaster* is about 30/My, a factor of ten less than the estimate for *C. elegans*. Large-scale "block" duplications of genomic regions, such as are seen in yeast (Seoighe and Wolfe 1999) and humans (Bailey et al. 2002), are not known in *Drosophila* (see Friedman and Hughes 2001).

The amplification of particular gene families has been associated with the development of field resistance to insecticides in both *Drosophila melanogaster* and culicine mosquitoes. Field populations of the former species are subject to selection for resistance to copper, which they meet as Bordeaux Mixture sprayed in vineyards against *Phylloxera vastatrix*. Copper ions and those of related heavy metals, such as zinc and cadmium, are bound (and rendered biologically inert) by metallothioneins. In some natural populations, the *Mtn* gene has been tandemly duplicated, leading to increased resistance to copper (Maroni et al. 1987).

Resistance to organophosphorous insecticides in *Culex pipiens* and *C. quinquifasciatus* is associated with a dramatic amplification of genes encoding esterases that detoxify these compounds (Mouchès et al. 1986). The amplification of the *esterase B* gene seems to have been a unique and recent event, in which a single gene is now a cluster of 250 or so tandemly repeated copies, organized as a contiguous chromosomally located block (Tomita et al. 1996). This amplified locus has now spread worldwide (Raymond et al. 1991, 1998).

Amplification of globin genes is seen in the chironomids, whose larvae live in hypoxic environments and use extracellular hemoglobins for oxygen transport. In these species, there are often more than 40 different globin genes in multigene clusters (e.g., Chen et al. 1995; Hankeln et al. 1998; for a review, see Vinogradov et al. 1992); in *D. melanogaster*, there is only one globin gene (Burmester and Hankeln 1999), as there is in *Gasterophilus intestinalis*, another species that lives in a hypoxic environment, the stomachs of horses (Dewilde et al. 1998).

Nonhomologous genetic exchange is a well-known mechanism for generating tandem duplications in *D. melanogaster* (Sturtevant 1925). Its frequency is quite high (Gelbart and Chovnick 1979) and probably results from exchange occurring between dispersed copies of identical transposable elements (Goldberg et al. 1983; Davis et al. 1987). In populations subject to strong positive selection, the chance duplication of genes whose products confer a selective advantage will quickly sweep through a population.

Regulatory Elements. A number of operationally different elements has been characterized as playing a role in the *cis*-regulation of transcription. These have been reviewed, for *D. melanogaster*, by Arnosti (2003). The transcription start sequence (TSS) is usually fewer than 400-bp 5′ to the ATG start codon (Ohler et al. 2002); it and its immediately contiguous sequences define the promoter, the region where the RNA polymerase binds and initiates transcription. In less than 30% of genes (in *D. melanogaster*), this region has a recognizable TATA box (consensus: TATAAA) at –30 with respect to the TSS (Ohler et al. 2002). The TATA box binds the TATA-binding protein, a component of the TFIID complex. The TSS itself is, in about 60% of genes, included within an initiator motif (INR) (consensus: TCA (G/T)T(C/T)), transcription beginning at the A residue. Some promoters contain a DPE motif (consensus: (A/G/T)(C/G)(A/T)(C/T)(A/C/G)(C/T)), located between +28 and +23 (relative to the TSS). It is usually considered that INR and DPE elements cooperate in binding TFIID in the absence of a TATA motif, although some promoters have both TATA and DPE motifs (Kutach and Kadonaga 2000).

By definition, enhancer elements bind transcription factors and influence transcription from an adjacent promoter, either positively or negatively. Regulatory regions may contain very many enhancers, functioning as complex *cis*-regulatory modules, and although these are classically 5′ to their cognate promoters, they are also found within introns or 3′ to the genes they regulate.

Insulator elements are sequences that function as barriers to *cis*-regulation. As a consequence of binding specific proteins, these sequences block the regulatory sequences of one gene influencing a neighbor (for a review, see Gerasimova and Corces 2001). These elements may act by partitioning the chromatin into autonomously functioning looped domains (see Gerasimova and Corces 2001) or as "enhancer decoys" (see Geyer 1997). Indeed, it is not clear that all insulators function by a common mechanism. In *D. melanogaster*, insulator elements have been well characterized at the *Bithorax* cluster (e.g., Hagstrom et al. 1996), at the *Hsp70A* locus (e.g., Kellum and Scheld 1992), at the *ftz-Scr* loci (Belozerov et al. 2003), and on the *gypsy* element, where they bind the *su(Hw)* protein (Gdula et al. 1996).

Introns and Variable Splicing. Introns in *D. melanogaster* are A + T rich and vary in length over a wide range, from about 40 bp to more than 3 Mb. There is, however, a peak in the frequency distribution of introns at 59–63 bp (Mount et al. 1992); only 32 introns are smaller than 48 bp; the smallest in *D. melanogaster* for which there is experimental evidence of 43 bp (*Acp98AB*). In the Release 3.1 sequence of *D. melanogaster*, 48,257 introns were annotated in 13,379 protein coding genes, an overall average of 3.6 introns/gene (i.e., the "average" gene has 4.6 exons). The record for intron number in *D. melanogaster* is the *Dscam* gene, encoding an axon guidance receptor, with 114 introns (Celotto and Graveley 2001). (The *dumpy* gene encodes the longest predicted protein, 23,054 amino acids; this is an extracellular matrix protein.) Most, if not all, genes have exons that are at least in part untranslated, at either or both of their 5′ and 3′ ends (Misra et al. 2002). A consequence of most genes having multiple exons is, of course, that there are opportunities for alternative splicing. The prediction of alternative splicing by computational means is not yet reliable, and its detection relies on EST or cDNA sequences. Misra et al. (2002) estimate that about 20% of genes in Release 3.1 have alternative transcripts; this is probably an underestimate. The *Dscam* gene has the potential to encode more than 38,000 different proteins (Celotto and Graveley 2001); 95 of its 114 exons are variably, and 20 constitutively, spliced (see Neves et al. 2004).

Although the great majority of introns in *D. melanogaster* are spliced by the U2 spliceosome, a few use the U12 spliceosome, in which the U1, U2, U4, and U6 snRNAs are replaced by U11, U12, U4atac, and U6atac snRNAs, respectively. These introns have AT/AC (rather than GU/AG) termini; the best studied is that in the *prospero* gene, in which a U2 intron is nested within a U12 intron (a "twintron"; Scamborova et al. 2004).

Large introns (i.e., those larger than 1 Kb) often contain regulatory sequences; for example, enhancers. It is a fallacy to believe that intron sequences evolve at a neutral rate. Interspecific comparisons in drosophilids (Bergman and Kreitman 2001; Bergman et al. 2002) show clustered blocks of highly conserved sequences in both introns and intergenic regions (see also Moses et al. 1990).

A consequence of long introns is that these genes take a considerable time to be transcribed. The rate of transcription, in *D. melanogaster* at 25°C, is 1.1 Kb/min (Thummel et al. 1990): a gene with a 70-Kb intron takes over 1 hour to transcribe. In early development, this may be far more than the cell-cycle time (see Rothe et al. 1992). It is thought that the time taken to transcribe these genes is a component of their regulation (see Gubb 1986).

Gene Structures: Variations on Themes. There is very considerable variety in the organization of genes. Polycistronic genes were an unexpected feature of eukaryotic genomes; yet in *D. melanogaster,* they are not uncommon and come in two major flavors. On the one hand are pairs of genes that are transcribed into a functional dicistronic mRNA, such as *stonedA* and *stonedB* and *Adh* and *Adhr* (Andrews et al. 1996; Brogna and Ashburner 1997). In these pairs, translation of the downstream gene is presumed to be by internal ribosome entry or by a mechanism involving partial disassembly of the ribosome and continued ribosome scanning by the 40S subunit. Thirty-one dicistronic genes of this class are predicted in Release 3.1 (Misra et al. 2002). On the other hand are genes that may encode a dicistronic primary transcript, but whose mRNAs are monocistronic. Then two (or more) mRNAs from a single gene will share 5′-sequences that may be untranslated (e.g., *sesB* and *Antp2;* Zhang et al. 1999). Alternatively, the shared 5′-region may encode the *N*-terminus of two proteins, which may be very different in function. The best-known cases of this in *Drosophila* are the *Su(var)3-9* gene, which encodes two proteins that have 80 identical *N*-terminal amino acids, but functions as a chromatin-binding protein and translation initiation factor (Krauss and Reuter 2000); and *Cha* and *VAChT,* in which case, the two proteins are functionally related, being a choline acetyl transferase and acetyl choline transporter, respectively (Kitamoto et al. 1998).

Genes with dicistronic primary, but monocistronic processed, transcripts represent a special case of "included" genes; that is, genes within introns of other genes. These also turn out to be surprisingly common in the genome of *D. melanogaster,* in which more than 7% of genes are within the introns of other genes (Ashburner et al. 1999; Misra et al. 2002). Generally, but by no means universally, intronic genes are transcribed in the opposite direction to the gene within which they are included and are transcribed independently (e.g., Henikoff and Eghtedarzadeh 1987).

Another unexpected discovery in *D. melanogaster* has been of genes whose transcripts are *trans*-spliced. Well known in trypanosomes and even nematodes, in which many transcripts may share a common *trans*-spliced leader sequence (for a review, see Blumenthal 1995), the single case of *trans*-splicing known to date in flies is that of *mod(mdg4).* This gene encodes a chromatin-binding protein with at least 22 different isoforms as a consequence of alternative splicing (Misra et al. 2002). The remarkable feature is that at least eight alternative 3′-exons are coded by the opposite strand to the common 5′-exons of this gene (Dorn et al.

2001; Labrador et al. 2001; Mongelard et al. 2002; Pirrotta 2002). Although these 3′-exons are normally contiguous with the common 5′-region of this gene, they can be *trans*-spliced even if moved to a distant genomic site. This gene is not *trans*-spliced in *An. gambiae;* indeed, the gene has been spectacularly rearranged in *An. gambiae* with respect to *Drosophila,* despite local synteny of the *mod(mdg4)* region in these species (Laborador and Corces 2003).

Although the great majority of *Drosophila* mRNAs use AUG as their initiating codon, there are a small number known in which this rule is broken, and translation begins with other codons (CUG, ACG, GUG) being read as methionine (see Ashburner et al. 2005). Some mRNAs have long 5′-UTR regions that include many small ATG-initiated open reading frames (e.g., there are 15 of these in the 1.7-Kb 5′-UTR of the P2 transcript of *Antp;* Oh et al. 1992; Ye et al. 1997). In these cases, initiation at the "correct" ATG is thought to be by internal ribosome entry (see Hart and Bienz 1996). Ye et al. (1997) point out that *Drosophila* genes with multiple small open reading frames in the upstream leaders of their mRNAs usually encode regulatory proteins and that their cap-independent translation may be a mechanism to dissociate the control of their translation from the regulation of protein synthesis as a whole.

Translational Read-Through and Selenocysteine. The genome of *D. melanogaster* encodes a single selenocysteinyl tRNA that can decode the normal termination codon UGA inserting the rare amino acid, SeC (Zhou et al. 1999). A consequence is the read-through of an apparent stop codon. It is the secondary structure of the mRNA—particularly the presence of the SECIS element in the 3′-UTR—that determines whether SeC incorporation or termination occurs at a UGA codon. From the point of view of genome annotation, SeC incorporation is indicated by an in-frame UGA stop codon within an otherwise intact open reading frame, and can be experimentally confirmed by ^{75}Se incorporation into the protein. Three selenoproteins are predicted in the genome of *D. melanogaster* (Castellano et al. 2001; Martin-Romero et al. 2001). The selenocysteinyl tRNA–specific translation elongation factor required for SeC incorporation has been identified in *D. melanogaster* (Tujebajeva et al. 2000).

Five genes have been identified in *D. melanogaster* with in-frame stop codons that are apparently read through by the translation machinery, although many more have been predicted (Sato et al. 2003). In some cases, the mechanism of read-through may be translational frameshifting. The only known example of this mechanism is the *Oda* gene in *D. melanogaster,* encoding the ornithine carboxylase antizyme. Here, a +1 translational frameshift is needed to translate the *Oda* mRNA into a functional protein (translational frameshifts are, of course, well known in retroelements). Polyamine levels regulate the efficiency of this frameshift in the *Oda* mRNA. This mechanism of regulating polyamine concentration is conserved from yeast to mammals (see Ivanov et al. 2000). In other cases (e.g., *headcase, kelch*), read-through is probably by tRNA suppression (Xue and Cooley 1993; Steneberg et al. 1998), or the stop codon may be bypassed by splicing (e.g., the female specific transcript of *transformer* (Boggs et al. 1987).

RNA Editing. First discovered in kinetoplastid transcripts of trypanosomes, the editing of the nucleotide sequence of gene transcripts is now known to be widespread. RNA editing involves the deamination of either cytosine residues (to create uridine) or adenosine residues (to create inosine) and is accomplished by specific RNA base deaminase (Gott and Emerson 2000; Bass 2001; Grauso et al. 2002; Hoopengardner et al. 2003; for reviews, see Bass 2001,

2002). RNA editing may alter the coding potential of a transcript. In *D. melanogaster,* at least 17 genes are known whose products are edited by an adenosine deaminase; most of these encode proteins that function as either voltage- or ion-gated ion channels or proteins involved in the synaptic release machinery (Hoopengardner et al. 2003; see also Stapleton et al. 2002b). Even though A-to-I editing creates codon changes in the proteins resulting from these transcripts, the enzyme required for editing is not vital, as knockout mutant flies are viable, although behaviorally abnormal (Palladino et al. 2000a,b). The conservation of edited sites among very different species of *Drosophila,* in the *para* (Na^+-channel) gene (Hanrahan et al. 2000) and between *Drosophila* and the lepidopteran *Heliothis virescens,* in the *Dalpha6* ACE-receptor subunit (Grauso et al. 2002), suggests that this phenomenon will be general in the Diptera. C-to-U editing has not, so far, been described from any invertebrate.

Antisense Transcripts. Antisense transcripts are known from a small number of genes in *D. melanogaster.* These transcripts include several from transposable elements (Contursi et al. 1993; Lankenau et al. 1994; Danilevskaya et al. 1999), an antisense transcript of *Trap100* in *D. pseudoobscura* (Noor et al. 2003), *Rnp4F,* and the Y-linked *Suppressor-of-Stellate* (see below). There is experimental evidence that antisense transcripts of the *I-element* play a role in the regulation of this non-LTR element's transcription (Jensen et al. 1999a,b, 2002; Malinsky et al. 2000).

The *Rnp4F* gene combines both RNA editing and regulation by an antisense transcript (Petschek et al. 1996, 1997; Peters et al. 2003). Transcription from one of the two promoters of the adjacent gene, *sas-10,* reads through the noncoding strand of *Rnp4F.* This transcript can form an RNA duplex with the *Rnp4F* transcript and target both transcripts for degradation and for promiscuous A-to-I editing (which contrasts with the specific A-to-I editing known in other edited transcripts of *Drosophila*). Although degradation could be by the RNAi pathway, it may also be the result of ribonucleases specific for inosine-containing nucleotides (see Scadden and Smith 2001).

The X-linked *Stellate* locus of *D. melanogaster* (it is absent in *D. simulans*) includes a repeat of genes that encode a protein that probably acts as a regulatory subunit of a protein kinase. In *X0* males, these genes are hyperactive, leading to the accumulation of this protein in primary spermatocytes, where it forms characteristic crystals whose morphology depends on the copy number of the *Stellate* gene on the X chromosome (Livak 1990, 1984; Palumbo et al. 1994; Bozzetti et al. 1995). The suppression of this gene in XY males requires the presence of a paralogous locus, *Su(Ste)* on the Y chromosome and the activities of other proteins, such as those encoded by *aubergine* (Schmidt et al. 1999), *spindle-E* (Aravin et al. 2001), and *armitage* (Tomari et al. 2004). The mechanism of suppression appears to be RNA interference via dsRNA formed from sense and antisense transcripts of *Su(Ste)* (Aravin et al. 2001; Gvozdev et al. 2003; Tritto et al. 2003). Both *aubergine* and *spindle-E* are known to be required for RNAi in the oocyte (Kennerdell et al. 2002).

Pseudogenes. A rigorous definition of pseudogenes is hard to formulate. However, in general, pseudogenes are nonfunctional paralogs of genes, originating by unequal genetic exchange, by the incorporation into the genome of a reverse transcript of an RNA, or by the "degeneration" associated with, for example, new Y chromosomes. In *Drosophila,* all three classes are known. The usual signatures of pseudogenes that have originated by unequal exchange are that they are in tandem with their "normal" paralog and, if this gene has introns, then these are retained in the pseudogene. The pseudogene's coding sequence is normally nonfunctional as a consequence of base substitutions and/or frameshift mutations. By contrast, retrotransposed pseudogenes usually lack the introns of their normal paralog, may retain a 3′-poly-(dA) run, and are located at distant genomic sites.

Unequal exchange-generating tandem duplication is a common genetic phenomenon (see the subsection on gene families above). One of the surprising results of whole-genome sequence analysis has been the extent to which tandem pairs (or more) of genes closely related in both sequence and structure are found. In many cases, both (or all) copies of these arrays appear to be, or are known to be, functional. Indeed, many were discovered genetically: for example, *knirps* and *knirps-related, polyhomeotic-distal* and *polyhomeotic-proximal, engrailed* and *invected.* These gene pairs may be functionally redundant (usually assessed by the fact that deletion of both is required for an obvious mutant phenotype). Even when functional redundancy is not seen by genetic techniques, the failure to do so may be because of developmental differences in expression between gene pairs, as in the case of *knirps* and *knirps-related* (Rothe et al. 1992). However, sequence divergence between tandem pairs of genes often results in complete functional divergence, as for *Adh* and *Adhr* (Ashburner 1998). Total inactivation of one member of a tandem gene array, to generate a truly nonfunctional pseudogene, is known in *Drosophila;* for example, the loss of the *CecropinA2* gene by partial deletion, in the *D. simulans* clade (Ramos-Onsins and Aguardé 1998). There is a cluster of ten *α-Esterase* genes in *D. melanogaster* spread over some 60 Kb. In this species, but not in *D. simulans* or *D. yakuba,* one of these clusters is a pseudogene, and is now evolving by neutral evolution (Robin et al. 2000). Substitution patterns consistent with neutral evolution are also seen in the *CecA2* pseudogene of *D. simulans* and in the larval cuticle protein pseudogenes of both *D. melanogaster* and *D. simulans* (Pritchard and Schaeffer 1997).

Processed pseudogenes, originating by the reverse transcription of an RNA and the insertion of this cDNA into the genome, are surprisingly rare in the genome of *Drosophila.* The first to be discovered were some tRNA pseudogenes (e.g., Sharp et al. 1981) and a pseudogene of *Adh* in *D. yakuba* and *D. teissieri* (Jeffs and Ashburner 1991). More detailed study of the latter suggests that this gene is indeed functional, having been "captured" by another gene and subverted to its control (Long et al. 1999). Processed *Adh* pseudogenes are also known in the *D. obscura* species group (Luque et al. 1997). Currie and Sullivan (1994) suggest that a phosphoglycerate mutase gene, *Pglm87,* may have originated by retrotransposition from its paralog, *Pglym78.* Other pairs of genes, one member of which may have a similar origin, include *Prat* and *Prat2* (Malmanche et al. 2003) and *Ntf-2* and *Ntf-2r* (Betrán and Long 2003). There is also evidence that the *Rhodopsin 4* gene of *D. virilis* evolved by retrotransposition (Neufeld et al. 1991). A *Histone-2B* pseudogene has been detected in *D. melanogaster* by Akhmanova and Hennig (1998).

Betrán et al. (2003) suggest that new genes originate quite frequently by retrotransposition in *Drosophila.* They have identified in the Release 2 sequence, 24 genes, presumably functional, that show the hallmarks of having originated in this way; half of these originated from genes on the X chromosome (see also Wang et al. 2000 and above).

The Release 3.1 annotation of the genome of *D. melanogaster* predicts only two processed pseudogenes (Misra et al. 2002). This is certainly an underestimate caused by the difficulty in annotating nonfunctional genes; Echols et al. (2002) predict 114 pseudogenes (of both classes) in the Release 2 sequence of this species, using a different computational method.

Using a K_a:K_s ratio of about 1 as a criterion for a pseudogene (suggestive, but by no means conclusive), Zdobnov et al. (2002) estimate between 439 and 1,319 pseudogenes (both classes) in *An. gambiae* and between 162 and 396 in *D. melanogaster.* The reason why processed pseudogenes are rare in these Diptera is not known and is somewhat surprising, given the abundance and ubiquity of retrotransposable elements that encode reverse transcriptases. Petrov and Hartl have argued that pseudogenes in *Drosophila* undergo a very high rate of DNA loss, due to deletion (Petrov and Hartl 1997; Petrov 2002) and that this accounts for their dearth.

Y Chromosome Genes. In many species of Diptera, but in by no means all, there is a marked differentiation between the X and Y chromosomes. Heteromorphic sex chromosomes have, presumably, arisen independently several times in the Diptera (see Rai and Black 1999). In *Drosophila,* the Y chromosome is always entirely heterochromatic and its only known function (other than perhaps encoding rRNA genes in some species) is for male fertility. In *D. melanogaster,* it is known to encode at least 16 different proteins, of which three are sperm-specific dyneins (Goldstein et al. 1982; Gepner and Hays 1993; Carvalho et al. 2000; Carvalho 2002). Assembly of the Y chromosome sequence has been difficult, because of the very repetitive nature of most of its sequence. However, even this assembly is now being achieved (Carvalho et al. 2001, 2003; Carvalho 2002). Studies, first with *D. hydei* and its relatives, showed that this chromosome is transcriptionally active in primary spermatocytes, forming large lampbrush loops (Hennig 1987). These loops are now known to be the sites of the protein coding genes. These genes are quite extraordinary: they have very large introns; one in *D. hydei* is estimated to be more than 3.6 Mb and contains both satellite DNA sequences and transposable elements. It is the transcription of these monsters (which will take almost 1 day) that forms the lampbrush loops (Kurek et al. 2000; Reugels et al. 2000). It is not known where these genes are located in the few drosophilids with *X0* males (e.g., *D. affinis*), or those few Diptera with female heterogamety; that is, the chironomid *Polypedilum nubifer* (Martin 1966) and the tephritid *Chrysotrypanea trifasciata* (Bush 1966).

Most of the Culicidae have homomorphic sex chromosomes with the Y carrying a male-determining locus (see Rai and Black 1999). The anophelines, however, have heteromorphic sex chromosomes with a heterochromatic Y. There is the scope here for comparative genomics to throw light on Y chromosome evolution. The Y chromosome of *An. gambiae* contains its own active transposable element, a cluster of some 12 copies of the *Ty1-copia* family retrotransposon *mtanga* (Rohr et al. 2002).

Neo-Y chromosomes also give us an opportunity to see what happens when a formerly functional chromosome or chromosome arm becomes genetically inert, or relatively so (Charlesworth and Charlesworth 2000). It is remarkable that the Y chromosomes of *D. melanogaster* and *D. pseudoobscura* are not homologous (Carvalho and Clark 2004). All of the Y-linked genes of *D. melanogaster* are autosomal in the latter species, whose Y chromosome has derived from an autosome (element *E*). The most detailed molecular study of *neo*-Y chromosome evolution is the *neo*-Y chromosome of *D. miranda.* This chromosome evolved as a consequence of a chromosome fusion between an ancestral Y and element *E* (equivalent to chromosome arm *2R* of *D. melanogaster*). This fusion is thought to have occurred about 2 Mya. Some genes expected to be present on this *neo*-Y, given its origin, are absent; for example, the *Amylase* genes (Steinemann and Steinemann 1999). Other genes, such as the larval cuticle protein gene cluster *Lcp1-4,* are massively rearranged and disrupted by the insertion of transposable elements (Steinemann and Steinemann 1992, 1995, 2000). The *neo*-Y chromosome of "americana americana" populations of *D. americana,* which has resulted from an X-4 chromosome fusion, is much younger than that of *D. miranda* (a few hundred thousand years), and has not yet begun to show obvious signs of degeneration (Charlesworth et al. 1997). The *Adh* gene, at least, of this chromosome does, however, show reduced nucleotide polymorphism, perhaps as the result of reduced genetic recombination (McAllister and Charlesworth 1999).

The "degeneration" of Y chromosomes may involve both the fixation of deleterious mutations and a slow rate of adaptive evolution due to an absence of recombination (Orr and Kim 1998; Charlesworth and Charlesworth 2000; see Muller 1918). The available data suggest that both mechanisms are at work (see Bachtrog and Charlesworth 2002). In a remarkably prescient suggestion, Darlington (1937: 331–332) pointed out that heterochromatic chromosomes, such as the Y, may accumulate "inert genes [that have] lost the capacity for reaction [i.e., function] [yet retain] the capacity for reproduction," that is, pseudogenes.

Genomic Organization

In recent years, much attention has been paid to the analysis of genomes at the level of genes; far less to genomic organization on a larger scale. The genomes of the Diptera present a wonderful opportunity for larger-scale analyses, as their genomes are often organized as well-structured polytene chromosomes. The stability, both during development and phylogenetically, of the polytene chromosome banding patterns is an enigma. Although we understand, at a general level, that these banding patterns reflect differences in chromatin compaction, we have very little idea how this is achieved. We do know that there is no obvious compositional bias in the DNA of bands vs. interbands, and all attempts to discover sequence elements that might correlate with band/interband boundaries have so far failed (M. Ashburner, unpublished). Nevertheless, there is the conviction that these boundaries must be reflected in the primary sequence.

Recent analyses of gene expression data, from microarray or chip experiments, have identified an unexpected organization of the genome of *D. melanogaster* (Boutanaev et al. 2002; Spellman and Rubin 2002; Ueda et al. 2002) and other species (Caron et al. 2001; Roy et al. 2002). The genome is organized into domains within which genes tend to be co-expressed. These domains are around 100 Kb in length and typically include 10–30 genes; between 20% and 30% of the genes appear to be organized in expression domains, which do not correlate with polytene chromosome banding patterns or any other recognized feature. It is too early to tell whether these domains have functional significance or are a simple consequence of an overarching structural organization of chromosomes. Comparative studies may help to distinguish these alternatives. In addition to these indications of high-order organization of chromosomes, there is—in *Drosophila* Kc tissue culture cells, at least—a correlation between gene expression and the time of replication; genes that replicate early in the S phase have a higher probability of being expressed than those that are replicated late (Schübeler et al. 2002).

Sex Determination

The great majority of Diptera reproduce sexually, although thelytokous parthenogenetic species have evolved in several different lineages, and paedogenesis has evolved in two tribes of Cecidomyiidae (for a review, see Ashburner 2000b). The genetic basis of sex determination seems, at first sight, to be highly variable in the Diptera (Nöthiger and Steinmann-Zwicky 1985). Male heterogamety is the rule in the Diptera, although female heterogamety is claimed for some chironomids (see Martin 1966) and is known in some strains of *Musca domestica* (see below). There is, however, a common underlying principle (Nöthiger and Steinmann-Zwicky 1985; Schütt and Nöthiger 2000). A "primary signal" regulates a "key gene" (*F*) that is activated only in females. The products of *F* activate one or a few "subordinate control genes," which themselves regulate the nature of the protein product from a "double-switch"—a gene that can produce either a male or a female specific isoform of a transcription factor, depending on whether the subordinate control genes have been activated by *F* (in a female) or not (in a male).

The primary signal for sex determination varies considerably within the Diptera and may even vary within a species (Davies and Roderick, Chapter 8). In the majority of Diptera that have been studied, there is a monogenic primary signal, but in *Drosophila,* the primary signal is the ratio of the number of X chromosomes to the number of autosomes (the *X:A* ratio); this ratio is 1.0 in females and 0.5 in males (Bridges 1925). The ratio is interpreted very early in development by each and every somatic cell through the combined effects of a small number of X-linked genes (so-called "numerator elements," whose transcription factor products are twice as abundant in female cells than in males), a single autosomal denominator element, and the transcription factor products of at least four genes that are active maternally. In females, these factors activate the *Sex-lethal* gene, whose product is an RNA splicing factor—this is the "key-gene" of *Drosophila.* The major subordinate control gene is *transformer.* This gene is transcribed in both males and females but is differentially spliced—only in females is a functional protein produced. The *Sxl* protein controls this splice; *transformer* also encodes a splicing factor. When present, the transcript of the double-switch gene, *doublesex,* is spliced to encode a female specific isoform of a transcription factor; when the *tra* protein is absent, then *dsx* produces a male-specific isoform of its protein (for a review, see Cline and Meyer 1996).

In groups as different as tipulids (*Nephrotoma crocata;* Ullerich et al. 1964), culicids (*Culex pipiens;* Gilchrest and Haldane 1947), simuliids (Rothfels and Nambiar 1981), phorids (Springer 1967), tephritids (Willhoeft and Franz 1996), muscids (Perje 1948), anthomyiids (Vosselman 1978), and calliphorids (Ullerich 1963; Ribbert 1967), there is a dominant male determiner (*M*), which may be carried on a heteromorphic Y chromosome or on a chromosome that is cytologically very similar to the X chromosome (homomorphic sex chromosomes). In the latter cases, the sex-determining region may be visible in polytene chromosomes; for example, as a thick ("heterochromatic") chromosome band (e.g., in *Chironomus thummi;* Hagele 1985; the Psychodid *Telmatoscopus albipunctatus;* Amabis 1977) or as other differences in banding (e.g., in *Culex pipiens;* Dennhofer 1975), or the Y chromosome may be marked by particular inversions (e.g., in the camptochironomids; Beermann 1955; Keyl 1962).

In *Megaselia scalaris* (Phoridae), sex is also determined by a dominant male-determining factor (Tokunaga 1955a,b; Springer 1967). This male-determining factor is unusual in that it can regularly move between nonhomologous chromosomes (Mainx 1964); these transposition events are probably premeiotic and can occur at frequencies of 0.08% to 0.3% (Burisch 1963; Mainx 1966) and may be mediated by transposable elements (Green 1980). The chromosomal region that is transferred between heterologs with the male-determining factor is small (Traut and Willhoeft 1990; Traut 1994). The process does not involve major translocations between chromosomes, as had been suggested, a conclusion reinforced by mapping the male-determiner in different chromosomes (Traut 1994; Willhoeft and Traut 1995). Molecular markers associated with the male-determining factor have now been identified, so there is the prospect that this fascinating genetic system will be understood at the molecular level (Willhoeft and Traut 1995; Traut and Wollert 1998).

Genetic studies of *Musca domestica,* which began with the rise in insecticide resistance in this species in the late 1940s, soon uncovered a fascinating variety of sex-determining systems (for reviews of the early literature, see Milani 1967; Milani et al. 1967). This species is clearly very labile, and its populations may well be evolving very quickly as a consequence of the selection imposed on them by human activities (Rubini et al. 1977). Until the late 1950s or so, all strains of *M. domestica* analyzed cytologically had a typical XX/XY chromosomal sex-determining system, both chromosomes being heterochromatic and virtually devoid of conventional genetic markers; indeed, both X0 and Y0 flies are viable and fertile (Rubini et al. 1972). Subsequently, however, there began to be reports of unusual populations; for example, of Y-linked inheritance of DDT resistance from Australia and sex-linked inheritance of mutants on autosomes in North America.

Strains derived from present-day natural populations of *Musca* fall into three major classes, according to their genetic sex-determining mechanism (for a review, see Dübendorfer et al. 1992). In "standard" strains, there is a conventional dominant male-determining factor, carried by the heterochromatic Y chromosome; males are XY, females are XX. In "autosomal *M*" strains, the *M* factor has translocated to an autosome—this may be any autosome (e.g., Milani and Franco 1959; Wagoner 1969) or even the X chromosome (Denholm et al. 1983). The *M* element appears to have different "strengths" according to its location, strength being determined by its ability to repress the activity of the *F* element (Schmidt et al. 1997b). In standard strains the Y chromosome carries at least two different *M* elements (Hediger et al. 1998a). Hediger et al. (1998b) suggest that these differences in *M* element according to position are not due to intrinsic differences in the elements themselves, but to differences in their activities according to their heterochromatic environments.

The third class of strains shows female, rather than male, heterogamety; in these strains, there is a dominant *F* element on an autosome, both males and females being homozygous for the *M* element (e.g., Malacrida et al. 1982). In addition, arrhenotokous females are often found in natural populations at a low frequency (e.g., McDonald et al. 1975). The autosomal *M* strains presumably result from the transposition of the *M* element from the Y chromosome to an autosome (see Hiroyoshi 1964). Whenever autosomaly located, the *M* element appears to be in the centric heterochromatin (Inoue and Hiroyoshi 1986). It should be noted that in *Musca,* transposition of the *M* element has not been seen under laboratory conditions, unlike the situation seen in *Megaselia* (Inoue and Hiroyoshi 1984). The genetic basis of strains that show female heterogamety has been interpreted as being due to a dominant constitutive mutation of the *F* element that makes it active despite the presence of an *M* element (see Dübendorfer et al. 1992). There is a close parallel to the constitutive *Sxl*^{M} mutations in *D. melanogaster* (see Cline and Meyer 1996), although *F* is not the *Musca Sxl* gene (Meise et al. 1998). In the laboratory, loss-of-function mutations of *F* have been recovered, leading to male development even in the absence of the *M* factor (Inoue and Hiroyoshi 1984; Schmidt et al. 1997a).

There is at least circumstantial evidence that this genetic differentiation is recent. No evidence of any unusual sex-determining system is found in populations from central Italy collected up to 1957; now these populations are entirely of the autosomal sex-determining class. Franco et al. (1982) suggest that these microevolutionary changes are a consequence of the massive use of insecticides for housefly control in the period since the Second World War.

It is clear that genes that determine sex may evolve very rapidly. Now that all of these genes have been characterized in *D. melanogaster,* there have been several attempts to clone the "homologous" genes from other Diptera, based on nucleic acid sequence similarity criteria. Many of these attempts have either failed or the genes, if isolated, do not play a role in sex determination, despite their similarities in sequence to their *Drosophila* homologs. Thus, for example, the *Drosophila Sex-lethal* homolog has been isolated from *Sciara* and several other Nematocera, *Ceratitis, Chrysomya, Megaselia,* and *Musca* (Müller-Holtkamp 1995; Meise et al. 1998; Saccone et al. 1998; Sievert et al. 2000; Ruiz et al. 2003; Serna et al. 2004). It is highly conserved in sequence with respect to the *Drosophila* gene, yet is not involved in sex determination. Within the Drosophilidae, the primary sex-determination genes in species as evolutionarily distant as *D. melanogaster* and *D. virilis* are functionally conserved (Bopp et al. 1996; Erickson and Cline 1998). Genes further down the sex-determination pathway, such as the double-switch *doublesex,* are apparently conserved in function as well as in sequence between families; for example, in *Megaselia* (Sievert et al. 1997; Kuhn et al. 2000), tephritids (Shearman and Frommer 1998), and *Musca* (Hediger et al. 2004).

Some of the genes in the *Drosophila* sex-determination pathway are hard to isolate from other Diptera because, even within the genus *Drosophila,* they are evolving exceptionally rapidly. The *transformer* gene is only 86% identical in amino acid sequence between *D. melanogaster* and *D. simulans* (species that diverged only 2 Mya) and only 31% identical between *D. melanogaster* and *D. hydei* (which diverged about 60 Mya) (O'Neil and Belote 1992; see also Kulathinal et al. 2003). By way of contrast, the *Adh* proteins of *D. melanogaster* and *D. hydei* are 82% identical in sequence.

It is of considerable interest that *transformer*-like proteins are implicated in sex-determination in both the medfly, *Ceratitis capitata,* and the honey bee. In the medfly, there is little similarity of the *tra* protein with that of *D. melanogaster,* yet the gene is differentially spiced in males and females, with, as in *Drosophila,* only females producing a full-length protein. Its splicing is not, however, under the control of the *Sxl* protein (which plays no role in sex-determination in this species). Yet, like *Sxl* in *Drosophila, tra* function in *Ceratitis* apparently involves an autoregulatory mechanism (Pane et al. 2002). A *tra*-like protein (the product of the *csd* gene) is the primary sex-determining signal in the honey bee, a species with a haplo-diplo mechanism of sex determination. In this case, the functional protein is a heterodimer of peptides encoded by two different alleles of the *csd* gene (Beye et al. 2003).

Evolution of Genes and Genomes

One major contribution of comparative genomic analysis will be to give some insight into the well-known phenomenon that different regions of genomes evolve at very different rates. It has been known for some time from DNA-DNA hybridization experiments using DNA from closely related species that a surprisingly high fraction of their genomes fail to re-anneal. For example, in the case of the very closely related species *D. melanogaster* and *D. simulans*, most of their genomes are only 3–4% divergent in sequence and readily form stable duplexes; yet 35% of their genomes do not, suggesting that this fraction differs by 25% or more in its nucleotide sequence; not surprisingly, this fraction of the genome is depleted for coding sequences (for a review, see Powell 1997). Yet even coding sequences may diverge very rapidly. Schmid and Tautz (1997) found, remarkably, that over one-third of a randomly selected set of cDNA clones from *D. melanogaster* did not cross-hybridize to DNA from *D. virilis*. Some of these cases were cloned from *D. yakuba* and shown to have a very high rate of amino acid substitution with respect to their *D. melanogaster* homologs. Some of the genes, at least, also show a high rate of intraspecific replacement polymorphism (Schmid et al. 1999).

Two general classes of genes show very rapid rates of molecular evolution: those involved in sex determination (see above) and male sexual characters, presumably because of the intensity of sexual selection (e.g., Civetta and Singh 1998; Wycoff et al. 2000); and those genes involved in defense against pathogens.

Both primary and secondary sexual structures are well known to evolve at very fast evolutionary rates (see Eberhard 1985). The same appears to be true of proteins that function during sexual behavior, at least in the drosophilids. One example is that of the proteins of the male seminal fluid, secreted by the accessory glands (for a review, see Wolfner 2002). These proteins are, on average, much more polymorphic than are others in *Drosophila* (e.g., Tsaur and Wu 1997, 1998; Begun et al. 2000). Genes represented in testes expressed sequence tag (EST) libraries of *D. melanogaster* are more divergent in amino acid sequence in *D. pseudoobscura* than are other genes (S. Richards, pers. comm.). Genes that are, somehow, involved in postzygotic species isolation also show much higher than average divergence among species. The first to be studied was *Odysseus*, a gene implicated in the sterility of male *D. simulans/ mauritiana* species hybrids. Although not sufficient to account for this sterility (Perez and Wu 1995), this gene, which encodes a homeodomain-containing protein, is evolving very rapidly: there are 15 amino acid substitutions in the homeodomains of *Odysseus* between *D. simulans* and *D. mauritiana* (Ting et al. 1998). Similarly, the *Hmr* gene—which, when mutant in *D. melanogaster*, can rescue to viability otherwise lethal hybrids of this species with its siblings (Hutter et al. 1990)—shows an extraordinary 20% amino acid difference between *D. melanogaster* and *D. simulans* (Barbash et al. 2003).

Many species of *Drosophila* (and perhaps mosquitoes; Rosen 1980) are infected by the rhabdovirus *sigma* (for a review, see Brun and Plus 1980). This has detrimental consequences in the wild and the laboratory (the affected flies are very sensitive to carbon dioxide). Natural populations of *D. melanogaster* are polymorphic for alleles at several genes that render them resistant to this virus. One of these, *ref(2)P* has been characterized. It encodes a PEST domain protein that is extraordinarily variable in coding sequence in natural populations, with more replacement than synonymous polymorphisms (Dru et al. 1993; Wayne et al. 1996). There is good evidence that this condition is an adaptive response to viral infection (Wyers et al. 1995).

Quite dramatic differences in gene organization can occur even between very closely related species. For example, two genes, one encoding a cytoplasmic dynein intermediate chain and the other encoding an annexin, are adjacent at the base of the X chromosome of *D. simulans.* In *D. melanogaster,* however, these genes have fused to form a novel gene, only expressed in testes and encoding an *N*-terminally truncated dynein protein. The testis-specific expression of this novel gene seems to have been formed fortuitously from non-promoter sequences of the annexin gene (Nurminsky et al. 1998).

The gain or loss of introns is a frequent evolutionary event. In nearly all drosophilids, the *Adh* gene has two introns interrupting the coding region, and these are always at precisely homologous sites; yet one group of species, *D. willistoni* and its relatives, has lost the second of these introns, and this is clearly a derived condition (Anderson et al. 1993). The subgenus *Sophophora* has lost the single intron characteristic of the *Gapdh* gene of other drosophilids (Wojtas et al. 1992). There is also evidence for several independent losses of introns in the *Amylase* genes of *Drosophila* (Da Lage et al. 1996). One of the most surprising discoveries is that of a polymorphism for two alleles, in *D. melanogaster,* of the *Rnp4F* gene, encoding a ribonucleoprotein; these two alleles (which are apparently both active) differ by the presence of two introns in one but not the other (Feiber et al. 2002). The most detailed study of intron gain/loss in the Diptera is that of Krzywinski and Besansky (2002) of the *white* gene. This gene may lack introns (e.g., in an undetermined tipulid), or possess one, two, three, or four introns in other groups. A phylogenetic study shows that there have been multiple independent events affecting intron presence within this gene in the Diptera (see also Llopart et al. 2002). We understand very little about how intron loss events occur, although it may come about as a homologous recombination between a gene and a processed pseudogene that may subsequently be lost from the genome (Fink 1987; Mourier and Jeffares 2003).

The Mitochondrial Genome

The mitochondrial genomes of eight species of Diptera have been completely sequenced: *An. gambiae* (Beard et al. 1993), *An. quadrimaculatus* (Cockburn et al. 1990), *Batrocera oleae* (Nardi et al. 2003), *Ceratitis capitata* (Spanos et al. 2000), *Chrysomya putoria* (Junqueira et al. 2001), *Cochliomya hominivorax* (Lessinger et al. 2000), *D. melanogaster* (Lewis et al. 1995), and *D. yakuba* (Clary and Wolstenholme 1985). They encode 13 proteins, 22 tRNAs, and two ribosomal RNAs (for a review, see Clary and Wolstenholme 1984).

These genomes are generally typical of the mitochondria of metazoa, variations in length (from 15.3 Kb to 19.5 Kb) being due to variations in the size of the noncoding D-loop region at the origin of replication of the mitochondrial DNA molecule. Indeed, this length may be polymorphic within a species. In species of *Drosophila,* at least, some populations have more than one class of mitochondrial DNA (Solignac et al. 1986; Tsujino et al. 2002), and individual flies may have two mitochondrial genomes (i.e., be heteroplasmic). The orientation of genes with respect to the origin of replication differs in the dipteran mitochondrial genome from that found in mammals (Goddard and Wolstenholme 1980). In the arthropods, transcription is divergent from the origin, whereas in mammals, all but one gene are transcribed from the same strand. The culicids show a derived condition with respect to the order of the genes encoding four of the mitochondrial tRNAs.

Genomes of Dipteran Symbionts

Wolbachia are α-Proteobacteria, first seen in the ovaries of the mosquito *Culex pipiens* nearly 80 years ago (see Hertig 1936), but it was only when they were recognized as the causative agent of cytoplasmic incompatibilities between different "cytotypes" of this species (Yen and Barr 1971) that the world sat up and noticed. Like their close relatives *Rickettsia, Wolbachia* are obligatory intracellular symbionts. We now know that *Wolbachia* are very widely spread in the insects and that they may be benign, pathogenic, or have interesting effects on reproduction—causing, for example, nonreciprocal cytoplasmic incompatibility (for reviews, see O'Neill et al. 1997; McGraw and O'Neill 2004). The genomes of several different *Wolbachia* strains are now being sequenced, and that of the *D. melanogaster w*Mel symbiont is published (Wu et al. 2004).

Although a dozen or so families of Diptera are hematophagous, only four families are blood-feeders throughout their life cycles: Glossinidae, Hippoboscidae, Streblidae, and Nycteribiidae. In common with other insects that rely on vertebrate blood for nutrition, these families all carry obligate endosymbionts, housed in special cells, and they rely on these for the provision of essential nutrients, such as the B-group vitamins (e.g., Nogge 1981). Only the symbionts of the Glossinidae are reasonably well known, and the genomic sequence of one, *Wigglesworthia glossinidia,* has been determined (Akman et al. 2002). Unlike *Wolbachia,* the evolution of *Wigglesworthia* is concordant with its host species, indicative of an ancient and stable association (Chen et al. 1999). In common with other intracellular organisms, *W. glossinidia* has a very reduced genome (only 698 Kb) and retains genes expected to be required for vitamin and cofactor biosynthesis, unlike the genomes of *Rickettsia prowazekii* and *Buchnera,* an obligate symbiont of aphids. Remarkably, *Wigglesworthia* also retains genes to synthesize a complete flagellar apparatus; this is unexpected in an intracellular symbiont, and Akman et al. (2002) speculate that there may be an extracellular phase associated with transfer from mother to progeny in the uterine gland of the fly.

Very ancient endosymbionts, such as mitochondria and chloroplasts, are well known to have transferred genes from their own genomes to the nuclear genomes of their hosts. This has not—yet—been seen for any dipteran endosymbiont but has been found in the Coleoptera, in which part of the genome of the *Wolbachia* of *Callosobruchus chinensis* has been transferred to the host's nuclear genome (Kondo et al. 2002).

The Somatic Genome

SOMATIC POLYPLOIDY

The larval tissues of the Diptera are the locus classicus of polytene chromosomes, first described by Balbiani in 1881 from the larval salivary glands of *Chironomus plumosus.* Polytene chromosomes are now known to be ubiquitous in many larval, and some adult, tissues of the Diptera. In his recent comprehensive review, Zhimulev (1996) lists their occurrence in 35 families, from Tipulidae to Glossinidae, although they are not always cytologically tractable, especially in species with a large number of dispersed repetitive elements (e.g.,

Ae. aegypti). Although typically studied in larval salivary glands, they are found in many other larval tissues (e.g., gut, Malphigian tubules, fat body, epidermal cells), as well as in pupal tissues (e.g., the footpad cells of Calliphoridae and Sarcophagidae, bristle-secreting cells in Tephritidae and Glossinidae) and ovarian nurse cells.

Polyteny is a special case of endopolyploidy in which the replicated chromosomes remain in intimate synapsis and are relatively lax. Typically, polytene nuclei are regular in shape, either circular or nearly so in cross section. This contrasts with the extraordinarily irregular shape of endopolyploid (but not polytene) nuclei in other insect orders (e.g., the silk-gland nuclei of Lepidoptera and Trichoptera; Ashburner 1980). The extent of polyploidy is highly variable, from 16 C (e.g., in some fat-body nuclei of *D. melanogaster*) to 32,768 C (i.e., 14 replications of a diploid nucleus) in the largest chromosomes of the salivary gland of *Chironomus tentans* (Zhimulev 1996: Table 12). Morphologically, there is considerable variation within and among different families in their polytene chromosomes. This is true not only with respect to the banding morphology—which, for example, is very diffuse in the Cecidomyiidae—but also with respect to the somatic synapsis of the paternal homologs, which is usually intimate but much less so in the Simuliidae.

The polytene chromosomes of the Diptera have been extraordinarily useful research tools. The colinearity of their banding patterns with the genetic map was established, in *D. melanogaster,* in the mid-1930s by C. B. Bridges, and this lead to their extensive use as what Dan Lindsley later called the "first physical maps of a genome." In *D. melanogaster,* the average DNA content of a polytene chromosome band is about 24 Kb, allowing high-resolution mapping of genes by either genetic methods or by in situ hybridization to polytene chromosomes of cloned genes or cDNAs (see Merriam et al. 1991). Epigenetic alterations to the banding patterns during development, seen as puffs, were suggested by Beermann (1956) to represent gene activity. The dramatic changes in the patterns of puffs in the salivary gland chromosomes of both *Chironomus* and *Drosophila,* associated with the beginning of metamorphosis, lead to detailed models of the control of gene expression by the ecdysteroid hormones (e.g., Ashburner et al. 1974).

The polytene chromosome banding patterns are species specific. In species in which the chromosomes are suitable for detailed study in several tissues, the general rule is that the basic banding patterns are identical, even between larval and adult stages (Redfern 1981). The only clear exception to this is in *Calliphora,* for which careful studies by Ribbert (1979) lead to the conclusion that the banding patterns are quite different in ovarian nurse cells and pupal trichogen cells.

Studies of natural populations of *D. pseudoobscura* by Sturtevant and Dobzhansky in the 1930s lead to the realization that differences in polytene chromosome banding patterns, due to major chromosomal mutations (typically paracentric inversions), could be used as phylogenetic markers, both for populations within a species and among species (for reviews, see Krimbas and Powell 1992). Extensive phylogenetic studies of various groups have been made with this technique, particularly in the Chironomidae, Culicidae, Simuliidae, and Drosophilidae. Although the phylogenetic trees derived with this method were classically considered to be unrooted, Green (1982) pointed out that this need not be so and that, with a suitable outgroup, rooted trees can be obtained. In cases for which we have parallel data from molecular evolutionary studies (e.g., in the *D. melanogaster* species group), the chromosome- and sequence-based phylogenies are concordant (see O'Grady et al. 2001).

DIFFERENTIAL AMPLIFICATION

Polytene chromosomes result from the endoreplication of diploid nuclei. It was realized early on in their study that not all of the genome was equally replicated, as was seen in the male *D. melanogaster,* whose Y chromosome is simply not visible in polytene nuclei (Painter 1933). In general, heterochromatic chromosomes (or heterochromatic chromosome regions) either fail to replicate or underreplicate in polytene nuclei. In many (e.g., Drosophilidae), but not all (e.g., Simuliidae, Chironomidae; but see Chubareva et al. 2003) groups, the polytene chromosomes unite in their centric heterochromatic regions to form a chromocenter. Gall et al. (1971) showed that in *Drosophila,* the satellite DNA sequences fail to replicate in polytene nuclei, remaining at their diploid levels; this is not so in at least the one species of chironomid studied, *Chironomus melanotus* (Steinemann 1978).

Many polytene chromosomes are characterized by constrictions, or weak points, at particular loci. The number of these points varies enormously between species; in *D. melanogaster,* for example, there are about ten major constrictions. They result from local underreplication (Spierer and Spierer 1984). This underreplication is under genetic control and a mutation has been characterized that both suppresses the condition and also allows considerable replication of some of the normally underreplicated pericentric heterochromatic sequences (Belyaeva et al. 1998).

Local overreplication of specific polytene chromosome loci was discovered in the neotropical sciarid *Rhynchosciara americana* by Pavan (see Pavan and da Cunha 1969; Glover et al. 1982; Lara et al. 1991; for a review, see Gerbi and Urnov 1996). Overreplication forms DNA puffs at particular salivary gland loci at the commencement of metamorphosis in the Sciaridae but not, as far as is known, in any other family. Overreplication is under ecdysteroid control, and the loci, like the RNA puffs characteristic of other families, are transcriptionally active and encode major components of the salivary gland secretion (e.g., Laicine et al. 1984).

Overreplication is a characteristic of the genes encoding the chorion proteins in *Drosophila.* In *D. melanogaster,* most of these are in one of two clusters (Spradling et al. 1980; Delidakis and Kafatos 1987, 1989), although other clusters are known (Claycomb et al. 2004). Both major chorion protein clusters overreplicate, that on the X chromosome about 16–20-fold, that on chromosome *3* about 60–80-fold; the overreplicated regions extend over about 40 Kb (for a review, see Royzman and Orr-Weaver 1998). Overreplication only occurs in the somatic follicle cells, which surround the oocyte and synthesize the structural proteins of the chorion (Spradling and Mahowald 1980). Amplification occurs as a consequence of the repeated firing of replication origins that are interspersed in each cluster (see Osheim et al. 1988). The sequence elements that control overreplication can function autonomously and are providing a useful model for the study of the control of replication (see Austin et al. 1999).

EPIGENETIC MODIFICATION

Methylation is one of the best-known epigenetic modifications of DNA. In vertebrates, methylation of cytosine residues, almost invariable at CpG doublets, is very common and in general is correlated with repression of the genome (see Bird and Wolffe 1999). The DNA of *Drosophila* shows no deviation from the expected frequency of CpG doublets; in contrast to vertebrate genomes, in which this frequency is much lower than expected (given the frequency of C and G residues), due to the risks of mutation should deamination of 5-MeC occur. It was, therefore, not unexpected that attempts to find 5-MeC in *D. melanogaster* and other Diptera, either chemically or by differences in the sensitivity of genomic DNA to restriction enzyme isoschizomers, were generally (Bird and Taggart 1980; Urieli-Shoval et al. 1982), but not always (Adams et al. 1979; Achwal et al. 1983, 1984) unsuccessful. More recently, Lyko et al. (2000a) and others have reinvestigated this problem and found a small amount of 5-MeC, mostly in CpT doublets rather than CpG, in embryonic but not adult DNA; less than 1% of the C residues of DNA from 1–2-h-old embryos are methylated. Gowher et al. (2000) have also detected low levels (<1% of C residues) of 5-MeC in adult *D. melanogaster* DNA.

If methylation of cytosine residues occurs, then the genome of *Drosophila* should encode a suitable methyltransferase. Indeed, a gene predicted to encode a protein with similarity to mammalian and yeast DNA (5-cytosine) methyltransferases has been identified in *D. melanogaster* (Hung et al. 1999; Tweedie et al. 1999; Lyko et al. 2000b). Lyko et al. (2000b) suggested that the methyltransferase homologue (*Mt2*) is carried by a transposon-like element. These researchers have artificially methylated CpG residues in *Drosophila* DNA in vivo (by expression of murine [5-cytosine] methyltransferases) and have shown that the process results in pupal lethality. In addition to *Mt2*, a gene encoding a protein that shares motifs with known 5-MeCpG binding proteins has also been identified (Tweedie et al. 1999), but there is some doubt as to whether it binds methylated DNA (Roder et al. 2000; Ballestar et al. 2001; see Marhold et al. 2002). The function of DNA methylation in *Drosophila* is unknown; indeed, blocking the activity of the methylase gene (*Mt2*) has no obvious phenotypic consequence (Kunert et al. 2003).

IMPRINTING

Imprinting is an epigenetic phenomenon in which the activity of genes, or the behavior of whole chromosomes, is influenced by their parental origin. Although now widely studied in mammals, genomic imprinting was in fact first discovered in a fly—in *Sciara* by Helen Crouse (Crouse 1960; Herrick and Seger 1999). In sciarids, for example *S. coprophila*, the somatic chromosome number is eight in females, seven in males; the former are XX, the latter X0. (Sciarids also have a set of germ-line–limited chromosomes, at least in some species or populations; see above.) Although female meiosis is normal, that in the male is decidedly not so; during meiosis I, there is no synapsis of homologs and the first anaphase is monopolar, with one set of somatic chromosomes plus the germ-line–limited chromosomes going to the only functional pole. There is genetic evidence that the somatic chromosomes retained are those the male has inherited from its mother. At the second anaphase, a bipolar spindle forms but is also functionally monopolar, as only one meiotic product is functional. The single X chromosome now divides and each product co-disjoins to the single functional pole, along with a single set of autosomes and germ-line–limited chromosomes. This is the first example of imprinting in *Sciara*—the two somatic sets are epigenetically different; only the maternal is transmitted in the sperm. All zygotes are triplo-X diploids. During early cleavage, not only are the germ-line–limited chromosomes eliminated from all but the future germ cell nuclei, but also either one or two X chromosomes (depending on maternal genotype) are eliminated from all somatic nuclei at the seventh or eighth cleavage divisions, as a consequence of incomplete sister chromatid separation and of remaining on the metaphase plate (de Saint Phalle and Sullivan 1996). Thus some embryos retain (somatically) one X chromosome (and are males), others two (and are females). Remarkably, only the paternal X chromosomes are eliminated. These chromosomes were, of course, maternal in the previous generation. Somehow these chromosomes have been epigenetically modified—imprinted—by their parental source. Nothing is yet known of the mechanism of imprinting in sciarids, although a controlling element has been identified genetically by Crouse (1979), and differences in histone acetylation between paternal and maternal chromosomes have been found (Goday and Ruiz 2002).

In *Drosophila*, there has been for some time evidence of parental origin effects on genes subject to position-effect inactivation (for reviews of the early data, see Spofford 1976; Lloyd et al. 1999; Lloyd 2000). Position-effect variegation is a phenomenon by means of which genes are epigenetically inactivated as a consequence of being placed next to a novel heterochromatic boundary (for genes normally located in the euchromatin) or a novel euchromatic boundary (for genes normally located in the heterochromatin; for a review, see Zhimulev 1998). Imprinting seems to be a general property of genes subject to position-effect variegation as a consequence of insertion into Y chromosome heterochromatin. Maggert and Golic (2002) studied the expression of a number of transgenes inserted into either the Y or autosomal heterochromatin; 22 of 23 Y chromosome transgenes displayed imprinting; only one of seven of those inserted into autosomal heterochromatin did so. Generally, transgenes inherited maternally showed higher expression than those inherited paternally. The conclusion must be that these transgenes are inheriting an imprint from their chromosomal environment, and that the Y chromosome and some (but not all) autosomal heterochromatic regions are normally imprinted in *Drosophila*. The mechanism of imprinting in insects remains a mystery.

Conclusions and Prospects

Which species, from a genomics perspective, now warrant serious genomic study? I argue that two general criteria can be used to answer this question. The first is a phylogenetic criterion: which species would give us the most information about the major evolutionary events in the 250-My history of the Diptera? This answer, of course, demands genomic studies of suitable outgroups, such as the Siphonoptera and Mecoptera (see Whiting, Chapter 1). I leave it for others to give a detailed answer to this question. My second criterion is that it would be very interesting to study those species that have unusual life styles or life histories. Here I will simply give an annotated list of some of my own choices: species that have evolved an oxygen-carrying hemoglobin (i.e., chironomids and the oestrid *Gasterophilus*), species that have evolved viviparity (e.g., tsetse flies, hippoboscids, streblids, nycteribiids), species that have evolved obligate paedogenesis (e.g., some primitive cecidomyiids and their closest sexual relatives), and species that have unusual chromosome behavior (e.g., *Sciara*). I would also include the cecidomyiid *Mayetiola destructor*, a pest of wheat, which shows a gene-for-gene relationship between its virulence genes and the resistance genes of its host plant (for a brief review, Ashburner 2000b). What is certain is that the study of dipteran genomes has much to teach us.

ACKNOWLEDGMENTS

My work is supported by a Programme Grant from the Medical Research Council, London. I wrote this review at the same time that I was updating the genome chapter for the revised edition of *Drosophila: A Laboratory Handbook* (Ashburner et al. 2005), and have used some of the same material for both. I thank José Ranz and Casey Bergman for their comments on the manuscript of this chapter. I also thank those who provided me with preprints or have allowed me to quote their unpublished work.

REFERENCES

Abad, J. P., M. Carmena, S. Baars, R. D. C. Saunders, D. M. Glover, P. Ludena, C. Sentis, C. Tyler-Smith, and A. Villasante. 1992. Dodeca satellite: A conserved G+C-rich satellite from the centromeric heterochromatin of *Drosophila melanogaster. Proc. Natl. Acad. Sci. USA* 89:4663–4667.

Abad, J. P., B. de Pablos, K. Osoegawa, P. J. de Jong, A. Martin-Gallardo, and A. Villasante. 2004a. Genomic analysis of *Drosophila melanogaster* telomeres: Full-length copies of *HeT-A* and *TART* elements at the telomeres. *Mol. Biol. Evol.* 21:1613–1619.

Abad, J. P., B. de Pablos, K. Osoegawa, P. J. de Jong, A. Martin-Gallardo, and A. Villasante. 2004b. TAHRE, a novel telomeric retrotransposon from *Drosophila melanogaster,* reveals the origin of *Drosophila* telomeres. *Mol. Biol. Evol.* 21:1620–1624.

Achwal, C. W., C. A. Iyer, and H. S. Chandra. 1983. Immunochemical evidence for the presence of 5mC, 6mA and 7mG in human, *Drosophila* and mealybug DNA. *FEBS. Lett.* 158:353–358.

Achwal, C. W., P. Ganguly, and H. S. Chandra. 1984. Estimation of the amount of 5-methylcytosine in *Drosophila melanogaster* DNA by amplified ELISA and photoacoustic spectroscopy. *EMBO J.* 3:263–266.

Adams, M. D., S. E. Celniker, R. A. Holt, C. A. Evans, J. D. Gocayne, P. G. Amanatides, S. E. Scherer, P. W. Li, R. A. Hoskins, R. F. Galle, R. A. George, S. E. Lewis, S. Richards, M. Ashburner, S. N. Henderson, G. G. Sutton, J. R. Wortman, M. D. Yandell, Q. Zhang, L. X. Chen, R. C. Brandon, Y. H. Rogers, R. G. Blazej, M. Champe, B. D. Pfeiffer, K. H. Wan, C. Doyle, E. G. Baxter, G. Helt, C. R. Nelson, G. L. Gabor Miklos, J. F. Abril, A. Agbayani, H. J. An, C. Andrews-Pfannkoch, D. Baldwin, R. M. Ballew, A. Basu, J. Baxendale, L. Bayraktaroglu, E. M. Beasley, K. Y. Beeson, P. V. Benos, B. P. Berman, D. Bhandari, S. Bolshakov, D. Borkova, M. R. Botchan, J. Bouck, P. Brokstein, P. Brottier, K. C. Burtis, D. A. Busam, H. Butler, E. Cadieu, A. Center, I. Chandra, J. M. Cherry, S. Cawley, C. Dahlke, L. B. Davenport, P. Davies, B. de Pablos, A. Delcher, Z. Deng, A. D. Mays, I. Dew, S. M. Dietz, K. Dodson, L. E. Doup, M. Downes, S. Dugan-Rocha, B. C. Dunkov, P. Dunn, K. J. Durbin, C. C. Evangelista, C. Ferraz, S. Ferriera, W. Fleischmann, C. Fosler, A. E. Gabrielian, N. S. Garg, W. M. Gelbart, K. Glasser, A. Glodek, F. Gong, J. H. Gorrell, Z. Gu, P. Guan, M. Harris, N. L. Harris, D. Harvey, T. J. Heiman, J. R. Hernandez, J. Houck, D. Hostin, K. A. Houston, T. J. Howland, M. H. Wei, C. Ibegwam, M. Jalali, F. Kalush, G. H. Karpen, Z. Ke, J. A. Kennison, K. A. Ketchum, B. E. Kimmel, C. D. Kodira, C. Kraft, S. Kravitz, D. Kulp, Z. Lai, P. Lasko, Y. Lei, A. A. Levitsky, J. Li, Z. Li, Y. Liang, X. Lin, X. Liu, B. Mattei, T. C. McIntosh, M. P. McLeod, D. McPherson, G. Merkulov, N. V. Milshina, C. Mobarry, J. Morris, A. Moshrefi, S. M. Mount, M. Moy, B. Murphy, L. Murphy, D. M. Muzny, D. L. Nelson, D. R. Nelson, K. A. Nelson, K. Nixon, D. R. Nusskern, J. M. Pacleb, M. Palazzolo, G. S. Pittman, S. Pan, J. Pollard, V. Puri, M. G. Reese, K. Reinert, K. Remington, R. D. Saunders, F. Scheeler, H. Shen, B. C. Shue, I. Siden-Kiamos, M. Simpson, M. P. Skupski, T. Smith, E. Spier, A. C. Spradling, M. Stapleton, R. Strong, E. Sun, R. Svirskas, C. Tector, R. Turner, E. Venter, A. H. Wang, X. Wang, Z. Y. Wang, D. A. Wassarman, G. M. Weinstock, J. Weissenbach, S. M. Williams, T. Woodage, K. C. Worley, D. Wu, S. Yang, Q. A. Yao, J. Ye, R. F. Yeh, J. S. Zaveri, M. Zhan, G. Zhang, Q. Zhao, L. Zheng, X. H. Zheng, F. N. Zhong, W. Zhong, X. Zhou, S. Zhu, X. Zhu, H. O. Smith, R. A. Gibbs, E. W. Myers, G. M. Rubin, and J. C. Venter. 2000. The genome sequence of *Drosophila melanogaster. Science* 287:2185–2195.

Adams, R. L. P., E. L. McKay, L. M. Craig, and R. H. Burdon. 1979. Methylation of mosquito DNA. *Biochim. Biophys. Acta* 563:72–81.

Akhmanova, A., and W. Hennig. 1998. *Drosophila melanogaster* histone *H2B* retropseudogene is inserted into a region rich in transposable elements. *Genome* 41:396–401.

Akman, L., A. Yamashita, H. Watanabe, K. Oshima, T. Shiba, M. Hattori, and S. Aksoy. 2002. Genome sequence of the endocellular obligate symbiont of tsetse flies, *Wigglesworthia glossinidia*. *Nat. Genet.* 32:402–407.

Amabis, J. M. 1977. Cytological evidence of male heterogamety in sex determination of *Telmatoscopus albipunctatus* (Diptera, Psychodidae). *Chromosoma* 62:133–138.

Anderson, C., E. A. Carew, and J. R. Powell. 1993. Evolution of the *Adh* locus in the *Drosophila willistoni* group: The loss of an intron, and a shift in codon usage. *Mol. Biol. Evol.* 10:605–618.

Andrews, J., M. Smith, J. Merakovsky, M. Coulson, F. Hannan, and L. E. Kelly. 1996. The *stoned* locus of *Drosophila melanogaster* produces a dicistronic transcript and encodes two distinct polypeptides. *Genetics* 143:1699–1711.

Aravin, A. A., N. M. Naumova, A. V. Tulin, V. V. Vagin, Y. M. Rozovsky, and V. A. Gvozdev. 2001. Double-stranded RNA-mediated silencing of genomic tandem repeats and transposable elements in the *D. melanogaster* germline. *Curr. Biol.* 11:1017–1027.

Aravin, A. A., M. Lagos-Quintana, A. Yalcin, M. Zavolan, D. Marks, B. Snyder, T. Gaasterland, J. Meyer, and T. Tuschl. 2003. The small RNA profile during *Drosophila melanogaster* development. *Devel. Cell.* 5:337–350.

Arkhipova, I. R., and M. Meselson. 2000. Transposable elements in sexual and ancient asexual taxa. *Proc. Natl. Acad. Sci. USA* 97:14473–14477.

Arkhipova, I. R., K. I. Pyatkov, M. Meselson, and M. B. Evgen'ev. 2003. Retroelements containing introns in diverse invertebrate taxa. *Nat. Genet.* 33:123–124.

Arnosti, D. N. 2003. Analysis and function of transcriptional regulatory elements: Insights from *Drosophila*. *Annu. Rev. Entomol.* 48:579–602.

Ashburner, M. 1980. Some aspects of the structure and function of the polytene chromosomes of the Diptera. In R. L. Blackman, G. M. Hewitt, and M. Ashburner, eds., *Insect Cytogenetics. Symposium of the Royal Entomological Society 10*, pp. 64–84. Oxford: Blackwell Scientific Publishing.

Ashburner, M. 1998. Speculations on the subject of alcohol dehydrogenase and its properties in *Drosophila* and other flies. *BioEssays* 20:949–954.

Ashburner, M. 2000a. A biologist's view of the *Drosophila* genome annotation assessment project. *Genome Res.* 10:391–393.

Ashburner, M. 2000b. Genetic systems in Palaearctic Diptera. In L. Papp and D. Darvas, eds., *Contributions to a Manual of Palaearctic Diptera. vol. 1. General and Applied Dipterology*, pp. 241–281. Budapest: Science Herald.

Ashburner, M., C. Chihara, P. Meltzer, and G. Richards. 1974. Temporal control of puffing activity in polytene chromosomes. *Cold Spring Harbor Symp. Quant. Biol.* 38:655–662.

Ashburner, M., S. Misra, J. Roote, S. E. Lewis, R. Blazej, T. Davis, C. Doyle, R. Galle, R. George, N. Harris, G. Hartzell, D. Harvey, L. Hong, K. Houston, R. Hoskins, G. Johnson, C. Martin, A. Moshrefi, M. Palazzolo, M. G. Reese, A. Spradling, G. Tsang, K. Wan, K. Whitelaw, B. Kimmel, S. Celniker, and G. M. Rubin. 1999. An exploration of the sequence of a 2.9-Mb region of the genome of *Drosophila melanogaster*. The *Adh* region. *Genetics* 153:179–219.

Ashburner, M., K. Golic, and S. H. Hawley. 2005. Drosophila: *A Laboratory Handbook*, 2nd ed. Cold Spring Harbor, N.Y.: Cold Spring Harbor Press.

Austin, R. J., T. L. Orr-Weaver, and S. P. Bell. 1999. *Drosophila* ORC specifically binds to ACE3, an origin of DNA replication control element. *Genes Dev.* 13:2639–2649.

Bachtrog, D., and B. Charlesworth. 2002. Reduced adaptation of a non-recombining *neo*-Y chromosome. *Nature* 416:323–327.

Bae, E., V. C. Calhoun, M. Levine, E. B. Lewis, and R. A. Drewell. 2002. Characterization of the intergenic RNA profile at *abdominal-A* and *abdominal-B* in the *Drosophila bithorax* complex. *Proc. Natl. Acad. Sci. USA* 99:16847–16852.

Bailey, J. A., Z. Gu, R. A. Clark, K. Reinert, R. V. Samonte, S. Schwartz, M. D. Adams, E. W. Myers, P. W. Li, and E. E. Eichler. 2002. Recent segmental duplications in the human genome. *Science* 297:1003–1007.

Balbiani, E. G. 1881. Sur la structure du noyeau des cellules salivaires chez les larves de *Chironomus*. *Zool. Anz.* IV:637–641, 662–666.

Ballestar, E., L. A. Pile, D. A. Wassarman, A. P. Wolffe, and P. A. Wade. 2001. A *Drosophila* MBD family member is a transcriptional corepressor associated with specific genes. *Europ. J. Biochem.* 68:5397–5406.

Barbash, D. A., D. F. Siino, A. M. Tarone, and J. Roote. 2003. A rapidly evolving MYB-related protein causes species isolation in *Drosophila. Proc. Natl. Acad. Sci. USA* 100:5302–5307.

Bass, B. L., ed. 2001. *RNA Editing.* Oxford: Oxford University Press.

Bass, B. L. 2002. RNA editing by adenosine deaminases that act on RNA. *A. Rev. Biochem.* 71:817–846.

Beard, C. B., D. M. Hamm, and F. H. Collins. 1993. The mitochondrial genome of the mosquito *Anopheles gambiae:* DNA sequence, genome organization, and comparisons with mitochondrial sequences of other insects. *Insect Mol. Biol.* 2:103–124.

Beermann, W. 1955. Geschlechtsbestimmung und Evolution der genetischen Y-Chromosomen bei *Chironomus. Biol. Zbl.* 74:525–544.

Beermann, W. 1956. Nuclear differentiation and functional morphology of chromosomes. *Cold Spring Harbor Symp. Quant. Biol.* 21:217–232.

Begun, D. J., and C. H. Langley. 2002. Proposal for sequencing the *Drosophila yakuba* and *D. simulans genomes.* A White Paper. Unpublished.

Begun, D. J., P. Whitley, B. L. Todd, H. M Waldrip-Dail, and A. G. Clark. 2000. Molecular population genetics of male accessory gland proteins in *Drosophila. Genetics* 156:1879–1888.

Bellen, H. J., R. W. Levis, G. Liao, Y. He, J. W. Carlson, G. Tsang, M. Evans-Holm, P. R. Hiesinger, K. L. Schulze, G. M. Rubin, R. A. Hoskins, and A. C. Spradling. 2004. The BDGP Gene Disruption Project: Single transposon insertions associated with 40% of *Drosophila* genes. *Genetics* 167:761–781.

Belozerov, V. E., P. Majumder, P. Shen, and H. N. Cai. 2003. A novel boundary element may facilitate independent gene regulation in the *Antennapedia* complex of *Drosophila. EMBO J.* 22:3113–3121.

Belyaeva, E. S., I. F. Zhimulev, E. I. Volkova, A. A. Alekseyenko, Y. M. Moshkin, and D. E. Koryakov. 1998. SuUR: A gene suppressing DNA underreplication in intercalary and pericentromeric heterochromatin of *Drosophila melanogaster* polytene chromosomes. *Proc. Natl. Acad. Sci. USA* 95:7532–7537.

Bennett, M. D., I. J. Leitch, H. J. Price, and J. S. Johnston. 2003. Comparisons with *Caenorhabditis* ~100 Mb and *Drosophila* ~175 Mb using flow cytometry show genome size in *Arabdiopsis* to be ~157 Mb and thus ~25% larger than the *Arabidopsis* Genome Initiative estimate of ~125 Mb. *Anns. Bot.* 91:547–557.

Benos, P. V., M. K. Gatt, M. Ashburner, L. Murphy, D. Harris, B. Barrell, C. Ferraz, S. Vidal, C. Brun, J. Demailles, E. Cadieu, S. Dreano, S. Gloux, V. Lelaure, S. Mottier, F. Galibert, D. Borkova, B. Minana, F. C. Kafatos, C. Louis, I. Siden-Kiamos, S. Bolshakov, G. Papagiannakis, L. Spanos, S. Cox, E. Madueno, B. de Pablos, J. Modolell, A. Peter, P. Schottler, M. Werner, F. Mourkioti, N. Beinert, G. Dowe, U. Schafer, H. Jackle, A. Bucheton, D. M. Callister, L. A. Campbell, A. Darlamitsou, N. S. Henderson, P. J. McMillan, C. Salles, E. A. Tait, P. Valenti, R. D. C. Saunders, and D. M. Glover. 2000. From sequence to chromosome: The tip of the X chromosome of *D. melanogaster. Science* 287:2220–2222.

Benos, P. V., M. K. Gatt, L. Murphy, D. Harris, B. Barrell, C. Ferraz, S. Vidal, C. Brun, J. Demaille, E. Cadieu, S. Dreano, S. Gloux, V. Lelaure, S. Mottier, F. Galibert, D. Borkova, B. Minana, F. C. Kafatos, S. Bolshakov, I. Siden-Kiamos, G. Papagiannakis, L. Spanos, C. Louis, E. Madueno, B. de Pablos, J. Modolell, A. Peter, P. Schottler, M. Werner, F. Mourkioti, N. Beinert, G. Dowe, U. Schafer, H. Jackle, A. Bucheton, D. Callister, L. Campbell, N. S. Henderson, P. J. McMillan, C. Salles, E. Tait, P. Valenti, R. D. Saunders, A. Billaud, L. Pachter, D. M. Glover, and M. Ashburner. 2001. From first base: The sequence of the tip of the X chromosome of *Drosophila melanogaster,* a comparison of two sequencing strategies. *Genome Res.* 11:710–730.

Bensaadi-Merchermek, N., C. Cagon, I. Desmons, J. C. Salvado, S. Kareme, D. D'Amico, and C. Mouchès. 1997. CM-gag, a transposable-like element reiterated in the genome of *Culex pipiens* mosquitoes, contains only a *gag* gene. *Genetica* 100:141–148.

Bergman, C. M., and M. Kreitman. 2001. Analysis of conserved noncoding DNA in *Drosophila* reveals similar constraints in intergenic and intronic sequences. *Genome Res.* 11:1335–1345.

Bergman, C. M., B. D. Pfeiffer, D. E. Rincon-Limas, R. A. Hoskins, A. Gnirke, C. J. Mungall, A. M. Wang, B. Kronmiller, J. Pacleb, S. Park, M. Stapleton, K. Wan, R. A. George, P. J. De Jong, J. Botas, G. M. Rubin, and S. E. Celniker. 2002. Assessing the impact of comparative genomic sequence data on the functional annotation of the *Drosophila* genome. *Genome Biol.* 3:RESEARCH 0086.1–86.20.

Besansky, N. J., and J. R. Powell. 1992. Reassociation kinetics of *Anopheles gambiae* (Diptera: Culicidae). DNA. *J. Med. Ent.* 29:125–128.

Betrán, E., and M. Long. 2003. Dntf-2r, a young *Drosophila* retroposed gene with specific male expression under positive Darwinian selection. *Genetics* 164:977–988.

Betrán, E., K. Thornton, and M. Long. 2003. Retroposed new genes out of the X in *Drosophila. Genome Res.* 12:1854–1859.

Bettencourt, B. R., and M. E. Feder. 2002. Rapid concerted evolution via gene conversion at the *Drosophila hsp70* genes. *J. Mol. Evol.* 54:569–586.

Beye, M., M. Hasselmann, M. K. Fondrk, R. E. Page, and S. W. Omholt. 2003. The gene *csd* is the primary signal for sexual development in the honeybee and encodes a SR-type protein. *Cell* 114:419–429.

Biessmann, H., and J. M. Mason. 1988. Progressive loss of DNA sequences from terminal chromosome deficiencies in *Drosophila melanogaster. EMBO J.* 7:1081–1086.

Biessmann, H., and J. M. Mason. 1997. Telomere maintenance without telomeres. *Chromosoma* 106:63–69.

Biessmann, H., J. M. Mason, K. Ferry, M. d'Hulst, K. Valgeirsdottir, K. L. Traverse, and M. L. Pardue. 1990. Addition of telomere-associated HeT DNA sequences "heals" broken chromosome ends in *Drosophila. Cell* 61:663–673.

Biessmann, H., L. E. Champion, M. O'Hair, K. Ikenaga, B. Kasravi, and J. M. Mason. 1992. Frequent transpositions of *Drosophila melanogaster HeT-A* transposable elements to receding chromosome ends. *EMBO J.* 11:4459–4469.

Biessmann, H., F. Kobeski, M. F. Walter, A. Kasravi, and C. W. Roth. 1998. DNA organization and length polymorphism at the 2L telomeric region of *Anopheles gambiae. Insect Mol. Biol.* 7:83–93.

Biessmann, H., M. Zurovcova, J. G. Yao, E. Lozovskaya, and M. F. Walter. 2000. A telomeric satellite in *Drosophila virilis* and its sibling species. *Chromosoma* 109:372–380.

Bird, A. P., and M. H. Taggart. 1980. Variable patterns of total DNA and rDNA methylation in animals. *Nucleic Acids Res.* 8:1485–1497.

Bird, A. P., and A. P. Wolffe. 1999. Methylation-induced repression—belts, braces and chromatin. *Cell* 91: 451–454.

Black, W. C. IV, and K. S. Rai. 1988. Genome evolution in mosquitoes: Intraspecific and interspecific variation in repetitive DNA amounts and organization. *Genet. Res. (Camb.)* 51:185–196.

Block, K. 1969. Chromosomal variation in Agromyzidae. II. *Phytomyza crassiseta* Zetterstedt—a parthenogenetic species. *Hereditas* 62:357–381.

Blumenthal, T. 1995. *Trans*-splicing and polycistronic transcription in *C. elegans. Trends Genet.* 11:132–136.

Boggs, R. T., P. Gregor, S. Idriss, J. M. Belote, and M. McKeon. 1987. Regulation of sexual differentiation in *Drosophila melanogaster* via alternative splicing of RNA from the transformer gene. *Cell* 50:739–747.

Boguski, M. S., T. M. Lowe, and C. M. Tolstoshev. 1993. dbEST—database for "expressed sequence tags." *Nat. Genet.* 4:332–333.

Bolshakov, V. N., P. Topalis, C. Blass, E. Kokoza, A. della Torre, F. C. Kafatos, and C. Louis. 2002. A comparative genomic analysis of two distant Diptera, the fruit fly, *Drosophila melanogaster,* and the malaria mosquito, *Anopheles gambiae. Genome Res.* 12:57–66.

Bonin, C. P., and R. S. Mann. 2004. A *piggyBac* transposon gene trap for the analysis of gene expression and function in *Drosophila. Genetics* 167:1801–1811.

Bopp, D., G. Calhoun, J. I. Horabin, M. Samuels, and P. Schedl. 1996. Sex-specific control of *Sex-lethal* is a conserved mechanism for sex determination in the genus *Drosophila. Development* 122:971–982.

Boutanaev, A. M., A. I. Kalmykova, Y. Y. Shevelyov, and D. I. Nurminsky. 2002. Large clusters of co-expressed genes in the *Drosophila* genome. *Nature* 420:666–669.

Bowen, N. J., and J. F. McDonald. 2001. *Drosophila* euchromatic LTR retrotransposons are much younger than the host species in which they reside. *Genome Res.* 11:1527–1540.

Bozzetti, M. P., S. Massari, P. Finelli, F. Meggio, L. A. Pinna, B. Boldyreff, O. G. Issinger, G. Palumbo, C. Ciriaco, and S. Bonaccorsi. 1995. The *Ste* locus, a component of the parasitic *cry-Ste* system of *Drosophila melanogaster,* encodes a protein that forms crystals in primary spermatocytes and mimics properties of the beta subunit of casein kinase 2. *Proc. Natl. Acad. Sci. USA* 92:6067–6071.

Brennecke, J., D. R. Hipfner, A. Stark, R. B. Russell, and S. M. Cohen. 2003. *bantam* encodes a developmentally regulated microRNA that controls cell proliferation and regulates the proapoptotic gene *hid* in *Drosophila. Cell* 113:25–36.

Bridges, C. B. 1925. Sex in relation to chromosomes and genes. *Am. Nat.* 59:127–137.

Brogna, S., and M. Ashburner. 1997. The *Adh*-related gene of *Drosophila melanogaster* is expressed as a functional dicistronic messenger RNA: Multigenic transcription in higher organisms. *EMBO J.* 16:2023–2031.

Brown, S. E., D. W. Severson, L. A. Smith, and D. L. Knudson. 2001. Integration of the *Aedes aegypti* mosquito genetic linkage and physical maps. *Genetics* 157:1299–1305.

Brun, G., and N. Plus. 1980. The viruses of *Drosophila*. In M. Ashburner and T. R. F. Wright, eds., *The Genetics and Biology of Drosophila*, 2nd ed., pp. 625–702. London and New York: Academic Press.

Burisch, E. 1963. Beiträge zur Genetik von *Megaselia scalaris* Loew Phoridae. *Z. Vererbungs.* 94:322–330.

Burmester, T., and T. Hankeln. 1999. A globin gene of *Drosophila melanogaster. Mol. Biol. Evol.* 16:1809–1811.

Bush, G. L. 1966. Female heterogamety in the family Tephritidae Acalyptratae, Diptera. *Am. Nat.* 100: 119–126.

Cáceres, M., J. M. Ranz, A. Barbadilla, M. Long, and A. Ruiz. 1999. Generation of a widespread *Drosophila* inversion by a transposable element. *Science* 285:415–418.

Cáceres, M., M. Puig, and A. Ruiz. 2001. Molecular characterization of two natural hotspots in the *Drosophila buzzatii* genome induced by transposon insertions. *Genome Res.* 11:1353–1364.

Caron, H., B. van Schaik, M. van der Mee, F. Baas, G. Riggins, P. van Shuis, M.-C. Hermus, R. van Asperen, K. Boon, P. A. Voûte, S. Heisterkamp, A. van Kampen, and R. Versteeg. 2001. The human transcriptome map: Clustering of highly expressed genes in chromosomal domains. *Science* 291:1289–1292.

Carvalho, A. B. 2002. Origin and evolution of the *Drosophila* Y chromosome. *Curr. Opin. Genet. Dev.* 12:664–668.

Carvalho, A. B., and A. G. Clark. 2004. Y chromosome of *D. pseudoobscura* is not homologous to the ancestral *Drosophila Y. Science.* Published online November 4, 2004.

Carvalho, A. B., B. P. Lazzaro, and A. G. Clark. 2000. Y chromosomal fertility factors *kl-2* and *kl-3* of *Drosophila melanogaster* encode dynein heavy chain polypeptides. *Proc. Natl. Acad. Sci. USA* 97:13239–13244.

Carvalho, A. B., B. A. Dobo, M. D. Vibranovski, and A. G. Clark. 2001. Identification of five new genes on the Y chromosome of *Drosophila melanogaster. Proc. Natl. Acad. Sci. USA* 98:13225–13230.

Carvalho, A. B., M. D. Vibranovski, J. W. Carlson, S. E. Celniker, R. A. Hoskins, G. M. Rubin, G. G. Sutton, M. D. Adams, E. M. Myers, and A. G. Clark. 2003. Y chromosome and other heterochromatic sequences of the *Drosophila melanogaster* genome: How far can we go? *Genetica* 117:227–237.

Casacuberta, E., and M.-L. Pardue. 2002. Coevolution of the telomeric retrotransposon across *Drosophila* species. *Genetics* 161:1113–1124.

Casals, F., M. Cáceres, and A. Ruiz. 2003. The Foldback-like transposon *Galileo* is involved in the generation of two different natural chromosomal inversions of *Drosophila buzzatii. Mol. Biol. Evol.* 20: 674–685.

Castellano, S., N. Morozova, M. Morey, M. J. Berry, F. Serras, M. Corominas, and R. Guigo. 2001. *In silico* identification of novel selenoproteins in the *Drosophila melanogaster* genome. *EMBO Rpts.* 2:697–702.

Caudy, A. A., M. Myers, G. J. Hannon, and S. M. Hammond. 2002. Fragile X-related protein and VIG associate with the RNA interference machinery. *Genes Dev.* 16:2491–2496.

Celniker, S. E., D. A. Wheeler, B. Kronmiller, J. W. Carlson, A. Halpern, S. Patel, M. Adams, M. Champe, S. P. Dugan, E. Frise, A. Hodgson, R. A. George, R. A. Hoskins, T. Laverty, D. M. Muzny, C. R. Nelson, J. M. Pacleb, S. Park, B. D. Pfeiffer, S. Richards, E. J. Sodergren, R. Svirskas, P. E. Tabor, K. Wan, M. Stapleton, G. G. Sutton, C. Venter, G. Weinstock, S. E. Scherer, E. W. Myers, R. A. Gibbs, and G. M. Rubin. 2002. Finishing a whole-genome shotgun: Release 3 of the *Drosophila melanogaster* euchromatic genome sequence. *Genome Biol.* 3:RESEARCH 0079-9.

Celotto, A. M., and B. R. Graveley. 2001. Alternative splicing of the *Drosophila Dscam* pre-mRNA is both temporally and spatially regulated. *Genetics* 159:599–608.

Charlesworth, B., and D. Charlesworth. 2000. The degeneration of Y chromosomes. *Proc. R. Soc. Lond. B* 355:1563–1572.

Charlesworth, B., D. Charlesworth, J. Hnilicka, A. Yu, and D. S. Guttman. 1997. Lack of degeneration of loci on the *neo*-Y chromosome of *Drosophila americana americana. Genetics* 145:989–1002.

Chen, X., L. Song, and S. Aksoy. 1999. Concordant evolution of a symbiont with its host insect species: Molecular phylogeny of genus *Glossina* and its bacteriome-associated endosymbiont *Wigglesworthia glossinidia. J. Mol. Evol.* 48:49–58.

Chen, Z.-Z., J. Martin, and B. T. O. Lee. 1995. Hemoglobins of *Kiefferulus,* sister genus of *Chironomus* (Diptera: Insecta): Evolution of the HbVIIB cluster. *J. Mol. Evol.* 41:909–919.

Chubareva, L. A. 2001. Triploidy in populations of three species of black flies of the genus *Prosimulium* Roubaud (Diptera, Simuliidae). *Tsitologia* 43:992–1000. [In Russian, English summary]

Chubareva, L. A., T. M. Grinchuk, and E. A. Kachvorian. 1974. Triploidy in natural populations of bloodsucking black flies. *Tsitologiia* 16:253–255. [In Russian, English summary]

Chubareva, L. A., N. A. Petrova, and E. A. Kachvorian. 2003. [Morphological diversity of centromere regions in polytene chromosomes of blackflies (Diptera, Simulidae).] *Tsitologia* 45:368–376. [In Russian, English summary]

Civetta, A., and R. S. Singh. 1998. Sex-related genes, directional sexual selection, and speciation. *Mol. Biol. Evol.* 15:901–909.

Clark, J. B., and M. G. Kidwell. 1997. A phylogenetic perspective on P transposable element evolution in *Drosophila. Proc. Natl. Acad. Sci. USA* 94:11428–11433.

Clary, D. O., and D. R. Wolstenholme. 1984. The *Drosophila* mitochondrial genome. *Oxford Surveys Eukaryotic Genes* 1:1–35.

Clary, D. O., and D. R. Wolstenholme. 1985. The mitochondrial DNA molecular of *Drosophilayakuba:* Nucleotide sequence, gene organization, and genetic code. *J. Mol. Evol.* 22:252–271.

Claycomb, J. M., M. Benasutti, G. Bosco, D. F. Fenger, and T. L. Orr-Weaver. 2004. Gene amplification as a developmental strategy: Isolation of two developmental amplicons in *Drosophila. Devel. Cell* 6: 145–155.

Cline, T. W., and B. J. Meyer. 1996. Vive la difference: Males vs females in flies and worms. *Annu. Rev. Genet.* 30:637–702.

Cockburn, A. F., S. E. Mitchell, and J. A. Seawright. 1990. Cloning of the mitochondrial genome of *Anopheles quadrimaculatus. Arch. Insect Biochem. Physiol.* 14:31–36.

Coelho, P. A., D. Nurminsky, D. Hartl, and C. E. Sunkel. 1996. Identification of Porto-1, a new repeated sequence that localises close to the centromere of chromosome 2 of *Drosophila melanogaster. Chromosoma* 105:211–222.

Cohen, E. H., and S. C. Bowman. 1979. Detection and location of three simple sequence DNAs in polytene chromosomes from *virilis* group species of *Drosophila. Chromosoma* 73:327–355.

Colby, C., and S. M. Williams. 1993. The distribution and spreading of rare variants in the histone multigene family in *Drosophila melanogaster. Genetics* 135:127–133.

Contursi, C., G. Minchiotti, and P. P. di Nocera. 1993. Functional dissection of two promoters that control sense and antisense transcription of *Drosophila melanogaster F* elements. *J. Mol. Biol.* 234:988–997.

Cordeiro, M., L. Wheeler, C. S. Lee, C. D. Kastritsis, and R. H. Richardson. 1975. Heterochromatic chromosomes and satellite DNAs of *Drosophila nasutoides. Chromosoma* 51:65–73.

Craig, N. L., R. Craigie, M. Gellert, and A. M. Lambowitz, eds. 2002. *Mobile DNA II.* Washington, D.C.: American Society of Microbiology.

Crouse, H. V. 1960. The controlling element in sex chromosome behavior in *Sciara. Genetics* 45:1429–1443.

Crouse, H. V. 1979. The controlling element in sex chromosome behavior in *Sciara. Chromosoma* 74: 219–239.

Csink, A. K., and S. Henikoff. 1998. Something for nothing: The evolution and utility of satellite repeats. *Trends Genet.* 14:200–204.

Currie, P. D., and D. T. Sullivan. 1994. Structure, expression and duplication of genes which encode phosphoglyceromutase of *Drosophila melanogaster. Genetics* 138:353–363.

Da Lage, J.-L., M. Wegnez, and M.-L. Cariou. 1996. Distribution and evolution of introns in *Drosophila* Amylase genes. *J. Mol. Evol.* 43:334–347.

Daniels, S. B., K. R. Peterson, L. D. Strausbaugh, M. G. Kidwell, and A. Chovnick. 1990. Evidence for horizontal transmission of the P transposable element between *Drosophila* species. *Genetics* 124:339–355.

Danilevskaya, O. N., K. L. Traverse, N. C. Hogan, P. G. DeBaryshe, and M.-L. Pardue. 1999. The two *Drosophila* telomeric transposable elements have very different patterns of transcription. *Mol. Cell. Biol.* 19: 873–881.

Darlington, C. D. 1937. *Recent Advances in Cytology,* 2nd ed. London: J. & A. Churchill.

Davidson, J. E., C. W. Schmidt, and N. Davidson. 1975. Interspersion of repeat and non-repeat DNA sequences in the *Drosophila melanogaster* genome. *Cell* 4:141–155.

Davis, P. S., M. W. Shen, and B. H. Judd. 1987. Asymmetrical pairings of transposons in and proximal to the *white* locus of *Drosophila* account for four classes of regularly occurring exchange products. *Proc. Natl. Acad. Sci. USA* 84:174–178.

de Carvalho, M. O., J. C. Silva, and E. L. S. Loreto. 2003. Analaysis of *P*-like transposable element sequences from the genome of *Anopheles gambiae. Insect Mol. Biol.* 13:55–63.

Deininger, P. L. 1989. SINEs: Short interspersed repeated DNA elements in higher eukaryotes. In D. E. Berg and M. M. Howe, eds., *Mobile DNA,* pp. 619–636. Washington, D.C.: American Society of Microbiology.

Delidakis, C., and F. C. Kafatos. 1987. Amplification of a chorion gene cluster in *Drosophila* is subject to multiple *cis*-regulatory elements and to long-range position effects. *J. Mol. Biol.* 197:11–26.

Delidakis, C., and F. C. Kafatos. 1989. Amplification enhancers and replication origins in the autosomal chorion gene cluster of *Drosophila. EMBO J.* 8:891–901.

Denholm, I., M. G. Franco, P. G. Rubini, and M. Vecchi. 1983. Identification of a male-determinant on the X chromosome of housefly (*Musca domestica* L.) populations in South-East England. *Genet. Res.* 42: 311–322.

Dennhofer, L. 1975. Die Speicheldrüsenchromosomen der Stechmücke *Culex pipiens L.* IV. Der chromosomale Geschlechtsdimorphismus. *Genetica* 45:163–175.

de Saint Phalle, B., and W. Sullivan. 1996. Incomplete sister chromatid separation is the mechanism of programmed chromosome elimination during early *Sciara coprophila* embryogenesis. *Development* 122: 3775–3784.

Dewilde, S., M. Blaxter, M. L. Van Hauwaert, K. Van Houte, A. Pesce, N. Griffon, L. Kiger, M. C. Marden, S. Vermeire, J. Vanfleteren, E. Esmans, and L. Moens. 1998. Structural, functional, and genetic characterization of *Gastrophilus* [sic] hemoglobin. *J. Biol. Chem.* 273:32467–32474.

Dimopoulos, G., G. K. Christophides, S. Meister, J. Schultz, K. P. White, C. Barillas-Mury, and F. C. Kafatos. 2002. Genome expression analysis of *Anopheles gambiae:* Responses to injury, bacterial challenge, and malaria infection. *Proc. Natl. Acad. Sci. USA* 99:8814–8819.

Djikeng, A., S. Shi, C. Tschudi, and E. Ullu. 2001. RNA interference in *Trypanosoma brucei:* Cloning of small interfering RNAs provides evidence for retroposon-derived 24–26 nucleotide RNAs. *RNA* 7: 1522–1530.

Dominski, Z., Y.-c. Yang, M. Purdy, and W. F. Marzluff. 2003. Cloning and characterization of the *Drosophila* U7 small nuclear RNA. *Proc. Natl. Acad. Sci. USA* 100:9422–9427.

Dorn, R., G. Reuter, and A. Loewendorf. 2001. Transgene analysis proves mRNA *trans*-splicing at the complex *mod(mdg4)* locus in *Drosophila. Proc. Natl. Acad. Sci. USA* 98:9724–9729.

Dru, P., F. Bras, S. Dezelee, P. Gay, A. M. Petitjean, A. M. Pierre-Deneubourg, D. Teninges, and D. Contamine. 1993. Unusual variability of *Drosophila melanogaster ref(2)P* protein which controls the multiplication of sigma rhabdovirus. *Genetics* 133:943–954.

Dübendorfer, A., D. Hilfiker-Kleiner, and R. Nöthiger. 1992. Sex determination mechanisms in dipteran insects: The case of *Musca domestica. Semin. Devel. Biol.* 3:349–356.

Eberhard, W. G. 1985. *Sexual Selection and Animal Genitalia.* Cambridge, Mass.: Harvard University Press.

Echols, N., P. Harrison, S. Balasubramanian, M. N. Luscombe, P. Bertone, Z. Zhang, and M. Gerstein. 2002. Comparative analysis of amino-acid and nucleotide composition in eukaryotic genomes, comparing genes and pseudogenes. *Nucleic Acids Res.* 30:2515–2523.

Eickbush, T. H. 2002. R2 and related site-specific non-long terminal repeat retrotransposons. In N. L. Craig, R. Craigie, M. Gellert, and A. M. Lambowitz, eds., *Mobile DNA II,* pp. 813–835. Washington, D.C.: American Society of Microbiology.

Eickbush, T. H., and H. S. Malik. 2002. Origins and evolution of retrotransposons. In N. L. Craig, R. Craigie, M. Gellert, and A. M. Lambowitz, eds., *Mobile DNA II,* pp. 1111–1144. Washington, D.C.: American Society of Microbiology.

Engels, W. R. 1989. *P* elements in *Drosophila melanogaster.* In D. E. Berg and M. M. Howe, eds., *Mobile DNA,* pp. 437–484. Washington, D.C.: American Society of Microbiology.

Enright, A., B. John, U. Gaul, T. Tuschl, C. Sander, and D. Marks. 2003. MicroRNA targets in *Drosophila. Genome Biol.* 5:R1.

Erickson, J. W., and T. W. Cline. 1998. Key aspects of the primary sex determination mechanism are conserved across the genus *Drosophila. Development* 125:3259–3268.

Evgen'ev, M. B., H. Zelentsova, N. Shostak, M. Kozitsina, V. Barskyi, D. H. Lankenau, and V. G. Corces. 1997. *Penelope,* a new family of transposable elements and its possible role in hybrid dysgenesis in *Drosophila virilis. Proc. Natl. Acad. Sci. USA* 94:196–201.

Evgen'ev, M. B., H. Zelentsova, L. Mnjoian, A. Poluectova, and M. G. Kidwell. 2000. Invasion of *Drosophila virilis* by the *Penelope* transposable element. *Chromosoma* 109:350–357.

Feiber, A. L., J. Rangarajan, and J. C. Vaugh. 2002. The evolution of single-copy *Drosophila* nuclear *4f-rnp* genes: Spliceosomal intron losses create polymorphic alleles. *J. Mol. Evol.* 55:401–413.

Feschotte, C., and C. Mouchès. 2000. Recent amplification of miniature inverted-repeat transposable elements in the vector mosquito *Culex pipiens:* Characterization of the *Mimo* family. *Gene* 250:109–116.

Feschotte, C., N. Fourrier, I. Desmons, and C. Mouchès. 2001. Birth of a retroposon: The *Twin* SINE family from the vector mosquito *Culex pipiens* may have originated from a dimeric tRNA precursor. *Mol. Biol. Evol.* 18:74–84.

Feschotte, C., X. Zhang, and S. R. Wessler. 2002. Miniature-inverted repeat transposable elements and their relationship to established DNA transposons. In N. L. Craig, R. Craigie, M. Gellert, and A. M. Lambowitz., eds., *Mobile DNA II,* pp. 1147–1158. Washington, D.C.: American Society of Microbiology.

Fink, G. R. 1987. Pseudogenes in yeast? *Cell* 10:5–6.

FlyBase. 2003. The FlyBase database of the *Drosophila* genome projects and community literature. *Nucleic Acids Res.* 31:172–175.

Foster, G. G., M. J. Whitten, C. Konovalov, J. T. A. Arnold, and G. Maffi. 1981. Autosomal genetic maps of the Australian sheep blowfly, *Lucilia cuprina dorsalis* R. D. (Diptera:Calliphoridae), and possible correlations with the linkage maps of *Musca domestica* L. and *Drosophila melanogaster* Mg. *Genet. Res. (Camb.)* 37:55–69.

Franco, M. G., P. G. Rubini, and M. Vechhi. 1982. Sex-determinants and their distribution in various populations of *Musca domestica* L. of Western Europe. *Genet. Res.* 40:279–293.

Franke, A., and B. S. Baker. 1999. The *roX1* and *roX2* RNAs are essential components of the compensasome, which mediates dosage compensation in *Drosophila melanogaster. Mol. Cell* 4:117–122.

Fraser, M. J. 2000. The TTAA-specific family of transposable elements: Identification, functional characterization, and utility for transformation of insects. In A. H. Handler and A. A. James, eds., *Insect Transgenesis,* pp. 249–270. Boca Raton, Fla.: CRC Press.

Friedman, R., and A. L. Hughes. 2001. Gene duplication and the structure of eukaryotic genomes. *Genome Res.* 11:373–381.

Gall, J. G., E. H. Cohen, and M. L. Polan. 1971. Repetitive DNA sequences in *Drosophila. Chromosoma* 33: 319–344.

Gdula, D. A., T. J. Gerasimova, and V. G. Corces. 1996. Genetic and molecular analysis of the gypsy chromatin insulator of *Drosophila. Proc. Natl. Acad. Sci. USA* 93:9378–9383.

Gelbart, W. M., and A. Chovnick. 1979. Spontaneous unequal exchange in the *rosy* region of *Drosophila melanogaster. Genetics* 92:849–859.

Gepner, J., and J. S. Hays. 1993. A fertility region of the Y chromosome of *Drosophila melanogaster* encodes a dynein microtubule motor. *Proc. Natl. Acad. Sci. USA* 90:10904–10906.

Gerasimova, T. I., and V. G. Corces. 2001. Chromatin insulators and boundaries: Effects on transcription and nuclear organization. *Annu. Rev. Genet.* 35:193–208.

Gerbi, S. A. 1986. Unusual chromosome movements in sciarid flies. In W. Hennig, ed., *Germ Line-Soma Differentiation. Results and Problems in Cell Differentiation 13,* pp. 71–104. Berlin and Heidelberg: Springer.

Gerbi, S. A., and F. D. Urnov. 1996. Differential DNA replication in insects. In M. L. DePamphilis, ed., *DNA Replication in Eukaryotic Cells,* pp. 947–970. Cold Spring Harbor, N.Y.: Cold Spring Harbor Press.

Geyer, P. K. 1997. The role of insulator elements in defining domains of gene expression. *Curr. Opin. Genet. Dev.* 7:242–248.

Gilchrest, B. M., and J. B. S. Haldane. 1947. Sex linkage and sex determination in a mosquito, *Culex molestus. Hereditas* 33:175–190.

Glover, D. M., A. Zaha, A. J. Stocker, R. V. Santelli, M. T. Pueyo, S. M. de Toledo, and F. J. S. Lara. 1982. Gene amplification in *Rhynchosciara angelae* salivary gland chromosomes. *Proc. Natl. Acad. Sci. USA* 79: 2947–2951.

Goday, C., and M. F. Ruiz. 2002. Differential acetylation of histones H3 and H4 in paternal and maternal germline chromosomes during development of sciarid flies. *J. Cell. Sci.* 115:4765–4775.

Goddard, J. M., and D. R. Wolstenholme. 1980. Origin and direction of replication in mitochondrial DNA molecules from the genus *Drosophila. Nucleic Acids Res.* 8:741–757.

Goldberg, M. L., R. Paro, and W. J. Gehring. 1982. Molecular cloning of the *white* locus region of *Drosophila melanogaster* using a large transposable element. *EMBO J.* 1:93–98.

Goldberg, M. L., J.-Y. Sheen, W. J. Gehring, and M. M. Green. 1983. Unequal crossing-over associated with asymmetrical synapsis between nomadic elements in the *Drosophila melanogaster* genome. *Proc. Natl. Acad. Sci. USA* 80:5017–5021.

Goldstein, L. S. B., R. W. Hardy, and D. L. Lindsley. 1982. Structural genes on the Y chromosome of *Drosophila melanogaster. Proc. Natl. Acad. Sci. USA* 79:7405–7409.

González, J., E. Betrán, M. Ashburner, and A. Ruiz. 2000. Molecular organization of the *Drosophila melanogaster Adh* chromosomal region in *D. repleta* and *D. buzzatii,* two distantly related species of the *Drosophila* genus. *Chromosome Res.* 8:375–385.

González, J., J. M. Ranz, and A. Ruiz. 2002. Chromosomal elements evolve at different rates in the *Drosophila* genome. *Genetics* 161:1137–1154.

Gopal, S., M. Schroeder, U. Pieper, A. Sczyrba, G. Aytekin-Kurban, S. Bekiranov, J. E. Fajardo, N. Eswar, R. Sanchez, A. Sali, and T. Gaasterland. 2001. Homology-based annotation yields 1,042 new candidate genes in the *Drosophila melanogaster* genome. *Nat. Genet.* 27:337–340.

Gott, J. M., and R. B. Emerson. 2000. Functions and mechanisms of RNA editing. *Annu. Rev. Genet.* 34:499–531.

Gowher, H., O. Leismann, and A. Jeltsch. 2000. DNA of *Drosophila melanogaster* contains 5-methylcytosine. *EMBO J.* 19:6918–6923.

Grauso, M., R. A. Reenan, E. Culetto, and D. B. Sattelle. 2002. Novel putative nicotinic acetylcholine receptor subunit genes, *Dalpha5, Dalpha6* and *Dalpha7,* in *Drosophila melanogaster* identify a new and highly conserved target of adenosine deaminase acting on RNA-mediated A-to-I pre-mRNA editing. *Genetics* 160:1519–1533.

Green, C. A. 1982. Cladistic analysis of mosquito chromosome data *Anopheles (Cellia) Myzomyia). J. Hered.* 73:2–11.

Green, M. M. 1980. Transposable elements in *Drosophila* and other Diptera. *Annu. Rev. Genet.* 14:109–120.

Green, P. 1997. Against a whole-genome shotgun. *Genome Res.* 7:410–417.

Gregory, T. R. 2001. Animal Genome Size Database. www.genomesize.com.

GtRDB. 2003. The Genomic tRNA Database. rna.wustl.edu/GtRDB.

Gu, Z., A. Cavalcanti, F.-C. Chen, P. Bouman, and W.-H. Li. 2002. Extent of gene duplication in the genomes of *Drosophila,* nematode and yeast. *Mol. Biol. Evol.* 19:256–262.

Gubb, D. 1986. Intron-delay and the precision of expression of homeotic gene products in *Drosophila. Dev. Genet.* 7:119–131.

Gvozdev, V. A., A. A. Aravin, Y. A. Abramov, M. S. Klenov, G. L. Kogan, S. A. Lavrov, N. M. Naumova, O. M. Olenkina, A. V. Tulin, and V. V. Vagin. 2003. *Stellate* repeats: Targets of silencing and modules causing *cis*-inactivation and *trans*-activation. *Genetica* 117:239–245.

Hagele, K. 1985. Identification of a polytene chromosome band containing a male sex determiner of *Chironomus thummi thummi. Chromosoma* 91:167–171.

Hagstrom, K., M. Muller, and P. Schedl. 1996. Fab-7 functions as a chromatin domain boundary to ensure proper segment specification by the *Drosophila bithorax* complex. *Genes Dev.* 10:3202–3215.

Handler, A. M. 2002. Use of the *piggyBac* transposon for germ-line transformation of insects. *Insect Biochem. Mol. Biol.* 32:1211–1220.

Handler, A. M., and R. A. Harrell. 1999. Germline transformation of *Drosophila melanogaster* with the *piggyBac* transposon vector. *Insect Mol. Biol.* 8:449–457.

Handler, A. H., and S. D. McCombs. 1999. The *piggyBac* transposon mediates germ-line transformation in the oriental fruit fly and closely related elements exist in its genome. *Insect Mol. Biol.* 9:605–612.

Hankeln, T., and E. R. Schmidt. 1990. New foldback transposable element TFB1 found in histone genes of the midge *Chironomus thummi. J. Mol. Biol.* 215:477–482.

Hankeln, T., C. Amid, B. Weich, J. Niessing, and E. R. Schmidt. 1998. Molecular evolution of the globin gene cluster E in two distantly related midges, *Chironomus pallidivittatus* and *C. thummi thummi. J. Mol. Evol.* 46:589–601.

Hanrahan, C. J., M. J. Palladino, B. Ganetzky, and R. A. Reenan. 2000. RNA editing of the *Drosophila para* Na^+ channel transcript. Evolutionary conservation and developmental regulation. *Genetics* 155:1149–1160.

Harden, N., and M. Ashburner. 1990. Characterization of the *FB-NOF* transposable element of *Drosophila melanogaster. Genetics* 126:387–400.

Hart, K., and M. Bienz. 1996. A test for cell autonomy, based on di-cistronic messenger translation. *Development* 122:747–751.

He, H., C. Rovira, S. Recco-Pimentel, C. Liao, and J.-E. Edström. 1995. Polymorphic SINEs in chironomids with DNA derived from the R2 insertion site. *J. Mol. Biol.* 245:34–42.

He, H., C. Liao, and J.-E. Edström. 1998. Centromere 3 specific tandem repeats from *Chironomus pallidivittatus. Chromosoma* 107:304–310.

Hediger, M., A.-D. Minet, M. Niessen, R. Schmidt, D. Hilfiker-Kleiner, S. Cakir, R. Nothiger, and A. Dubendorfer. 1998a. The male-determining activity on the Y chromosome of the housefly *Musca domestica* L. consists of separable elements. *Genetics* 150:651–661.

Hediger, M., M. Niessen, J. Muller-Navia, R. Nothiger, and A. Dubendorfer. 1998b. Distribution of heterochromatin on the mitotic chromosomes of *Musca domestica* L. in relation to the activity of male-determining factors. *Chromosoma* 107:267–271.

Hediger, M., G. Burghardt, C. Siegenthaler, N. Buser, D. Hilfiker-Kleiner, A. Dübendorfer, and D. Hopp. 2004. Sex determination in *Drosophila melanogaster* and *Musca domestica* converges at the level of the terminal regulator *doublesex*. *Devel. Genes Evol.* 214:29–42.

Heinrich, J. C., X. Li, R. A. Henry, N. Haack, L. Stringfellow, A. C. Heath, and M. J. Scott. 2001. Germ-line transformation of the Australian sheep blowfly *Lucilia cuprina*. *Insect Mol. Biol.* 10:1–10.

Henikoff, S., and M. K. Eghtedarzadeh. 1987. Conserved arrangement of nested genes at the *Drosophila* Gart locus. *Genetics* 117:711–725.

Henikoff, S., K. Ahmad, and H. S. Malik. 2001. The centromere paradox: Stable inheritance with rapidly evolving DNA. *Science* 293:1098–1102.

Hennig, W. 1987. The Y chromosomal lampbrush loops of *Drosophila*. In W. Hennig, ed., *Structure and Function of Eukaryotic Chromosomes. Results and Problems in Cell Differentiation 14*, pp. 133–146. Berlin and Heidelberg: Springer.

Herrick, G., and J. Seger. 1999. Imprinting and paternal genome elimination in insects. In R. Ohlsson, ed., *Imprinting: An Interdisciplinary Approach. Results and Problems in Cell Differentiation 25*, pp. 41–71. Berlin and Heidelberg: Springer.

Hertig, M. 1936. The rickettsia, *Wolbachia pipientis* gen. et sp. n. and associated inclusions of the mosquito *Culex pipiens*. *Parasitology* 28:453–486.

Hild, M., B. Beckmann, S. A. Haas, B. Koch, V. Solovyev, C. Busold, K. Fellenberg, M. Boutros, M. Vington, F. Sauer, J. D. Hoheisel, and R. Paro. 2003. An integrated gene annotation and transcriptional profiling approach towards full gene content of the *Drosophila* genome. *Genome Biol.* 5:R3.

Hiroyoshi, T. 1964. Sex-limited inheritance and abnormal sex ratios in the housefly. *Genetics* 50:373–385.

Holmquist, G. 1975. Organization and evolution of *Drosophila virilis* heterochromatin. *Nature* 257:503–506.

Holt, R. A., G. M. Subramanian, A. Halpern, G. G. Sutton, R. Charlab, D. R. Nusskern, P. Wincker, A. G. Clark, J. M. Ribeiro, R. Wides, S. L. Salzberg, B. Loftus, M. Yandell, W. H. Majoros, D. B. Rusch, Z. Lai, C. L. Kraft, J. F. Abril, V. Anthouard, P. Arensburger, P. W. Atkinson, H. Baden, V. de Berardinis, D. Baldwin, V. Benes, J. Biedler, C. Blass, R. Bolanos, D. Boscus, M. Barnstead, S. Cai, A. Center, K. Chatuverdi, G. K. Christophides, M. A. Chrystal, M. Clamp, A. Cravchik, V. Curwen, A. Dana, A. Delcher, I. Dew, C. A. Evans, M. Flanigan, A. Grundschober-Freimoser, L. Friedli, Z. Gu, P. Guan, R. Guigo, M. E. Hillenmeyer, S. L. Hladun, J. R. Hogan, Y. S. Hong, J. Hoover, O. Jaillon, Z. Ke, C. Kodira, E. Kokoza, A. Koutsos, I. Letunic, A. Levitsky, Y. Liang, J. J. Lin, N. F. Lobo, J. R. Lopez, J. A. Malek, T. C. McIntosh, S. Meister, J. Miller, C. Mobarry, E. Mongin, S. D. Murphy, D. A. O'Brochta, C. Pfannkoch, R. Qi, M. A. Regier, K. Remington, H. Shao, M. V. Sharakhova, C. D. Sitter, J. Shetty, T. J. Smith, R. Strong, J. Sun, D. Thomasova, L. Q. Ton, P. Topalis, Z. Tu, M. F. Unger, B. Walenz, A. Wang, J. Wang, M. Wang, X. Wang, K. J. Woodford, J. R. Wortman, M. Wu, A. Yao, E. M. Zdobnov, H. Zhang, Q. Zhao, S. Zhao, S. C. Zhu, I. Zhimulev, M. Coluzzi, A. della Torre, C. W. Roth, C. Louis, F. Kalush, R. J. Mural, E. W. Myers, M. D. Adams, H. O. Smith, S. Broder, M. J. Gardner, C. M. Fraser, E. Birney, P. Bork, P. T. Brey, J. C. Venter, J. Weissenbach, F. C. Kafatos, F. H. Collins, and S. L. Hoffman. 2002. The genome sequence of the malaria mosquito *Anopheles gambiae*. *Science* 298:129–149.

Holyoake, A. J., and M. G. Kidwell. 2003. *Vege* and *Mar*: Two novel *hAT* MITE families in *Drosophila willistoni*. *Mol. Biol. Evol.* 20:163–167.

Hoopengardner, B., T. Bhalla, C. Staber, and R. Reenan. 2003. Nervous system targets of RNA editing identified by comparative genomics. *Science* 301:832–836.

Horn, C., N. Offen, S. Nystedt, U. Hacker, and E. A. Wimmer. 2003. *piggyBac*-based insertional mutagenesis and enhancer detection as a tool for functional insect genomics. *Genetics* 163:647–661.

Hoskins, R. A., C. R. Nelson, B. P. Berman, T. R. Laverty, R. A. George, L. Ciesiolka, M. Naeemuddin, A. D. Arenson, J. Durbin, R. G. David, P. E. Tabor, M. R. Bailey, D. R. DeShazo, J. Catanese, A. Mammoser, K. Osoegawa, P. J. de Jong, S. E. Celniker, R. A. Gibbs, G. M. Rubin, and S. E. Scherer. 2000. A BAC-based physical map of the major autosomes of *Drosophila melanogaster*. *Science* 287:2271–2274.

Hoskins, R. A., C. D. Smith, J. W. Carlson, A. B. Carvalho, A. Halpern, J. S. Kaminker, C. Kennedy, C. J. Mungall, B. A. Sullivan, G. G. Sutton, J. C. Yasuhara, B. T. Wakimoto, E. W. Myers, S. E. Celniker, G. M. Rubin, and G. H. Karpen. 2002. Heterochromatic sequences in a *Drosophila* whole-genome shotgun assembly. *Genome Biol.* 3:RESEARCH 0085-5.

Houck, M. A., J. B. Clark, K. R. Peterson, and M. G. Kidwell. 1991. Possible horizontal transfer of *Drosophila* genes by the mite *Proctolaelaps regalis. Science* 253:1125–1129.

Hung, M. S., N. Karthikeyan, B. Huang, H. C. Koo, J. Kiger, and C. K. J. Shen. 1999. *Drosophila* proteins related to vertebrate DNA 5-cytosine methyltransferases. *Proc. Natl. Acad. Sci. USA* 96:11940–11945.

Hutter, P., J. Roote, and M. Ashburner. 1990. A genetic basis for the inviability of hybrids between sibling species of *Drosophila. Genetics* 124:909–920.

Iida, T., and S. Kobayashi. 1998. Essential role of mitochondrially encoded large rRNA for germ-line formation in *Drosophila* embryos. *Proc. Natl. Acad. Sci. USA* 95:11274–11278.

Inoue, H., and T. Hiroyoshi. 1986. A maternal-effect sex-transforming mutant of the housefly, *Musca domestica* L. *Genetics* 112:469–482.

Ivanov, I. P., R. F. Gesteland, and J. F. Atkins. 2000. Antizyme expression: A subversion of triplet decoding, which is remarkably conserved by evolution, is a sensor for an autoregulatory circuit. *Nucleic Acids Res.* 28:3185–3196.

Jakubczak, J., W. D. Burke, and T. H. Eickbush. 1991. Retrotransposable elements *R1* and *R2* interrupt the rDNA genes of most insects. *Proc. Natl. Acad. Sci. USA* 88:3295–3299.

Jeffs, P., and M. Ashburner. 1991. Processed pseudogenes in *Drosophila. Proc. R. Soc. Lond. B* 244:151–159.

Jensen, R. C., Y. Wang, S. B. Hardin, and W. E. Stumph. 1998. The proximal sequence element PSE plays a major role in establishing the RNA polymerase specificity of *Drosophila* U-snRNA genes. *Nucleic Acids Res.* 26:616–622.

Jensen, S., M. P. Gassama, and T. Heidmann. 1999a. Cosuppression of I transposon activity in *Drosophila* by I-containing sense and antisense transgenes. *Genetics* 153:1767–1774.

Jensen, S., M. P. Gassama, and T. Heidmann. 1999b. Taming of transposable elements by homology-dependent gene silencing. *Nat. Genet.* 21:209–212.

Jensen, S., M. P. Gassama, X. Dramard, and T. Heidmann. 2002. Regulation of I-transposon activity in *Drosophila:* Evidence for cosuppression of nonhomologous transgenes and possible role of ancestral I-related pericentromeric elements. *Genetics* 162:1197–1209.

Jiang, N., Z. Bao, X. Zhang, H. Hirochika, S. R. Eddy, S. R. McCouch, and S. R. Wessler. 2003. An active DNA transposon family in rice. *Nature* 421:163–167.

Jordan, I. K., L. V. Matyunia, and J. F. McDonald. 1999. Evidence for the recent horizontal transfer of long terminal repeat retrotransposons. *Proc. Natl. Acad. Sci. USA* 96:12621–12625.

Junqueira, A. C. M., A. C. Lessinger, T. T. Torres, A. L. Vettore, F. R. da Silva, P. Arruda, and A. M. L. Azeredo-Espin. 2001. The complete mitochondrial genome of the *Chrysomya putoria.* Genbank:AF352790.

Kaminker, J. S., C. M. Bergman, B. Kronmiller, J. Carlson, R. Svirskas, S. Patel, E. Frise, D. A. Wheeler, S. E. Lewis, G. M. Rubin, M. Ashburner, and S. E. Celniker. 2002. The transposable elements of the *Drosophila melanogaster* euchromatin: A genomics perspective. *Genome Biol.* 3:RESEARCH 0084-4.

Kapitonov, V. V., and J. Jurka. 2001. Rolling-circle transposons in eukaryotes. *Proc. Natl. Acad. Sci. USA* 98:8714–8719.

Kapitonov, V. V., and J. Jurka. 2002. Looper1_DM, a family of Looper/*piggyBac* DNA transposons in *D. melanogaster. RepBase Reps.* 2(3):6.

Kapitonov, V. V., and J. Jurka. 2003. Molecular paleontology of transposable elements in the *Drosophila melanogaster* genome. *Proc. Natl. Acad. Sci. USA* 100:6569–6574.

Kapitonov, V. V., and J. Jurka. 2004. A reference collection of *A. gambiae* repetitive elements. RepBase Update, Release 1.1.5. orka.girinst.org/server/archive/RepBase9.02/RepBase9.02.embl/angrep.ref.

Karpen, G. H., and R. C. Allshire. 1997. The case for epigenetic effects on centromere identity and function. *Trends Genet.* 13:489–496.

Karpen, G. H., and A. C. Spradling. 1992. Analysis of subtelomeric heterochromatin in the *Drosophila* minichromsoome *Dp1187* by single *P* element insertional mutagenesis. *Genetics* 132:737–753.

Kellum, R., and P. Schedl. 1992. A group of *scs* elements function as domain boundaries in an enhancer-blocking assay. *Mol. Cell. Biol.* 12:2424–2431.

Kennerdell, J. R., S. Yamaguchi, and R. W. Carthew. 2002. RNAi is activated during *Drosophila* oocyte maturation in a manner dependent on *aubergine* and *spindle-E. Genes Dev.* 16:1884–1889.

Keyl, H.-G. 1962. Chromosomenevolution bei *Chironomus.* II. Chromosomenumbauten und phylogenetische Beziehungen der Arten. *Chromosoma* 13:464–514.

Keyl, H.-G. 1965. Duplikationen von Untereinheite der chromosomalen DNS während der Evolution von *Chironomus thummi. Chromosoma* 17:139–180.

Kitamoto, T., W. Wang, and P. M. Salvaterra. 1998. Structure and organization of the *Drosophila* cholinergic locus. *J. Biol. Chem.* 273:2706–2713.

Knudson, D. L., L. Zheng, S. W. Gordon, S. E. Brown, and F. C. Kafatos. 1996. Genome organization of vectors. In B. J. Beaty and W. C. Marquardt, eds., *The Biology of Disease Vectors,* pp. 175–214. Niwot, Colo.: University of Colorado Press.

Knudson, D. L., S. E. Brown, and D. W. Severson. 2002. Culicine genomics. *Insect Biochem. Mol. Biol.* 32: 1193–1197.

Kogan, G. L., A. V. Tulin, A. A. Aravin, Y. A. Abramov, A. I. Kalymkova, C. Maisonhaute, and V. A. Gvozdev. 2003. The *GATE* retrotransposon of *Drosophila melanogaster:* Mobility in heterochromatin and aspects of expression in germline tissues. *Mol. Gen. Genomics* 269:234–242.

Kondo, N., N. Nikoh, N. Ijichi, M. Shimada, and T. Fukatsu. 2002. Genome fragment of *Wolbachia* endosymbiont transferred to the X chromosome of host insect. *Proc. Natl. Acad. Sci. USA* 99:14280–14285.

Krauss, V., and G. Reuter. 2000. Two genes become one. The genes encoding heterochromatin protein SU VAR. 3–9 and translation initiation factor subunit eIF-2gamma are joined to a dicistronic unit in holometabolic insects. *Genetics* 156:1157–1167.

Kress, H., K. Bechler, U. Swida, and S. Maletz. 2001. Evolution of 5S rRNA gene families in *Drosophila. Chromosome Res.* 9:403–415.

Krimbas, C. B., and J. R. Powell, eds. 1992. *Inversion Polymorphism.* Boca Raton, Fla.: CRC Press.

Krzywinski, J., and N. J. Besansky. 2002. Frequent intron loss in the *white* gene: A cautionary tale for phylogeneticists. *Mol. Biol. Evol.* 19:362–366.

Kuhn, S., V. Sievert, and W. Traut. 2000. The sex-determination gene *doublesex* in the fly *Megaselia scalaris:* Conserved structure and sex-specific splicing. *Genome* 43:1011–1020.

Kulathinal, R. J., L. Skwarek, R. A. Morton, and R. S. Singh. 2003. Rapid evolution of the sex-determining gene, *transformer:* Structural diversity and rate heterogeneity among sibling species of *Drosophila. Mol. Biol. Evol.* 20:441–452.

Kumar, A., and K. S. Rai. 1990. Intraspecific variation in nuclear DNA content among world populations of a mosquito, *Aedes albopictus* Skuse. *Theoret. Appl. Genet.* 79:748–752.

Kunert, N., J. Marhold, J. Stanke, D. Stach, and F. Lyko. 2003. A Dnmt2-like protein mediates DNA methylation in *Drosophila. Development* 130:5083–5090.

Kurek, R., A. M. Reugels, U. Lammermann, and H. Bünemann. 2000. Molecular aspects of intron evolution in dynein encoding mega-genes on the heterochromatic Y chromosomes of *Drosophila* species. *Genetica* 109:113–123.

Kutach, A. K., and J. T. Kadonaga. 2000. The downstream promoter element DPE appears to be as widely used as the TATA box in *Drosophila* core promoters. *Mol. Cell. Biol.* 20:4754–4764.

Labrador, M., and V. G. Corces. 2003. Extensive exon reshuffling over evolutionary time coupled with *trans*-splicing in *Drosophila. Genome Res.* 13:2220–2228.

Labrador, M., F. Mongelard, P. Plata-Rengifo, E. M. Baxter, V. G. Corces, and T. I. Gerasimova. 2001. Protein encoding by both DNA strands. *Nature* 409:1000.

Lachaise, D., M.-L. Cariou, J. R. David, F. Lemeunier, L. Tsacas, and M. Ashburner. 1988. Historical biogeography of the *Drosophila melanogaster* species subgroup. *Evol. Biol.* 22:159–225.

Lagos-Quintana, M., R. Rauhut, W. Lendeckel, and T. Tuschl. 2001. Identification of novel genes coding for small expressed RNAs. *Science* 294:853–858.

Lai, L. C., P. Tomancak, R. W. Williams, and G. M. Rubin. 2003. Computational identification of *Drosophila* microRNA genes. *Genome Biol.* 4:R42.

Laicine, E. M., M. A. R. Alves, J. C. de Almeida, E. Rizzo, W. C. Albernaz, and H. Sauia. 1984. Development of DNA puffs and patterns of polypeptide synthesis in the salivary glands of *Bradysia hygida. Chromosoma* 89:280–284.

Laird, C. D., and B. J. McCarthy. 1969. Molecular characterization of the *Drosophila* genome. *Genetics* 63: 865–882.

Lankenau, S., V. G. Corces, and D. H. Lankenau. 1994. The *Drosophila* micropia retrotransposon encodes a testis-specific antisense RNA complementary to reverse transcriptase. *Mol. Cell. Biol.* 14:1764–1775.

Lara, F. J. S., A. J. Stocker, and J. M. Amabis. 1991. DNA sequence amplification in sciarid flies: Results and perspectives. *Braz. J. Med. Biol. Res.* 24:233–248.

Le, M.-H., D. Duricka, and G. H. Karpen. 1995. Islands of complex DNA are widespread in *Drosophila melanogaster* centric heterochromatin. *Genetics* 141:283–303.

Leblanc, P., S. Desset, B. Dastugue, and C. Vaury. 1997. Invertebrate retroviruses: *ZAM,* a new candidate in *Drosophila melanogaster. EMBO J.* 16:7521–7531.

Leblanc, P., S. Desset, F. Giorgi, A. R. Taddei, A. M. Fausto, M. Mazzini, B. Dastugue, and C. Vaury. 2000. Life cycle of an endogenous retrovirus, ZAM, in *Drosophila melanogaster. J. Virol.* 74:10658–10669.

Lecher, P., A. Bucheton, and A. Pelisson. 1997. Expression of the *Drosophila* retrovirus *Gypsy* as ultrastructurally detectable particles in the ovaries of flies carrying a permissive *flamenco* allele. *J. Genet. Virol.* 68:510–516.

Lee, S. H., J. B. Clark, and M. G. Kidwell. 1999. A *P* element-homologous sequence in the house fly, *Musca domestica. Insect Mol. Biol.* 8:491–500.

Lehane, M. J., S. Aksoy, W. Gibson, A. Kerhornou, M. Berriman, J. Hamilton, M. B. Soares, M. F. Bonaldo, S. Lehane, and N. Hall. 2003. Adult midgut expressed sequence tags from the tsetse fly *Glossina morsitans morsitans* and expression analysis of putative immune response genes. *Genome Biol.* 4:R63.

Lessinger, A. C., A. C. Martins Junqueira, T. A. Lemos, E. L. Kemper, F. R. Da Silva, A. L. Vettor, P. Arruda, and A. M. L. Azeredo-Espin. 2000. The mitochondrial genome of the primary screwworm fly *Cochliomyia hominivorax* (Diptera: Calliphoridae). *Insect Mol. Biol.* 9:521–529.

Levis, R. W., R. Ganesan, K. Houtchens, L. A. Tolar, and F. M. Sheen. 1993. Transposons in place of telomeric repeats at a *Drosophila* telomere. *Cell* 17:1083–1093.

Lewis, D. L., C. L. Farr, and L. S. Kaguni. 1995. *Drosophila melanogaster* mitochondrial DNA: Completion of the nucleotide sequence and evolutionary comparisons. *Insect. Mol. Biol.* 4:263–278.

Lewis, S. E., M. Ashburner, and M. G. Reese. 2000. Annotating eukaryote genomes. *Curr. Opin. Struct. Biol.* 10:349–354.

Lewis, S. E., S. M. Searle, N. Harris, M. Gibson, V. Lyer, J. Richter, C. Wiel, L. Bayraktaroglir, E. Birney, M. A. Crosby, J. S. Kaminker, B. B. Matthews, S. E. Prochnik, C. D. Smithy, J. L. Tupy, G. M. Rubin, S. Misra, C. J. Mungall, and M. E. Clamp. 2002. Apollo: A sequence annotation editor. *Genome Biol.* 3: RESEARCH 0082-2.

Liao, C., C. Rovira, H. He, and J.-E. Edström. 1998. Site-specific insertion of a SINE-like element, *Cp1,* into centromeric tandem repeats from *Chironomus pallidivittatus. J. Mol. Biol.* 280:811–820.

Liao, G.-C., E. J. Rehm, and G. M. Rubin. 2000. Insertion site preferences of the *P* transposable element in *Drosophila melanogaster. Proc. Natl. Acad. Sci. USA* 97:3347–3351.

Lifton, R. P., M. L. Goldberg, R. M. Karp, and D. S. Hogness. 1978. The organization of the histone genes in *Drosophila melanogaster:* Functional and evolutionary implications. *Cold Spring Harbor Symp. Quant. Biol.* 42:1047–1051.

Livak, K. J. 1984. Organization and mapping of a sequence on the *Drosophila melanogaster* X and Y chromosomes that is transcribed during spermatogenesis. *Genetics* 107:611–634.

Livak, K. J. 1990. Detailed structure of the *Drosophila melanogaster Stellate* genes and their transcripts. *Genetics* 124:303–316.

Llopart, A., J. M. Comeron, F. G. Brunet, D. Lachaise, and M. Long. 2002. Intron presence-absence polymorphism in *Drosophila* driven by positive Darwinian selection. *Proc. Natl. Acad. Sci. USA* 99:8121–8126.

Lloyd, V. K. 2000. Parental imprinting in *Drosophila. Genetica* 109:35–44.

Lloyd, V. K., D. A. Sinclair, and T. A. Grigliatti. 1999. Genomic imprinting and position-effect variegation in *Drosophila melanogaster. Genetics* 151:1501–1516.

Lobo, N., X. Li, and M. J. Fraser. 1999. Transposition of the *piggyBac* element in embryos of *Drosophila melanogaster, Aedes aegypti* and *Trichoplusia ni. Mol. Gen. Genet.* 261:803–810.

Lobo, N., F. A. Hua-Van, X. Li, B. Nolen, and M. J. Fraser. 2002. Germ line transformation of the yellow fever mosquito, *Aedes aegypti,* mediated by transpositional insertion of a *piggyBac* vector. *Insect Mol. Biol.* 11:133–139.

Locke, J., L. T. Howard, N. Aippersbach, L. Podemski, and R. B. Hodgetts. 1999. The characterization of *DINE-1,* a short, interspersed repetitive element present on chromosome *4* and in the centric heterochromatin of *Drosophila melanogaster. Chromosoma* 108:356–366.

Lohe, A. R., and P. A. Roberts. 1988. Evolution of satellite DNA sequences in *Drosophila.* In R. Verma, ed., *Heterochromatin: Molecular and Structural Aspects,* pp. 148–186. New York: Cambridge University Press.

Lohe, A. R., A. J. Hilliker, and P. A. Roberts. 1993. Mapping simple repeated DNA sequences in heterochromatin of *Drosophila melanogaster. Genetics* 134:1149–1174.

Long, M., W. Wang, and J. Zhang. 1999. Origin of new genes and source of *N*-terminal domain of the chimerical gene, *jingwei,* in *Drosophila. Gene* 238:135–141.

López, C. C., L. Nielsen, and J.-E. Edström. 1996. Terminal long tandem repeats in chromosomes from *Chironomus pallidivitatus. Mol. Cell. Biol.* 16:3285–3290.

López, C. C., E. Rodriguez, J. Díez, J.-E. Edström, and G. Morcillo. 1999. Histochemical localization of reverse transcriptase in polytene chromosomes of chironomids. *Chromosoma* 108:302–307.

Lovering, R., N. Harden, and M. Ashburner. 1991. The molecular structure of *TE146* and its derivatives in *Drosophila melanogaster. Genetics* 128:357–372.

Lozovskaya, E. R., D. A. Petrov, and D. L. Hartl. 1993. A combined molecular and cytogenetic approach to genome evolution in *Drosophila* using large-fragment DNA cloning. *Chromosoma* 102:253–266.

Luque, T., G. Marfany, and R. Gonzalez-Duarte. 1997. Characterization and molecular analysis of *Adh* retrosequences in species of the *Drosophila obscura* group. *Mol. Biol. Evol.* 14:1316–1325.

Lyko, F., B. H. Ramsahoye, H. Kashevsky, M. Tudor, M. A. Mastrangelo, T. L. Orr-Weaver, and R. Jaenisch. 1999. Mammalian cytosine-5 methyltransferases cause genomic DNA methylation and lethality in *Drosophila. Nat. Genet.* 23:363–366.

Lyko, F., B. H. Ramsahoye, and R. Jaenisch. 2000a. DNA methylation in *Drosophila melanogaster. Nature* 408:538–540.

Lyko, F., A. J. Whittaker, T. Orr-Weaver, and R. Jaenisch. 2000b. The putative *Drosophila* methyltransferase gene *dDnmts2* is contained in a transposon-like element and is expressed specifically in ovaries. *Mech. Dev.* 95:215–217.

Lynch, M., and J. S. Conery. 2000. The evolutionary fate and consequences of duplicate genes. *Science* 290:1151–1155.

Lyttle, T. W., and D. S. Haymer. 1992. The role of the transposable element *hobo* in the origin of endemic inversions in wild populations of *Drosophila melanogaster. Genetica* 86:113–126.

Lyubomirskaya, N. V., J. M. Smirnova, O. V. Razorenova, N. N. Karpova, S. A. Surkov, S. N. Avedisov, A. I. Kim, and Y. Y. Ilyin. 2001. Two variants of the *Drosophila melanogaster* retrotransposon *gypsy mdg4:* Structural and functional differences, and distribution in fly stocks. *Mol. Genet. Genomics* 265:367–374.

Ma, R. Z., W. C. Black, and J. C. Reese. 1992. Genome size and organization in an aphid *Schizaphis graminum. J. Insect Physiol.* 38:161–165.

Madueño, E., G. Papagiannakis, G. A. Rimmington, R. D. C. Saunders, C. Savakis, I. Sidén-Kiamos, G. Skavdis, L. Spanos, J. Trenear, P. Adam, M. Ashburner, P. Benos, V. N. Bolshakov, D. Coulson, D. M. Glover, S. Herrmann, F. C. Kafatos, C. Louis, T. Majerus, and J. Modolell. 1995. A physical map of the X chromosome of *Drosophila melanogaster:* Cosmid contigs and sequence tagged sites. *Genetics* 139:1631–1647.

Maggert, K. A., and K. G. Golic. 2002. The Y chromosome of *Drosophila melanogaster* exhibits chromosome-wide imprinting. *Genetics* 162:1245–1258.

Mainx, F. 1964. The genetics of *Megaselia scalaris* Loew (Phoridae): A new type of sex determination in Diptera. *Am. Nat.* 98:415–430.

Mainx, F. 1966. Die Geschlechtsbestimmung bei *Megaselia scalaris* Loew (Phoridae). *Z. Vererbungsl.* 98:49–60.

Malacrida, A., G. Gasperi, G. F. Biscaldi, R. Milani, and M. G. Franco. 1982. Identificazione di un determinante di femminilata *F* sul IV gruppo di concatenazione di *Musca domestica* L. *Atti. Assoc. Genet. Ital.* 28:243–244.

Malacrida, A., G. Gasperi, G. F. Biscaldi, and R. Milani. 1985. Significato funzionale di blocchi di geni concatenati in *Musca domestica* e altri ditteri. *Atti del secondo Congresso nazionale della Societa italiana di Ecologia,* pp. 307–310. Parma: Edizione Zara.

Malacrida, A., G. Gasperi, G. F. Biscaldi, and R. Milani. 1986. Persistence of linkage relationships among enzyme loci in some dipteran species. *Atti. Assoc. Genet. Ital.* 32:121–122.

Malik, H. S., and S. Henikoff. 2001. Adaptive evolution of Cid, a centromere-specific histone in *Drosophila. Genetics* 157:1293–1298.

Malik, H. S., S. Henikoff, and T. H. Eickbush. 2000. Poised for contagion: Evolutionary origins of the infectious abilities of invertebrate retroviruses. *Genome Res.* 10:1307–1318.

Malik, H. S., D. Vermaak, and S. Henikoff. 2002. Recurrent evolution of DNA-binding motifs in the *Drosophila* centromeric histone. *Proc. Natl. Acad. Sci. USA* 99:1449–1454.

Malinsky, S., A. Bucheton, and I. Busseau. 2000. New insights on homology-dependent silencing of *I* factor activity by transgenes containing ORF1 in *Drosophila melanogaster. Genetics* 156:1147–1155.

Marhold, J., M. Zbylut, D.-H. Lankenau, M. Li, D. Gerlich, E. Ballestar, B. M. Mechler, and F. Lyko. 2002. Stage-specific chromosomal association of *Drosophila* dMBD2/3 during gene activation. *Chromosoma* 111:13–21.

Marin, I., and A. Fontdevila. 1995. Characterization of Gandalf, a new inverted-repeat transposable element of *Drosophila koepferae. Mol. Gen. Genet.* 248:423–433.

Maroni, G., J. Wise, J. E. Young, and E. Otto 1987. Metallothionein gene duplications and metal tolerance in natural populations of *Drosophila melanogaster. Genetics* 117:739–744.

Martin, J. 1966. Female heterogamety in *Polypedilum nubifer* (Diptera, Nematocera). *Am. Nat.* 100:157–159.

Martinez, J.-L., J.-E. Edström, G. Morcillo, and J. L. Díez. 2001. Telomeres in *Chironomus thummi* are characterized by different subfamilies of complex DNA repeats. *Chromosoma* 110:221–227.

Martinho, R. G., P. S. Kunwar, J. Casanova, and R. Lehmann. 2004. A noncoding RNA is required for the repression of RNAPolII-dependent transcription in primordial germ cells. *Curr. Biol.* 14:159–165.

Martin-Romero, F. J., G. V. Kryukov, A. V. Lobanov, B. A. Carlson, B. J. Lee, V. N. Gladyshev, and D. L. Hatfield. 2001. Selenium metabolism in *Drosophila:* selenoproteins, selenoprotein mRNA expression, fertility, and mortality. *J. Biol. Chem.* 276:29798–29804.

Matthews, T. C., and L. E. Munstermann. 1994. Chromosomal repatterning and linkage group conservation in mosquito karyotypic evolution. *Evolution* 48:146–154.

Matuszewski, B. 1982. *Diptera 1. Cecidomyiidae. Animal Cytogenetics. Vol. 3. Insecta 3.* Berlin: Gebruder Borntraeger.

McAllister, B. F., and B. Charlesworth. 1999. Reduced sequence variability on the *neo*-Y chromosome of *Drosophila americana americana. Genetics* 153:221–233.

McDonald, I. C., D. E. Overland, R. A. Leopold, M. E. Degrugillier, P. B. Morgan, and H. C Hofmann. 1975. Genetics of house flies. Variability studies with North Dakota, Texas, and Florida populations. *J. Hered.* 66:137–140.

McGraw, E. A., and S. L. O'Neill. 2004. *Wolbachia pipientis:* Intracellular infection and pathogenesis in *Drosophila. Curr. Opin. Microbiol.* 7:67–70.

McKee, B. D. 1996. The license to pair: Identification of meiotic pairing sites. *Chromosoma* 105:135–141.

McKee, B. D., and G. H. Karpen. 1990. *Drosophila* ribosomal RNA genes function as a X-Y pairing site during meiosis. *Cell* 61:61–72.

McNamara-Schroeder, K. J., R. F. Hennessey, G. A. Harding, R. C. Jensen, and W. E. Stumph. 2001. The *Drosophila* U1 and U6 gene proximal sequence elements act as important determinants of the RNA polymerase specificity of small nuclear RNA gene promoters in vitro and in vivo. *J. Biol. Chem.* 276: 31786–31792.

Meise, M., D. Hilfiker-Kleiner, A. Dübendorfer, C. Bruner, R. Nöthiger, and D. Bopp. 1998. *Sex-lethal,* the master sex-determining gene in *Drosophila,* is not sex-specifically regulated in *Musca domestica. Development* 125:1487–1494.

Meller, V. H., and M. I. Kuroda. 2002. Sex and the single chromosome. *Adv. Genet.* 46:1–24.

Meller, V. H., G. Roman, M. I. Kuroda, and R. L. Davis. 1997. *roX1* paints the X chromosome of male *Drosophila* and is regulated by the dosage compensation system. *Cell* 88:445–457.

Melnikova, L., and P. Georgiev. 2002. *Enhancer of terminal gene conversion,* a new mutation in *Drosophila melanogaster* that induces telomere elongation by gene conversion. *Genetics* 162:1301–1312.

Merriam, J. R., M. Ashburner, D. L. Hartl, and F. C. Kafatos. 1991. Toward cloning and mapping the genome of *Drosophila. Science* 254:221–225.

Milani, R. 1967. The genetics of *Musca domestica* and of other muscoid flies. In J. W. Wright and R. Pal, eds., *Genetics of Insect Vectors of Disease,* pp. 315–369. Amsterdam: Elsevier.

Milani, R., and M. G. Franco. 1959. Comportamento ereditario della resistenza al DDT in incroci tra il ceppo Orlando-R e ceppi *kdr* e kdr^+ di *Musca domestica* L. *Symp. Genet. Biol. Ital.* 6:249–268.

Milani, R., P. G. Rubini, and M. G. Franco. 1967. Sex-determination in the housefly. *Genet. Agraria* 21: 385–411.

Miller, W. J., A. Nagel, J. Bachmann, and L. Bachmann. 2000. Evolutionary dynamics of the *SGM* transposon family in the *Drosophila obscura* species group. *Mol. Biol. Evol.* 17:1597–1609.

Misra, S., M. A. Crosby, C. J. Mungall, B. B. Matthews, K. S. Campbell, P. Hradecky, Y. Huang, J. S. Kaminker, G. H. Millburn, S. E. Prochnik, C. D. Smith, J. L. Tupy, E. J. Whitfied, L. Bayraktaroglu, B. P. Berman,

B. R. Bettencourt, S. E. Celniker, A. D. de Grey, R. A. Drysdale, N. L. Harris, J. Richter, S. Russo, A. J. Schroeder, S. Q. Shu, M. Stapleton, C. Yamada, M. Ashburner, W. M. Gelbart, G. M. Rubin, and S. E. Lewis. 2002. Annotation of the *Drosophila melanogaster* euchromatic genome: A systematic review. *Genome Biol.* 3:RESEARCH 0083-3.

Mongelard, F., M. Labrador, E. M. Baxter, T. I. Gerasimova, and V. G. Corces. 2002. *Trans*-splicing as a novel mechanism to explain interallelic complementation in *Drosophila. Genetics* 160:1481–1487.

Morcillo, G., J. L. Diez, M. E. Carbajal, and R. M. Tanguay. 1993. Hsp90 associates with specific heat shock puffs Hsro in polytene chromosomes of *Drosophila* and *Chironomus. Chromosoma* 102:648–659.

Moses, K., U. Heberlein, and M. Ashburner. 1990. The *Adh* gene promoters of *Drosophilamelanogaster* and *Drosophila orena* are functionally conserved and share features of sequence structure and nuclease-protected sites. *Mol. Cell. Biol.* 10:539–548.

Mouchès, C., N. Pasteur, J. B. Berge, O. Hyrien, M. Raymond, B. R. de Saint Vincent, M. de Silvestris, and G. P. Georghiou. 1986. Amplification of an esterase gene is responsible for insecticide resistance in a California *Culex* mosquito. *Science* 233:778–780.

Mount, S. M., C. Burks, G. Hertz, G. D. Stormo, O. White, and C. Fields. 1992. Splicing signals in *Drosophila:* intron size, information content, and consensus sequences. *Nucleic Acids Res.* 20:4255–4262.

Mourier, T., and D. C. Jeffares. 2003. Eukaryotic intron loss. *Science* 300:1393.

Muller, H. J. 1918. Genetic variability, twin hybrids, and constant hybrids in a case of balanced lethal factors. *Genetics* 3:422–499.

Muller, H. J. 1940. Bearings of the "*Drosophila*" work on systematics. In J. Huxley, ed., *The New Systematics*, pp. 185–268. Oxford: Clarendon Press.

Müller-Holtkamp, F. 1995. The *Sex-lethal* gene homologue in *Chrysomya rufifacies* is highly conserved in sequence and exon-intron organization. *J. Mol. Evol.* 41:467–477.

Mungall, C. J., S. Misra, B. P. Berman, J. Carlson, E. Frise, N. Harris, B. Marshall, S. Shu, J. S. Kaminker, S. E. Prochnik, C. D. Smith, E. Smith, J. L. Tupy, C. Wiel, G. M. Rubin, and S. E. Lewis. 2002. An integrated computational pipeline and database to support whole-genome sequence annotation. *Genome Biol.* 3:RESEARCH 0081-1.

Murphy, T. D., and G. H. Karpen. 1995. Localization of centromere function in a *Drosophilamelanogaster* minichromosome. *Cell* 82:599–609.

Myers, E. W., G. G. Sutton, A. L. Delcher, I. M. Dew, D. P. Fasulo, M. J. Flanigan, S. A. Kravitz, C. M. Mobarry, K. H. Reinert, K. A. Remington, E. L. Anson, R. A. Bolanos, H. H. Chou, C. M. Jordan, A. L. Halpern, S. Lonardi, E. M. Beasley, R. C. Brandon, L. Chen, P. J. Dunn, Z. Lai, Y. Liang, D. R. Nusskern, M. Zhan, Q. Zhang, X. Zheng, G. M. Rubin, M. D. Adams, and J. C. Venter. 2000. A whole-genome assembly of *Drosophila. Science* 287:2196–2204.

Nakamura, A., R. Amikura, M. Mukai, S. Kobayashi, and P. F. Lasko. 1996. Requirement for a noncoding RNA in *Drosophila* polar granules for germ cell establishment. *Science* 274:2075–2079.

Nardi, F., A. Carapelli, R. Dallai, and F. Frati. 2003. The mitochondrial genome of the olive fly *Bactrocera oleae:* Two haplotypes from distant geographic locations. *Insect Mol. Biol.* 12:605–611.

Nei, M., and A. L. Hughes. 1992. Balanced polymorphism and evolution by the birth-and-death process in the MHC loci. In K. Tsuji, M. Aizawa, and T. Sasuki, eds., *11th Histocompatibility Workshop and Conference*, pp. 27–38. Oxford: Oxford University Press.

Nei, M., I. B. Rogozin, and H. Piontkivska. 2000. Purifying selection and birth-and-death evolution in the ubiquitin gene family. *Proc. Natl. Acad. Sci. USA* 97:10866–10871.

Neufeld, T. P., R. W. Carthew, and G. M. Rubin. 1991. Evolution of gene position: Chromosomal arrangement and sequence comparison of the *Drosophila melanogaster* and *Drosophila virilis sina* and *Rh4* genes. *Proc. Natl. Acad. Sci. USA* 88:10203–10207.

Neves, G., J. Zucker, M. Daly, and A. Chess. 2004. Stochastic yet biased expression of multiple *Dscam* splice variants by individual cells. *Nat. Genet.* 36:240–246.

Nogge, G. 1981. Significance of symbionts for the maintenance of an optimal nutritional state for successful reproduction in haematophagous arthropods. *Parasitology* 82:101–104.

Noor, M. A. F., P. Michalak, and D. Donze. 2003. Characterization of a male-predominant antisense transcript underexpressed in hybrids of *Drosophila pseudoobscura* and *D. persimilis. Genetics* 165:1823–1830.

Nöthiger, R., and M. Steinmann-Zwicky. 1985. A single principle for sex determination in insects. *Cold Spring Harbor Symp. Quant. Biol.* 50:615–621.

Nurminsky, D. I., M. V. Nurminskaya, D. de Aguiar, and D. L. Hartl. 1998. Selective sweep of a newly evolved sperm-specific gene in *Drosophila. Nature* 396:573–575.

O'Brochta, D. A., W. D. Warren, K. J. Saville, and P. W. Atkinson. 1996. Hermes, a functional non-Drosophilid insect gene vector from *Musca domestica. Genetics* 142:907–914.

O'Grady, P. M., R. H. Baker, C. M. Durando, W. T. Etges, and R. DeSalle. 2001. Polytene chromosomes as indicators of phylogeny in several species groups of *Drosphila. BMC Evol. Biol.* 1:6.

Oh, S. K., M. P. Scott, and P. Sarnow. 1992. Homeotic gene *Antennapedia* mRNA contains 5′-noncoding sequences that confer translational initiation by internal ribosome binding. *Genes Dev.* 6:1643–1653.

Ohler, U., G. C. Liao, H. Niemann, and G. M. Rubin. 2002. Computational analysis of core promoters in the *Drosophila* genome. *Genome Biol.* 3:RESEARCH 0087.

Okazaki, S., T. Tsuchida, H. Maekawa, H. Ishikawa, and H. Fuyiwara. 1993. Identification of a pentanucleotide telomeric sequence $TTAGG_n$ in the silkworm *Bombyx mori* and in other insects. *Mol. Cell. Biol.* 13:1424–1432.

O'Neil, M. T., and J. M. Belote. 1992. Interspecific comparison of the *transformer* gene of *Drosophila* reveals an unusually high degree of evolutionary divergence. *Genetics* 131:113–128.

O'Neill, S. L., A. A. Hoffman, and J. H. Werren, eds. 1997. *Influential Passengers. Inherited Microorganisms and Arthropod Reproduction.* Oxford: Oxford University Press.

Orr, H. A., and Y. Kim. 1998. An adaptive hypothesis for the evolution of the Y chromosome. *Genetics* 150:1693–1698.

Osheim, Y. N., O. L. Miller, Jr., and A. L. Beyer. 1988. Visualization of *Drosophila melanogaster* chorion genes undergoing amplification. *Mol. Cell. Biol.* 8:2811–2821.

Painter, T. S. 1933. A new method for the study of chromosome aberrations and the plotting of chromosome maps. *Science* 78:585–586.

Palladino, M. J., L. P. Keegan, M. A. O'Connell, and R. A. Reenan. 2000a. dADAR, a *Drosophila* double-stranded RNA-specific adenosine deaminase is highly developmentally regulated and is itself a target for RNA editing. *RNA N.Y.* 6:1004–1018.

Palladino, M. J., L P. Keegan, M. A. O'Connell, and R. A. Reenan. 2000b. A-to-I pre-mRNA editing in *Drosophila* is primarily involved in adult nervous system function and integrity. *Cell* 102:437–449.

Palumbo, G., S. Bonaccorsi, L. G. Robbins, and S. Pimpinelli. 1994. Genetic analysis of *Stellate* elements of *Drosophila melanogaster. Genetics* 138:1181–1197.

Pane, A., M. Salvemini, P. Delli Bovi, C. Polito, and G. Saccone. 2002. The *transformer* gene in *Ceratitis capitata* provides a gentic basis for selecting and remembering the sexual fate. *Development* 129:3715–3725.

Pantazidis, A., M. Labrador, and A. Fontdevila. 1999. The retrotransposon *Osvaldo* from *Drosophila buzzatii* displays all structural features of a functional retrovirus. *Mol. Biol. Evol.* 16:909–921.

Papusheva, E., M. C. Gruhl, E. Berezikov, T. Groudieva, S. V. Scherbik, J. Martin, A. Blinov, and G. Bergtrom. 2004. The evolution of SINEs and LINEs in the genus *Chironomus* Diptera. *J. Mol. Evol.* 58:269–279.

Pardue, M. L., and P. G. DeBaryshe. 2003. Retrotransposons provide an evolutionarily robust non-telomerase mechanism to maintain telomeres. *Annu. Rev. Genet.* 37:485–511.

Pasquinelli, A. E. 2002. MicroRNAs: Deviants no longer. *Trends Genet.* 18:171–173.

Pavan, C., and A. B. da Cunha. 1969. Gene amplification in ontogeny and phylogeny of animals. *Genetics* 61(Suppl. 1):289–304.

Pelisson, A., S. U. Song, N. Prud'homme, P. A. Smith, A. Bucheton, and V. G. Corces. 1994. *Gypsy* transposition correlates with the production of a retroviral envelope-like protein under tissue-specific control of the *Drosophila flamenco* gene. *EMBO J.* 13:4401–4411.

Pelisson, A., L. Mejlumian, V. Robert, C. Terzian, and A. Bucheton. 2002. *Drosophila* germline invasion by the endogenous retrovirus *gypsy:* Involvement of the viral *env* gene. *Insect Biochem. Mol. Biol.* 32: 1249–1256.

Perez, D. E., and C.-I. Wu. 1995. Further characterization of the *Odysseus* locus of hybrid sterility in *Drosophila:* One gene is not enough. *Genetics* 140:201–206.

Perje, A. M. 1948. Studies on spermatogenesis in *Musca domestica. Hereditas* 34:209–232.

Perkins, H. D., and A. J. Howells. 1992. Genomic sequences with homology to the *P* element of *Drosophila melanogaster* occur in the blowfly *Lucilia cuprina. Proc. Natl. Acad. Sci. USA* 89:10753–10757.

Peter A., P. Schöttler, M. Werner, N. Beinert, G. Dowe, P. Burkert, F. Mourkioti, L. Dentzer, Y. He, P. Deak, P. V. Benos, M. K. Gatt, L. Murphy, D. Harris, B. Barrell, C. Ferraz, S. Vidal, C. Brun, J. Demaille,

E. Cadieu, S. Dreano, S. Gloux, V. Lelaure, S. Mottier, F. Galibert, D. Borkova, B. Miñana, F. C. Kafatos, S. Bolshakov, I. Sidén-Kiamos, G. Papagiannakis, L. Spanos, C. Louis, E. Madueño, B. de Pablos, J. Modolell, A. Bucheton, D. Callister, L. Campbell, N. S. Henderson, P. J. McMillan, C. Salles, E. Tait, P. Valenti, R. D. Saunders, A. Billaud, L. Pachter, R. Klapper, W. Janning, D. M. Glover, M. Ashburner, H. J. Bellen, H. Jäckle, and U. Schäfer. 2002. Mapping and identification of essential gene functions on the X chromosome of *Drosophila. EMBO Rpts.* 3:34–38.

Peters, N. T., J. A. Rohrbach, B. A. Zalewski, C. M. Byrkett, and J. C. Vaughn. 2003. RNA editing and regulation of *Drosophila 4f-rnp* expression by *sas-10* antisense readthrough mRNA transcripts. *RNA* 9: 698–710.

Petrov, D. A. 2002. DNA loss and evolution of genome size in *Drosophila. Genetica* 115:81–91.

Petrov, D. A., and D. L. Hartl. 1997. Trash DNA is what gets thrown away: High rate of DNA loss in *Drosophila. Gene* 205:279–289.

Petschek, J. P., M. J. Mermer, M. R. Scheckelhoff, A. A. Simone, and J. C. Vaughn. 1996. RNA editing in *Drosophila 4f-rnp* gene nuclear transcripts by multiple A-to-G conversions. *J. Mol. Biol.* 259:885–890.

Petschek, J. P., M. R. Scheckelhoff, M. J. Mermer, and J. C. Vaughn. 1997. RNA editing and alternative splicing generate mRNA transcript diversity from the *Drosophila 4f-rnp* locus. *Gene* 204:267–276.

Pinsker, W., E. Haring, S. Hagemann, and W. J. Miller. 2001. The evolutionary life history of *P* transposons: From horizontal invaders to domesticated neogenes. *Chromosoma* 110:148–158.

Piontkivska, H., A. P. Rooney, and M. Nei. 2002. Purifying selection and birth-and-death evolution in the histone H4 gene family. *Mol. Biol. Evol.* 19:689–697.

Pirrotta, V. 2002. *Trans*-splicing in *Drosophila. BioEssays* 24:988–991.

Platero, J. S., A. K. Csink, A. Quintanilla, and S. Henikoff. 1998. Changes in chromosomal localization of heterochromatin-binding proteins during the cell cycle in *Drosophila. J. Cell Biol.* 140:1297–1306.

Potter, S., M. Truet, M. Phillips, and A. Maher. 1980. Eukaryotic transposable elements with inverted terminal repeats. *Cell* 20:639–647.

Potter, S., M. Truet, M. Phillips, and A. Maher. 1982. DNA sequence of a foldback transposable element in *Drosophila. Nature* 297:201–204.

Powell, J. R. 1997. *Progress and Prospects in Evolutionary Biology. The* Drosophila *Model.* New York and Oxford: Oxford University Press.

Prasanth, K. V., T. K. Rajendra, A. K. Lal, and S. C. Lakhotia. 2000. Omega speckles—A novel class of nuclear speckles containing hnRNPs associated with noncoding hsr-omega RNA in *Drosophila. J. Cell Sci.* 113: 3485–3497.

Pritchard, J. K., and S. W. Schaeffer. 1997. Polymorphism and divergence at a *Drosophila* pseudogene. *Genetics* 147:199–208.

Prud'homme, N., M. Gans, M. Masson, C. Terzian, and A. Bucheton. 1995. *Flamenco,* a gene controlling the *gypsy* retrovirus of *Drosophila melanogaster. Genetics* 139:697–711.

Quesneville, H., D. Nouaud, and D. Anxolabéhère. 2003. Detection of new transposable element families in *Drosophila melanogaster* and *Anopheles gambiae* genomes. *J. Mol. Evol.* 57:S50–S59.

Rai, K. S., and W. C. Black IV. 1999. Mosquito genomes: Structure, organization, and evolution. *Adv. Genet.* 41:1–33.

Rajewsky, N., and N. D. Socci. 2004. Computational identification of microRNA targets. *Devel. Biol.* 267: 529–535.

Ramos-Onsins, S., and M. Aguadé. 1998. Molecular evolution of the *Cecropin* multigene family in *Drosophila:* Functional genes vs. pseudogenes. *Genetics* 150:157–171.

Ranz, J. M., C. Segarro, and A. Ruiz. 1997. Chromosomal homology and molecular organization of Muller's elements *D* and *E* in the *Drosophila repleta* species group. *Genetics* 145:281–295.

Ranz, J. M., F. Casalls, and A. Ruiz. 2001. How malleable is the euchromatic genome? Extreme rate of chromosomal rearrangement in the genus *Drosophila. Genome Res.* 11:230–239.

Ranz, J. M., J. González, F. Casals, and A. Ruiz. 2003. Low occurrence of gene transposition events during the evolution of the genus *Drosophila. Evolution* 57:1325–1335.

Rao, P. N., and K. S. Rai. 1987. Inter- and intraspecific variation in nuclear DNA content in *Aedes* mosquitoes. *Heredity* 59:253–258.

Raymond, M., A. Callaghan, P. Fort, and N. Pasteur. 1991. Worldwide migration of amplified insecticide resistance genes in mosquitoes. *Nature* 350:151–153.

Raymond, M., C. Chevillon, T. Guillemaud, T. Lenormand, and N. Pasteur. 1998. An overview of the evolution of overproduced esterases in the mosquito *Culex pipiens. Phil. Trans. R. Soc. Lond. B* 353: 1707–1711.

Redfern, C. P. F. 1981. Homologous banding patterns in the polytene chromosomes from the larval salivary glands and ovarian nurse cells of *Anopheles stephensi* Liston (Diptera:Culicidae). *Chromosoma* 83: 221–240.

Reinhart, B. J., and D. P. Bartel. 2002. Small RNAs correspond to centromere heterochromatic repeats. *Science* 297:1831.

Ren, X., L. Eisenhour, C. Hong, Y. Lee, and B. D. Mckee. 1997. Roles of rDNA spacer and transcription unit-sequences in X-Y meiotic pairing in *Drosophila melanogaster* males. *Chromosoma* 106:29–36.

Reugels, A. M., R. Kurek, U. Lammermann, and H. Bünemann. 2000. Mega-introns in the dynein gene *Dh7(Y)* on the heterochromatic Y chromosome give rise to the *giant* threads loops in primary spermatocytes of *Drosophila hydei. Genetics* 154:759–769.

Ribbert, D. 1967. Die Polytanchromosomen bei Borstenbildungszellen von *Calliphora erythrocephala. Chromosoma* 21:296–344.

Ribbert, D. 1979. Chromosomes and puffing in experimentally induced polytene chromosomes of *Calliphora erythrocephala. Chromosoma* 74:269–298.

Rio, D. C. 2002. *P* transposable elements in *Drosophila melanogaster.* In N. L. Craig, R. Craigie, M. Gellert, and A. M. Lambowitz, eds., *Mobile DNA II,* pp. 484–518. Washington, D.C.: American Society of Microbiology.

Robertson, H. M. 2002. Evolution of DNA transposons in eukaryotes. In N. L. Craig, R. Craigie, M. Gellert, and A. M. Lambowitz, eds., *Mobile DNA II,* pp. 1093–1110. Washington, D.C.: American Society of Microbiology.

Robin, G. C. de, R. J. Russell, D. J. Cutler, and J. G. Oakshott. 2000. The evolution of an alpha-esterase pseudogene inactivated in the *Drosophila melanogaster* lineage. *Mol. Biol. Evol.* 17:563–575.

Roder, K., M. S. Hung, T. L. Lee, T. Y. Lin, H. Xiao, K. I. Isobe, J. L. Juang, and C. K. J. Shen. 2000. Transcriptional repression by *Drosophila* methyl-CpG-binding proteins. *Mol. Cell. Biol.* 20:7401–7409.

Roethele, J. B., J. Romero-Severson, and J. L. Feder. 2001. Evidence for broad-scale conservation of linkage map relationships between *Rhagoletis pomonella* Diptera: Tephritidae and *Drosophila melanogaster* Diptera: Drosophilidae. *Ann. Entomol. Soc. Am.* 94:936–947.

Rohr, C. J. B., H. Ranson, X. Wang, and N. J. Bezansky. 2002. Structure and evolution of *mtanga,* a retrotransposon actively expressed on the Y chromosome of the African malaria vector *Anopheles gambiae. Mol. Biol. Evol.* 19:149–162.

Rosen, L. 1980. Carbon dioxide sensitivity in mosquitoes infected with sigma, vesicular stomatitis and other rhabdoviruses. *Science* 207:989–991.

Rosén, M., I. Kamnert, and J.-E. Edström. 2002. Extrachromosomal RN-DNA complex containing long telomeric repeats in chironomids. *Insect Mol. Biol.* 11:167–174.

Ross, R., T. Hankeln, and E. R. Schmidt. 1997. Complex evolution of tandem-reptitive DNA in *Chironomus thummi* species group. *J. Mol. Evol.* 44:321–326.

Roth, C. W., F. Kobeski, M. F. Walter, and H. Biessmann. 1997. Chromosome end elongation by recombination in the mosquito *Anopheles gambiae. Mol. Cell. Biol.* 17:5176–5183.

Rothe, M., M. Pehl, H. Taubert, and H. Jäckle. 1992. Loss of gene function through rapid mitotic cycles in the *Drosophila* embryo. *Nature* 359:156–159.

Rothfels, K. H., and R. Nambiar. 1981. A cytological study of natural hybrids between *Prosimulium multidentatum* and *P. magnum* with notes on sex determination in the Simuliidae (Diptera). *Chromosoma* 82:673–691.

Rovira, C., W. Beermann, and J.-E. Edström. 1993. A repetitive DNA sequence associated with the centromeres of *Chironomus pallidivitatus. Nucleic Acids Res.* 21:1775–1781.

Roy, P. J., J. M. Stuart, J. Lund, and S. K. Kim. 2002. Chromosomal clustering of muscle-expressed genes in *Caenorhabditis elegans. Nature* 418:975–978.

Royzman, I., and T. L. Orr-Weaver. 1998. S phase and differential DNA replication during *Drosophila* oogenesis. *Genes Cells* 3:767–776.

Rubin, G. M., M. D. Yandell, J. R. Wortman, G. L. Gabor Miklos, C. R. Nelson, I. K. Hariharan, M. E. Fortini, P. W. Li, R. Apweiler, W. Fleischmann, J. M. Cherry, S. Henikoff, M. P Skupski, S. Misra, M. Ashburner,

E. Birney, M. S. Boguski, T. Brody, P. Brokstein, S. E. Celniker, S. A. Chervitz, D. Coates, A. Cravchik, A. Gabrielian, R. F. Galle, W. M. Gelbart, R. A. George, L. S. Goldstein, F. Gong, P. Guan, N. L. Harris, B. A. Hay, R. A. Hoskins, J. Li, Z. Li, R. O. Hynes, S. J. Jones, P. M. Kuehl, B. Lemaitre, J. T. Littleton, D. K. Morrison, C. Mungall, P. H. O'Farrell, O. K. Pickeral, C. Shue, L. B. Vosshall, J. Zhang, Q. Zhao, X. H. Zheng, F. Zhong, W. Zhong, R. Gibbs, J. C. Venter, M. D. Adams, and S. Lewis. 2000a. Comparative genomics of the eukaryotes. *Science* 287:2204–2215.

Rubin, G. M., L. Hong, P. Brokstein, M. Evans-Holm, E. Frise, M. Stapleton, and D. A. Harvey. 2000b. A *Drosophila* complementary DNA resource. *Science* 287:2222–2224.

Rubini, P. G., M. G. Franco, and S. Vanossi-Este. 1972. Polymorphisms for heterosomes and autosomal sex-determinants in *Musca domestica* L. *Atti. Cong. Naz. Ital. Entomol.* 9:341–352.

Rubini, P. G., C. van Heemert, and M. G. Franco. 1977. Rapporto sessi anomolo in una popolazione naturale di *Musca domestica* L. e sue basi genetiche. *Genet. Agraria* 31:21–38.

Ruiz, M. F., C. Goday, P. Gonzalez, and L. Sanchez. 2003. Molecular analysis and developmental expression of the *Sex-lethal* gene of *Sciara ocellaris* (Diptera Order, Nematocera Suborder). *Gene Expr. Patterns* 3:341–346.

Saccone, G., I. Peluso, D. Artiaca, E. Giordano, D. Bopp, and L. C. Polito. 1998. The *Ceratitiscapitata* homologue of the *Drosophila* sex-determining gene *Sex-lethal* is structurally conserved, but not sex-specifically regulated. *Development* 125:1495–1500.

Saigo, K., L. Milstein, and C. A. Thomas. 1981. The organization of the *Drosophila melanogaster* histone genes. *Cold Spring Harbor Symp. Quant. Biol.* 45:815–827.

Sarkar, A., C. Sim, Y. S. Hong, J. R. Hogan, M. J. Fraser, H. M. Robertson, and F. H. Collins. 2003a. Molecular evolutionary analysis of the widespread *piggyBac* transposon family and related "domesticated" sequences. *Mol. Genet. Genomics* 270:173–180.

Sarkar, A., R. Sengupta, J. Krzywinski, X. Wang, C. Roth, and F. C. Collins. 2003b. *P* elements are found in the genomes of the nematoceran insects of the genus *Anopheles. Insect Biochem. Mol. Biol.* 33:381–387.

Sato, M., H. Umeki, R. Saito, A. Kanai, and M. Tomita. 2003. Computational analysis of stop codon readthrough in *D. melanogaster. Bioinformatics* 19:1371–1380.

Scadden, A. D. J., and C. W. J. Smith. 2001. RNAi is antagonized by AAEI hyper-editing. *EMBO Rpts.* 2:1107–1111.

Scamborova, P., A. Wong, and J. A. Steitz. 2004. An intronic enhancer regulates splicing of the twintron of *Drosophila melanogaster prospero* pre-mRNA by different spliceosomes. *Mol. Cell. Biol.* 24:1855–1869.

Schmid, C. W., J. E. Manning, and N. Davidson. 1975. Inverted repeat sequences in the *Drosophila* genome. *Cell* 5:159–172.

Schmid, K., and T. Tautz. 1997. A screen for fast-evolving genes in *Drosophila. Proc. Natl. Acad. Sci. USA* 94:9746–9750.

Schmid, K., L. Nigro, C. F. Aquadro, and D. Tautz. 1999. Large number of replacement polymorphisms in rapidly evolving genes of *Drosophila:* Implications for genome-wide surveys of DNA polymorphism. *Genetics* 153:1717–1729.

Schmidt, A., G. Palumbo, M. P. Bozzetti, P. Tritto, S. Pimpinelli, and U. Schafer. 1999. Genetic and molecular characterization of *sting,* a gene involved in crystal formation and meiotic drive in the male germ line of *Drosophila melanogaster. Genetics* 151:749–760.

Schmidt, R., M. Hediger, R. Nöthiger, and A. Dübendorfer. 1997a. The mutation *masculinizer (man)* defines a sex-determining gene with maternal and zygotic functions in *Musca domestica* L. *Genetics* 145:173–183.

Schmidt, R., M. Hediger, S. Roth, R. Nöthiger, and A. Dübendorfer. 1997b. The Y-chromosomal and autosomal male-determining *M* factors of *Musca domestica* are equivalent. *Genetics* 147:271–280.

Schübeler, D., D. Scalzo, C. Kooperberg, B. van Steensel, J. Delrow, and M. Groudine. 2002. Genome-wide DNA replication profile for *Drosophila melanogaster:* A link between transcription and replication timing. *Nat. Genet.* 32:438–442.

Schütt, C., and R. Nöthiger. 2000. Structure, function and evolution of sex determining systems in dipteran insects. *Development* 127:667–677.

Segarra, C., and M. Aguadé. 1992. Molecular organization of the X chromosome in different species of the *obscura* group of *Drosophila. Genetics* 130:513–521.

Segarra, C., E. R. Lozovskaya, G. Ribó, M. Aguadé, and D. L. Hartl. 1995. P1 clones from *Drosophila melanogaster* as markers to study the chromosomal evolution of Muller's *A* element in the species of the *obscura* group of *Drosophila. Chromosoma* 104:129–136.

Segarra, C., G. Ribó, and M. Aguadé. 1996. Differentiation of Muller's chromosomal elements *D* and *E* in the *obscura* group of *Drosophila. Genetics* 144:139–146.

Seoighe, C., and K. H. Wolfe. 1999. Updated map of duplicated regions in the yeast genome. *Gene* 238:253–261.

Serna, E., E. Gorab, M. Fernanda Ruiz, C. Goday, J. M. Eirín-López, and L. Sánchez. 2004. The gene *Sex-lethal* of the Sciaridae family (order Diptera, suborder Nematocera) and its phylogeny in dipteran insects. *Genetics* 168:907–921.

Severson, D. W., S. E. Brown, and D. L. Knudson. 2001. Genetic and physical mapping in mosquitoes: Molecular approaches. *Annu. Rev. Entomol.* 46:183–219.

Severson, D. W., D. L. Knudson, M. B. Soares, and N. J. Loftus. 2004. *Aedes aegypti* genomics. *Insect Biochem. Mol. Biol.* 34:715–721.

Sharp, S., D. de Franco, M. Silberklang, H. A. Hosbach, T. Schmidt, E. Kubli, J. P. Gergen, P. C. Wensink, and D. Söll. 1981. The initiator tRNA genes of *Drosophila melanogaster:* Evidence for a tRNA pseudogene. *Nucleic Acids Res.* 9:5867–5882.

Shearman, D. C., and M. Frommer. 1998. The *Batroceras tryoni* homologue of the *Drosophilamelanogaster* sex-determining gene *doublesex. Insect Mol. Biol. Biochem.* 7:355–366.

Sheen, F.-M., and R. W. Levis. 1994. Transposition of the LINE-like retrotransposon *TART* to *Drosophila* chromosome termini. *Proc. Natl. Acad. Sci. USA* 91:12510–12514.

Sievert, V., S. Kuhn, and W. Traut. 1997. Expression of the sex determining cascade genes *Sex-lethal* and *doublesex* in the phorid fly *Megaselia scalaris. Genome* 40:211–214.

Sievert, V., S. Kuhn, A. Paululat, and W. Traut. 2000. Sequence conservation and expression of the *Sex-lethal* homologue in the fly *Megaselia scalaris. Genome* 43:382–390.

Silber, J., C. Bazin, F. Lemeunier, S. Aulard, and M. Volovitch. 1989. Distribution and conservation of the *foldback* transposable element in *Drosophila. J. Mol. Evol.* 28:220–224.

Silva, J. C., and M. G. Kidwell. 2000. Horizontal transfer and selection in the evolution of *P* elements. *Mol. Biol. Evol.* 17:1542–1557.

Siriaco, G. M., G. Cenci, A. Haoudi, L. E. Champion, C. Zhou, M. Gatti, and J. M. Mason. 2002. *Telomere elongation (Tel),* a new mutation in *Drosophila melanogaster* that produces long telomeres. *Genetics* 160:235–245.

Smith, G. P. 1974. Unequal crossingover and the evolution of multigene families. *Cold Spring Harbor Symp. Quant. Biol.* 38:507–513.

Smoller, D. A., D. Petrov, and D. L. Hartl. 1991. Characterization of bacteriophage P1 library containing inserts of *Drosophila* DNA of 75–100 kilobase range. *Chromosoma* 100:487–494.

Solignac, M., M. Monnerot, and J.-C. Mounolou. 1986. Mitochondrial DNA evolution in the *melanogaster* species subgroup of *Drosophila. J. Mol. Evol.* 23:31–40.

Spanos, L., G. Koutroumbas, M. Kotsyfakis, and C. Louis. 2000. The mitochondrial genome of the mediterranean fruit fly, *Ceratitis capitata. Insect Mol. Biol.* 9:139–144.

Spellman, P. T., and G. M. Rubin. 2002. Evidence for large domains of similarly expressed genes in the *Drosophila* genome. *J. Biol.* 1:5–8.

Spierer, A., and P. Spierer. 1984. Similar level of polyteny in bands and interbands of *Drosophila* giant chromosomes. *Nature* 307:176–178.

Spofford, J. B. 1976. Position-effect variegation in *Drosophila.* In M. Ashburner and E. Novitski, eds., *The Genetics and Biology of Drosophila,* vol. 1c, pp. 955–1018. London: Academic Press.

Spradling, A. C., and A. P. Mahowald. 1980. Amplification of genes for chorion proteins during oogenesis in *Drosophila melanogaster. Proc. Natl. Acad. Sci. USA* 77:1096–1100.

Spradling, A. C., M. E. Digan, A. P. Mahowald, M. Scott, and E. A. Craig. 1980. Two clusters of genes for major chorion proteins of *Drosophila melanogaster. Cell* 19:905–914.

Spradling, A. C., D. Stern, I. Kiss, J. Roote, T. Laverty, and G. M. Rubin. 1995. Gene disruptions using *P* transposable elements. *Proc. Natl. Acad. Sci. USA* 92:10824–10830.

Springer, R. 1967. Konkurrenzversuche mit den drei Geschlechtsbestimmungstypen von *Megaselia scalaris* Loew (Phoridae). *Mol. Gen. Genet.* 99:125–132.

Staiber, W. 1991. Structural homologies between germ line–limited and soma chromosomes in *Acricotopus lucidus* (Diptera, Chironomidae). *J. Hered.* 82:247–249.

Staiber, W. 2002. Isolation of a new germ line–specific repetitive DNA family in *Acricotopus* by microdissection of polytenized germ line–limited chromosome sections from a permanent larval salivary gland preparation. *Cytogen. Genome Res.* 98:210–215.

Staiber, W., and C. Schiffkowki. 2000. Structural evolution of the germ-line–limited chromosomes in *Acricotopus. Chromosoma* 109:343–349.

Staiber, W., I. Wech, and A. Preiss. 1997. Isolation and chromosomal localization of a germ line–specific highly repetitive DNA family in *Acricotopus lucidus* (Diptera, Chironomidae). *Chromosoma* 106:267–275.

Stapleton, M., G. Liao, P. Brokstein, L. Hong, P. Carninci, T. Shiraki, Y. Hayashizaki, M. Champe, J. Pacleb, K. Wan, C. Yu, J. Carlson, R. George, S. E. Celniker, and G. M. Rubin. 2002a. The *Drosophila* gene collection: Identification of putative full-length cDNAs for 70% of *D. melanogaster* genes. *Genome Res.* 12:1294–1300.

Stapleton, M., J. Carlson, P. Brokstein, C. Yu, M. Champe, R. George, H. Guarin, B. Kronmiller, J. Pacleb, S. Park, K. Wan, G. M. Rubin, and S. E. Celniker. 2002b. A *Drosophila* full-length cDNA resource. *Genome Biol.* 3:RESEARCH 0080-0.

Stapleton, W., S. Das, and B. D. McKee. 2001. A role of the *Drosophila homeless* gene in repression of *Stellate* in male meiosis. *Chromosoma* 110:228–240.

Stark, A., J. Brennecke, R. B. Russell, and S. M. Cohen. 2003. Identification of *Drosophila* microRNA targets. *PLoS Biol.* 1:397–409.

Stein, L. D., Z. Bao, T. Blumenthal, M. R. Brent, N. Chen, A. Chinwalla, L. Clarke, C. Clee, A. Coghlan, A. Coulson, P. D'Eustachio, D. H. A. Fitch, L. A. Fulton, R. E. Fulton, S. Griffiths-Jones, T. W. Harris, L. W. Hillier, R. Kamarth, P. E. Kuwabara, E. R. Mardis, M. A. Marra, T. L. Miner, P. Minx, J. C. Mullikin, R. W. Plumb, J. Rogers, J. E. Schein, M. Sohrmann, J. Spieth, J. E. Stajich, C. Wei, D. Willey, R. K. Wilson, R. Durnin, and R. H. Waterston. 2003. The genome sequence of *Caenorhabditis briggsae:* A platform for comparative genomics. *PloS Biol.* 1:166–192.

Steinemann, M. 1978. Co-replication of satellite DNA of *Chironomus melanotus* with mainband DNA during polytenization. *Chromosoma* 66:127–139.

Steinemann, M., and S. Steinemann. 1992. Degenerating Y chromosome of *Drosophila miranda:* A trap for retrotransposons. *Proc. Natl. Acad. Sci. USA* 89:7591–7595.

Steinemann, M., and S. Steinemann. 1995. A duplication including the Y allele of *Lcp2* and the *TRIM* retrotransposon at the *Lcp* locus on the degenerating *neo*-Y chromosome of *Drosophila miranda:* Molecular structure and mechanisms by which it may have arisen. *Genetics* 134:497–505.

Steinemann, M., and S. Steinemann. 2000. Common mechanisms of Y chromosome evolution. *Genetica* 109:105–111.

Steinemann, S., and M. Steinemann. 1999. The *amylase* gene cluster on the evolving sex chromosomes of *Drosophila miranda. Genetics* 151:151–161.

Steneberg, P., C. Englund, J. Kronhamn, T. A. Weaver, and C. Samakovlis. 1998. Translational readthrough in the *hdc* mRNA generates a novel branching inhibitor in the *Drosophila* trachea. *Genes Dev.* 12: 956–967.

Sturtevant, A. H. 1925. The effects of unequal crossing over at the *Bar* locus in *Drosophila. Genetics* 10: 117–147.

Sturtevant, A. H., and E. Novitski. 1941. The homologies of the chromosome elements in the genus *Drosophila. Genetics* 26:517–541.

Sullivan, B. A., M. D. Blower, and G. H. Karpen. 2001. Determining centromere identity: Cyclical stories and forking paths. *Nat. Rev. Genet.* 2:584–596.

Sun, X., H. D. Le, J. M. Wahlstrom, and G. H. Karpen. 2003. Sequence analysis of a functional *Drosophila* centromere. *Genome Res.* 13:182–194.

Thompson-Stewart, D., G. H. Karpen, and A. C. Spradling. 1994. A transposable element can drive the concerted evolution of tandemly repetitious DNA. *Proc. Natl. Acad. Sci. USA* 91:9042–9046.

Thummel, C. S., K. C. Burtis, and D. S. Hogness. 1990. Spatial and temporal patterns of E74 transcription during *Drosophila* development. *Cell* 61:101–111.

Ting, C.-T., S.-C. Chern, M.-L. Wu, and C.-T. Wu. 1998. A rapidly evolving homeobox at the site of a hybrid sterility gene. *Science* 282:1501–1504.

Tokunaga, C. 1955a. A possible explanation concerning the sex determination mechanisms in *Aphiochaeta xanthina* Spieser. *Kobe College Studies* 4:65–75.

Tokunaga, C. 1955b. The presence of male determining factor in *Aphiochaeta xanthina* Spieser. *Kobe College Studies* 2:1–32.

Tomari, Y., T. Du, B. Haley, D. S. Schwartz, R. Bennett, H. A. Cook, B. S. Koppetsch, W. E. Theurkauf, and P. Zamore. 2004. RISC assembly defects in the *Drosophila* RNAi mutant *armitage. Cell* 116:831–841.

Tomita, T., Y. Kono, and T. Shimada. 1996. Chromosomal localization of amplified esterase genes in insecticide resistant *Culex* mosquitoes. *Insect Biochem. Mol. Biol.* 26:853–857.

Torok, T., M. Gorjanacz, P. J. Bryant, and I. Kiss. 2000. Prod is a novel DNA-binding protein that binds to the 1.686 g/cm^3 10 bp satellite repeat of *Drosophila melanogaster. Nucleic Acids Res.* 28:3551–3557.

Traut, W. 1994. Sex determination in the fly *Megaselia scalaris,* a model system for primary steps of sex chromosome evolution. *Genetics* 136:1097–1104.

Traut, W., and U. Willhoeft. 1990. A jumping sex determining factor in the fly *Megaselia scalaris. Chromosoma* 99:407–412.

Traut, W., and B. Wollert. 1998. An X/Y DNA segment forms an early stage of sex chromosome differentiation in the fly *Megaselia scalaris. Genome* 41:289–294.

Tritto, P., V. Specchia, L. Fanti, M. Berloco, R. D'Alessandro, S. Pimpinelli, G. Palumbo, and M. P. Bozzetti. 2003. Structure, regulation and evolution of the crystal-Stellate system of *Drosophila. Genetica* 117:247–257.

Troiano, G. 1978. Triploidy in the natural population of the psychodine moth fly *Psychoda parthenogenetica* Tonnoir (Diptera: Psychodidae). *Caryologia* 31:225–232.

Trotta, C. R., and J. Abelson. 1999. tRNA splicing: An RNA world add-on or an ancient reaction? In R. F. Gesterland, T. R. Cech, and J. F. Atkins, eds., *The RNA World,* 2nd ed., pp. 561–584. Cold Spring Harbor, N.Y.: Cold Spring Harbor Press.

Truet, M. A., R. S. Jones, and S. S. Potter. 1981. Unusual structure of the *FB* family of transposable elements in *Drosophila. Cell* 24:753–763.

Tsaur, S. C., and C.-I. Wu. 1997. Positive selection and the molecular evolution of a gene of male reproduction, *Acp26Aa* of *Drosophila. Mol. Biol. Evol.* 14:544–549.

Tsaur, S. C., and C.-I. Wu. 1998. Positive selection driving the evolution of a gene of male reproduction, *Acp26Aa,* of *Drosophila:* II. Divergence versus polymorphism. *Mol. Biol. Evol.* 15:1040–1046.

Tsujino, F., A. Kosemura, K. Inohira, T. Hara, Y. F. Otsuka, M. K. Obara, and E. T. Matsuura. 2002. Evolution of the A + T-rich region of the mitochondrial DNA in the *melanogaster* species group of *Drosophila. J. Mol. Evol.* 55:573–583.

Tu, Z. 1997. Three novel families of miniature inverted-repeat transposable elements are associated with genes in the yellow fever mosquito, *Aedes aegypti. Proc. Natl. Acad. Sci. USA* 94:7475–7480.

Tu, Z. 1999. Genomic and evolutionary analysis of *Feilai,* a diverse family of highly reiterated SINEs in the yellow fever mosquito, *Aedes aegypti. Mol. Biol. Evol.* 16:760–772.

Tu, Z. 2000. Molecular and evolutionary analysis of two divergent subfamilies of a novel miniature inverted repeat transposable element in the yellow fever mosquito, *Aedes aegypti. Mol. Biol. Evol.* 17:1313–1325.

Tu, Z. 2001a. *Maque,* a family of extremely short interspersed repetitive elements: Characterization, possible mechanisms of transposition, and evolutionary implications. *Gene* 263:247–253.

Tu, Z. 2001b. Eight novel families of miniature inverted-repeat transposable elements in the African malaria mosquito, *Anopheles gambiae. Proc. Natl. Acad. Sci. USA* 98:1699–1704.

Tu, Z., and C. Coates. 2004. Mosquito transposable elements. *Insect Biochem. Mol. Biol.* 34:631–644.

Tu, Z., and S. P. Orphanidis. 2001. *Microuli,* a family of miniature subterminal inverted-repeat transposable elements (MITEs): Transposition without terminal inverted repeats. *Mol. Biol. Evol.* 18:893–895.

Tujebajeva, R. M., P. R. Copeland, X.-M. Xu, B. A. Carlson, J. W. Harney, D. M. Driscoll, D. L. Hatfield, and M. L. Berry. 2000. Decoding apparatus for eukaryotic selenocysteine insertion. *EMBO Rpts.* 1:158–163.

Tweedie, S., H. H. Ng, A. L. Barlow, B. M. Turner, B. Hendrich, and A. Bird. 1999. Vestiges of a DNA methylation system in *Drosophila melanogaster? Nat. Genet.* 23:389–390.

Tycowski, K. T., and J. A. Steitz. 2001. Non-coding snoRNA host genes in *Drosophila:* Expression strategies for modification guide snoRNAs. *Europ. J. Cell Biol.* 80:119–125.

Ueda, H. R., A. Matsumoto, M. Kawamura, M. Iino, T. Tanimura, and S. Hashimoto. 2002. Genome-wide transcriptional orchestration of circadian rhythms in *Drosophila. J. Biol. Chem.* 277:14048–14052.

Ullerich, F.-H. 1963. Geschlechtschromosomen und Geschlechtsbestimmung bei einigen Calliphorinen (Calliphoridae, Diptera). *Chromosoma* 14:45–110.

Ullerich, F.-H., H. Bauer, and R. Dietz. 1964. Geschlechtsbestimmung bei Tipuliden (Nematocera, Diptera). *Chromosoma* 15:591–605.

Urieli-Shoval, S., Y. Gruenbaum, J. Sedat, and A. Razin. 1982. The absence of detectable methylated bases in *Drosophila melanogaster* DNA. *FEBS Letts.* 146:148–152.

Venter, J. C., M. D. Adams, C. G. Sutton, A. R. Kerlavage. H. O. Smith, and M. Hunkapiller. 1998. Shotgun sequencing of the human genome. *Science* 280:1540–1542.

Vieira, J., C. P. Vieira, D. L. Hartl, and E. R. Lozovskaya. 1997. A framework physical map of *Drosophila virilis* based on P1 clones: Applications in genome evolution. *Chromosoma* 106:99–107.

Vieira, V., C. Nardon, C. Arpin, D. Lepetit, and C. Biemont. 2002. Evolution of genome size in *Drosophila.* Is the invader's genome being invaded by transposable elements? *Mol. Biol. Evol.* 19:1154–1161.

Viera, C., G. Piganeau, and C. Biemont. 2000. High copy numbers of multiple transposable element families in an Australian population of *Drosophila simulans. Genet. Res. (Camb.)* 76:117–119.

Vinogradov S. N., D. A. Walz, and B. Pohajdak. 1992. Organization of non-vertebrate *globin* genes. *Comp. Biochem. Physiol. B* 103:759–773.

Vosselman, L. 1978. Sex determination of the onion fly, *Hylemya antiqua* Meig. I. Sex chromosome polymorphism, gynandromorphism and X-polysomy. *Chromosoma* 67:201–218.

Wagoner, D. E. 1969. Presence of male-determining factors found on three autosomes in the housefly, *Musca domestica. Nature* 223:187–188.

Wang, W., J. Zhang, C. Alvarez, A. Llopart, and M. Long. 2000. The origin of the *jingwei* gene and the complex modular structure of its parental genes, *yellow emperor,* in *Drosophilamelanogaster. Mol. Biol. Evol.* 17:1294–1301.

Warren, A. M., and J. M. Crampton. 1991. The *Aedes aegypti* genome: Complexity and organization. *Genet. Res. (Camb.)* 58:225–232.

Warren, W. D., P. W. Atkinson, and D. A. O'Brochta. 1994. The *Hermes* transposable element from the house fly, *Musca domestica,* is a short inverted repeat-type element of the *hobo, Ac,* and *Tam3 hAT* element family. *Genet. Res.* 64:87–97.

Wayne, M. L., D. Contamine, and M. L. Kreitman. 1996. Molecular population genetics of *ref(2)P,* a locus which confers viral resistance in *Drosophila. Mol. Biol. Evol.* 13:191–199.

Wessler, S. R., T. E. Bureau, and S. E. White. 1995. LTR-transposons and MITES: Important players in the evolution of plant genomes. *Curr. Opin. Genet. Dev.* 5:814–821.

White, M. J. D. 1973. *Animal Cytology and Evolution,* 3d ed. Cambridge: Cambridge University Press.

Wilder, J., and H. Hollocher. 2001. Mobile elements and the genesis of microsatellites in Dipterans. *Mol. Biol. Evol.* 18:384–392.

Willhoeft, U., and G. Franz. 1996. Identification of the sex-determining region of the *Ceratitis capitata* Y chromosome by deletion mapping. *Genetics* 144:737–745.

Willhoeft, U., and W. Traut. 1995. The sex-determining region in the *Megaselia scalaris* Diptera. Y chromosome. *Chromosome Res.* 3:59–65.

Wojtas, K. M., L. von Kaim, J. R. Weaver, and D. T. Sullivan. 1992. The evolution of duplicate glycerate-4-dehydrogenase genes in *Drosophila. Genetics* 132:789–797.

Wolfner, M. F. 2002. The gifts that keep on giving: Physiological functions and evolutionary dynamics of male seminal proteins in *Drosophila. Heredity* 88:85–93.

Wu, C.-I., T. W. Lyttle, M. L. Wu, and G. F. Lin. 1988. Association between a satellite DNA sequence and the Responder of *Segregation Distorter* in *Drosophila melanogaster. Cell* 54:179–189.

Wu, M., L. V. Sun, J. Vamathevan, M. Riegler, R. Deboy, J. C. Brownlie, E. A. McGraw, W. Martin, C. Esser, N. Ahmadinejad, C. Wiegand, R. Madupu, M. J. Beanan, L. M. Brinkac, S. C. Daugherty, A. S. Durkin, J. F. Kolonay, W. C. Nelson, Y. Mohamoud, P. Lee, K. Berry, M. B.Young, T. Utterback, J. Weidman, W. C. Nierman, I. T. Paulsen, K. E. Nelson, H. Tettelin, S. L. O'Neill, and J. A. Eisen. 2004. Phylogenomics of the reproductive parasite *Wolbachia pipientis w*Mel: A streamlined genome overrun by mobile genetic elements. *PloS Biol.* 2:327–341.

Wycoff, G. J., W. Wang, and C.-I. Wu. 2000. Rapid evolution of male reproductive genes in the descent of man. *Nature* 403:304–309.

Wyers, F., A. M. Petitjean, P. Dru, P. Gay, and D. Contamine. 1995. Localization of domains within the *Drosophila ref(2)P* protein involved in the intracellular control of sigma rhabdovirus multiplication. *J. Virol.* 69:4463–4470.

Xu, P., S. Y. Vernooy, M. Guo, and B. A. Hay. 2003. The *Drosophila* microRNA Mir-14 suppresses cell death and is required for normal fat metabolism. *Curr. Biol.* 13:790–795.

Xue, F., and L. Cooley. 1993. *kelch* encodes a component of intercellular bridges in *Drosophila* egg chambers. *Cell* 72:681–693.

Ye, X., P. Fong, N. Iizuka, D. Choate, and D. R. Cavener. 1997. *Ultrabithorax* and *Antennapedia* 5′ untranslated regions promote developmentally regulated internal translation initiation. *Mol. Cell. Biol.* 17: 1714–1721.

Yeates, D. K., and B. M. Wiegmann. 1999. Congruence and controversy: Toward a higher-level phylogeny of Diptera. *Annu. Rev. Entomol.* 44:397–428.

Yen, J. H., and A. R. Barr. 1971. New hypothesis of the cause of cytoplasmic incompatability in *Culex pipiens. Nature* 232:657–658.

Yoshiyama, M., Z. Tu, Y. Kainoh, H. Honda, T. Shono, and K. Kimura. 2001. Possible horizontal transfer of a transposable element from host to parasitoid. *Mol. Biol. Evol.* 18:1952–1958.

Yuan, G., C. Klambt, J. P. Bachellerie, J. Brosius, and A. Huttenhofer. 2003. RNomics in *Drosophila melanogaster:* Identification of 66 candidates for novel non-messenger RNAs. *Nucleic Acids Res.* 31:2495–2507.

Zacharias, H. 1986. Tissue-specific schedule of selective replication in *Drosophila nasutoides. Roux's Arch. Dev. Biol.* 195:378–388.

Zacharias, H. 1990. Heteropycnosis of an underreplicating chromosome. *Chromosoma* 99:24–35.

Zdobnov, E. M., C. von Mering, I. Letunic, D. Torrents, M. Suyama, R. R. Copley, G. K. Christophides, D. Thomasova, R. A. Holt, G. M. Subramanian, H. M. Mueller, G. Dimopoulos, J. H. Law, M. A. Wells, E. Birney, R. Charlab, A. L. Halpern, E. Kokoza, C. L. Kraft, Z. Lai, S. Lewis, C. Louis, C. Barillas-Mury, D. Nusskern, G. M. Rubin, S. L. Salzberg, G. G. Sutton, P. Topalis, R. Wides, P. Wincker, M.Yandell, F. H. Collins, J. Ribeiro, W. M. Gelbart, F. C. Kafatos, and P. Bork. 2002. Comparative genome and proteome analysis of *Anopheles gambiae* and *Drosophila melanogaster. Science* 298:149–159.

Zhang, Y. Q., J. Roote, S. Brogna, A. W. Davis, D. A. Barbash, D. Nash, and M. Ashburner. 1999. *Stress sensitive B* encodes an adenine nucleotide translocase in *Drosophilamelanogaster. Genetics* 153:891–903.

Zhimulev, I. F. 1996. Morphology and structure of polytene chromosomes. *Adv. Genet.* 34:1–497.

Zhimulev, I. F. 1998. Polytene chromosomes, heterochromatin, and position effect variegation. *Adv. Genet.* 37:1–566.

Zhou, X., S. I. Park, M. E. Moustafa, B. A. Carlson, P. F. Crain, A. M. Diamond, D. L. Hatfield, and B. J. Lee. 1999. Selenium metabolism in *Drosophila.* Characterization of the selenocysteine tRNA population. *J. Biol. Chem.* 274:18729–18734.

CHAPTER FIVE

Evolutionary Developmental Biology of the Diptera: The "Model Clade" Approach

Rob DeSalle

For the past century, thousands of biologists have relied on the model organism research paradigm to fuel their research programs. This paradigm, to say the least, has been incredibly useful and productive. The model organism paradigm, however, has suffered from the obvious limitation that taxonomic sampling has focused on only a few organisms. Although researchers working on model organisms have been able to examine the genetic, developmental, physiological, and morphological phenomena inherent to their model organisms, their ability to extend their systems beyond the model organism has been limited. More recent efforts by a broad community of biologists to include nonmodel organisms in comparative biology and evolutionary biology studies have expanded the number of organisms involved, but to date, little attention has been given to a systematic approach to incorporating developmental information. Some examples of the importance of a focused taxonomic approach to evolutionary developmental biology are summarized in DeSalle (2002), Wilkins (2002), and Carroll (2003).

A dipteran—*Drosophila melanogaster*—has been used as one of those model organisms. It was "constructed" as a model organism (Kohler 1994) in the early part of the twentieth century. Its utility as a tool in biology arises as a result of several advantages that it conferred on researchers. *Drosophila* has also been a major player in the development of the nascent field of evolutionary developmental biology (EDB). The three most recent treatments of this field (Carroll et al. 2001; Davidson 2001; Wilkins 2002) all rely heavily on this model organism to describe the emerging EDB paradigm in modern evolutionary biology. In addition, this dipteran species, because of its model organism status, was the third eukaryotic genome to be fully sequenced. The use of this fly as a model organism in EDB makes great sense, as it has the following important characteristics that make it very amenable to developmental biology study: (1) its whole genome is sequenced; (2) it has a long and detailed tradition of genetic analysis; (3) it is an easy organism to culture and to manipulate experimentally; (4) its life cycle is well understood and easily manipulated; and (5) its embryos are large enough to manipulate and observe. In addition, the *Drosophila* community is an excellent, established research community in which data, ideas, and technologies are freely shared (see some of the sites given in Box 1). Until recently, the utility of *Drosophila* as a tool for macroevolutionary studies and systematics was limited (DeSalle and Grimaldi 1993; Powell and DeSalle 1995; DeSalle et al. 1996; Powell 1998). Aside from some beautiful work on polytene chromosomes and some comparative molecular biology across the family Drosophilidae, focus

Box 1. A Beginner's Guide to *D. melanogaster* Development

A. General *Drosophila* Development Websites

The WWW Virtual Library: Model Organisms — www.ceolas.org/VL/mo

The WWW Virtual Library: *Drosophila* — www.ceolas.org/fly

biol.net Drosophila — biol.net/Drosophila.htm

FlyBase — flybase.bio.indiana.edu

The Interactive Fly — sdb.bio.purdue.edu/fly/aimain/1aahome.htm

GadFly: Genome Annotation Database of Drosophila — www.fruitfly.org/annot

J-Fly; Japanese Drosophila resource — jfly.nibb.ac.jp

FLYMOVE — flymove.uni-muenster.de:8020

FlyView — flyview.uni-muenster.de:8010

Drosophila developmental gene expression time course site — quantgen.med.yale.edu

B. Texts

The Making of a Fly (P. Lawrence; Blackwell Scientific Publishing, 1992)

The Development of Drosophila melanogaster (2 vols., eds. M. Bate and A. Martinez-Arias; Cold Spring Harbor Press, 1993)

Biology of Drosophila (ed. M. Demerec; Cold Spring Harbor Press, 1994).

The Embryonic Development of Drosophila melanogaster (J. A. Campos-Ortega and V. Hartenstein; Springer-Verlag, 1985)

Imaginal Discs (L. I. Held; Cambridge University Press, 2002)

The genome of Drosophila melanogaster (D. L. Lindsley and G. G. Zimm; Academic Press, 1992)

Drosophila, *a Practical Approach* (ed. D. B. Roberts; IRL Press, 1986)

Drosophila, *a Laboratory Manual* (M. Ashburner; Cold Spring Harbor Press, 1989)

Drosophila melanogaster: *Practical Uses in Cell and Molecular Biology* (ed. L. S. Goldstein and E. A. Fryberg; Academic Press, 1994)

Drosophila (B. Shorrocks; Ginn & Co., 1972)

C. Drosophila Genome Sequence Sites

Berkeley — www.fruitfly.org

European Drosophila Genome Project (EDGP) — edgp-dev.ebi.ac.uk

Celera private site — public.celera.com/cds/login.cfm

National Centre for Biotechnology Information — www.ncbi.nlm.nih.gov/PMGifs/Genomes/7227.html

Mosquito Genomics WWW Server — klab.agsci.colostate.edu/

D. Specific Websites

Drosophila researcher websites — www.ceolas.org/fly/labs.html

FlyView — pbio07.uni-muenster.de

Drosophila-related EST sequences — www.tigem.it/LOCAL/drosophila/dros.html

Protocols — ceolas.org/VL/fly/protocols.html

Tracheal development — www.biozentrum.unibas.ch/affolter/trachea

Malphigian tubule development — www.mblab.gla.ac.uk/tubules/tubules-www.html

Gene Networks Database — www.csa.ru:82/Inst/gorb_dep/inbios/genet/s0sgnt.htm

flyEx; a database of segmentation gene expression in *Drosophila* — www.csa.ru/flyex
FlyTrap — www.fly-trap.org
Virtual fly brain — www.neurofly.de
Fly nervous system atlas — brain.biologie.uni-freiburg.de/Atlas/text/atlasFi.html
Fly motor axon — www.its.caltech.edu/~zinnlab/motoraxons/fma%20home%20page.html
FlyBrain Project — flybrain.neurobio.arizona.edu
FlyMove: Fly development movies — flyview.uni-muenster.de:8080
DB Cinema main page; Fly Morph-o-genesis — sdb.bio.purdue.edu/dbcinema/kaufman/kaufman.html
Body part viewer — flybase.bio.indiana.edu/.bin/fbimage
Drosophila anatomy and images — flybase.bio.indiana.edu/anatomy

on the Drosophilidae in particular and the Diptera in general as evolutionary tools has primarily been at the population level.

In this chapter, I summarize the utility of expanding the *Drosophila* model system approach to the order Diptera in the EDB paradigm in general and explore the future approaches that will be applied to an expansion of applications to other Diptera. The prevailing paradigm in EDB has been to focus on an interesting morphological problem using what can only be described as a patchy taxonomic sampling. The research paradigm essentially uses model organisms as data points, as well as sources of technology to produce data for other organisms.

Drosophila and Dipteran Development: A User's Guide

A comprehensive description of the developmental biology of *D. melanogaster* would take volumes (see Bate and Martinez-Arias 1993 and Held 2002 as examples of comprehensive volumes on the subject). Instead I present a "user's guide" to *Drosophila* development (Box 1) and a cursory review of the historical foundation of modern developmental approaches relevant to expanding developmental studies in the Diptera.

The developmental biology of the dipteran *D. melanogaster* is, to make an understatement, well understood. During the past century, hundreds of biologists undertook the genetic and developmental dissection of this organism and were able to coordinate detailed embryonic studies with exquisite genetic studies. Two events occurred in the last quarter of the twentieth century that greatly accelerated the discovery process in drosophilid development. The first culminated in 1981 with the publication by Nuesslein-Volhard and Wieschaus (1980) of a paper in *Nature* summarizing an ambitious saturation of the *D. melanogaster* genome with mutations for early embryonic effects. In concert with Lewis's (1978) and Kaufman's (Scott et al. 1982) work on the homeotic loci in the Bithorax (BX-C) and Antennapedia (ANT-C) complexes, respectively, these studies laid the foundation for coupling genetic analysis with embryonic development. The second event was the fusion of genetics with molecular biology, allowing the introduction of genes into *Drosophila*. Important transgenic systems such as *P-element* transformation, enhancer trap, and UAS-GAL4 overexpression were developed, leading to the design of ectopic gain-of-function screens. These systems were crucial for extending the fine-tuned analysis from larval patterning in the embryo to adult patterning in the imaginal discs. The third was the development and refining of several tools for the visualization of gene activity in specific body parts of the embryo, larva, and adult fly. These techniques include whole-mount antibody staining of embryos, in situ hybridization of nucleic acid probes to developing embryos, enhancer trap methods with colorimetric assay systems to localize phenotypic effect of enhancer trap, among others.

Nuesslein-Volhard and Wieschaus (1980) used three steps to discover a broad range of developmental effects in *Drosophila:* (1) saturation mutagenesis, (2) detailed screen for alterations in embryonic morphology, and (3) focus on early embryogenesis. Until they developed their screen, most embryonic lethals were ignored because most researchers were interested only in adult morphologies. It was also assumed that the majority of embryonic lethal loci were developmentally insignificant housekeeping genes, and that genetic deletion of early developmental genes would not result in informative embryonic phenotypes. Nuesslein-Volhard and Wieschaus (1980) reasoned that a significant fraction of embryonic lethals might hold important information for embryonic development in *Drosophila* and could also be discovered because of specific, detectable, and informative patterning defects. Thus they paid special attention to embryos that were nonviable in early stages. Their approach allowed them to discover and classify many early embryonic patterning genes. What ensued after their study and the coupling of other developmental systems, such as the homeotic systems with molecular genetics, is one of the best-understood developmental systems in biology. Their work resulted in a hierarchical organization of developmental gene action that revolutionized the way developmental biologists thought about *Drosophila* development in particular, and organismal development in general. In particular, the now well-known arrangement of maternal affect, gap, pair rule, and segment polarity gene interactions was suggested and refined from this pioneering work.

The technical advancement of approaches that assisted in the visualization of developmental gene products in embryonic and adult tissues in *D. melanogaster* has been equally impressive. To date, a virtual Noah's Ark of embryos has been examined using these visualization techniques. The approach is simple: a probe (either an in situ nucleic acid probe or an antibody probe) for a gene product (either mRNA or protein) is labeled as a marker. The probe is then used to localize the tissue or tissues in which gene expression is active for the probe. In the case of labeled nucleic acid probes, hybridization of the probe with target mRNA indicates the presence of transcriptional activity of the probe. In the case of antibody probes, reaction of the antibody with target proteins indicates translational activity of the probed gene.

Starting with radioactively labeled probes for in situ hybridization studies in the 1980s and leading to the refinement of antibody staining approaches, the spatial information generated by these methods is stunning. An amazing example of the kind of information that can be obtained can be found at www.flyex.ams.sunysb.edu (John Reinitz home page) or www.csa.ru:82/Inst/gorb_dep/inbios/genet/s0patt.htm (GeNet/SegNet website). Because the antibodies and in situ probes in general cross-react and cross-hybridize, respectively, use of these approaches can be extended outside the model organisms. And indeed, some interesting and exciting data have arisen from this approach. Today the best access to this information is through the consortium developed by the *Drosophila* community over the past few decades (see Box 1).

The *Drosophila*-Diptera Approach in Practice

Two of the earliest and most prominent EDB study systems for animals involved examination of the expression patterns of *engrailed* (Patel 1994) and *distalless* (Panganiban et al. 1997) in a wide variety of animals. These systems examined expression patterns of developmental genes over broad evolutionary distances (across phyla) and established that interesting inferences about the evolution of body plan could be made when placed into a phylogenetic context. The choice of these two systems was reasonable, given that evolutionary change of segmentation patterns and limb formation is best examined across phyla. Other significant firsts, focusing more on insects, include the use of Ubx/AbdA antibody staining to observe segmental identity (Kelsh et al. 1994), examination of mesoderm formation using twist and snail expression patterns (Sommer and Tautz 1994), and *hunchback* expression patterns to examine gap pattern formation (Treier et al. 1989).

More recently, several studies have examined the role of developmental genes in the establishment of the evolution of development in dipteran divergence. These study systems include the evolution of bristle formation, pigmentation in the body and wings, wing vein evolution, and, surprisingly, the evolution of sex determination and of early development. The latter are surprising in that these two aspects of the developmental biology of this order of insects are expected to be highly conserved across insects. Here I present short reviews of these six dipteran EDB systems (Fig. 5.1).

EVOLUTION OF EARLY DEVELOPMENTAL GENES IN DIPTERA

By far the most interesting aspect of stem lineage evolution during dipteran divergence to examine is the establishment of segmental identity and of the anteroposterior embryonic body axis in developing embryos. The most extensive analysis of the role of early development in insects has been carried out in *D. melanogaster,* but as pointed out by several researchers, this dipteran has a highly derived and highly specialized developmental program. To establish the extent of derivation of the *Drosophila* developmental program, several studies have begun to examine the developmental programs of more basal dipteran taxa.

The *bicoid* gene is located in the HOM-C of *Drosophila* and is one of three genes (*bicoid, zerknüllt,* and *fushi tarazu*) in this complex that have evolved more rapidly than average in the cluster. This gene is responsible for establishing the boundaries of the larval head and thorax (Rushlow and Levine 1990; Pankratz and Jackle 1993). The function of the *bicoid* and *zerknüllt* genes are well known and characterized in *Drosophila.* The *zerknüllt* gene has undergone a very recent duplication to produce *zen1* and *zen2* in *D. melanogaster.* Searches for a *bicoid* homolog in organisms other than cyclorrhaphan diptera—such lower cyclorrhaphan dipterans as *Megaselia abdita* (Stauber et al. 2000) or muscoid Diptera (Shaw et al. 2001), and such Schizophora as *D. melanogaster* (Stauber et al. 1999)—have been undertaken with no success. However, a *zerknüllt/Hox 3* homolog has been found in a wide variety of Lower Diptera and other orders of insects.

In addition, the timing of expression of these two genes introduces some interesting phylogenetic patterns of divergence that only a thorough comparative approach can decipher. For instance, the expression dynamics of *bicoid* in the anterior tip of the blastoderm embryo are well conserved throughout the Cyclorrhapha (Stauber et al. 2002). But for *zerknüllt,* the

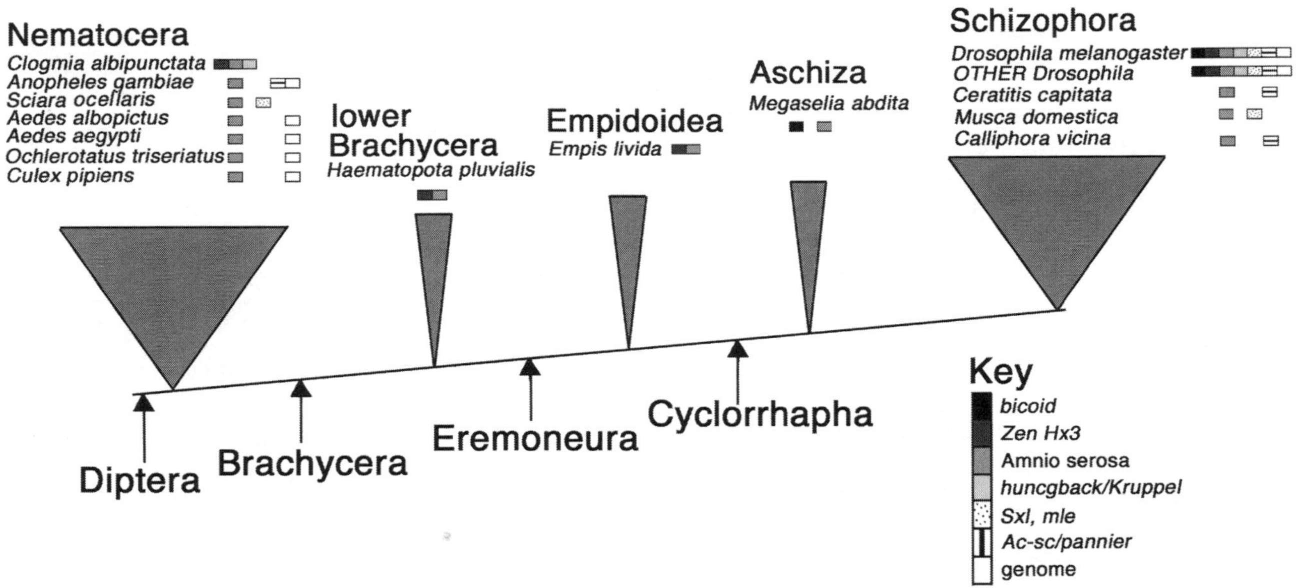

FIGURE 5.1. Reduced schematic representation of dipteran phylogeny with the taxa discussed in this chapter mapped onto the phylogeny. Major groups within the Diptera are also indicated along the backbone of the phylogeny. The kinds of studies utilized in discussion of dipteran EDB in this chapter are also indicated next to the species name. The key at the bottom of the figure indicates which EDB system was examined for the indicated taxon.

conservation of expression patterns is more complex. *Zerknüllt* has both a maternal and zygotic distribution in *Drosophila,* but even in the basal Cyclorrhaphan, *M. abdita,* the maternal component of its expression patterns is missing (Stauber et al. 2002). The pattern seen in most insects for *zerknüllt* is a zygotic expression domain on the dorsal side of the embryo that is required for establishment of extra-embryonic tissue fate. Because the highly derived state of *D. melanogaster* is at odds with the broader understanding of the expression patterns and genomic presence or absence of these genes, a more complete taxonomic sampling of dipteran taxa for the presence or absence of the *bicoid* gene in the genome, an establishment of orthology for *bicoid* and *zerknüllt* to *Hox3* genes, and a better understanding of the expression timing and domains of these genes was undertaken by Stauber et al. (1999, 2000, 2002).

It is clear that *bicoid* orthologs do not exist outside the Cyclorrhapha (Stauber et al. 2002), or if they do, they have diverged sufficiently to have lost their ancestral function. *Zerknüllt* orthologs do, however, exist in lower dipteran genomes, as evidenced by their presence in *Empis livida* (Empidoidea), *Haematopota pluvialis* (a lower brachyceran), and *Clogmia albopunctata* (a lower dipteran). Furthermore, the expression patterns of this *zerknüllt* ortholog in these lower flies show patterns reminiscent of both *zerknüllt* and *bicoid.* Stauber et al. (2002), using the phylogenetic placement of these flies, suggest that the ancestral genomic state in all Diptera was the presence of a *zerknüllt/Hox3* ortholog that duplicated in the ancestor of cyclorrhaphan flies. Subsequent divergence of these duplicated genes resulted in the loss of maternal expression in the *zerknüllt* paralog and the loss of zygotic expression in the *bicoid* paralog.

LONG AND SHORT GERM-BANDS

Krause (1939) and Sander (1976) first characterized the early events of embryonic development in insects as being long, short, and intermediate banded. In the early development of all insects, the germ-band stage consists of the same basic segmental organization of the developing embryo, with a head with three gnathal segments (mandibular, maxillary, and labial) and a procephalic region. This head region is followed by three thoracic segments and from eight to 11 abdominal segments. "Long germ bandedness" refers to the pattern of laying down the entire segmental pattern with the onset of gastrulation, whereas short germ banded insects generate body segments during a growth phase after formation of the blastoderm. Intermediate germ banded insects show a mixture of short and long germ patterns. For instance, some intermediate banded insects have segments determined by the blastoderm stage as far posterior as the thorax and even some of the more anterior abdominal segments. Segments posterior to these in intermediate banded insects are established after gastrulation in a processional fashion.

Germ bandedness has been an interesting embryonic morphology to examine using the EDB approach (Patel et al. 1992, 1994; Tautz 1992; Patel 1994; Tautz et al. 1994; Tautz and Sommer 1995; DeSalle et al. 1996; Davis and Patel 2002) because of its interesting distribution in the various orders of insects. A detailed examination of dipteran taxa for early segmentation gene patterns promises to unravel some of the processes involved in determining germ-band organization in insects. Although almost all dipteran taxa are considered to be long germ banded in their early embryonic development, there are two interesting divergences of developmental program when comparing Lower Diptera and Brachycera (Rohr et al. 1999). First, although Lower Diptera are derived with respect to being long germ banded, they retain the ancestral insect embryological character of forming separate extra-embryonic membranes, amnion, and serosa (Sander 1976), which are rudimentary and do not separate in the derived Brachycera (Rohr et al. 1999). Basal Brachycera do, however, develop an amnion (Schmidt-Ott 2000). Rohr et al. (1999) also first suggested that the ancestral expression of *hunchback* in the serosa was lost in the cyclorraphan stem lineage, indicating a molecular genetic correlate with amioserosa evolution. Second, the embryonic head region in Cyclorrhapha develops through an involuted stage, leading to acephalic larva with a cephalopharyngeal head skeleton, whereas the lower dipteran head anlagen do not involute. Because the developmental differences between some Brachycera and Lower Diptera are so distinct, an examination of the prepatterning of the embryos of taxa in these two major dipteran "groups" has shed light on the germ-band organization issue in insects.

Sander (1996) summarizes several of the studies on early development of dipteran taxa examined using EDB approaches. The general consensus concerning these studies is that patterns of divergence and conservation are both observed. In particular, the prepatterning of the early embryo appears to be highly conserved in the various dipteran species examined (Schmidt-Ott et al. 1994; Curtis et al. 1995). However, Rohr et al. (1999) examined the early acting segmentation genes *hunchback* and *Kruppel* and the pair rule gene *even-skipped* in *Clogmia albipuncatatus* and found that this lower dipteran showed features of both short and long germ-band embryogenesis. The germ bandedness of this lower dipteran therefore can be characterized as intermediate (Rohr et al. 1999), adding more information to how germ bandedness can be interpreted in a phylogenetic context. As the phylogenetic distribution of the embryological character germ bandedness (with its character states long, short, and intermediate) suggests homoplasy, the dissection of this character using EDB approaches appears to be the most efficient and easily interpreted method.

WING VEINS

Stark et al. (1999) reviewed the developmental and systematic literature on wing veins in Diptera to place the emerging developmental and genetic information on wing vein formation (Garcia-Bellido and de Cellis 1992; Sturtevant and Bier 1995) in insects into a phylogenetic systematic context. Because wing veins have been an important source of information in anatomical studies in diptera for over a century, Stark et al. (1999) surveyed the extent of variation in wing vein characteristics in Diptera and the current state of knowledge about the molecular basis for wing vein formation in *Drosophila*. Specific statements about the role of developmental genes in the formation of wing veins in dipteran taxa were not made in this review, but some guidelines for how EDB approaches could interpret the history of wing vein change in dipteran taxa were made. Their conclusion was that discovering the molecular basis of morphological change in dipteran wing vein change, although somewhat of a fishing expedition, could be attempted with the "phyletic phenocopy" (DeSalle and Carew 1992) paradigm in mind and using the emerging molecular and developmental biology probes for wing vein development.

In an attempt to make predictions about specific wing vein morphology (the crossvein), Marcus (2001) suggested that signaling processes play crucial roles in the development of wing vein morphology. Specifically, Marcus (2001) suggests that there are three stages in the development of crossveins that are relevant to control of morphology of these wing veins and that in the first stage, the number and placement of crossveins is determined by signaling along the proximodistal axis of the wing. During the second stage, a signaling system on the dorsal wing epithelium of the wing communicates with the ventral epithelium of the wing to prepattern the wing. Finally, in the third stage, the exact location of crossveins is defined by signaling interactions in signal transduction pathways. Marcus (2001) makes some very specific predictions of the kinds of genes and developmental process that might be involved in these developmental interactions. For instance, the cdc42 gene product and the Jun-N-terminal Kinase transduction pathway are implicated in the first stage, a BMP-like signaling pathway is implicated in the second stage, and a EGF-receptor Map Kinase signal transduction pathway and a Notch-Delta signaling interaction are implicated in the third stage. Because molecular markers and tools are available for the examination of these pathways and developmental gene interactions, Marcus (2001) has made some specific suggestions as to an approach to understanding crossvein changes in insect evolution in general and dipteran taxa in particular.

PIGMENTATION IN DIPTERA

The developmental biology of pigment formation in insect embryos has also become an exciting and well-studied area. Wright (1987) defined many of the biochemical interactions that occur in developing dipteran tissues that produce pigmentation differences. More recently, True and Carroll and colleagues (True and Carroll 1998; True et al. 1999; Carroll et al. 2001; Wittkopp et al. 2002a,b, 2003) have established the fine detail of the biochemical and developmental pathways of melanin deposition in developing embryos and imaginal discs in *Drosophila*. They conclude that pigmentation developmental processes are relatively simple, involving only a few genetic elements (tyrosine hydroxylase, dopa decarboxylase, and ebony proteins), and in particular, that complex hormonal regulation is not a factor (True et al. 1999; Kopp and True 2002). The wide variety of wing vein pigmentation variation in drosophilid species (O'Grady and DeSalle 2000) can possibly now be placed in the context of the genetic and developmental interactions in these flies. In addition, the level of melanin production in the developing *Drosophila* body is dependant on the interaction of two proteins (the Yellow protein and the Ebony protein), indicating a simple genetic developmental pathway for the pigmentation effects in dipteran insects as well.

THE EDB OF SEX DETERMINATION AND DOSAGE COMPENSATION IN DIPTERA

For recent reviews of the molecular mechanisms involved in sex determination in animals, see Cline and Meyer (1996) and Marin and Baker (1998). A brief summary of the major events is as follows (see also Ashburner, Chapter 4). Sex determination in *Drosophila* is triggered by the ratio of X to autosomes (A). If a ratio of X:A of 1.0 or greater is detected, then the *Sxl* gene is turned on and the SXL protein is produced. The SXL protein acts as a splicing protein, which processes the RNA produced by the *transformer* (*tra*) gene to produce active TRA protein. TRA protein together with the protein product from the *transformer 2* gene (TRA-2) determines the female specific splicing of the RNA from the double switch gene *doublesex* (*dsx*) and the *fruitless* gene (*fru*) to produce DSX-F and FRU-F proteins, of which, the DSX-F protein results in female differentiation. If the male specific ratio is detected, the

Sxl gene is switched off and the cascade described above does not occur; instead, male specific splicing of the *dsx* gene and the *fruitless* genes occurs (DSX-M and FRU-M, respectively). The male *dsx* spliced transcript produces male DSX-M, which results in male differentiation in most somatic tissues and the male version of FRU-M is required for male sexual differentiation in the central nervous system.

In the context of these molecular mechanisms, there are a large number of ways dipteran taxa determine sex. These modes of sex determination in Diptera have been examined in detail by White (1972), Blackman (1995), and Marin and Baker (1998), and include male determining dominant factor, genotye of mother, X:A balance, or a combination of these. In addition, the sex chromosome makeup ranges from the drosophilid constitution of X and Y chromosomes to other families with Z, W, and V chromosomes and some families with homomorphic chromosome constitution. Even within a genus, highly variable determination systems exist (see data for *Musca* in Marin and Baker 1998: Table 1).

Examination of sex determination factors at the molecular level within the genus *Drosophila* (Erickson and Cline 1998) indicated that across three species in the two subgenera in this genus, there is no variation in the sex determination pathway. Specifically, one of the important "X:A numerator genes" that detects the ratio of sex chromosomes to autosomes in *Drosophila* (and hence determines sex of the fly) is called *sisterless A* (*sisA*). Sex is determined by controlling the transcription of *Sxl*. Examination of the *sisA* gene in *D. virilis* and *D. pseudoobscura* indicated that *sisA* and *Sxl* are equally tightly coupled across the genus, indicating that the same primary sex determination mechanism exists across the genus.

Schutt and Nothinger (2000) have also examined the role of *Sxl* and other genes in sex determination in dipterans. Cloning and functional analysis of several of the genes from the pathway outlined above in *Musca domestica* and other dipteran insects indicate that the same basic strategy for implementing sex determination in these other dipteran taxa exists. The pathway as deciphered by Schutt and Nothinger (2000) is: (1) a basic genetic signal that is different in males and females is detected by the sex determining system, (2) a key gene responds to this basic genetic signal that produces a gene product, (3) the gene product in step 2 reacts with a double switch gene that implements the selection between the two alternative sexual programs. Although this basic pathway is conserved across all diptera, the molecular players in the pathway have diverged. In particular, *Sxl*, which has a major role in triggering the sexual differentiation of drosophilids, appears to not have a function in sex determination in any of the dipteran genera examined. However, *dsx*, the double switch gene that acts at the end of the pathway, does seem to have conserved function in the sex determination cascade.

Dosage compensation effects in dipteran taxa also follow this pattern of divergence. Ruiz et al. (2000) have examined the role of the *maleless* (*mle*) gene in dosage compensation in *Sciara occelaris*, a lower dipteran. Whereas the *mle* gene implements dosage compensation in drosophilids, it appears to have no such function in *S. occelaris* and it would not be surprising to find other genera in the Diptera that lack a role for *mle* in dosage compensation.

The divergence of the role of the specific gene products involved in these two important determinative processes in dipteran development indicates an unexpected plasticity in the role of genes in these pathways. Marin and Baker (1998) have demonstrated plasticity at higher taxonomic levels in the genes that are used to determine sex. But the demonstration that a similar plasticity works at the lower taxonomic levels is surprising and indicates an extremely rapid rate of evolution in sex determining cascades in comparison to other regulatory pathways.

BRISTLES

Bristles have been a major source of information for taxonomists and systematists in understanding dipteran biology. Wheeler (1981: 4) points out that the systematics of the Drosophilidae has relied heavily on "whether a certain bristle is present or absent or if it is directed forward or backward." The molecular genetic basis for bristle development in *Drosophila* is sufficiently well understood that some researchers have undertaken a detailed examination of bristle evolution in the Diptera. In an initial review article, the extent and phylogenetic pattern of variation in bristle development and positioning in dipteran taxa were undertaken (Simpson et al. 1999; see also Simpson 2002). This review pointed out that ancestral dipteran taxa have randomly distributed but uniformly spaced bristles, whereas the more derived taxa tend to have bristles aligned into longitudinal rows. With respect to larger bristles, which reside on the scutum, Simpson et al. (1999) suggest that the innovation of four rows of bristles occurred in the cyclorrhaphous Brachycera. Within the cyclorrhaphous Brachycera, most species have a variation on this major theme of four rows of bristles on the scutum. Based on the current state of knowledge about bristle formation, Simpson et al. (1999) suggest that these character systems are most likely controlled by the regulation of genes in the achaete-scute complex of genes.

Wulbeck and Simpson (1999, 2000, 2002), Pistillo et al. (2002), and Skaer et al. (2002) have examined this hypothesis by obtaining homologs of genes in the achaete-scute complex, as well as the transcriptional activator pannier and other members of the *Iriquois* gene family from several Lower Diptera. In the first of three studies, Wulbeck and Simpson (2000) isolated the ac-sc complex homologs from *Ceratitis capitata,* a species of acalypterate Schizophora. The bristle patterns in this species of fly have diverged slightly from the patterns seen in the highly derived Drosophilidae, and molecular dissection of three genes from the ac-sc complex—*scute, lethal of scute,* and *asense*—is highly conserved at the amino acid level over the 100 My of divergence of these groups of flies. In addition, the expression levels and spatial distribution of expression patterns for these genes as well as for the transcriptional activator pannier are conserved, providing evidence for broad conservation of many *cis*-regulatory interactions involved in the development and final disposition of bristles on the notum across a relatively large phylogenetic distance. More importantly from a systematic standpoint, the dissection of the molecular basis for these bristle characters indicates that the origin for the stereotyped bristle patterns on the notum of Schizophora has a common molecular developmental basis. Similar molecular examination of the expression of the *scute* gene of *Calliphora vicina* suggests that the pattern of longitudinal stripes has a common molecular mechanism for the higher Diptera. Wulbeck and Simpson (2002) have extended this approach to the mosquito *Anopheles gambiae* by dissecting the interaction of *pannier* gene expression and several ac-sc complex genes. Specifically, the *An. gambiae pannier* gene product is spatially distributed in identical positions on the dorsal medial notum at sites where sensory organs emerge. The distribution of pannier in *An. gambiae* is identical to the distribution to the *An. gambiae* ac-sc gene product ASH, which is also the pattern of pannier and ASH in higher Diptera. These results suggest that activation of ASH by pannier in *An. gambiae* has been conserved over the divergence of Lower Diptera and Brachycera.

Expanding Technology in the Study of Diptera: Microarrays, RNAi, Broad Spectrum Gene Transfer, and Transformation and Genomics

One of the more interesting and important approaches in developmental biology is the examination of the phenotypic effect of mutagenizing or altering a gene in the genome. In fact, this approach has become part of the developmental analysis paradigm. When a gene product is suspected of producing a phenotype, the best way to show the effect is to create a mutant or mutagenized lines and examine their phenotypes. Genes can be mutagenized in a targeted manner by using screens for enhancing or suppressing mutation, which does not require transgenics. The alternative approach of creating genetic constructs in *D. melanogaster* by transformation or genetic transfer with a cloned altered gene has been used for a couple of decades in *Drosophila.* This approach takes advantage of genetic elements, such as transposable elements, to implement the transfer. Coupled to the genetic applications are a large number of pest control applications.

For insects and Diptera, the approach was at first limited with respect to branching out into organisms other than *D. melanogaster,* because many of the transformation vectors were nonfunctional in even very closely related *Drosophila* species (Rubin and Spradling 1982; O'Brochta and Atkinson 1996). Some success in transformation approaches has been achieved using *Drosophila*-specific transposable elements (Loukeris et al. 1995; Lozovskya et al. 1995; Lohe and Hartl 1996; O'Brochta and Atkinson 1996), but more recently, to broaden taxonomic efficacy, researchers have developed and refined two approaches that have the potential to cover large phylogenetic distances. The first is the use of broad spectrum transformation vectors, such as Sindbis virus (Higgs et al. 1999; Lewis et al. 1999; Olson et al. 2000; Cheng et al. 2001), baculoviruses (Fraser et al. 1985; Lobo et al. 1999; Oppenheimer et al. 1999), and *piggyBacs* (Berghammer et al. 1999; Horn and Wimmer 2000; Peloquin et al. 2000, 2002; Tamura et al. 2000; Hacker et al. 2003; Horn et al. 2003). These broad spectrum vehicles can implement gene transfer among organisms relatively distantly related to *D. melanogaster.*

A second set of approaches that promises to have an impact on creating controlled genetic "constructs" in dipteran species as well as a wide range of other organisms is the knockout, or more appropriately, "knockdown" of the function of genes in developmentally important pathways in RNA interference or RNAi. Guo and Kemphues (1995) first used the approach in *Ceanorhabditis elegans* and it has since been applied to a broad spectrum of organisms (Kennerdell and Carthew 1998; Misquitta and Paterson 1999; Hughes and Kaufman 2000). RNAi refers to the production of hypomorphic phenotypes due to the introduction of homologous double stranded RNA (dsRNA) into the mRNA pool. The dsRNA initiates the building of a gene-specific RNA degradation multi-enzyme complex involving RNA-dependent RNA polymerase and dsRNA cleavage activities (for reviews, see Denli and Hannon 2003; Wimmer 2003; Dykxhoorn et al. 2004). Wimmer (2003: 226) suggests that "targeted expression systems and RNAi knockout in non-model organisms will make possible comparative phenotypic analyses of gene functions in diverse insect species."

Perhaps one of the most exciting technologies recently developed is the use of microarray approaches to understand gene expression in a developmental context. White, Rosbash, Greenspan, and colleagues (White et al. 1999; Furlong et al. 2001; McDonald and Rosbash 2001; Toma et al. 2002) have been the most prominent workers in this area, and to date, have examined three major areas of interest in dipteran developmental biology. The first area concerns the profiling of gene expression during metamophosis using a *D. melanogaster* gene chip containing several thousand *D. melanogaster* gene sequences (6,240 gene sequences that included more than 4,500 unique cDNA expressed sequence tag [EST] clones along with a number of ecdysone-regulated control genes having predictable expression patterns; White et al. 1999: 2179). White et al. (1999) examined the number and kinds of genes expressed during stages that span two pulses of ecdysone during the development of *D. melanogaster* and demonstrated the feasibility of examining differential expression of genes on the scale of the genome. In three other studies, the genome-wide expression of genes in early development of the mesoderm (Furlong et al. 2001) and in the regulation of circadian rhythm (McDonald and Rosbash 2001; Toma et al. 2002) were characterized. These approaches are landmark studies in that they show the efficacy of the whole genome approach to gene expression analysis using developmental stages. The applicability of the approach to other dipteran species is still to be shown, but the promise of this approach is very high as evidenced by a recent study of *D. melanogaster* group flies (Rifkin et al. 2003), in which several species of drosophilids were examined for temporal patterns of gene expression using microarray technology.

More recently, the *An. gambiae* genome has been completed (Holt et al. 2002). Along with the *D. melanogaster* completed genome and plans to complete several *Drosophila* genomes in the next five years as well as some other lower dipteran genomes like *Aedes aegypti, Ae. albopictus, Ochlerotatus* (formerly *Aedes*) *triseriatus,* and *Culex pipiens* (see www.klab.agsci.colostate.edu), the number of dipteran genomes sequenced over the next few years will only grow larger. Although the implications for the sequencing of the *An. gambiae* genome in medicine and human health are obvious (Aultman et al. 2002; Jasny et al. 2002; Morel et al. 2002), the implications for the utility of the full genomes of this dipteran in understanding development of insects have also been recognized (Kaufman et al. 2002; Zdobnov et al. 2002). Even though previous studies of dipteran regulatory region evolution have demonstrated relatively rapid primary sequence change in these regions (Ludwig et al. 1998, 2000), the comparative genomics of *Drosophila* and *Anopheles* might lead to the discovery of new regulatory regions and new candidate *cis*-regulatory sequences important in developmental processes using the phylogenetic shadowing approach (Boffelli et al. 2003).

Conclusion

In this chapter, I have examined clades of organisms as models for EDB and for broader genomic analyses. Because much of the data collected for EDB concerns the localization of spatial and temporal gene expression, one of the requirements for establishing a clade of organisms as a model is their culturability and ease of manipulation of the organisms in the clade. The model clade should also be part of a major organismal radiation and show a broad range of morphological variation. The clade should also either have a strong tradition of systematic analysis or the potential for unraveling the systematic relationships of the members of the clade. During the establishment of EDB as a prevailing paradigm in modern evolutionary biology, the approach has been to focus on interesting evolutionary and morphological questions. From a systematic standpoint, these studies offer an interesting and important approach to establishing hypotheses about the targeted organisms. In particular, focus on the major questions in EDB using *Drosophila* have been on what could be called "deep morphology" and are in essence studies that attempt to hang the ornamentation of developmental biology on the already established tree of life, or more appropriately, on only specific limbs of the tree of life.

In particular, the following have been the major focus of study in the context of *Drosophila* biology or have used *Drosophila* as a knowledge base for approaching the specific developmental question: (1) the origin of limbs, (2) the origin of wings, (3) the origin of long and short germ band form of embryonic development, (4) the origin of mouthparts, (5) the origin of arthropod eyes, (6) the origin of bristles and sensilla, and (7) the origin of segments in arthropods. Although all of these studies are elegant and interesting, they utilize a deep phylogenetic sampling and approach problems of deep morphology with very limited taxonomic sampling. From the vantage of systematics, the EDB approach can be made more useful with more detailed and richer taxonomic sampling (Bang et al. 2000, 2002). The Diptera are a morphologically diverse group of organisms that affords a much tighter phylogenetic arrangement of taxa and hence could be considered a model clade for future EDB studies.

ACKNOWLEDGMENTS

I thank the Lewis and Dorothy Culman Program for Molecular Systematics Studies at the American Museum of Natural History. The two anonymous reviewers substantially improved the manuscript. Special thanks to Brian Weigmann and David Yeates for their support and encouragement during the writing of this chapter.

REFERENCES

Aultman, K. S., M. Gottlieb, M. Y. Giovanni, and A. S. Fauci. 2002. *Anopheles gambiae* genome: Completing the malaria triad. *Science* 298:13–14.

Bang, R., R. DeSalle, and W. Wheeler. 2000. Transformationalism, taxism, and developmental biology in systematics. *Syst. Biol.* 49:19–27.

Bang, R., T. R. Schultz, and R. DeSalle. 2002. Development, homology and systematics. *EXS 2002* 92: 175–186.

Bate, M., and A. Martinez-Arias. 1993. *The Development of* Drosophila melanogaster, vols. I and II. Plainview, N.Y.: Cold Spring Harbor Press.

Berghammer, A. J., M. Klingler, and E. A. Wimmer. 1999. A universal marker for transgenic insects. *Nature* 402:370–371.

Blackman, R. L. 1995. Sex determination in insects. In S. R. Leather and J. Hardie, eds., *Insect Reproduction,* pp. 57–94. Boca Raton, Fla.: CRC Press.

Boffelli, D., K. McAuliffe, D. Ovcharenko, K. D. Lewis, I. Ovcharenko, I. L. Pachter, and E. M. Rubin. 2003. Phylogenetic shadowing of primate sequences to find functional regions of the human genome. *Science* 299:1391–1394.

Carroll, S. B. 2003. Genetics and the making of *Homo sapiens. Nature* 422:849–857.

Carroll, S. B., A. Kopp, J. R. True, and P. Wittkopp. 2001. The development and evolution of pigmentation patterns in *Drosophila. Dev. Biol.* 235:288–295.

Cheng, L. L., L. C. Bartholomay, K. E. Olson, C. Lowenberger, J. Vizioli, S. Higgs, B. J. Beaty, and B. M. Christensen. 2001. Characterization of an endogenous gene expressed in *Aedes aegypti* using an orally infectious recombinant Sindbis virus. *J. Insect Sci.* 1:1–8.

Cline, T. W., and B. J. Meyer. 1996. Vive la difference: Males vs. females in flies vs. worms. *Annu. Rev. Genet.* 30:637–701.

Curtis, D., J. Apfeld, and R. Lehmann. 1995. Nannos is an evolutionarily conserved organizer of anterior-posterior polarity. *Development* 121:1899–1910.

Davidson, E. H. 2001. *Genomic Regulatory Systems. Development and Evolution.* San Diego: Academic Press.

Davis, G. K., and N. H. Patel. 2002. Short, long and beyond: Molecular and embryological aproaches to insect segmentation. *Annu. Rev. Entomol.* 47:669–699.

Denli, A. M., and G. J. Hannon. 2003. RNAi: An ever-growing puzzle. *Trends Biochem. Sci.* 28:196–201.

DeSalle, R. 2002. Bold yet austere. *Bioessays* 24:865–866.

DeSalle, R., and E. Carew. 1992. Phyletic phenocopy and the role of developmental genes in morphological evolution in the Drosophilidae. *J. Evol. Biol.* 5:363–374.

DeSalle, R., and D. Grimaldi. 1993. Phylogenetic pattern and developmental process in *Drosophila. Syst. Biol.* 42:458–475.

DeSalle, R., D. Agosti, M. Whiting, B. Perez-Sweeney, J. Remsen, R. Baker, J. Bonacum, and R. Bang. 1996. Crossroads, milestones, and landmarks in insect development and evolution: Implications for systematics. *Aliso* 14:305–321.

Dykxhoorn, D. M., C. D. Novina, and P. M. Sharp. 2004. Killing the messenger: Short RNAs that silence gene expression. *Nat. Rev. Mol. Cell Biol.* 4:457–467.

Erickson, J. W., and T. W. Cline. 1998. Key aspects of the primary sex determination mechanism are conserved across the genus *Drosophila. Development* 125:3259–3268.

Fraser, M., J. Brusca, G. Smith, and M. Summers. 1985. Transposon mediated mutagenisis of a baculovirus. *Virology* 145:356–361.

Furlong, E. E. M., E. C. Andersen, B. Null, K. P. White, and M. P. Scott. 2001. Patterns of gene expression during *Drosophila* mesoderm development. *Science* 293:1629–1633.

Garcia-Bellido, A., and J. F. de Cellis. 1992. Developmental genetics of wing veination pattern of *Drosophila. Annu. Rev. Genet.* 26:277–304.

Guo, S., and K. J. Kemphues. 1995. *par-1,* a gene required for establishing polarity in *C. elegans* embryos, encodes a putative Ser/Thr kinase that is asymmetrically distributed. *Cell* 81:611–620.

Hacker, U., S. Nystedt, M. P. Barmchi, C. Horn, and E. A. Wimmer. 2003. *Piggybac*-based insertional mutagenesis in the presence of stably integrated *P elements* in *Drosophila. Proc. Natl. Acad. Sci. USA* 100: 7720–7725.

Held, L. I. 2002. *Imaginal Discs: The Genetic and Cellular Logic of Pattern Formation.* London: Cambridge University Press.

Higgs, S., C. T. Oray, K. M. Myles, K. E. Olson, and B. J. Beaty. 1999. Infecting larval arthropods with a chimeric, double subgenomic Sindbis virus vector to express genes of interest. *Biotechniques* 27:908–911.

Holt, R. A., G. M. Subramanian, A. Halpern, G. G. Sutton, R. Charlab, D. R. Nusskern, P. Wincker, A. G. Clark, J. M. C. Ribeiro, R. Wides, S. L. Salzberg, B. Loftus, M. Yandell, W. H. Majoros, D. B. Rusch, Z. Lai, C. L. Kraft, J. F. Abril, V. Anthouard, P. Arensburger, P. W. Atkinson, H. Baden, V. de Berardinis, D. Baldwin, V. Benes, J. Biedler, C. Blass, R. Bolanos, D. Boscus, M. Barnstead, S. Cai, A. Center, K. Chatuverdi, G. K. Christophides, M. A. Chrystal, M. Clamp, A. Cravchik, V. Curwen, A. Dana, A. Delcher, I. Dew, C. A. Evans, M. Flanigan, A. Grundschober-Freimoser, L. Friedli, Z. Gu, P. Guan, R. Guigo, M. E. Hillenmeyer, S. L. Hladun, J. R. Hogan, Y. S. Hong, J. Hoover, O. Jaillon, Z. Ke, C. Kodira, E. Kokoza, A. Koutsos, I. Letunic, A. Levitsky, Y. Liang, J.-J. Lin, N. F. Lobo, J. R. Lopez, J. A. Malek, T. C. McIntosh, S. Meister, J. Miller, C. Mobarry, E. Mongin, S. D. Murphy, D. A. O'Brochta, C. Pfannkoch, R. Qi, M. A. Regier, K. Remington, H. Shao, M. V. Sharakhova, C. D. Sitter, J. Shetty, T. J. Smith, R. Strong, J. Sun, D. Thomasova, L. Q. Ton, P. Topalis, Z. Tu, M. F. Unger, B. Walenz, A. Wang, J. Wang, M. Wang, X. Wang, K. J. Woodford, J. R. Wortman, M. Wu, A. Yao, E. M. Zdobnov, H. Zhang, Q. Zhao, S. Zhao, S. C. Zhu, I. Zhimulev, M. Coluzzi, A. della Torre, C. W. Roth, C. Louis, F. Kalush, R. J. Mural, E. W. Myers, M. D. Adams, H. O. Smith, S. Broder, M. J. Gardner, C. M. Fraser, E. Birney, P. Bork, P. T. Brey, J. C. Venter, J. Weissenbach, F. C. Kafatos, F. H. Collins, and S. L. Hoffman. 2002. The genome sequence of the malaria mosquito *Anopheles gambiae. Science* 298:129–149.

Horn, C., and E. A. Wimmer. 2000. A versatile vector set for animal transgenesis. *Dev. Genes Evol.* 210: 630–637.

Horn, C., N. Offen, S. Nystedt, U. Hacker, and E. A. Wimmer. 2003. *piggyBac*-based insertional mutagenesis and enhancer detection as a tool for functional insect genomics. *Genetics* 163:647–661.

Hughes, C. L., and T. C. Kaufman. 2000. RNAi analysis of *Deformed, proboscipedia* and *Sex combs reduced* in the milkweed bug *Oncopeltus fasciatus:* Novel roles for Hox genes in the hemipteran head. *Development* 127:3683–3694.

Jasny, B. R., S. Orla, M. L. Roberts, and M. Enserink. 2002. Taking a genomic bite of the malaria mosquito. *Science* 298:77–78.

Kaufman T. C., D. W. Severson, and G. E. Robinson. 2002. The *Anopheles* genome and comparative insect genomics. *Science* 298:97–98.

Kelsh, R., R. O. Weinzierl, R. A. White, and M. Akam. 1994. Homeotic gene expression in the locust *Schistocerca:* An antibody that detects conserved epitopes in Ultrabithorax and abdominal-A proteins. *Dev. Genet.* 15:19–31.

Kennerdell, J. R., and R. W. Carthew. 1998. Use of dsRNA-mediated genetic interference to demonstrate that frizzled and frizzled 2 act in the Wingless pathway. *Cell* 95:1017–1026.

Kohler, R. 1994. *Lords of the Fly:* Drosophila *Genetics and the Experimental Life.* Chicago: University of Chicago Press.

Kopp A., and J. R. True. 2002. Evolution of male sexual characters in the oriental *Drosophila melanogaster* species group. *Evol. Dev.* 4:278–291.

Krause G. 1939. Die Eitypen der Insekten. *Biol. Zentralbl.* 59:495–536.

Lewis, D. L., M. A. de Camillis, C. R. Brunetti, G. Halder, V. A. Kassner, J. E. Selegue, S. Higgs, and S. B. Carroll. 1999. Ectopic gene expression and homeotic transformations in arthropods using recombinant Sindbis viruses. *Curr. Biol.* 9:1279–1287.

Lewis, E. B. 1978. A gene complex controlling segmentation in *Drosophila. Nature* 276:565–570.

Lobo, N., X. Li, and M. Fraser. 1999. Transposition of the *piggyBac* element in embryos of *Drosophila melangaster, Aedes aegypti* and *Trichoplusia. Mol. Gen. Genet.* 261:803–810.

Lohe, A. R., and D. L. Hartl. 1996. Germline transformation of *Drosophila virilis* with the transposable element mariner. *Genetics* 143:365–374.

Loukeris, T. G., I. Livadaras, B. Arca, S. Zabalou, and C. Savakis. 1995. Gene transfer into the medfly, *Ceratitis capitata,* with a *Drosophila hydei* transposable element. *Science* 270:2002–2005.

Lozovskaya, E. R., D. I. Nurminsky, D. L. Hartl, and D. T. Sullivan. 1995. Germline transformation of *Drosophila virilis* mediated by the transposable element hobo. *Genetics* 142:173–177.

Ludwig, M. Z., N. H. Patel, and M. Kreitman. 1998. Functional analysis of eve stripe 2 enhancer evolution in *Drosophila:* Rules governing conservation and change. *Development* 125:949–958.

Ludwig, M. Z., C. Bergman, N. H. Patel, and M. Kreitman. 2000. Evidence for stabilizing selection in a eukaryotic enhancer element. *Nature* 403:564–567.

Marcus, J. M. 2001. The development and evolution of crossveins in insect wings. *J. Anat.* 199:211–216.

Marin, I., and B. S. Baker. 1998. The evolutionary dynamics of sex determination. *Science* 281:1990–1994.

McDonald, M. J., and M. Rosbash. 2001. Microarray analysis and organization of circadian gene expression in *Drosophila. Cell* 107:567–578.

Misquitta, L., and B. M. Paterson. 1999. Targeted disruption of gene function in *Drosophila* by RNA interference (RNA-i): A role for nautilus in embryonic somatic muscle formation. *Proc. Natl. Acad. Sci. USA* 96:1451–1456.

Morel C. M., Y. T. Tour, È, B. Dobrokhotov, and A. M. J. Oduola. 2002. The mosquito genome—A breakthrough for public health. *Science* 298:79–80.

Nuesslein-Volhard, C., and E. Wieschaus. 1980. Mutations affecting segment number and polarity in *Drosophila. Nature* 287:795–801.

O'Brochta, D. A., and P. W. Atkinson. 1996. Transposable elements and gene transformation in non-drosophilid insects. *Insect Biochem. Mol. Biol.* 26:739–753.

O'Grady, P. M., and R. DeSalle. 2000. How the fruit fly changed (some of) its spots. *Curr. Biol.* 10:75–77.

Olson, K. E., K. M. Myles, R. C. Seabaugh, S. Higgs, J. O. Carlson, and B. J. Beaty. 2000. Development of a Sindbis virus expression system that efficiently expresses green fluorescent protein in midguts of *Aedes aegyti* following *per os* infection. *Insect Mol. Biol.* 9:57–65.

Oppenheimer, D. I., A. M. Mac Nicol, and N. H. Patel. 1999. Functional conservation of the wingless engrailed interaction as shown by a widely applicable baculovirus misexpression system. *Curr. Biol.* 25:1288–1296.

Panganiban, G. S., M. Irvine, C. Lowe, H. Roehl, L. S. Corely, B. Sherbon, J. K. Grenier, J. F. Fallon, J. Kimble, M. Walker, G. A. Wray, B. J. Swalla, M. Q. Martindale, and S. B. Carroll. 1997. The origin and evolution of animal appendages. *Proc. Natl. Acad. Sci. USA* 94:5126–5166.

Pankratz, M. J., and H. Jackle. 1993. Blastoderm segmentation. In M. Bate and A. Martinez-Arias, eds., *The Development of* Drosophila melanogaster, vol. I. Plainview, N.Y.: Cold Spring Harbor Press.

Patel, N. H. 1994. Developmental evolution: Insights from studies of insect segmentation. *Science* 266: 581–590.

Patel, N. H., E. E. Ball, and C. S. Goodman. 1992. Changing role of *even-skipped* during the evolution of insect pattern formation. *Nature* 357:339–342.

Patel, N. H., B. G. Condron, and K. Zinn. 1994. Pair-rule expression patterns of *even-skipped* are found in both short- and long-germ beetles. *Nature* 367:429–434.

Peloquin, J. J., S. Thibault, R. Staten, and T. Miller. 2000. Germline transformation of pink bollworm (Lepidoptera: Gelechidae) mediated by the *piggyBac* transposable element. *Insect Mol. Biol.* 9:323–333.

Peloquin. J. J., C. R. Lauzon, S. Potter, and T. A. Miller. 2002. Transformed bacterial symbionts reintroduced to and detected in host gut. *Curr. Microbiol.* 45:41–45.

Pistllo, D., N. Skaer, and P. Simpson. 2002. Scute expression in *Calliphora vicinia* reveals an ancestral pattern of longitudinal stripes on the thorax of higher Diptera. *Development* 129:563–572.

Powell, J. R. 1997. *Progress and Prospects in Evolutionary Biology: The* Drosophila *Model.* London: Oxford University Press.

Powell, J. R., and R. DeSalle. 1995. *Drosophila* molecular phylogenies and their uses. *Evol. Biol.* 28:87–138.

Rifkin, S. A., J. Kim, and K. P. White. 2003. Evolution of gene expression in the *Drosophila melanogaster* subgroup. *Nat. Genet.* 33:138–144.

Rohr, K. B., D. Tautz, and K. Sander. 1999. Segmentation gene expression in the moth midge *Clogmia albipunctata* (Diptera, Psychodidae) and other primitive dipterans. *Dev. Genes Evol.* 209:145–154.

Rubin, G. M., and A. C. Spradling. 1982. Genetic transformation of *Drosophila* with transposable element vectors. *Science* 218:348–353.

Ruiz, M. F., M. Esteban, C. Donoro, C. Goday, and L. Sanchez. 2000. Evolution of dosage compensation in Diptera: The *maleless* gene implements dosage compensation in *Drosophila* (Brahcycera suborder) but its homolog in *Sciara* (Nematocera suborder) appears to play no role in dosage compensation. *Genetics* 156:1853–1865.

Rushlow, C., and M. Levine. 1990. The role of *zerknüllt* gene in dorsal-ventral pattern formation in *Drosophila. Adv. Genet.* 27:277–307.

Sander, K. 1976. Specification of the basic body pattern in insect embryogenesis. *Adv. Insect Physiol.* 12: 125–238.

Sander, K. 1996. Variants of embryonic patterning mechanisms in insects—Hymenoptera and Diptera. *Semin. Cell Dev. Biol.* 7:573–582.

Schmidt-Ott, U. 2000. The amnioserosa is an apomorphic character of cyclorrhaphan flies. *Dev. Genes Evol.* 210:373–376.

Schmidt-Ott, U., M. Gonzalez-Gaitan, H. Jackle, and G. M. Technau. 1994. Number, identity, and sequence of the *Drosophila* head segments as revealed by neural elements and their deletion patterns in mutants. *Proc. Natl. Acad. Sci. USA* 91:8363–8367.

Schutt, C., and R. Nothinger. 2000. Structure, function and evolution of sex-determining systems in Dipteran insects. *Development* 127:667–677.

Scott, M. P., A. J. Weiner, T. I. Hazelrigg, B. A. Polisky, V. Pirrotta, F. Scalenghe, and T. C. Kaufman. 1982. The molecular organization of the *Antennapedia* locus of *Drosophila. Genetics* 105:345–356.

Shaw, P. J., A. Salameh, A. P. McGregor, S. Bala, and G. A. Dover. 2001. Divergent structure and function of the bicoid gene in Muscoidea fly species. *Evol. Dev.* 3:251–262.

Simpson, P. 2002. Evolution of development in closely related species of flies and worms. *Nat. Rev. Genet.* 3:907–910.

Simpson, P., R. Woehl, and K. Usui. 1999. The development and evolution of bristle patterns in Diptera. *Development* 126:1349–1364.

Skaer, N., D. Pistillo, J.-M. Gibert, P. Lio, C. Wolbeck, and P. Simpson. 2002. Gene duplication at the achaeteñscute complex and morphological complexity of the peripheral nervous system in Diptera. *Trends Genet.* 18:399–405.

Sommer, R. J., and D. Tautz. 1994. Expression patterns of twist and snail in *Tribolium* (Coleoptera) suggest a homologous formation of mesoderm in long and short germ band insects. *Dev. Genet.* 15: 32–37.

Stark, J., J. Bonacum, J. Remsen, and R. DeSalle. 1999. The evolution and development of dipteran wing veins: A systematic approach. *Annu. Rev. Entomol.* 44:97–129.

Stauber, M., H. Jackle, and U. Schmidt-Ott. 1999. The anterior determinant *bicoid* of *Drosophila* is a derived *Hox* class 3 gene. *Proc. Natl. Acad. Sci. USA* 96:3786–3789.

Stauber, M., H. Taubert, and U. Schmidt-Ott. 2000. Function of *bicoid* and *hunchback* homologs in the basal cyclorrhaphan fly *Megaselia* (Phoridae). *Proc. Natl. Acad. Sci. USA* 97:10844–10849.

Stauber, M., A. Prell, and U. Schmidt-Ott. 2002. A single *Hox3* gene with composite bicoid and zerknüllt expression characteristics in non-Cyclorrhaphan flies. *Proc. Natl. Acad. Sci. USA* 99:274–279.

Sturtevant, M. A., and E. Bier. 1995. Analysis of the genetic hierarchy guiding wing vein development in *Drosophila. Development* 121:785–801.

Tamura,T., C. Thibert, R. Corrine, T. Kanda, E. Abraham, M. Kamba, N. Komoto, J. Thomas, B. Mauchamp, G. Chavency, P. Sirk, M. Fraser, J. Pruddhomme, and P. Couble. 2000. Germline transformation of the silkworm *Bombyx mori* L. using a *piggyBac* transposon-derived vector. *Nat. Biotech.* 18:81–84.

Tautz, D. 1992. Redundancies, development and the flow of information. *BioEssays* 14:263–266.

Tautz, D., and R. Sommer. 1995. Evolution of segmentation genes in insects. *Trends Genet.* 11:23–27.

Tautz, D., M. Friedrich, and R. Schröder. 1994. Insect embryogenesis—What is ancestral and what is derived? In M. Akam, P. Holland, P. Ingham, and G. Wray, eds., *Development 1994 Supplement*, pp. 193–199. Cambridge: The Company of Biologists Limited.

Toma, D. P., K. P. White, J. Hirsch, and R. J. Greenspan. 2002. Identification of genes involved in *Drosophila melanogaster* geotaxis, a complex behavioral trait. *Nat. Genet.* 31:349–353.

Treier, M., C. Pfeifle, and D. Tautz. 1989. Comparison of the gap segmentation gene *hunchback* between *Drosophila melanogaster* and *Drosophila virilis* reveals novel modes of evolutionary change. *EMBO J.* 8:1517–1525.

True, J. R., and S. B. Carroll. 1998. Genetic analysis of melanin patterning in *Drosophila. Dev. Biol.* 198: 199–207.

True, J. R., K. A. Edwards, D. Yamamoto, and S. B. Carroll. 1999. *Drosophila* wing melanin patterns form by vein-dependent elaboration of enzymatic prepatterns. *Curr. Biol.* 9:1382–1391.

Wheeler, M. R. 1981. The Drosophilidae: A taxonomic overview. In M. Ashburner, H. L. Carson, and J. N. Thompson, Jr., eds., *The Genetics and Biology of* Drosophila, vol. 3a., pp. 1–97. New York: Academic Press.

White, K. P., S. A. Rifkin, P. Hurban, and D. S. Hogness. 1999. Microarray analysis of *Drosophila* development during metamorphosis. *Science* 286:2179–2184.

White, M. J. D. 1972. *Animal Cytology and Evolution.* Cambridge: Cambridge University Press.

Wilkins, A. 2002. *The Evolution of Developmental Pathways.* Sunderland, Mass.: Sinauer.

Wimmer, E. A. 2003. Innovations: Applications of insect transgenesis. *Nat. Rev. Genet.* 4:225–232.

Wittkopp, P. J., J. R. True, and S. B. Carroll. 2002a. Reciprocal functions of the *Drosophila* Yellow and Ebony proteins in the development and evolution of pigment patterns. *Development* 129:1849–1858.

Wittkopp, P. J., K. Vaccaro, and S. B. Carroll. 2002b. Evolution of Yellow gene regulation and pigmentation in *Drosophila. Curr. Biol.* 12:1547–1556.

Wittkopp, P. J., B. L. Williams, J. E. Selegue, and S. B. Carroll. 2003. *Drosophila* pigmentation evolution: Divergent genotypes underlying convergent phenotypes. *Proc. Natl. Acad. Sci. USA* 100:1808–1813.

Wright, T. 1987. The genetics of biogenic amine metabolism, sclerotization, and melanization in *Drosophila melanogaster. Adv. Genet.* 24:127–222.

Wulbeck, C., and P. Simpson. 1999. The development and evolution of bristle patterns in Diptera. *Development* 126:1349–1364.

Wulbeck, C., and P. Simpson. 2000. Expression of achaete-scute homologues in discrete proneural clusters on the developing notum of the medfly *Ceratitis capitata*, suggests a common origin for the stereotyped bristle patterns of higher Diptera. *Development* 127:1411–1420.

Wulbeck, C., and P. Simpson. 2002. The expression of pannier and achaete-scute homologues in a mosquito suggests an ancient role of *pannier* as a selector gene in the regulation of the dorsal body pattern. *Development* 129:3861–3871.

Zdobnov, E. M., C. von Mering, I. Letunic, D. Torrents, M. Suyama, R. R. Copley, G. K. Christophides, D. Thomasova, R. A. Holt, G. M. Subramanian, H.-M. Mueller, G. Dimopoulos, J. H. Law, M. A. Wells, E. Birney, R. Charlab, A. L. Halpern, E. Kokoza, C. L. Kraft, Z. Lai, S. Lewis, C. Louis, C. Barillas-Mury, D. Nusskern, G. M. Rubin, S. L. Salzberg, G. G. Sutton, P. Topalis, R. Wides, P. Wincker, M. Yandell, F. H. Collins, J. Ribeiro, W. M. Gelbart, F. C. Kafatos, and P. Bork. 2002. Comparative genome and proteome analysis of *Anopheles gambiae* and *Drosophila melanogaster. Science* 298:149–159.

CHAPTER SIX

Transposable Elements and the Evolution of Dipteran Genomes

Margaret G. Kidwell

Transposable elements (TEs) are mobile DNA sequences that have been identified in a broad range of species in all kingdoms of life. Although originally discovered around 50 years ago (McClintock 1948) in plants, it was not until about 20 years later that mobile elements were unambiguously identified in other eukaryotic species such as *Drosophila melanogaster* (e.g., Finnegan et al. 1978). In the past 20 to 30 years, this model organism has played a leading role in advancing our knowledge of eukaryotic TEs. However, their study in a broad range of dipteran species awaited the routine use of sophisticated molecular techniques in the 1990s. The potential applicability of TEs for control of dipteran pest species, such as mosquitoes, has been an important factor in spurring this research, which has increased dramatically during the past decade, as shown in Table 6.1. The acceleration in nondrosophilid TE research is expected to continue following the recent sequencing of the genome of the malaria mosquito *Anopheles gambiae* (Holt et al. 2002). Comparative analyses of the genomic complements of TEs in *An. gambiae* and *D. melanogaster* are now producing a wealth of new information and insights (see Ashburner, Chapter 4).

The purpose of this chapter is to review the importance of research on Diptera in understanding the evolution and impact of TEs in natural populations. In addition, techniques that use dipteran TEs in basic and applied research are described and evaluated.

Classification of TEs

Most TEs can be assigned to two main classes, according to their mechanism of transposition. Class I elements encode a reverse transcriptase (RT) and employ an RNA-mediated mode of transposition. Class II elements, the transposons (sensu strictu), use a DNA-based mode of transposition. Both autonomous and nonautonomous members are found in most TE families of both classes. Autonomous elements are able to catalyze their own transposition, whereas nonautonomous elements depend for their mobility on autonomous elements from the same family.

CLASS I ELEMENTS

Class I elements are members of the retroelements that include both retrotransposons and retroviruses (Malik et al. 2000). Retroelements use an RNA-mediated mode of transposition and autonomous members encode an RT. Class I elements are amplified by the transposition process and have a high potential for increase in copy number. The vast majority of Class I

TABLE 6.1. Number of Citations to Transposable Element Studies in Various Dipteran Taxa[a]

Taxon	Before 1990	1990–2002	Total
Drosophila melanogaster	274	579	853
Drosophila other than *melanogaster*	186	470	656
Mosquitoes			
Anopheles spp.	2	28	30
Culex spp.	0	5	5
Aedes spp.	1	19	20
Tephritid fruit flies	0	16	16
Chironomid midges	2	10	12
Musca domestica (housefly)	0	12	12
Lucilia cuprina (sheep blowfly)	0	5	5
Mayetiola destructor (Hessian fly)	0	3	3
Other	0	4	4
TOTAL	465	1151	1616

[a] Included in the Pub Med database, 30 June 2002.

elements can be assigned to two subclasses, the LTR retrotransposons that are characterized by direct, long terminal repeats (LTRs), and the non-LTR retrotransposons (or retroposons) that lack terminal repeats. Here these elements are referred to as LTR retrotransposons and non-LTR retrotransposons, respectively. A third clade, the *Penelope*-like elements (PLEs), is clearly phylogenetically distinct from the LTR and non-LTR retrotransposons (Arkhipova et al. 2003) and is represented in Diptera by *Penelope* elements in *D. virilis* and other species of the *virilis* group (Evgen'ev et al. 1997). PLEs have unusual structural features and the ability to retain introns (Arkhipova et al. 2003).

LTR Retrotransposons. About 20 years ago, the *copia*-like elements of *D. melanogaster* were identified as one clearly distinct class of middle repetitive DNA (Rubin 1983). At that time, at least seven repeated sequence families, *copia, 412, 297, mdg1, mdg3, B104,* and *gypsy,* were identified. These elements ranged in size from 5 to 8.5 kb in length, with copy numbers varying from 10 to 100 per genome. Today, based on phylogenetic analysis of their RT sequences, four main lineages of LTR retrotransposons and two closely related clades of viruses are recognized (Eickbush and Malik 2002). Three out of four of these LTR retrotransposon lineages, *Ty1-copia, BEL,* and *Ty-3 gypsy,* are represented in dipteran species (see below for details).

Non-LTR Retrotransposons. Long interspersed nuclear elements (*LINE*s) and short interspersed nuclear elements (*SINE*s) were first identified as two basic structural types of interspersed repetitive elements (Singer 1982). Subsequently, *LINE*-like elements were found to encode autonomous members and are now recognized as representing one of the two major types of autonomous retroelements, LTR retrotransposons and non-LTR retrotransposons. *SINE*s only include nonautonomous elements.

Non-LTR retrotransposons (or *LINE*-like elements) are subdivided into five major groups: *R2, L1, RTE, I,* and *Jockey* (Eickbush and Malik 2002). Three of these clades, *R2, I,* and *Jockey,* are represented in dipteran species (see below for details).

CLASS II ELEMENTS

Class II elements are divided into three broad subclasses: (1) elements that transpose by a cut-and-paste mechanism and in which a DDE (aspartic acid, glutamic acid, conserved amino acid triad) signature is present, exemplified by *mariner* in *D. mauritiana;* (2) elements that transpose by a cut-and-paste mechanism and in which a DDE signature is absent, exemplified by the *P* and *hobo* elements in *Drosophila;* and (3) elements that transpose using a rolling circle (RC) transposition mechanism similar to that found in prokaryotes (Kapitonov and Jurka 2001). The first two subclasses are quite prevalent in Diptera. The miniature inverted repeat transposable elements (MITEs) are now considered nonautonomous members of the Class II elements (Le et al. 2000; Turcotte et al. 2001).

Characterization and Distribution of Transposable Elements in Diptera

CLASS I ELEMENTS

LTR Retrotransposons. Three lineages of LTR retrotransposons have been identified, belonging to the *Ty1-copia, gypsy,* and *BEL* groups.

The Ty1-copia group of LTR retrotransposons. The *copia* element of *D. melanogaster* was one of the first LTR elements to be identified in the 1970s. Along with the *Ty1* element of the yeast *Saccharomyces cerevisiae,* it gave its name to the *Ty-copia* group of retroelements. Based on the divergence of their RT sequences, this group represents the most ancient lineage of LTR retrotransposons (Eickbush and Malik 2002). *Ty-copia* elements contain open reading frames similar to the *gag* and *pol* genes of retroviruses. *Copia* elements are broadly distributed in dipteran species and are one of the most intensively studied elements in *D. melanogaster.* These elements are normally inherited vertically, but one case of recent horizontal transfer, from *D. melanogaster* to *D. willistoni,* has been well documented (Jordan et al. 1999).

Gypsy LTR retrotransposons with viruslike properties. It has been known for some time that retroviruses and LTR retrotransposable elements share a number of structural genes, such as *gag* and *pol.* In addition, all retroviruses but only a few LTR retrotransposons have an envelope (*env*) gene. About 20 years ago, there were two opposing theories about the relationship between retroviruses and LTR retrotransposable elements. One school of thought (e.g., Finnegan 1983) argued that retroviruses came first and LTR elements evolved from them. Recent phylogenetic analyses suggest that LTR retrotransposons can acquire additional reading frames that enable them to mediate infection, providing a definitive answer to this old question. Moreover, this acquisition took place a number of independent times. One of the most well-studied examples is provided by the *gypsy* and related LTR retrotransposons (the insect errantiviruses) that have acquired their envelope gene from a class of insect baculoviruses (Malik et al. 2000). Errantiviruses with dipteran hosts include *17.6, 297, gypsy/mdg4, Zam, Idefix Tirant,* and *412* in many *Drosophila* species, *Tv1* in *D. virilis,* and the *tom* element in *D. ananassae.* The envelope protein present in the *Osvaldo* TE of *D. buzzatii* and *D. koepferae* represents a viral acquisition that was independent of that of the *gypsy* group (Pantazidis et al. 1999; Malik et al. 2000).

The *BEL* group of LTR retrotransposons. The *BEL* lineage was named after an abundant, active element in *D. melanogaster* (Bell et al. 1985). Based on the RT domain, the *BEL* lineage is second only to the *Ty1/copia* lineage in terms of age (Eickbush and Malik 2002). Members of this lineage in *D. melanogaster* include the *BEL, Roo,* and *Mazi* families (Bowen and McDonald 2001).

Non-LTR Retrotransposons. Three clades of non-LTR retrotransposons, *R2, I,* and *Jockey,* are represented in dipteran species. The *R2* elements are the oldest and most broadly distributed in the Diptera. All elements in this group insert into specific locations in their host genome. The *I* clade includes the *R1* elements from *Drosophila* species, the *RT* elements of *Anopheles,* the *LOA* element from *D. silvestris,* the *Lian* and *Mosqu1* elements of *Aedes aegypti,* the *TRIM* element of *D. miranda,* the *Bilboa* element of *D. subobscura,* and the *I* elements of *D. melanogaster* and *D. teissieri.* The *Jockey* clade includes the *D. melanogaster Jockey, G, Doc, F, BS,* and *TART* elements, as well as *JuanA* from *Aedes* species and *JuanC* from *Culex* species, and *T1* and *Q* from *An. gambiae.*

R1 and *R2* non-LTR elements. In the 1970s, sequences referred to as type I and type II insertions were identified within various copies of cloned *rRNA* genes in *D. melanogaster* (Glover and Hogness 1977; Wellauer and Dawid 1977). Their identity as a subclass of retroelements with properties very different from those of the LTR retrotransposons was not recognized immediately. Only later was it recognized that the *Drosophila R1* and *R2* elements were the first TEs to be isolated by recombinant DNA techniques (Eickbush 2002). Identification, and characterization of *R1* and *R2* homologs in the silk moth, *Bombyx mori,* allowed further progress in understanding the origin, structure, and distribution of the *R1* and *R2* elements (Eickbush 2002). *R1* and *R2* elements appear to have occupied a specialized niche within the 28S rRNA region of the genomes of virtually all arthropods throughout their 500-My history (Eickbush 2002). These elements have the ability to turn off rRNA synthesis from a promoter located more than 6 kb upstream of their location in the rDNA unit. Despite their similarities in behavior, the *R1* and *R2* element families are phylogenetically diverged from one another (Eickbush and Malik 2002). The *R1* elements are members of the *I* lineage of non-LTR retrotransposons, whereas the *R2* elements belong to the *R2* lineage (Eickbush and Malik 2002).

I elements in *D. melanogaster.* *I* elements are *LINE*-like elements whose activity was found to be responsible for the inducer-reactive (I-R) system of hybrid dysgenesis in *D. melanogaster* (Bucheton et al. 1984). More than a decade before the characterization of *I* elements at the molecular level, observations were first made of genetic abnormalities that occurred when particular strains of flies, called inducer (I) and reactive (R), were crossed together (Picard et al. 1972). The pattern of occurrence and type of abnormality were similar, but not identical, to those produced by the P-M system of hybrid dysgenesis caused by the *P* element (see below).

SINEs. In contrast to mammals in which *SINE* elements are highly abundant, very few TEs of this type have been reported in *Drosophila* and other Diptera. Exceptions include the highly repetitive *SINEs Feilai* in *Ae. aegypti* (Tu 1999), *SINE200* (Holt et al. 2002), and the *Maque* element (Tu 2001b) in *An. gambiae,* as well as the *DINE-1* element of *D. melanogaster* (Locke et al. 1999) and the *Cp1* element of *Chironomus pallidivittatus* (Liao et al. 1998).

CLASS II ELEMENTS

IS630-Tc1-mariner *Super Family.* The eukaryotic DNA transposon families *Tc1* and *mariner,* and the bacterial *IS630* element and its relatives in prokaryotes and ciliates make up a superfamily of prokaryotic and eukaryotic TEs based on overall sequence similarities and common

TA dinucleotide insertion target (Capy et al. 1996; Shao et al. 2001). *Tc1/mariner* elements are small, only about 1.3–2.4 kb in length, and contain a single gene encoding a transposase. The typical DDE or DDD (aspartic acid conserved amino acid triad) motif found in most transposases and integrases is contained in the transposase protein. Inverted repeats are found at the element termini. Integration into the host genome always occurs at a TA host sequence that is replicated to flank both element termini on insertion.

The discovery of the *mariner* element (Jacobson et al. 1986) grew out of earlier observations of an unstable allele, white peach, at the *white* locus of *D. mauritiana*. The vast majority of *mariner* and *mariner*-like elements are nonautonomous. However, an intact, autonomous element, *Mos1*, was also discovered in *D. mauritiana*, enabling the later development of this element for germ-line transformation of a broad array of heterologous species (see later).

P Elements. *P* elements are one of the most intensively studied families of TEs in terms of their molecular and genetic properties and their evolutionary history. The first observed manifestations of the activity of these elements were male recombination, mutation, and sterility in interstrain crosses of *D. melanogaster*. *P* elements are the causative agent of P-M hybrid dysgenesis (Bingham et al. 1982). They were transferred horizontally from *D. willistoni* to *D. melanogaster* and their invasion into the latter species during the middle of the twentieth century was tracked in natural populations (Kidwell 1983; Anxolabéhère et al. 1988). Canonical *P* elements are restricted to a few New World species, but diverged subfamilies have a much more widespread distribution in Diptera and beyond. The majority of elements are nonautonomous. They are abundant in species of the subgenus *Sophophora* and have been identified sporadically in the subgenus *Drosophila*. Outside of *Drosophila*, *P* elements have been found in *Liodrosophila*, *Lordiphosa*, and *Scaptomyza* (Anxolabéhère et al. 1985; Lansman et al. 1985; Daniels et al. 1990b; Haring et al. 2000). *P* elements have also been identified in the blowfly, *Lucilia cuprina* (Perkins and Howells 1992), the house fly (*Musca domestica*), and anopheline mosquitoes (Sarkar et al. 2003), and are abundant in *An. gambiae* (Oliveira de Carvalho et al. 2004). *P* elements are perhaps best known for their applications in the genetic engineering of *Drosophila* (see below).

The hAT Elements. The *hAT* superfamily of autonomous transposons obtained its name from the *hobo* element (first identified in *Drosophila*), *Ac* in maize, and *Tam3* in the snapdragon *Anthirinhum majus*. These elements typically encode short terminal inverted repeats and duplicate 8 bp of host target site DNA upon transposition.

Within Diptera, a number of *hAT* superfamily members have been identified. These include *Homer* and the nonautonomous element *HLE* from the tephritid fly *Batrocera tryoni* (Pinkerton et al. 1999), *Hermes* from the housefly *Musca domestica* (Warren et al. 1994), *Hermit* from the blowfly *Lucilia cuprina* (Coates et al. 1996), and *hopper* elements from the tephritid *Batrocera dorsalis* (Handler and Gomez 1997; Handler 2003). Partial *hAT*-like sequences have also been derived from the Australian bush fly *Musca vetustissima* (Warren et al. 1995).

Within *Drosophila*, the *hobo* element (McGinnis et al. 1983) is the only *hAT* element so far identified. A Southern hybridization screen of *hobo* elements from 134 taxa within the genus *Drosophila* (Daniels et al. 1990a) showed that *hobo* elements were restricted to the *melanogaster* and *montium* subgroups of the *melanogaster* species group. Only *D. melanogaster* and its sibling species, *D. simulans* and *D. mauritiana,* contained potentially complete *hobo* elements.

MITEs. Within Diptera, MITEs appear to be abundant in the mosquito species *An. gambiae, Ae. aegypti,* and *Culex pipiens.* Ten MITE families were reported in *An. gambiae* (Besansky et al. 1996; Tu 2001a), and one in *An. stephensi* (Luckhart and Rosenberg 1999). Twelve families have been described in *Ae. aegypti* (Tu 1997, 2000, 2001b), and three families in *C. pipiens* (Feschotte and Mouches 2000a; Feschotte et al. 2002).

Although abundant in mosquitoes, MITEs are unusually rare in *Drosophila.* Possible exceptions are the *Dm-mPogo* and *SGM-IS* elements (Feschotte et al. 2002). *Dm-mPogo* has dinucleotide (TA) target site duplications, and perfect 21-bp (or imperfect 25-bp) long terminal inverted repeats and is essentially identical to the sequence of the autonomous *Pogo* element. *Dm-mPogo* may therefore be a severely truncated nonautonomous version of the *Pogo* element family (Kapitonov and Jurka 2002). Likewise, *SGM-IS* sequences have recently been identified as members of the *Mini-me* element family, common among *Drosophila* (Wilder and Hollocher 2001). Two true MITE families, *Vege* and *Mar,* have, however, recently been identified in *D. willistoni* (Holyoake and Kidwell 2003). These elements appear to be associated with the *hAT* super family of TEs.

Foldback *Elements. Foldback* (FB) elements were first discovered in *D. melanogaster* (Truett et al. 1981). These elements are very broadly distributed in *Drosophila* and other dipteran species. They have unusually long inverted terminal repeats that under certain conditions can fold back to form branched DNA structures. These repeats are made up of two domains, an outer tandemly arranged subrepeat and an inner nonrepetitive region. FB elements can potentially encode three proteins, but at this time, their function is uncertain.

In *Drosophila,* FB elements are responsible for diverse chromosomal rearrangements. They are also frequently associated with sites of DNA breakage. Both properties serve to scramble host genomes. The genome reshuffling associated with an FB-mediated inversion in *D. buzzatii* (see below) provides a dramatic natural example of TE-induced genomic rearrangements.

Contributions of Diptera to TE Studies

A number of *Drosophila* species have provided excellent experimental systems for the study of TEs. The early emphasis on *Drosophila* can be partially attributed to the wealth of information that was available on the classical genetics of a number of these species. In addition, the giant polytene salivary gland chromosomes of *Drosophila* larvae allow the estimation of TE numbers and locations, using the technique of in situ hybridization. In the 40 years preceding the development of molecular methods of analysis in the late 1960s and 1970s, a number of independent observations of highly mutable *Drosophila* loci were reported (Crow 1988).

HYBRID DYSGENESIS

The term "hybrid dysgenesis" describes a group of abnormal traits, such as high mutation rates and sterility, that were first observed independently in France, Australia, and the United States in the early 1970s (Picard et al. 1972; Kidwell and Kidwell 1975; Sved 1976). When two interacting strains of *Drosophila* are crossed in reciprocal combinations, hybrid dysgenesis is usually observed in only one direction. The reciprocal cross produces normal progeny. The causes of hybrid dysgenesis have subsequently been traced to the activation of several independent families of TEs, including the *P, I,* and *hobo* elements in *D. melanogaster* and the *Penelope, Ulysses, Helena, Paris,* and *Telemac* elements in *D. virilis.* More recently, some of these elements have been developed for use in genetically manipulating genes, both within and between species, as described below.

Hybrid dysgenic traits include high frequencies of offspring inviability, male recombination (which does not normally occur in *Drosophila*), mutations, chromosomal aberrations, and transmission ratio distortions (Kidwell et al. 1977). These traits result from instability of the germ-line in F_1 hybrids. The DNA in the soma, the body cells, is usually not affected.

In addition to *Drosophila,* several potential examples of hybrid dysgenesis have been reported in other dipterans, including the fly *Chironomus* (Hagele and Oschmann 1987) and the Mediterranean fruit fly, *Ceratis capitata* (Torti et al. 1994), but few of these have been clearly linked to the activation of TEs. It is not clear why hybrid dysgenesis has so far been observed almost exclusively in *Drosophila.*

HORIZONTAL TRANSFER

Although it has been recognized for some time that horizontal transfer is common in bacteria and has played an important role in the evolution of the eukaryotic cell, the recent occurrence of this phenomenon in eukaryotes is less well known. In fact, several dipteran TEs provide some of the best examples of eukaryotic horizontal transfer, including the *Drosophila P* element, whose invasion of natural populations was actually documented during the middle of the twentieth century (Kidwell 1994). In addition, there is evidence that the *P* element has transferred multiple times during its past evolutionary history (Silva and Kidwell 2000). Moreover, invasions of both *hobo* (another Class II transposon) and the *I* elements (a non-LTR retroelement) occurred in *D. melanogaster* a little earlier than did the *P* element invasion. Such a coincidence is likely to be related to the spread of *D. melanogaster* into the New World only about 200 years ago (Engels 1992). Members of another transposon family, *mariner,* originally identified in *D. mauritiana,* have subsequently been shown to have potentially transferred horizontally across a broad range of taxonomic boundaries (Robertson 1997).

Evidence has also been presented for the horizontal transfer and recent invasion into new species of other dipteran TEs. These include the *Penelope* element into *D. virilis* (Evgen'ev et al. 2000a), the retroelement *copia,* from *D. willistoni* to *D. melanogaster* (Jordan et al. 1999), and *copia*-like retrotransposons in *Anopheles* species in Thailand (Rongnoparut et al. 1998).

TES IN *DROSOPHILA MELANOGASTER*

The first dipteran genome to be sequenced was that of *D. melanogaster* (Adams et al. 2000). Initial estimates of the abundance of TEs were applicable to only the euchromatic fraction and ignored the heterochromatic third of the genome, in which TEs are especially abundant. Soon after the release of the initial sequence, 54 TE families were identified in the sequenced portion of the *D. melanogaster* genome (Rizzon et al. 2002), representing only 2% of the euchromatic fraction. However, a subsequent study provided evidence that TEs are at least three times more abundant than the earlier estimates, accounting for at least 6% and 60% of the euchromatic and heterochromatic portions of the genome, respectively (Kapitonov and Jurka 2003).

Twenty-eight families of LTR retrotransposons were identified in the *D. melanogaster* genome, representing three out of four of the identified superfamilies; namely, the *copia, gypsy,* and *BEL* groups (Bowen and McDonald 2001). Members of seven different clades of non-LTR retrotransposons were also identified (Kapitonov and Jurka 2003), with the majority of families belonging to the *Jockey* clade (e.g., *Doc,* the *F* element). Also represented are the *R2* clade (e.g., *R2Dm*) and the *I* clade (e.g., *R1Dm,* the *I* element). Telomeres in *D. melanogaster* are composed of multiple copies of two non-LTR retrotransposons families, *HeT-A* and *TART* (the latter a member of the *Jockey* clade), instead of the short DNA repeats generated by telomerase in most organisms. However, the two telomeric transposable elements have very different patterns of transcription (Danilevskaya et al. 1999), suggesting that they are not closely related. Surprisingly, the members of LTR retrotransposon families show less than 1% divergence from one another (Bowen and McDonald 2001), strongly suggesting recent transposition of these elements well after this species diverged from its closest relative, *D. simulans,* and a high frequency of element loss over evolutionary time. The ten families of DNA transposons found in *D. melanogaster* include *P, hobo* (see the earlier section on hybrid dysgenesis), *Hoppel (1360), Bari, I, Foldback, HB1, Hopper, Pogo S,* and *Vivi* (Bartolome et al. 2002; Rizzon et al. 2002).

The relationship between the density of TEs and recombination rates in various parts of the *D. melanogaster* genome has been examined (Rizzon et al. 2002). Although the density of LTR and non-LTR retrotransposons was high in regions with low recombination, no clear-cut association between TE density and recombination rate was demonstrated. However, the density of DNA transposons was significantly negatively correlated with recombination rate. This suggests, but does not prove, that transposons may be subject to negative selection (Rizzon et al. 2002).

TES IN *DROSOPHILA* OTHER THAN *MELANOGASTER*

In addition to *D. melanogaster,* TEs have been extensively studied in many *Drosophila* species. However, in the absence of sequence data, no detailed comparisons among them are currently possible. Different *Drosophila* species groups appear to be associated with different suites of elements, and there is wide variation in the TE compositions of even sibling species. For example, it has been known for some time that the amount of dispersed, middle repetitive DNA in *D. simulans* is only about one-third of that of its sibling species *D. melanogaster* (Dowsett and Young 1982). Consistent with this, *D. melanogaster* has a higher average chromosomal insertion site per genome than *D. simulans* for 29 out of 34 TE families examined (Vieira et al. 1999). Although space constraints do not allow systematic coverage of the TE complements of other *Drosophila* taxa, pertinent information about these is included in several sections of this chapter.

TEs in Mosquitoes. TEs have been characterized widely in mosquitoes, being found in the anopheline, culicine, and *Aedes* groups.

Anopheline mosquitoes. During the past decade, particularly the past five years, spectacular progress has been made in identifying and describing TEs in species of the *An. gambiae* complex (see Table 6.2 for a summary). The recent sequencing of the *An. gambiae* genome

(Holt et al. 2002) is providing opportunities for analysis that have previously only been possible in *D. melanogaster.* Initially about 40 different TE families were identified (Holt et al. 2002) in the *An. gambiae* genome, but subsequently more comprehensive search methods indicate that the actual number of families was considerably larger. For example, 12 types of non-LTR retrotransposons were initially identified in the *An. gambiae* genome (Holt et al. 2002), but subsequently, Biedler and Tu (2003) found over 100 families of the same type using a reiterative and comprehensive strategy. This level of diversity is unprecedented in any other species that has been examined.

In *Anopheles* species, TE family copy number appears to be relatively low compared to that of *Ae. aegypti,* consistent with differences in genome size. Although it is too early to make any major generalizations, *An. gambiae* TEs tend to be located "within islands of short-period interspersed repetitive DNA in a sea of long-period interspersed, mostly unique sequence DNA" (Hill et al. 2001: 215). Chromosomal paracentric inversions are a characteristic of *An. arabiensis,* a member of the *An. gambiae* complex. Identification of a TE named *Odysseus* at the distal breakpoint of one of these inversions suggested that formation of the inversion might have been related to TE activity (Mathiopoulos et al. 1998).

Because of the urgent need to provide improved methods of control of malaria vectors, considerable effort has been put into developing efficient transformation systems for anopheline mosquitoes during the past decade. Consequently, *An. stephensi* has recently been successfully transformed by the *Minos* element from *D. hydei* (Catteruccia et al. 2000a) and both *An. stephensi* and *An. gambiae* have been transformed by the *piggyBac* element from the Lepidopteran species *Trichoplusia ni* (Grossman et al. 2001; Nolan et al. 2002).

Aedes mosquitoes. Two families of LTR retrotransposons and at least 11 families of non-LTR retrotransposons have been identified in *Ae. aegypti* (Table 6.2). Approximately 59,000 copies of the *SINE* family *Feilai* are present in this species, equivalent to 2% of the entire genome (Tu 1999). Class II elements are represented by many families that are often present in high copy numbers (Table 6.2). *Ae. aegypti* has now been successfully transformed with three different TEs, *mariner* (Coates et al. 1998), *Hermes* (Jasinskiene et al. 1998), and *piggyBac* (Kokoza et al. 2001).

Culicine mosquitoes. Although successful transformation with a *Hermes*-based TE has been recently achieved (Allen et al. 2001), studies of endogenous TEs in culicine mosquitoes have been relatively few and have so far been restricted to *C. pipiens.* Many full-length copies of *Juan-C,* a *LINE*-like element, have been described in this species (Agarwal et al. 1993). These elements are most similar in their sequence to the *Juan-A* elements identified in *Aedes* species (Mouches et al. 1992), but they also have significant sequence similarity with *Drosophila LINEs*, such as *Jockey.* Also found in *C. pipiens* are the *CM-gag* elements that are similar in genetic organization to the *Het-A* elements of *D. melanogaster* (Bensaadi-Merchermek et al. 1997), and *Mimo,* a family of MITEs that appear to have recently amplified in this vector species (Feschotte and Mouches 2000b).

TEs in Tephritid Fruit Flies. Similar to mosquitoes, interest in mobile elements in the tephritid fruit flies has been spurred by the search for efficient mechanisms for genetic transformation of these pests of fruit crops. Although the first successful transformation of the Mediterranean fruit fly *Ceratitis capitata* occurred in 1995 (see section below), a number of

TABLE 6.2. Endogenous TEs in *Anopheles, Aedes,* and *Culex* Mosquitoes

	Anopheles			*Aedes*			*Culex*		
Element	*Species*	Name (Copy Number)	Reference	*Species*	Name (Copy Number)	Reference	*Species*	Name (Copy Number)	Reference
Class I									
LTR	*gambiae*	*Ozymandias* (20)	Lander et al. 2001	*aegypti*	*Mosqcopi* (<10) *Mosqninja* (ND)	J. Tu, pers. comm.			
	gambiae	*Moose*	Biessmann et al. 1999						
	gambiae	*Mtanga-Y*	Rohr et al. 2002						
	gambiae	8 families	Holt et al. 2002						
	gambiae	10 *mdg-1* families	Tubio et al. 2004						
Non-LTR	*gambiae*	*T1Ag* (100)	Besansky 1990	*aegypti*	*Lian* (6)	Tu et al. 1998	*pipiens*	*Juan-C* (many)	Agarwal et al. 1993
		Q (20) (63)	Besansky et al. 1994		*Mosqul* (14)	Tu 1999	*pipiens*	*CM-gag*	Bensaadi-Merchermek et al. 1997
		Vash (20) *Guildenstern* *JuanAg*	Lander et al. 2001	*aegypti, albopictus, polynesiensis*	*Juan-A*	Mouches et al. 1992			
		RT1 in 28S rRNA	Besansky et al. 1992	*aegypti*	Eight uncharacterized families	J. Tu, pers. comm.			
	gambiae	Twelve families	Holt et al. 2002						
	gambiae	More than 100 families	Biedler and Tu 2003						
	arabiensis	*RT2* in 28S rRNA	Besansky et al. 1992						
SINEs	*gambiae*	*Sine200* (2389)	Holt et al. 2002	*aegypti*	*Felai*	Tu 1999			
Other	*gambiae*	*Maque* (220)	Tu 2001b						

Class II									
IS630-Tc1-mariner superfamily	*gambiae*	Three families (11–21)	Grossman et al. 1999	*aegypti*	*Zebedee*	Warren et al. 1997			
	gambiae	*Crusoe* (25)	Lander et al. 2001	*aegypti*	*MsqTc3* (100)	Shao et al. 2001			
	albimanus	*Quetzal* (11)	Ke et al. 1996	Three *Aedes* spp.	*ItmD37E1*	Shao et al. 2001			
	albimanus	Unnamed (few)	Liu et al. 1999						
	gambiae	*mariner* (65)	Mukabayire and Besansky 1996						
	gambiae	*Ikirara*	Romans et al. 1998						
	gambiae	*ItmD37E1*	Shao and Tu 2001						
	Six species	*P*	Sarkar et al. 2003; Oliveira de Carvalho et al. 2004						
	gambiae	Thirteen families	Holt et al. 2002						
MITEs	*gambiae*	*Pegasus* (90) *Joey* (1,120)	Besansky et al. 1996; Tu 2001a	*aegypti*	Three families (2,100–2,700)	Tu 1997	*pipiens*	*Mimo* (1000)	Feschotte and Mouches 2000b
	gambiae	Eight families (14–1,340)	Tu 2001	*aegypti*	Two *Pony* A and B (8,400 and 9,900)	Tu 2000	*pipiens*	*Milord* (3000)	Feschotte et al. 2002
	stephensi	*NOS*	Luckhart and Rosenberg 1999	*aegypti*	*Microuli* (3000)	Tu and Orphanidis 2001	*pipiens*	*Nemo* *Mikado* (1500) *Mirza*	Feschotte et al. 2002
	gambiae	Seven families	Holt et al. 2002	*aegypti*	Seven uncharacterized families	J. Tu, pers. comm.			

subsequent studies of the basic biology of endogenous elements in this and other fruit fly species have been completed.

In one of the earliest genetic investigations of endogenous TEs in nondrosophilid fruit flies, a syndrome of abnormal traits resembling hybrid dysgenesis in *D. melanogaster* was reported in *C. capitata* (Torti et al. 1997). It was concluded that the complex patterns observed were probably the result of interactions between more than one different mobile element systems. Full-length *mariner* elements with high sequence similarity were identified in *C. capitata, Ceratitis rosa,* and *Trirhithrum coffeae.* However, further characterization of *mariner* sequences in *C. rosa* and *T. coffeae* revealed wide copy number variation and no functional elements (Torti et al. 1998). The *mariner* elements amplified from these closely related tephritid species appear to belong to the mellifera subfamily (Torti et al. 1998). A second unrelated *mariner* element was subsequently identified in the Natal fruit fly, *C. rosa.* This and related elements in other tephritid species were found to represent a new basal subfamily of *mariner* elements, the rosa subfamily (Gomulski et al. 2001). The unusual distribution of these elements implicated horizontal transfer in addition to vertical transmission. Representatives of five distinct types of *mariner* family elements were identified in *B. tryoni* and its sibling species, *Bactrocera neohumeralis* (Green and Frommer 2001).

The functionality of the *D. melanogaster hobo* element was tested in several tephritid species using transient embryonic excision assays (Handler and Gomez 1996). A permissive state for *hobo* mobility was indicated in *Anastrepha suspensa, Bactrocera dorsalis, Bactrocera cucurbitae, C. capitata,* and *Toxotrypana curcicauda. Hobo*-related sequences were identified in all species examined except for *T. curcicauda,* some probably originating from horizontal transfer events. Two types of *hAT* elements, *Homer* and *Homer*-like elements (HLE) were later identified in the Queensland fruit fly *B. tryoni* (Pinkerton et al. 1999).

Endogenous copies of the *piggyBac* TE were found in the Oriental fruit fly, *B. dorsalis,* but not in other tephritid species (Handler and McCombs 2000). The *piggyBac* element was originally discovered by its ability to transpose into infecting bacuolovirus (Fraser et al. 1985) and was isolated from the cabbage looper moth *Trichoplusia ni.* The occurrence of *piggyBac* elements as endogenous TEs in *B. dorsalis* is of particular interest because of its implications for natural horizontal transfer. It also has considerable potential importance for the assessment of safety of insect transformation using TE vectors.

TEs in Midges. A number of both non-LTR and DNA element families have been described in *Chironomus* species. Among the non-LTR elements are *NLRCth1* from *Chironomus thummi* (Blinov et al. 1993), and *NLRCt2* from *C. tentans* (Blinov et al. 1997). A *SINE*-like dispersed element, *Cp1,* was reported to have a site-specific insertion pattern in centromeric tandem repeats of *C. pallidivittatus* (Liao et al. 1998). Among the transposons described in *C. thummi* are *MEC* (Blinov et al. 1991), *TFB1,* an FB element found in histone genes (Hankeln and Schmidt 1990), and *TECth1,* found in the 3′ flanking region of a Balbiani ring gene (Wobus et al. 1990). A family of tandem repetitive DNA sequences, the *Cla* elements, is present in the genomes of several *Chironomus* species. Interspersed clusters of these elements are widely distributed all over the chromosomes in the subspecies *C. thummi thummi,* whereas they seem to be limited to the centromeric regions in the closely related subspecies *C. t. piger.* This differential distribution is due to transposition of *Cla* elements along with flanking DNA in the *C. t. thummi* genome (Hankeln et al. 1994).

TEs in Other Dipteran Species. Members of three families of transposons have been identified in the house fly, *Musca domestica.* These include *Hermes,* an active element member of the *hAT* superfamily (Warren et al. 1994), and deleted, nonautonomous elements of the *P* (Lee et al. 1999) and *mariner* (Yoshiyama et al. 2000) families. In addition, a related species *Musca vetustissima,* the Australian bush fly, contains a sequence related to transposons of the *hobo, Ac,* and *Tam3* superfamily (Warren et al. 1995). The house fly has been successfully transformed by the *piggyBac* TE (Hediger et al. 2001).

Hermit, a member of the *hAT* family of TEs, was identified in the Australian sheep blowfly, *Lucilia cuprina* (Coates et al. 1996), as well as two *P* elements that were highly diverged from those of *D. melanogaster* (Perkins and Howells 1992).

Mariner-like elements were identified in the Hessian fly, *Mayetiola destructor* (Say), which is an agriculturally important pest of wheat, *Triticum aestivum* L., in the United States and other parts of the world (Shukle and Russell 1995). The distribution of sequences was predominantly, but not exclusively, in paracentromeric regions (Russell and Shukle 1997).

Applications of Dipteran TEs

TRANSFORMATION SYSTEMS AND TRANSGENESIS

One of the most important applications of dipteran TEs has been their development as transformation vectors. Most of the basic pioneering research on stable germ-line transformation was accomplished with *D. melanogaster.* This has elicited immense amounts of research on fundamental analyses of gene structure, function, and gene regulation and has provided a broad understanding of how flies develop (Lawrence 1992). Furthermore, this new technology has made it possible to identify, isolate, and clone many new genes. Closely following the discovery of the molecular basis of P-M hybrid dysgenesis in 1982, the *P* element was used in the first successful germ-line transformation of *D. melanogaster* (Rubin and Spradling 1982). Early experiments demonstrated that a complete *P* element within a plasmid could transpose into the genome of an early embryo. Subsequently, using a helper plasmid carrying a transposase gene, defective *P* elements having intact termini, and carrying an eye-color marker, were also shown to transpose successfully. This was the basis for the first binary transformation system in *Drosophila.* All transposase-mediated systems developed since then use a similar system, in which vector and helper plasmids are coinjected into preblastoderm embryos prior to pole cell formation. Integrations are usually stable in the absence of a transposase source, but can be remobilized by either injecting an appropriate transposase-producing DNA or by crossing to strains that carry a stable source of transposase (Bingham et al. 1982; Robertson et al. 1988).

In the period immediately following the first transformation of *Drosophila,* there were high hopes that *P* elements could be used successfully as vectors for transformation of a broad array of animal species. However, despite considerable efforts, it was eventually determined that the host range for successful *P* element transformation was disappointingly narrow. Although successful transformation of *D. simulans* and other species of *Drosophila* was achieved, embryonic transient mobility assays indicated that *P* element activity decreased as a function of the relatedness of a species to *D. melanogaster,* with no mobility evident outside the Drosophilidae (Handler and O'Brochta 1991; O'Brochta et al. 1991). Subsequently it has been speculated that the narrow host range of the *P* element can be attributed to its dependence on host factors necessary for transposition.

During the past six years, four new TE-based vector systems have been successfully developed for stable germ-line transformation of nondrosophilid insects. These systems are based on the *Mos-1* (active *mariner*) element from *D. mauritiana,* the *Hermes* element from *M. domestica,* the *Minos* element from *D. hydei,* and the *piggyBac* element from *Trichoplusia ni* (Table 6.3). In addition to *Drosophila* species, successful transformation of mosquitoes, tephritid fruit flies, and other dipteran species has been achieved (Table 6.3). In fact, insect transformation can now be considered routine (Handler 2001). Despite the huge progress made, there probably is still room for improvement in the development of more robust genetic tools for efficient gene transfer (Atkinson et al. 2001).

TRANSPOSABLE ELEMENT MUTAGENESIS AND GENE TAGGING

The first example of gene tagging in Diptera was the use of a retrotransposon to isolate the white locus in *D. melanogaster* (Bingham et al. 1981). Shortly thereafter, dysgenesis-induced mutations at the white locus were used to identify the molecular basis of P-M hybrid dysgenesis (Bingham et al. 1982). Later, the development and use of single genetically marked *P* elements simplified the identification and recovery of induced mutations (Cooley et al. 1988). About the same time, the identification of single *P* elements carrying the transposase gene at a defined chromosomal location (Robertson et al. 1988) made it possible to identify transposition events using genetic crosses rather than by embryo injection. This facilitated the removal of the transposase and subsequent stabilization of the new insertion. Single *P* element mutagenesis has made possible the identification of several thousand lethal *P* element insertions in distinct genes and has played an important role in the *Drosophila* Genome Project (Spradling et al. 1995).

ENHANCER TRAPS

Another important application of the *P* element has been the development of the method of enhancer trapping using the weak *P* element promoter fused to the B-galactosidase (*lacZ*) gene of *Escherichia coli* (O'Kane and Gehring 1987). After insertion of the *P-lacZ* elements into the *D. melanogaster* genome, it was found that the expression of *lacZ* often corresponded in tissue or temporal specificity to that of a gene near the insertion site because of adjacent enhancers acting on the weak *P* element promoter. Subsequently, large-scale screens were initiated that identified novel genes and provided markers for specific cell or tissue types in development.

GENE REPLACEMENT

A method of gene replacement was developed that uses the double-strand gap that is left as the result of the *P* element cut-and-paste transposition process (Gloor et al. 1991). The gap repair process can use ectopic templates to copy new information near the site of a *P* element-induced break. It was observed that this repair is quite efficient and exhibits a strong *cis* preference. This system was the first of its kind to be developed in a metazoan organism and is used for gene targeting in *Drosophila* (Lankenau and Gloor 1998). The method is particularly useful for large genes that cannot be subjected to normal *P* element-mediated transformation studies (Rio 2002).

TABLE 6.3. List of Successful TE Transformations

Species Transformed	Element	Reference
Drosophila		
D. melanogaster	*P* from *D. melanogaster*	Rubin and Spradling 1982
	hobo from *D. melanogaster*	Blackman et al. 1989
	Mos1 from *D. mauritiana*	Garza et al. 1991
	Minos from *D. hydei*	Loukeris et al. 1995
	piggyBac from *Trichoplusia ni*	Handler and Harrell 1999
	Penelope from *D. virilis*	Evgen'ev et al. unpub. data
D. hawaiiensis	*P* from *D. melanogaster*	Brennan et al. 1984
D. simulans	*P* from *D. melanogaster*	Scavarda and Hartl 1984
D. virilis	*hobo* from *D. melanogaster*	Lozovskaya et al. 1996
	Mos1 from *D. mauritiana*	Lohe and Hartl 1996
Mosquitoes		
Aedes aegypti	*Mos1* from *D. mauritiana*	Coates et al. 1998
	Hermes from *Musca domestica*	Jasinskiene et al. 1998
	piggyBac from *T. ni*	Kokoza et al. 2001
Culex quinquefasciatus	*Hermes* from *Musca domestica*	Allen et al. 2001
Anopheles stephensi	*Minos* from *D. hydei*	Catteruccia et al. 2000b
	piggyBac from *T. ni*	Nolan et al. 2002
An. gambiae	*piggyBac* from *T. ni*	Grossman et al. 2001
An. albimanus	*piggyBac* from *T. ni*	Perera et al. 2002
Tephritid fruit flies		
Ceratitis capitata	*MInos* from *D. hydei*	Loukeris et al. 1995
	piggyBac from *T. ni*	Handler et al. 1998
	Hermes from *M. domestica*	Michel et al. 2001
Bactrocera dorsalis	*piggyBac* from *T. ni*	Handler and McCombs 2000
B. tryoni	*hobo* from *D. melanogaster*[a]	Raphael et al. 2004
Anastrepha suspensa	*piggyBac* from *T. ni*	Handler and Harrell 2001
Other Diptera		
Stomoxys calcitrans	*Hermes* from *M. domestica*	O'Brochta et al. 2000
Musca domestica	*piggyBac* from *T. ni*	Hediger et al. 2001
Lucilia cuprina	*piggyBac* from *T. ni*	Heinrich et al. 2002
Cochliomyia hominivorax	*piggyBac* from *T. ni*	Allen et al. 2004

[a] Aberrant integration.

TRANSPOSABLE ELEMENTS AS MARKERS IN EVOLUTIONARY STUDIES

*SINE*s are tRNA-derived retroelements that are present in well over 10,000 total copies in many eukaryotic genomes. The enormous number of *SINE* amplifications per organism makes them important evolutionary agents for shaping the diversity of genomes, and the irreversible, independent nature of their insertion allows them to be used for diagnosing common ancestry among host taxa with extreme confidence. In humans, the *Alu* family of *SINE*s has been used very successfully as markers in population genetic analysis. This success depends on the fact that recently active *SINE*s can produce a high level of polymorphism and their transposition does not involve excision. *SINE*s potentially constitute a very useful system of genetic markers, but in *D. melanogaster,* their frequency is very low. This potential is being exploited in *Ae. aegypti* using the *Felai* family of tRNA-related *SINE*s (Tu 1999). Approximately 60,000 copies of *Feilai* are present in *Ae. aegypti,* equivalent to 2% of the entire genome. This system is very easy to use, in addition to being highly efficient, and represents a powerful new tool for use in systematic biology.

In addition to *SINEs*, *mariner* elements have been successfully used to study genome diversity within and between the Indian biotypes of the Asian rice gall midge, *Orseolia oryzae*, a major pest of rice (Behura et al. 2001).

The Role of Transposable Elements in Fly Evolution

Until recently, broad acceptance of the selfish DNA hypothesis often led to dismissal of transposable elements (TEs) as selfish entities having little relevance for the evolution of their host organisms (Kidwell and Lisch 2000). Now, thanks to many studies in Diptera and other eukaryotes, a more complex and enlightened view is emerging that recognizes the relationship between TEs and their hosts as flexible and opportunistic, rather than rigid. The parasitic-mutualistic continuum provides a useful framework organizing the multiple possibilities of host-element relationships (Kidwell and Lisch 2000; Kidwell and Lisch 2001). Although TEs are expected to tend toward the parasitic end of such a continuum, during periods of invasion of a new host species, relationships that are more mutualistic can be expected to evolve, often over extended periods. The properties that have lead to the frequent labeling of TEs as "junk DNA" may have provided genomes with opportunities for further evolution.

One of the most important ways that TEs contribute to the evolution of their host organisms is through the production of multiple types of genomic variation (Kidwell and Lisch 1997, 2002). TEs provide sources of variation in two main ways. First, because of their activity, they induce new variation directly through mutations and genomic rearrangements. However, due to natural and artificial selection, the standing variation in populations reflects only a fraction of the original variation induced by these elements. Second, the co-option of TE sequences and enzymatic machinery by their host organisms provides another way that TEs contribute to variation over evolutionary time. Once again, such co-opted sequences and mechanisms are likely to be increasingly difficult to identify with the passage of time. Dipteran species provide a number of examples of both these types of TE-induced variation.

HOST VARIATION PRODUCED DIRECTLY BY TE ACTIVITY

TE-Induced Mutations. Insertions by TEs into coding regions often disrupt gene function, a property that lead to the discovery of *P* elements (Bingham et al. 1982). A series of null mutations produced by hybrid dysgenesis at the *white* locus in *Drosophila* were shown to result from *P* element insertions into the host gene (Rubin et al. 1982). Because of their low organismal fitness, such mutations tend to be relatively uncommon in natural populations of Diptera. However, *P* elements in *D. melanogaster* provide some of the most detailed and comprehensive examples of the potency of TE insertions in generating new quantitative variation. In a series of large-scale experiments, Mackay and colleagues have shown that *P* element insertions induced by hybrid dysgenesis can provide statistically significant changes in both the mean and variance of a wide range of quantitative genetic traits, including those affecting fitness, morphology, and behavior (e.g., Lyman et al. 1996; Currie et al. 1998).

In addition to their ability to disrupt or alter gene function in various ways upon insertion, TEs that move by a cut-and-paste mechanism often produce further variation when they excise. This is because after repair, the original sequence is seldom restored precisely as it existed previously. There is evidence in many eukaryotes that transposon excision from a given site can generate a high degree of variation in DNA sequence and phenotype. The excision process may result in either the addition of new sequences or deletion of host sequences, flanking the insertion site. The genesis of microsatellite DNAs by *Mini-me* elements in many dipteran species provides an interesting example of the addition of new sequences (Wilder and Hollocher 2001). In *Drosophila,* precise excision of *P* elements results from double-strand gap repair using the wild type homolog to restore the excision site (Engels et al. 1990). In this case, precise excision is a direct expression of the relatively frequent presence of wild type homologues for use as repair templates.

TE-Induced Genomic Rearrangements. The ability of TEs to induce chromosomal rearrangements, such as deletions, duplications, inversions, and reciprocal translocations, provides the potential for both small- and large-scale reorganization of genomes. In fact, a study of mobilization rates of nine TEs in *D. melanogaster* indicated that most changes in restriction patterns were consistent with rearrangements, rather than with true transposition (Dominguez and Albornoz 1996). Furthermore, clustering of movement across transposon families was observed, suggesting that transposition of different families may not be independent.

Two mechanisms are considered most likely to be responsible for TE-induced karyotypic changes. The best-known mechanism is ectopic recombination, in which homologous recombination occurs between two TE copies present in a genome. A second mechanism for inducing genomic rearrangements is alternative transposition of Class II elements, first described in *Drosophila* (Gray 2000). In contrast to the traditional mode of cut-and-paste transposition that involves the synapsis of complementary ends of a single TE, alternative transposition involves the synapsis of complementary ends of different TEs on the same, or different, chromosomes. The hybrid element produced excises and reinserts in a new genomic location, as occurs in traditional transposition. This mode of transposition provides the potential for producing many different types of chromosomal rearrangements (Engels and Preston 1984; Gray 2000).

Polymorphic inversions are very common in *Drosophila,* and evidence for the implication of TEs has been found for a number of these. For example, many breakpoints in hybrid dysgenesis–induced chromosomal rearrangements occurred at, or very near to, the sites of *P* and *hobo* element insertions (Engels and Preston 1984; Lim and Simmons 1994). Over half of the newly induced chromosomal rearrangements induced by hybrid dysgenesis in *D. virilis* were also found to contain *Penelope* and *Ulysses* TEs in the same chromosomal subsections as their breakpoints (Evgen'ev et al. 2000b). Many of these breakpoints also coincide with chromosomal locations of *Penelope* and *Ulysses* insertions in the parental strains and with breakpoints of inversions previously established for other species of the *virilis* group. These observations suggest that TEs may have played an important role in the evolution of the *virilis* species group. In *D. buzzatii,* the breakpoints of the cosmopolitan inversion 2j contain large insertions corresponding to a TE (Caceres et al. 1999). The inversion arose by ectopic recombination between two copies of the transposon located in opposite orientations. Target site duplications generated upon insertion were apparently exchanged during the inversion event. In the mosquito *An. arabiensis,* the *Odysseus* TE was recently discovered by detailed molecular analysis to lie at the distal breakpoint of a polymorphic inversion (Mathiopoulos et al. 1998). TEs are more likely to be found associated with endemic inversions than cosmopolitan ones, because of the tendency of TEs to be lost over time from inversion breakpoints (Engels and Preston 1984; Mathiopoulos et al. 1998).

RECRUITMENT OF TRANSPOSABLE ELEMENTS FOR NEW HOST FUNCTIONS

In contrast to the prevailing selfish DNA paradigm of the past 20 years, a minority opinion has maintained that at least some TE sequences have become "useful" for some functions of their hosts. This idea has been expressed in varying terminology. A number of examples of TE-induced variation have been cited as exaptations (i.e., features that have previously evolved for functions other than those now used, or for no function at all, but which have been co-opted for a new use). The term "molecular domestication" has been applied more recently to various TE sequences in *Drosophila.*

An example of the variety of ways in which host/element relationships can be interpreted is providing by the *R2* elements, described earlier. Their stable, long-term presence at low copy number in the genomes of all arthropods for 500 My may be explained by their ability to behave as extremely efficient parasites. Under this scenario, the elements might be maintained by their ability to efficiently monitor and maintain low copy numbers and to occupy a genomic niche that never threatens the fitness of their hosts because of the multicopy nature of rDNA genes (Eickbush and Malik 2002). However, an alternative hypothesis that postulates that these elements are maintained by positive selection, because of their ability to provide a function advantageous to their hosts, cannot be ruled out at this time.

HET-A *and* TART *Elements in Diptera.* In *Drosophila* and other Diptera, one of the most compelling examples of host recruitment of TEs is seen at the ends of chromosomes. In place of the telomeres found in all other eukaryotes, flies have recruited two families of non-LTR retrotransposons, *HET-A* and *TART,* to stabilize chromosome ends (Biessmann et al. 1992; Levis et al. 1993). However, in contrast to the *HET-A* and *TART* elements, the nature of the relationship between the majority of TEs and their host organisms has not yet been completely worked out.

Molecular Domestication. The stationary *P* element–related gene clusters of *D. guanche, D. madeirensis,* and *D. subobscura* provide an interesting example of molecular domestication (Miller et al. 1992, 1997). Each cluster unit consists of a *cis*-regulating section composed of different insertion sequences followed by the first three exons of a *P* element that encodes a 66-kDa "repressor-like" protein. In contrast to this normal repressor function, these stationary *P* element repeats appear to have evolved the function of transcription factors (Miller et al. 1995). Remarkably, the *D. guanche* P-protein produces an enhancer-like effect, rather than repressing canonical *P* element activity in transgenic *D. melanogaster* (W. Miller, pers. comm.). The insertion sequence, which gave rise to the de novo A-type promoter of this *P*-gene cluster, has recently been identified (Miller et al. 2000). This insertion belongs to a new MITE-like TE family, designated *SGM,* which is related to poorly characterized *IS* elements of other *obscura* group species.

Evidence that molecular domestication of a TE family may recur in a host lineage is provided by a truncated immobile *P* sequence cloned from *D. tscasi,* a species of the *D. montium* subgroup (Nouaud and Anxolabéhère 1997). This truncated element produces a polyadenylated RNA that also has a coding capacity for a 66-kDa "repressor-like" protein. This domestication event has lead to the evolution of a new promoter and a new intron (Nouaud et al. 1999). Although the immobile *obscura* and *montium P* sequences were derived from the same ancestral mobile *P* element family, the structures of the flanking regions of these sequences indicate that they were produced by two completely independent evolutionary events.

Transposable Elements in Host Gene Regulation. There are several ways that TEs can affect gene regulation. Either a new insertion can disrupt an existing host regulatory element, or it can contribute its own regulatory sequences to a host gene in which it has been inserted in an appropriate location. Inserted sequences may also provide the potential for future evolution of regulatory sequences. The preference of a number of TEs for insertion in genic regions makes these elements good candidates for co-option as regulatory elements. Good examples are MITEs in the yellow fever mosquito, *Ae. aegypti* (Tu 2000).

Another excellent example of a TE insertion in the regulatory region of a gene is provided by the *gypsy* element that inserted into the 5′ upstream region of the *Yellow* gene in *Drosophila.* This causes a loss of expression of the *Yellow* gene in specific tissues (Corces and Geyer 1991). The loss of expression in some tissues, and not others, is the result of the interaction of the element, tissue-specific enhancers upstream of the element with specific host factors.

Changes in *cis*-regulatory regions of duplicated genes may be more important for the evolution and divergence of functional and morphological characters than are mutations in coding sequences. Only recently has evidence started to accumulate to support this hypothesis. In *Drosophila,* for instance, the three-homeotic genes *paired* (*prd*), *gooseberry* (*gsb*), and *gooseberry neuro* (*gsbn*) have evolved from a single ancestral gene, following gene duplication. They now have distinct developmental functions during embryogenesis (Li and Noll 1994).

Heterochromatin. Heterochromatin represents a significant portion of the eukaryotic genome. As much as one-third of the *D. melanogaster* genome, for instance, is made up of heterochromatin (Bartolome et al. 2002). Heterochromatin is rich in satellite DNA and TEs and relatively poor in host gene sequences. Recombination within heterochromatin is markedly reduced, and it has been suggested that TEs are often overrepresented in heterochromatin, because selection has preferentially removed elements from euchromatic sites where ectopic recombination would select for their loss (Charlesworth et al. 1994). Although one analysis of release 2 of the draft *D. melanogaster* genome sequence suggests that TEs constitute 18% of the heterochromatic fraction, compared with only 2% of the euchromatic fraction of this genome (Bartolome et al. 2002), a subsequent analysis estimates these fractions to be at least as high as 60% for heterochromatin and 6% for euchromatin (Kapitonov and Jurka 2003). Even though there is also validation of the interpretation of differential selection in the medfly (Torti et al. 2000), there are several lines of evidence indicating that other processes are involved as well. First, differential selection due to differences in recombination frequency would not be expected to produce the pattern seen within *D. melanogaster* heterochromatin (Pimpinelli et al. 1995), in which each of several classes of elements inhabit a specific domain, as recombination is suppressed throughout heterochromatin. Second, some TEs, such as the *I* elements of *D. melanogaster,* clearly target heterochromatin (Dimitri et al. 1997). De novo insertion of *I* elements into a heterochromatic gene was an order of magnitude higher in frequency than that into a euchromatic gene on the same chromosome. Third, there is evidence that the presence of transposons themselves can contribute to the formation of heterochromatin. For example, there is compelling evidence that the first step in Y chromosome degeneration in *D. miranda* is driven by the accumulation of TEs, especially retrotransposons (Steinemann and Steinemann 1998). An enrichment of these elements along an evolving Y chromosome could account for the switch from a euchromatic to a heterochromatic chromatin structure.

Conclusion

During the past 30 years, Diptera has played an important role in the emerging field of TE biology. Although in the early years, the use of *Drosophila* predominated in studies of this type, there has more recently been an accelerating trend toward the description, analysis, and application of TEs in other dipteran species. Interest in the development of genetic transformation systems for pest species, notably mosquitoes and tephritid fruit flies, has provided the major impetus for the study of TEs in nondrosophilids. Some TEs have good potential for use as genetic markers. In addition to their important applications, TEs are increasingly being viewed as important contributors to the evolution of dipteran genomes. The recent availability of the genome sequences of *D. melanogaster* and *An. gambiae* promises a wealth of interesting comparative information on the evolution of these sequences in dipteran lineages that diverged more than 200 Mya.

REFERENCES

Adams, M. D., S. E. Celniker, R. A. Holt, C. A. Evans, J. D. Gocayne, P. G. Amanatides, S. E. Scherer, P. W. Li, R. A. Hoskins, R. F. Galle, R. A. George, S. E. Lewis, S. Richards, M. Ashburner, S. N. Henderson, G. G. Sutton, J. R. Wortman, M. D. Yandell, Q. Zhang, L. X. Chen, R. C. Brandon, Y. H. Rogers, R. G. Blazej, M. Champe, B. D. Pfeiffer, K. H. Wan, C. Doyle, E. G. Baxter, G. Helt, C. R. Nelson, G. L. Gabor Miklos, J. F. Abril, A. Agbayani, H. J. An, C. Andrews-Pfannkoch, D. Baldwin, R. M. Ballew, A. Basu, J. Baxendale, L. Bayraktaroglu, E. M. Beasley, K. Y. Beeson, P. V. Benos, B. P. Berman, D. Bhandari, S. Bolshakov, D. Borkova, M. R. Botchan, J. Bouck, P. Brokstein, P. Brottier, K. C. Burtis, D. A. Busam, H. Butler, E. Cadieu, A. Center, I. Chandra, J. M. Cherry, S. Cawley, C. Dahlke, L. B. Davenport, P. Davies, B. de Pablos, A. Delcher, Z. Deng, A. D. Mays, I. Dew, S. M. Dietz, K. Dodson, L. E. Doup, M. Downes, S. Dugan-Rocha, B. C. Dunkov, P. Dunn, K. J. Durbin, C. C. Evangelista, C. Ferraz, S. Ferriera, W. Fleischmann, C. Fosler, A. E. Gabrielian, N. S. Garg, W. M. Gelbart, K. Glasser, A. Glodek, F. Gong, J. H. Gorrell, Z. Gu, P. Guan, M. Harris, N. L. Harris, D. Harvey, T. J. Heiman, J. R. Hernandez, J. Houck, D. Hostin, K. A. Houston, T. J. Howland, M. H. Wei, C. Ibegwam, M. Jalali, F. Kalush, G. H. Karpen, Z. Ke, J. A. Kennison, K. A. Ketchum, B. E. Kimmel, C. D. Kodira, C. Kraft, S. Kravitz, D. Kulp, Z. Lai, P. Lasko, Y. Lei, A. A. Levitsky, J. Li, Z. Li, Y. Liang, X. Lin, X. Liu, B. Mattei, T. C. McIntosh, M. P. McLeod, D. McPherson, G. Merkulov, N. V. Milshina, C. Mobarry, J. Morris, A. Moshrefi, S. M. Mount, M. Moy, B. Murphy, L. Murphy, D. M. Muzny, D. L. Nelson, D. R. Nelson, K. A. Nelson, K. Nixon, D. R. Nusskern, J. M. Pacleb, M. Palazzolo, G. S. Pittman, S. Pan, J. Pollard, V. Puri, M. G. Reese, K. Reinert, K. Remington, R. D. Saunders, F. Scheeler, H. Shen, B. C. Shue, I. Siden-Kiamos, M. Simpson, M. P. Skupski, T. Smith, E. Spier, A. C. Spradling, M. Stapleton, R. Strong, E. Sun, R. Svirskas, C. Tector, R. Turner, E. Venter, A. H. Wang, X. Wang, Z. Y. Wang, D. A. Wassarman, G. M. Weinstock, J. Weissenbach, S. M. Williams, T. Woodage, K. C. Worley, D. Wu, S. Yang, Q. A. Yao, J. Ye, R. F. Yeh, J. S. Zaveri, M. Zhan, G. Zhang, Q. Zhao, L. Zheng, X. H. Zheng, F. N. Zhong, W. Zhong, X. Zhou, S. Zhu, X. Zhu, H. O. Smith, R. A. Gibbs, E. W. Myers, G. M. Rubin, and J. C. Venter. 2000. The genome sequence of *Drosophila melanogaster*. *Science* 287:2185–2195.

Agarwal, M., N. Bensaadi, J. C. Salvado, K. Campbell, and C. Mouches. 1993. Characterization and genetic organization of full-length copies of a LINE retroposon family dispersed in the genome of *Culex pipiens* mosquitoes. *Insect Biochem. Mol. Biol.* 23:621–629.

Allen, M. L., D. A. O'Brochta, P. W. Atkinson, and C. S. Levesque. 2001. Stable, germ-line transformation of *Culex quinquefasciatus* (Diptera: Culicidae). *J. Med. Entomol.* 38:701–710.

Allen, M. L., A. M. Handler, D. R. Berkebile, and S. R. Skoda. 2004. *piggyBac* transformation of the New World screwworm, *Cochliomyia hominivorax,* produces multiple distinct mutant strains. *Med. Vet. Entomol.* 18:1–9.

Anxolabéhère, D., D. Nouaud, and G. Périquet. 1985. Séquences homologues a l'élément *P* chez des espèces de *Drosophila* du groupe *obscura* et chez *Scaptomyza pallida (Drosophilidae). Génét. Sél. Evol.* 17: 579–584.

Anxolabéhère, D., M. G. Kidwell, and G. Périquet. 1988. Molecular characteristics of diverse populations are consistent with the hypothesis of a recent invasion of *Drosophila melanogaster* by mobile *P* elements. *Mol. Biol. Evol.* 5:252–269.

Arkhipova, I. R., K. I. Pyatkov, M. Meselson, and M. B. Evgen'ev. 2003. Retroelements containing introns in diverse invertebrate taxa. *Nat. Genet.* 33:123–124.

Atkinson, P. W., A. C. Pinkerton, and D. A. O'Brochta. 2001. Genetic transformation systems in insects. *Annu. Rev. Entomol.* 46:317–346.

Bartolome, C., X. Maside, and B. Charlesworth. 2002. On the abundance and distribution of TEs in the genome of *Drosophila melanogaster. Mol. Biol. Evol.* 19:926–937.

Behura, S. K., S. Nair, and M. Mohan. 2001. Polymorphisms flanking the mariner integration sites in the rice gall midge (*Orseolia oryzae* Wood-Mason) genome are biotype-specific. *Genome* 44:947–954.

Bell, J. R., A. M. Bogardus, T. Schmidt, and M. Pellegrini. 1985. A new *copia*-like transposable element found in a *Drosophila* rDNA gene unit. *Nucleic Acids Res.* 13:3861–3871.

Bensaadi-Merchermek, N., C. Cagnon, I. Desmons, J. C. Salvado, S. Karama, F. D'Amico, and C. Mouches. 1997. *CM-gag,* a transposable-like element reiterated in the genome of *Culex pipiens* mosquitoes, contains only a *gag* gene. *Genetica* 100:141–148.

Besansky, N. J. 1990. A retrotransposable element from the mosquito *Anopheles gambiae. Mol. Cell. Biol.* 10:863–871.

Besansky, N. J., S. M. Paskewitz, D. M. Hamm, and F. H. Collins. 1992. Distinct families of site-specific retrotransposons occupy identical positions in the rRNA genes of *Anopheles gambiae. Mol. Cell. Biol.* 12: 5102–5110.

Besansky, N. J., J. A. Bedell, and O. Mukabayire. 1994. Q: A new retrotransposon from the mosquito *Anopheles gambiae. Insect Mol. Biol.* 3:49–56.

Besansky, N. J., O. Mukabayire, J. A. Bedell, and H. Lusz. 1996. *Pegasus,* a small terminal inverted repeat transposable element found in the *white* gene of *Anopheles gambiae. Genetica* 98:119–129.

Biedler, J., and Z. Tu. 2003. Non-LTR retrotransposons in the African malaria mosquito, *Anopheles gambiae:* Unprecedented diversity and evidence of recent activity. *Mol. Biol. Evol.* 20:1811–1825.

Biessmann, H., K. Valgeirsdottir, A. Lofsky, C. Chin, B. Ginther, R. W. Levis, and M. L. Pardue. 1992. *HeT-A,* a transposable element specifically involved in "healing" broken chromosome ends in *Drosophila melanogaster. Mol. Cell. Biol.* 12:3910–3918.

Biessmann, H., M. F. Walter, D. Le, S. Chuan, and J. G. Yao. 1999. *Moose,* a new family of LTR-retrotransposons in the mosquito *Anopheles gambiae. Insect Mol. Biol.* 8:201–212.

Bingham, P. M., R. Levis, and G. M. Rubin. 1981. Cloning of DNA sequences from the *white* locus of *D. melanogaster* by a novel and general method. *Cell* 25:693–704.

Bingham, P. M., M. G. Kidwell, and G. M. Rubin. 1982. The molecular basis of P-M hybrid dysgenesis: The role of the *P* element, a *P*-strain-specific transposon family. *Cell* 29:995–1004.

Blackman, R. K., M. M. Koehler, R. Grimaila, and W. M. Gelbart. 1989. Identification of a fully-functional *hobo* transposable element and its use for germ-line transformation of *Drosophila. EMBO J.* 8: 211–217.

Blinov, A. G., Y. V. Sobanov, E. K. Gaidamakova, S. S. Bogachev, N. N. Kolesnikov, M. A. Filippova, and I. I. Kiknadze. 1991. *MEC:* A transposable element from *Chironomus thummi* (Diptera). *Mol. Gen. Genet.* 229:152–154.

Blinov, A. G., Y. V. Sobanov, S. S. Bogachev, A. P. Donchenko, and M. A. Filippova. 1993. The *Chironomus thummi* genome contains a non-LTR retrotransposon. *Mol. Gen. Genet.* 237:412–420.

Blinov, A. G., Y. V. Sobanov, S. V. Scherbik, and K. G. Aimanova. 1997. The *Chironomus (Camptochironomus) tentans* genome contains two non-LTR retrotransposons. *Genome* 40:143–150.

Bowen, N. J., and J. F. McDonald. 2001. *Drosophila* euchromatic LTR retrotransposons are much younger than the host species in which they reside. *Genome Res.* 11:1527–1540.

Brennan, M. D., R. G. Rowan, and W. J. Dickinson. 1984. Introduction of a functional *P* element into the germ line of *Drosophila hawaiiensis. Cell* 38:147–151.

Bucheton, A., R. Paro, H. M. Sang, A. Pelisson, and D. J. Finnegan. 1984. The molecular basis of I-R hybrid dysgenesis: Identification, cloning and properties of the I factor. *Cell* 38:153–163.

Caceres, M., J. M. Ranz, A. Barbadilla, M. Long, and A. Ruiz. 1999. Generation of a widespread *Drosophila* inversion by a transposable element. *Science* 285:415–418.

Capy, P., R. Vitalis, T. Langin, D. Higuet, and C. Bazin. 1996. Relationships between transposable elements based upon the integrase-transposase domains: Is there a common ancestor? *J. Mol. Evol.* 42:359–368.

Catteruccia, F., T. Nolan, C. Blass, H. M. Muller, A. Crisanti, F. C. Kafatos, and T. G. Loukeris. 2000a. Toward *Anopheles* transformation: *Minos* element activity in anopheline cells and embryos. *Proc. Natl. Acad. Sci. USA* 97:2157–2162.

Catteruccia, F., T. Nolan, T. G. Loukeris, C. Blass, C. Savakis, F. C. Kafatos, and A. Crisanti. 2000b. Stable germline transformation of the malaria mosquito *Anopheles stephensi. Nature* 405:959–962.

Charlesworth, B., P. Sniegowski, and W. Stephan. 1994. The evolutionary dynamics of repetitive DNA in eukaryotes. *Nature* 371:215–220.

Coates, C. J., K. N. Johnson, H. D. Perkins, A. J. Howells, D. A. O'Brochta, and P. W. Atkinson. 1996. The *hermit* transposable element of the Australian sheep blowfly, *Lucilia cuprina,* belongs to the *hAT* family of transposable elements. *Genetica* 97:23–31.

Coates, C. J., N. Jasinskiene, L. Miyashiro, and A. A. James. 1998. *Mariner* transposition and transformation of the yellow fever mosquito, *Aedes aegypti. Proc. Natl. Acad. Sci. USA* 95:3748–3751.

Cooley, L., R. Kelley, and A. Spradling. 1988. Insertional mutagenesis of the *Drosophila* genome with single P elements. *Science* 239:1121–1128.

Corces, V. G., and P. K. Geyer. 1991. Interactions of retrotransposons with the host genome. *Trends in Genet.* 7:86–90.

Crow, J. F. 1988. The genesis of dysgenesis. *Genetics* 120:315–318.

Currie, D. B., T. F. Mackay, and L. Partridge. 1998. Pervasive effects of *P* element mutagenesis on body size in *Drosophila melanogaster. Genet. Res.* 72:19–24.

Daniels, S. B., A. Chovnick, and I. A. Boussy. 1990a. Distribution of *hobo* transposable elements in the genus *Drosophila. Mol. Biol. Evol.* 7:589–606.

Daniels, S. B., K. R. Peterson, L. D. Strausbaugh, M. G. Kidwell, and A. Chovnick. 1990b. Evidence for horizontal transmission of the *P* transposable element between *Drosophila* species. *Genetics* 124:339–355.

Danilevskaya, O. N., K. L. Traverse, N. C. Hogan, P. G. DeBaryshe, and M. L. Pardue. 1999. The two *Drosophila* telomeric transposable elements have very different patterns of transcription. *Mol. Cell. Biol.* 19:873–881.

Dimitri, P., B. Arca, L. Berghella, and E. Mei. 1997. High genetic instability of heterochromatin after transposition of the *LINE*-like I factor in *Drosophila melanogaster. Proc. Natl. Acad. Sci. USA* 94:8052–8057.

Dominguez, A., and J. Albornoz. 1996. Rates of movement of transposable elements in *Drosophila melanogaster. Mol. Gen. Genet.* 251:130–138.

Dowsett, A. P., and M. W. Young. 1982. Differing levels of dispersed repetitive DNA among closely related species of *Drosophila. Proc. Natl. Acad. Sci. USA* 79:4570–4574.

Eickbush, T. H. 2002. *R2* and related site-specific non-long terminal repeat retrotransposons. In N. L. Craig, R. Craigie, M. Gellert, and A. M. Lambowitz, eds., *Mobile DNA II,* pp. 813–835. Washington, D.C.: ASM Press.

Eickbush, T. H., and H. S. Malik. 2002. Origin and evolution of retrotransposons. In N. L. Craig, R. Craigie, M. Gellert, and A. M. Lambowitz, eds., *Mobile DNA II,* pp. 1111–1144. Washington, D.C.: ASM Press.

Engels, W. R. 1992. The origin of *P* elements in *Drosophila melanogaster. Bioessays* 14:681–686.

Engels, W. R., and C. R. Preston. 1984. Formation of chromosome rearrangements by *P* factors in *Drosophila. Genetics* 107:657–678.

Engels, W. R., D. M. Johnson-Schlitz, W. B. Eggleston, and J. Sved. 1990. High-frequency *P* element loss in *Drosophila* is homolog dependent. *Cell* 62:515–525.

Evgen'ev, M. B., H. Zelentsova, N. Shostak, M. Kozitsina, V. Barsky, D.-H. Lankenau, and V. G. Corces. 1997. *Penelope,* a new family of transposable elements and its possible role in hybrid dysgenesis in *Drosophila virilis. Proc. Natl. Acad. Sci. USA* 94:196–201.

Evgen'ev, M. B., H. Zelentsova, L. Mnjoian, H. Poluectova, and M. G. Kidwell. 2000a. Invasion of *Drosophila virilis* by the *Penelope* transposable element. *Chromosoma* 109:350–357.

Evgen'ev, M. B., H. Zelentsova, H. Poluectova, G. T. Lyozin, V. Veleikodvorskaja, K. I. Pyatkov, L. A. Zhivotovsky, and M. G. Kidwell. 2000b. Mobile elements and chromosomal evolution in the *virilis* group of *Drosophila. Proc. Natl. Acad. Sci. USA* 97:11337–11342.

Feschotte, C., and C. Mouches. 2000a. Evidence that a family of miniature inverted-repeat transposable elements (MITEs) from the *Arabidopsis thaliana* genome has arisen from a *pogo*-like DNA transposon. *Mol. Biol. Evol.* 17:730–737.

Feschotte, C., and C. Mouches. 2000b. Recent amplification of miniature inverted-repeat transposable elements in the vector mosquito *Culex pipiens:* Characterization of the *Mimo* family. *Gene* 250:109–116.

Feschotte, C., X. Zhang, and S. R. Wessler. 2002. Miniature inverted repeat transposable elements and their relationship to established DNA transposons. In N. L. Craig, R. Craigie, M. Gellert, and A. M. Lambowitz, eds., *Mobile DNA II,* pp. 1147–1158. Washington, D.C.: ASM Press.

Finnegan, D. J. 1983. Retroviruses and transposable elements—Which came first? *Nature* 302:105–106.

Finnegan, D. J., G. M. Rubin, M. W. Young, and D. S. Hogness. 1978. Repeated gene families in *Drosophila melanogaster. Cold Spring Harbor Symp. Quant. Biol.* 42:1053–1063.

Fraser, M. J., J. S. Brusca, G. E. Smith, and M. D. Summers. 1985. Transposon-mediated mutagenesis of a baculovirus. *Virology* 145:356–361.

Garza, D., M. Medhora, A. Koga, and D. L. Hartl. 1991. Introduction of the transposable element *mariner* into the germline of *Drosophila melanogaster. Genetics* 128:303–310.

Gloor, G. B., N. A. Nassif, D. M. Johnson-Schlitz, C. R. Preston, and W. R. Engels. 1991. Targeted gene replacement in *Drosophila* via *P* element–induced gap repair. *Science* 253:1110–1117.

Glover, D. M., and D. S. Hogness. 1977. A novel arrangement of the 18S and 28S sequences in a repeating unit of *Drosophila melanogaster* rDNA. *Cell* 10:167–176.

Gomulski, L. M., C. Torti, M. Bonizzoni, D. Moralli, E. Raimondi, P. Capy, G. Gasperi, and A. R. Malacrida. 2001. A new basal subfamily of mariner elements in *Ceratitis rosa* and other tephritid flies. *J. Mol. Evol.* 53:597–606.

Gray, Y. H. 2000. It takes two transposons to tango: Transposable-element–mediated chromosomal rearrangements. *Trends Genet.* 16:461–468.

Green, C. L., and M. Frommer. 2001. The genome of the Queensland fruit fly *Bactrocera tryoni* contains multiple representatives of the *mariner* family of transposable elements. *Insect Mol. Biol.* 10:371–386.

Grossman, G. L., A. J. Cornel, C. S. Rafferty, H. M. Robertson, and F. H. Collins. 1999. *Tsessebe, Topi* and *Tiang:* Three distinct *Tc1*-like transposable elements in the malaria vector, *Anopheles gambiae. Genetica* 105:69–80.

Grossman, G. L., C. S. Rafferty, J. R. Clayton, T. K. Stevens, O. Mukabayire, and M. Q. Benedict. 2001. Germline transformation of the malaria vector, *Anopheles gambiae,* with the *piggyBac* transposable element. *Insect Mol. Biol.* 10:597–604.

Hagele, K., and B. Oschmann. 1987. Non-reciprocal gonadal dysgenesis in hybrids of the chironomid midge *Chironomus thummi.* III. Germ line specific abnormalities. *Chromosoma* 96:50–54.

Handler, A. M. 2001. A current perspective on insect gene transformation. *Insect Biochem. Mol. Biol.* 31: 111–128.

Handler, A. M. 2003. Isolation and analysis of a new hopper *hAT* transposon from the *Bactrocera dorsalis* white eye strain. *Genetica* 118:17–24.

Handler, A. M., and S. P. Gomez. 1996. The *hobo* transposable element excises and has related elements in tephritid species. *Genetics* 143:1339–1347.

Handler, A. M., and S. P. Gomez. 1997. A new *hobo, Ac, Tam3* transposable element, hopper, from *Bactrocera dorsalis* is distantly related to *hobo* and *Ac. Gene* 185:133–135.

Handler, A. M., and R. A. Harrell, 2nd. 1999. Germline transformation of *Drosophila melanogaster* with the *piggyBac* transposon vector. *Insect Mol. Biol.* 8:449–457.

Handler, A. M., and R. A. Harrell, 2nd. 2001. Transformation of the Caribbean fruit fly, *Anastrepha suspensa,* with a *piggyBac* vector marked with polyubiquitin-regulated GFP. *Insect Biochem. Mol. Biol.* 31:199–205.

Handler, A. M., and S. D. McCombs. 2000. The *piggyBac* transposon mediates germ-line transformation in the Oriental fruit fly and closely related elements exist in its genome. *Insect Mol. Biol.* 9:605–612.

Handler, A. M., and D. A. O'Brochta. 1991. Prospects for gene transformation in insects. *Annu. Rev. Entomol.* 36:159–183.

Handler, A. M., S. D. McCombs, M. J. Fraser, and S. H. Saul. 1998. The lepidopteran transposon vector, *piggyBac,* mediates germ-line transformation in the Mediterranean fruit fly. *Proc. Natl. Acad. Sci. USA* 95:7520–7525.

Hankeln, T., and E. R. Schmidt. 1990. New foldback transposable element *TFB1* found in histone genes of the midge *Chironomus thummi. J. Mol. Biol.* 215:477–482.

Hankeln, T., A. Rohwedder, B. Weich, and E. R. Schmidt. 1994. Transposition of minisatellite-like DNA in *Chironomus* midges. *Genome* 37:542–549.

Haring, E., S. Hagemann, and W. Pinsker. 2000. Ancient and recent horizontal invasions of drosophilids by *P* elements. *J. Mol. Evol.* 51:577–586.

Hediger, M., M. Niessen, E. A. Wimmer, A. Dubendorfer, and D. Bopp. 2001. Genetic transformation of the housefly *Musca domestica* with the lepidopteran derived transposon *piggyBac. Insect Mol. Biol.* 10: 113–119.

Heinrich, J. C., X. Li, R. A. Henry, N. Haack, L. Stringfellow, A. C. Heath, and M. J. Scott. 2002. Germ-line transformation of the Australian sheep blowfly *Lucilia cuprina. Insect Mol. Biol.* 11:1–10.

Hill, S. R., S. S. Leung, N. L. Quercia, D. Vasiliauskas, J. Yu, I. Pasic, D. Leung, A. Tran, and P. Romans. 2001. Ikirara insertions reveal five new *Anopheles gambiae* transposable elements in islands of repetitious sequence. *J. Mol. Evol.* 52:215–231.

Holt, R. A., G. M. Subramanian, A. Halpern, G. G. Sutton, R. Charlab, D. R. Nusskern, P. Wincker, A. G. Clark, J. M. Ribeiro, R. Wides, S. L. Salzberg, B. Loftus, M. Yandell, W. H. Majoros, D. B. Rusch, Z. Lai, C. L. Kraft, J. F. Abril, V. Anthouard, P. Arensburger, P. W. Atkinson, H. Baden, V. de Berardinis, D. Baldwin, V. Benes, J. Biedler, C. Blass, R. Bolanos, D. Boscus, M. Barnstead, S. Cai, A. Center, K. Chatuverdi, G. K. Christophides, M. A. Chrystal, M. Clamp, A. Cravchik, V. Curwen, A. Dana, A. Delcher, I. Dew, C. A. Evans, M. Flanigan, A. Grundschober-Freimoser, L. Friedli, Z. Gu, P. Guan, R. Guigo, M. E. Hillenmeyer, S. L. Hladun, J. R. Hogan, Y. S. Hong, J. Hoover, O. Jaillon, Z. Ke, C. Kodira, E. Kokoza, A. Koutsos, I. Letunic, A. Levitsky, Y. Liang, J. J. Lin, N. F. Lobo, J. R. Lopez, J. A. Malek, T. C. McIntosh, S. Meister, J. Miller, C. Mobarry, E. Mongin, S. D. Murphy, D. A. O'Brochta, C. Pfannkoch, R. Qi, M. A. Regier, K. Remington, H. Shao, M. V. Sharakhova, C. D. Sitter, J. Shetty, T. J. Smith, R. Strong, J. Sun, D. Thomasova, L. Q. Ton, P. Topalis, Z. Tu, M. F. Unger, B. Walenz, A. Wang, J. Wang, M. Wang, X. Wang, K. J. Woodford, J. R. Wortman, M. Wu, A. Yao, E. M. Zdobnov, H. Zhang, Q. Zhao, S. Zhao, S. C. Zhu, I. Zhimulev, M. Coluzzi, A. della Torre, C. W. Roth, C. Louis, F. Kalush, R. J. Mural, E. W. Myers, M. D. Adams, H. O. Smith, S. Broder, M. J. Gardner, C. M. Fraser, E. Birney, P. Bork, P. T. Brey, J. C. Venter, J. Weissenbach, F. C. Kafatos, F. H. Collins, and S. L. Hoffman. 2002. The genome sequence of the malaria mosquito *Anopheles gambiae. Science* 298:129–149.

Holyoake, A. J., and M. G. Kidwell. 2003. Vege and Mar: Two novel *hAT* MITE families from *Drosophila willistoni. Mol. Biol. Evol.* 20:163–167.

Jacobson, J. W., M. M. Medhora, and D. L. Hartl. 1986. Molecular structure of a somatically unstable transposable element in *Drosophila. Proc. Natl. Acad. Sci. USA* 83:8684–8688.

Jasinskiene, N., C. J. Coates, M. Q. Benedict, A. J. Cornel, C. S. Rafferty, A. A. James, and F. H. Collins. 1998. Stable transformation of the yellow fever mosquito, *Aedes aegypti,* with the *Hermes* element from the housefly. *Proc. Natl. Acad. Sci. USA* 95:3743–3747.

Jordan, I. K., L. V. Matyunina, and J. F. McDonald. 1999. Evidence for the recent horizontal transfer of long terminal repeat retrotransposon. *Proc. Natl. Acad. Sci. USA* 96:12621–12625.

Kapitonov, V. V., and J. Jurka. 2001. Rolling-circle transposons in eukaryotes. *Proc. Natl. Acad. Sci. USA* 98:8714–8719.

Kapitonov, V. V., and J. Jurka. 2002. POGON1, a "bona fide" family of nonautonomous DNA transposons. *Repbase Reports* 2:7.

Kapitonov, V. V., and J. Jurka. 2003. Molecular paleontology of transposable elements in the *Drosophila melanogaster* genome. *Proc. Natl. Acad. Sci. USA* 100:6569–6574.

Ke, Z., G. L. Grossman, A. J. Cornel, and F. H. Collins. 1996. Quetzal: A transposon of the *Tc1* family in the mosquito *Anopheles albimanus. Genetica* 98:141–147.

Kidwell, M. G. 1983. Evolution of hybrid dysgenesis determinants in *Drosophila melanogaster. Proc. Natl. Acad. Sci. USA* 80:1655–1659.

Kidwell, M. G. 1994. The Wilhelmine E. Key 1991 Invitational Lecture. The evolutionary history of the *P* family of transposable elements. *J. Hered.* 85:339–346.

Kidwell, M. G., and J. F. Kidwell. 1975. Cytoplasm-chromosome interactions in *Drosophila melanogaster. Nature* 253:755–756.

Kidwell, M. G., and D. Lisch. 1997. Transposable elements as sources of variation in animals and plants. *Proc. Natl. Acad. Sci. USA* 94:7704–7711.

Kidwell, M. G., and D. R. Lisch. 2000. Transposable elements and host genome evolution. *Trends Ecol. Evol.* 15:95–99.

Kidwell, M. G., and D. R. Lisch. 2001. Perspective: Transposable elements, parasitic DNA, and genome evolution. *Evolution Int. J. Org. Evolution* 55:1–24.

Kidwell, M. G., and D. R. Lisch. 2002. Transposable elements as sources of genomic variation. In N. L. Craig, R. Craigie, M. Gellert, and A. M. Lambowitz, eds., *Mobile DNA II,* pp. 59–90. Washington, D.C.: ASM Press.

Kidwell, M. G., J. F. Kidwell, and J. A. Sved. 1977. Hybrid dysgenesis in *Drosophila melanogaster:* A syndrome of aberrant traits including mutation, sterility and male recombination. *Genetics* 36:813–833.

Kokoza, V., A. Ahmed, E. A. Wimmer, and A. S. Raikhel. 2001. Efficient transformation of the yellow fever mosquito *Aedes aegypti* using the *piggyBac* transposable element vector pBac[3xP3-EGFP afm]. *Insect Biochem. Mol. Biol.* 31:1137–1143.

Lankenau, D. H., and G. B. Gloor. 1998. In vivo gap repair in *Drosophila:* A one-way street with many destinations. *Bioessays* 20:317–327.

Lansman, R. A., S. N. Stacey, T. A. Grigliatti, and H. W. Brock. 1985. Sequences homologous to the *P* mobile element of *Drosophila melanogaster* are widely distributed in the subgenus Sophophora. *Nature* 318: 561–563.

Lawrence, P. A. 1992. *The Making of a Fly: The Genetics of Animal Design.* London: Blackwell Scientific.

Le, Q. H., S. Wright, Z. Yu, and T. Bureau. 2000. Transposon diversity in *Arabidopsis thaliana. Proc. Natl. Acad. Sci. USA* 97:7376–7381.

Lee, S. H., J. B. Clark, and M. G. Kidwell. 1999. A *P* element-homologous sequence in the house fly, *Musca domestica. Insect Mol. Biol.* 8:491–500.

Levis, R. W., R. Ganesan, K. Houtchens, L. A. Tolar, and F. M. Sheen. 1993. Transposons in place of telomeric repeats at a *Drosophila* telomere. *Cell* 75:1083–1093.

Li, X., and M. Noll. 1994. Evolution of distinct developmental functions of three *Drosophila* genes by acquisition of different *cis*-regulatory regions. *Nature* 367:83–87.

Liao, C., C. Rovira, H. He, and J. E. Edstrom. 1998. Site-specific insertion of a *SINE*-like element, *Cp1,* into centromeric tandem repeats from *Chironomus pallidivittatus. J. Mol. Biol.* 280:811–821.

Lim, J. K., and M. J. Simmons. 1994. Gross chromosome rearrangements mediated by transposable elements in *Drosophila melanogaster. Bioessays* 16:269–275.

Liu, K., P. Lizardi, J. Yu, and M. H. Rodriguez. 1999. Evidence for an inactive transposable *mariner*-like element in *Anopheles albimanus. J. Am. Mosq. Control Assoc.* 15:463–467.

Locke, J., L. T. Howard, N. Aippersbach, L. Podemski, and R. B. Hodgetts. 1999. The characterization of *DINE*-1, a short, interspersed repetitive element present on chromosome and in the centric heterochromatin of *Drosophila melanogaster. Chromosoma* 108:356–366.

Lohe, A. R., and D. L. Hartl. 1996. Germline transformation of *Drosophila virilis* with the transposable element *mariner. Genetics* 143:365–374.

Loukeris, T. G., B. Arca, I. Livadaras, G. Dialektaki, and C. Savakis. 1995. Introduction of the transposable element *Minos* into the germ line of *Drosophila melanogaster. Proc. Natl. Acad. Sci. USA.* 92: 9485–9489.

Lozovskaya, E. R., D. I. Nurminsky, D. L. Hartl, and D. T. Sullivan. 1996. Germline transformation of *Drosophila virilis* mediated by the transposable element *hobo. Genetics* 142:173–177.

Luckhart, S., and R. Rosenberg. 1999. Gene structure and polymorphism of an invertebrate nitric oxide synthase gene. *Gene* 232:25–34.

Lyman, R. F., F. Lawrence, S. V. Nuzhdin, and T. F. Mackay. 1996. Effects of single *P*-element insertions on bristle number and viability in *Drosophila melanogaster. Genetics* 143:277–292.

Malik, H. S., S. Henikoff, and T. H. Eickbush. 2000. Poised for contagion: Evolutionary origins of the infectious abilities of invertebrate retroviruses. *Genome Res.* 10:1307–1318.

Mathiopoulos, K. D., A. della Torre, V. Predazzi, V. Petrarca, and M. Coluzzi. 1998. Cloning of inversion breakpoints in the *Anopheles gambiae* complex traces a transposable element at the inversion junction. *Proc. Natl. Acad. Sci. USA* 95:12444–12449.

McClintock, B. 1948. Mutable loci in maize. *Carnegie Inst. Wash. Yrbk.* 47:155–169.

McGinnis, W., A. W. Shermoen, and S. K. Beckendorf. 1983. A transposable element inserted just 5′ to a *Drosophila* glue protein gene alters gene expression and chromatin structure. *Cell* 34:75–84.

Michel, K., A. Stamenova, A. C. Pinkerton, G. Franz, A. S. Robinson, A. Gariou-Papalexiou, A. Zacharopoulou, D. A. O'Brochta, and P. W. Atkinson. 2001. *Hermes*-mediated germ-line transformation of the Mediterranean fruit fly *Ceratitis capitata. Insect Mol. Biol.* 10:155–162.

Miller, W. J., S. Hagemann, E. Reiter, and W. Pinsker. 1992. *P*-element homologous sequences are tandemly repeated in the genome of *Drosophila guanche. Proc. Natl. Acad. Sci. USA* 89:4018–4022.

Miller, W. J., N. Paricio, S. Hagemann, M. J. Martinez-Sebastian, W. Pinsker, and R. de Frutos. 1995. Structure and expression of clustered *P* element homologues in *Drosophila subobscura* and *Drosophila guanche. Gene* 156:167–174.

Miller, W. J., J. F. McDonald, and W. Pinsker. 1997. Molecular domestication of mobile elements. *Genetica* 100:261–270.

Miller, W. J., A. Nagel, J. Bachmann, and L. Bachmann. 2000. Evolutionary dynamics of the *SGM* transposon family in the *Drosophila obscura* species group. *Mol. Biol. Evol.* 17:1597–1609.

Mouches, C., N. Bensaadi, and J. C. Salvado. 1992. Characterization of a *LINE* retroposon dispersed in the genome of three non-sibling *Aedes* mosquito species. *Gene* 120:183–190.

Mukabayire, O., and N. J. Besansky. 1996. Distribution of *T1, Q, Pegasus* and *mariner* transposable elements on the polytene chromosomes of PEST, a standard strain of *Anopheles gambiae. Chromosoma* 104: 585–595.

Nolan, T., T. M. Bower, A. E. Brown, A. Crisanti, and F. Catteruccia. 2002. *piggyBac*-mediated germline transformation of the malaria mosquito *Anopheles stephensi* using the red fluorescent protein dsRED as a selectable marker. *J. Biol. Chem.* 277:8759–8762.

Nouaud, D., and D. Anxolabéhère. 1997. *P* element domestication: A stationary truncated *P* element may encode a 66-kDa repressor-like protein in the *Drosophila montium* species subgroup. *Mol. Biol. Evol.* 14:1132–1144.

Nouaud, D., B. Boeda, L. Levy, and D. Anxolabéhère. 1999. A *P* element has induced intron formation in *Drosophila. Mol. Biol. Evol.* 16:1503–1510.

O'Brochta, D. A., S. P. Gomez, and A. M. Handler. 1991. *P* element excision in *Drosophila melanogaster* and related drosophilids. *Mol. Gen. Genet.* 225:387–394.

O'Brochta, D. A., P. W. Atkinson, and M. J. Lehane. 2000. Transformation of *Stomoxys calcitrans* with a *Hermes* gene vector. *Insect Mol. Biol.* 9:531–538.

O'Kane, C. J., and W. J. Gehring. 1987. Detection in situ of genomic regulatory elements in *Drosophila. Proc. Natl. Acad. Sci. USA* 84:9123–9127.

Oliveira de Carvalho, M., J. C. Silva, and E. L. Loreto. 2004. Analyses of *P*-like transposable element sequences from the genome of *Anopheles gambiae. Insect Mol. Biol.* 13:55–63.

Pantazidis, A., M. Labrador, and A. Fontdevila. 1999. The retrotransposon *Osvaldo* from *Drosophila buzzatii* displays all structural features of a functional retrovirus. *Mol. Biol. Evol.* 16:909–921.

Perera, O. P., I. R. Harrell, and A. M. Handler. 2002. Germ-line transformation of the South American malaria vector, *Anopheles albimanus,* with a *piggyBac/EGFP* transposon vector is routine and highly efficient. *Insect Mol. Biol.* 11:291–297.

Perkins, H. D., and A. J. Howells. 1992. Genomic sequences with homology to the *P* element of *Drosophila melanogaster* occur in the blowfly *Lucilia cuprina. Proc. Natl. Acad. Sci. USA* 89:10753–10757.

Picard, G., A. Bucheton, J. M. Lavige, and A. Fleuriet. 1972. [A sterility phenomenon of nonmendelian determinism in *Drosophila melanogaster.*] *C. R. Acad. Sci. Hebd. Seances Acad. Sci. D* 275:933–936. [In French]

Pimpinelli, S., M. Berloco, L. Fanti, P. Dimitri, S. Bonaccorsi, E. Marchetti, R. Caizzi, C. Caggese, and M. Gatti. 1995. Transposable elements are stable structural components of *Drosophila melanogaster* heterochromatin. *Proc. Natl. Acad. Sci. USA* 92:3804–3808.

Pinkerton, A. C., S. Whyard, H. A. Mende, C. J. Coates, D. A. O'Brochta, and P. W. Atkinson. 1999. The Queensland fruit fly, *Bactrocera tryoni,* contains multiple members of the *hAT* family of transposable elements. *Insect Mol. Biol.* 8:423–434.

Raphael, K. A., S. Whyard, D. Shearman, X. An, and M. Frommer. 2004. *Bactrocera tryoni* and closely related pest tephritids—Molecular analysis and prospects for transgenic control strategies. *Insect Biochem. Mol. Biol.* 34:167–176.

Rio, D. C. 2002. *P* transposable elements in *Drosophila melanogaster.* In N. L. Craig, R. Craigie, M. Gellert, and A. M. Lambowitz, eds., *Mobile DNA II,* pp. 484–518. Washington, D.C.: ASM Press.

Rizzon, C., G. Marais, M. Gouy, and C. Biemont. 2002. Recombination rate and the distribution of transposable elements in the *Drosophila melanogaster* genome. *Genome Res.* 12:400–407.

Robertson, H. M. 1997. Multiple *mariner* transposons in flatworms and hydras are related to those of insects. *J. Hered.* 88:195–201.

Robertson, H. M., C. R. Preston, R. W. Phillis, D. M. Johnson-Schlitz, W. K. Benz, and W. R. Engels. 1988. A stable genomic source of *P* element transposase in *Drosophila melanogaster. Genetics* 118:461–470.

Rohr, C. J., H. Ranson, X. Wang, and N. J. Besansky. 2002. Structure and evolution of *mtanga,* a retrotransposon actively expressed on the Y chromosome of the African malaria vector *Anopheles gambiae. Mol. Biol. Evol.* 19:149–162.

Romans, P., R. K. Bhattacharyya, and A. Colavita. 1998. *Ikirara,* a novel transposon family from the malaria vector mosquito, *Anopheles gambiae. Insect Mol. Biol.* 7:1–10.

Rongnoparut, P., N. Sirichotpakorn, R. Rattanarithikul, and K. J. Linthicum. 1998. Sequence heterogeneity in *copia*-like retrotransposons in *Anopheles* (Diptera: Culicidae) in Thailand. *J. Med. Entomol.* 35: 771–777.

Rubin, G. M. 1983. Dispersed repetitive DNAs in *Drosophila.* In J. A. Shapiro, ed., *Mobile Genetic Elements,* pp. 329–361. New York: Academic Press.

Rubin, G. M., and A. C. Spradling. 1982. Genetic transformation of *Drosophila* with transposable element vectors. *Science* 218:348–353.

Rubin, G. M., M. G. Kidwell, and P. M. Bingham. 1982. The molecular basis of P-M hybrid dysgenesis: The nature of induced mutations. *Cell* 29:987–994.

Russell, V. W., and R. H. Shukle. 1997. Molecular and cytological analysis of a *mariner* transposon from Hessian fly. *J. Hered.* 88:72–76.

Sarkar, A., R. Sengupta, J. Krzywinski, X. Wang, C. Roth, and F. H. Collins. 2003. *P* elements are found in the genomes of nematoceran insects of the genus *Anopheles. Insect Biochem. Mol. Biol.* 33:381–387.

Scavarda, N. J., and D. L. Hartl. 1984. Interspecific DNA transformation in *Drosophila. Proc. Natl. Acad. Sci. USA* 81:7515–7519.

Shao, H., and Z. Tu. 2001. Expanding the diversity of the *IS630-Tc1-mariner* superfamily: Discovery of a unique DD37E transposon and reclassification of the DD37D and DD39D transposons. *Genetics* 159:1103–1115.

Shao, H., Y. Qi, and Z. Tu. 2001. *MsqTc3,* a *Tc3*-like transposon in the yellow fever mosquito *Aedes aegypti. Insect Mol. Biol.* 10:421–425.

Shukle, R. H., and V. W. Russell. 1995. *Mariner* transposase-like sequences from the Hessian fly, *Mayetiola destructor. J. Hered.* 86:364–368.

Silva, J. C., and M. G. Kidwell. 2000. Horizontal transfer and selection in the evolution of *P* elements. *Mol. Biol. Evol.* 17:1542–1557.

Singer, M. F. 1982. *SINE*s and *LINE*s: Highly repeated short and long interspersed sequences in mammalian genomes. *Cell* 28:433–434.

Spradling, A. C., D. M. Stern, I. Kiss, J. Roote, T. Laverty, and G. M. Rubin. 1995. Gene disruptions using *P* transposable elements: An integral component of the *Drosophila* genome project. *Proc. Natl. Acad. Sci. USA* 92:10824–10830.

Steinemann, M., and S. Steinemann. 1998. Enigma of Y chromosome degeneration: *Neo*-Y and *neo*-X chromosomes of *Drosophila miranda* a model for sex chromosome evolution. *Genetica* 103:409–420.

Sved, J. A. 1976. Hybrid dysgenesis in *Drosophila melanogaster:* A possible explanation in terms of spatial organization of chromosomes. *Aus. J. Biol. Sci.* 29:375–388.

Torti, C., A. R. Malacrida, G. Yannopoulos, C. Louis, and G. Gapari. 1994. Hybrid dysgenesis-like phenomena in the medfly, *Ceratis capitata* (Diptera, Tephritidae). *J. Hered.* 85:92–99.

Torti, C., L. M. Gomulski, A. R. Malacrida, P. Capy, and G. Gasperi. 1997. Genetic and molecular investigations on the endogenous mobile elements of non-drosophilid fruitflies. *Genetica* 100:119–129.

Torti, C., L. M. Gomulski, A. R. Malacrida, P. Capy, and G. Gasperi. 1998. Characterization and evolution of *mariner* elements from closely related species of fruit flies (Diptera: Tephritidae). *J. Mol. Evol.* 46: 288–298.

Torti, C., L. M. Gomulski, D. Moralli, E. Raimondi, H. M. Robertson, P. Capy, G. Gasperi, and A. R. Malacrida. 2000. Evolution of different subfamilies of *mariner* elements within the medfly genome inferred from abundance and chromosomal distribution. *Chromosoma* 108:523–532.

Truett, M. A., R. S. Jones, and S. S. Potter. 1981. Unusual structure of the *FB* family of transposable elements in *Drosophila. Cell* 24:753–763.

Tu, Z. 1997. Three novel families of miniature inverted-repeat transposable elements are associated with genes of the yellow fever mosquito, *Aedes aegypti. Proc. Natl. Acad. Sci. USA* 94:7475–7480.

Tu, Z. 1999. Genomic and evolutionary analysis of *Feilai,* a diverse family of highly reiterated *SINEs* in the yellow fever mosquito, *Aedes aegypti. Mol. Biol. Evol.* 16:760–772.

Tu, Z. 2000. Molecular and evolutionary analysis of two divergent subfamilies of a novel miniature inverted repeat transposable element in the yellow fever mosquito, *Aedes aegypti. Mol. Biol. Evol.* 17:1313–1325.

Tu, Z. 2001a. Eight novel families of miniature inverted repeat transposable elements in the African malaria mosquito, *Anopheles gambiae. Proc. Natl. Acad. Sci. USA* 98:1699–1704.

Tu, Z. 2001b. *Maque,* a family of extremely short interspersed repetitive elements: Characterization, possible mechanism of transposition, and evolutionary implications. *Gene* 263:247–253.

Tu, Z., and S. P. Orphanidis. 2001. *Microuli,* a family of miniature subterminal inverted-repeat transposable elements (MSITEs): Transposition without terminal inverted repeats. *Mol. Biol. Evol.* 18:893–895.

Tu, Z., J. Isoe, and J. A. Guzova. 1998. Structural, genomic, and phylogenetic analysis of *Lian,* a novel family of non-LTR retrotransposons in the yellow fever mosquito, *Aedes aegypti. Mol. Biol. Evol.* 15:837–853.

Tubio, J. M., J. C. Costas, and H. F. Naveira. 2004. Evolution of the *mdg1* lineage of the *Ty3/gypsy* group of LTR retrotransposons in *Anopheles gambiae. Gene* 330:123–131.

Turcotte, K., S. Srinivasan, and T. Bureau. 2001. Survey of transposable elements from rice genomic sequences. *Plant J.* 25:169–179.

Vieira, C., D. Lepetit, S. Dumont, and C. Biemont. 1999. Wake up of transposable elements following *Drosophila* simulans worldwide colonization. *Mol. Biol. Evol.* 16:1251–1255.

Warren, A. M., M. A. Hughes, and J. M. Crampton. 1997. *Zebedee:* A novel *copia-Ty1* family of transposable elements in the genome of the medically important mosquito *Aedes aegypti. Mol. Gen. Genet.* 254: 505–513.

Warren, W. D., P. W. Atkinson, and D. A. O'Brochta. 1994. The *Hermes* transposable element from the house fly, *Musca domestica,* is a short inverted repeat-type element of the *hobo, Ac,* and *Tam3* (*hAT*) element family. *Genet. Res.* 64:87–97.

Warren, W. D., P. W. Atkinson, and D. A. O'Brochta. 1995. The Australian bushfly *Musca vetustissima* contains a sequence related to transposons of the *hobo, Ac* and *Tam3* family. *Gene* 154:133–134.

Wellauer, P. K., and I. B. Dawid. 1977. The structural organization of ribosomal DNA in *Drosophila melanogaster. Cell* 10:193–212.

Wilder, J., and H. Hollocher. 2001. Mobile elements and the genesis of microsatellites in dipterans. *Mol. Biol. Evol.* 18:384–392.

Wobus, U., H. Baumlein, S. S. Bogachev, I. V. Borisevich, R. Panitz, and N. N. Kolesnikov. 1990. A new transposable element in *Chironomus thummi. Mol. Gen. Genet.* 222:311–316.

Yoshiyama, M., H. Honda, T. Shono, and K. Kimura. 2000. Survey of *mariner*-like elements in the housefly, *Musca domestica. Genetica* 108:81–86.

CHAPTER SEVEN

Evolution and Development of the Dipteran Nervous System

David J. Merritt

The Diptera are an ancient group; hence, an understanding of evolution and development of their nervous system will open windows into 250 My of evolutionary history (Labandeira, Chapter 9). Among insects, the dipteran nervous system is the best understood, primarily because the fruit fly, *Drosophila melanogaster,* is the foremost invertebrate genetic model organism. Studies of *Drosophila* have provided extraordinary insights into biological processes at the levels of genes, molecules, cells, tissues, and behavior. In many of these investigations, a question that has been often posed is: how representative is *Drosophila?* To answer this question, comparisons with other Diptera can establish an evolutionary perspective within the order. Unfortunately, the number of such comparative studies is limited, and the genetic techniques available in *Drosophila* cannot yet be extended to other species. In this review, the majority of developmental information comes from *Drosophila* and where possible, I draw on comparative studies to provide an evolutionary perspective.

Central Nervous System Morphology

The nervous system of insects is composed of the supraoesophageal ganglion (the "brain"), which sits above the suboesophageal ganglion as a single fused mass (Strausfeld 1976), and the ventral nerve cord (VNC), made up of the suboesophageal ganglion and the thoracic and abdominal ganglia (Fig. 7.1A). The segmentally originating units of the central nervous system (CNS) are termed "neuromeres." They show various degrees of fusion into ganglia; for example, the brain represents a fusion of embryonic neuromeres whose segmental identity has been traced by expression patterns of segmental genes and the positions of nerve roots (Schmidt-Ott et al. 1994). Neuromeres and ganglia are composed of central neuropil surrounded by the cell bodies of neurons. The neuropil represents the synaptic region, made up of finely interwoven axons, dendrites, and synapses. The axon tracts are made up of bundled arrays of axons that may extend through connectives into adjacent ganglia or may diverge from tracts to branch in the neuropil.

BRAIN

Comprehensive atlases of the brains of two species of fly, *Musca domestica* (Strausfeld 1976) and *D. melanogaster* (Power 1943), have been published. As in all insects, the brain is subdivided into protocerebrum, deutocerebrum, and tritocerebrum. Protocerebral structures

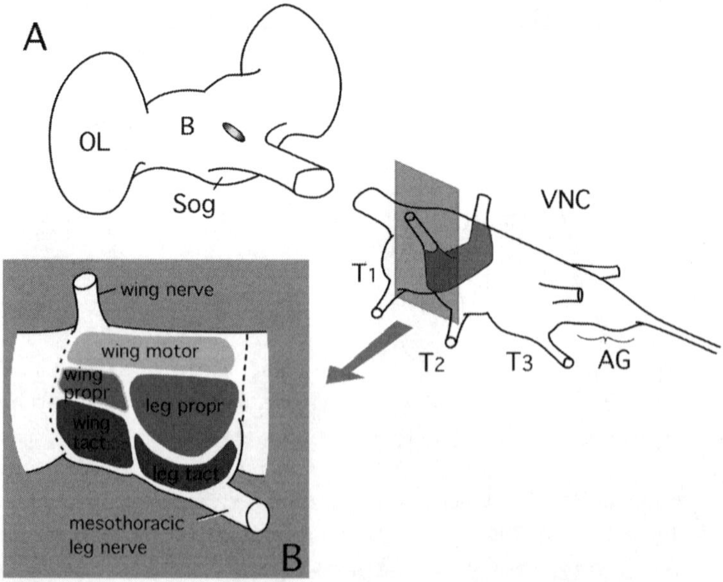

FIGURE 7.1. Diagram of the nervous system of a cyclorrhaphan fly. (A) The central brain, B, is flanked on either side by the optic lobes, OL. The suboesophageal ganglion, Sog, is also located in the head. The thoracic and abdominal ganglia compose the ventral nerve cord, VNC. Thoracic neuromeres, T1–3, are expanded to accommodate the leg neuropils. The abdominal ganglia, AG, are fused with the thoracic ganglia. (B) Lateral view of the mesothoracic neuromere, showing the functional subdivision of neuropils according to the location of the sensory neurons or motor axons.

include the pars intercerebralis, the antennal lobes, the central complex, and the mushroom bodies (corpora pedunculata). The mushroom bodies are coming under intense scrutiny in *Drosophila* because they are the seat of olfactory learning and higher behaviors, such as context generalization and visually mediated choice behaviors (Liu et al. 1999; Waddell and Quinn 2001; Heisenberg 2003). Clusters of olfactory glomeruli—the ball-like projection neuropils of olfactory sensilla—are located anteriorly in the protocerebrum, ventral to the mushroom bodies, forming the antennal lobe (Stocker 1994). They are the site of olfactory sensory neurons' first synapses with interneurons. The optic lobes, another site of sensory integration, lie on either side of the protocerebrum (Fig. 7.1A). In adult Diptera, they represent a considerable proportion of the cephalic nervous system, reflecting the importance of vision in fly behavior. The optic lobe neuropils are: the outer lamina, lying immediately beneath the compound eye; the medulla, lying beneath the lamina; and the lobula complex, lying deeper still (Fig. 7.2A). The lamina and medulla are very highly ordered neuropils carrying a retinotopic mapping of input from the photoreceptor neurons called "retinula cells."

An extreme example of evolutionary change in the nervous system accompanying external morphological change is seen in the stalk-eyed flies (Diopsidae), whose compound eyes are located on long stalks (Buschbeck and Hoy 1998). The stretch in head width has led to the lamina, medulla, lobula, and lobula plate lying within the eye stalks close to the compound

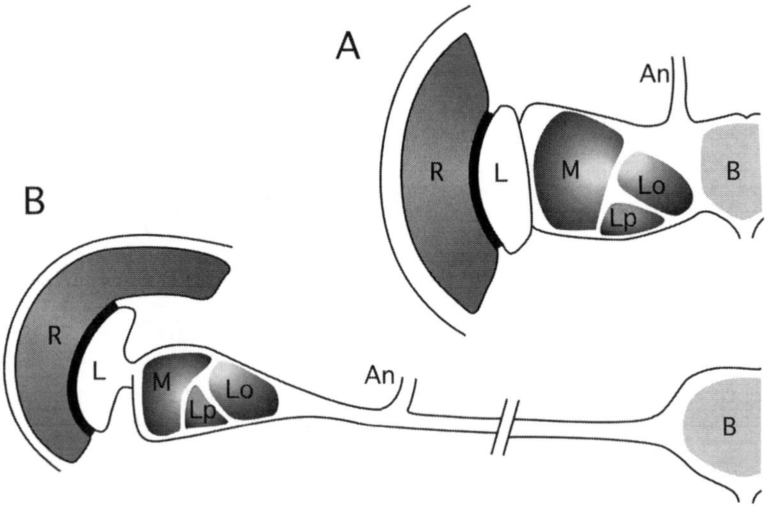

FIGURE 7.2. Dorsal view of (A) brain neuropils in a typical cyclorrhaphan fly compared to (B) a stalk-eyed fly (Diopsidae), in which the optic neuropils lie beneath the retina of the compound eye. Abbreviations: An, antennal nerve; B, brain; L, lamina; Lo, lobula; Lp, lobula plate; M, medulla; R, retina.

eye, and a unique, elongate optic nerve connecting each eye complex to the brain lying on the midline (Fig. 7.2B). Reduction in the numbers of some neural classes suggests there is a cost associated with the wide eye separation (Buschbeck and Hoy 1998).

VNC

The VNC includes the suboesophageal ganglion, the thoracic ganglia, and the abdominal ganglia. The thoracic and abdominal neuromeres can be fused into ganglia. In general, a string of isolated ganglia posterior to the suboesophageal ganglion is regarded as the primitive condition of VNC organization (Bullock and Horridge 1965). Variation in the degree of ganglion fusion in flies was investigated in detail by Brandt (1879). A subsequent systematic examination of neuromere fusion in adult Brachycera by Yeates et al. (2002) revealed that a consistent fusion pattern is usually seen within each family, and the degree of fusion is not affected by body shape. The basal configuration, seen in Lower dipteran outgroups, is a pattern of minimum fusion except for the joining of thoracic neuromeres 1 and 2 and joining of thoracic neuromere 3 with abdominal neuromere 1 (Fig. 7.3). The derived features are: (1) fusion of the thoracic and the anterior abdominal neuromeres, and (2) progressive fusion of the posterior neuromeres into a terminal abdominal ganglion. Neuromere fusion is seen in its most complete form in the joining of all thoracic and abdominal neuromeres into a synganglion in calyptrate flies. However, complete fusion appears to have evolved at least four times within the Eremoneura (Fig. 7.3). The suggested advantages of neuromere fusion include reduced neural conduction times, the elimination of relay interneurons, increased availability of sensory inputs to arrays of neurons (Leise 1991), and enhancement of sensorimotor coordination in the centers of locomotion (Yeates et al. 2002).

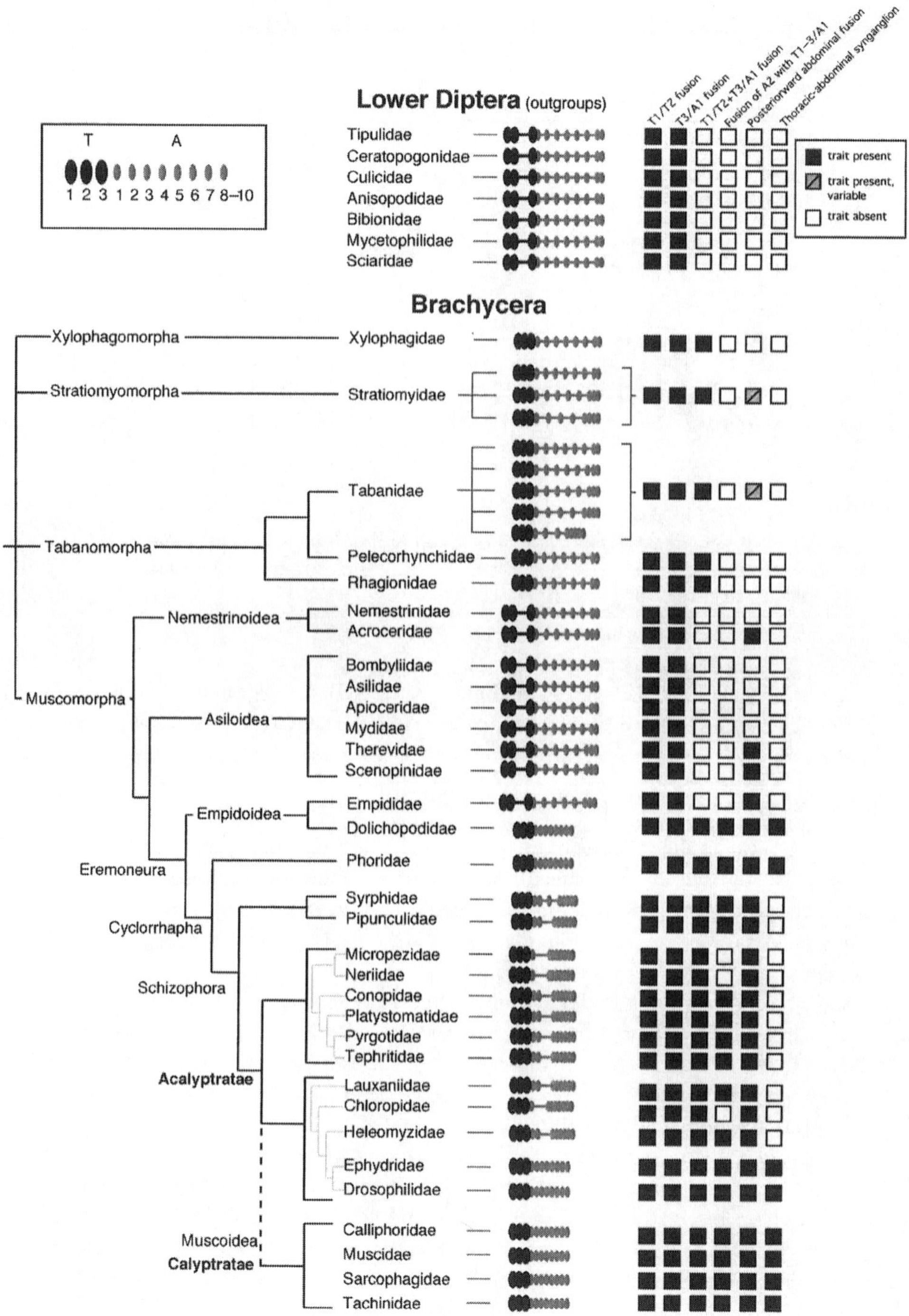

Lower Diptera (outgroups)
T A
1 2 3 1 2 3 4 5 6 7 8–10
T1/T2 fusion
T3/A1 fusion
T1/T2+T3/A1 fusion
Fusion of A2 with T1–3/A1
Posteriorward abdominal fusion
Thoracic-abdominal syngangion
trait present
trait present, variable
trait absent
Tipulidae
Ceratopogonidae
Culicidae
Anisopodidae
Bibionidae
Mycetophilidae
Sciaridae
Brachycera
Xylophagomorpha
Xylophagidae
Stratiomyomorpha
Stratiomyidae
Tabanidae
Tabanomorpha
Pelecorhynchidae
Rhagionidae
Nemestrinoidea
Nemestrinidae
Acroceridae
Bombyliidae
Asilidae
Muscomorpha
Asiloidea
Apioceridae
Mydidae
Therevidae
Scenopinidae
Empidoidea
Empididae
Dolichopodidae
Eremoneura
Phoridae
Syrphidae
Cyclorrhapha
Pipunculidae
Micropezidae
Neriidae
Schizophora
Conopidae
Platystomatidae
Pyrgotidae
Tephritidae
Acalyptratae
Lauxaniidae
Chloropidae
Heleomyzidae
Ephydridae
Drosophilidae
Calliphoridae
Muscoidea
Muscidae
Calyptratae
Sarcophagidae
Tachinidae

Larval VNC structure has not been as comprehensively sampled. Lower brachyceran larvae tend to show partial fusion of the larval VNC into a "5-limbed chain" or "12-limbed chain" (Melzer et al. 1995) that is difficult to reconcile with patterns of fusion in adults within the same families (Yeates et al. 2002). In Cyclorrhapha, both larvae and adults show the same pattern of complete VNC fusion (Melzer et al. 1995). The degree of fusion seen in embryos and its correspondence with larval and adult structure have received less attention. A preliminary study of the VNC of late-stage embryos of members of Stratiomyidae, Calliphoridae, and Tephritidae shows that they closely resemble the *Drosophila* embryonic VNC; that is, they possess an elongate trunklike cord with a pair of longitudinal connectives on either side of the midline and an anterior and posterior commissure within each neuromere (R. Herbert, J. Rice, and D. Merritt, pers. obs.). The larva of the tsetse fly, *Glossina pallidipes*, brooded within the female until the third instar, tends to retain embryonic features; namely, a reduced degree of VNC condensation (Truman 1990). Furthermore, the embryonic neuroblasts that normally become quiescent in the first instar remain active through the larval stages, indicating a general retardation of VNC development.

A number of studies have recognized that the individual neuromeres of flies, including the fused synganglion of the cyclorrhaphan flies, *Drosophila melanogaster*, and *Phormia regina* (Calliphoridae), follow the generic insect ground plan arrangement of longitudinal tracts, commissures, and neuropils (Murphey et al. 1989a; Merritt and Murphey 1992; Boyan and Ball 1993; arthropod ground plan reviewed by Leise 1991). Many neuropils have distinct boundaries and some can be identified as the site of sensorimotor integration in relation to discrete functions. For example, three pairs of ventral bulges of the thoracic neuromeres of flies have been termed the "leg neuropils" because they are the projection region of the sensory input from the legs. Equivalent neuropil bulges are not recognizable in the abdominal neuromeres because of the greatly reduced sensory input and motor output in the abdomen compared with the legs (Fig. 7.1A). Similarly, the axons from the rich complement of wing sensory receptors project into the wing neuropil (Fig. 7.1B), an expanded region of the T3 neuromere anterior to the T3 leg neuropils (Merritt and Murphey 1992). Typically, motor axons tend to have their axonal branches in the dorsal region of the ganglion. For example, the dendrites of the *Drosophila* flight motor neurons branch in the dorsal neuropil of the mesothoracic neuromere of the synganglion, where they are exposed to inputs from the halteres (Trimarchi and Schneiderman 1994; Consoulas et al. 2002). Homologous flight motor neurons can be identified in Calliphoridae based on neuronal morphology and soma position (Trimarchi and Schneiderman 1994). However, the large-diameter giant fiber neurons found in diverse cyclorrhaphan flies may not represent homologous neurons (King and Valentino 1983).

Different regions of each neuromere receive axons from sensory neurons of different sensory modalities. Proprioceptive sensory input, such as that derived from hair plate and

FIGURE 7.3. Diagrammatic representations of ventral nerve cord fusion in Diptera. The left column is a schematic representation of the nerve cord with fusion of neuromeres indicated by abutting ovals (thoracic and abdominal neuromeres shown in different shades according to inset, with T indicating thoracic neuromeres and A, abdominal neuromeres). The right-hand column shows ventral nerve cord fusion interpreted in the form of character states. Derivation of character states from ventral nerve cord fusion is described in Yeates et al. (2002). Reprinted with permission from Yeates et al. (2002).

campaniform sensilla, projects into the intermediate neuropil regions (Murphey et al. 1989a; Merritt and Murphey 1992), whereas input from tactile and gustatory receptors projects into a ventral layer, termed the "ventral association center." A further principle of organization is that sensory input from ventral appendages (legs) terminates posteriorly and ventrally in the neuromere, whereas input from dorsal appendages (wings and halteres) is found anteriorly in the neuromere (Fig. 7.1B). These principles of organization are conserved within Diptera; for example, receptors of the same type in homologous locations in *Phormia regina* (Calliphoridae) and *D. melanogaster* have very similar axonal projections (Murphey et al. 1989a). Signs of modality-specific layering are present in the embryonic and larval VNC of *Drosophila* (Schrader and Merritt 2000).

DEVELOPMENT

Many reviews have been written about specific aspects of neural development of *D. melanogaster,* including a number of chapters in the two-volume set, "The Development of *Drosophila melanogaster*" (Bate and Martinez-Arias 1993). Rather than reiterate these studies, I will give a brief overview. The brain and VNC is derived from neurectoderm defined during the early stages of embryonic development. In the VNC, all 30 neural precursor cells (neuroblasts) per hemisegment have been identified by their position (Broadus et al. 1995). Furthermore, the lineages and locations of motor neurons, interneurons, neurosecretory cells, and glia arising from each neuroblast have been identified in detail (Schmid et al. 1999). Together, the four classes comprise approximately 400 cells per abdominal hemisegment and approximately 500 per thoracic hemisegment (Schmid et al. 1999). Few comparative studies of embryonic development of the CNS have been carried out among Diptera; however, comparisons between the grasshopper and *Drosophila* indicate a high degree of conservation of neuroblasts and neuronal lineages (Thomas et al. 1984; Boyan and Ball 1993), as well as transcription factor expression (e.g., Hayward et al. 1995), indicating that the developmental pattern of the insect CNS is highly conserved.

One fundamental difference between hemimetabolous and holometabolous insects is the dissociation of CNS development in Holometabola into two phases associated with the larva and adult. By comparison with Hemimetabola, they have different body forms and a different repertoire of behaviors associated with their divergent habitats, seen in its most extreme form in higher Diptera. Surprisingly, the CNS is one of the more stable organs through metamorphosis. Although extensive remodeling occurs, primarily through addition of neurons in the larva (comprehensively reviewed by Consoulas et al. 2000; Truman et al. 1993; Truman 1996; Tissot and Stocker 2000), the fundamental arrangement does not change. Studies of metamorphic processes in *Drosophila* have revealed several principles. First, all adult motor neurons are present in the embryo. Some innervate larval muscles then undergo regression and regrowth of neurites to innervate adult muscles. Others do not send axons outside the CNS until metamorphosis, when the appropriate muscles form (Consoulas et al. 2002). Second, a set of neuroblasts undergoes mitotic arrest in the late embryo and re-enters mitosis in the larva to give rise to adult neurons, predominately interneurons. A small proportion of the abdominal neuroblasts reactivates in larvae, whereas the majority of thoracic neuroblasts do so, adding 4,000 or so neurons to each neuromere. Considerable expansion of the optic lobes also takes place (Truman and Bate 1988). Third, persisting neurons undergo considerable remodeling of their axonal arborizations and synaptic contacts to accommodate the modified adult motor and sensory systems. Fourth, some larval neurons die by a process of apoptosis.

Peripheral Nervous System

The peripheral nervous system (PNS) includes components of both sensory and motor systems. The sensory neurons originate in the peripheral ectoderm of the embryo and send their axons into the developing CNS. It is beyond the scope of this review to cover in detail the variety of sensillum types in Diptera. The structure of adult insect sensilla (Keil and Steinbrecht 1984; Steinbrecht 1984) and the sensilla of immature insects (Zacharuk and Shields 1991) have been comprehensively reviewed. I briefly mention the main types of sensilla, outline the nature of their axonal projections in the CNS, and present cases in which comparative studies have provided insights into PNS development or evolution.

TYPES OF SENSILLA IN LARVAE AND ADULTS

With the possible exception of olfactory receptors (see below), the cells that make up a sensillum are related by their direct descent from a single cell, a sense organ precursor (SOP). Sensilla may be innervated by one or many neurons, as well as several support cells derived from the SOP. The term "organule," coined by Lawrence (1966), is useful because of indications that the lineage relationships of sensilla are also present in evolutionarily related, noninnervated structures, such as dermal silk glands (see below) and the scales of Lepidoptera (Galant et al. 1998).

The external sensilla include tactile, gustatory, hair plate, campaniform, and olfactory sensilla. A single sheath cell (thecogen cell) wraps the neurons and secretes an extracellular electron-dense substance called the "dendritic sheath." At least two additional support cells are present in tactile and gustatory hairs, the tormogen and trichogen cells. These cells have dual roles: secretion of cuticular parts during development, and maintenance of ionic environment in the functioning sensillum.

Tactile Sensilla. Tactile sensilla are found either widely dispersed on the body or arranged in rows on the legs, wing margins, thorax, and other body parts of adults. They are innervated by a single ciliated sensory neuron. The sensory neuron terminates at the base of the hair shaft in a transducing structure. In the thoracic and abdominal neuromeres, the tactile neurons project their axons into the ventral association center (Murphey et al. 1989a). Frequently, the axons of the tactile hairs are somatotopically mapped into the CNS, the pattern of axonal terminations reflecting the positions of the sensilla on the body wall (Murphey et al. 1989b).

In Lower dipteran flies, the notum is covered with relatively unpatterned, randomly spaced tactile bristles of equal size. More derived Diptera show a progressive tendency toward location of the bristles in rows, and the differentiation of bristles into two classes: smaller microchaetes and larger macrochaetes. There is a tendency for the macrochaetes to become reduced in number and restricted to precise locations, indicating that the alignment of bristles into rows and stereotyped positioning of macrochaetes are derived features (Simpson et al. 1999). Comparing the patterns of large, thoracic, tactile sensilla (macrochaetes) in *Drosophila* with higher flies, Garcia-Bellido (1981) found that certain taxa show loss or gain of macrochaetes.

To derive the bristle patterns of the full range of flies, one must turn to the most complex pattern, termed the "prepattern," first recognized by Stern (1954), and subsequently examined by Simpson and coworkers. In Calyptrata, the macrochaetes are restricted to four well-defined rows on either side of the midline, and in most species, the number of macrochaetes is stereotyped. In Acalyptrata, the number of macrochaetes is reduced, although they are confined to positions that are recognizably within the borders of the four-row prepattern seen in the calyptrates.

The positions of microchaetes and macrochaetes have been examined in relation to the expression patterns of the proneural genes required for SOP formation and their regulatory genes. The *Drosophila achaete-scute* complex comprises four paralogous genes with very similar sequences. Although they have different functions, they also show a degree of functional overlap (discussed by Wülbeck and Simpson 2000). At least three of the four genes are present in the tephritid fly, *Ceratitis capitata.* If the fourth member, *achaete,* is definitely missing in *Ceratitis,* it indicates that a relatively recent duplication event has given rise to this gene in *Drosophila* (Wülbeck and Simpson 2000). Stripelike patterns of *scute* orthologs are seen in *Ceratitis* and *Calliphora,* similar to the *Drosophila* pattern. This expression pattern is suggested to be the precursor to individually specified macrochaetes, supporting the prepattern hypothesis (Pistillo et al. 2002). In the mosquito, *Anopheles gambiae,* the *achaete-scute* complex appears to be composed of two genes: an *asense* gene and a *lethal of scute*-like gene, suggesting that further duplications took place some time after the separation of Nematocera and Brachycera (Wülbeck and Simpson 2000).

Gustatory Sensilla. Gustatory sensilla are found at many locations on the bodies of adult flies (reviewed by Stocker 1994; Pollack and Balakrishnan 1997). They are concentrated on the mouthparts, the legs, the wing margins, and may be found on the female terminalia. They have been the subject of electrophysiological investigations in a wide range of species (Dethier 1976; Stocker 1994), and more recently, a family of genes encoding gustatory receptor molecules has been isolated from *Drosophila* (Dunipace et al. 2001; Scott et al. 2001). Usually three or four chemoreceptive neurons are associated with each sensillum, accessing the external environment through a pore at the tip of the hair. A mechanoreceptive dendrite is inserted at the base of the hair, so these receptors can be termed multimodal. Interestingly, the mechanoreceptive axon terminates in a different region of neuropil from the multiple gustatory axons associated with the same sense organ, indicating that, although they are clonally derived, the neurons do not necessarily show the same axonal projection characteristics (Murphey et al. 1989a). Although the gustatory sensilla on the wing margin are evenly spaced and increase in number with wing size in Drosophilidae (Dickinson et al. 1997), the gustatory sensilla on the legs of *Drosophila* are individually specified, like the notal macrochaetes (Held 2002).

Hair Plate and Campaniform Sensilla. Hair plates are mechanoreceptive sensilla found in clusters at the joints between the segments of the body or appendages. They resemble tactile sensilla with a shortened hair shaft and, like tactile sensilla, are each innervated by a single mechanosensory neuron. Campaniform sensilla, like hair plates, are concentrated in foci rather than being widely distributed. Morphologically, they are composed of a low-profile dome in a socket. They, too, have a single mechanosenory neuron terminating in a transducing structure at the base of the cuticular dome. Electrophysiological examinations show they respond to stresses in the cuticle. Both classes are proprioceptors, providing internal feedback on limb and body position and loading. Both generally have large-diameter axons, indicative of rapid conduction velocities, and widely ramifying axonal arborizations in the CNS. Comparison of the distributions of leg sensilla in *D. melanogaster* and two species of calliphorid flies shows high conservation of the location, number of constitutive sensilla, and morphological characters of hair plates and campaniform sensilla among cyclorrhaphan flies (Bryant 1978; Gnatzy et al. 1987; Merritt and Murphey 1992). The axonal projection characteristics are also conserved, the neurons of homologous sense organs in *Drosophila* and *Phormia* showing the same branching characteristics (Merritt and Murphey 1992). An investigation of the distribution of hair plate and campaniform sensilla on the coxa and trochanter throughout Diptera revealed that the number of component sensilla in identified clusters increases isometrically according to the animal's size in noncyclorrhaphan flies, whereas, with a few exceptions, in Cyclorrhapha, the number is low and invariant—no matter the insect's size (Frantsevich and Gladun 2002). Similarly, a fixed, size-independent number of isolated campaniform sensilla is present on the wing blade of acalyptrate and calyptrate flies; however, position and number are variable in more basal Diptera (Dickinson and Palka 1987; Dickinson et al. 1997). The evolutionary tendency toward size-independent stereotypy of the number and location of several different types of sensilla (campaniform sensilla, hair plate sensilla, and macrochaetes) in Cyclorrhapha hints at an evolutionary transition in the way these sensilla are patterned. The Cyclorrhapha appear to have abandoned classical spacing mechanisms (Held 2002) that result in sensillum numbers increasing proportionally as the size of the competent tissue patch increases with growth. Rather, the stereotypy suggests that a flexible spacing mechanism became a fixed mechanism at some point in evolutionary history, and in their subsequent radiation, the Cyclorrhapha have maintained the resulting pattern.

Olfactory and Antennal Sensilla. Olfactory sensilla are a class of external sensilla found on the antennae and maxillary palps of adult Diptera (Stocker 1994), and possibly on the female genitalia (Merritt and Rice 1984). They are generally innervated by several neurons, each with finely branched dendrites that ramify within an external cuticular component, with many fine pores in the cuticular wall. In *D. melanogaster,* olfactory sensilla are present on the surface of the funiculus and the maxillary palps (Venkatesh and Singh 1984; Stocker 1994). The axons of antennal olfactory neurons of *D. melanogaster* project into well-defined glomeruli within the antennal lobe. The maxillary palp olfactory afferents enter the CNS in the labial nerve and pass through the suboesophageal ganglion into the antennal lobes (Stocker et al. 1990). A total of 57 odor receptor–encoding candidate genes have been identified in *D. melanogaster* and 79 in *An. gambiae,* based on genome analysis (Hill et al. 2002). Interestingly, only one odorant receptor type shows high sequence conservation between the two species. An outstanding feature of comparison of the two genomes is the expansion of subfamilies of odor receptor types unique to each lineage, perhaps due to the ecological and physiological specializations associated with host finding in *Drosophila* (fruit-associated) vs. *Anopheles* (vertebrate-associated) (Hill et al. 2002).

Additional sensory modalities found on the antennae or palps of adult Diptera include thermoreceptive sensilla, CO_2 receptors, and chordotonal organs. Combinations of behavioral analysis, mutant analysis, and ablation techniques have confirmed that a hygroreceptive sensitivity is located in the arista and a thermoreceptive sensitivity in the funiculus of adult *D. melanogaster* (Sayeed and Benzer 1996). A sensillum with either thermoreceptive (Foelix et al. 1989) or hygroreceptive (Sayeed and Benzer 1996) function is present at the base of the arista of *Drosophila, Musca,* and *Calliphora.* Highly lamellate dendrites within a sensillum are suggestive of a thermoreceptive function in insects (Steinbrecht 1984). Such sensilla are found on the antennae of mosquitoes and midges (Boo and Richards 1975; Cribb 1996), and maxillary palps of adult midges (Chu-Wang et al. 1975).

External sensilla of the larvae of Diptera are at times difficult to classify into the categories listed above. For example, the sensilla of soft-bodied cyclorrhaphan larvae have reduced external structures, and behavioral and electrophysiological determinations of their function are lacking. In another example, the function of putative chemoreceptive sensilla on larval mosquito antennae has yet to be proven (Clements 1999). The antennal sensilla of larval Cyclorrhapha have been reviewed by Cobb (1999). The dorsal organ is the main olfactory organ, determined in *Drosophila* by electrophysiological recordings and by abolition of olfactory orientation after toxin expression in the olfactory cells (Heimbeck et al. 1999; Oppliger et al. 2000). Nicastro et al. (1998) carried out a Diptera-wide investigation of antennal shape and their component sensilla in larvae. They found that the "sensory cone" structure—the most conspicuous component of the cyclorrhaphan dorsal organ—is common to almost all Diptera, the ground plan being composed of seven olfactory sensillum units, each with two to three dendrites. Various synapomorphies and plesiomorphies of the sensory cone and a "peg organ" were identified and mapped onto a cladogram of the Diptera (Nicastro et al. 1998).

Chordotonal Sensilla. Chordotonal sensilla are stretch-sensitive units that are suspended at either end between cuticular attachment points, such as the joints of appendages. In adult flies, chordotonal sensilla are found within the legs, at the base of the antenna, and along the body wall (Field and Matheson 1998). They may be composed of a single sensillum or be compound organs, such as Johnston's organ found at the base of the adult antenna. At its most elaborate form in Culicidae and Chironomidae, it is composed of 25,000 to 30,000 neurons per antenna (Boo and Richards 1975). The dense innervation of the antennal pedicel and associated complex cuticular structures provide sensitivity to airborne vibrations. An interesting innovation is the evolution of a hearing organ in the anterior thorax of ormiine flies (Tachinidae). These orthopteran parasites detect their host by sound. A proprioceptive chordotonal organ found in the anterior thorax of all flies has been modified through an increase in the sensillum complement and development of a tympanum to provide the audio spectral sensitivity to locate their hosts (Lakes-Harlan and Heller 1992; Robert et al. 1992). Insects appear to have a conserved pattern of chordotonal sensilla, with similarities between Orthoptera, Diptera, and Lepidoptera in their distribution in larvae, as discussed by Wong and Merritt (2002).

Multiterminal Sensilla. Multiterminal (multiple dendrite [md]) neurons are characterized by widely ramifying dendrites associated with the epidermis or internal organs and are generally considered to be proprioceptors (Finlayson 1976). In Diptera, little is known of their distribution and function outside of *D. melanogaster.* A number of roles have been suggested for the md neurons of larvae but few have been proven. A sensitivity to punctate stimuli is possible in the soft-bodied larvae of Cyclorrhapha, such as *Drosophila,* because some md axons project into the same neuropil as tactile receptors (Schrader and Merritt 2000; Grueber et al. 2002). The larval md neurons also mediate the thermal nociceptive response (Tracey et al. 2003) and could act as general thermoreceptors (Liu et al. 2003). Each abdominal and thoracic segment of *Drosophila* bears a "dorsal bipolar dendrite" md neuron (dbd) that is located in a characteristic dorsolateral position and is attached to a clonally related ligament-like horizontal support cell (Bodmer et al. 1989; Jones et al. 1995). A neuron with dbd-like neuronal morphology and support cell is also present in Calliphoridae, Tephritidae, and Stratiomyidae (R. Herbert, J. Rice, and D. Merritt, pers. obs.) and is also likely to be present in other insects (Osborne and Finlayson 1962; Osborne 1963), possibly indicating a conserved, proprioceptive function.

External Glands as Modifications of Sensilla. It has been suggested that some external glands could be derived from sensilla (Wigglesworth 1953). Male empidid flies possess swollen foretarsi containing a concentration of silk-producing glands that are used to wrap nuptial gifts. An ultrastructural investigation revealed that the silk-producing structures are most likely modified contact chemoreceptive sensilla that have lost their neural component and developed the secretory function (Young and Merritt 2003). Tergal glands of tephritid flies are noninnervated, sensillum-like structures that resemble adjacent sensory hairs (Evans 1967), indicating that glands may have independently arisen from sensilla in a number of taxa. These examples show the extraordinary evolutionary flexibility of the organules that make up the PNS.

SENSILLUM DEVELOPMENT

The development of sensilla has been well studied in *D. melanogaster* (Ghysen and Dambly-Chaudière 1992; Jan and Jan 1993, 1994). A cluster of ectodermal cells takes on the potential to become neural precursors due to the expression of a proneural gene in all the cells of the cluster. Known proneural genes include the *achaete scute* complex, required for formation of tactile, gustatory, campaniform, and hair plate sensilla (Leyns et al. 1989; Held 1990; Dominguez and Campuzano 1993); *atonal,* required for chordotonal organs (Jarman et al. 1993) and a subset of olfactory sensilla; and *amos,* required for a subset of md (including dbd) and subset of olfactory sensilla (Goulding et al. 2000; Huang et al. 2000). The cells of the proneural cluster interact through a process of lateral inhibition, resulting in one cell becoming a SOP. The SOP then undergoes a series of specific, oriented divisions to give rise to the support cells and associated sensory neurons.

In tactile receptors, the trichogen and tormogen cells form the template upon which cuticle is laid down. In some cases, one of these two outer support cells degenerates in the adult. In the calliphorid, *Calliphora vicina,* the trichogen cell of the thoracic macrochaetes degenerates (Keil 1978). In *D. melanogaster,* both the tormogen and trichogen cells degenerate after formation of the interommatidial bristles (Perry 1968). The lineage of gustatory sensilla requires numerous extra divisions to give rise to the multiple chemosensory neurons (reviewed by Pollack and Balakrishnan 1997). All *achaete-scute*–dependent sensilla have a single thecogen cell wrapping a single neuron or multiple neurons.

The constituent cells of the chordotonal sensillum are a cap cell, an ectodermal attachment cell, a scolopale cell, a ligament cell, and one or two neurons (Brewster and Bodmer 1995). Considering lineage, gene expression, and morphology, a number of cellular homologies are apparent between the external sense organs and chordotonal organs; for example, the sheath cells (thecogen and scolopale, respectively) and neurons show a number of homologies in structure (Merritt 1997) and stimulus transduction mechanisms (Eberl et al. 2000).

The olfactory sensilla of *Drosophila* are composed of three support cells—trichogen, tormogen and thecogen—plus one to four neurons (Venkatesh and Singh 1984). By comparison, the sensilla basiconica of *C. erythrocephala* (Calliphoridae) have two neurons and four support cells. One cell, the tormogen cell, degenerates in the late pupa (Kuhbandner 1984). Rodrigues and coworkers (Ray and Rodrigues 1995; Sen et al. 2003) claim that development of olfactory sensilla in *Drosophila* is divergent from the normal clonal mode of sensillum development, more closely resembling the recruitment process seen in the retinula cells of the eye (see below). According to lineage reconstructions using immunostaining and enhancer trap lines, a founder cell recruits a cluster of three secondary progenitor cells that divide further to give rise to additional support cells and neurons. Founder cell formation requires the action of the proneural genes *atonal* or *amos* (Gupta and Rodrigues 1997; Goulding et al. 2000). As Keil (1997) points out, this process is fundamentally different from sensillum development in other sensillum types of *Drosophila,* and indeed from olfactory sensillum development in the silkmoth *Antheraea,* which displays a conventional clonal lineage.

The dissociation of larval and adult development seen in the CNS of Diptera is also seen in the PNS development. A major difference is that although the CNS retains a recognizable structure through metamorphosis, the sensory system is substantially replaced. However, it is apparent that persistent sensory elements of the larval PNS are important in establishing an axon scaffold that the adult sensory neurons follow into the CNS (Garcia-Alonso 1999), just as the axons of newly added sensilla at each moult in hemimetabolous insects follow pre-existing pathways. Although pre-existing nerves are not necessarily essential for new axons to find the CNS (Tix et al. 1989; Kunes et al. 1993), errors of axon growth within the CNS are seen in the absence of the persistent neurons of the abdominal and thoracic segments of the *Drosophila* larva (Williams and Shepherd 2002).

A subset of leg sensory neurons also persist through metamorphosis. Sensory structures called "Keilin's organs" (KO), located on the thorax of cyclorrhaphan larvae at the point of connection of the imaginal disc with the hypodermis, are suggested to be vestiges of ancestral larval legs (Keilin 1915). In both *D. melanogaster* and *Phormia regina* (Calliphoridae), the KO are connected by long dendrites to neurons whose somata are drawn inward into the developing leg imaginal disc (Lakes-Harlan et al. 1991a,b). Their axons provide the pathway for the guidance of adult sensory axons. Some of the persistent sensilla of *Phormia* re-emerge from metamorphosis as adult tarsal sensilla (Lakes-Harlan et al. 1991a,b; Fig. 7.4). Positional homologs of KO are present in the larval thoracic segments of the phantom midge *Chaoborus crystallinus* (Melzer et al. 1999). Although neuronal persistence has not been investigated, the somata of these sensory neurons are less extensively internalized than in the Cyclorrhapha, perhaps indicating an intermediate stage in the trend to the extreme internalization of the sensory neurons of imaginal discs seen in Cyclorrhapha (Fig. 7.4).

In the examples above, a subset of sensory neurons of the thoracic body wall, leg rudiments, and abdomen persist through metamorphosis, the remaining neurons degenerating. The antenna is a sensory appendage in which the sensory neurons do not persist through metamorphosis, at least in Cyclorrhapha. The pathfinding role of larval axons is implemented in the pupa when adult afferent axons grow into the CNS along the pre-existing larval anten-

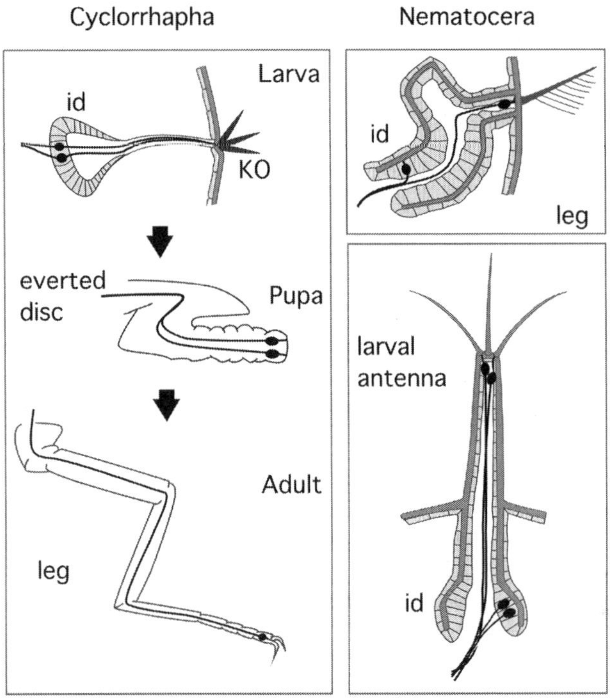

FIGURE 7.4. Diagram of persistent sensory neurons associated with the Keilin's organs, leg imaginal discs, and adult legs. In Cyclorrhapha (left), the somata of sensory neurons innervating Keilin's organs, KO, are located in the epithelium of the deeply invaginated imaginal discs, id. After imaginal disc eversion in the pupa, the persisting sensory neurons are located in the tip of the adult leg. By comparison, the leg imaginal discs of larval Lower Diptera are located immediately beneath the corresponding larval structures; in this case, the leg rudiments of the phantom midge, *Chaoborus crystallinus*. The larval antenna of *C. crystallinus* is also depicted, showing the continuity of the larval antenna and the imaginal tissue that will give rise to the adult antenna. Lower Diptera diagrams adapted from Melzer et al. (1999). Cyclorrhapha diagrams adapted from Lakes-Harlan et al. (1991b).

nal nerve before the larval afferents degenerate (Tissot et al. 1997). Correspondingly, the larval and adult antennal lobes represent the same neuropil at different stages of the life history (Tissot et al. 1997; Python and Stocker 2002). Once again, an intermediate stage appears to be present in the nematoceran *Chaoborus crystallinus,* in which imaginal tissue forms as a basal ingrowth of the larval antenna (Fig. 7.5). Larval sensory structures may persist as the tip of the future adult structure (Melzer et al. 1999), as occurs in Lepidoptera (Svacha 1992). In mosquitoes, the antennal imaginal discs are further internalized, lying within the larval head directly below and connected to the larval antennae (Imms 1908). The Cyclorrhapha appear to have undergone a relocation of imaginal tissue from the larval antenna base to the

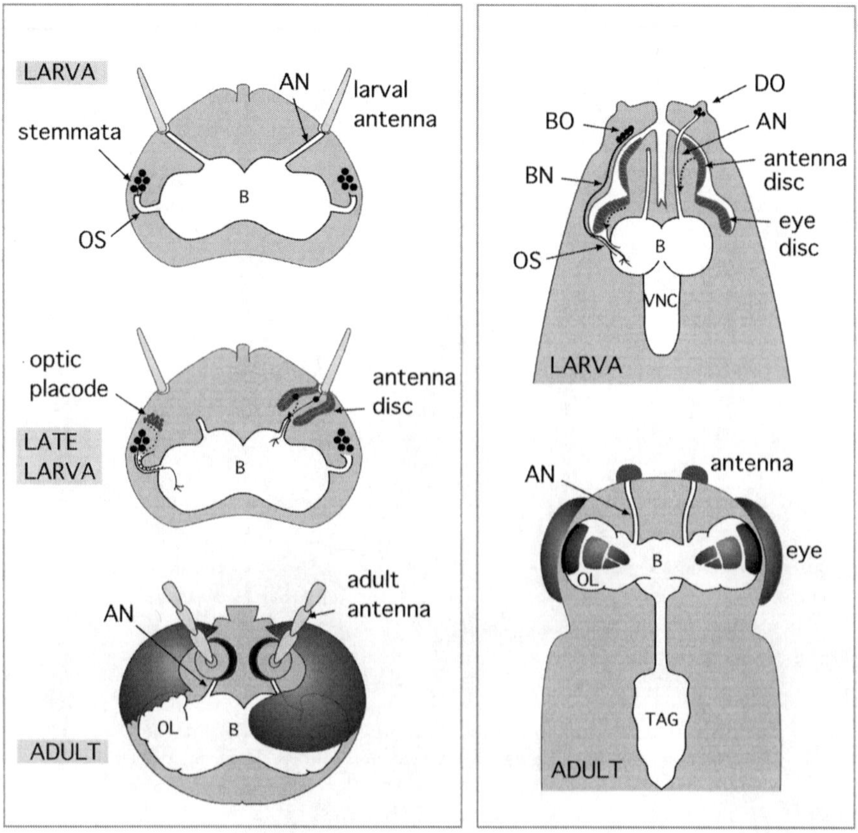

FIGURE 7.5. Representation of the larva-to-adult transition in eyes and antennae of a Lower dipteran, such as a mosquito or midge (left), and a cyclorrhaphan, such as *Drosophila* (right). In *Drosophila,* the larval Bolwig's organ, BO, is the light receptor. The eye component of the eye-antenna imaginal disc lies internally, away from the Bolwig's organ. Outgrowing retinal axons in the late larva follow the Bolwig's nerve, BN, into the optic lobes (dotted line) along the optic stalk, OS, before the Bolwig's organ neurons degenerate in the pupa. The larval antenna is the dorsal organ, DO. The olfactory axons from the developing adult antenna, derived from the antenna disc, follow the larval antennal nerve, AN, into the brain, B (right side). By the adult stage, the larval olfactory and visual neurons have degenerated and the imaginal discs everted. In the Lower dipteran example, the photoreceptor axons derived from larval stemmata (black circles) send axons into the optic lobe, OL. In the last instar, the optic placode (gray circles) develops anterior to the stemmata. The axons of newly added photoreceptor neurons putatively follow the pre-existing optic stalk into the optic lobes. The antennal imaginal disc is depicted as an invagination of the base of the larval antenna. The adult olfactory axons also putatively follow the pre-existing antennal nerve into the olfactory lobes. The terminal abdominal ganglion, TAG, formed by the fusion of all thoracic and abdominal ganglia, is found in many Cyclorrhapha.

"dorsal pouch" that gives rise to the eye-antennal imaginal disc (Jürgens and Hartenstein 1993), probably associated with the evolution of larval head involution.

The extensive replacement of hypodermis occurring in the higher Diptera has led to the concept of the imaginal discs of Holometabola as undifferentiated embryonic tissue that completely regenerates the external body plan at metamorphosis. Instead, Svacha (1992) has argued that imaginal discs are not undifferentiated and "embryonic"; rather, they are continuous parts of the larval body plan, and they represent sequential homologs of embryonic and adult structures. The foregoing studies of the sensory system through metamorphosis support the sequential homology by indicating that (1) in some segments, sensory neurons persist from the homologous larval structure to the adult; (2) in cases in which sensory neurons do not persist, adult sensory afferents utilize pre-existing nerves to enter the CNS; and (3) intermediate forms of imaginal disc are recognizable in Lower Diptera, which can explain the evolutionary derivation of extremely invaginated imaginal discs.

VISUAL SYSTEM

Adult flies possess compound eyes and ocelli located between the compound eyes. The compound eyes are composed of many hexagonal ommatidia. Eight photoreceptors, retinula cells R1–8, are clustered beneath each ommatidium. The archetypal arrangement of retinula cells in insects is a symmetrical arrangement of R1–6 beneath each ommatidium. Their rhabdomeres (the terminal microvillar component of the R cells) abut to form a single, central, closed rhabdom. Lower Diptera generally show this basal, symmetrical pattern of retinula cells and a closed rhabdom beneath the ommatidium (see Melzer and Paulus 1991). However, Brachycera possess an open rhabdom, derived from the basal, closed structure (Shaw 1989). The rhabdomeres of R1–6 are arranged around a cavity containing a central rhadomere made up of R7 and R8. Furthermore, the R3 cell takes an eccentric location, forming a trapezoid arrangement of retinula cells. The orientation of the trapezoid is flipped 180°, depending on whether the ommatidium lies above or below an equatorial line running through each eye. The trapezoid arrangement and open rhadoms are seen only in Brachycera (Shaw 1989, 1990), indicative of an evolutionary progression in Diptera.

The axonal projections of retinula cells also show evolutionary changes. The ground plan arrangement in insects is for the R1–6 axons of a single ommatidium to project into a single cartridge in the lamina, representing the same point in space as the ommatidium, thus forming a retinotopic map in the brain. This arrangement is modified in Diptera. In Lower Diptera, the photoreceptor axons branch in the lamina, sending collaterals into adjacent cartridges as well as their corresponding cartridge (neural superposition). The result is that each ommatidium sends input to 18 surrounding cartridges (Melzer et al. 1997), seen as a way of increasing low-light sensitivity (Nilsson and Ro 1994). In Brachycera, the collaterals are pruned, but adjacency of innervation is retained: the axons of R1–6 from a single ommatidium project into adjacent lamina cartridges (Shaw 1990). This arrangement results in the retinula cells from adjacent ommatidia that see the same point in space, due to ommatidial lens curvature projecting to the same cartridge, proposed as a way of increasing visual acuity. The arrangement represents the endpoint in Diptera of a progressive evolution of axonal projection characteristics due to differing visual requirements (Melzer et al. 1997).

Evolutionary changes in synaptogenesis in the adult dipteran eye have also been closely studied. Within the lamina, the identity of cells is the same across the families; however, progressive alterations in axonal arborization patterns and synaptic partners are apparent (reviewed by Shaw 1989). Synaptic relationships have shown progressive changes over evolutionary time, tending toward greater complexity. Six characteristics were found by Shaw and Meinertzhagen (1986) to vary across the Muscomorpha and were used to construct a phylogenetic tree that showed strong accord with the accepted phylogeny of Diptera with some possible alternative placements (Buschbeck 2000).

VISUAL SYSTEM DEVELOPMENT

Visual units of flies develop in a fundamentally different mode to sensilla. The photoreceptors and support cells of each ommatidium are not clonally derived; rather, they are recruited by inductive processes from a pool of surrounding cells (Wolff and Ready 1993). *D. melanogaster* is typical of Cyclorrhapha, with the eyes and antennae forming from deeply invaginated imaginal discs that undergo extensive differentiation in the third instar larva and extensive morphogenesis in the pupa. The cellular and genetic processes involved in formation of the adult compound eye have been described in detail (Dickson and Hafen 1993; Wolff and Ready 1993; Frankfort and Mardon 2002). Differentiation begins as a morphogenetic furrow sweeps across the disc, leaving behind groups of neurons. The R8 photoreceptor is the first to differentiate, followed by R2–7 as they are progressively recruited from surrounding undifferentiated cells behind the morphogenetic furrow. In contrast to the wealth of knowledge of ommatidium development in *Drosophila*, very little is known of the process in more basal Diptera.

HOMOLOGY OF LARVAL AND ADULT EYES

The evolutionary and developmental relationship between larval and adult eyes of Diptera is contentious because of the extreme morphological changes occurring at metamorphosis. The dipteran visual system is presumably originally derived from the hemimetabolous pattern, in which ommatidia are progressively added to the compound eyes at each moult, usually from a proliferative zone at the dorsal eye border (Bodenstein 1953). Lower Dipteran larvae rarely possess compound eyes. They have stemmata, considered by Paulus (1979) to be modified ommatidia. In the mosquito, *Aedes aegypti*, the neonate initially possesses a small number of stemmata. Through subsequent instars, the number of stemmata remains constant, and an epidermal thickening (optic placode) develops anterior and adjacent to the stemmata (Fig. 7.5). In the fourth instar, a mitotic wave moves from posterior to anterior across the placode, leaving in its wake the differentiated ommatidia of the compound eye that becomes fully functional in the adult, by which time the stemmata have degenerated (White 1961). Thus larval-to-adult eye development in the mosquito can be interpreted as a modification of the hemimetabolous pattern through (1) reduction and specialization of the larval component of the eye primordium (=stemmata), along with (2) delayed development of the adult component. Comparing this process with eye-antenna imaginal disc development in Cyclorrhapha, the optic placode is the Lower dipteran equivalent of the eye imaginal disc that has not yet been relocated and internalized, and the mitotic wave is the forerunner of the cyclorrhaphan morphogenetic furrow.

The photoreceptors of cyclorrhaphan larvae are termed "Bolwig's organs" (BO) (Bolwig 1946). Like the stemmata of larval Lower Diptera, the BO of Cyclorrhapha degenerate in the pupa and do not contribute to the adult compound eye. The BO are recognized by some authors as homologs of Lower dipteran stemmata (Paulus 1989; Wolff and Ready 1993); however, alternatives have been suggested (Tomlinson and Ready 1987; see Meinertzhagen and Hanson 1993). Several lines of developmental evidence support a sequential homology between BO and the adult compound eye. First, in *D. melanogaster*, the BO and adult eyes derive from the same continuous region of the embryonic head ectoderm (Daniel et al. 1999; Namba and Minden 1999). The developing BO is drawn inward at head involution, whereas the future imaginal disc cells invaginate to form a pouch in the late embryo that later forms the eye-antenna imaginal disc. Second, axons of the compound eye photoreceptors follow the BO photoreceptor axons into the optic lobe (Fig. 7.5), entering by the same nerve (Steller et al. 1987) and projecting into the lamina neuropil that was initially established by the BO axons. Furthermore, the photoreceptor units of BOs and compound eyes show similar gene expression and development (Daniel et al. 1999), as well as expressing a similar suite of rhodopsin genes (Pollock and Benzer 1988). Although more extensive studies are required, it appears that larval stemmata and adult compound eyes are temporal variants of homologous units in Diptera and that stemmata and BO are homologous units.

Conclusion

One of the reasons for the success of Diptera is the ability of the larva to exploit a wide range of habitats commonly associated with high moisture content. Consequently, dipteran larvae usually show little resemblance to the adult form. Although some researchers regard the relatively simple larval stages of Diptera as showing primitive features, observations of nervous system development lead to the conclusion that they are highly adapted, derived stages. By comparison with hemimetabolous insects, evolution has modified and simplified the larval stages. In the CNS, neuroblasts become quiescent late in embryogenesis and formation of the full suite of adult neurons is delayed. The larval PNS is pared down compared to the adult and is substantially replaced at metamorphosis. However, persistent sensory neurons guide the newly formed adult sensory afferents into the CNS. Indeed, by tracing the axonal pathways taken by adult sensory neurons developing in the imaginal discs and following persistent sensory neurons through metamorphosis, it is possible to establish identities between larval and adult structures in the Dipteran body. The phylogenetic viewpoint of the origin of imaginal discs as structures that possess sensory elements and whose final differentiation is delayed until metamorphosis dispels the notion that imaginal discs are undifferentiated, embryonic tissue.

A question posed at the beginning of this review was: how representative is *Drosophila* as a model organism? From the viewpoint of nervous system development and evolution, *Drosophila* appears to be typical of the Cyclorrhapha. This highly successful group has many features in common; namely, a similar degree of CNS fusion, the development of extremely internalized imaginal discs, and extreme modification of the larval head as a result of embryonic head involution. The Lower Brachycera and Lower Diptera do not show the same degree of internalization of imaginal discs and head remodeling. Given the availability of the genome of the mosquito, *Anopheles,* and the cyclorrhaphan fly, *Drosophila,* the opportunity is now available to link changes in gene expression with these radical evolutionary changes in the body plan and the nervous system.

ACKNOWLEDGMENTS

Thanks to David Yeates for his support and patience during the writing of the chapter, and to Eldon Ball for his comments on the manuscript.

REFERENCES

Bate, M., and M. Martinez Arias. 1993. *The Development of* Drosophila melanogaster. Plainview, N.Y.: Cold Spring Harbor Laboratory Press.

Bodenstein, D. 1953. Postembryonic development. In K. D. Roeder, ed., *Insect Physiology*, pp. 822–865. New York: John Wiley and Sons.

Bodmer, R., R. Caretto, and Y. N. Jan. 1989. Neurogenesis of the peripheral nervous system in *Drosophila* embryos: DNA replication patterns and cell lineages. *Neuron* 3:21–32.

Bolwig, N. 1946. Senses and sense organs of the anterior end of the housefly larva. *Vidensk. Meddr. dansk naturh. Foren.* 109:81–217.

Boo, K. S., and A. G. Richards. 1975. Fine structure of the scolopidia in the Johnston's organ of male *Aedes aegypti* (L.) (Diptera: Culicidae). *Int. J. Insect Morphol. Embryol.* 4:549–566.

Boyan, G. S., and E. E. Ball. 1993. The grasshopper, *Drosophila* and neuronal homology (advantages of the insect nervous system for the neuroscientist). *Prog. Neurobiol.* 41:657–682.

Brandt, E. 1879. Vergleichend-anatomische Untersuchungen über das Nervensystem der Zweiflügler (Diptera). *Horae Soc. Entomol. Russ.* 15:84–101.

Brewster, R., and R. Bodmer. 1995. Origin and specification of type II sensory neurons in *Drosophila. Development* 121:2923–2936.

Broadus, J., J. B. Skeath, E. P. Spana, T. Bossing, G. Technau, and C. Q. Doe. 1995. New neuroblast markers and the origin of the aCC/pCC neurons in the *Drosophila* central nervous system. *Mech. Dev.* 53: 393–402.

Bryant, P. J. 1978. Pattern formation in imaginal discs. In M. Ashburner and T. R. F. Wright, eds., *The Genetics and Biology of* Drosophila, vol. 2c, pp. 230–336. New York: Academic Press.

Bullock, T. H., and G. A. Horridge. 1965. *Structure and Function in the Nervous System of Invertebrates.* San Francisco: W. H. Freeman.

Buschbeck, E. K. 2000. Neurobiological constraints and fly systematics: How different types of neural characters can contribute to a higher level dipteran phylogeny. *Evolution* 54:888–898.

Buschbeck, E. K., and R. R. Hoy. 1998. Visual system of the stalk-eyed fly, *Cyrtodiopsis quinqueguttata* (Diopsidae, Diptera): An anatomical investigation of unusual eyes. *J. Neurobiol.* 37:449–468.

Chu-Wang, I., R. C. Axtell, and D. L. Kline. 1975. Antennal and palpal sensilla of the sand fly *Culicoides furens* (Poey) (Diptera: Ceratopogonidae). *Int. J. Insect Morphol. Embryol.* 4:131–149.

Clements, A. N. 1999. *The Biology of Mosquitoes.* Vol. 2: *Sensory Reception and Behaviour.* Oxon: CABI.

Cobb, M. 1999. What and how do maggots smell? *Biol. Rev. Camb. Philos. Soc.* 74:425–459.

Consoulas, C., C. Duch, R. J. Bayline, and R. B. Levine. 2000. Behavioral transformations during metamorphosis: Remodeling of neural and motor systems. *Brain Res. Bull.* 53:571–583.

Consoulas, C., L. L. Restifo, and R. B. Levine. 2002. Dendritic remodeling and growth of motoneurons during metamorphosis of *Drosophila melanogaster. J. Neurosci.* 22:4906–4917.

Cribb, B. W. 1996. Antennal sensilla of the female biting midge: *Forcipomyia* (Lasiohelea) *townsvillensis* (Taylor) (Diptera: Ceratopogonidae). *Int. J. Insect Morphol. Embryol.* 25:405–425.

Daniel, A., K. Dumstrei, J. A. Lengyel, and V. Hartenstein. 1999. The control of cell fate in the embryonic visual system by *atonal, tailless* and EGFR signaling. *Development* 126:2945–2954.

Dethier, V. G. 1976. *The Hungry Fly : A Physiological Study of the Behavior Associated with Feeding.* Cambridge, Mass.: Harvard University Press.

Dickinson, M. H., and J. Palka. 1987. Physiological properties, time of development, and central projection are correlated in the wing mechanoreceptors of *Drosophila. J. Neurosci.* 7:4201–4208.

Dickinson, M., S. Hannaford, and J. Palka. 1997. The evolution of insect wings and their sensory apparatus. *Brain. Behav. Evol.* 50:13–24.

Dickson, B., and E. Hafen. 1993. Genetic dissection of eye development in *Drosophila.* In M. Bate and A. Martinez-Arias, eds., *The Development of* Drosophila melanogaster, vol. 2, pp. 1327–1362. Plainview, N.Y.: Cold Spring Harbor Laboratory Press.

Dominguez, M., and S. Campuzano. 1993. *asense,* a member of the *Drosophila achaete-scute* complex, is a proneural and neural differentiation gene. *EMBO J.* 12:2049–2060.

Dunipace, L., S. Meister, C. McNealy, and H. Amrein. 2001. Spatially restricted expression of candidate taste receptors in the *Drosophila* gustatory system. *Curr. Biol.* 11:822–835.

Eberl, D. F., R. W. Hardy, and M. J. Kernan. 2000. Genetically similar transduction mechanisms for touch and hearing in *Drosophila*. *J. Neurosci.* 20:5981–5988.

Evans, J. J. T. 1967. The integument of the Queensland fruit fly *Dacus tryoni* (Frogg.). 1, The tergal glands. *Z. Zellforsch. Mikrosk. Anat.* 81:18–33.

Field, L. H., and T. Matheson. 1998. Chordotonal organs of insects. *Adv. Insect Physiol.* 27:1–228.

Finlayson, L. 1976. Abdominal and thoracic receptors in insects, centipedes and scorpions. In P. Mill, ed., *Structure and Function of Proprioceptors in the Invertebrates,* pp. 153–212. London: Chapman and Hall.

Foelix, R. F., R. F. Stocker, and R. A. Steinbrecht. 1989. Fine structure of a sensory organ in the arista of *Drosophila melanogaster* and some other Dipterans. *Cell Tissue Res.* 258:277–287.

Frankfort, B. J., and G. Mardon. 2002. R8 development in the *Drosophila* eye: A paradigm for neural selection and differentiation. *Development* 129:1295–1306.

Frantsevich, L., and D. Gladun. 2002. Evolution of the middle leg basal articulations in flies (Diptera). *Acta Zool.* 83:125–147.

Galant, R., J. B. Skeath, S. Paddock, D. L. Lewis, and S. B. Carroll. 1998. Expression pattern of a butterfly *achaete-scute* homolog reveals the homology of butterfly wing scales and insect sensory bristles. *Curr. Biol.* 8:807–813.

Garcia-Alonso, L. A. 1999. Postembryonic sensory axon guidance in *Drosophila. Cell. Mol. Life Sci.* 55: 1386–1398.

Garcia-Bellido, A. 1981. From the gene to the pattern: Chaeta differentiation. In C. Looyd and D. Rees, eds., *Cellular Controls in Differentiation,* pp. 281–304. London: Academic Press.

Ghysen, A., and C. Dambly-Chaudière. 1992. Development of the peripheral nervous system in *Drosophila.* In M. Shankland and E. R. Macagno, eds., *Determinants of Neuronal Identity,* pp. 225–292. San Diego: Academic Press.

Gnatzy, W., U. Grunert, and M. Bender. 1987. Campaniform sensilla of *Calliphora vicina* (Insecta, Diptera) I. Topography. *Zoomorphology* 106:312–319.

Goulding, S. E., P. zur Lage, and A. P. Jarman. 2000. *Amos,* a proneural gene for *Drosophila* olfactory sense organs that is regulated by *lozenge. Neuron* 25:69–78.

Grueber, W. B., L. Y. Jan, and Y. N. Jan. 2002. Tiling of the *Drosophila* epidermis by multidendritic sensory neurons. *Development* 129:2867–2878.

Gupta, B. P., and V. Rodrigues. 1997. *atonal* is a proneural gene for a subset of olfactory sense organs in *Drosophila. Genes Cells* 2:225–233.

Hayward, D. C., N. H. Patel, E. J. Rehm, C. S. Goodman, and E. E. Ball. 1995. Sequence and expression of grasshopper *Antennapedia:* Comparison to *Drosophila. Dev. Biol.* 172:452–465.

Heimbeck, G., V. Bugnon, N. Gendre, C. Haberlin, and R. F. Stocker. 1999. Smell and taste perception in *Drosophila melanogaster* larva: Toxin expression studies in chemosensory neurons. *J. Neurosci.* 19: 6599–6609.

Heisenberg, M. 2003. Mushroom body memoir: From maps to models. *Nat. Rev. Neurosci.* 4:266–275.

Held, L. I. 1990. Arrangement of bristles as a function of bristle number on a leg segment in *Drosophila melanogaster. Roux Arch. Dev. Biol.* 199:48–62.

Held, L. I. 2002. *Imaginal Discs: The Genetic and Cellular Logic of Pattern Formation.* Cambridge: Cambridge University Press.

Hill, C. A., A. N. Fox, R. J. Pitts, L. B. Kent, P. L. Tan, M. A. Chrystal, A. Cravchik, F. H. Collins, H. M. Robertson, and L. J. Zwiebel. 2002. G protein-coupled receptors in *Anopheles gambiae. Science* 298:176–178.

Huang, M. L., C. H. Hsu, and C. T. Chien. 2000. The proneural gene *amos* promotes multiple dendritic neuron formation in the *Drosophila* peripheral nervous system. *Neuron* 25:57–67.

Imms, A. 1908. On the larval and pupal stages of *Anapheles maculipennis,* Meigen. *Parasitology* 1:103–133.

Jan, Y. N., and L. Y. Jan. 1993. The peripheral nervous system. In M. Bate and A. Martinez-Arias, eds., *The Development of* Drosophila melanogaster, vol. 2, pp. 1207–1244. Plainview, N.Y.: Cold Spring Harbor Laboratory Press.

Jan, Y. N., and L. Y. Jan. 1994. Genetic control of cell fate specification in *Drosophila* peripheral nervous system. *Annu. Rev. Genet.* 28:373–393.

Jarman, A. P., Y. Grau, L. Y. Jan, and Y. N. Jan. 1993. *atonal* is a proneural gene that directs chordotonal organ formation in the *Drosophila* peripheral nervous system. *Cell* 73:1307–1321.

Jones, B. W., R. D. Fetter, G. Tear, and C. S. Goodman. 1995. *glial cells missing:* A genetic switch that controls glial versus neuronal fate. *Cell* 82:1013–1023.

Jürgens, G., and V. Hartenstein. 1993. The terminal regions of the body pattern. In M. Bate and A. Martinez-Arias, eds., *The Development of* Drosophila melanogaster, vol. 1, pp. 687–746. Plainview, N.Y.: Cold Spring Harbor Laboratory Press.

Keil, T. A. 1978. Die Makrochaeten auf dem Thorax von *Calliphora vicina* Robineau-Desvoidy (Calliphoridae, Diptera). *Zoomorphologie* 90:151–180.

Keil, T. A. 1997. Functional morphology of insect mechanoreceptors. *Microsc. Res. Tech.* 39:506–531.

Keil, T., and R. Steinbrecht. 1984. Mechanosensitive and olfactory sensilla of insects. In R. King and H. Akai, eds., *Insect Ultrastructure,* vol. 2, pp. 477–516. New York: Plenum Press.

Keilin, D. 1915. Recherches sur les larves des Diptères cyclorrhaphes. *Bull. Sci. Fr. Bel.* 49:14–198.

King, D. G., and K. L. Valentino. 1983. On neuronal homology: A comparison of similar axons in *Musca, Sarcophaga,* and *Drosophila* (Diptera: Schizophora). *J. Comp. Neurol.* 219:1–9.

Kuhbandner, B. 1984. Ultrastructure and ontogeny of the hair sensilla on the funicle of *Calliphora erythrocephala* (Insecta, Diptera). I. The basiconical sensilla. *Zoomorphology* 104:373–385.

Kunes, S., C. Wilson, and H. Steller. 1993. Independent guidance of retinal axons in the developing visual system of *Drosophila. J. Neurosci.* 13:752–767.

Lakes-Harlan, R., and K.-G. Heller. 1992. Ultrasound-sensitive ears in a parasitoid fly. *Naturwissenschaften* 79:224–226.

Lakes-Harlan, R., G. S. Pollack, and D. J. Merritt. 1991a. From embryo to adult: Anatomy and development of a leg sensory organ in *Phormia regina* Meigen (Insecta: Diptera). I. Anatomy and physiology of a larval "leg" sensory organ. *J. Comp. Neurol.* 308:188–199.

Lakes-Harlan, R., G. S. Pollack, and D. J. Merritt. 1991b. From embryo to adult: Anatomy and development of a leg sensory organ in *Phormia regina,* Meigen (Insecta: Diptera). II. Development and persistence of sensory neurons. *J. Comp. Neurol.* 308:200–208.

Lawrence, P. A. 1966. Development and determination of hairs and bristles in the milkweed bug, *Oncopeltus fasciatus* (Lygaeidae, Hemiptera). *J. Cell Sci.* 1:475–498.

Leise, E. M. 1991. Evolutionary trends in invertebrate ganglionic structure. *Sem. Neurosci.* 3:369–377.

Leyns, L., C. Dambly-Chaudière, and A. Ghysen. 1989. Two different sets of *cis* elements regulate *scute* to establish two different sensory patterns. *Roux Arch. Dev. Biol.* 198:227–232.

Liu, L., R. Wolf, R. Ernst, and M. Heisenberg. 1999. Context generalization in *Drosophila* visual learning requires the mushroom bodies. *Nature* 400:753–756.

Liu, L., O. Yermolaieva, W. A. Johnson, F. M. Abboud, and M. J. Welsh. 2003. Identification and function of thermosensory neurons in *Drosophila* larvae. *Nat. Neurosci.* 6:267–273.

Meinertzhagen, I., and T. Hanson. 1993. The development of the optic lobe. In M. Bate and A. Martinez-Arias, eds., *The Development of* Drosophila melanogaster, vol. 2, pp. 1362–1491. Plainview, N.Y.: Cold Spring Harbor Laboratory Press.

Melzer, R. R., A. Panzinger, F. Reckel, and U. Smola. 1995. Central nervous system of brachyceran larvae (Insecta, Diptera). *Zool. Anz.* 234:113–123.

Melzer, R. R., and H. F. Paulus. 1991. Morphology of the visual system of *Chaoborus crystallinus* (Diptera, Chaoboridae). I. Larval compound eyes and stemmata. *Zoomorphology* 110:227–238.

Melzer, R. R., T. Zimmerman, and U. Smola. 1997. Modification of dispersal patterns of branched photoreceptor axons and the evolution of neural superposition. *Cell. Mol. Life Sci.* 53:242–247.

Melzer, R. R., J. Sprenger, D. Nicastro, and U. Smola. 1999. Larva-adult relationships in an ancestral dipteran: A re-examination of sensillar pathways across the antenna and leg anlagen of *Chaoborus crystallinus* (DeGeer, 1776; Chaoboridae). *Dev. Genes Evol.* 209:103–112.

Merritt, D. J. 1997. Transformation of external sensilla to chordotonal sensilla in the *cut* mutant of *Drosophila* assessed by single cell marking in the embryo and larva. *Microsc. Res. Tech.* 39:492–505.

Merritt, D. J., and R. K. Murphey. 1992. Projections of leg proprioceptors within the CNS of the fly *Phormia* in relation to the generalized insect ganglion. *J. Comp. Neurol.* 322:16–34.

Merritt, D. J., and M. J. Rice. 1984. Innervation of the cercal sensilla on the ovipositor of the Australian sheep blowfly (*Lucilia cuprina*). *Physiol. Entomol.* 9:39–47.

Murphey, R. K., D. Possidente, G. Pollack, and D. J. Merritt. 1989a. Modality-specific axonal projections in the CNS of the flies *Phormia* and *Drosophila. J. Comp. Neurol.* 290:185–200.

Murphey, R. K., D. R. Possidente, P. Vandervorst, and A. Ghysen. 1989b. Compartments and topography of leg afferent projections in *Drosophila. J. Neurosci.* 9:3209–3217.

Namba, R., and J. S. Minden. 1999. Fate mapping of *Drosophila* embryonic mitotic domain 20 reveals that the larval visual system is derived from a subdomain of a few cells. *Dev. Biol.* 212:465–476.

Nicastro, D., R. R. Melzer, H. Hruschka, and U. Smola. 1998. Evolution of small sense organs: Sensilla on the larval antennae traced back to the origin of the Diptera. *Naturwissenschaften* 85:501–505.

Nilsson, D.-E., and A.-I. Ro. 1994. Did neural pooling for night vision lead to the evolution of neural superposition eyes? *J. Comp. Physiol. A* 175:289–302.

Oppliger, F. Y., P. M. Guerin, and M. Vlimant. 2000. Neurophysiological and behavioural evidence for an olfactory function for the dorsal organ and a gustatory one for the terminal organ in *Drosophila melanogaster* larvae. *J. Insect Physiol.* 46:135–144.

Osborne, M. P. 1963. The sensory neurones and sensilla in the abdomen and thorax of the blowfly larva. *Q. J. Microsc. Sci.* 104:227–241.

Osborne, M., and L. Finlayson. 1962. The structure and topography of stretch receptors in representatives of seven orders of insects. *Q. J. Microsc. Sci.* 103:227–242.

Paulus, H. F. 1979. Eye structure and monophyly of Arthropoda. In A. P. Gupta, ed., *Arthropod Phylogeny,* pp. 199–383. New York: Van Nostrand Reinhold.

Paulus, H. 1989. Das Homologisierin in der Feinstruktur-Forschung: Das Bolwig-Organ der höheren Dipteren und seine Homologisierung mit Stemmata und Ommatidien eines ursprünglichen Fazettenauges der Mandibulata. *Zool. Beitr. N. F.* 32:437–478.

Perry, M. M. 1968. Further studies on the development of the eye of *Drosophila melanogaster.* II. The interommatidial bristles. *J. Morphol.* 124:249–262.

Pistillo, D., N. Skaer, and P. Simpson. 2002. *scute* expression in *Calliphora vicina* reveals an ancestral pattern of longitudinal stripes on the thorax of higher Diptera. *Development* 129:563–572.

Pollack, G. S., and R. Balakrishnan. 1997. Taste sensilla of flies: Function, central neuronal projections, and development. *Microsc. Res. Tech.* 39:532–546.

Pollock, J., and S. Benzer. 1988. Transcript localization of four opsin genes in the three visual organs od *Drosophila;* RH2 is ocellus specific. *Nature* 323:779–782.

Power, M. 1943. The brain of *Drosophila melanogaster. J. Morphol.* 72:517–559.

Python, F., and R. F. Stocker. 2002. Adult-like complexity of the larval antennal lobe of *D. melanogaster* despite markedly low numbers of odorant receptor neurons. *J. Comp. Neurol.* 445:374–387.

Ray, K., and V. Rodrigues. 1995. Cellular events during development of the olfactory sense organs in *Drosophila melanogaster. Dev. Biol.* 167:426–438.

Robert, D., J. Amoroso, and R. R. Hoy. 1992. The evolutionary convergence of hearing in a parasitoid fly and its cricket host. *Science* 258:1135–1137.

Sayeed, O., and S. Benzer. 1996. Behavioral genetics of thermosensation and hygrosensation in *Drosophila. Proc. Natl. Acad. Sci. USA* 93:6079–6084.

Schmid, A., A. Chiba, and C. Q. Doe. 1999. Clonal analysis of *Drosophila* embryonic neuroblasts: Neural cell types, axon projections and muscle targets. *Development* 126:4653–4689.

Schmidt-Ott, U., M. Gonzalezgaitan, H. Jackle, and G. M. Technau. 1994. Number, identity, and sequence of the *Drosophila* head segments as revealed by neural elements and their deletion patterns in mutants. *Proc. Natl. Acad. Sci. USA* 91:8363–8367.

Schrader, S., and D. J. Merritt. 2000. Central projections of *Drosophila* sensory neurons in the transition from embryo to larva. *J. Comp. Neurol.* 425:34–44.

Scott, K., R. Brady, A. Cravchik, P. Morozov, A. Rzhetsky, C. Zuker, and R. Axel. 2001. A chemosensory gene family encoding candidate gustatory and olfactory receptors in *Drosophila. Cell* 104:661–673.

Sen, A., G. V. Reddy, and V. Rodrigues. 2003. Combinatorial expression of Prospero, Seven-up, and Elav identifies progenitor cell types during sense-organ differentiation in the *Drosophila* antenna. *Dev. Biol.* 254:79–92.

Shaw, S. 1989. The retina-lamina pathway in insects, particularly Diptera, viewed from an evolutionary perspective. In D. Stavenga and R. Hardie, eds., *Facets of Vision,* pp. 187–212. Berlin: Springer-Verlag.

Shaw, S. 1990. The photoreceptor axon projection and its evaluation in the neural superposition eyes of some primitive brachyceran Diptera. *Brain. Behav. Evol.* 35:107–125.

Shaw, S., and I. Meinertzhagen. 1986. Evolutionary progression at synaptic connections made by identified homologous neurones. *Proc. Natl. Acad. Sci. USA* 83:7961–7965.

Simpson, P., R. Woehl, and K. Usui. 1999. The development and evolution of bristle patterns in Diptera. *Development* 126:1349–1364.

Steinbrecht, R. 1984. Arthropoda: Chemo-, hygro-, and thermoreceptors. In J. Bereiter-Hahn, A. Matoltsy, and K. Richards, eds., *Biology of the Integument*, vol. 1, pp. 523–553. Berlin: Springer-Verlag.

Steller, H., K. F. Fischbach, and G. M. Rubin. 1987. *disconnected:* A locus required for neuronal pathway formation in the visual system of *Drosophila. Cell* 50:1139–1153.

Stern, C. 1954. Two or three bristles. *Am. Scient.* 42:213–247.

Stocker, R. F. 1994. The organization of the chemosensory system in *Drosophila melanogaster:* A review. *Cell Tissue Res.* 275:3–26.

Stocker, R. F., M. C. Lienhard, A. Borst, and K. F. Fischbach. 1990. Neuronal architecture of the antennal lobe in *Drosophila melanogaster. Cell Tissue Res.* 262:9–34.

Strausfeld, N. J. 1976. *Atlas of an Insect Brain.* Berlin: Springer-Verlag.

Svacha, P. 1992. What are and what are not imaginal discs: Reevaluation of some basic concepts (Insecta, Holometabola). *Dev. Biol.* 154:101–117.

Thomas, J. B., M. J. Bastiani, M. Bate, and C. S. Goodman. 1984. From grasshopper to *Drosophila:* A common plan for neuronal development. *Nature* 310:203–207.

Tissot, M., and R. F. Stocker. 2000. Metamorphosis in *Drosophila* and other insects: The fate of neurons throughout the stages. *Prog. Neurobiol.* 62:89–111.

Tissot, M., N. Gendre, A. Hawken, K. F. Stortkuhl, and R. F. Stocker. 1997. Larval chemosensory projections and invasion of adult afferents in the antennal lobe of *Drosophila. J. Neurobiol.* 32:281–297.

Tix, S., M. Bate, and G. M. Technau. 1989. Pre-existing neuronal pathways in the leg imaginal discs of *Drosophila. Development* 107:855–862.

Tomlinson, A., and D. F. Ready. 1987. Neuronal differentiation in the *Drosophila* ommatidium. *Dev. Biol.* 120:366–376.

Tracey, W. D., R. I. Wilson, G. Laurent, and S. Benzer. 2003. *painless,* a *Drosophila* gene essential for nociception. *Cell* 113:261–273.

Trimarchi, J. R., and A. M. Schneiderman. 1994. The motor neurons innervating the direct flight muscles of *Drosophila melanogaster* are morphologically specialized. *J. Comp. Neurol.* 340:427–443.

Truman, J. W. 1990. Metamorphosis of the central nervous system of *Drosophila. J. Neurobiol.* 21: 1072–1084.

Truman, J. W. 1996. Metamorphosis of the insect nervous system. In L. Gilbert, J. Tata and B.G. Atkinson, eds., *Metamorphosis: Postembryonic Reprogramming of Gene Expression in Amphibian and Insect Cells,* pp. 283–320. San Diego: Academic Press.

Truman, J. W., and M. Bate. 1988. Spatial and temporal patterns of neurogenesis in the central nervous system of *Drosophila melanogaster. Dev. Biol.* 125:145–157.

Truman, J., B. Taylor, and T. Awad. 1993. The formation of the adult nervous system. In M. Bate and A. Martinex-Arias, eds., *The Development of* Drosophila melanogaster, vol. 2, pp. 1245–1275. Plainview, N.Y.: Cold Spring Harbor Laboratory Press.

Venkatesh, S., and R. N. Singh. 1984. Sensilla on the third antennal segment of *Drosophila melanogaster* Meigen (Diptera: Drosophilidae). *Int. J. Insect Morphol. Embryol.* 13:51–63.

Waddell, S., and W. G. Quinn. 2001. Flies, genes, and learning. *Annu. Rev. Neurosci.* 24:1283–1309.

White, R. 1961. Analysis of the development of the compound eye in the mosquito *Aedes aegypti. J. Exp. Zool.* 1148:223–240.

Wigglesworth, V. 1953. The origin of sensory neurones in an insect, *Rhodnius prolixus* (Hemiptera). *Quart. J. Microsc. Sci.* 94:93–112.

Williams, D. W., and D. Shepherd. 2002. Persistent larval sensory neurones are required for the normal development of the adult sensory afferent projections in *Drosophila. Development* 129:617–624.

Wolff, T., and D. Ready. 1993. Pattern formation in the *Drosophila* retina. In M. Bate and A. Martinez-Arias, eds., *The Development of* Drosophila melanogaster, vol. 2, pp. 1277–1325. Plainview, N.Y.: Cold Spring Harbor Laboratory Press.

Wong, D. C., and D. J. Merritt. 2002. The role of homeotic genes in determining the segmental pattern of chordotonal organs in *Drosophila. Int. J. Dev. Biol.* 46:475–481.

Wülbeck, C., and P. Simpson. 2000. Expression of *achaete-scute* homologues in discrete proneural clusters on the developing notum of the medfly *Ceratitis capitata,* suggests a common origin for the stereotyped bristle patterns of higher Diptera. *Development* 127:1411–1420.

Yeates, D. K., D. J. Merritt, and C. H. Baker. 2002. The adult ventral nerve cord as a phylogenetic character in brachyceran Diptera. *Org. Divers. Evol.* 2:89–96.
Young, J., and D. Merritt. 2003. The ultrastructure and function of the silk-producing basitarsus in the Hilarini (Diptera: Empididae). *Arthropod Struct. Dev.* 32:157–165.
Zacharuk, R. Y., and V. D. Shields. 1991. Sensilla of immature insects. *Annu. Rev. Entomol.* 36:331–354.

CHAPTER EIGHT

Dipteran Sex Chromosomes in Evolutionary Developmental Biology

Neil Davies and George K. Roderick

Sex chromosomes are involved in many fundamental biological phenomena, such as sex determination (Bull 1983; Charlesworth 1996; Marin and Baker 1998), imprinting (Crouse 1960), dosage compensation (Cline and Meyer 1996; Chatterjee 1998), recombination suppression (Korol et al. 1994; Otto and Lenormand 2002), meiotic drive (Hurst and Pomiankowski 1991), sexual selection (Hastings 1994), sexual dimorphism (Rice 1984), and speciation (Haldane 1922b; Coyne 1992; Sperling 1994). Although these phenomena are clearly interrelated at some level, it is far from certain that any unified theory is possible and to what extent such a theory might explain them. Understanding the evolutionary forces that have led to the differentiation of sex chromosomes, however, will certainly shed light on many of these important aspects of biology. In particular, sex chromosome evolution is an attractive system for studying the assembly and maintenance of developmental pathways and gene networks. The sex chromosomes of Diptera offer a particularly fertile ground for such investigations due to their diversity, tractability, and economic significance.

Sex chromosomes are conventionally distinguished as X and Y (or Z and W, when females are the heterogametic sex). It is generally believed that the sex chromosomes were originally homologous but began to differentiate once they evolved the genes for sex determination, recombination became restricted, and selection favored alleles advantageous in one sex but disadvantageous in the other. Eventually, recombination became suppressed along the entire length of the two sex chromosomes (Charlesworth 1996; Rice 1996). Many sex chromosome systems appear to have evolved very long ago, making it difficult to study their origins. For example, female heterogamety arose in the common ancestor of the Trichoptera and Lepidoptera and has remained fixed in those groups ever since. Similarly, there are few exceptions to XY sex determination in mammals (Cattanach et al. 1971; Fredga et al. 1976; Fredga 1994), with the differences among marsupials and eutherian mammals being only subtle (Cooper et al. 1977). Cytogenetic evidence, however, shows that reptiles and Diptera offer a wealth of sex chromosome variation for comparative analysis (White 1973; Bull 1983). Of the two groups, Diptera have significant advantages because they represent a phylogenetically cohesive group that includes the model organism *Drosophila melanogaster,* and because the sterile insect technique (Calkins et al. 1994) to control dipteran pests and disease vectors has provided economic motives for understanding dipteran sex chromosome biology (Shearman 2002).

Dipteran sex chromosomes, and particularly their role in sex determination, represent a potential model system for the study of gene network evolution (Carroll et al. 2001; Mac-

Carthy et al. 2003). Importantly, evolutionary developmental biologists can take advantage of (1) the detailed knowledge of *Drosophila* sex determination gene network architecture and its pattern of gene expression through time, and (2) the evolutionary diversification of sex determination in Diptera (Schutt and Nothiger 2000; Saccone et al. 2002; Shearman 2002). Here we review (and anticipate) sex chromosome research that, although rooted in the *Drosophila* model organism, exploits the diversity of sex chromosome structure and function across the Diptera model clade (DeSalle, Chapter 5).

Sex Determination

The origin of a two-gender system, with its ecological and behavioral consequences, is a rich field of evolutionary study (Maynard Smith 1978; Krebs and Davies 1991; Hurst and Peck 1996). Evolutionary developmental biology has focused on sex determination to understand the organization of developmental genetic pathways because it has two clear-cut outcomes (Marx 1995a) and because aspects of sex determination appear to evolve very rapidly, whereas other elements are strongly conserved (Blackman 1995; Marin and Baker 1998). For such a fundamental property, the mechanics of sex determination are surprisingly diverse (Bull 1983; Marx 1995b; Mittwoch 1996). This might reflect the fact that sex determination is unusually tolerant of mutations because the failure of a developmental pathway to function merely results in the default sex (i.e., still a perfectly viable individual), but what are the evolutionary drivers responsible for the diversification of some developmental pathways and what has constrained the evolution of others?

Gender is sometimes mediated through environmental cues, such as temperature. This is the case in many reptiles (Charnov and Bull 1977), but relatively few dipterans—*Aedes stimulans* and species of *Heteropeza* providing some examples (Nothiger and Steinemann-Zwicky 1985). In most Diptera, sex is determined genetically, and early studies sought to compare sex determination in *D. melanogaster* with other model organisms. For example, the ratio of X chromosomes to autosomes decides gender in *Drosophila* and the nematode *Caenorhabditis elegans* (Mittwoch 1996), but the molecular genetic pathways responsible are quite different (Parkhurst and Meneely 1994; Cline and Meyer 1996). More recently, attention has focused on whether *Drosophila* is even very representative of other insects. Several authors have compared the *Drosophila* paradigm with other insect species, especially dipterans (for recent reviews, see Schutt and Nothiger 2000; Saccone et al. 2002; Shearman 2002). There is no need to detail dipteran sex determination pathways again here, but a few of the major findings are discussed.

In *Drosophila,* the primary sex determining signal is the ratio of X chromosomes to autosomes, X:A (see Schutt and Nothiger 2000), which sets the activity of a gene called *sex-lethal* (*sxl*) that with the nonsex-specific gene product of *transformer-2* directs the sex-specific splicing of *transformer* (*tra*) to produce a functional TRA protein in females. This in turn causes female-specific splicing of the double-switch gene *doublesex* (dsx_F), which leads to female differentiation. The *Drosophila* sex determination cascade seems to be conserved within the genus *Drosophila* (Erickson and Cline 1998), and evolutionarily, it might be "locked in" because *sxl* controls three different pathways: somatic sex determination, dosage compensation, and oogenesis. Whereas in other dipterans, mutations above *dsx* simply change the sex of the individual produced, in *Drosophila,* a mutation can cause sex-specific lethality or sterility (Nothiger and Steinemann-Zwicky 1985). The divergence of nondrosophilid dipteran sex determination would seem to support this hypothesis. Evidence from the medfly, *Ceratitis capitata* (Saccone et al. 2002), *Bactrocera tryoni* (Shearman 2002), *Megaselia scalaris* (Kuhn et al. 2000), and the housefly, *Musca domestica* (Hilfiker-Kleiner et al. 1994; Schutt and Nothiger 2000) shows that the *Drosophila* system is only partially conserved across the Diptera. In fact, most dipteran sex determination mechanisms (in Tephritidae, Muscidae, Chironomidae, Culicidae, and Calliphoridae) are based on a male-determining factor, M, linked to a heteromorphic Y chromosome or one of the homomorphic chromosomes; it is thus parsimonious to presume that this is the ancestral system and the *Drosophila* mechanism is a more derived state (Saccone et al. 2002).

The common evolutionary origin of sex determination systems and the well-documented *Drosophila* example prompted MacCarthy et al. (2003) to use it as a model to investigate the evolutionary constraints and flexibility of the sex determination regulatory system. Using the gene network modeling approach (reviewed in DeJong 2002), they compared the known *Drosophila* network to a population of randomly generated networks. The results revealed a large population of functionally equivalent sex determination networks containing subsets of networks that are closely related mutationally. Many functional networks are mutually accessible through single network mutations, suggesting considerable flexibility in network architecture. Interestingly, the *Drosophila* network turns out to be the most parsimonious network (fewest network interactions) and occurs in a subset of the random network population that is particularly diverse. In other words, a single mutation in the *Drosophila* network can access many viable pattern variations, meaning it has considerable evolutionary potential. For example, a viable network is accessible by one mutation in *Drosophila tra* that adds a positive autoregulatory interaction to the gene. Data on the *tra* homolog in *C. capitata* support such an ancestral relationship to *Drosophila* (Pane et al. 2002), suggesting that the approach used by MacCarthy et al. (2003) could also help to generate hypotheses for empirical investigation.

Although *Drosophila* might turn out to be somewhat unusual among Diptera, the pattern of diversification evident across the dipteran clade supports the hypothesis that developmental pathways, at least for sex determination and perhaps in general, evolve from the bottom up (Wilkins 1995). In Diptera, changes at the top of the hierarchy seem more likely than those lower down (Schutt and Nothiger 2000). The evolution of the X:A counting mechanism in *Drosophila*, presumably from an ancestral M factor system, is one example. The M factor system common to most dipterans also harbors a great deal of diversity that can be divided into two broad groups (Shearman 2002). In the first group, M is fixed to the heteromorphic Y chromosome, as is the case in many tephritids. In the second group, the location and role of the M factor appears to change rapidly, as in *Megaselia scalaris* (Traut 1994). Although these groups might be convenient, they are certainly artificial; *Musca domestica* falls into both categories, with variation even among intraspecific populations. In standard housefly strains, sex determination is the XX/XY system with the Y carrying the M factor. In other strains, however, M can be found on autosomes, and in still others, the M factor is homozygous and sex is determined by a female-determining (FD) gene (Hilfiker-Kleiner et al. 1994; Schutt and Nothiger 2000).

Further support for the "bottom-up theory" comes from the comparison of gene structure and function between species. These data are demonstrating a trend toward the conservation of basal genes, like *doublesex* (*dsx*), compared to genes higher up in the sex determination cascade, such as *sxl*. Indeed, *dsx* appears to be conserved even between *Drosophila* and *C. elegans* (Raymond et al. 1998) and related genes have also been found in humans (Raymond et al. 2000), birds (Shan et al. 2000), and fish (Huang et al. 2002). Interestingly, evidence indicates that although in Diptera, *dsx* follows male-specific binding unless regulated by female-specific splice-site activation (i.e., male by default), the situation might be reversed in Lepidoptera; in the silk moth, *Bombyx mori*, it seems that male-specific *dsx* splicing needs to be regulated (Saccone et al. 2002). As Lepidoptera have homogametic (ZZ) females, it might be worthwhile investigating *dsx* in ZZ Diptera (e.g., tephritines).

Shearman (2002) developed the theory proposed by Wilkins (1995) that sex determination cascades have evolved by the hierarchical addition of genes to the top of the cascade. In this model, any gene exerting negative control over the primary signal of the sex determining cascade could become the new primary signal itself. The process is driven by selection to maintain a balanced (1:1) sex ratio (Fisher 1930) and involves a switch in the heterogametic sex as a new hierarchical level is added to the regulatory cascade. Consequently, it predicts that more derived systems will be more complex and intermediate states should show switches in the heterogametic sex. Shearman also hypothesizes an important role for transposable elements. For example, the dominant M factor might have been inserted into ancestral dipteran chromosomes homologous with the *Drosophila* dot chromosomes. The dot chromosomes would then have grown by accumulating heterochromatin and transposable elements. This would explain why, unlike *Drosophila*, other dipterans, such as *C. capitata*, *B. tryoni*, and *Musca domestica*, have relatively few genes on the sex chromosomes—except for the male determining factor (Shearman 2002). There are a few difficulties with this theory, such as the discovery of fertility genes on the Y chromosome of the medfly (Willhoeft and Franz 1996). The question of sex chromosome evolution (usually the degeneration of the Y chromosome) is a whole topic in itself that we consider next.

Sex Chromosome Degeneration

Many taxa have morphologically distinct (heteromorphic) sex chromosomes, yet in other groups, the sex chromosomes are undifferentiated and isomorphic (Ohno 1967). Usually, the Y or W chromosomes are reduced (Morell 1994), but in houseflies, both the X and Y chromosomes are degenerate (Bull 1983). In some reptiles, the Y chromosome is larger than the X (Gorman and Atkins 1966; Cole and Lowe 1967). The evolution of heteromorphic sex chromosomes has been the subject of various hypotheses (Jablonka and Lamb 1990). Sex chromosomes are assumed to have been regular autosomes that happened to evolve the genes for sex determination. The differentiation of the sex chromosomes might have been initiated by structural changes, such as inversions or translocations (Ohno 1967). Alternatively, the first step in sex chromosome differentiation may be the delayed replication of the Y or W chromosomes, as suggested by studies on the primitive sex chromosomes of shrews (Raman and Nanda 1986), frogs (Schempp and Schmid 1981), and snakes (Ray-Chauduri et al. 1972). Raman and Nanda (1986) proposed that such differences in replication time lead to heterochromatinization and the resulting divergence of the sex chromosomes. Studies in mammals have shown conclusively that the X and Y chromosomes are homologous (Lahn et al. 2001), although there are still competing theories to explain the degeneration of the

Y chromosome, including: Muller's ratchet, background selection, the Hill-Robertson effect with weak selection, and the hitchhiking of deleterious alleles by favorable mutations (for a review, see Charlesworth and Charlesworth 2000).

The evolution of heteromorphic sex chromosomes in Diptera has, of course, focused on *Drosophila.* Several authors have recently reviewed the degeneration of Y chromosomes and the *Drosophila* model provides some of the most important evidence (Rice 1996; Charlesworth and Charlesworth 2000; Carvalho 2002; Rice 2002). Sixteen genes have so far been identified on the *Drosophila* Y chromosome; they appear to have been acquired from the autosomes and have mainly male-related functions, such as sperm motility (Carvalho 2002, and references therein). The lack of homology between genes or repetitive sequences among X and Y in *Drosophila* has led to the suggestion that, unlike in mammals, the drosophilid Y is not a degenerate X chromosome (Lohe et al. 1993) but may have originated from the B chromosomes (Hackstein et al. 1996). Carvalho (2002) thus recommends caution when using the *Drosophila* Y as evidence to support general theories of Y chromosome degeneration. He argues that *Drosophila* might be rather unusual in this respect and speculates that the ancestral dipteran M factor Y chromosomal sex determination might have led to XO sex determination in early *Drosophila* and from there a "Y chromosome was reinvented from B-chromosomes" (see Carvalho 2002: 665, Fig. 1).

Despite this caveat, studies of sex chromosome (Y) degeneration in Diptera remain focused on *Drosophila.* One of the most difficult issues is how the process is initiated, and *Drosophila* provides a promising study system due to the evolutionarily recent centromeric fusion of the X chromosome and an autosome in some *Drosophila,* such as *D. americana* (see McAllister 2003) and *D. miranda* (Steinemann and Steinemann 2000 and references therein). The result is neo-X and neo-Y chromosomes, which provide an opportunity to study the early stages of sex chromosome differentiation (Steinemann and Steinemann 2000; Bachtrog and Charlesworth 2002; Bachtrog 2003; McAllister 2003; Yi et al. 2003). The neo-Y chromosome appears to experience a severe reduction in effective population size and begins to diverge rapidly. Evidence suggests that degeneration can occur in a few million generations (Charlesworth 2002).

Male or Female Heterogamety?

Does it matter in which sex degeneration of the Y or W chromosome occurs? The null hypothesis would predict the degeneration of the Y chromosome to occur equally often in either males or females. At first sight, it appears that the heterogametic sex choice is made early in the evolution of a lineage and remains fixed. Birds, snakes, and butterflies have heterogametic (ZW) females, whereas the heterogametic (XY) sex is male in mammals and some lower vertebrates (Bull 1983). Most orders of insect are fixed for either male or female heterogamety (Table 8.1). As described earlier, however, some theories for the evolution of sex determination from the "bottom up" (Wilkins 1995) predict switches from male to female heterogamety and back again (Shearman 2002). The data, therefore, suggest that these new primary switches are not acquired very rapidly. Interestingly, there is a statistically significant bias toward male heterogamety among insects, with only three out of 16 insect orders having species with heterogametic females (Bull 1983). The Diptera are unique in having both male and female heterogamety; nevertheless, most dipteran families have male heterogamety (Table 8.2).

TABLE 8.1. The Heterogametic Sex Insects

	Heterogametic Sex	
Taxon	Male	Female
Exopterygota		
Ephemeroptera	X	
Odonata	X	
Orthoptera	X	
Dermaptera	X	
Isoptera	X	
Embioptera	X	
Plecoptera	X	
Psocoptera	X	
Hemiptera	X	
Endopterygota		
Mecoptera	X	
Tricoptera		X
Lepidoptera		X
Diptera	X	X
Siphonaptera	X	
Neuroptera	X	
Coleoptera	X	

SOURCE: Bull 1983.

Bull (1983) asked whether female heterogamety rarely evolved or was just harder to maintain. A possible mechanism for biasing the evolution of male vs. female heterogamety was proposed by Kraak and De Looze (1993) for vertebrates. They suggested that heterogametic female genetic sex determination (ZW) is most likely to evolve from environmental sex determination when there is a female size advantage. Conversely, heterogametic male genetic sex determination (XY) will tend to evolve when there is male size advantage. Their hypothesis assumes that the original sex determining genetic element accelerates growth. Data exist to suggest that growth rate is indeed an important factor in sex determination (Mittwoch 1971, 1986). Kraak and De Looze (1993) only apply their size-advantage hypothesis to

TABLE 8.2. The Heterogametic Sex: Diptera

	Heterogametic Sex	
Taxon	Male	Female
Tipulidae	X	
Culicidae	X	
Chironomidae	X	X
Simulidae	X	
Tephritidae	X	X
Drosophilidae	X	
Muscidae	X	X
Calliphoridae	X	
Phoridae	X	
Anthomyidae	X	

SOURCE: Bull 1983.

vertebrates and they expect different evolutionary processes to operate among invertebrates for reasons that are not fully explained. Somewhat pessimistically, they state that no unified theory of sex determination is possible.

There are other differences that depend on (or maybe help determine) which sex is heterogametic. For example, the sex chromosomes behave differently (for a review, see Jablonka and Lamb 1988). In taxa with male heterogamety, sex chromosomes undergo more dramatic structural changes at meiosis, which Jablonka and Lamb (1990) have suggested results from the larger size of oocytes compared to spermatocytes. The difference in gamete size between the two sexes means that meiosis is a much longer process in oocytes than in spermatocytes. Genes on the X or Z chromosomes, therefore, must be kept active to sustain metabolic activities. Indeed, in both XX and ZW oocytes, the sex chromosomes appear to be active during early meiotic prophase (Jablonka and Lamb 1988). In spermatocytes, however, the sex chromosomes can be deactivated for the relatively brief meiosis. In species with heterogametic males, the X and Y chromosomes become heterochromatic and they are genetically inactive (Jablonka and Lamb 1988)—although cases of meiotic drive in many dipteran species argue that at least some X-linked gene products are present. When males are the homogametic sex, it may still be necessary for both Z chromosomes to be active, even if their gene products are not needed, because crossing over can only occur among Z chromosomes in the homogametic sex. Heterochromatic (inactivated) chromosomes do not cross-over and so the Z chromosomes of homogametic males may remain active in order for recombination to take place (Jablonka and Lamb 1990).

It follows that if ZW chromosomes must be active in oocytes because Z-linked gene products are needed, then W chromosomes must also be uncondensed to avoid pairing failure (Jablonka and Lamb 1990). Jablonka and Lamb (1990) acknowledged, however, that their hypothesis may be wrong because many groups with heterogametic females also have meroistic ovaries. In meroistic taxa, nurse cells supply essential metabolites to the oocyte (King and Buning 1985). We do not know whether the euchromatic oocyte Z chromosome persists despite the subsequent evolution of nurse cells. At least in some species, the Z chromosomes, although diffuse, do not seem to be very active (Telfer 1975). Relatively little is known about the evolution of meroistic ovaries, which occur sporadically among arthropods (most dipterans are meroistic). Female heterogamety probably evolved in tephritids subsequent to the evolution of nurse cells. If so, the active Z chromosome in oocytes presumably evolved from a male heterogametic system with nurse cells.

If gene products are (or were initially) needed in oocytes, the routes to heteromorphism will be more limited in taxa with female heterogamety because crossing over may still occur among the sex chromosomes. It is thus possible that W chromosome degeneration will be less likely than Y chromosome degeneration (Jablonka and Lamb 1990). If correct, one might expect Z and W chromosomes to be less heteromorphic than X and Y chromosomes and so female heterogamety may be more susceptible to switches to male heterogamety. If so, we should see male heterogamety evolving from female heterogamety more frequently than the reverse, and the size differential among sex chromosomes should be greater for taxa with male heterogamety (indeed, there do not seem to be many ZO taxa, the most extreme case of heteromorphy). To our knowledge, there has not been a systematic comparison of rates of Y vs. W degeneration.

Haldane's Rules

The evolution of genetic sex determination and the subsequent divergence (degeneration) of sex chromosomes are related to some fascinating and still somewhat puzzling biological phenomena; notably, imprinting (Crouse 1960), dosage compensation (Cline and Meyer 1996; Chatterjee 1998), recombination suppression (Korol et al. 1994; Otto and Lenormand 2002), meiotic drive (Hurst and Pomiankowski 1991), sexual selection (Hastings 1994), sexual dimorphism (Rice 1984), and speciation (Haldane 1922b; Coyne 1992; Sperling 1994). The mechanisms, evolutionary origins, and consequences of these phenomena are much less well understood than sex determination and sex chromosome degeneration, yet they may all be interrelated. The phylogenetic distribution of these phenomena and the mechanisms underlying them, however, requires much investigation. A comprehensive treatment of all these phenomena and the possible role of sex chromosomes would take a whole volume in itself. Here we limit ourselves to a brief review of those that seem most strongly influenced by the sex chromosomes, due to the importance of heterogamety in their expression. The differential manifestation of biological phenomena in one sex in relation to heterogamety was first formalized by J. B. S. Haldane and we title this section in his honor.

ACHIASMATE MEIOSIS

The evolution of recombination dimorphism on sex chromosomes has received a great deal of attention, and interestingly, this dimorphism appears sometimes to extend to the autosomes. Diptera provide some of the most dramatic examples, with no recombination occurring during male gametogenesis in *Drosophila* (Morgan 1914; Gethmann 1988). In Lepidoptera also, although the W chromosome is euchromatinized (avoiding pairing failure), there is no recombination between the W and Z chromosomes or any of the autosomes in heterogametic females (Turner and Sheppard 1975; Nokkala 1987). Copepods (female heterogamety and heteromorphic sex chromosomes) also show achiasmate meiosis in females (White 1973). Furthermore, achiasmate meiosis is perhaps only the extreme of a more general phenomenon, called "heterochiasmy" by Lenormand (2003), whereby recombination rates differ among males and females (Korol et al. 1994). As the above examples suggest, the suppression of meiotic recombination rates appears to disproportionately affect the heterogametic sex, a phenomenon first noted by Haldane (1922a) and Huxley (1928), and consequently known as the "Haldane-Huxley rule" (Bell 1982).

Explanations for the Haldane-Huxley rule might involve the sex chromosomes. For example, achiasmy and heterochiasmy could represent a pleiotropic consequence of selection against recombination among the sex chromosomes. Alternatively, they might be due to the Y (or W) chromosome evolving in the sex that had lower or no recombination. Both these explanations, however, are unsatisfactory in various ways. For example, pleiotropy does not explain why heterochiasmy should be maintained once sex chromosome heteromorphy (hence, lack of recombination) is well established even in the most extreme case of XO species (Lenormand 2003). Furthermore, there are several exceptions to the Haldane-Huxley rule (Singer et al. 2002). Birds have female heterogamety and heteromorphic sex chromosomes, yet there appears to be recombination in both sexes (Rahn and Solari 1986; Solari et al. 1988), and crossing-over in marsupials is suppressed in the homogametic sex (Bennett et al. 1986; Hayman et al. 1988).

In Diptera, data from mosquitoes also throw doubt on the link between heterogamety and achiasmate meiosis (Baker and Rabbani 1970), and studies on tephritids are intriguingly contradictory. In female-heterogametic tephritines, for example, achiasmate meiosis occurs in the ZZ males, not the ZW females (Bush 1966). This might suggest that the ancestral state of Diptera is achiasmate meiosis in males that has been retained despite the evolution of female heterogamety in the Tephritinae. However, Berlocher (1993) concluded from a study of allozyme linkage disequilibria that there was no recombination in the heterogametic (although isomorphic) females of *Rhagoletis suavis*. In other words, recombination suppression switched with the heterogametic sex. An interesting phenomenon, the Haldane-Huxley rule remains to be fully tested, rejected, or explained.

HYBRID STERILITY AND INVIABILITY

In addition to identifying the potential pattern of recombination suppression, Haldane also proposed that when one sex of an interspecific hybrid is inviable, sterile, or absent, it is usually the heterogametic sex (Haldane 1922b). The observation, which has become known as "Haldane's rule," is one of the only regularities in speciation. Explaining Haldane's rule is thus crucial to understanding the process of biological diversification and has contributed greatly to our understanding of speciation; particularly, the genetic aspects of this process (Coyne and Orr 1989; Wu et al. 1996; Turelli and Orr 2000).

A key genetic observation has provided the context for many studies: the genes having the greatest effect on hybrid sterility and inviability are X-linked, the so-called "large X effect" (Coyne and Orr 1989). Several explanations for the large X effect are possible, given that varying recombination rates, sex-linked inheritance (e.g., the effective population size of sex-linked genes), and sex-biased mutation rates are likely to cause important differences in the molecular evolution of X chromosomes, Y chromosomes, and autosomes (Charlesworth et al. 1987; Miyata et al. 1990). Interestingly, however, the higher mutation rates shown in males of some groups, such as mammals, birds, and fish (see Kirkpatrick and Hall 2004 and references therein), have not been found in *Drosophila* (Bauer and Aquadro 1997). Nevertheless, the X chromosome in *Drosophila simulans* and *D. mauritiana* harbors a disproportionate number of genes that cause hybrid male sterility (Tao et al. 2003), whereas the Y chromosome of *D. melanogaster* contributes disproportionate amounts of genetic variance to lifetime male fitness (Chippendale and Rice 2001). Furthermore, Reinhold (1998) showed that male reproductive display traits in mammals and insects (predominantly *Drosophila*) are X-linked. Similar patterns have been shown for Z chromosomes in Lepidoptera (Sperling 1994). One potential explanation is that sexually antagonistic genes will tend to accumulate on the sex chromosomes (Rice 1984), and this might also underlie Haldane's rule.

Whatever the role of sexual antagonism, two theories have emerged that directly explain both fertility and viability manifestations of Haldane's rule. The first is known as the "dominance theory" and is based on the supposition that most of the genes causing incompatibility are recessive and thus fully expressed by the hemizygous X in the heterogametic sex (Muller 1942; Davies and Pomiankowski 1995; Turelli and Orr 1995). The second is not a Haldane effect, as it only applies when males are the heterogametic sex. The faster-male evolution theory (due either to stronger sexual selection on males or the sensitivity of spermatogenesis) explains why hybrid male sterility appears to be more abundant than female sterility or male inviability in male-heterogametic taxa, such as *Drosophila* and mammals (Wu et al. 1996).

Much of the research into the genetic basis of Haldane's rule has come from studies of *Drosophila* (Coyne and Orr 1989; Coyne 1997). Furthermore, Haldane's rule is one of the best examples of how hypotheses formulated from *Drosophila* can be expanded and tested by harnessing the diversity of sex chromosomes in the Diptera. Mosquitoes of the genus *Anopheles* have degenerate Y chromosomes, as do *Drosophila,* but *Aedes* mosquitoes have single-locus sex determination, with two functional sex chromosomes in both sexes. This enabled Presgraves and Orr (1998) to test both the faster-male and dominance theories of Haldane's rule in a group lacking a hemizygous sex (*Aedes*). Furthermore, they were able to compare this with the "more typical" situation demonstrated by closely related *Anopheles* (and more distant *Drosophila*). Presgraves and Orr reported three crucial results: (1) Male-only hybrid sterility is common in *Aedes* compared to female-only sterility, supporting the faster-male theory; (2) a larger proportion of *Anopheles* than *Aedes* hybridizations show Haldane's rule for sterility, supporting the dominance theory; and (3) *Anopheles* have many cases of hybrid inviability but *Aedes* do not, supporting the dominance theory. Thus these findings in mosquitoes corroborate hypotheses previously generated from *Drosophila* that Haldane's rule results from a composite of the faster-male and dominance theories.

DOSAGE COMPENSATION

In some groups and circumstances, it appears to be important to maintain an equal quantity and quality of X-linked gene products in both the sexes, despite the heterogametic sex having only a single copy of X-linked genes due to the degenerate nature of the Y chromosome (Bull 1983; Charlesworth 1996). The evolution of dosage compensation is poorly understood (Chatterjee 1998); indeed, it is not even known how common it is. According to Parkhurst and Meneely (1994: 930), dosage compensation "is widespread among animals and nearly universal among animals with heteromorphic sex chromosomes." But Jablonka and Lamb (1990) pointed out that dosage compensation is found only in species with male heterogamety (Cock 1964; Johnson and Turner 1979; Baverstock et al. 1982; Ohno 1983). Although this rule (if it turns out to be a rule) cannot be assigned to Haldane, it is one he might very well have come up with.

It has been proposed that dosage compensation followed from the evolution of X chromosome inactivation during meiosis, which appears to be common taxonomically, including in *Drosophila* (McKee and Handel 1993). The ability to regulate X chromosome activity arose by selection to avoid pairing failure, and the mechanism was then applied for dosage compensation (Lifschytz and Lindsley 1972; Lyon 1974a,b). The reduced chiasma formation in female marsupials (Bennett et al. 1986; Hayman et al. 1988) may be related to their deactivation of the paternal X chromosome. Jablonka and Lamb (1988) suggested that the apparent absence of dosage compensation in ZW taxa results because Z chromosomes, unlike X chromosomes, are not condensed at meiosis. Consequently, these groups never acquired the machinery for dosage compensation along mammalian lines (Ohno 1967; Johnson and Turner 1979; Baverstock et al. 1982). More recently, Wu and Xu (2003) proposed the sexual antagonism and X inactivation (SAXI) hypothesis to explain germline X inactivation in mammals, *Drosophila,* and nematodes. The data from these groups indicate that late spermatogenic genes are redistributed from the X to autosomes, leading to eventual germline X inactivation because the demasculinized X chromosome accumulates genes that are beneficial to females but costly to males; that is, sexually antagonistic genes (Rice 1984). Again Diptera provide the potential testing ground for this hypothesis. Wu and Xu (2003) predict that such groups as mosquitoes with incipient sex chromosomes that have not (yet) evolved germline X inactivation (McKee and Handel 1993) will show a moderate redistribution of spermatogenic genes.

Whatever the evolutionary origins of dosage compensation, its mechanism seems to differ greatly among mammals, flies, and nematodes (Akhtar 2003). In marsupials, the paternal X chromosome is inactivated (Sharman 1971). X inactivation also occurs in eutherian mammals, but in the embryo, it is random with respect to its parental origin (Sharman 1971; Cooper et al. 1977). In nematodes and *Drosophila,* there is no X inactivation, and the transcription rate of the X chromosome is reduced in homogametic nematode females (Parkhurst and Meneely 1994), but increased for X-linked genes in heterogametic drosophilid males (Bashaw and Baker 1996; Cline and Meyer 1996; Lucchesi 1996). Remarkably, in crickets, there is evidence of both mammalian-style X inactivation (Rao and Arora 1979) and *Drosophila*-style X-expression adjustment (Rao and Ali 1982).

Reminiscent of the sex determination pathway, dosage compensation in Diptera appears to be controlled by a mixture of ancient, conserved genes and evolutionary plastic pathways. The five genes (constituting the so-called "compensasome" gene complex) that control dosage compensation in *Drosophila* have also been found in mammals (Marin 2003), yet even within Diptera, dosage compensation gene networks seem to have diverged. In *Sciara ocellaris* (Nematocera), for example, dosage compensation is achieved by hypertranscription of the X chromosome, as in *Drosophila,* but the molecular pathway appears to use different proteins (Ruiz et al. 2000). Apparently there is no dosage compensation in *M. domestica* because there are few genes on the X chromosome (Shearman 2002). There also appear to be few genes on the X chromosome in other nondrosophilid Diptera, such as the *C. capitata* (for a reviewed, see Rossler et al. 1994). This would not explain, however, why Lepidoptera, with many genes on the Z chromosome (Sperling 1994; Presgraves 2002), are apparently able to function without dosage compensation. Jablonka and Lamb (1990) predict that dosage compensation does in fact occur in groups with female heterogamety, but that regulation will be posttranscriptional on a gene-by-gene basis rather than at the whole-chromosome level. Recent evidence suggests that dosage compensation might indeed occur in birds, although not through Z inactivation (see McQueen 2001).

The third "pseudo-Haldane rule" may be proved wrong. If true, sex determination pathways in birds and butterflies might be constrained as they are in *Drosophila* by a dual role in sex determination and dosage compensation. Whether or not Haldane's rules continue to stand the test of time, they certainly help to motivate future studies, particularly in Diptera.

Future Directions

The knowledge generated so far on dipteran sex chromosomes suggests that further investigation is likely to be richly rewarded. In particular, we would argue that more phylogenetically controlled studies are needed in groups with apparently high levels of sex chromosome variation. The importance of mosquitoes has been discussed earlier. Other dipterans of interest include *Megaselia* flies, which appear to have very primitive sex chromosomes with a single male-determining factor marking the Y chromosome. *Megaselia scalaris*, for example, has three isomorphic chromosome pairs, and the male-determining factor moves at a slow rate with respect to other chromosomes, creating new Y chromosomes in the process (Traut 1994). Chironomid flies have cases of ZW sex determination, and it has been suggested that one species, *Polypedilum nubifer*, may still be in the process of changing from XY to ZW, because some males appear to be heterogametic and the sex chromosomes are not highly heteromorphic (Martin 1966). Sex chromosome diversity in tephritid fruit flies is particularly remarkable, including species with isomorphic sex chromosomes, as well as heteromorphic ZW and XY systems, although the majority seem to have XY males (Bush 1966; Frias 1992). The subfamily Tephritinae might prove to be the most diverse insect group of all.

The study of tephritid cytology has a long history (Metz 1916; Keuneke 1924), but it was in the 1960s that Guy Bush conducted the first major investigation and review of tephritid sex chromosome morphology. Bush (1966) found that most tephritids, like *Drosophila* and other dipterans (Bull 1983), had XY sex chromosomes. There was, however, one major exception. The subfamily Tephritinae contain not only XO and XY species but also—indeed, the majority—have either nondifferentiated (isomorphic) sex chromosomes or show female heterogamety. Whereas other tephritids (the Dacinae and Trypetinae) are fruit feeders, the Tephritinae are either gall-formers or feed on flower heads (generally of Asteraceae and less commonly, Acanthaceae, Goodeniaceae, Verbenaceae, and Lamiaceae). It is intriguing but probably coincidental that the switch from fruit feeding coincided with the evolution of female heterogametic species.

Frias (1992) has reviewed subsequent studies of tephritid sex determination. The five species examined from the Dacinae have heterogametic males. More variation has been found in the Trypetinae, for which 41 species have been typed. Three species are X_1X_2Y, many are XY, and the speciose genus *Rhagoletis* are mostly isomorphic. Genetic data suggest that *Rhagoletis suavis* may have heterogametic females (Berlocher 1993) but they are cytologically isomorphic (Bush 1966). Intraspecific variation in sex chromosome morphology has also been found in *Anastrepha* (Solferini and Morgante 1987, 1990). Heteromorphic ZW sex chromosomes are still unknown for tephritids outside of the Tephritinae. Frias (1992) lists the heterogametic sex for 15 species of tephritines from ten genera (Table 8.3). They include species with XY, XO, ZW, and isomorphic sex chromosomes. Why have so many modes of sex determination evolved in this subfamily but apparently not in other tephritids or other insects?

The monophyly of the tephritines seems robust in terms of DNA sequences (Han and McPheron 1994), morphology, and host plant use (Foote et al. 1993). Within the Tephritinae, however, classification above the genus level is controversial (Hardy and Drew 1996). Of particular interest in terms of their sex chromosomes are the closely related tribes Dithrycini and Tephritini. The tribe Dithrycini contains both XY and ZW species; indeed, XY and ZW species occur in the same subtribe, Cecidocharina. Tephritines therefore, like mosquitoes, could provide important data with respect to speciation genetics. Read and Nee (1991) showed that Haldane's rule is phylogenetically insignificant. Hybridization of female-heterogametic tephritids could provide crucial data in establishing the significance of Haldane's rule in a controlled phylogenetic context.

TABLE 8.3. Sex Chromosomes in the Tephritinae

Tribe/Subtribe	Species	Sex Chromosomes
Tribe Eutretini	*Rachiptera limbata*	ZW
Tribe Tephritini	*Acinia lucata*	ZW
	A. mallochi	ZW
	Dyseuaresta impluvita	Isomorphic
	Spathulina arincae	XO
	Trypanaresta marisolae	Isomorphic
	Trupanea chrysanthemifolii	Isomorphic
	T. foliosi	Isomorphic
	T. footei	Isomorphic
	T. thuriferae	Isomorphic
Tribe Dithrycini		
Subtribe Cecidocharina	*Cecidocharella borrichia*	ZW
	Procecidochares utilis	XY
Subtribe Oedaspidina	*Chrysotrypanea trifasciata*	ZW
	Chrysotrypanea sp.	ZW
Increte sedis Tephritini	*Tephritis* sp.	ZW

SOURCE: Frias 1992.

Conclusion

Patterns of genetic sex determination, the suppression of recombination, the degeneration of sex chromosomes, dosage compensation, achiasmate meiosis, and Haldane's rule may yet be explained by a unified theory of sex chromosome evolution. If such a theory exists to be discovered, the Diptera are probably the best place to start looking for it.

ACKNOWLEDGMENTS

Thanks to the Pacific Education and Research Laboratories (PEARL) program at the University of California–Berkeley. The manuscript benefited greatly from the suggestions of anonymous reviewers. We are grateful to Bruce McPheron for input to an earlier incarnation of this chapter, and to David Yeates and Brian Weigmann for their tenacity in putting together this volume.

REFERENCES

Akhtar, A. 2003. Dosage compensation: An intertwined world of RNA and chromatin remodelling. *Curr. Opin. Genet. Dev.* 13:161–169.

Bachtrog, D. 2003. Protein evolution and codon usage bias on the neo-sex chromosomes of *Drosophila miranda*. *Genetics* 165:1221–1232.

Bachtrog, D., and B. Charlesworth. 2002. Reduced adaptation of a non-recombining neo-Y chromosome. *Nature* 416:323–326.

Baker, R., and M. Rabbani. 1970. Complete linkage in females of *Culex tritaeniorhynchus* mosquitoes. *Heredity* 61:59–61.

Bashaw, G., and B. Baker. 1996. Dosage compensation and chromatin structure in *Drosophila*. *Curr. Opin. Genet. Dev.* 6:496–501.

Bauer, V., and C. Aquadro. 1997. Rates of DNA sequence evolution are not sex-biased in *Drosophila melanogaster* and *D. simulans*. *Mol. Biol. Evol.* 14:1252–1257.

Baverstock, P., M. Adams, R. Polkinghorne, and M. Gelder. 1982. A sex-linked enzyme in birds—Z-chromosome conservation but no dosage compensation. *Nature* 296:763–766.

Bell, G. 1982. *The Masterpiece of Nature.* Berkeley: University of California Press.

Bennett, J., D. Hayman, and R. Hope. 1986. Novel sex differences in linkage values and meiotic chromosome behaviour in a marsupial. *Nature* 323:59–60.

Berlocher, S. 1993. Gametic disequilibrium between allozyme loci and sex chromosomes in the genus *Rhagoletis. J. Hered.* 84:431–437.

Blackman, R. L. 1995. Sex determination in insects. In S. R. Leather and J. Hardie, eds., *Insect Reproduction,* pp. 57–94. Boca Raton, Fla.: CRC Press.

Bull, J. 1983. *Evolution of Sex Determining Mechanisms.* Menlo Park, Calif.: Benjamin/Cummings Publishing.

Bush, G. 1966. Female heterogamety in the family Tephritidae (Acalyptratae, Diptera). *Am. Nat.* 100:119–126.

Calkins, C. O., W. Klassen, and P. Liedo, eds. 1994. *Fruit Flies and the Sterile Insect Technique.* Boca Raton, Fla.: CRC Press.

Carroll, S. B., J. K. Grenier, and S. D. Weatherbee. 2001. *From DNA to Diversity: Molecular Genetics and the Evolution of Animal Design.* Malden, Mass.: Blackwell Science.

Carvalho, A. 2002. Origin and evolution of the *Drosophila* Y chromosome. *Curr. Opin. Genet. Dev.* 12: 664–668.

Cattanach, B., C. Pollard, and S. Hawkes. 1971. Sex-reversed mice: XX and XO males. *Cytogenetics* 10:318–337.

Charlesworth, B. 1996. The evolution of chromosomal sex determination and dosage compensation. *Curr. Biol.* 6:149–162.

Charlesworth, B. 2002. The evolution of chromosomal sex determination. In D. Chadwick and J. Goode, eds., *Genetics and Biology of Sex Determination. Novartis Foundation Symposium,* pp. 207–224. New York: John Wiley & Sons.

Charlesworth, B., and D. Charlesworth. 2000. The degeneration of Y chromosomes. *Philos. Trans. R. Soc. B* 355:1563–1572.

Charlesworth, B., J. A. Coyne, and N. Barton. 1987. The relative rates of evolution of sex-chromosomes and autosomes. *Am. Nat.* 130:113–146.

Charnov, E., and J. Bull. 1977. When is sex environmentally determined? *Nature* 266:828–830.

Chatterjee, R. N. 1998. *Mechanism and evolutionary origins of gene dosage compensation.* In R. N. Chatterjee and L. Sanchez, eds., *Genome Analysis in Eukaryotes: Developmental and Evolutionary Aspects,* pp. 167–214. New Dehli: Springer/Narosha.

Chippendale, A., and W. Rice. 2001. Y chromosome polymorphism is a strong determinant of male fitness in *Drosophila melanogaster. Proc. Natl. Acad. Sci. USA* 98:5677–5682.

Cline, T. W., and B. J. Meyer. 1996. Vive la difference: Males vs females in flies vs worms. *Annu. Rev. Genet.* 30:637–702.

Cock, A. 1964. Dosage compensation and sex-chromatin in non-mammals. *Genet. Res.* 5:354–365.

Cole, C., and C. Lowe. 1967. Sex chromosomes in lizards. *Science* 155:1028–1029.

Cooper, D., P. Johnston, G. Sharman, and J. Vanderberg. 1977. The control of gene activity on eutherian and metatherian X chromosomes—A comparison. In J. Calaby and C. Tyndale-Biscoe, eds., *Reproduction and Evolution,* pp. 81–87. Canberra: Australian Academy of Science.

Coyne, J. A. 1992. Genetics and speciation. *Nature* 355:511–515.

Coyne, J. A. 1997. Patterns of speciation in *Drosophila* revisited. *Evolution* 51:295–303.

Coyne, J., and H. Orr. 1989. Two rules of speciation. In D. Otte and J. Endler, eds., *Speciation and Its Consequences,* pp. 180–207. Sunderland, Mass.: Sinauer.

Crouse, H. 1960. The controlling element in sex chromosome behavior in *Sciara. Genetics* 45:1429–1443.

Davies, N., and A. Pomiankowski. 1995. Haldane's rule: Old theories are the best. *Trends Ecol. Evol.* 10:350–351.

DeJong, H. 2002. Modeling and simulation of genetic regulatory systems: A literature review. *J. Comp. Biol.* 9:67–103.

Erickson, J. W., and T. W. Cline. 1998. Key aspects of the primary sex determination mechanism are conserved across the genus *Drosophila. Development* 125:3259–3268.

Fisher, R. 1930. *The Genetical Theory of Natural Selection.* Oxford: Clarendon Press.

Foote, R., F. Blanc, and A. Norrbom. 1993. *Handbook of the Fruit Flies (Diptera: Tephritidae) of America North of Mexico.* Ithaca and London: Comstock Publishing Associates.

Fredga, K. 1994. Bizarre mammalian sex-determining mechanisms. In R. Short and E. Balaban, eds., *The Difference Between the Sexes*, pp. 419–431. Cambridge: Cambridge University Press.

Fredga, K., H. Gropp, H. Winking, and F. Frank. 1976. Fertile XX- and XY-type females in the wood lemming *Myopus shisticolor*. *Nature* 261:225–227.

Frias, L. 1992. Aspetos de la biología evolutiva de especies de Tephritidae (Diptera) de distribución chilena. *Acta Ent. Chilena* 17:67–79.

Gethmann, R. 1988. Crossing over in males of higher Diptera (Brachycera). *J. Hered.* 79:344–350.

Gorman, G., and L. Atkins. 1966. Chromosome heteromorphism in some male lizards of the genus *Anolis*. *Am. Nat.* 100:579–583.

Hackstein, J., R. Hochstenbach, and E. Hauschteck-Jungen. 1996. Is the Y chromosome of *Drosophila* an evolved supernumary chromosome? *BioEssays* 18:317–323.

Haldane, J. 1922a. The part played by recurrent mutation in evolution. *Am. Nat.* 67:5–9.

Haldane, J. 1922b. Sex-ratio and unisexual sterility in hybrid animals. *J. Genet.* 12:101–109.

Han, H.-Y., and B. McPheron. 1994. Phylogenetic study of selected Tephritid flies (Insecta: Dipetera: Tephritidae) using partial sequences of the nuclear 18S ribosomal DNA. *Biochem. Syst. Ecol.* 22:447–457.

Hardy, D. E., and R. A. I. Drew. 1996. Revision of the Australian Tephritini (Diptera: Tephritidae). *Invert. Tax.* 10:213–405.

Hastings. 1994. Manifestations of sexual selection may depend on the genetics basis of sex determination. *Proc. R. Soc. B* 258:83–87.

Hayman, D., H. Moore, and E. Evans. 1988. Further evidence for novel differences in chiasma distribution in marsupials. *Heredity* 61:455–458.

Hilfiker-Kleiner, D., A. Dubendorfer, A. Hilfiker, and R. Nothiger. 1994. Genetic control of sex determination in the germ line and soma of the housefly *Musca domestica*. *Development* 9:2531–2538.

Huang, X., H. Cheng, Y. Guo, L. Liu, J. Gui, and R. Zhou. 2002. A conserved family of double-sex related genes from fishes. *J. Exp. Zool.* 294:63–67.

Hurst, L., and J. Peck. 1996. Recent advances in understanding of the evolution and maintenance of sex. *Trends Ecol. Evol.* 11:46–52.

Hurst, L., and A. Pomiankowski. 1991. Causes of sex ratio bias may account for unisexual sterility in hybrids: A new explanation of Haldane's rule and related phenomena. *Genetics* 128:841–858.

Huxley, J. S. 1928. Sexual difference of linkage *Grammarus chereuxi*. *J. Genet.* 20:145–156.

Jablonka, E., and M. Lamb. 1988. Meiotic pairing constraints and the activity of sex chromosomes. *J. Theoret. Biol.* 133:23–36.

Jablonka, E., and M. Lamb. 1990. The evolution of heteromorphic sex chromosomes. *Biol. Rev.* 65:249–227.

Johnson, M., and J. Turner. 1979. Absence of dosage compensation for sex-linked enzymes in butterflies (*Heliconius*). *Heredity* 43:71–77.

Keuneke, W. 1924. Über die Spermatogenese einiger Dipteren. *Z. Wiss. Biol.* 1:357–412.

King, R., and J. Buning. 1985. The origin and functioning of insect oocytes and nurse cells. In G. Kerkut and L. Gilbert, eds., *Comprehensive Insect Physiology, Biochemistry, and Pharmacology*, pp. 37–82. Oxford: Pergamon Press.

Kirkpatrick, M., and D. Hall. 2004. Male-biased mutation, sex linkage, and the rate of adaptive evolution. *Evolution* 58:437–440.

Korol, A., I. A. Preigel, and S. I. Preigel. 1994. *Recombination, Variability, and Evolution*. London: Chapman and Hall.

Kraak, S., and E. De Looze. 1993. A new hypothesis on the evolution of sex determination in vertebrates; big females ZW, big males XY. *Netherlands J. Zool.* 43:260–273.

Krebs, J., and N. Davies, eds. 1991. *Behavioural Ecology*. Oxford: Blackwell Scientific Publishing.

Kuhn, S., V. Sievert, and W. Traut. 2000. The sex-determining gene *doublesex* in the fly *Megaselia scalaris*: Conserved structure and sex-specific splicing. *Genome* 43:1011–1020.

Lahn, B., N. Pearson, and K. Jegalian. 2001. The human Y chromosome, in the light of evolution. *Nat. Rev. Genet.* 2:207–216.

Lenormand, T. 2003. The evolution of sex dimorphism in recombination. *Genetics* 163:811–822.

Lifschytz, E., and D. Lindsley. 1972. The role of X-chromosome inactivation during spermatogenesis. *Proc. Natl. Acad. Sci. USA* 69:182–186.

Lohe, A., A. Hilfiker, and P. Roberts. 1993. Mapping simple repeated DNA sequences in heterochromatin of *Drosophila melanogaster*. *Genetics* 134:1149–1174.

Lucchesi, J. 1996. Dosage compensation in *Drosophila* and the "complex" world of transcriptional regulation. *BioEssays* 18:541–547.

Lyon, M. 1974a. Evolution of X-chromosome inactivation in mammals. *Nature* 250:651–653.

Lyon, M. 1974b. Mehanisms and evolutionary origins of variable X-chromosome activity in mammals. *Proc. R. Soc. B* 187:243–268.

MacCarthy, T., R. Seymour, and A. Pomiankowski. 2003. The evolutionary potential of the *Drosophila* sex determination gene network. *J. Theoret. Biol.* 225:461–468.

Marin, I. 2003. Evolution of chromatin-remodeling complexes: Comparative genomics reveals the ancient origin of "novel" compensasome genes. *J. Mol. Evol.* 56:527–539.

Marin, I., and B. Baker. 1998. The evolutionary dynamics of sex determination. *Science* 281:1990–1994.

Martin, J. 1966. Female heterogamety in *Polypedilum nubifer* (Diptera). *Am. Nat.* 100:157–158.

Marx, J. 1995a. Snaring the genes that divide the sexes for mammals. *Science* 269:1824–1825.

Marx, J. 1995b. Tracing how the sexes develop. *Science* 269:1822–1825.

Maynard Smith, J. 1978. *The Evolution of Sex.* Cambridge: Cambridge University Press.

McAllister, B. F. 2003. Sequence differentiation associated with an inversion on the neo-X chromosome of *Drosophila americana. Genetics* 165:1317–1328.

McKee, B. D., and M. A. Handel. 1993. Sex chromosome, recombination, and chromatin conformation. *Chromosoma* 102:71–80.

McQueen, H. 2001. Dosage compensation: Do birds do it as well? *Trends Genet.* 18:25–28.

Metz, C. 1916. Chromosomes studies on the Diptera. II. The paired association of chromosomes in the Diptera, and its significance. *J. Exp. Zool.* 21:213–279.

Mittwoch, U. 1971. Sex determination in birds and mammals. *Nature* 231:432–434.

Mittwoch, U. 1986. Males, females and hermaphrodites. *Ann. Human Genet.* 50:103–121.

Mittwoch, U. 1996. Sex-determining mechanisms in animals. *Trends Ecol. Evol.* 11:63–67.

Miyata, T., K. Kuma, N. Iwabe, H. Hayashida, and T. Yasunaga. 1990. Different rates of evolution of autosome-, X chromosome-, and Y chromosome-linked genes: Hypothesis of male-driven molecular evolution. In N. Takahata and J. Crow, eds., *Population Biology of Genes and Molecules,* pp. 341–357. Tokyo: Baifukan.

Morell, V. 1994. Rise and fall of the Y chromosome. *Science* 263:171–172.

Morgan, T. H. 1914. No crossing over the male *Drosophila* of genes in the second and third pair of chromosomes. *Biol. Bull.* 26:195–204.

Muller, H. 1942. Isolating mechanisms, evolution, and temperature. *Biol. Sym.* 6:71–125.

Nokkala, S. 1987. Cytological characteristics of chromosome behaviour during female meiosis in *Sphinx ligustri* L. (Sphingidae, Lepidoptera). *Hereditas* 106:169–179.

Nothiger, R., and M. Steinemann-Zwicky. 1985. A single principle for sex determination in insects. *Cold Spring Harbor Symp.* 50:615–621.

Ohno, S. 1967. *Sex Chromosomes and Sex-linked Genes.* Berlin: Springer-Verlag.

Ohno, S. 1983. *Phylogeny of the X chromosome of man.* In A. Sandberg, ed. *Cytogenetics of the Mammalian X Chromosome,* pp. 1–19. New York: Alan R. Liss.

Otto, S., and T. Lenormand. 2002. Resolving the paradox of sex and recombination. *Nat. Rev. Genet.* 3:252–261.

Pane, A., M. Salvemini, P. Bovi, L. Polito, and G. Saccone. 2002. The transformer gene in *Ceratitis capitata* provides a genetic basis for selecting and remembering the sexual fate. *Development* 129:3715–3725.

Parkhurst, S., and P. Meneely. 1994. Sex determination and dosage compensation: Lessons from flies and worms. *Science* 264:924.

Presgraves, D. C. 2002. Patterns of postzygotic isolation in Lepidoptera. *Evolution* 56:1168–1183.

Presgraves, D. C., and H. A. Orr. 1998. Haldane's rule in taxa lacking a hemizygous X. *Science* 282:952–954.

Rahn, M., and A. Solari. 1986. Recombination nodules in the oocytes of the chicken, *Gallus domesticus. Cytogenet. Cell Genet.* 43:187–193.

Raman, R., and I. Nanda. 1986. Mammalian sex chromosomes. I. Cytological changes in the chiasmatic sex chromosomes of the male musk shrew, *Suncus murinus. Chromosoma* 93:367–374.

Rao, S., and S. Ali. 1982. Insect sex chromosomes. IV. A presumptive hyperactivation of the male X chromosome in *Acheta domesticus* (L.). *Chromosoma* 86:325–339.

Rao, S., and P. Arora. 1979. Insect sex chromosomes III. Differential susceptibility of homologous X chromosomes of *Gryllotalpa fossor* to 3H-Urd-induced aberrations. *Chromosoma* 74:241–252.

Ray-Chauduri, S., L. Singh, and T. Sharma. 1972. Evolution of sex chromosomes and formation of W-chromatin in snakes. *Chromosoma* 33:239–251.

Raymond, C., C. Shamu, M. Shen, B. Seifert, B. Hirsch, J. Hodgkin, and D. Zarkower. 1998. Evidence for evolutionary conservation of sex determining genes. *Nature* 391:691–695.

Raymond, C., M. Murphy, M. O'Sullivan, V. J. Bardwell, and D. Zarkower. 2000. *Dmrt1,* a gene related to worm and fly sexual regulators, is required for mammalian testis differentiation. *Genes Devel.* 14: 2587–2595.

Read, A., and S. Nee. 1991. Is Haldane's Rule significant? *Evolution* 45:1707–1709.

Reinhold, K. 1998. Sex linage among genes controlling sexually selected traits. *Behav. Ecol. Sociobiol.* 44:1–7.

Rice, W. 1984. Sex chromosomes and the evolution of sexual dimorphism. *Evolution* 38:735–742.

Rice, W. 1996. Evolution of the Y sex chromosome in animals. *Bioscience* 46:331–343.

Rice, W. 2002. Experimental tests of the adaptive significance of sexual recombination. *Nat. Rev. Genet.* 3:241–251.

Rossler, Y., A. Malacrida, and M. C. Zapater. 1994. Mutants chromosomes and genetic maps in the Mediterranean fruit fly. In C. O. Calkins, W. Klassen, and P. Liedo, eds., *Fruit Flies and the Sterile Insect Technique.* Boca Raton, Fla.: CRC Press.

Ruiz, M. F., M. Esteban, C. Donoro, C. Goday, and L. Sanchez. 2000. Evolution of dosage compensation in Diptera: The *maleless* gene implements dosage compensation in *Drosophila* (Brahcycera suborder) but its homolog in Sciara (Nematocera suborder) appears to play no role in dosage compensation. *Genetics* 156:1853–1865.

Saccone, G., A. Pane, and L. Polito. 2002. Sex determination in flies, fruitflies, and butterflies. *Genetica* 116:15–23.

Schempp, W., and M. Schmid. 1981. Chromosome banding in Amphibia. VI. BrdU-replication patterns in *Anura* and demonstration of XX/XY sex chromosomes in *Rana esculenta. Chromosoma* 83:697–710.

Schutt, C., and R. Nothiger. 2000. Structure, function and evolution of sex-determining systems in Dipteran insects. *Development* 127:667–677.

Shan, Z., I. Nanda, Y. Wang, M. Schmid, A. Vortkamp, and T. Haaf. 2000. Sex-specific expression of an evolutionary conserved male regulatory gene, *dmrt1,* in birds. *Cytogenet. Cell Genet.* 89:252–257.

Sharman, G. 1971. Late DNA replication in the paternally derived X chromosome of female kangaroos. *Nature* 230:231–232.

Shearman, D. 2002. The evolution of sex determination systems in dipteran insects other than *Drosophila. Genetica* 116:25–43.

Singer, A., H. Perlman, Y. Yan, C. Walker, G. Corley-Smith, B. Brandhorst, and J. Postlethwait. 2002. Sex-specific recombination rates in zebrafish (*Danio rerio*). *Genetics* 160:649–657.

Solari, A., N. Fechheimer, and J. Bitgood. 1988. Pairing of ZW gonosomes and the localized recombination nodule in two Z-autosome translocations in *Gallus domesticus. Cytogenet. Cell Genet.* 48:130–136.

Solferini, V., and J. Morgante. 1987. Karyotype study of eight species of Anastrepha (Diptera: Tephritidae). *Caryologia* 40:229–241.

Solferini, V., and J. Morgante. 1990. X1X1X2X2:X1X2Y mechanism of sex determination in *Anestrepha bistrigata* and *A. serpentina* (Diptera: Tephritidae). *Rev. Brazil. Genet.* 13:201–208.

Sperling, F. 1994. Sex-linked genes and species differences in Lepidoptera. *Can. Entomol.* 126:807–818.

Steinemann, M., and S. Steinemann. 2000. Common mechanisms of Y chromosome evolution. *Genetica* 109:105–111.

Tao, Y., S. Chen, D. Hartl, and C. Laurie. 2003. Genetic dissection of hybrid incompatibilities between *Drosophila simulans* and *D. mauritiana.* I. Differential accumulation of hybrid male sterility effects on the X and autosomes. *Genetics* 164:1383–1397.

Telfer, W. 1975. Development and physiology of the oocyte-nurse cell syncytium. *Adv. Insect Physiol.* 11: 223–319.

Traut, W. 1994. Sex determination in the fly *Megaselia scalaris,* a model system for primary steps of sex chromosome evolution. *Genetics* 136:1097–1104.

Turelli, M., and H. A. Orr. 1995. The dominance theory of Haldane's rule. *Genetics* 140:389–402.

Turelli, M., and H. A. Orr. 2000. Dominance, epistasis, and the genetics of postzygotic isolation. *Genetics* 154:1663–1679.

Turner, J., and P. Sheppard. 1975. Absence of crossing-over in female butterflies (Heliconius). *Heredity* 34:265–269.

White, M. 1973. *Animal Cytology and Evolution.* Cambridge: Cambridge University Press.
Wilkins, A. 1995. Moving up the hierarchy: A hypothesis on the evolution of a genetic sex determination pathway. *BioEssays* 17:71–77.
Willhoeft, U., and G. Franz. 1996. Identification of the sex-determining region of the *Ceratitis capitata* Y chromosome by deletion mapping. *Genetics* 144:737–745.
Wu, C.-I., and E. Y. Xu. 2003. Sexual antagonism and X inactivation—The SAXI hypothesis. *Trends Genet.* 19:243–247.
Wu, C.-I., N. Johnson, and M. Palopoli. 1996. Haldane's rule and its legacy: Why are there so many sterile males? *Trends Ecol. Evol.* 11:281–284.
Yi, S., D. Bachtrog, and B. Charlesworth. 2003. A survey of chromosomal and nucleotide sequence variation in *Drosophila miranda. Genetics* 164:1369–1381.

PART III

Evolutionary Ecology and Biogeography

CHAPTER NINE

Fossil History and Evolutionary Ecology of Diptera and Their Associations with Plants

Conrad C. Labandeira

The Diptera are one of the four major and dominant orders of holometabolous insects whose larval and adult stages have invaded virtually every available terrestrial and freshwater habitat on the planet. In these continental environments, dipterans consume every nutritionally rewarding food resource and are able to thrive in settings as varied as petroleum-rich pools, highly saline playa lakes, and urine soaked soils (Séguy 1950; Oldroyd 1964; Skevington and Dang 2002). Although the origin of the Diptera has been controversially assigned to both sides of the catastrophic Permo-Triassic mass extinction 252 Mya (Amorim and Silva 2002), their major plant-associated diversification commenced during the later Triassic, when there was exploitation of aquatic ecosystems by larvae, followed by a major expansion from the mid-Jurassic to Early Cretaceous of terrestrial adults on seed plants, and most recently, during the Late Cretaceous to Paleogene, by terrestrial phytophagous larvae endophytically colonizing varied tissues of vascular plants. Each of these three diversification events was associated with an invasion of novel food resources and the establishment of myriad associations among disparate dipteran and plant clades. These events have resulted in major dietary trajectories that have defined the present ecological breadth of the order.

Even though the Diptera have a much lower percentage of species that interact with plants when contrasted to the other major holometabolous clades (Strong et al. 1984), they do possess ecologically a similar breadth of plant associational originations compared to the Coleoptera, Lepidoptera, and Hymenoptera. This deficiency in taxonomic diversity has been compensated by numerous independent origins of phytophagy in dipteran clades, resulting in the colonization of a wide variety of plant taxa through life habits as varied as pollinivory, galling, seed predation, and leaf mining. For example, almost all leaf mining originated once within the basal Lepidoptera and at most a few times in the basal Hymenoptera (Pschorn-Walcher and Altenhofer 1989; Powell et al. 1999), where it is a plesiotypic life-habit. By contrast, leaf mining originated numerous times within the Coleoptera (Hespenheide 1991; Connor and Taverner 1997) and is distributed across many apomorphic clades. The dipteran pattern of leaf mining is most similar to the Coleoptera, where it is distributed throughout the order, and has originated independently at least 20 times among 16 family-level clades (Connor and Taverner 1997; Table 9.1). Notably, many of the family-level taxa of dipteran leaf miners contain subordinate taxa that have exploited other types of endophytic phytophagy —galling, seed predation, and boring—and have replicated similarly broad patterns of plant colonization (Roskam 1992; Dempewolf 2001).

TABLE 9.1. Endophytic Phytophagy of Recent Diptera

Family, Higher Clade	Example	Reference
Leaf- and Stem Mining		
Chironomidae (Culicomorpha)	*Chironomus* on *Nuphar* (Nympheaceae)	Petit 1990; Berg 1950; Velde and Hiddink 1987
Ceratopogonidae (Culicomorpha)	*Dasyhelea* on *Salvinia* (Salviniaeae)	A. Borkent, pers. comm.
Tipulidae (Tipulomorpha)	*Dicranomyia* on *Cyrtandra* (Gesneriaceae)	Swezey 1915; Walshe 1951
Sciaridae (Bibionomorpha)	*Lycoriella* on *Codonopis* (Campanulaceae)	Sasakawa 1997
Phoridae (Phoroidea)	*Megaselia* on *Cicer* (Fabaceae)	Cakar and Disney 1991
Dolichopodidae (Empidoidea)	*Oligochaetus* on *Eleocharis* (Cyperaceae)	Frohne 1939
Syrphidae (Syrphoidea)	*Cheilosia* on *Umbilicus* (Crassulaceae)	Rotheray 1988
Psilidae (Diopsoidea)	*Chamaepsila* on *Daucus* (Apiaceae)	Oldroyd 1964; Cole 1969
Tephritidae (Tephritoidea)	*Euleia* on *Pastinaca* (Apiaceae)	Hering 1967
Lauxaniidae (Lauxanioidea)	*Lauxania* (?) on *Opuntia* (Cactaceae)	Hering 1951
Agromyzidae (Opomyzoidea)	*Tropicomyia* on *Angiopteris* (Marattiaceae)	Spencer 1990
Chloropidae (Carnoidea)	*Oscinella* on *Avena* (Poaceae)	Oldroyd 1964
Drosophilidae (Ephydroidea)	*Scaptomyza* on *Stellaria* (Caryophyllaceae)	Hering 1927; Collinge and Louda 1988
Ephydridae (Ephydroidea)	*Hydriella* on *Potamogeton* (Potamogetoniaceae)	Hering 1962; Keiper et al. 2002
Scathophagidae (Muscoidea)	*Cordilura* on *Carex* (Cyperaceae)	Hering 1923; Neff and Wallace 1969
Anthomyiidae (Muscoidea)	*Pegomyia* on *Rumex* (Polygonaceae)	Hering 1962; Roberts 1971
Galling		
Cecidomyiidae	*Asphondylaa* on *Annona* (Annonaceae)	Gagné 1994
Platypezidae (Platypezoidea)	*Agathomyia* on *Ganoderma* (Polyporaceae fungus)	Dreger-Jauffret and Shorthouse 1992
Lonchaeidae (Tephritoidea)	*Dasiops* on *Cynodon* (Poaceae)	Kolomoets et al. 1989
Tephritidae (Tephritoidea)	*Euribia* on *Cirsium* (Asteraceae)	Meyer 1987
Lauxaniidae (Lauxanioidea)	*Calliopum* on *Viola* (Violaceae)	Collin 1948; Kolomoets et al. 1989
Fergusoniidae (Opomyzoidea)	*Fergusonina* on *Eucalyptus* (Myrtaceae)	Skevington and Dang 2002
Agromyzidae (Opomyzoidea)	*Melanagromyza* on *Sarothamnus* (Asteraceae)	Houard 1903; Ionescu and Neasçu 1969
Chloropidae (Carnoidea)	*Lipara* on *Phragmites* (Poaceae)	Docteurs van Leeuwen 1949; Meyer 1987
Anthomyiidae (Muscoidea)	*Craspedochaeta* on *Athyrium* (Aspleniaceae)	Meyer 1987
Wood Boring		
Cecidomyiidae (Bibionomorpha)	*Helicomyia* on *Salix* (Salicaceae)	Barnes 1951
Pantophthalmidae (Stratiomyioidea)	*Pantophthalmus* on *Erythrina* (Fabaceae)	Thrope 1934; Oldroyd 1964
Asilidae (Asiloidea)	*Andrenosoma* on unknown dicot	Melin 1923; Teskey 1976
Syrphidae (Syrphoidea)	*Temnostoma* on *Betula* (Betulaceae)	Heiss 1938; Krivosheina and Mamaev 1962
Psilidae (Diopsoidea)	*Chyliza* on *Ulmus* (Ulmaceae)	Pechuman 1943
Agromyzidae (Opomyzoidea)	*Phytobia* on *Callitris* (Cupressaceae)	Süss 1979; Solomon 1995

TABLE 9.1. *Continued*

Family, Higher Clade	Example	Reference
Seed Predation		
Sciaridae (Bibionomorpha)	*Bradysia* on *Quercus* (Fagaceae)	Keen 1958
Cecidomyiidae (Bibionomorpha)	*Contarinia* on *Pseudotsuga* (Pinaceae)	Ebel et al. 1975; Prévost 1990
Phoridae (Phoroidea)	*Megaselia* on unknown vascular plant	Grebenshchikova and Naumov 1985
Lonchaeidae (Tephritoidea)	*Earomyia* on *Abies* (Pinaceae)	McAlpine 1956; Morge and Nanu 1981
Otitdae (Tephritoidea)	*Eumetopiella* on *Echinochloa* (Poaceae)	Valley et al. 1969
Tephritidae (Tephritoidea)	*Urophora* on *Acamptopappus* (Asteraceae)	Goeden et al. 1995
Piophilidae (Tephritoidea)	*Mycetaulus* on *Pinus* (Pinaceae)	Keen 1958
Pallopteridae (Tephritoidea)	*Palloptera* on *Abies* (Pinaceae)	Keen 1958
Chamaemyiidae (Lauxanioidea)	*Leucopia* on *Pinus* (Pinaceae)	Keen 1958
Lauxaniidae (Lauxanioidea)	*Minettia* on *Abies* (Pinaceae)	Keen 1958
Chloropidae (Carnoidea)	*Elachiptera* on *Pinus* (Pinaceae)	Keen 1958
Chyromyiidae (Ephydroidea)	*Chyromyia* on *Picea* (Pinaceae)	Keen 1958
Sphaeroceridae (Sphaeroceroidea)	*Leptocera* on *Acer* (Sapindaceae)	Keen 1958
Anthomyiidae (Muscoidea)	*Chiastochaeta* on *Trollius* (Ranunculaceae)	Pellmyr 1992; Michelsen 1988
Muscidae (Muscoidea)	*Hylemyia* on *Abies* (Pinaceae)	Keen 1958
Tachinidae (Oestroidea)	*Mycophasia* on *Quercus* (Fagaceae)	Keen 1958

Given this ecological wealth of larval and adult associations with plants, what is the historical pattern of interactions between plants and their dipteran associates? Two basic approaches have been used to address this fundamental query (Labandeira 2002a). The first employs a more indirect, phylogenetically based methodology whereby there is the mapping of various life-habit features, such as functional feeding group or dietary guild membership, on a phylogram that is independently established by morphological, molecular, or combined data (e.g., Matile 1997). Such features are typically ecological attributes that are structurally defined at particular character states or clade nodes, or as more encompassing life-habit assignments to particular clade branches. The second, perhaps more direct, approach employs data from the plant, dipteran, and interaction fossil records using five types of associational evidence—together with a sixth, taxonomic uniformitarianism—that can provide weak to strong inferences for particular types of interactions (Labandeira 1998a, 2002a; Fig. 9.1). Ideally a combination of mutually reinforcing neontological and paleontological approaches provides the most complete account of the history of plant-dipteran associations. Nevertheless, these six types of evidence occur idiosyncratically in time and space, and reliance on a broad spectrum of data is not always possible (Fig. 9.1). In this context, dipterans present particular limitations; namely, their widespread strategy of fluid feeding in terrestrial environments that are not encountered in other mandibulate, phytophagous insect groups. Thus, any determination of the fossil associational record of the Diptera requires mention of the major ways in which dipterans interact with plants.

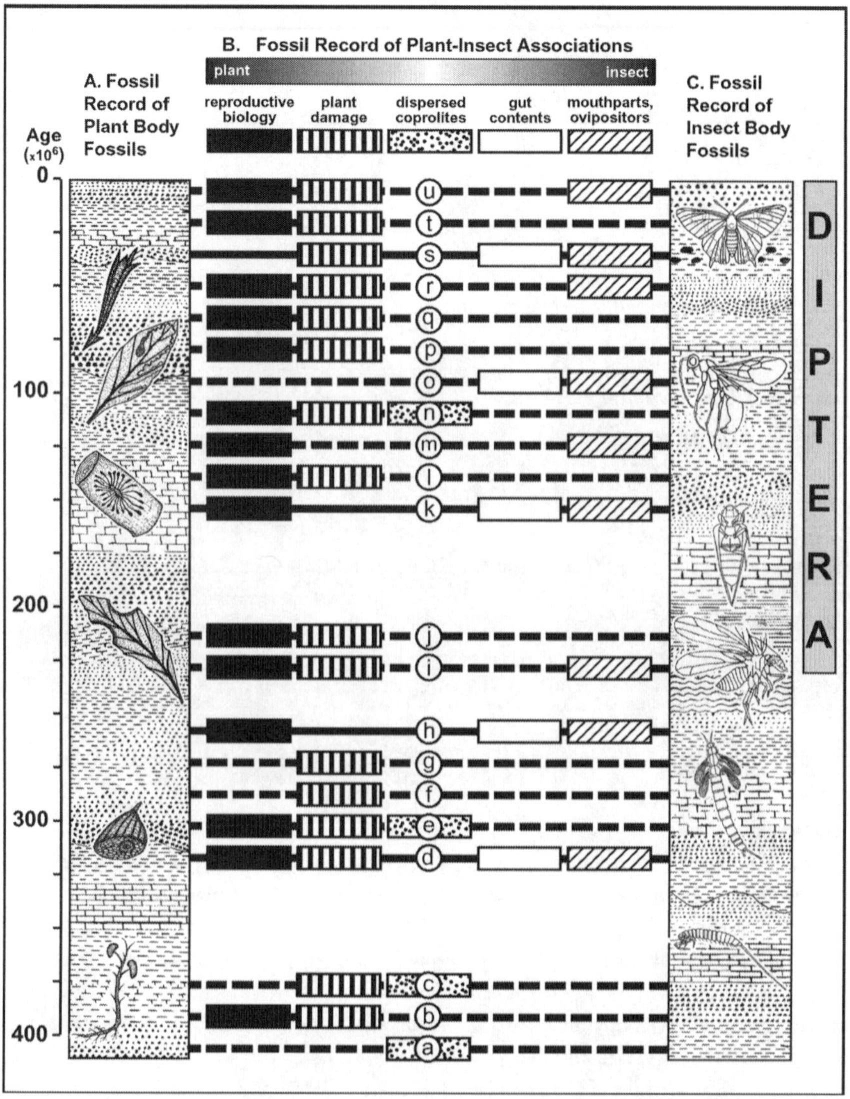

FIGURE 9.1. Types of evidence for plants, insects, and their associations for 21 biotas (a–u in figure), of which the Diptera represent the upper 12 biotas that are younger than 250 My (right vertical bar). The biotas depicted are a small but representative sample selected from the record of fossil Lagerstätten and are documented from the literature. The fossil record of plant-insect associations (center panel with links) consists of one to four of the five major categories of evidence for plant-insect associations that have been used to establish direct (solid lines) to indirect links (dashed lines) between plants and insects. These types of evidence range from those centered primarily on the plant (reproductive biology), to those concentrating on both plant and insect (dispersed coprolites), to those focusing on insect structures (mouthparts and ovipositors). These biotas are not a complete inventory of all major fossil Lagerstätten but represent many of the best-case examples, placed in geochronological order and approximately pegged to their absolute ages (left). For stratigraphic, geographic, and age-related details for each biota, see Labandeira (2002a). Reproduced from Labandeira (2002a) with permission of Blackwell Scientific Publishing.

Contemporary Interactions of Diptera and Plants

The most pervasive and direct way that dipterans interact with plants is by their mouthparts; namely, through the sequestering, processing, and consumption of food. Ovipositors also are important structures as insertion devices for eggs into one or more plant tissues in certain clades (Skevington and Dang 2002). Legs, particularly those of the prothorax, have an ancillary function in some taxa through the manipulation of pollen and other food. In addition to the salient role that mouthpart structure provides for making interpretations between dipterans and their plant associates (Chaudonneret 1990; Labandeira 1997), two other ecological categories are important. One is the functional feeding group, an approach initially developed in freshwater aquatic ecosystems (Cummins and Klug 1979) to distinguish among four principal modes of processing and consumption of solid or particulate plant matter (Merritt and Cummins 1996). The conceptual framework of the functional feeding group (Root 1973; Simberloff and Dayan 1991) is especially applicable to aquatic dipteran larvae, although it was extended to the terrestrial realm by recognition of distinctive types of phytophagous feeding (Root 1973; Strong et al. 1984). The last ecological category is the dietary guild, conceptually formalized by Root (1967), which is an alternative way of characterizing the ecological relationships between dipterans and their food. In summary, mouthpart types circumscribe *the particular apparatus* that is used to consume food; functional feeding groups describe *how* food is consumed; and dietary guilds define *what type* of food is consumed. In this chapter, I discuss functional feeding groups and broad dietary guilds, although the locus of action and functional explanation of the feeding process involves an initial consideration of mouthpart structure.

MOUTHPARTS

Of all insect orders, the Diptera has the greatest diversity of mouthpart types. The evidence for this comes from larval and adult archetypal classes (Chaudonneret 1990), phenetic classifications (Metcalf 1929; Labandeira 1997), comparative studies (Elzinga and Broce 1986; McKeever and French 1999), and phylogenetic evaluations (Sinclair 1992; Fang et al. 1999). Amid this structural diversity, only two major uses of mouthparts are conspicuously absent among the Diptera compared to other holometabolans. First, with the sole exception of a terrestrial Lower Dipteran larva feeding externally on an angiosperm (Cylindrotomidae; Cameron 1918), there are apparently no known instances of typical mandibulate mouthparts occurring either in a dipteran larval or adult stage that are used for terrestrial external feeding on live foliar tissue analogous to that of herbivorous Coleoptera or Lepidoptera. Second, there are no demonstrable examples of piercing and sucking by adult dipterans through fasciculate stylets for plant sap, such as those for the Palaeodictyopteroidea and Hemipteroidea (Labandeira and Phillips 1996; Dixon 1998). These absences aside, both larval and adult mouthparts have been modified independently into structural ensembles that provide for a broad spectrum of phytophagous functional feeding groups. For larvae, these include filtering, scraping, gathering, and collecting of food in aquatic habitats; and leaf mining, galling, boring, and seed predation in terrestrial situations (Krivosheina 1969; Skevington and Dang 2002). For terrestrial adults, the dominant feeding modes are terrestrial pollinivory and nectarivory, occasionally combined with hematophagy and insectivory (Séguy 1957; Downes and Dahlem 1987; Foster 1995).

Larvae. Larval dipteran mouthparts are considerably complex. The larvae of almost all taxa are considered the primary trophic stage whereby food is assimilated for the increase of body mass that is eventually transferred to the adult stage through pupation (Séguy 1950; Oldroyd 1964). For the Diptera, there was the transformation from transversely adducting/abducting, chewing mandibles involved in solid and particulate feeding among plesiomorphic Lower dipteran lineages to a somewhat gradational series of obliquely protracting/retracting, piercing-and-sucking structures for fluid feeding in orthorrhaphous lineages (Cook 1949; Schremmer 1951). This trend culminates in the highly stereotyped and falcate mouthhooks that move in a vertical plane within cyclorrhaphan taxa (Teskey 1981). A parallel trend also occurs for the Cecidomyiidae (Otter 1938). Additionally, the consolidation of the internal cephalic endoskeleton and certain mouthpart elements into the cephalopharyngeal complex has gradually rendered cyclorrhaphan larvae externally "acephalous," as virtually all head and most mouthpart elements have been retracted within their thoraces (Matsuda 1965; Krivosheina 1969).

For most Lower dipteran clades, aquatic larval mouthparts have been modified extensively into elaborate, often setose or pilose, prominent structures for processing suspended particulate and substrate-attached plant and animal material (Merritt and Cummins 1996; Wichard et al. 2002). Several recurring themes are evident in this radiation of larval mouthpart types. First is the modification of the labrum and mandible into a variously deployed fan with long setae or hairs for passive filtering of particulate matter from water. Clades employing this filtering mechanism include most taxa of the Culicidae (labral and mandibular brushes of mosquitoes), Simuliidae (labral fans of black flies), and occasional independent originations among other aquatic families (Kramer 1954; Craig 1977; Dahl et al. 1988). The mouthparts of other aquatic Lower dipteran larvae are typically less elaborate and emphasize mandibular or maxillary modification involved in scraping, gathering, or shredding of live plants or detritus that are encrusted or otherwise attached to benthic substrates (Gouin 1959). The principal organs for extraction of anchored food are robust, occluding mandibles and maxillary elements, which bear toothlike projections or stiff setae.

Lower brachyceran larvae bear externally conspicuous, decurved, unornamented mandibles that may articulate with similarly prominent and parallel extensions of the maxilla, such as the stipes. This shift in mouthpart position and function from the Lower dipteran condition has been associated with a dependence on fluid feeding from a wide variety of organismal sources. Acquisition of fluid feeding by dominantly terrestrial Lower brachyceran clades are best expressed by extraoral digestion associated with carnivory (Roberts 1969). In some lineages, such as the Rhagionidae (snipe flies) and Athericidae, this type of feeding is enhanced by ventral grooves or canals occurring along the length of raptorial mandibles that result in channelized fluid flow, particularly when left and right mandibles are occluded (Schremmer 1951; Tsacas 1963; Nagatomi et al. 1999). Lower brachyceran larvae are overwhelmingly parasitoids or predators; occasional exceptions include such taxa as the saprophagous Stratiomyiidae (soldier flies) and phytophagous Therevidae (stiletto flies) (Roberts 1969).

In cyclorrhaphous larvae, the mandibles are transformed into rather uniform, scythe-shaped mouthhooks. Mouthhooks are frequently used for predation or saprophagy in many clades, where they are typically unornamented, bear a smooth surface, and lack accessory teeth (Okada 1963; Roberts 1971). Modifications occur in many taxa, particularly schizophoran clades that feed on live plants, which are characterized by supernumerary accessory teeth. For leaf-mining larvae, frequently there are multiple adjacent teeth (Thomas 1938; Beri 1983), arranged as a serrate curved series analogous to circular sawblade mandibles of leaf-mining Lepidoptera and Coleoptera. In other phytophagous forms, such as the seed-predating larvae of the Anthomyiidae (root maggot flies), there is fusion of the right and left mandibles into a single penetrative organ. By contrast, agromyzid cambium borers have only two subapical teeth that operate in unison on left and right mandibles that are conjoined into a single structure (Hanson and Benjamin 1967). For the Otitidae (picturewinged flies), phytophagous species have mouthhooks bearing accessory teeth, whereas saprophagous taxa lack such dentition (Allen and Foote 1975; Valley et al. 1969).

Adults. In adults, two basic mouthpart conditions have resulted in a diversity of mouthpart types responsible for accessing liquid food. This diversity is categorized by two major functional types; namely, (1) a modified labium containing a terminal padlike labellum, typically with an extensive pseudotracheal network, for adsorbing fluids; and (2) the presence of elongate mandibles, maxillary laciniae, a labrum, and a hypopharynx modified either singly or in a fasciculate arrangement to form a styletal penetration mechanism. These two major mouthpart types (Labandeira 1997), the labellate and stylate conditions respectively, constitute the major mechanical ways that dipterans access fluid and particulate food. A common modification is a prolongation of the labellate condition into an elongate and largely tubular proboscis, accompanied by considerable reduction of the terminal labellar pad and stylets, used for the consumption of hidden nectar. Although most dipteran species possess both a labellum and various styliform elements, the distinction between labellate and stylate mouthparts is a functional one that centers on the principal mode of accessing fluids and pollen: labellate mouthparts emphasize or are solely dominated by an expanded and fleshy labellum accompanied by a capillary-based pseudotracheal network, whereas stylate mouthparts are overwhelmingly controlled by a stylet-within-a-tube apparatus. A third, much less common type is the vestibulate condition, which occurs in some bibionomorph taxa, characterized by an anterior labrum, lateral and separate labella, and a ventral projection of the postgenal region that forms a comparatively short "vestibule" of surrounding mouthparts (Gagné 1994). Finally, in some nominally anthophilous clades, there is the occasional reduction of the proboscis and constituent mouthparts into a vestigial organ lacking any capacity for feeding (Paramonov 1953).

Lower dipteran mouthparts are either labellate, stylate, or vestibulate (Peterson 1916; Hoyt 1952; Gagné 1994), all of which occasionally are used for consumption of nectar (Downes 1958; Gagné and Boldt 1989). The commonest type of feeding is penetration of tissues by two to six stylets or similar structures for consumption of blood from vertebrate hosts or the fluidized contents of insect prey. Styletal functionality and number are often sexually dimorphic within a species, allowing females a fuller complement of tissue-piercing stylets for obtaining a blood meal, which is generally necessary for ovarian follicle development (Lehane 1991). However, consumption of carbohydrate- or even lipid-rich fluids is also present and is typically required for sustaining flight in adults by supplementing blood diets in females or supplanting them in males (Haslett 1989). Examples of unusually elongate, nectaring proboscides—often several times the length of the head—occur among flower-associated Lower dipterans and include the tipulid (crane fly) *Elephantomyia*, the blepharocerid (net-winged midge) *Aspistomyia*, the mycetophilid (fungus gnat) *Gnoriste*, and the sciarid (black fungus gnat) *Eugnoriste* (Frey 1913; Peterson 1916; Porsch 1958; Steffan 1981). Nectar consumption has been established for many Lower dipteran taxa, including the Tipulidae (crane flies; Knab 1910; May 1979); Cecidomyiidae (gall midges; Feil 1992; Ollerton 1996b), Sciaridae (root gnats; Webber 1981), and Mycetophilidae (Mesler et al. 1980; Breckon and Ortíz 1983), and also includes species in typically blood-feeding clades, such as Ceratopogonidae (biting midges; Downes 1955; Saunders 1959) and Culicidae (Larsen 1948; Thien 1969). Additionally, some of these taxa are implicated in pollination mutualisms (Thien 1969; Breckon and Ortíz 1983).

Among Lower brachyceran lineages, there is a tendency toward the elongation of mouthparts, which may be either stylate and responsible for insectivory or hematophagy, or labellate and involved in nectarivory, pollinivory, or imbibition of animal-derived surface fluids (Zaitzev 1984). Those that are nectarivores, pollinivores, or consume mixtures of both floral products (Gilbert 1981), typically have elongate to occasionally very long proboscides—a condition that occurs among many taxa of the pangioniine Tabanidae (horse flies); Vermileonidae (worm lions); Pelecorhynchidae; Rhagionidae; Nemestrinoidea; and a few asiloid lineages, especially the Bombyliidae (bee flies), Hilarimorphidae, Apioceridae (flower-loving flies), Mydidae (mydas flies), and Scenopinidae (window flies) (Nagatomi and Soroida 1985; Liu and Nagatomi 1995). The trophic division into tissue-penetrating zoophagous taxa vs. those that imbibe such surface fluids as nectar is associated with proboscis structure, including overall length and terminal labellar modifications, but most importantly, styletal modification, particularly the length, rigidity, and dentition of the mandibular and maxillary stylets. Nectarivores in many orthorrhaphous clades lack mandibular stylets. However, for the Empididae (dance flies), zoophagous taxa possess functional maxillary stylets, whereas in nectarivorous species, the maxillary stylets are rudimentary and extend much less of the length of the proboscis, leaving only the labrum-epipharyngeal and hypopharyngeal elements that are modified into nonstylate structures, the latter with a broad preterminal duct for secreting salivary fluids (Bletchly 1954; Nagatomi and Soroida 1985). In fact, many taxa of obligately floricolous taxa, such as the Nemestrinidae (tanglevein flies), Acroceridae (small-headed flies), Vermileonidae, Apioceridae, and Bombyliidae, have diminutive maxillary stylets or lack them altogether and possess nonrigid and shortened maxillary stylets. This condition leaves the proboscis as a closed tube consisting of an upper labrum and lower labium, with only the hypopharynx extending almost to the labellum and providing enzymatic secretions for predigestion or transport of nectar or pollen (Peterson 1916). In some lineages, this is done by possession of diminutive mandibular or maxillary stylets, whereas in others, it is accomplished by positioning these mouthpart elements outside of the proximal portion of the labral-labial tube, below where the proboscis is attached to the clypeal tormae (Szucsich and Krenn 2000). Such taxa as *Fallenia* of the Nemestrinidae and *Oligoneura* of the Acroceridae possess elongate tubular proboscides that lack functional mandibles, although they bear flexible, thin laciniae incapable of penetrating tissue (Becher 1882; Nagatomi and Soroida 1985). In an undetermined species of the nemestrinid *Prosoeca*, Vogel (1954) showed the absence of maxillary laciniae at proboscis midsection, which consisted only of a labrum and membranously attached lower labrum that was supported by cuticular invaginations. This genus and other long-proboscid nemestrinids typically nectar deep-throated flowers of the Cape Floral Province of South Africa and are ecological equivalents of sphingid moths elsewhere (Nilsson 1988). Proboscides of these taxa bear modifications of the labellum that are significant for the initial processing of fluids, including a small relative size when compared to the more typical, expansive labellum (Peterson 1916). In addition, there is arrangement of the left and right labellar lobes into a distinctive elongate and bifid structure (Nitzulescu 1927; Dusa 1968; Stuckenberg 1996), the proximal part of which bears pseudotracheae that are altered into dentate rather than spinose structures for achieving closure of channels that debouche into the food canal (Zaitzev 1982). Finally, such taxa as *Atriadops* and *Nycterimyia* of the Nemestrinidae lack functional mouthparts and undoubtedly are nontrophic (Nagatomi and Soroida 1985).

Labellate mouthparts in cyclorrhaphans are typically shorter and more compact than the frequently elongate proboscides of many Lower brachyceran groups. Unlike the proboscides of Lower brachyceran lineages, which are continuous with the head capsule and tend to dangle (Matsuda 1965), the proboscides of most cyclorrhaphan species are suspended by a membranous region and divided into three functional parts: the basiproboscis (rostrum), medioproboscis (haustellum), and distiproboscis (labellum), each of which is defined by internal muscles but also shares muscles with the other regions (Graham-Smith 1930; Lall and Davies 1971). This division allows for the cyclorrhaphan proboscis to be uniquely designed for protraction or retraction into the head cavity by contraction of muscles that either deploy or withdraw a lever-based folding mechanism (Graham-Smith 1930; Downes 1963). Once protracted, the labellum is used to sponge surface fluids through the capillary action of its numerous branching channels, or pseudotracheae, which collect fluids into a centrally located oral tube. In some cases, the proboscis can be protracted completely, yet the labellum simultaneously can be retracted sufficiently so that the oral opening is adpressed directly to the food surface, with suction provided by the oral tube and not by labellar adpression (Müller 1883). Most labella are designed to collect fluids on exposed surfaces, such as oozing organismal exudates, deliquescing macrofungal tissues, fluidized carrion and dung, and commonly, nectar (Baumgartner 1986; Skevington and Dang 2002). Significant modifications of the labellum nevertheless exist, including exceptional elongation in some clades, which retain the distinctive cyclorrhaphan retraction mechanism, such as some Conopidae (thick-headed flies; Smith 1967) and Syrphidae (hover flies; Schiemenz 1957). Another significant modification has occurred among cyclorrhaphous lineages, in which there has been major structural alteration of the labellum and its pseudotracheal denticles into a newly constituted organ for rasping and penetrating plant tissues or animal skin in hematophagous taxa. This is accomplished by the presence of numerous abrading denticles (Elzinga and Broce 1986) or by fewer stylate teeth (Jobling 1933). Alternatively, in other lineages, the labellum has been converted into a device for entrapment, transport, and consumption of pollen (Elvers 1980; Pont 1993; Ssymank and Gilbert 1993), where such structures also are used for the consumption of pollen-nectar slurries.

FUNCTIONAL FEEDING GROUPS

There are three major classes of herbivore functional feeding groups among the Diptera, defined by distinctive modes of accessing and consuming plant products. These feeding types reflect a combination of (1) differing larval versus adult nutritional requirements, (2) the presence of a basic mouthpart structure and subordinate modifications that are essential for securing certain types of food, and (3) the historical record of evolutionary radiations into major feeding niches.

Aquatic Feeding. Larval Diptera have invaded virtually every freshwater environment. By virtue of a wide repertoire of mouthpart types, they process solid and particulate food as diverse is planktic diatoms, surface algal films, attached benthic plants, detrital drift trapped in the neuston zone, and foliar tissues of submerged plants (Nessimian et al. 1999; Alverson and Courtney 2002; Wichard et al. 2002). Four major functional feeding groups describe this exploitation of food resources: filtering, scraping, gathering, and collecting. Filtering is the more or less passive entrapment of suspended particulate matter through the action of mouthparts that act as a sieve. Both dead and live plant (and animal) material is strained from the ambient water and consumed, especially by the Culicidae and Simuliidae (Dahl et al. 1988). A second feeding strategy is the use of biting mouthparts, particularly mandibles and maxillary lobes, to scrape plant material, such as algal films, protist colonial accumulations, and other substrate-attached food, especially in lotic environments, such as streams. This method of feeding is used by the Lower dipteran Blepharoceridae and Thaumaleidae (solitary midges), the cyclorrhaphous Canaceidae (beach flies), and the cyclorrhaphous Ephydridae–Parydrinae (shore flies) (Merritt and Cummins 1996). By contrast, the third theme, collecting, results in the active removal of suspended and particulate material through the creation of vortices by mouthpart action and a pharyngeal pump, with or without the assistance of food-entrapping silken webs (Wichard et al. 2002). Collectors include most taxa in the Lower dipteran Ptychopteridae (phantom crane flies); Dixidae (dixid midges); five subfamilies of Chironomidae, Psychodidae (moth flies); the Lower brachyceran Stratiomyidae; and the cyclorrhaphous Syrphidae and some Ephydridae (Merritt and Cummins 1996; Nessimian et al. 1999). For collectors, Schremmer (1951) discussed the process employed by the mandibular-maxillary apparatus of stratiomyiid larvae for sequestering food. Lastly, shredders typically feed on submerged dead or live vascular plant vegetation, which they pulverize with their mouthparts; clades with species possessing this feeding habit are the Lower dipteran Tipulidae and Cylindrotomidae, and the cyclorrhaphous Scathophagidae (dung flies) (Merritt and Cummins 1996). Although these four basic aquatic feeding strategies occur among other groups of insects, nowhere are they as collectively well established as in the Diptera.

Nectarivory and Pollinivory. Adult Diptera are very important pollinators worldwide and have developed interactions with seed plants, primarily angiosperms, that are categorized into ten basic types of flowers or analogous structures. This is attributable to their role as obligate fluid and small-particle feeders as adults, resulting from their possession of a wide variety of labellate and even stylate mouthpart modifications, which have allowed many taxa to become preeminent nectarivores and pollinivores. This is accomplished by their use of scent as a stimulative in approaching flowers at short distances and by their alighting on blossoms of particular colors (typically white, yellow, or green; less commonly bluish in the case of flowers with deeper-seated nectaries; and brown, maroon, or purple for carrion flowers) (Bänziger 1991; Pellmyr et al. 1991; Wacht et al. 2000). A unique repertoire of sensory cues, in conjunction with modified mouthparts and the presence of frequently large to holoptic eyes with stereoscopic and probably color vision in advanced forms (Grinfel'd 1955; Proctor et al. 1996), has allowed the exploitation of a stupendous variety of flower types and angiosperm taxa. This phenomenon has occurred in all ecosystems (Kevan and Baker 1983; Lack and

Kevan 1984) and has resulted in dipterans becoming the only important pollinators in some extreme environments (Hagerup 1951; Totland 1993).

The two overwhelming nutritive rewards that dipterans pursue on the reproductive structures of vascular plants are nectar and pollen. Nectar is a liquid rich in carbohydrates, lipids, and proteins (Fahn 1979; Baker and Baker 1983) and is produced by glands from vegetative tissues, such as those on the petiole axils of certain ferns (Bonnier 1879). It is overwhelmingly associated with the flowers of angiosperms (Endress 1994; Vogel 1997, 1998). The pollination-drop fluids that occur in the strobilar micropyles of gymnosperms had an origin separate from the nectarial glands of ferns and angiosperms. However, the composition of these secreted fluids are nutritionally similar, and all are consumed by insects (Baker and Baker 1983; Owens et al. 1998).

For dipterans, the consumption of nectar consists of three major types of feeding strategies, as evidenced by mouthpart structure and behavioral studies. The first, ranging taxonomically from Tipulidae to the Cyclorrhapha, is sponging by species with compact, short proboscides that terminate in an expansive, fleshy labellum with pseudotracheal channels for the capillary uptake of fluid (Graham-Smith 1930; Elzinga and Broce 1986). This type occurs sporadically among many groups of dipterans but is best developed in advanced cyclorrhaphan lineages. The second type of nectar consumption is from taxa with a tubular proboscis, often elongate and bearing a relatively small terminal labellum, and housing a comparatively capacious food canal involved in the active siphoning of nectar (Schumacher and Hoffmann 1982; Szucsich and Krenn 2000). This type of feeding principally is found among Lower brachyceran lineages and reaches its greatest development in long-proboscid forms that are able to reach nectar at the base of deep-throated flowers (Manning and Goldblatt 1997). The third type of nectar feeding is done by nominally hematophagous or insectivorous taxa with fasciculate stylate mouthparts and is developed in both females and males. This type of nectarivory occurs among Lower dipteran taxa and is known from typically blood-feeding families, such as Psychodidae (Grensted 1947), Ceratopogonidae (Downes 1978), Culicidae (Thien 1969; Brantjes and Leemans 1976), Simuliidae (Smart 1943), Chironomidae (Schlee 1977; Beardsell and Bernhardt 1983), and Tabanidae (Lall and Davies 1971), frequenting plants as varied as orchids, ivy, and jacks-in-the-pulpit.

For pollinivory, there are four basic feeding strategies, characterized by both mouthpart morphology and stereotyped behaviors for extracting pollen contents. First is the direct ingestion of pollen by taxa with tubular mouthparts, typically housing large-diameter food canals; principally the Acroceridae, Bombyliidae, Syrphidae, and to a lesser extent, Empididae (Schuhmacher and Hoffman 1982; Szucsich and Krenn, 2000; Wacht et al. 2000). Although typically entomophilous pollen is consumed by these taxa, anemophilous types also are accessed, particularly by syrphids, and include pollen from grasses, sedges, *Plantago,* and catkins of trees (Sonderstrom and Calderón 1971; Stelleman 1981; Leereveld 1982; Ssymank and Gilbert 1993). Consumption of anemophilous pollen also occurred among other insect lineages during the mid-Mesozoic (Rasnitsyn and Krassilov 1996) and may be relevant for the origin of dipteran pollination both during and prior to the ecological radiation of angiosperms (Labandeira 2000). Second is extraoral digestion by certain Drosophilidae (pomace flies), such as *Drosophila flavohirta,* which collect pollen on their proboscis. In *D. flavohirta,* collection is followed by extraoral mastication and predigestion of the molded pollen clump, then grain rupture, ending in imbibition of the protoplast contents (Nicholson 1994). This feeding mechanism is functionally and behaviorally identical to that of the Neotropical butterflies *Heliconius* (Gilbert 1972) and *Parides* (DeVries 1979), which represent two independent originations within the Lepidoptera. The third mechanism consists of the punch-and-suck procedure initially described for thrips (Kirk 1984), which is known to occur in the ceratopogonid midge *Atrichopogon* (Downes 1955, 1971; Meillon and Wirth 1989) and in the empidid fly *Anthalia* (Downes and Smith 1969). Fourth, and most poorly known, is the collection and transport of pollen along the pseudotracheal channels of muscoid flies. In some taxa, there is use of adjacent, modified teeth or blades to scarify or otherwise modify pollen grains for eventual freeing of their protoplasts, and evidently some muscid taxa can accommodate particular pollen sizes and shapes (Elvers 1980; Elzinga and Broce 1986). This type of pollinivory occurs among the Lower dipteran Scatopsidae (minute black scavenger flies), but principally among the cyclorrhaphous Muscidae (house flies), Anthomyiidae, and Scathophagidae (Elvers 1980; Pont 1993; Larson et al. 2001). All four types of pollen consumption are often associated with the imbibition of such foods as nectar or honeydew, thus offering relatively balanced diets consisting of mixtures of proteins, lipids, and carbohydrates (Allen 1929; Haslett 1989). With the exception of the last type of pollinivory, it is notable that these mechanisms occur in other, nondipteran insect taxa, although they are poorly documented in the fossil record.

Historically the associations between flies and the flowers they pollinate have been characterized variously as "the fly pollination syndrome," "myiophily," "sapromyiophily," or similar attributions (Armstrong 1979; Faegri and van der Pijl 1980; Proctor et al. 1996). However, considerable evidence now indicates that such "syndromes" are not as objective as once thought, predominantly because specialization in pollination is the exception rather than the rule (Ollerton 1996a; Waser et al. 1996; Johnson and Steiner 2000). This realization calls into question the applicability of "pollination syndromes" that were based on presumed, repeated structural and behavioral correlates regarding particular pollinator and associated plant taxa (see Kugler 1938, 1955). For example, the "sapromyiophilous syndrome" includes a variety of basic floral types listed in Table 9.2. Additionally, there is geographical evidence for major shifts in pollinator spectra for plant species with particular floral structures that were once thought to constitute a well-defined syndrome (Lippok and Renner 1997). Because recent studies that emphasize the lability of pollination systems are based on both extrinsic and intrinsic factors and affect a variety of pollinator taxa and visited plants, the term "syndrome" is not used here. This view, nevertheless, acknowledges that there are indeed a limited number of taxonomically broad but recurring associations between plants and dipterans (Johnson and Steiner 2000), examples of which are presented in Table 9.2. This shift in understanding is based on recent empirical data and has implications for hypotheses regarding the role of insects, particularly the Diptera, during the early radiation of angiosperms (Crepet 1996; Grimaldi 1999; Dilcher 2000).

The variety of seed-plant reproductive structures that are nectared, pollinated, or otherwise visited by dipterans ranges from advanced seed-plant clades, such as cycads and gnetaleans, to basal angiosperms, to catkin-bearing and nominally wind-pollinated groups, such as sycamores and grasses, and ends with the comparatively derived angiosperms that include sunflowers and orchids (Dilcher 2000; Labandeira 2000). Table 9.2 provides documentation for the ten basic types of recent floral structures or their nonangiospermous equivalent that offer rewards or provide lures to major dipteran taxa, ordinated by their inferred sequence of appearance in the fossil record. These data are based on an extensive literature of fly pollination and visitation studies that include field observational data, functional-morphological assessments of both floral structure and insect mouthpart features, analyses of gut and body vestiture contents, and other examinations of the pollination mutualisms (Faegri and van der Pijl 1980; Proctor et al. 1996). It can be inferred from Table 9.2 that dipterans are major visitors and pollinators for major groups of seed plants, second only to hymenopterans, and that they probably had major associations with seed plants during the mid-Mesozoic, including certain Gnetales, Cycadales, and possibly Coniferales that produced ovular pollination drops initially designed for wind pollen entrapment (Rothwell 1977; Poort et al. 1996). These nutritious secretions of gymnosperms undoubtedly were secondarily co-opted by insects as an alternative source for nectar, as judged by extant dipterans that are the most common visitors of modern Ephedraceae (Mormon tea family; Meeuse et al. 1990), Gnetaceae (van der Pijl 1953; Kato and Inoue 1994) and Welwitschiaceae (Wetschnig and Depisch 1999). Dipteran visitors are much less common for the Cycadaceae and Zamiaceae (cycads; Breckon and Ortíz 1983) and virtually absent in the conifers. An additional potentially important food resource are the prepollen and pollen of various gymnosperm pollen organs, which could serve as food, sources of desired volatile scents, and gathering sites resulting from plant thermogenesis (Terry 2001; entry 1 in Table 9.2).

The other nine basic types of reproductive structures frequented by dipterans are angiosperms that constitute a broad range of floral arrangement, scent production, size, shape, color, thermogenesis presence, insect rewards, and overall specialization. In approximate order of geochronologic appearance (Table 9.2), they are: small, bowl-shaped, clustered flowers from ANITA-grade angiosperms (entry 2); florets of anemophilous catkins (entry 3); midge flowers (entry 4); nontrap carrion flowers (entry 5); medium-depth connate flowers (entry 6); brush flowers (entry 7); trap carrion flowers (entry 8); long-proboscid fly flowers (entry 9); and orchids (entry 10). Collectively these fly-frequented flowers span an enormous breadth of pollination mechanisms, phylogenetic diversity, geographic range, ecological habitat, and geologic timing of origination. Some floral types are plesiomorphic (ANITA-grade basal groups and anemophilous trees), whereas others are quite varied in their phylogenetic representation (medium depth connate flowers, brush flowers), and others consist of highly advanced, apomorphic taxa (many long-proboscid fly flowers and orchids). Table 9.2 strongly suggests that each of these ten assemblages represents a phylogenetically motley collection of both pollinators and floral hosts that have achieved some structural consistency resulting from ecological convergence within similar habitats in time and space. (For reasons of space, these floral types will not be detailed here; Table 9.2 provides a categorization and entry into the relevant literature.)

Endophytic Phytophagy. Phytophagy is defined as the consumption of live plant tissues. For Diptera, phytophagy primarily involves the consumption of (1) particulate plankton or substrate-attached plants, such as diatoms or filamentous algae by aquatic larvae; (2) external plant products, such as nectar or pollen by terrestrial adults; and (3) internal plant tissues of all plant organs that are consumed from within by terrestrial larvae. For the latter, four major functional feeding groups are principally found in the Diptera: mining, galling, boring, and seed predation (Nartshuk 1985; Table 9.1). Root feeding by larvae, such as those of the

TABLE 9.2. Principal Types of Seed-Plant Reproductive Structures Visited or Pollinated by Diptera

Reproductive Structure Type and Key Features	Examples of Visited or Pollinated Plant Taxa	Examples of Major Dipteran Visitors or Pollinators	Rewards or Lures	Geochronologic Appearance[a]	Reference
1. Ovules and pollen organs of nonangiospermous seed plants. Exposed or hidden ovules secreting nectarlike, micropylar fluids; pollen or prepollen produced by organs dioecidously or moneciously.	Ephedraceae (*Ephedra*), Gnetaceae (*Gnetum*), Welwitschiaceae (*Welwitschia*), Cycadaceae (*Cycas*), Zamiaceae (*Zamia, Macrozamia, Encephalartos*); Cheirolepidaceae (*Classopollis*)	Culicidae, Mycetophilidae; Syrphidae, Anthomyiidae, Calliphoridae, Chloropidae, Muscidae, Otitidae, Sacrophagidae, Tachinidae	Pollination drops, prepollen, pollen; endogenous heat; oviposition sites	Late Triassic to Early Cretaceous, depending on the clade	Pijl 1953; Breckon and Ortíz 1983; Meeuse et al. 1990; Kato and Inoue 1994; Wetschnig and Depisch 1999; Labandeira 2000, this chapter; Terry 2001
2. ANITA[b]-grade basal angiosperms. Small, bowl-shaped, white to yellowish, actinomorphic flowers; exposed sexual organs; perianth of separate sepals and petals, often clustered in inflorescenses.	Amborellaceae, Nympheaceae, Illiciaceae, Trimeniaceae, Austrobaileyaceae; Degeneriaceae, Monimiaceae, Sauraraceae; Saxigragaceae	Syrphidae; Muscoidea, especially Anthomyiidae, Faniidae, Muscidae, and Lauxaniidae	Deceit (e.g., pseudo nectaries); scent (musky or sweet); limited nectar	Early Cretaceous: (Hauterivian to Aptian); Late Cretaceous	Thien, 1982; White and Thien 1985; Lloyd and Wells 1992; Feil 1992; Dilcher 2000; Endress and Igersheim 2000; Labandeira 2000
3. Florets of anemophilous male catkins. Pendant clusters of small staminate florets; brown to greenish; occurring in amentiferous trees or herbs such as graminoids.	Hamamelidaceae, Platanaceae; Cyperaceae, Poaceae Plataginaceae	Syrphidae; ?Bombyliidae	Pollen (nectar absent)	Woody trees: mid-Cretaceous (Albian to Cenomanian); Herbaceous forms: Late Paleocene to Middle Miocene	Clifford 1962; Sonderstrom and Calderón 1971; Stelleman 1981; Leereveld 1982; Ssymank and Gilbert 1993
4. Midge flowers. Small, white, purplish or green; hermaphroditic; mostly zygomorphic flowers with hidden rewards; often bearing staminodes; occasional petal modifications for pollination.	Apiaceae, Caprifoliaceae, Piperaceae, Rosaceae, Piperaceae, and especially Ranunculaceae, Sterculiaceae (Byttnerereae)	Culicidae, Mycetophilidae, Sciaridae, Cecidomyiidae, Ceratopogonidae (Forcipo, myiinae), Simuliidae, Phoridae; Milichiidae, a few other cyclorrhaphan taxa of small-bodied flies	Nectar, pollen, scent, mating site	Late Cretaceous (Coniacian) to Early Eocene[c]	Downes 1955, 1958; Young 1984; Young et al. 1986; Ollerton 1996b; Hurtado-Mejía and Vélez-Angel 1999

5. Nontrap carrion flowers.[c] Purplish to brownish, fetid, solitary, broad, deep-tubular flowers; moist surface appearance, translucent light "windows" below perianthtube; no trap mechanism.	Araceae (*Arisarum*), Aristolochaceae (*Asarum*), Asclepidaceae (*Stapelia*)		Deceit; scent; surface texture; endogenous heat	Late Cretaceous (Senonian) to Middle Eocene	Vogel 1978a; Koach and Galil 1986; Meeuse and Raskin 1988; Kite et al. 1998
6. Medium-depth connate flowers. Actinomorphic, sympetalous flowers bearing 1–2-cm deep throats; bearing concealed nectar rewards next to receptacle; mostly herbaceous.	Asteraceae, Balsaminaceae, Boraginaceae, Brassicaceae, Campanulaceae, Caryophyllaceae, Caryophyllaceae, Ericaceae, Gentianaceae, Lythraceae, Rosaceae, Valerianaceae	Dolichopodidae, Syrphidae, Phoridae, Bombyliidae; Conopidae, other cyclorrhaphous taxa with medium-length proboscides	Nectar; pollen (infrequently)	Latest Cretaceous (Maastrichtian) to Middle Eocene	Christy 1922; Kugler 1938; Knight 1968; Holloway 1976; Disney 1980; Schuhmacher and Hoffman 1982
7. Brush flowers. Red to whitish, small narrow diameter tubular flowers; prominently exerted stigmas and stamens, arranged into dense, pendant, prominent inflorescences.	Apiaceae, Asteraceae, Geraniaceae, Lamiaceae, Lecythidaceae, Mimosaceae, Myrtaceae, Papilionaceae, Proteaceae	Tipuidae, Bibionidae, Bombyliidae; also labellage flies with short proboscides	Nectar	Latest Cretaceous (Maastrichtian) to Late Miocene	Yeo 1972; Lack and Kevin 1984; Proctor et al. 1996
8. Trap carrion flowers.[c] As in 5 above, but flowers protogynous and modification of inner funelliform perianth into a lure and trap with decurved trichomes and trap-door for confining small insects.	Araceae (*Amorphophallus, Arisaema, Arum*), Aristolochiaceae (*Aristolochia*), Asclepidaceae (*Ceropega*), Hydnoraceae, Orchidaceae, (*Bulbophyllum, Pterostylis*), Solanaceae (*Trechonaetes*), Taccaceae	Schizophoran flies: Calliphoridae, Scathophagidae, Sacrophagidae, Muscidae. To a lesser extent: Certopogonidae, Psycodidae, Simuliidae, Mycetophilidae, Phoridae, Psycodidae, Simuliidae, Mycetophilidae, Phoridae, Drosophilidae, Sphaeroceridae, Sepsidae, Milichiidae, Chloropidae	Deceit (including pseudocopulation); brood sites; scent; surface texture; endogenous heat	Middle Eocene to Late Oligocene	Ridley 1890; Knoll 1926; Sargent 1934; Grensted 1947; Pijl 1953; Vogel 1954; Yeo 1962; Brantjes 1980; Beaman et al. 1988; Bänziger 1991; Hall and Brown 1993

continued

TABLE 9.2. *Continued*

Reproductive Structure Type and Key Features	Examples of Visited or Pollinated Plant Taxa	Examples of Major Dipteran Visitors or Pollinators	Rewards or Lures	Geochronologic Appearance[a]	Reference
9. Long-proboscid fly flowers. Reddish to bluish, deep-throated, tubular flowers bearing nectar at the corolla base and typically lacking a petal landing platform.	Amarillidaceae, Geraniaceae, Iridaceae, Lamiaceae, Orchidaceae, Polemoniaceae, Scrophulariaceae	Acroceridae, Apioceridae, Bombyliidae, Empididae, Nemestrinidae, Tabanidae (Pangioniinae), Vermelionidae	Nectar, pollen	Probably Paleogene	Nitzulescu 1927; Vogel 1954; Cazier 1963; Grant and Grant 1965; Coscarón and Phillip 1979; Nagatomi and Soroida 1985; Nilsson 1988; Johnson and Steiner 2000; Stuckenberg 2000; Goldblatt and Manning 2000
10. Orchids. Distinctive monocot flowers; six perianth segments, the lowermost (labellum) an alighting platform, bearing a nectariferous spur; central column, pollinia.	Orchidaceae (especially Cypripedioideae)	Mycetophilidae, Sciaridae, Syrphidae	Deceit (including pseudocopulation. and pseudonecaries); scent; nectar	Miocene	Darwin 1862; Vogel 1978b; Dodson 1966; Mesler et al. 1980; Schick 1982; Osche 1983; Beardsell and Bernhardt 1983; Ackerman 1986; Chase and Peacor 1987; Nilsson 1992; Johnson and Steiner 1997

[a] In approximate order of plant temporal appearance, based on fossil occurrences from Collinson et al. (1993), recent updates, and sister-group comparisons.
[b] The acronym ANITA refers to the first five plant families in the column at right.
[c] Also referred to, in part, as "fungus gnat flowers" (see Vogel 1973, 1978a).

Bibionidae (March flies), Anthomyiidae, and Psilidae (rust flies), have been considered a fifth, distinctive feeding type (Skevington and Dang 2002), but are not considered further here and lack a demonstrable fossil record. Dipteran endophytic phytophages are overwhelmingly host specific, and their fluid-feeding larvae are phylogenetically clustered in particular subclades within the order.

Of these four functional feeding groups, those with the highest family-level representation are leaf miners and seed predators, each representing 16 families (Table 9.1). By contrast, gallers represent about half the number of mining and seed predator families, with nine families, and borers are even less diverse, with six families. These higher-level patterns, however, may obscure underlying speciosity, particularly as galling Cecidomyiidae (Gagné 1994) and leaf-mining Agromyzidae (leafminer flies: Spencer 1990) contain the greatest number of known species of endophytic Diptera. Of the 31 endophytic family-level taxa that are present in these four functional feeding groups, there is a substantial amount of taxonomic commonality. No family has representatives for all four functional feeding groups. However, the Cecidomyiidae, Agromyzidae, Lauxaniidae, Chloropidae, Tephritidae (fruit flies), and Anthomyiidae are found in three; and the Sciaridae, Syrphidae, Phoridae, Lonchaeidae, and Psilidae are represented in two. Although there are sporadic occurrences of phytophagy across the Diptera, the two major clusters of feeding on live plants clearly are the Cecidomyiidae, and the higher Cyclorrhapha; in particular, the Schizophora. These two lineages underwent significant evolutionary radiations characterized by extensive larval consumption of internal plant tissues, mostly as specialists on particular plant species. Notably these two taxonomic isolates of dipteran larval phytophagy occur within a broad swath of animal predation, parasitoidism, parasitism, saprophagy, and microvory (Fig. 9.2). Additionally, within these superfamilies, the phytophagous habit typically does not extend to closely related clades, including sister-families, which have more eclectic feeding habits that span the wider gamut of feeding on animals, fungi, or decomposing plant tissues. An example is the origin of endophytic phytophagy in the Cecidomyiidae from fungivorous ancestors (Bissett and Borkent 1988).

It appears that the frequent presence of the same family-level clades at lower taxonomic units in two to four of the endophytic functional feeding groups may be a consequence of tissue-transferable adaptations to phytophagy. Thus once phytophagy has been established within plant tissues, feeding mechanisms and dietary requirements can be carried to other organs or tissues within the same host plant, and eventually to other host plants (Buhr 1955; Rotheray 1988; Dempewolf 2001). Such endophytic colonization of closely—or even more distantly—related taxa might be a major cause for the elevation of species-level diversity in cyclorrhaphous lineages.

Mining. Leaf mining is the consumption of live, internal foliar or herbaceous stem tissues by insects and mites (Hering 1951; Connor and Taverner 1997; Walter and Proctor 1999). Some endophytic structures, such as those produced by the cecidomyiid *Monoarthropus* on *Buxus* (Buxaceae), resemble blotch mines, even though they only induce the production of callus tissue and lack distinctive cecidogenic tissues typical of galls. Such structures contain a stationary maggot that typically feeds on vascular sap and historically have been considered blister galls (Bronner 1992). Nevertheless, leaf miners have the most impressive phylogenetic range of any dipteran phytophagous functional feeding group. They occur sporadically among the Lower dipteran clades of Culicomorpha (Chironomidae, Ceratopogonidae);

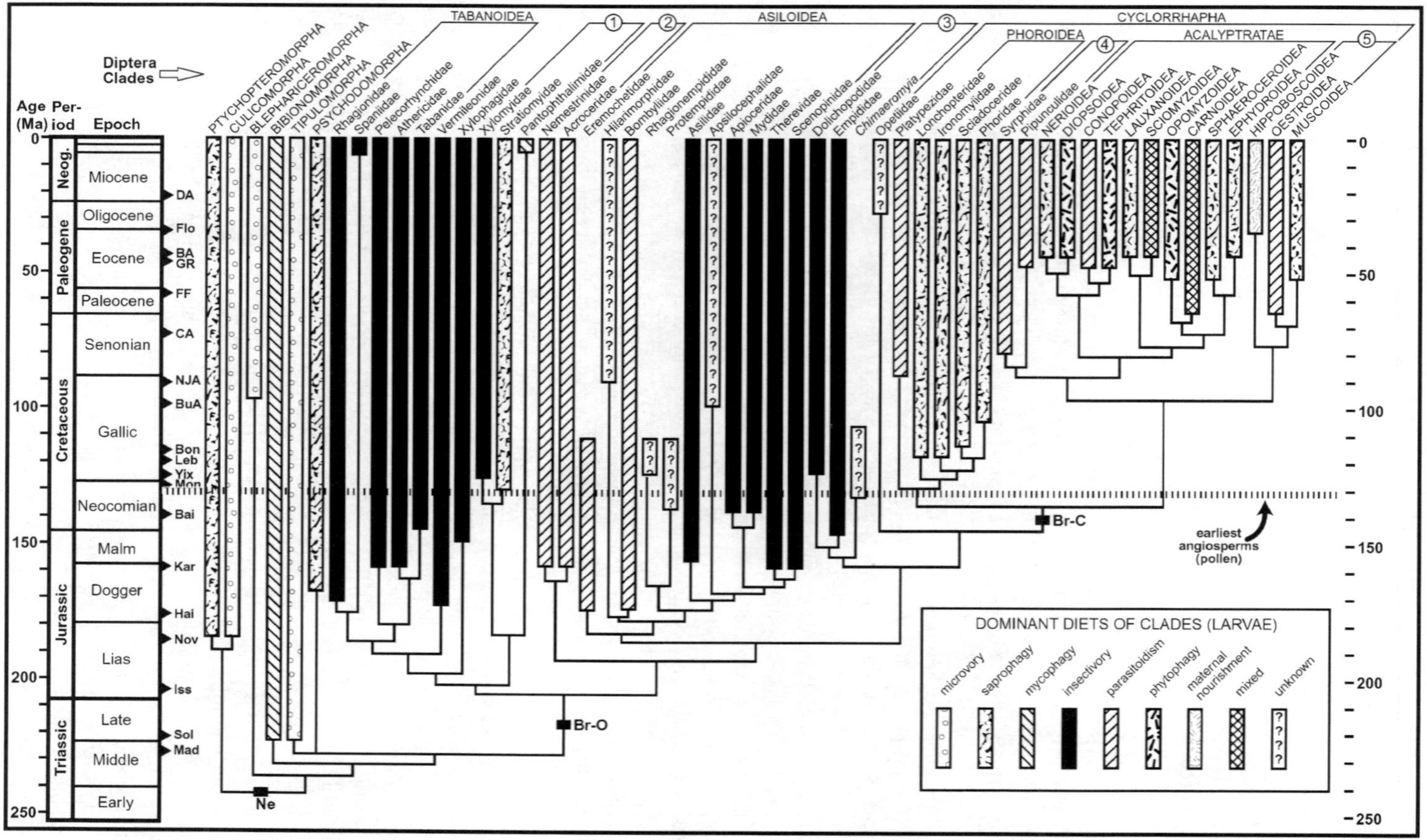
Age (Ma)
Period
Epoch
Diptera Clades
Neog.
Paleogene
Cretaceous
Jurassic
Triassic
Miocene
Oligocene
Eocene
Paleocene
Senonian
Gallic
Neocomian
Malm
Dogger
Lias
Late
Middle
Early
DA
Flo
BA
GR
FF
CA
NJA
BuA
Bon
Leb
Yix
Mon
Bai
Kar
Hai
Nov
Iss
Sol
Mad
0
50
100
150
200
250
TABANOIDEA
ASILOIDEA
CYCLORRHAPHA
PHOROIDEA
ACALYPTRATAE
1
2
3
4
5
PTYCHOPTEROMORPHA
CULICOMORPHA
BLEPHARICEROMORPHA
BIBIONOMORPHA
TIPULOMORPHA
PSYCHODOMORPHA
Rhagionidae
Spaniidae
Pelecorhynchidae
Athericidae
Tabanidae
Vermileonidae
Xylophagidae
Xylomyidae
Stratiomyidae
Pantophthalmidae
Nemestrinidae
Acroceridae
Eremochetidae
Hilarimorphidae
Bombyliidae
Rhagionempididae
Protempididae
Asilidae
Apsilocephalidae
Apioceridae
Mydidae
Therevidae
Scenopinidae
Dolichopodidae
Empididae
Chimaeromyia
Opetiidae
Platypezidae
Lonchopteridae
Ironomyiidae
Sciadoceridae
Phoridae
Syrphidae
Pipunculidae
NERIOIDEA
DIOPSOIDEA
CONOPOIDEA
TEPHRITOIDEA
LAUXANOIDEA
SCIOMYZOIDEA
OPOMYZOIDEA
CARNOIDEA
SPHAEROCEROIDEA
EPHYDROIDEA
HIPPOBOSCOIDEA
OESTROIDEA
MUSCOIDEA
Ne
Br-O
Br-C
earliest angiosperms (pollen)
DOMINANT DIETS OF CLADES (LARVAE)
microvory
saprophagy
mycophagy
insectivory
parasitoidism
phytophagy
maternal nourishment
mixed
unknown

Tipulomorpha (Tipulidae); Bibionomorpha (Sciaridae); the Dolichopodidae for orthorrhaphous clades; the Phoridae and Syrphidae among "lower" cyclorrhaphous clades; and the Diopsoidea (Psilidae), Tephritoidea (Tephritidae), Lauxanioidea (Lauxaniidae), Opomyzoidea (Agromyzidae), Carnoidea (Chloropidae), Ephydroidea (Drosophilidae, Ephydridae), and Muscoidea (Scathophagidae, Anthomyiidae) among the "higher" cyclorrhaphous clades (Table 9.1). Notably, with the limited exception of the Dolichopodidae, there are no leaf-mining taxa among larval orthorrhaphous clades; instead, these taxa engage extensively in predation and parasitoidism. The spectrum of dipteran leaf mining includes at least 25 independent originations and has favored herbaceous angiosperm taxa, unlike other holometabolous leaf-mining clades that have much fewer originations and have targeted woody species. The host-plant taxa for miners include aquatic herbaceous hosts, especially among nematocerous miners, and grasses, sedges, herbaceous dicots, and (less commonly) woody dicots for cyclorrhaphous leaf miners, such as the Agromyzidae (Spencer 1990; Scheffer and Wiegmann 2000). Given the distinctive features of dipteran mines, such as the occurrence of frass trails along one or the other side of the active mine and the presence of an often distinctive slit at the mine terminal chamber (Hering 1951; Scheirs et al. 1997), as well as their occurrence in aquatic plants, there should be a high likelihood of encountering dipteran mines in the fossil record.

Galling. A second type of endophytic phytophagy involves insect metabolic control of a plant host's capacity for producing normal tissues. This developmental usurpation results in the formation of newly created anomalous and localized tissues that are organized into galls

FIGURE 9.2. Dominant dietary trajectories for larvae of major dipteran clades. Diet-type assignments were determined by the greatest relative number of species possessing a particular diet within the clade under consideration. For dietary designations, microvory is the consumption of particulate matter suspended in aqueous media; saprophagy refers to consumption variously of all nonliving fluidized substances; phytophagy includes leaf mining, galling, wood boring, and seed predation, in which there is consumption of live plant tissue; maternal nourishment refers to larviparous taxa that nourish larvae internally after they have hatched within the female; and a mixed diet is a combination of two or more of the nine categories without the predominance of any one diet at the level of taxonomic analysis. Larvae are unknown for the Apsilocephalidae, Hilarimorphidae, and Opetiidae. Durations are plotted using the range-through method; first and last appearances are depicted at midpoints of relevant time units; geochronology modified after Gradstein and Ogg (1996). Major sources for phylogeny, dietary habits, and geochronology are Clausen (1940), Séguy (1950), Downes (1958), Oldroyd (1964), Cole (1969), McAlpine et al. (1981, 1987), Slansky and Rodriguez (1987), Evenhuis (1989, 1994), Proctor et al. (1996), Grimaldi and Cumming (1999), Stuckenberg (1999, 2001), Yeates and Wiegmann (1999), Krzeminski and Evenhuis (2000), Wiegmann et al. (2000), Amorim and Silva (2002). Abbreviations: 1, Stratiomyioidea; 2, Nemestrinoidea; 3, Empidoidea; 4, Syrphioidea; 5, Pupipara; DA, Dominican amber (Dominican Republic); Flo, Florissant (Colorado, United States); BA, Baltic amber (Baltic Sea region); CA, California (United States); GR, Green River (Colorado, United States); FF, Fur Formation (Denmark); NJA, New Jersey amber (New Jersey, United States); BuA, Burmese amber (northern Myanmar); Bon, Bon Tsagan (Mongolia); Leb, Lebanese amber; Yix, Yixian Formation (Liaoning Province, China); Mon, Montsec (Spain); Bai, Baissa (Transbaikalia, Russia); Kar, Karatau (southern Kazakhstan); Hai, Haiffangou Formation (Liaoning and Hebei Provinces, China); Nov, Novospasskoe (Transbaikalia, Russia); Iss, Issyk-Kul (northeastern Kyrgyzstan); Sol, Solite Quarry (Virginia and North Carolina, United States); and Mad, Madygen (western Kyrgyzstan). Clade designations at base of cladograms are Ne, nematocerous Lower Diptera, probably paraphyletic; Br-O, orthorrhaphous Lower brachyceran Diptera, paraphyletic; and Br-C, cyclorrhaphous Diptera, probably monophyletic. Clades lacking fossil records are depicted to be minimally Pleistocene in age for purposes of inclusion.

for the benefit of a larval dipteran. One or more larvae are typically is ensconced within the gall, a teratological and typically rigid structure. For the Diptera, the most taxonomically diverse and ecologically dominant gallers are the Cecidomyiidae, which have a worldwide occurrence on a wide variety of dicot and monocot angiosperms, conifers, and pteridophytes (Gagné 1994). This group experienced an evolutionary radiation during the Late Cretaceous and Paleogene based on body fossils from several amber deposits (Pike 1995; Arillo and Nel 2000; Rasnitsyn and Ross 2000) and probable fossil leaf galls (Srivastava et al. 2000; Labandeira pers. obs.). Gallers from basal cyclorrhaphan clades include sporadically the Platypezidae (flatfooted flies), albeit on polypores, and Lonchaeidae (Kolomoets et al. 1989; Dreger-Jauffret and Shorthouse 1992). Acalyptrate and muscoid gallers from the same families as leaf miners, such as Tephritidae, Chloropidae, Agromyzidae, Fergusoniidae, Lauxaniidae, and Anthomyiidae, also contribute to the worldwide galling fauna, particularly on herbaceous angiosperms (Meyer 1987; Kolomoets et al. 1989; Table 9.1). In terms of family-level diversity, galling dipterans are less diverse than those of leaf miners or seed predators, although they share many groups in common with the three other functional feeding groups. A taxonomically eclectic range of plant hosts are galled by dipterans, albeit angiosperms are highly targeted, particularly advanced dicots, such as the Vitaceae (grape family), Rosaceae, Asteraceae, and monocots, such as the Poaceae.

Boring. Boring is defined as the tunneling into live trees, whether it consists of the consumption of such live tissues as cambium or of such nominally dead tissues as secondary xylem (Teskey 1976; Solomon 1995). True wood borers probably involve protistan and fungal mutualisms that colonize tunnel plant detritus, and subsequently are cropped by the occupant. Larval wood borers possess gut endosymbionts that allow for digestion of cellulosic and other refractory constituents (Inoue et al. 2000). According to this more restrictive definition, borers do not include saprophages on decomposing wood, nor do they include predators that inhabit the tunnels of borers. Borers are the least taxonomically diverse of the four ecological groupings of endophytic herbivores, representing only six families, but these originate from all major dipteran clades, thus indicating more of an opportunistic colonization rather than evolutionary radiations typical of the three other functional feeding groups (but see below). For the Diptera, borers occur in Lower dipteran clades (Cecidomyiidae), Lower brachyceran clades (Asilidae: robber flies; Panthophthalmidae), the latter representing the largest known dipteran larva (Oldroyd 1964), and sporadically across the cyclorrhaphous brachyceran clades (Syrphidae, Psilidae, and especially Agromyzidae) (Teskey 1976; Süss 1979; Table 9.1). It is the Agromyzidae, however, that represents the only significant evolutionary radiation of dipteran borers in woody plants, particularly the genus *Phytobia,* which has extensively colonized the cambial tissue of conifer and dicot angiosperm trees. With the exception of cambium borers, dipterans have never significantly radiated into the wood-boring niche.

Seed Predation. Dipteran seed predation is the consumption by larvae of endosperm or embryonic tissues, causing the death of a diploid individual. Seed predation is distributed throughout insects and has originated numerous times in such groups as the Hemiptera, Coleoptera, and Hymenoptera (Janzen 1971). For family-level dipteran clades, seed predation is as diverse as leaf mining and is represented by 16 families (Keen 1958; Grebenshchikova and Naumov 1985; Turgeon et al. 1994). Seed predators share six out of the 16 families in common with leaf miners; namely, the Sciaridae, Phoridae, Tephritidae, Chloropidae, Lauxaniidae, and Anthomyiidae (Table 9.1); those families that are shared with gallers and borers are the Cecidomyiidae, Tephritidae, Chloropidae, Agromyzidae, Lauxaniidae, Lonchaeidae, and Anthomyiidae. Seed predation, like other types of endophytic phytophagy, is confined to Lower dipteran clades (Cecidomyiidae and Sciaridae) and especially cyclorrhaphous clades. Within the latter group, there is representation among the Phoridae and various acalyptrate groups, especially the Tephritoidea (Tephritidae, Lonchaeidae, Piophilidae [skipper flies], Pallopteridae [flutter flies], and Otitidae), but also Lauxanioidea (Chamaemyiidae [aphid flies] and Lauxaniidae), Sphaeroceroidea (Sphaeroceridae), Carnoidea (Chloropidae), and Ephydroidea (Chyromyiidae); additional occurrences are found among the calyptrate dipterans (Muscidae, Anthomyiidae, and Tachinidae). Unlike leaf miners, wood borers, and—to a lesser extent—gallers, seed predators have extensively invaded conifers as a food source (Keen 1958). Remarkably, the only dipteran seed-predating families without representatives on gymnosperm seeds are the Tephritidae and Otitidae, which have an abundance of asteraceous and poaceous hosts. It is unclear if any of these conifer-dipteran associations are ancient (cf. Farrell 1998), although the appropriate tests have not been made.

DIETARY GUILDS

In a major clade of insects as diverse as the Diptera, general larval and adult dietary trajectories can be explored at the family or superfamily level without sacrificing undue generality at higher ranks or being swamped by numerous subpatterns of dietary exceptions at lower ranks (Figs. 9.2, 9.3). The distribution of phytophagy and other feeding types primarily assumes a uniformitarian mapping of dietary data from known extant species to their immediate fossil lineages. This approach is appropriate for slowly evolving and more recent lineages, but requires confirmation for more rapidly evolving and older lineages. Supplemental verification is provided primarily by fossil damage data, but also by functional morphological analysis. The primary purpose of using midlevel taxonomic units is to avoid inundation from data representing complex dietary shifts from geographically restricted and often highly specialized clades representing, for example, genera and species. This approach, in which 53 extant and fossil clades representing all of the Diptera were examined, indicates that there are highly patterned dietary trajectories for both larvae and adults from the most plesiomorphous to the most apomorphous taxa within the order.

Larvae. For larvae, aquatic Lower dipteran clades are engaged overwhelmingly in the consumption of particulate suspended matter and substrate-attached filamentous algae and other plants, surface films, and unattached detritus and live tissues. Live and dead plant material alike consumed, often without distinction by a given species. Within this milieu, two major types of diets are present. The first is microvory, in which dead organic and live protistan and algal matter is consumed as particles from the water column by filterers; the second comprises herbivory, saprophagy, and mycophagy, in which bacterial or algal films, leaf detritus, or fungi occurring either subaqueously or in wet subaerial environments are extracted by scrapers, gatherers, and shredders (Krivosheina 1969; Nessimian et al. 1999; Alverson and Courtney 2002; Fig. 9.2). In this obligately aquatic existence, there is a tendency toward terrestriality that is evident from the more derived Lower dipteran clades, followed by basal brachyceran clades whose larvae inhabit terrestrial environments, albeit often in wet or moist habitats. A preeminent theme among basal brachyceran taxa is the trophic dominance

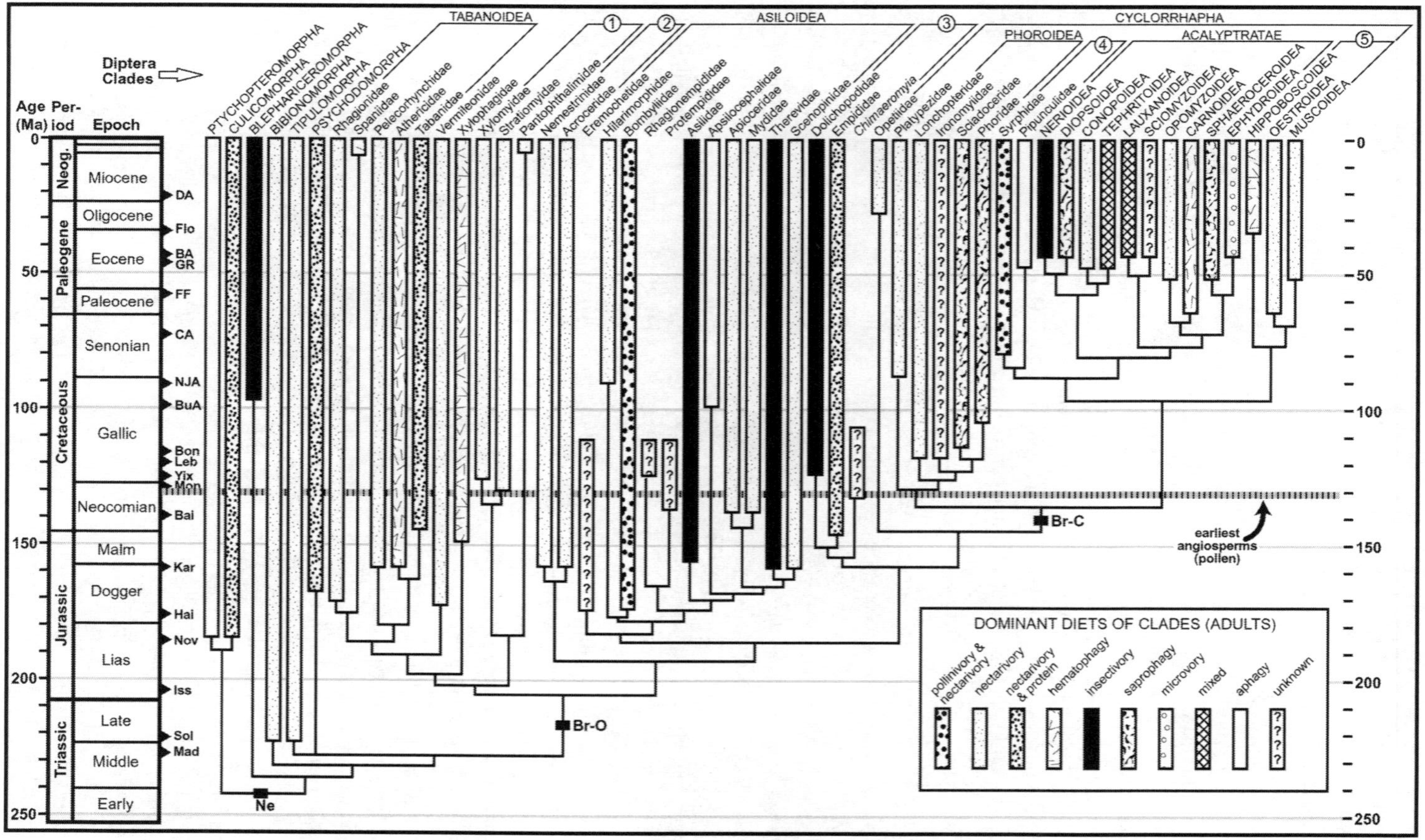

Age (Ma)
Period
Epoch
Diptera Clades
Neog.
Paleogene
Cretaceous
Jurassic
Triassic
Miocene
Oligocene
Eocene
Paleocene
Senonian
Gallic
Neocomian
Malm
Dogger
Lias
Late
Middle
Early
DA
Flo
BA
GR
FF
CA
NJA
BuA
Bon
Leb
Yix
Mon
Bai
Kar
Hai
Nov
Iss
Sol
Mad
0
50
100
150
200
250
PTYCHOPTEROMORPHA
CULICOMORPHA
BLEPHARICEROMORPHA
BIBIONOMORPHA
TIPULOMORPHA
PSYCHODOMORPHA
Rhagionidae
Spaniidae
Pelecorhynchidae
Athericidae
Tabanidae
Vermileonidae
Xylophagidae
Xylomyidae
Stratiomyidae
Pantophthalmidae
Nemestrinidae
Acroceridae
Eremochaetidae
Hilarimorphidae
Bombyliidae
Rhagionempididae
Protempididae
Asilidae
Apsilocephalidae
Apioceridae
Mydidae
Therevidae
Scenopinidae
Dolichopodidae
Empididae
Chimaeromyia
Opetiidae
Platypezidae
Lonchopteridae
Ironomyiidae
Sciadoceridae
Phoridae
Syrphidae
Pipunculidae
NERIOIDEA
DIOPSOIDEA
CONOPOIDEA
TEPHRITOIDEA
LAUXANOIDEA
SCIOMYZOIDEA
OPOMYZOIDEA
CARNOIDEA
SPHAEROCEROIDEA
EPHYDROIDEA
HIPPOBOSCOIDEA
OESTROIDEA
MUSCOIDEA
TABANOIDEA
1
2
ASILOIDEA
3
CYCLORRHAPHA
PHOROIDEA
4
ACALYPTRATAE
5
Ne
Br-O
Br-C
earliest angiosperms (pollen)
DOMINANT DIETS OF CLADES (ADULTS)
pollinivory & nectarivory
nectarivory
nectarivory & protein
hematophagy
insectivory
saprophagy
microvory
mixed
aphagy
unknown

of carnivorous forms, principally predatory and parasitoid forms (Fig. 9.2). Carnivorous taxa almost exclusively characterize the Tabanoidea, Nemestrinoidea, Asiloidea, and Empidoidea. The only major exceptions are the highly saprophagous Stratiomyiidae and the mycophagous (wood-boring) Pantophthalmidae (Thorpe 1934). By contrast, larval cyclorrhaphous larvae have predominantly abandoned carnivory in favor of exploiting decomposing animal, plant, and fungal tissues, as well as true phytophagy (Oldroyd 1964). Either ancestrally retained or secondarily acquired parasitoidism is present in the Platypezidae, Syrphoidea, Conopidae, and Oestroidea, the latter three of which are phylogenetically embedded among surrounding clades of saprophages and phytophages (Fig. 9.2). In several lineages of the Pupipara, active larval feeding has been abandoned and replaced by maternal nourishment for a larva that is retained within the female (Oldroyd 1964).

Adults. For adults, the distribution of diets across clades is highly patterned, but in different modes than their associated larvae. Lower dipteran adults are predominantly penetrative or surface fluid feeders, frequently expressed as nectarivory, which sometimes is combined with the consumption of pollen or vertebrate blood (Fig. 9.3). Although hematophagous taxa are economically important, they do not constitute the dominant dietary component by species of any nematocerous superfamily, although the Culicomorpha is characterized by a high percentage of blood-feeding taxa. Most Lower brachyceran taxa also have nectarivorous or related fluid diets (e.g., Watanabe and Kamimura 1975), although a few basal clades have shifted to predominantly hematophagy; particularly the Spaniidae, Athericidae, Xylophagidae, and certain lineages within the Tabanidae. A distinctive pattern in the more advanced orthorrhaphous clades is the emergence of adult insectivory, typified by the Asilidae, Therevidae, and Dolichopodidae. Minor trends include incorporation of other foods along with the consumption of nectar, such as pollen in the Bombyliidae or insect protein in the Empididae. For cyclorrhaphous adults, there is continuation of fluid feeding such as nectarivory in the basalmost clades, but also a distinct shift favoring saprophagy sporadically between some Phoroidea and major schizophoran clades. The Syrphoidea have acquired pollinivorous (Syrphidae), as well as aphagous (Pipunculidae, bigheaded flies), habits. Notably, the schizophoran calyptrates display the widest variety of diets of any larger dipteran clade, and include microvory, insectivory, nectarivory, saprophagy, hematophagy, and mixtures of multiple food types that are singularly not characterizable at the superfamily level of analysis. The Pupipara are dominated by hematophagy and nectarivory, the latter possibly as a

FIGURE 9.3. Dominant dietary trajectories for adults of major dipteran clades. Focus is on Lower brachyceran taxa of family level in lower-case letters; taxa in upper-case letters are suprafamilial ranks of Lower Diptera and cyclorrhaphous Brachycera. Diet-type assignments were determined by the greatest relative number of species possessing a particular diet within the clade under consideration. For dietary designations, nectarivory includes sources of floral and extrafloral nectar, including those of nonangiospermous plants, and imbibition of surface fluids; hematophagy consists of feeding on blood and other sources of animal dermal exudates, such as lymph and sebaceous secretions; saprophagy refers to consumption variously of all nonliving fluidized substances; microvory is the consumption of particulate matter suspended in aqueous media; and a mixed diet is a combination of two or more of the ten categories without the predominance of any one diet, at the level of taxonomic analysis. Taxonomic focus, methods of plotting clade durations, references, abbreviations, major clade designations, and conventions for depicting clades with absent fossil records are the same as for Fig. 9.2.

retained diet type from more basal cyclorrhaphous clades. Aphagy occurs in some Pupipara and sporadically among Lower dipteran clades.

Use of Fossil Data

Fossil and related data for the study of the interactions between plants and dipterans come in a variety of forms. Four approaches are used for seeking these data. Two of these approaches—phylogenetic and associational—are used in this study to ferret out the types of interactions that particular dipteran clades have had with plants. Within the associational approach, several types of evidence are available in the fossil record. Associational, phylogenetic, functional morphological, and taxic approaches can provide mutually reinforcing ways of examining issues, such as the timing of the origin and presence of convergent plant-dipteran interactions in the fossil record.

APPROACHES

Each of these four approaches has varying strengths and weaknesses, but when combined, they can provide a robust, multidisciplinary approach to the study of the dipteran fossil record and their relationships with plants. Of these, associational data have been the most underappreciated but historically have been the most direct source of information regarding the interactions between plants and insects (Labandeira 2002a; Figs. 9.1, 9.4). By contrast, phylogenetic approaches, expressed as cladistic studies, have been crucial in establishing the basic framework of organismic relationships upon which associational and functional morphological data can be contrasted. The functional morphological approach can supply intrinsic insights into how plants and insects interact, particularly regarding trophic specializations and the likely presence of entomophilous pollination mechanisms. Both plant- and insect-based functional morphology overlap somewhat with associational data. In this context, functional morphological and associational data can be mapped onto cladograms to investigate the evolutionary history of phenomena such as mouthpart type, functional feeding group, or dietary convergences and parallelisms, or alternatively, the role of convergence in the history of dipteran feeding strategies.

The taxic approach forms a fourth, integral approach for examining the evolutionary dynamics of fossil insect and plant diversity. It provides valuable data on clade extinction and origination, taxon longevities, and turnover rates, which are unavailable through other approaches. The taxic approach describes the evolutionary dynamics of clades and works best at lower taxonomic levels; the genus is the preferred level of analysis. Recently taxic approaches have been used in insect tropical ecology, where it is termed "the higher-taxon approach" (Gaston and Williams 1993). In several studies (Williams and Gaston 1994; Lee 1997; Balmford et al. 2000), the higher taxon approach has been used to address questions of regional conservation status, species richness, latitudinal diversity patterns, hotspot endemicity, trophic diversity, and other synecological issues where species- or genus-level identifications are unavailable for clades as diverse as vascular plants, insect herbivores, and butterflies. Interestingly, in extant insect faunas with known species identification and composition, family-level diversity has a very high statistical correlation with underlying species-level diversity (Balmford et al. 2000), thus allowing for extensive use of family-level data to infer species-level patterns. Given the particularly rich body-fossil record of Diptera compared

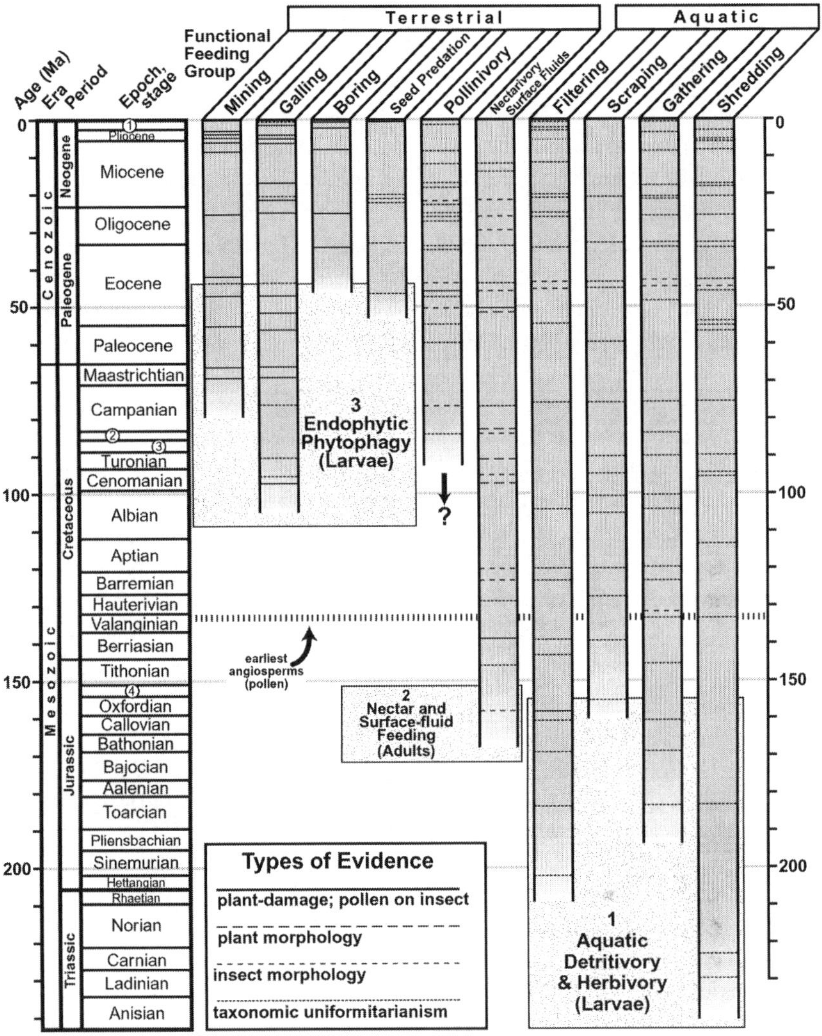

FIGURE 9.4. Geochronological distribution of major functional feeding groups for the Diptera. Each occurrence is represented by a horizontal line, which may be solid, dashed, or dotted, depending on the type of evidence (see Labandeira 2002a: Fig. 1). The three major phases of associations with plants are indicated. Geochronology abbreviations: 1, Pleistocene; 2, Santonian; 3, Coniacian; and 4, Kimmeridgian. Time scale modified after Gradstein and Ogg (1996).

with most other insect orders, the taxic approach can reveal important evolutionary dynamics of particular lineages (Jarzembowski and Ross 1996).

TYPES OF ASSOCIATIONAL EVIDENCE

The fossil record of plant-insect associations actually encompasses three, rather distinct, fossil records (Labandeira 2002a). The first is the body-fossil record of plants; the second, the body-fossil record of insects; and third is the trace-fossil, or associational, record of plants and their interacting insects (Fig. 9.1). In relevant deposits of exceptional preservation, any combination of these records may be present. In the broadest sense of the associational record, there are five distinctive types of evidence that are arrayed along a continuum that emphasizes plants at one pole and insects at the other (Labandeira 2002a). The indirect evidence from the plant and insect poles of this continuum is based on aspects of body fossils with ecological implications—for example, nectaries and mouthparts—that are not technically trace fossils but are part of the body-fossil record. The first type of evidence is from vascular plants, and centers on their reproductive biology, such as size and surface structure of pollen, presence and location of nectaries, and specialized morphology of floral elements, all of which can reveal significant information regarding insect pollination. Proceeding from the plant end of this continuum, the second type of evidence is the occurrence of insect-mediated, live plant damage, which reveals stereotypy in the pattern of insect consumption of tissues and records the histological response of plants to particular types of induced trauma. The center of the spectrum consists of the dispersed coprolite record (fossilized fecal pellets). When encountered as exceptionally preserved permineralization, dispersed coprolites provide valuable species-level and histological data on the identity and processing of consumed food, insect diet specificity, and food-web flow patterns through the ecosystem from producers to carnivores. For the previous three types of evidence, preservation of insects is lacking: only their reciprocal impact on plant reproductive structure or their indirect effects of plant consumption—damage and coprolites—is present. The fourth type of evidence is the occurrence of gut contents, frequently revealing the specific nature of the consumed plant and the consuming insect (Krassilov and Rasnitsyn 1996; Afonin 2000). Fifth and finally, at the opposite pole of plant reproductive structure and merging with functional morphology, is the structure of insect mouthparts and ovipositors. Mouthpart data can provide constraints on membership in a functional feeding group or inclusion in a particular dietary guild. For dipterans, dispersed coprolites and gut contents have not been sought for in the fossil record, attributable to their overwhelmingly fluid-feeding habits and consequent lack of identifiable structures in preserved fecal pellets or food boluses. Nevertheless, the gut contents and coprolites of particle feeders, such as adult pollinivores or larval consumers of diatoms, are undoubtedly present as identifiable trace fossils. To these five types of trace-fossil and structural data is added the taxonomic affiliation of the fossil insect, particularly if it is resolvable to an extant taxon whose dietary ecology is sufficiently established.

COMPLEMENTARITY

Phylogenetic, associational, functional morphologic, and taxic data collectively contribute in a mutually supplemental way for understanding issues in the plant/dipteran fossil record. One example is the role that angiosperm origins played in the contemporaneous and subsequent diversification of insects during the Early Cretaceous (Labandeira and Sepkoski 1993; Grimaldi 1999; Gorelick 2001), of which Diptera are a relevant anthophilous clade during this time interval (Dilcher 2000; Labandeira 2000; Thien et al. 2000; Larson et al. 2001). Several sources of largely independent data have been used as evidence for or against the coordinated early angiosperm/insect diversification hypothesis (Crepet 1985; Grimaldi 1999). Although there is previous mention in the literature that the rise of Early Cretaceous angiosperms is connected with insect diversification (Regal 1977; Burger 1981), the first evidence to suggest that there was not an associated diversification, reflected in higher-level taxa, was that of Labandeira and Sepkoski (1993), who examined family-level diversity of insects across this interval. Jarzembowski and Ross (1996) confirmed these results, using an independently compiled dataset. This conclusion was seemingly not supported by phylogenetic relationships from cladistic analyses by Grimaldi (1999), based primarily on mid-Early Cretaceous and younger amber insect faunas, which have their oldest occurrences approximately contemporaneous with the earliest angiosperms. Thus Grimaldi's study largely did not consider earlier Early Cretaceous to late mid-Jurassic preangiospermous compression faunas. An implied inconsistency was suggested between Labandeira and Sepkoski's (1993) family-level taxa providing evidence for an earlier bout of coadaptation between insects and nonangiospermous but advanced seed plants, and alternatively, independently gleaned cladistic data of Grimaldi (1999), indicating interclade and presumably intraclade diversification contemporaneous with early angiosperm expansion. Interestingly, these two approaches are mutually supported when examined more closely. The most parsimonious and mutually congruent interpretation is one in which family-level taxic data indicate an earlier phase of diversification with advanced seed plants, followed by a slackening of insect diversification at that level during the early rise of angiosperms, during which insect diversification was at a lower rank, suggested by the cladistic data (Grimaldi and Cumming 1999). This interpretation also is supported by a diverse associational data from a variety of plant and insect clades, including orthorrhaphous lineages, indicating that plant-insect interactions, such as nectarivory and pollinivory, were present during the preangiospermous mid-Mesozoic (Lloyd and Wells 1992; Kato and Inoue 1994; Labandeira 2000). For other functional feeding groups, particularly those involving endophytic phytophagy, there is additional associational evidence for other Holometabola, including leaf mining (Rozefelds 1988; Anderson and Anderson 1989), galling (Grauvogel-Stamm and Kelber 1996; Ash 1997), wood boring (Walker 1938; Zhou and Zhang 1989; Jarzembowski 1990), and seed predation (Nishida and Hayashi 1996; Falder et al. 1998). Additionally, the functional morphology of gymnospermous reproductive structures and their pollen, such as those of cheirolepidaceous Coniferales, Cycadales, Bennettitales, Gnetales, and some advanced seed-fern lineages, buttress the conclusion that nectarivory and pollinivory probably were present (Labandeira 1998b,c, 2000). This hypothesis is advanced by additional associational and functional morphological data demonstrating that clusters of cheirolepidaceous pollen occurred on the heads and bases of elongate proboscides of Lower brachyceran dipterans from the Baissa locality of Transbaikalian Russia (Fig. 9.5M), dated approximately at 130 Ma (Zherikhin et al. 1999).

The Paleobiological Record

The paleobiological record is the only direct source of information for past trophic habits of dipterans and their associated plants. However, for dipterans, documentation of their role in plant interactions is particularly challenging. Dipterans are not fluid-feeders as adults and thus do not produce definable coprolites with identifiable solid contents, nor are they piercer-and-suckers of plants that would reveal distinctive response tissue surrounding stylet insertion marks (Labandeira and Phillips 1996), both of which are crucial evidence for other insect groups. Consequently several coordinated approaches are necessary to infer the fossil associations between plants and dipterans, including associational evidence and the mapping

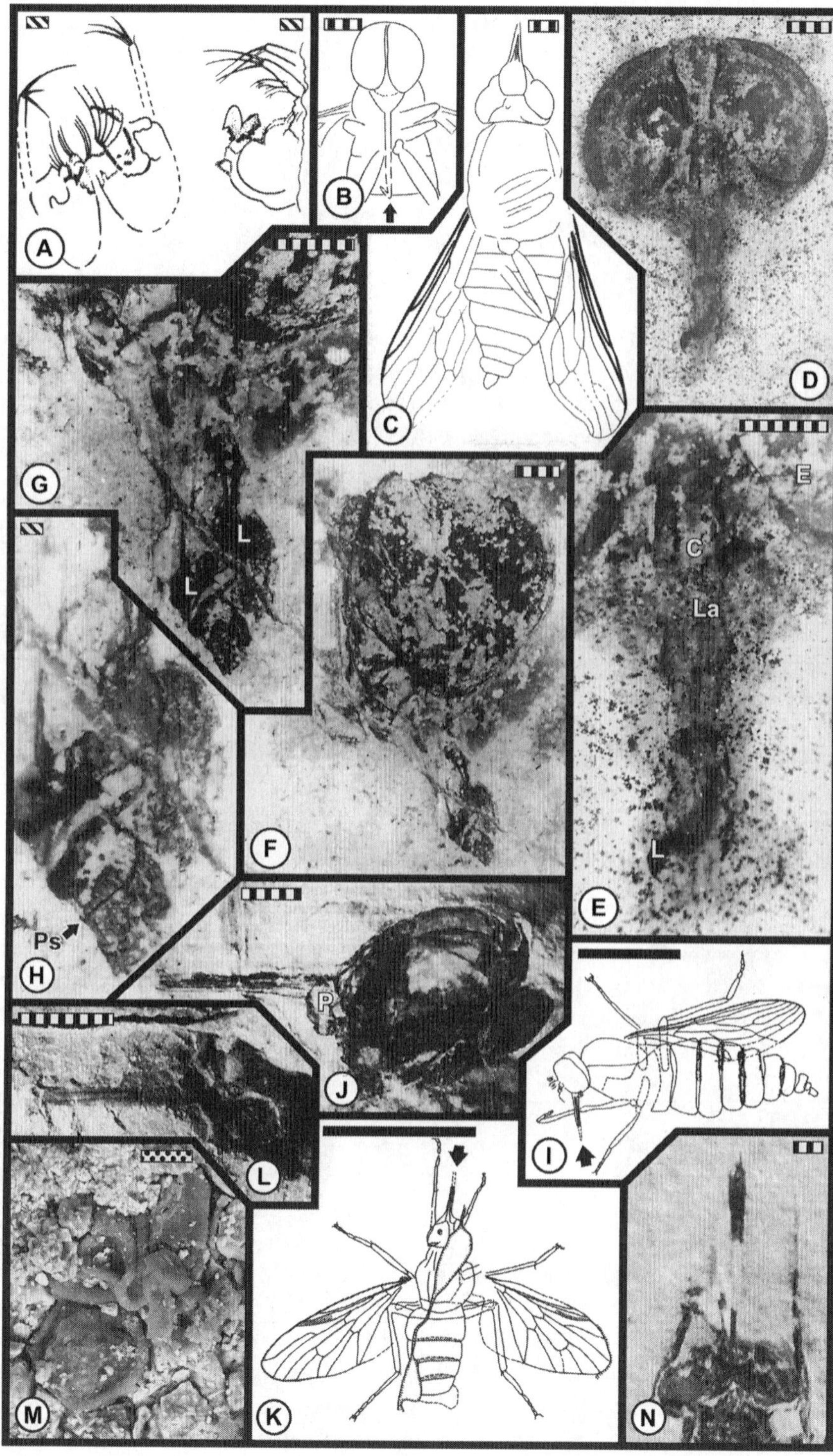

A
B
C
D
G
L
L
E
C
La
L
F
E
H
Ps
P
J
I
L
M
K
N

of larval and adult dietary data on dipteran phylogeny, calibrated by fossil occurrences. Nevertheless, major types of data can provide a well-supported account and highlight critical gaps in our knowledge of the history of dipteran feeding habits.

WHAT IS THE PRIMITIVE DIET OF DIPTERA?

The nature of the primitive adult dipteran diet hinges on the issue of what is the most plesiomorphic condition for mouthparts in the Diptera. The three previously discussed options—(1) the stylate, (2) the labellate, and (3) the vestibulate mouthpart conditions (Chaudonneret 1990; Gagné 1994; Labandeira 1997)—are not mutually exclusive within Lower dipteran lineages and in fact many taxa often represent structural emphases on suites of characters from more than one condition. The vestibulate condition is phylogenetically limited (Gagné and Boldt 1989), confined to bibionomorph taxa (Gagné 1994). The stylate condition is dietarily associated with penetrative fluid feeding, especially on vertebrate blood or on insect hemolymph, whereas labellate forms tend to imbibe sugar- or (less frequently) lipid- or protein-rich surface fluids. This is a distinction that typically is made among closely related species or is intraspecifically based on gender, given the nutritional requirements for oögenesis and active flight (Haslett 1989; Foster 1995; but see Lane and Anderson 1983).

FIGURE 9.5. Evidence for surface fluid feeding from late mid-Jurassic to Early Cretaceous Diptera, indicating aquatic detritivory or phytophagy (A), and associations with preangiospermous seed plants (B–M). (A) Head capsule and associated appendages of the chaoborid *Hypsocorethra toficola* Kalugina and Kovalev (1985) from Uda, in Transbaikalia, Russia. The age of this deposit is late Callovian or early Oxfordian (PIN 3053/916 at left, PIN 3053/866 at right). From Kalugina and Kovalev (1985). (B) Thorax, head, and proboscis of the nemestrinid *Protonemestrius rohdendorfi* Mostovski (1998), from the Karabastau Formation (Oxfordian) of Karatau, eastern Kazakhstan (PIN 2784/73). Arrow points to flexure in distal proboscis. From Mostovski (1998). (C) Body of *Protonemestrius martynovi* (Rohdendorf), from the same locality as (B). Note proboscis and indication of upturned distalmost wing veins (PIN 2239/2198); from Rohdendorf (1968). (D) Frontal view of the head and proboscis of the apiocerid *Protapiocera* sp., from the probably Hauterivian Zaza Formation of Baissa, Transbaikalia, Russia (PIN 3064/9930). Note large compound eyes and extensive proboscis. (E) Close-up of proboscis in (D), showing mesial aspect of compound eyes, E, clypeal region, C, at base of proboscis surrounded laterally by opaque black bars that are probable tormae, and associated labrum, La. The semicircular area L represents a labellar lobe at proboscis terminus. (F) Right lateral view of head of an unassigned apiocerid, also from Baissa as in (D). Note heart-shaped labellum at the end of the proboscis (PIN 4210/6380). (G) Enlargement of proboscis in (F), showing labellar lobes, L. (H) Higher magnification of labellum in (G), showing detail of labellar lobes and possible pseudotracheae, Ps. (I) The nemestrinid *Florinemestrius pulcherrimus* Ren (1998), from the Barremian Yixian Formation of Laiyang, China (NGMC LB97009). From Ren (1998). Note upturned distalmost wing veins; arrow indicates extension of proboscis. (J) Close-up of head and proboscis in (I), showing probable maxillary palp, P, and individual mouthpart elements constituting the proboscis. (K) The lower brachyceran *Palaeopangonius eupterus* Ren (1998), from the same deposit as (I) (NGMC LB97018), From Ren (1998). Arrow shows position of proboscis. (L) Higher magnification of head and proboscis in (K), showing individual mouthpart elements. (M) SEM micrograph displaying a cluster of five or six cheirolepidaceous pollen grains on the head of a proboscid specimen of *Paroikus* sp., from Baissa, same locality as (D) (PIN 2904/1560). *Paroikus* may be a rhagioempidid or a basal therevoid. (N) An enigmatic stratiomyomorph brachyceran, the cratomyiid *Cratomyia macrorrhyncha* Mazzarolo and Amorim from the Santana Formation of the Araripe Basin, Ceará, Brazil (DBRP-0051). From Mazzarolo and Amorim (2000). Abbreviations: DBRP, Department of Biology, University of São Paulo at Ribeirão Preto; NGMC, National Geological Museum, Beijing; PIN, Paleontological Institute of the Russian Academy of Sciences, Moscow. Scale bars: solid, 1 cm; striped, 0.1 cm; backslashed, 0.01 cm; dotted, 1 micrometer.

Additionally there are examples of intraspecific, gender-based variation of mouthpart structure and correlated diet among Lower brachycarans and Lower dipterans (Tetley 1918; Downes 1978). This dietary dichotomy among adult females undoubtedly was present among the earliest Lower dipterans during the Middle Triassic and probably extends to the origin of the Diptera in the Early Triassic (Downes and Dahlem 1987; Krzeminski 1992). Given the requirement of both protein and carbohydrate consumption in adult Lower dipterans and the associated occurrence of both piercing-and-sucking and sponging mouthpart types, it may be that both feeding strategies are primitive (Downes 1971; Kneipert 1980). In lieu of knowledge of well-resolved mouthpart structure of the earliest known dipterans, it is difficult to determine whether nectarivory or hematophagy was the primitive feeding strategy for the Diptera (Downes and Dahlem 1987; Szadziewski 1988). However, if a carbohydrate food source was essential, it probably consisted of sugary secretions from vascular plants either in the form of pollination drops, extrastrobilar nectaries, or honeydew produced as a byproduct of homopteran sap-feeding on vegetative tissues (Bonnier 1879; Porsch 1958; Fahn 1979; Dixon 1998). The various modes of pollinivory are phylogenetically derived, such as that of the Bombyliidae, Syrphidae, and Muscidae (Holloway 1976; Nicholson 1994; Wacht et al. 2000; Szucsich and Krenn 2000), and are a secondary source of protein through the consumption of nutritive protoplasts. One basal lineage of the Ceratopogonidae, the Forcipomyiinae, contains some species of *Atrichopogon* that are punch-and-suck pollinivores (Borkent 1995), whose oldest occurrence is Late Paleocene (60 Mya), also indicating that pollinivory is a derived condition for the family.

ASSOCIATIONS BETWEEN DIPTERA AND PLANTS: TYPE AND TIMING

Ever since the Late Carboniferous, insects have independently and iteratively evolved similar feeding mechanisms and life-habits to consume the basic types of live plant tissues and their products, ranging from unicellular algae to oak trees. These repeated convergences and parallelisms in feeding mechanisms, especially regarding replacements by holometabolous insects after the end-Permian mass extinction, have provided the basic trophic structure in freshwater and terrestrial ecosystems. It is in this context that early dipterans became major associates of plants (Fig. 9.4), as shown by their current wide breadth of larval and adult mouthpart design and in their fine-scale trophic partitioning of live plant tissue.

Aquatic Feeding. The earliest dipteran adults were exploiting terrestrial food resources of insects, primarily blood and sugary fluids, as their conspecific larvae invaded aquatic habitats and consumed food resources in lakes and streams (Miller and Labandeira 2002). This colonization is the first penetration of larval Diptera into a novel mode of herbivory (Fig. 9.4). Based on the taxonomic assignments of Mesozoic Lower dipteran fossils to extant clades with known modes of aquatic feeding, larval head capsules with mouthparts, and sedimentary trace fossils, the collective evidence indicates that the earliest occurrences of filtering, scraping, gathering, and shredding (Merritt and Cummins 1996; Wichard et al. 2002) ranged from the Middle Triassic to the mid-Jurassic (Figs. 9.4, 9.5A). Of these, the earliest dipteran functional feeding group was shredding of the Middle Triassic and the most recent is scraping, first recorded during the Middle Jurassic. Based on this Early to mid-Mesozoic distribution of aquatic herbivore and saprophage feeding types, there is a strong presumption that lotic and lentic freshwater environments were significantly partitioned by dipteran and other insect taxa (Sinichenkova and Zherikhin 1996; Sinichenkova 2001; Miller and Labandeira 2002). Herbivorous diets included algae, either in the form of films on water or benthic substrate surfaces, or as particulate items in the water column, such as diatoms or other microscopic plants (Ponomarenko 1996). Additionally, submerged pteridiphytic and gymnospermous plants were present (Fraser et al. 1996). One note of caution, however, is that the larvae of primitive clades, including some Culicomorpha, inhabit marginal, quiescent, spatially circumscribed aquatic habitats, such as watery films on rock faces, waterlogged tree holes, humus, and small pools in forests (Borkent, pers. comm.), which suggests an alternative or possibly earlier environment from which there was the eventual colonization of lakes and streams.

Nectarivory and Pollinivory. The second major phase of dipteran association with plants was the consumption of externally accessible plant products, such as nectarlike fluids and pollen during the earlier Mesozoic. The diversity of nectarivory and pollinivory based on mouthpart type and behavior circumscribe the feeding strategies of dipterans on flowering plants today. For some of these modern taxa, it is difficult to know when these types of feeding originated, based on fossil mouthpart morphology alone (Labandeira 1997). For example, although the distally expanded labellum is documented throughout the Cretaceous and Cenozoic, knowledge of pseudotracheal teeth structure, which is important to determine if a certain type of pollinivory is present, is often absent due to either poor preservation or to lack of investigator interest (but see Fig. 9.5H). By contrast, elongate, probably flexible proboscides with comparatively diminutive terminal labella are known from the late mid-Jurassic to Early Cretaceous (Mostovski 1998; Ren 1998; Fig. 9.5B–L) and were undoubtedly involved in imbibition of nectarlike fluids. Additionally there is the occurrence of mid-Cretaceous forcipomyiine Ceratopogonidae, whose modern descendants use the punch-and-suck technique to extract pollen contents (Downes 1955; Borkent 1995, 2000). Other primitive lineages of ceratopogonids were likely pollinators on midge flowers (Table 9.2), as evidenced by hirsute bodies, elongated mouthparts, and requirements of sugar meals for oviposition in extant descendants (Borkent 1995). Additionally, tipulids with elongate proboscides co-occur with tubular flowers during the mid-Cretaceous (Rayner and Waters 1991). Interestingly, other orthorrhaphous taxa, such as the stratiomyomorph *Cratomyia* from the mid-Cretaceous Santana locality of northeastern Brazil (Mazzarolo and Amorim 2000) have an elongate and apparently fasciculate proboscis, but with unknown feeding habits (Fig. 9.5N). The presence of nectarivory by mostly hematophagous Lower dipterans with fasciculate stylate mouthparts also extends minimally to the Cretaceous, and probably is Late Triassic to Early Jurassic in origin. However, it is most likely that small bibionomorph dipterans possessed significantly expanded labella with simple pseudotracheae that were among the earliest angiosperm pollinators. Most early flowers were simple, actinomorphic, helically borne on an axis, and consisted initially of prominently elevated stigmas, visually attractive stamens and closed carpels that were more likely bisexual than unisexual (Crepet et al. 1991; Dilcher 2000; Endress 2001). This condition probably represents the primitive, fly-pollinated angiosperm flower. Subsequently, as elements were added to the receptacle (Sun et al. 2002), flowers became more bowl-shaped, carpels were sunk into the receptacle, and the stigma was receptive earlier than adjacent pollen maturation. The earliest floral rewards were nectar, perhaps pollen (but see Crepet et al. 1991), and scent may have been critical as well (Pellmyr et al. 1991; Azuma et al. 1999). Although the first angiosperm pollinators were small flies and beetles of similar size, each likely was attracted to different types of flowers and major taxa (Thien et al. 1983; Dilcher 2000). Later, during the diversification of the angiosperms, floral structures became more variable and elaborate, nectaries were anatomically deployed in various positions within the flower, and more ingenious pollination mechanisms developed, producing a great diversity of dipteran nectarivore and pollinivore types that evolved to exploit these and other food rewards (Zaitzev 1992; Gandolfo et al. 1998; Grimaldi and Cumming 1999; Friis et al. 2000). By the Paleogene, a modern spectrum of nectarivorous and pollinivorous dipterans and associated angiosperms were present, as exemplified by long-proboscate Bombyliidae from the Middle Eocene Green River Formation of northwestern Colorado (Fig. 9.6; also see Crepet 1979), and Nemestrinidae and other dipterans from the Late Eocene Florissant Formation of central Colorado (Bequaert and Carpenter 1936; Hull 1973).

Endophytic Phytophagy. The third major phase of dipteran phytophagy was the consumption of internal tissues that commenced during the Late Cretaceous and continued into the Paleogene (Fig. 9.4). Evidence for this colonization of plants originates principally from Cretaceous occurrences of cecidomyiid body fossils and insect-mediated damage in the fossil record, particularly leaf mines (Fig. 9.7), galls (Fig. 9.8A–E), and borings (Fig. 9.8G–H) on coniferous and dicotyledonous hosts, especially for the Paleogene (Süss and Müller-Stoll 1980; Lang et al. 1995; Srivastava et al. 2000). Documentation for dipteran seed predation from ascribable fossil damage is absent but is represented by body-fossil evidence from several relevant clades (Fig. 9.4). In addition to the fossil record, phylogenetic and other studies of some modern clades have provided general times of origin for some of these feeding types that extend to the Paleogene or perhaps Late Cretaceous (Michelsen 1988; Spencer 1990; Gilbert et al. 1994; Dempewolf 2001). Within the 31 families that have made the shift to significant phytophagy, the greatest number of species occurs among the predominantly galling Cecidomyiidae (Fig. 9.8A–E), leaf-mining Agromyzidae (Fig. 9.7), and the phytophagously more varied Tephritidae. Of these, the Cecidomyiidae probably represents a more ancient, Cretaceous colonization of vascular plants, although the appropriate analyses have not been

FIGURE 9.6. Unassigned Bombyliidae showing the variety of preservation of heads and proboscides in a typical Cenozoic fine-grained shale Fossil-Lagerstätten. All material is from the Green River insect fauna and is of lower mid-Eocene age; it originates from the Piceance Creek Basin, from the Green River Formation, near Rifle, in west-central Colorado, United States. Specimens are from the Kohls Collection at the National Museum of Natural History (USNM) in Washington, D.C. Arrows indicate position of the proboscis that is figured in each subsequent, higher-magnification photograph. (A) Typical habitus of bombyliid specimens, with outstretched wings (USNM 520438). (B) Enlargement of (A), showing proboscis and maxillary palps at its base. (C) Left lateral view of specimen with unusually positioned legs (USNM 520439). (D) Higher magnification of (C), showing proboscis, antennae, and compound eye. (E) Oblique dorsolateral view of specimen (USNM 520440). (F) Enlargement of (E), showing size of proboscis with regard to prothoracic legs. (G) Species with a truncate head (USNM 520441). (H) Close-up of (G) showing head shape and proboscis. (I) Specimen with head twisted with respect to thorax and abdomen (USNM 520442). (J) Detail of head, antennae, and proboscis in (I). (K) Specimen in position similar to (A) (USNM 520443). (L) Magnified head, proboscis, and prothoracic legs in (K), showing individual elements in proboscis and perhaps an associated maxillary palp. Dark longitudinal lines positioned medially along the proboscis are probably the food tube walls. Scale bars: solid, 1 cm; striped, 0.1 cm.

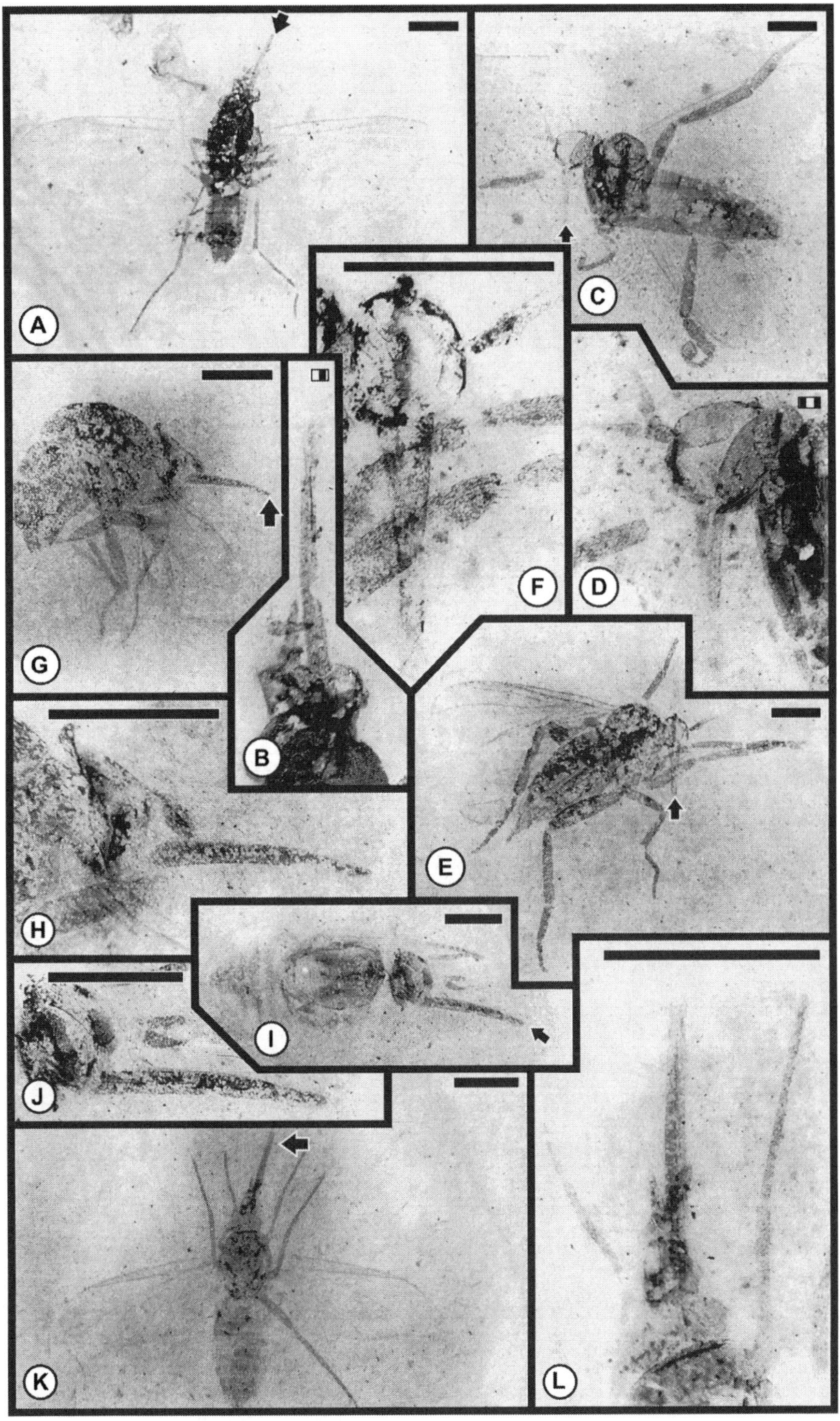
A
C
F
D
G
B
E
H
I
J
K
L

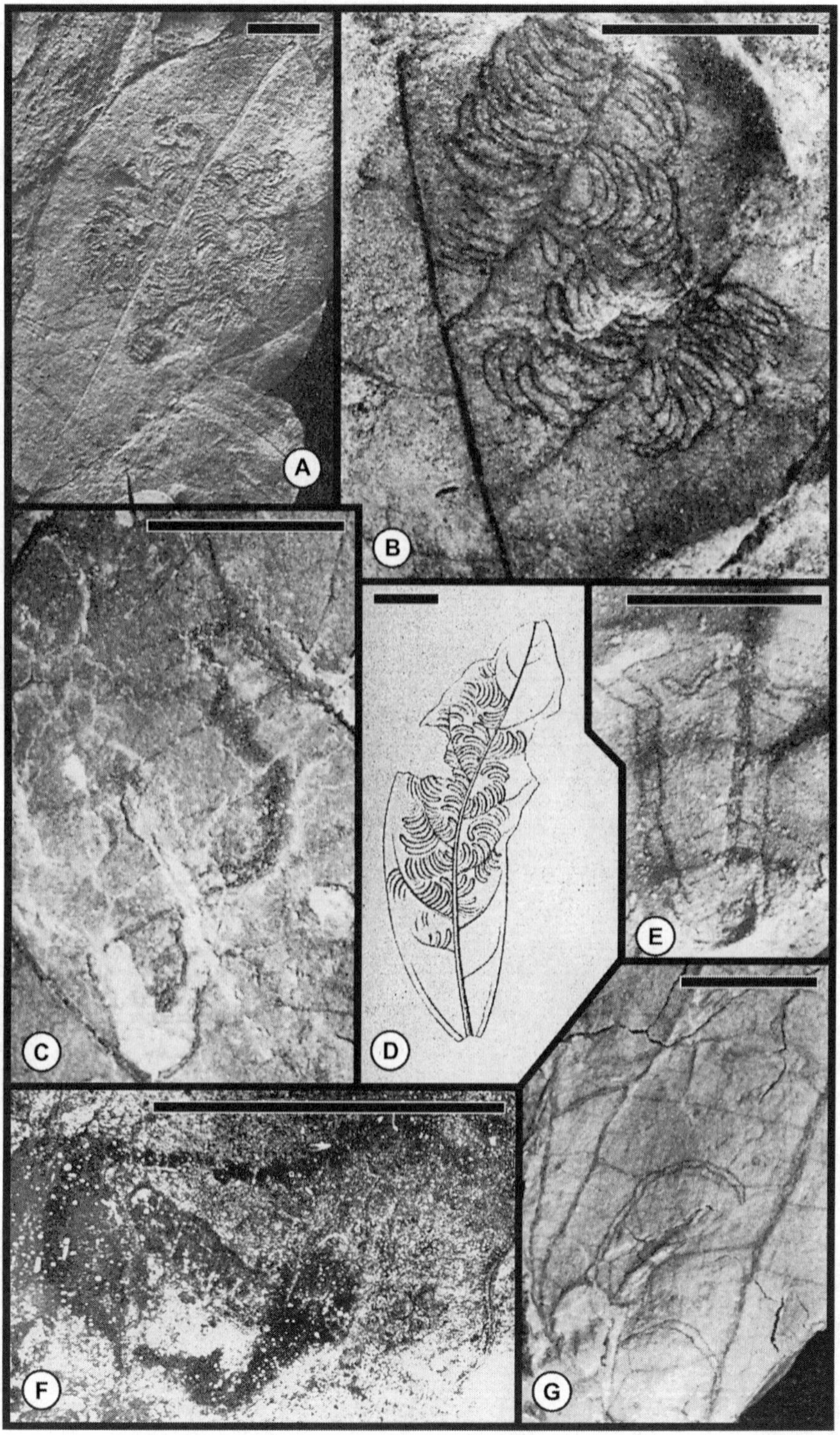
A
B
C
D
E
F
G

made of whether the most basal lineages occur on gymnosperms or angiosperms or both (cf. Farrell 1998). Agromyzids, however, appear to have radiated onto angiosperms, although cambium-boring genera, such as the very diverse *Phytobia*, occur on many conifers and may be phylogenetically basal within the family (Dempewolf 2001; Fig. 9.8G–H). These radiations onto plant tissues, unlike other phytophagous, holometabolous taxa, are geochronologically late and apparently targeted angiosperms. Endophytic phytophagous clades from the Coleoptera, Hymenoptera, and Lepidoptera occur earlier in the Mesozoic and colonized gymnospermous as well as basal angiospermous host plants (Crepet 1974; Rozefelds 1988; Jarzembowski 1990; Labandeira et al. 1994; Nishida and Hayashi 1996; Krassilov et al. 1997; Falder et al. 1998; Farrell 1998). This relative delay in phytophagy is explained by the Late Cretaceous origin of cyclorrhaphous lineages (Grimaldi and Cumming 1999), taxa in which the bulk of endophytic tissue consumption occurs, although the speciose Cecidomyiidae likely represented an earlier, independent, Cretaceous association with plants.

DIETS OF DOMINANT CLADES

An alternative to the functional feeding group mode of documenting the occurrence of phytophagy within the Diptera is an examination of dietary trajectories for particular clades within the order (Figs. 9.2, 9.3). This approach was discussed previously from the viewpoint of extant lineages, but is addressed below in a paleobiological context.

Larvae. During the Triassic, Lower dipteran clades such as the Culicomorpha, Ptychopteromorpha, and Psychodomorpha invaded freshwater habitats by pursuing four primary dietary strategies (see Ponomarenko 1996). The first was microvory, in which filterers consumed unicellular or finely particulate plant material suspended in the water column (Alverson and Courtney 2002). The second was herbivory, in which live plant material was eaten (Arens 1989; Nessimian et al. 1999). Saprophagy, the third type of feeding, consisted of the consumption of dead plant material in subaqueous but also in wet terrestrial habitats of algal films and waterlogged soils (Caspers and Wagner 1980). Finally, mycophagy, the consumption of live fungal or other saprobic organisms often associated with degraded plant detritus, also was a major type of feeding (Wichard et al. 2002). Herbivory, saprophagy, and mycophagy are found among larvae bearing mandibulate mouthparts engaged in scraping, gathering, and shredding (Merritt and Cummins 1996), and sometimes these qualitative trophic distinctions are not reflected in actual diets. By contrast, terrestrial phytophagy in Lower dipteran

FIGURE 9.7. Published examples of dipteran leaf mining (Agromyzidae) from the fossil record. (A) "*Phytopus*" *antiquus* Heyden on *Salix abbreviata* Göppert, from the upper Oligocene of Rott, in Siebengebirge, Germany. From Sittig (1927). (B) Mine on *Cinnamomum* sp. (Lauraceae), from the Upper Miocene (Tortonian) near Sarajevo, Bosnia. From Berger (1949). (C) Mine on unknown dicot, from the mid-Eocene Branksome Sand Formation of Hampshire, England. From Lang et al. (1995). (D) One of the oldest illustrations of a fossil leaf mine, an earlier reproduction of "*Phytopus*" *antiquus* Heyden shown in (A). From Heyden (1862). (E) Mine on unknown dicot, from the mid-Eocene Branksome Sand Formation of Hampshire, England. From Lang et al. (1995). (F) *Phytomyza lethe* Hering, ensconced between primary veins of an undetermined dicot, from the Upper Miocene (Messinian) Dysodil beds of Randecker Maar, Germany. From Hering (1930). (G) Unassigned agromyzid mine on *Platanus schimperi* (Heer) Saporta and Marion, from the Upper Paleocene Reading Beds of Berkshire, England. From Crane and Jarzembowski (1980). Scale bars: 1 cm.

A
B
C
D
E
F
G
H

clades predominantly occurs among the Bibionomorpha, including the comparatively older but dietarily diverse and dominantly gallicolous Cecidomyiidae (Gagné 1994; Amorim and Silva 2002). Phytophagy also occurs in related families, such as the Sciaridae, which are characterized by rare seed predation and leaf mining, although such feeding strategies are probably derived and originated during the Cenozoic.

In distinct contrast to typically aquatic Lower dipteran larvae, those of the Lower brachyceran lineages are predominantly predaceous, as demonstrated in the Tabanoidea, Vermileonidae, Xylophagidae, "higher" Asiloidea, and Empidoidea. Alternatively, other Lower brachyceran clades pursued parasitoid life habits, exemplified by the Nemestrinoidea, Eremochaetidae (based on fossil ovipositor structure), and Bombyliidae of the "lower" Asiloidea (Eggleton and Belshaw 1992). The dramatic expansion of predatory and parasitoid carnivory during the mid-Jurassic to mid-Cretaceous had a major effect on food-web structure of terrestrial ecosystems through the introduction of two novel types of insect carnivory that evidently did not exist prior to the Jurassic (Labandeira 2002b). Nevertheless, those taxa with larval parasitoid life-habits have adult stages that overwhelmingly consume nectar or similar surface fluids (Figs. 9.2, 9.3) and possibly had a major effect through these associations on preangiospermous seed plants. Among this radiation of larval parasitoidism there were occasional instances of plant consumption in certain clades, indicated by wood boring and mycophagy in the Pantophthalmidae and Asilidae, and leaf mining in the Dolichopodidae (Frohne 1939; Teskey 1976).

The dietary pattern of major cyclorrhaphous brachyceran clades is significantly different from those of their Lower brachyceran precursors. Among extant representatives of basal cyclorrhaphous clades that originated during the mid-Cretaceous, fungal galling is expressed in the Platypezidae, seed predation and leaf mining in the Phoridae, and leaf mining and wood boring in the Syrphidae (Grebenshchikova and Naumov 1985; Rotheray 1988; Cakar and Disney 1991; Dreger-Jauffret and Shorthouse 1992), although there is no evidence that these current associations extend to the Cretaceous. By contrast, apomorphic, cyclorrhaphan groups of the Schizophora and Pupipara literally host a vast variety of phytophagous clades,

---

FIGURE 9.8. Published examples of dipteran galling (Cecidomyiidae, A–E) and wood boring (Agromyzidae, G–H) from the fossil record. (A) Probable *Cecidomyia salicis* Schrank on *Salix amygdalina* L. (Salicaceae), from the Pleistocene Stegodon beds at Yagi-Higashiei, Japan. From (Miki (1937). (B) *Thecodiplosis clarkiensis* Lewis on *Taxodium* sp. (Cupressaceae) from the mid-Miocene Clarkia flora of northern Idaho. From Lewis (1985). This gall is very similar to extant *Taxodiomyia cupressiananassa* (Osten Sacken) on *Taxodium distichum* (L.) Richard (Gagné, pers. comm. 1997). (C) *Mikiola pontiensis* Villalta on *Fagus pristina* Saporta (Fagaceae), from the Upper Miocene of Cerdaña, Lérida, Spain. From Villalta (1957); Diéguez et al. (1996). (D) Gall comparable to *Dasyneura ruebsaameni* Kieffer on *Carpinus* sp. (Betulaceae), from the Upper Pliocene of Willershausen, Harz, Germany. From Straus (1977). (E) *Sequoiomyia kraeuseli* Möhn on *Sequoia langsdorfi* Brongniart (Cupressaceae), from the Upper Miocene Braunkohle of the Rhineland, Germany. From Möhn (1960); Gagné (1968). (F) Oblique tangential section of tunnels of *Palaeophytobia prunorum* Süss and Müller-Stoll on *Pruninium gummosum* (Platen) Süss and Müller-Stoll (Rosaceae), from the Upper Eocene of Yellowstone National Park, Wyoming, United States. From Süss and Müller-Stoll (1980). (G) Tangential section of *Palaeophytobia platani* Süss and Müller-Stoll on *Platanoxylon* sp. (Platanaceae), from the Upper Miocene of Mikófalva, Heves, Hungary. From Süss and Müller-Stoll (1975). (H) Same as in (F), but a radial section of the tunnel. From Süss (1980). Scale bars: solid, 1 cm; striped, 0.1 cm; dotted, 0.01 cm.

including all four major functional feeding groups as well as root feeders (Skevington and Dang 2002). This radiation of 20 family-level clades, especially the Agromyzidae and Tephritidae, probably began during the early Late Cretaceous and continued into the Paleocene, concomitant with the ecological expansion of modern angiosperm lineages. A second round of diversification occurred during the mid- to Late Miocene, synchronous with the ecological expansion of grasslands in arid midlatitudinal belts. Many taxa of Agromyzidae, Chloropidae, and Lonchaeidae, among others, are dependent on grasses (Ionescu and Neasçu 1969; Spencer 1990). A similar situation may exist for the Tephritidae and their frequent asteraceous hosts (McPheron and Steck 1996).

Adults. Adult feeding patterns are more complicated, owing to gender-based dietary segregation into protein- and carbohydrate-rich diets. Nevertheless, there are distinctive adult dietary trajectories throughout the Diptera, commencing during the mid-Triassic to Lower Jurassic, among a wide variety of Lower dipteran clades. These clades exhibit the consumption of nectarivory + protein, nectarivory + hematophagy, nectarivory and other similar fluids, insectivory, and aphagy (Fig. 9.3). This spectrum of feeding persists among more plesiomorphic, orthorrhaphous clades, which radiated during the mid-Jurassic to mid-Cretaceous and consisted of the Tabanoidea, Xylophagoidea, Stratiomyioidea, and Nemestrinoidea. This trend is supplemented in more apomorphic, Lower brachyceran groups, occasionally by pollinivory in the Bombyliidae, and a strong tendency toward obligate insectivory in the Asilidae, Therevidae, and Dolichopodidae. Interestingly, the extremely eclectic diets of adult cyclorrhaphous clades parallel those of their larval stages (Fig. 9.2), particularly the Acalyptratae and Pupipara, which diversified during the Late Cretaceous and Paleogene (McAlpine 1970; Grimaldi and Cumming 1999). Adult cyclorrhaphans, in contrast to orthorrhaphans, have a more eclectic dietary spectrum and consume virtually all available foods, including fluidized plant tissues (Oldroyd 1964). The most salient pattern in adult dipteran phytophagy is the consumption of various mixtures of nectar, pollen, blood, and fluidized insect contents, which typically support both the requirements of ovarian development and the energetics of powered flight. This provides prima facie evidence that a combination of sugars and protein is the essential dietary requirements for many extant dipterans and probably was important during their early evolution.

ADVANCED GYMNOSPERMS, ADULT DIPTERA, AND ANGIOSPERMS

Direct evidence for nectarivory in fossil dipterans is rarely encountered. Although the best direct evidence for food consumption in fossil insects is their gut contents, in most cases, such data are unavailable for dipterans because of their fluid-feeding habits. Ingested plant sap, nectar, blood, or other fluid diets lack any identifying features and probably occur as nondiagnostic intestinal films in fossils. Consequently, more indirect forms of evidence are needed, such as mouthpart structure, putative host-plant reproductive features, and features of pollen, such as size, shape, surface ornamentation, dispersal pattern, and especially occurrence on the proboscides and bodies of suspect nectarivores and pollinivores (Labandeira 2000; Fig. 9.5M), which has been documented on modern dipterans (Stelleman 1981; Grimaldi 1988; Nicholson 1994). Although pollen has been described in the intestines of modern pollinivorous dipterans (Wilson and Lieux 1972; Leereveld 1982), gut contents have not been documented in fossil taxa, even though they occur in other insect groups from the Middle Carboniferous to the Recent (Kukalová-Peck 1987; Rasnitsyn and Krassilov 1996; Afonin 2000; Labandeira 2000). Lower brachyceran taxa bearing long, apparently flexible proboscides of the families Nemestrinidae, Apioceridae, Tabanidae–Pangioniinae, and Therevidae have been described from preangiospermous biotas. These taxa typically have proboscides as long as the axial length of the head and some approach the length of their respective bodies (Ren 1998; Mostovski and Martínez-Delclòs 2000; Fig. 9.7B,D,I–L), which are similar but shorter than the long proboscides of modern confamilial descendants that nectar deep-throated flowers (Manning and Goldblatt 1997; Goldblatt and Manning 2000; Stuckenberg 2000). An alternative interpretation of these structures is that they were constructed for hematophagy (Grimaldi 1999), although these proboscides exhibit pronounced flexibility without brittle fracture, evidenced by a recurved 135° bend in orientation shown for a Jurassic nemestrinid by Mostovski (1998; Fig. 9.5B). Such an example of pronounced proboscis deflection is inconsistent with a fascicle of sclerotized, relatively rigid stylets that are essential for penetrating the dermal tissues of animal hosts (Nagatomi and Soroida 1985). The presence of pollen clusters attached to the head region or adjacent to the proboscis bases of these insects (Fig. 9.5M), in addition to the extraordinary length, flexibility, and development of a reduced and terminal labellum of these proboscides (Fig. 9.5D–H), indicates that fluid feeding on sugary plant secretions, with or without pollination, was their most likely food source. Finally, similarly constructed and elongate proboscides of extant descendants of these families are nectarivores, including the subfamily Pangioniinae of the nominally hematophagous Tabanidae (Coscarón and Phillip 1979; Goldblatt and Manning 2000; but see Mitter 1918).

A recent interpretation of these anthophilous insects involves an assemblage from the Yixian Formation of Liaoning Province in northeastern China, formerly considered of Late Jurassic (Tithonian) age (Ren 1998). Ren proposed that the presence of long-proboscid dipterans from this deposit indicates that they were consuming angiospermous nectar, and therefore the geochronological range of angiosperms should be extended downward by about 20 My to the Tithonian Stage (Ren 1998). This view, however, is not tenable for two major reasons. The first is that the age of the Yixian Formation historically has been controversial, and currently has been established as Early Cretaceous, equivalent to the mid-Barremian Stage, with a $^{40}Ar/^{39}Ar$ radioisotopic date of 124.6 ± 0.25 Ma (Swisher et al. 1999), corroborated by dinoflagellate (Smith et al. 1995), pollen (Li and Liu 1994), and other data (Barrett 2000). The best explanation for this discordance in dates is that the Yixian biota was a "refugium for relicts" (Luo 1999), and represents an assemblage of typically Late Jurassic groups, including insects, birds, dinosaurs, pterosaurs, and primitive mammals that persisted regionally into the Early Cretaceous in northeastern China. Second, despite extensive worldwide searching, there is no evidence for angiosperm pollen prior to the mid-Valanginian Stage (Crane et al. 1995; Wing and Boucher 1998); the earliest known megafloral material is coeval with the Yixian dipteran fossils and has been documented in stratigraphically adjacent Barremian beds from the same formation (Sun et al. 2002). These angiosperms, representing two species of *Archaeofructus,* are the basalmost angiosperms known from megafloral evidence, a conclusion also based on a molecular and morphological cladistic analysis of *Archaeofructus,* modern ANITA-grade taxa, and other primitive angiosperms (Sun et al. 2002). *Archaeofructus* curiously is an emergent aquatic form with highly dissected leaves and plesiomorphous reproductive characters, including absence of a perianth, and may have been insect pollinated (Sun et al. 2002). The ecological context of *Archaeofructus* within the

Yixian Biota is one of a diverse assemblage of advanced seed plants, including diverse gnetaleans (Guo and Wu 2000; Sun et al. 2002), many of which, like their extant descendants, were producers of pollination drops from their ovular micropyles. Similar assemblages containing both advanced seed plants and Lower brachyceran dipterans with long proboscides have been documented for the older, Barremian Las Hoyas Biota of Spain (Mostovski and Martínez-Delclòs 2000), the likely Hauterivian Baissa Biota in Transbaikalian Russia (Zherikhin et al. 1999), and the uppermost mid-Jurassic Karatau Biota in southeastern Kazakhstan (Rohdendorf 1968; Mostovski 1998). For the Karatau Biota, a recent stratigraphic and paleobotanical revision by Kirichkova and Doludenko (1996) indicates that it is mostly of upper Callovian in age and thus uppermost mid-Jurassic. Interestingly, a nemestrinid taxon with a vestigial proboscis, presumably nonfeeding, occurs in the Late Jurassic (Tithonian) Solnhofen Biota of Germany (Ansorge and Mostovski 2000).

Rather than posit an extension of angiosperms into the Late Jurassic that is inconsistent with fossil evidence, a more reasonable hypothesis is afforded by observations of modern anthophilous fluid-feeding insects and by recent advances in knowledge about the reproductive biology of mid-Mesozoic seed plants. This alternate hypothesis holds that the occurrences of long-proboscate, nonhematophagous flies documented by Ren (1998), Mostovski (1998), and others—in floras either lacking angiosperms entirely or having very few angiosperms that lack reproductive features inconsistent with long-tongued pollinators—instead indicate associations with gymnospermous seed plants that secreted nectarlike substances (Labandeira 2000). In this context, the presence of clusters of cheirolepidaceous *Classopollis* pollen preserved on the head of *Paroikus*, a probable basal therevoid from Baissa (Fig. 9.5M) indicates that feeding on fluids produced by a conifer strobilus was a likely source of nutrition. This conclusion is consistent with the moderately elongate, labellate mouthpart structure of *Paroikus*, and it suggests that pollination may have been a mutualism with an advanced seed plant, a status similar to that proposed by Zaitzev (1982) for fossil Eremochaetidae. Accordingly, the pollination drop, produced by secretion of sugary fluids in the micropyles and adjacent ovular surfaces of gymnospermous reproductive structures, was the primary reward (Thien et al. 1983; Kato and Inoue 1994; Labandeira 2000). Interestingly, *Classopollis* pollen, whose source plant was the coniferous family Cheirolepidiaceae from typically arid regions (Watson 1988), was consumed by prophalangopsid grasshoppers and xyelid sawflies at Karatau (Krassilov et al. 1997). Additionally, *Classopollis* pollen has been attributed by some palynologists as insect vectored because of its large size, lack of sacci that would have provided aerial buoyancy, peculiar surface ornamentation resembling some extant flowering-plant taxa reminiscent of self-incompatibility systems, asteraceous-like exine ultrastructure, and the frequent presence of mutually attached pollen grains suggestive of insect-vectored pollen (Courtinat 1980; Pocock et al. 1990). The female cones pollinated by *Classopollis* are poorly known (Watson 1988), but one taxon is sufficiently well preserved that the reproductive biology can be reasonably approximated. In particular, the cone of *Frenelopsis alata* exhibits microstructural detail that indicates a pollination process atypical for any modern conifer, described as a uniquely specialized pollination system (Kvacek 2000) or reminiscent of certain angiosperms (Pocock et al. 1990).

This presumed type of external fluid feeding by proboscate dipterans occurs in extant angiosperms, as well as in gymnosperms that produce pollination drops. Notably, all of the three principal lineages of the Gnetales—the Gnetaceae, Ephedraceae, and Welwitschiaceae—

secrete pollination drops (Moussel 1980; Kato and Inoue 1994). These carbohydrate- and sometimes lipid-rich secretions are avidly consumed by fluid-feeding dipterans (McMillan 1958; Meeuse et al. 1990; Owens et al. 1998), as well as by other insects, such as moths, beetles, and thrips. Pollination drops also occur in the second major extant clade of gymnospermous seed plants, the cycads (Poort et al. 1996), and provide nutritional rewards for fungus gnats (Breckon and Ortíz 1983) and other insects (Pellmyr et al. 1991), albeit there is less evidence than that for the Gnetales. Among the more speciose conifers, the third major gymnospermous clade, many genera secrete observable pollination drops (Tomlinson et al. 1991; Owens et al. 1995), but evidently insects are rarely attracted even though copious pollen is produced that is consumed by adult insects, including dipterans (Burdick 1961; Holloway 1976). The fourth major seed-plant group of today, the currently monotypic *Ginkgo*, also produces ovular secretions as pollination drops (Gifford and Foster 1989), but there are few observations of fluid consumption by flies or other insects. The principal conclusion from fly-involved consumption of nectarlike secretions in modern gymnospermous seed plants is that it probably was a precursor to the imbibition of stigmatic exudates by small flies in basal angiosperms (Thien et al. 1983; Dilcher 2000). This shift from gymnospermous to angiospermous sources of sugary fluid reward, in turn, was elaborated by a more pervasive type of nectar provision by specialized nectar-bearing tissue. This was followed by the establishment of more persistent or faithful pollination associations found in more advanced angiosperms (Crepet 1985, 1996). Eventually these types of nectar production became standardized in nectarial gland position and type that dominate most plant-pollinator associations of today.

Role of Diptera in Ecosystem Evolution

As a major holometabolous order of exceptional diversity, the Diptera has modestly contributed to the evolution of plant-insect associations in the freshwater and terrestrial realms. A more important role was accomplished by the mid-Mesozoic addition of larval and adult parasites and parasitoids, which supplemented and eventually supplanted the role of predators as the dominant carnivores on land (Labandeira 2002b). Predator-based food webs are typically lean at top and enlarge downward to lower trophic levels by virtue of their eclectic choice of prey. By contrast, food webs rich in parasitoids and parasites are characterized by carnivores with high host specificities, frequently monospecific associations, and expand toward the top of the trophic scale. This effect often is magnified by the presence of additional relationships such as hyperparasitoids and cleptoparasites. The Hymenoptera largely launched the Mesozoic parasitoid/parasite revolution during the mid-Jurassic, but the Diptera also contributed to an expansion of this novel type of carnivory (Labandeira 2002b; Fig. 9.2). This shift in carnivore feeding overwhelmingly was concentrated at the larval stage, and eventually surpassed adult predation in trophic impact. As a consequence, the adults were able to consume food important for mating and oviposition during their relatively brief lives. For many holometabolous clades whose larvae are parasitoids or parasites, adults are nectarivores, less frequently pollinivores, or a combination of both, or otherwise consume other energy-rich surface fluids (Burgin and Hunter 1997), illustrated by a comparison of Figs. 9.1 and 9.2. It is most probable that these carbohydrate- and lipid-rich fluid sources were provided dominantly by mid-Mesozoic seed plants, prior to the ecological expansion of angiosperms during the Early Cretaceous (Labandeira 2000; Gorelick 2001). Thus the mid-Mesozoic shift toward parasitoid and parasite modes of carnivory was probably associated with adult feeding on nectar or other plant-derived equivalents. The result was an ecosystem-wide effect on the associations between many holometabolous insects and seed plants. The expansion of fluid feeding on plant reproductive products also is seen in the increase of mid-Mesozoic mouthpart types involved in external fluid feeding (Labandeira 1997), including the Diptera.

Additional ecosystem penetration involved the occupation of several ephemeral, terrestrial habitats during the Mesozoic by larval and adult Diptera. These habitats included especially the microenvironments of carrion, dung, and macrofungi, which provided alternative sources of fluid nutrition through the production of putrefying or deliquescing fluid from decomposing tissues or organic matter (Hammer 1941; Hackman and Meinander 1979). An interesting parallel development was the origination in several clades of angiosperms whose flowers mimicked in sight, smell, and taste the same microhabitat features, thereby co-opting the same dipteran visitors that frequented carrion, dung, and macrofungi (Lendner 1935; Vogel 1978a; Meeuse and Raskin 1988). Although these three decomposer habitats are ancient and undoubtedly extended to the Late Paleozoic, they required extraordinary coordination of overall life cycle and larval developmental timing to colonize relatively evanescent food resources. Such developmental histories probably were accomplished by earlier Mesozoic holometabolous insects with highly honed life cycles, of which cyclorrhaphous dipterans are prominent examples (Denno and Cothran 1976). An important ecosystem consequence was the important role that occupation of such habitats during the later Mesozoic had in nutrient recycling by enlarging the saprobic decomposer subcycle of terrestrial food webs (McAlpine 1970; Chin and Gill 1996).

More temporally persistent microhabitats are the varied internal tissues of vascular plants, particularly in more aseasonal regions. Precluded by mouthpart type from the external foliage feeding of most all other insect groups, terrestrial larval dipterans were preferentially channeled into endophytic modes of phytophagy. It is difficult to assess the contribution of larval dipteran herbivory to the structuring of modern-aspect terrestrial ecosystems when compared to the distribution of other holometabolous larval clades with endophytic feeding habits in the same niches. In this context the role of dipterans appears to be less impressive than that of the coleopterans, lepidopterans, and possibly hymenopterans. As a clade, the Diptera may be the least phytophagous of the four holometabolous orders, and geochronologically the last one to colonize vascular plants. Because dipteran endophytic phytophagy—leaf mining, galling, boring, seed predation, and root feeding—appeared relatively late in seed-plant evolution, their role is best characterized as enlarging the endophytic colonization of plants previously established by the other three holometabolous orders of the earlier Mesozoic (Crepet 1974; Labandeira et al. 1994; Farrell 1998; Labandeira 1998c). For example, endophytic herbivorous clades, such as the cyclorrhaphan Agromyzidae, appeared late relative to other holometabolous clades occupying the same plant-host niche (Hennig 1965; Dempewolf 2001). Within the Agromyzidae, leaf mining is primitive, occurring probably during the Late Cretaceous (Labandeira, pers. observ.), but other endophytic subclades, such as seed predators and cambium borers, are derived (Dempewolf 2001) and likely have their origins during the Cenozoic. It does appear, however, that endophytic dipteran phytophages had a potentially important effect on the expansion of ecosystem trophic webs during the mid-Cenozoic, coincident with the emergence of plant hosts such as herbaceous Asteraceae and Poaceae (Crepet and Feldman 1991; Pfretzschner 1998).

Finally, the presence of aquatic larval dipterans, especially Lower dipteran clades, in lentic and lotic freshwater ecosystems is important. With the colonization of fresh water by numerous other insect clades, commencing during the Early Permian (Miller and Labandeira 2002), and including the Odonata, Ephemeroptera, Plecoptera, Trichoptera, Coleoptera, Raphidioptera, and Diptera, there was the establishment of basic trophic webs by the mid-Jurassic (Ponomarenko 1996; Sinichenkova and Zherikhin 1996). These early lake and river ecosystems consisted of detritivores, herbivores, and predators, in which dipterans played significant roles (Sinichenkova and Zherikhin 1996). This is based on larval body fossils, as well as fossil occurrences of family-level clades with extant aquatic larvae of known feeding habits (Merritt and Cummins 1996; Miller and Labandeira 2002; Fig. 9.5A). The trophic complexity of these ecosystems increased during the Late Cretaceous, although it is unknown what result a regional K-T extinction of insect herbivores (Labandeira et al. 2002) had on their trophic structure. By the mid-Eocene, both lentic and lotic ecosystems were structured similar to that of today.

Conclusions

The broadest spectrum of available evidence should be used to infer the ecological history of an insect group as diverse as the Diptera. Uniformitarian and nonuniformitarian fossil evidence, such as dipteran-mediated plant associations, data indicating pollen or nectar consumption by reference to mouthpart or plant-reproductive structures, or calibration by body fossils of dipteran herbivore phylogenies, are all-important for determining the dipteran colonization of plants. The presence of fluid feeding in both the larval and adult stages nevertheless has limited the abundance of evidence gleaned from the fossil record. In addition to data from the associational record and phylogenetic analyses, functional morphologic and taxic approaches are relevant for understanding the totality of the ecological and evolutionary history between dipterans and their plant associates. This ecumenical theme advances not only alternative approaches, but also provides mutually supportive ways of arriving at a more complete ecological and evolutionary account.

Researchers have minimally used these classes of evidence to infer the ecological relationships between extinct plants and dipterans. More needs to be done in employing all the evidence and approaches for ferreting out this increasingly complex history. It is well known that dipterans have a wide variety of associations in extant fresh water and terrestrial ecosystems. Particularly on land, these associations include a significant role in detritivory among such habitats as carrion, dung, macrofungi, and rotting wood, as well as a major presence as carnivores on other arthropods in the form of predators, parasitoids, and parasites. Although a minority of dipteran taxa are herbivorous, particular clades have ecologically radiated as endophytic herbivores on vascular plant tissues and have become consumers of nectar, pollen, and other external plant fluids. Much of this pattern needs to be fleshed out in the fossil record. Advances in biological and paleobiological methods are increasing our knowledge of the geochronology, spatial distribution, and phylogenetic pattern of several clades of endophytic phytophages during the Late Cretaceous to mid-Cenozoic. The major challenge is to understand the role of dipterans in early angiosperm-dominated ecosystems during the

Cretaceous and especially in earlier preangiospermous ecosystems. Of particular interest is the specific role of adult dipterans with respect to gymnospermous plants whose ecologies are poorly known and subsequently became extinct or currently have relict distributions. This issue involves greater understanding of the fossil compression record from Late Triassic to Early Cretaceous biotas. It will also require forging collaborations among paleoentomologists, paleobotanists, and researchers studying modern plant-insect associations. Although the terrestrial, mid-Mesozoic compression fossil record is said to be inferior to the amber fossil record that commences relatively late in the history of the Diptera, the earlier fossils nevertheless are significantly older and potentially very informative. It is this older fossil record that will provide surprises in our understanding of the establishment of such dipteran associations as nectarivory, pollinivory, and perhaps pollination mutualisms on seed plants long before the advent of angiosperms.

ACKNOWLEDGMENTS

I extend my thanks to Brian Wiegmann and David Yeates for inviting me to contribute to this volume. Finnegan Marsh graciously formatted Figs. 9.1–9.8 into superb, copy-ready figures. I am grateful to Leonard Thien and Peter Bernhardt for providing prepublication data on the insect pollinators of modern *Amborella*. This manuscript has benefited from discussions with A. Borkent, M. Irwin, M. Mostovski, and M. von Tschirnhaus. I thank Art Borkent and Ray Gagné for helpful reviews. This is publication 94 of the Evolution of Terrestrial Ecosystems Consortium of the National Museum of Natural History.

REFERENCES

Ackerman, J. D. 1986. Mechanisms and evolution of food-deceptive pollination systems in orchids. *Lindleyana* 1:108–113.

Afonin, S. A. 2000. Pollen grains of the genus *Cladaitina* extracted from the gut of the Early Permian insect *Tillyardembia* (Grylloblattida). *Paleontol. J.* 34:575–579.

Allen, E. J., and B. A. Foote. 1975. Biology and immature stages of *Tritoxa incurva* (Diptera: Otitidae). *Proc. Entomol. Soc. Wash.* 77:246–257.

Allen, H. W. 1929. An annotated list of the Tachinidae of Mississippi. *Ann. Entomol. Soc. Am.* 22:676–690.

Alverson, A. J., and G. W. Courtney. 2002. Temporal patterns of diatom ingestion by larval net-winged midges (Diptera: Blephariceridae: *Blepharicera*). *Freshw. Biol.* 47:2087–2097.

Amorim, D. S., and V. C. Silva. 2002. How far advanced was Diptera in the Pangaea? *Ann. Soc. Entomol. France* 38:177–200.

Anderson, J. M., and H. M. Anderson. 1989. *Palaeoflora of Southern Africa: Molteno Formation (Triassic). Vol. 2: Gymnosperms, Excluding Dicroidium.* Rotterdam: Balkema.

Ansorge, J., and M. B. Mostovski. 2000. Redescription of *Prohirmoneura jurassica* Handlirsch 1906 (Diptera: Nemestrinidae) from the Lower Tithonian lithographic limestone of Eichstätt (Bavaria). *Neues Jb. Geol. Paläont. Mn.* 2000:125–243.

Arens, W. 1989. Comparative functional morphology of the mouthparts of stream animals feeding on epilithic algae. *Arch. Hydrobiol. Suppl.* 83:253–354.

Arillo, A., and A. Nel. 2000. Two new fossil cecidomyiid flies from the Lower Cretaceous amber of Álava (Spain) (Diptera, Cecidomyiidae). *Bull. Soc. Entomol. France* 105:285–288.

Armstrong, J. A. 1979. Biotic pollination mechanisms in the Australian flora—A review. *New Zealand J. Bot.* 17:467–508.

Ash, S. 1997. Evidence of arthropod-plant interactions in the Upper Triassic of the southwestern United States. *Lethaia* 29:237–248.

Azuma, H., L. B. Thien, and S. Kawano. 1999. Floral scents, leaf volatiles and thermogenic flowers in Magnoliaceae. *Pl. Sp. Biol.* 14:121–127.

Baker, H. G., and I. Baker. 1983. A brief historical review of the chemistry of floral nectar. In B. Bentley and T. Elias, eds., *The Biology of Nectarie,* pp. 126–152. New York: Columbia University Press.

Balmford, A., A. J. L. Lyon, and R. M. Lang. 2000. Testing the higher-taxon approach to conservation planning in a megadiverse group: The macrofungi. *Biol. Conserv.* 93:209–217.

Bänziger, H. 1991. Stench and fragrance: Unique pollination lure of Thailand's largest flower, *Rafflesia kerri* Meijer. *Nat. Hist. Bull. Siam* 39:19–52.

Barnes, H. F. 1951. *Gall Midges of Economic Importance. V. Gall Midges of Trees.* London: Crosby, Lockwood and Son.

Barrett, P. M. 2000. Evolutionary consequences of dating the Yixian Formation. *Trends Ecol. Evol.* 15: 99–103.

Baumgartner, D. L. 1986. Notes on distribution and bionomics of the dung fly, *Scathophaga tropicalis* (Diptera: Scathophagidae), in Peru. *Entomol. News* 97:208–214.

Beaman, R. S., P. J. Decker, and J. H. Beaman. 1988. Pollination of *Rafflesia* (Rafflesiaceae). *Am. J. Bot.* 75:1148–1162.

Beardsell, D. V., and P. Bernhardt. 1983. Pollination biology of Australian terrestrial orchids. In E. G. Williams, R. B. Knox, J. H. Gilbert, and P. Bernhardt, eds., *Pollination '82,* pp. 166–183. Parkville, Victoria: University of Melbourne Press.

Becher, E. 1882. Kenntnis der Mundtheile der Dipteren. *Denk. Akad. Wiss., Math.-Naturwiss. Klasse* 45: 123–162.

Bequaert, J. C., and F. M. Carpenter. 1936. The Nemestrinidae of the Miocene of Florissant, Colorado, and their relations to the Recent fauna. *J. Paleontol.* 10:395–409.

Berg, C. O. 1950. Biology of certain Chironomidae reared from *Potamogeton. Ecol. Mongr.* 20:84–99.

Berger, W. 1949. Lebensspuren schmarotzender Insekten an jungtertiären Laubblättern. *Sitz. Öster. Akad. Wiss.* 158:789–792.

Beri, S. K. 1983. Comparative morphological studies on the cephalopharyngeal skeleton of larval leaf-mining flies (Agromyzidae: Diptera). *Indian J. For.* 6:301–308.

Bissett, J., and A. Borkent. 1988. Ambrosia galls: The significance of fungal nutrition in the evolution of the Cecidomyiidae (Diptera). In K. A. Pirozynski and D. L. Hawksworth, eds., *Coevolution of Fungi with Plants and Animals,* pp. 203–225. Toronto: Academic Press.

Bletchly, J. D. 1954. The mouth-parts of the dance fly, *Empis livida* L. (Diptera, Empididae). *Proc. Zool. Soc. London* 124:317–334.

Bonnier, G. 1879. *Les Nectaires. Étude Critique, Anatomique et Physiologique.* Paris: G. Masson.

Borkent, A. 1995. *Biting Midges in the Cretaceous Amber of North America (Diptera: Ceratopogonidae).* Leiden: Backhuys.

Borkent, A. 2000. Biting midges (Ceratopogonidae: Diptera) from Lower Cretaceous Lebanese amber with a discussion of the diversity and patterns found in other ambers. In D. Grimaldi, ed., *Studies on Fossil Amber, with Particular Reference to the Cretaceous of New Jersey,* pp. 355–451. Leiden: Backhuys.

Brantjes, N. B. M. 1980. Flower morphology of *Aristolochia* species and the consequences for pollination. *Acta Bot. Neerl.* 29:212–213.

Brantjes, N. B. M., and J. A. A. M. Leemans. 1976. *Silene otites* (Caryophyllaceae) pollinated by nocturnal Lepidoptera and mosquitoes. *Acta. Bot. Neerl.* 25:281–295.

Breckon, G., and V. N. Ortíz. 1983. Pollination of *Zamia pumila* by fungus-gnats. *Am. J. Bot.* 70(5,2): 106–107.

Bronner, R. 1992. The role of nutritive cells in the nutrition of cynipids and cecidomyiids. In J. D. Shorthouse and O. Rohfritsch, eds., *Biology of Insect-Induced Galls,* pp. 118–140. New York: Oxford University Press.

Buhr, H. 1955. Minierfliegen als Gallenerzeuger. *Nachrich. Bayer. Entomol.* 4:81–87.

Burdick, D. J. 1961. A taxonomic and biological study of the genus *Xyela* Dalman in North America. *Univ. Calif. Publ. Entomol.* 17:281–353.

Burger, W. V. 1981. Why are there so many kinds of flowering plants? *Bioscience* 31:345–393.

Burgin, S. G., and F. F. Hunter. 1997. Evidence of honeydew feeding in black flies (Diptera: Simuliidae). *Canad. Entomol.* 129:859–869.

Cakar, L., and R. H. L. Disney. 1991. A leaf-mining scuttle fly (Diptera: Phoridae) as a pest of chickpeas (Leguminosae) in Yugoslavia. *Bull. Entomol. Res.* 81:21–24.

Cameron, A. E. 1918. Life-history of the leaf-eating crane-fly, *Cylindrotoma splendens*, Doane. *Ann. Entomol. Soc. Am.* 11:67–89.

Caspers, N., and R. Wagner. 1980. Emergenz-Untersuchungen an einem Mittelgebirgsbach bei Bonn. II. Psychodien Emergenz 1976/1977. *Arch. Hydrobiol.* 88:73–95.

Cazier M. A. 1963. The description and bionomics of a new species of *Apiocera*, with notes on other species (Diptera: Apioceridae). *Wasmann J. Biol.* 21:205–234.

Chase, M. W., and D. R. Peacor. 1987. Crystals of calcium oxalate hydrate on the perianth of *Stelis* Sw. *Lindleyana* 2:91–94.

Chaudonneret, J. 1990. *Les Pièces Buccales des Insectes.* Dijon: Berthier.

Chin, K., and B. D. Gill. 1996. Dinosaurs, dung beetles, and conifers: Participants in a Cretaceous food web. *Palaios* 11:280–285.

Christy, M. 1922. The pollination of the British primulas. *J. Linn. Soc.* 46:105–139.

Clausen, C. P. 1940. *Entomophagous Insects.* New York: McGraw-Hill.

Clifford, H. T. 1962. Insect pollination of *Plantago lanceolata* L. *Nature* 193:196.

Cole, F. R. 1969. *The Flies of Western North America.* Berkeley: University of California Press.

Collin, J. E. 1948. A short synopsis of the British Sapromyzidae (Diptera). *Trans. Roy. Entomol. Soc. London* 99:225–242.

Collinge, S. K., and S. M. Louda. 1988. Herbivory by leaf miners in response to experimental shading of a native crucifer. *Oecologia* 75:559–566.

Collinson, M. E., M. C. Boulter, and P. L. Holmes. 1993. Magnoliophyta ("Angiospermae"). In M. J. Benton, ed., *The Fossil Record Record 2*, pp. 809–841. London: Chapman and Hall.

Connor, E. F., and M. P. Taverner. 1997. The evolution and adaptive significance of the leaf-mining habit. *Oikos* 79:6–25.

Cook, E. F. 1949. The evolution of the head in the larvae of the Diptera. *Microentomology* 14:1–57.

Coscarón, S., and C. B. Phillip. 1979. A revision of Mycteromyiini ("Genus *Mycteromyia*" of authors), a new tribe of neotropical horse flies (Diptera, Tabanidae). *Proc. Calif. Acad. Sci.* (4) 41:427–452.

Courtinat, B. 1980. Structure d'adhérence des grains de pollen en tétrade du genre *Classopollis* Pflug, 1953, de l'Hettangien de Saint-Fromont, Manche (France). *Géobios* 13:209–229.

Craig, D. A. 1977. Mouthparts and feeding behaviour of Tahitian larval Simuliidae (Diptera: Nematocera). *Quaest. Entomol.* 13:195–218.

Crane, P. R., and E. A. Jarzembowski. 1980. Insect leaf mines from the Palaeocene of southern England. *J. Nat. Hist.* 14:629–636.

Crane, P. R., E. M. Friis, and K. R. Pedersen. 1995. The origin and early diversification of angiosperms. *Nature* 374:27–33.

Crepet, W. L. 1974. Investigations of North American cycadeoids: The reproductive biology of *Cycadeoidea. Palaeontographica A* 148:144–169.

Crepet, W. L. 1979. Some aspects of the pollination biology of middle Eocene angiosperms. *Rev. Palaeobot. Palynol.* 27:213–238.

Crepet, W. L. 1985. Advanced (constant) insect pollination mechanisms: Pattern of evolution and implications vis-à-vis angiosperm diversity. *Ann. Missouri Bot. Gard.* 71:607–630.

Crepet, W. L. 1996. Timing in the evolution of derived floral characters: Upper Cretaceous (Turonian) taxa with tricolpate and tricolpate-derived pollen. *Rev. Palaeobot. Palynol.* 90:339–359.

Crepet, W. L., and G. D. Feldman. 1991. The earliest remains of grasses in the fossil record. *Am. J. Bot.* 78:1010–1014.

Crepet, W. L., E. M. Friis, and K. C. Nixon. 1991. Fossil evidence for the evolution of biotic pollination. *Phil. Trans. R. Soc. Lond. B* 333:187–195.

Cummins, K. W., and M. J. Klug. 1979. Feeding ecology of stream invertebrates. *Annu. Rev. Ecol. Syst.* 10:147–172.

Dahl, C., L.-E. Widahl, and C. Nilsson. 1988. Functional analysis of the suspension feeding system in mosquitoes (Diptera: Culicidae). *Ann. Entomol. Soc. Am.* 81:105–127.

Darwin, C. 1862. *On the Various Contrivances by which British and Foreign Orchids are Fertilised by Insects.* London: Murray.

Dempewolf, M. 2001. *Larvalmorphologie und Phylogenie der Agromyzidae (Diptera).* Dissertation. Amsterdam: Universität Bielefeld.

Denno, R. F., and W. R. Cothran. 1976. Competitive interactions and ecological strategies of sarcophagid and calliphorid flies inhabiting rabbit carrion. *Ann. Entomol. Soc. Am.* 69:109–113.

DeVries, P. J. 1979. Pollen-feeding rainforest *Parides* and *Battus* butterflies in Costa Rica. *Biotropica* 11: 237–238.

Diéguez, C., J. L. Nieves-Aldrey, and E. Barrón. 1996. Fossil galls (zoocecids) from the Upper Miocene of la Cerdaña (Lérida, Spain). *Rev. Palaeobot. Palynol.* 94:329–343.

Dilcher, D. L. 2000. Toward a new synthesis: Major evolutionary trends in the angiosperm fossil record. In F. J Ayala, W. M. Fitch, and M. T. Clegg, eds., *Variation and Evolution in Plants and Microorganisms: Toward a New Synthesis 50 Years after Stebbins,* pp. 255–270. Washington, D.C.: National Academy Press.

Disney, R. H. L. 1980. Records of flower visiting by scuttle flies (Diptera: Phoridae) in the British Isles. *Naturalist* 105:45–50.

Dixon, A. F. G. 1998. *Aphid Ecology.* London: Chapman and Hall.

Docteurs van Leeuwin, W. 1949. Gallen kweken. *Natura* 41:1–4.

Dodson, C. H. 1966. Studies in orchid pollination—*Cypripedium, Phragmipedium* and allied genera. *Am. Orch. Soc. Bull.* 35:125–128.

Downes, J. A. 1955. The food habits and description of *Atrichopogon polllinivorus* sp. n. (Diptera: Ceratopogonidae). *Trans. R. Entomol. Soc. Lond.* 106:439–453.

Downes, J. A. 1958. The feeding habits of biting flies and their significance in classification. *Annu. Rev. Entomol.* 3:249–266.

Downes, J. A. 1971. The ecology of blood-sucking Diptera: An evolutionary perspective. In A. M. Fallis, ed., *Ecology and Physiology of Parasites,* pp. 232–258. Toronto: University of Toronto Press.

Downes, J. A. 1978. Feeding and mating in the insectivorous Ceratopogonidae (Diptera). *Mem. Entomol. Soc. Canada* 104:1–62.

Downes, J. A., and S. M. Smith. 1969. New or little known feeding habits in Empididae (Diptera). *Can. Entomol.* 101:404–408.

Downes, W. L. 1963. A reinterpretation of certain head structures in calyptrate Diptera. *Proc. Entomol. Soc. Wash.* 65:293–298.

Downes, W. L., and G. A. Dahlem. 1987. Keys to the evolution of Diptera: Role of Homoptera. *Environ. Entomol.* 16:847–854.

Dreger-Jauffret, F., and J. D. Shorthouse. 1992. Diversity of gall-inducing insects and their galls. In J. D. Shorthouse and O. Rohfritsch, eds., *Biology of Insect-Induced Galls,* pp. 8–33. New York: Oxford University Press.

Dusa, L. 1968. Contributi la studiul aparatului bucal la Bombyliidae *Stud. Univ. Babes-Bolyai, Biol.* 1:79–83.

Ebel, B. H., T. H. Flavell, L. E. Drake, H. O. Yates III, and G. L. DeBarr. 1975. Southern pine seed and cone insects. U.S. Department of Agriculture General Technical Report SE-8, pp. 1–42. Washington, D.C.: U.S. Government Printing Office.

Eggleton, P., and R. Belshaw. 1992. Insect parasitoids: An evolutionary overview. *Phil. Trans. R. Entomol. Soc. Lond. B* 337:1–20.

Elvers, I. 1980. Pollen eating *Thricops* flies (Diptera, Muscidae) on *Arrhenantherum pubescens* and some other grasses. *Bot. Not.* 133:49–52.

Elzinga, R. J., and A. B. Broce. 1986. Labellar modifications of Muscomorpha flies (Diptera). *Ann. Entomol. Soc. Am.* 79:150–209.

Endress, P. K. 1994. *Diversity and Evolutionary Biology of Tropical Flowers.* Cambridge: Cambridge University Press.

Endress, P. K. 2001. The flowers in extant basal angiosperms and inferences on ancestral flowers. *Int. J. Plant Sci.* 162:1111–1140.

Endress, P. K., and A. Igersheim. 2000. The reproductive structures of the basal angiosperm *Amborella trichopoda* (Amborellaceae). *Int. J. Plant Sci.* 161(6 Suppl):S237–S248.

Evenhuis, N. L., ed. 1989. Catalog of the Diptera of the Australasian and Oceanian regions. *Bishop Mus. Spec. Publ.* 86:1–1152.

Evenhuis, N. L. 1994. *Catalog of the Fossil Flies of the World (Insecta: Diptera).* Leiden: Backhuys.

Faegri, K., and L. van der Pijl. 1980. *The Principles of Pollination Ecology,* 3d rev. ed. Oxford: Pergamon.

Fahn, A. 1979. *Secretory Tissues in Plants.* London: Academic Press.

Falder, A. B., G. W. Rothwell, G. Mapes, R. H. Mapes, and L. R. Doguzhaeva. 1998. *Pityostrobus milleri* sp. nov., a pinaceous cone from the Lower Cretaceous (Aptian) of southwestern Russia. *Rev. Paleobot. Palynol.* 103:253–261.

Fang, Q. Q., S. McKeever, and F. E. French. 1999. Cladistic analysis of tabanids (Diptera: Tabanidae) using microscopic characters of the mouthparts. *Mem. Entomol. Int.* 14:355–366.

Farrell, B. 1998. "Inordinate fondness" explained: Why are there so many beetles? *Science* 281:555–559.

Feil, J. P. 1992. Reproductive ecology of dioecious *Siparuna* (Monimiaceae) in Ecuador—A case of gall midge pollination. *Bot. J. Linn. Soc.* 110:171–203.

Foster, W. A. 1995. Mosquito sugar feeding and reproductive energetics. *Annu. Rev. Entomol.* 40:443–474.

Fraser, N. B., D. A. Grimaldi, P. C. Olsen, and B. Axsmith. 1996. A Triassic Lagerstätte from eastern North America. *Nature* 380:615–620.

Frey, R. 1913. Über die Mundteile der Mycetophiden, Sciariden, und Cecidomyiiden. *Acta Soc. Fauna Flora Fenn.* 37:1–53.

Friis, E. M., K. R. Pedersen, and P. R. Crane. 2000. Reproductive structure and organization of basal angiosperms from the Early Cretaceous (Barremian or Aptian) of western Portugal. *Int. J. Plant Sci.* 161(6 Suppl.):S169–S182.

Frohne, W. C. 1939. Biology of certain subaquatic flies reared from emergent water plants. *Pap. Mich. Acad. Sci. Arts Lett.* 24:139–147.

Gagné, R. J. 1968. Revision of the gall midges of bald cypress (Diptera: Cecidomyiidae) *Entomol. News* 79:269–274.

Gagné, R. J. 1994. *The Gall Midges of the Neotropical Region.* Ithaca, N.Y.: Cornell University Press.

Gagné, R. J., and P. E. Boldt. 1989. A new species of *Neolasiptera* (Diptera: Cecidomyiidae) from *Baccharis* (Asteraceae) in southern United States and the Dominican Republic. *Proc. Entomol. Soc. Wash.* 91: 169–174.

Gandolfo, M. A., K. C. Nixon, and W. L. Crepet. 1998. *Tylerianthus crossmanensis* gen. et sp. nov. (aff. Hydrangeaceae) from the Upper Cretaceous of New Jersey. *Am. J. Bot.* 85:376–386.

Gaston, K. J., and P. H. Williams. 1993. Mapping the world's species—The higher taxon approach. *Biodiv. Lett.* 1:2–8.

Gifford, E. M., and A. S. Foster. 1989. *Morphology and Evolution of Vascular Plants,* 3d ed. New York: W. H. Freeman.

Gilbert, F. S. 1981. Foraging ecology of hoverflies: Morphology of the mouthparts in relation to feeding on nectar and pollen in some common urban species. *Ecol. Entomol.* 6:245–262.

Gilbert, F. S., G. Rotheray, P. Emerson, and R. Zafar. 1994. The evolution of feeding strategies. In P. Eggleton and R. I. Vane-Wright, eds., *Phylogenetics and Ecology,* pp. 323–343. London: Academic Press.

Gilbert, L. E. 1972. Pollen feeding and reproductive biology. *Proc. Natl. Acad. Sci. USA* 69:1403–1407.

Goeden, R. D., D. H. Headrick, and J. A. Teerink. 1995. Life history and description of immature stages of *Urophora timberlakei* Blanc and Foote (Diptera: Tephritidae) on native Asteraceae in southern California. *Proc. Entomol. Soc. Wash.* 97:779–790.

Goldblatt, P., and J. C. Manning. 2000. The long-proboscid fly pollination system in southern Africa. *Ann. Missouri Bot. Gard.* 87:146–170.

Gorelick, R. 2001. Did insect pollination cause increased seed plant diversity. *Biol. J. Linn. Soc.* 74:407–427.

Gouin, F. J. 1959. Morphology of the larval head of some Chironomidae (Diptera, Nematocera). *Smithson. Misc. Coll.* 137:175–201.

Gradstein, F., and J. Ogg. 1996. A Phanerozoic time scale. *Episodes* 19:3–5.

Graham-Smith, G. S. 1930. Further observations on the anatomy and function of the proboscis of the blowfly, *Calliphora erthrocephala* L. *Parasitology* 22:47–115.

Grant, V., and K. A. Grant. 1965. *Flower Pollination in the Phlox Family.* New York: Columbia University Press.

Grauvogel-Stamm, L., and K. P. Kelber. 1996. Plant-insect interactions and coevolution during the Triassic in western Europe. *Paleont. Lomb.* 5:5–23.

Grebenshchikova, V. P., and F. V. Naumov. 1985. Ecological features of Diptera in communities of cone-inhabiting insects. In O. A. Skarlato, ed., *Systematics of Diptera (Insecta)—Ecological and Morphological Principles,* pp. 18–22. New Delhi: Amerind Publishing.

Grensted, L. W. 1947. Diptera in the spathes of *Arum maculatum. Entomol. Mon. Mag.* 83:1–3.

Grimaldi, D. 1988. Bee flies and bluets: *Bombylius* (Diptera: Bombyliidae) flower-constant on the distylous species, *Hedyotis caerulea* (Rubiaceae), and the manner of foraging. *J. Nat. Hist.* 22:1–10.

Grimaldi, D. 1999. The co-radiations of pollinating insects and angiosperms in the Cretaceous. *Ann. Missouri Bot. Gard.* 86:373–406.

Grimaldi, D., and J. Cumming. 1999. Brachyceran Diptera in Cretaceous ambers and Mesozoic diversification of the Eremoneura. *Bull. Am. Mus. Nat. Hist.* 239:1–124.

Grinfel'd, E. K. 1955. [Diptera feeding on nectar and pollen and their role in the pollination of plants.] [*J. Leningrad Univ.*] 10:15–25. [In Russian]

Guo, S.-X., and X.-W. Wu. 2000. *Ephedrites* from latest Jurassic Yixian Formation in western Liaoning, northeast China. *Acta Paleont. Sinica* 39:81–91. [In Chinese and English]

Hackman, W., and M. Meinander. 1979. Diptera feeding as larvae on macrofungi in Finland. *Ann. Zool. Fenn.* 16:50–83.

Hagerup, O. 1951. Pollination in the Faroes—In spite of rain and poverty in insects. *Biol. Meddel.* 18:1–48.

Hall, D. W., and B. V. Brown. 1993. Pollination of *Aristolochia littoralis* (Aristolochiales: Aristolochiaceae) by males of *Megaselia* spp. (Diptera: Phoridae). *Ann. Entomol. Soc. Am.* 86:609–613.

Hammer, O. 1941. Biological and ecological investigations of flies associated with pasturing cattle and their excrement. *Vidensk. Medd. Dansk Naturhist. For.* 105:141–393.

Hanson, J. B., and D. M. Benjamin. 1967. Biology of *Phytobia setosa,* a cambium miner of sugar maple. *J. Econ. Entomol.* 60:1351–1355.

Haslett, J. R. 1989. Adult feeding by holometabolous insects: Pollen and nectar as complementary nutrient sources for *Rhingia campestris* (Diptera Syrphidae). *Oecologia* 81:361–363.

Heiss, E. M. 1938. A classification of the larvae and puparia of the Syrphidae of Illinois, exclusive of aquatic forms. *Ill. Biol. Monogr.* 16:1–142.

Hennig, W. 1965. Die Acalyptratae des baltischen Bernsteins. *Stutt. Beit. Naturk.* 145:1–215.

Hering, E. M. 1923. Minien-studien III. *Deut. Entomol. Z.*, N.F. 1923:188–206.

Hering, E. M. 1927. Die Minenfauna der canarischen Inseln. *Zool. Jb.* 53:405–486.

Hering, E. M. 1930. Eine Agromyziden-Mine aus dem Tertiär (Dipt. Agromyz.). *Deutsche Entomol. Z.* 4:63–65.

Hering, E. M. 1951. *Biology of the Leaf Miners.* S'Gravenhage, Netherlands: W. Junk.

Hering, E. M. 1962. Neue Blattminen-Studien II. *Deut. Entomol. Z.*, N.F. 9:30–65.

Hering, E. M. 1967. Blattminen der Insel Hvar (Col., Dipt., Hym., Lep.). *Deut. Entomol. Z.*, N.F. 14:1–80.

Hespenheide, H. A. 1991. Bionomics of leaf-mining insects. *Annu. Rev. Entomol.* 36:535–560.

Heyden, C. von. 1862. Gliederthiere aus der Braunkohle des Niederrheins, der Wetterau und der Röhn. *Palaeontographica* 10:62–82.

Holloway, B. A. 1976. Pollen-feeding in hover-flies (Diptera: Syrphidae). *New Zealand J. Zool.* 3:339–350.

Houard, C. 1903. Recherches anatomiques sur les galles de tiges: Pleurocécidies. *Bull. Sci. France Belgique* 38:140–419.

Hoyt, C. P. 1952. The evolution of the mouth parts of adult Diptera. *Microentomology* 17:61–125.

Hull, F. M. 1973. *Bee Flies of the World: The Genera of the Family Bombyliidae.* Washington, D.C.: Smithsonian Institution Press.

Hurtado-Mejía, J. G., and R. Vélez-Angel. 1999. Mosquitos (Diptera, Ceratopogonidae) asociados al cacao (*Theobroma cacao* L.) en Fredonia y Támesis (Antoquia, Colombia). *Rev. Brasil. Entomol.* 43:99–104.

Inoue, T., O. Kitade, T. Yoshimura, and I. Yamaoka. 2000. Symbiotic associations with protists. In T. Abe, D. E. Bignell, and M. Higashi, eds., *Termites: Evolution, Sociality, Symbioses, Ecology,* pp. 275–288. Dordrecht: Kluwer.

Ionescu, M. A., and P. Neasçu. 1969. *Phytomyza taraxacocecis* Hrtg., specie de Dipter galicol, noua in fauna romaniei. *Ann. Univ. Bucuresti* 18:77–81.

Janzen, D. H. 1971. Seed predation by animals. *Annu. Rev. Ecol. Syst.* 2:465–492.

Jarzembowski, E. A. 1990. A boring beetle from the Wealden of the Weald. In A. J. Boucot, ed., *Evolutionary Paleobiology of Behavior and Coevolution,* pp. 373–376. Amsterdam: Elsevier.

Jarzembowski, E. A., and A. J. Ross. 1996. Insect origination and extinction in the Phanerozoic. In M. B. Hart, ed., *Biotic Recovery from Mass Extinction Events. Geol. Soc. Spec. Publ.* 102:65–78.

Jobling, B. 1933. A revision of the structure of the head, mouthparts and salivary glands of *Glossina palpalis* Rob.-Desv. *Parasitology* 24:449–490.

Johnson, S. D., and K. E. Steiner. 1997. Long-tongued fly pollination and evolution of floral spur length in the *Disa draconis* complex (Orchidaceae). *Evolution* 51:45–53.

Johnson, S. D., and K. E. Steiner. 2000. Generalization versus specialization in plant pollination systems. *Trends Ecol. Evol.* 15:140–143.

Kalugina, N. S., and V. G. Kovalev. 1985. *Dipterous Insects from the Jurassic of Siberia.* Moscow: Academy of Sciences.

Kato, M., and T. Inoue. 1994. Origin of insect pollination. *Nature* 368:195.

Keen, F. P. 1958. Cone and seed insects of western forest trees. *U.S. Dept. Agricult. Tech. Bull.* 1169:1–168.

Keiper, J. B., D. L. Deonier, J. Jiannino, M. Sanford, and W. E. Walton. 2002. Biology, immature stages, and redescriptions of *Hydrellia personata* Deonier (Diptera: Ephydridae), a *Lemna* miner. *Proc. Entomol. Soc. Wash.* 104:458–467.

Kevan, P. G., and H. G. Baker. 1983. Insects as flower visitors and pollinators. *Annu. Rev. Entomol.* 28: 407–453.

Kirichkova, A. I., and M. P. Doludenko. 1996. [New data on phytostratigraphy of the Jurassic deposits of Kazakhstan.] *Stratigr. Geol. Korrelyatsiya* 4:35–52. [In Russian]

Kirk, W. D. J. 1984. Pollen feeding in thrips (Insecta: Thysanoptera). *J. Zool.* 204:107–117.

Kite, G. C., W. L. A. Hetterscheid, M. J. Lewis, P. C. Boyce, J. Ollerton, E. Cocklin, A. Diaz, and M. S. J. Simmonds. 1998. Inflorescence odours and pollinators of *Arum* and *Amorphophallus* (Araceae). In S. J. Owens and P. J. Rudall, eds., *Reproductive Biology,* pp. 295–315. Kew: Royal Botanic Gardens.

Knab, F. 1910. The feeding habits of *Elephantomyia* (Diptera: Tipulidae). *Proc. Entomol. Soc. Wash.* 12:61–66.

Kneipert, F.-W. 1980. Blood-feeding and nectar-feeding in adult Tabanidae (Diptera). *Oecologia* 46:125–129.

Knight, G. H. 1968. Observations on the behaviour of *Bombylius major* L. and *B. discolor* Mik. (Dipt., Bombyliidae) in the Midlands. *Entomol. Mongr. Mag.* 103:177–181.

Knoll, E. 1926. Die *Arum*-Blütenstände und ihre Besucher. *Abhand. Zool.-Bot. Ges. Wien* 12:379–481.

Koach, J., and J. Galil. 1986. The breeding system of *Arisarum vulgare* Targ.-Tozz. *Israel J. Bot.* 35:79–90.

Kolomoets, T. P., B. M. Mamaev, M. L. Zerova, E. P. Narchuk, V. M. Ermolenko, and L. A. Diakonchuk. 1989. [*Insect Gall Makers—Diptera.*] Kiev: Izdatelstvo Naukova Dumka. [In Russian]

Kramer, S. 1954. The musculature of the head of the phantom crane fly larva *Bittacomorpha* (Diptera Ptychopteridae). *J. Morphol.* 94:409–437.

Krassilov, V. A., and A. P. Rasnitsyn. 1996. Pollen in the guts of Permian insects: First evidence of pollinivory and its evolutionary significance. *Lethaia* 29:369–372.

Krassilov, V. A., V. V. Zherikhin, and A. P. Rasnitsyn. 1997. *Classopollis* in the guts of Jurassic insects. *Palaeontology* 40:1095–1101.

Krivosheina, N. P. 1969. *Ontogenesis and Evolution of the Two-Winged Insects.* Washington, D.C.: U.S. Department of Agriculture and National Science Foundation. [Translated from Russian]

Krivosheina, N. P., and B. M. Mamaev. 1962. Larvae of the European species of the genus *Temnostoma* (Diptera, Syrphidae). *Entomol. Rev.* 41:570–575.

Krzeminski, W. 1992. Triassic and Lower Jurassic stage of Diptera evolution. *Mitt. Schweiz. Entomol. Ges.* 65:39–59.

Krzeminski, W., and N. L. Evenhuis. 2000. Review of Diptera palaeontological records. In L. Papp and B. Darvas, eds., *Contributions to a Manual of Palaearctic Diptera,* vol. 1, pp. 535–564. Budapest: Science Herald.

Kugler, H. 1938. Sind *Veronica chamaedrys* L. und *Circaea lutetiana* L. Schwebfliegenblumen? *Bot. Archiv* 39:147–165.

Kugler, H. 1955. Zum Problem der Dipterenblumen. *Öster. Bot. Z.* 102:529–541.

Kukalová-Peck, J. 1987. New Carboniferous Diplura, Monura, Thysanura, the hexapod ground plan, and the role of thoracic side lobes in the origin of wings (Insecta). *Canad. J. Zool.* 65:2327–2345.

Kvacek, J. 2000. *Frenelopsis alata* and its microsporangiate and ovuliferous reproductive structures from the Cenomanian of Bohemia (Czech Republic, Central Europe). *Rev. Palaeobot. Palynol.* 112:51–78.

Labandeira, C. C. 1997. Insect mouthparts: Ascertaining the paleobiology of insect feeding strategies. *Annu. Rev. Ecol. Syst.* 28:153–193.

Labandeira, C. C. 1998a. Early history of arthropod and vascular plant associations. *Annu. Rev. Earth Planet. Sci.* 26:329–377.

Labandeira, C. C. 1998b. How old is the flower and the fly? *Science* 280:57–59.

Labandeira, C. C. 1998c. The role of insects in Late Jurassic to Middle Cretaceous ecosystems. In S. G. Lucas, J. I. Kirkland, and J. W. Estep, eds., *Lower and Middle Cretaceous Terrestrial Ecosystems. New Mexico Mus. Nat. Hist. Sci. Bull.* 14:105–124.

Labandeira, C. C. 2000. Paleobiology of pollination and its precursors. In R. A. Gastaldo and W. A. DiMichele, eds., *Phanerozoic Terrestrial Ecosystems. Paleont. Soc. Pap.* 6:233–269.

Labandeira, C. C. 2002a. The history of associations between plants and animals. In C. M. Herrera and O. Pellmyr, eds., *Plant-Animal Interactions: An Evolutionary Approach,* pp. 26–74, 248–261. London: Blackwell Scientific Publishing.

Labandeira, C. C. 2002b. Paleobiology of predators, parasitoids, and parasites: Death and accommodation in the fossil record of continental invertebrates. *Paleontol. Soc. Pap.* 8:211–250.

Labandeira, C. C., and T. L. Phillips. 1996. Insect fluid-feeding on Upper Pennsylvanian tree ferns (Palaeodictyoptera, Marattiales) and the early history of the piercing-and-sucking functional feeding group. *Ann. Entomol. Soc. Am.* 89:157–183.

Labandeira, C. C., and J. J. Sepkoski, Jr. 1993. Insect diversity in the fossil record. *Science* 261:310–315.

Labandeira, C. C., D. L. Dilcher, D. R. Davis, and D. L. Wagner. 1994. Ninety-seven million years of angiosperm-insect association: Paleobiological insights into the meaning of coevolution. *Proc. Natl. Acad. Sci. USA* 91:12278–12282.

Labandeira, C. C., K. R. Johnson, and P. Wilf. 2002. Impact of the terminal Cretaceous event on plant-insect associations. *Proc. Natl. Acad. Sci. USA* 99:2061–2066.

Lack, A. J., and P. G. Kevan. 1984. On the reproductive biology of a canopy tree, *Syzygium sygygioides* (Myrtaceae), in a rain forest in Sulawesi, Indonesia. *Biotropica* 16:31–36.

Lall, S. B., and D. M. Davies. 1971. An intergeneric comparison of cephalic structures in tabanids (Diptera) in relation to feeding habits. *J. Med. Entom.* 8:700–706.

Lane, R. S., and J. R. Anderson. 1983. Multiple autogeny and burrow oviposition by a marine horse fly (Diptera: Tabanidae). *J. Med. Entomol.* 20:212–213.

Lang, P. J., A. C. Scott, and J. Stephenson. 1995. Evidence of plant-arthropod interactions from the Eocene Branksome Sand Formation, Bournemouth, England: Introduction and description of leaf mines. *Tert. Res.* 15:145–174.

Larsen, E. B. 1948. Observations on the activity of some culicids. *Entomol. Meddel.* 25:263–277.

Larson, B. M. H., P. G. Kevan, and D. W. Inouye. 2001. Flies and flowers: Taxonomic diversity of anthophiles and pollinators. *Can. Entomol.* 133:439–465.

Lee, M. S. Y. 1997. Documenting present and past biodiversity: Conservation meets palaeontology. *Trends Ecol. Evol.* 12:132–133.

Leereveld, H. 1982. Anthecological relations between reputedly anemophilous flowers and syrphid flies. III. Worldwide survey of crop and intestine contents of certain anthophilous syrphid flies. *Tijd. Entomol.* 125:25–35.

Lehane, M. J. 1991. *Biology of Blood-Sucking Insects.* London: HarperCollins.

Lendner, A. 1935. Plantes fétides et insectes coprophiles. *Rev. Hort. Suisse* 8:49–54.

Lewis, S. E. 1985. Miocene insects from the Clarkia deposits of northern Idaho. In C. Smiley, ed., *Late Cenozoic History of the Pacific Northwest,* pp. 245–264. Washington, D.C.: American Association for the Advancement of Science.

Li, W.-B., and Z.-S. Liu. 1994. The Cretaceous palynofloras and their bearing on stratigraphic correlation in China. *Cret. Res.* 15:333–365.

Lippok, B., and S. S. Renner. 1997. Pollination of *Nuphar* (Nympheaceae) in Europe: Flies and bees rather than *Donacia* beetles. *Plant Syst. Evol.* 207:273–283.

Liu, N., and A. Nagatomi. 1995. The mouthpart structure of Scenopinidae (Diptera). *Japan. J. Entomol.* 63:181–202.

Lloyd, D. G., and M. S. Wells. 1992. Reproductive biology of a primitive angiosperm, *Pseudowintera colorata* (Winteraceae), and the evolution of pollination systems in the Anthophyta. *Pl. Syst. Evol.* 181:77–95.

Luo, Z. 1999. A refugium for relicts. *Nature* 400:23–24.

Manning, J. C., and P. Goldblatt. 1997. The *Moegistorhynchus longirostris* (Diptera: Nemestrinidae) pollination guild: Long-tubed flowers and a specialized long-proboscid fly pollination system in southern Africa. *Plant Syst. Evol.* 206:51–69.

Matile, L. 1997. Phylogeny and evolution of the larval diet in the Sciaroidea (Diptera, Bibionomorpha) since the Mesozoic. In P. Grandcolas, ed., *The Origin of Biodiversity in Insects: Phylogenetic Tests of Evolutionary Scenarios. Mem. Natl. Mus. Nat. Hist.* 173:273–303.

Matsuda, R. 1965. Morpohology and evolution of the insect head. *Mem. Am. Entomol. Inst.* 4:1–334.

May, E. M. 1979. Adaptations of crane fly mouthparts for nectar feeding. *J. Kansas Entomol. Soc.* 52:564.

Mazzarolo, L. A., and D. S. Amorim. 2000. *Cratomyia macrorrhyncha,* a Lower Cretaceous brachyceran fossil from the Santana Formation, Brazil, representing a new species, genus and family of the Stratiomyomorpha (Diptera). *Insect Syst. Evol.* 31:91–102.

McAlpine, J. F. 1956. Cone-feeding lonchaeids of the genus *Earomyia* Zett., with descriptions of five new species from western North America (Diptera: Lonchaeidae). *Can. Entomol.* 88:178–196.

McAlpine, J. F. 1970. First record of calypterate flies in the Mesozoic Era (Diptera; Calliphoridae). *Can. Entomol.* 102:342–346.

McAlpine, J. F., B. V. Peterson, G. E. Shewell, H. J. Teskey, J. R. Vockeroth, and D. M. Wood, eds. 1981. *Manual of Nearctic Diptera,* vol. 1. *Res. Branch Agric. Can. Mongr.* 27:1–674.

McAlpine, J. F., B. V. Peterson, G. E. Shewell, H. J. Teskey, J. R. Vockeroth, and D. M. Wood, eds. 1987. *Manual of Nearctic Diptera,* vol. 2. *Res. Branch Agric. Can. Mongr.* 28:1–674.

McKeever, S., and F. E. French. 1999. Comparative study of adult mouthparts of fifty-two species representing thirty genera of Tabanidae (Diptera). *Mem. Entomol. Int.* 14:327–353.

McMillan, J. R. 1958. The role of the micropyle in the pollination of *Pinus. Bot. Gaz.* 120:109–117.

McPheron, B. A., and G. J. Steck, eds. 1996. *Fruit Fly Pests: A World Assessment of Their Biology and Management.* Delray Beach, Fla.: St. Lucie Press.

Meeuse, A. D. J., A. H. de Meijer, O. W. P. Mohr, and S. M. Wellinga. 1990. Entomophily in the dioecious gymnosperm *Ephedra aphylla* Forsk. (=*E. alte* C. A. Mey.), with some notes on *Ephedra campylopoda* C. A. Mey. III. Further anthecological studies and relative importance of entomophily. *Israel J. Bot.* 39:113–123.

Meeuse, B. J. D., and I. Raskin. 1988. Sexual reproduction in the arum lily family, with emphasis on thermogenicity. *Sex Pl. Reprod.* 1:3–5.

Meillon, B. D., and W. W. Wirth. 1989. A new pollen feeding *Atrichopogon* midge from Madagascar, with notes on closely related subsaharan species (Diptera: Ceratopogonidae). *Revue Fran. Entomol.* 11:85–89.

Melin, D. 1923. Contributions to the knowledge of the biology, metamorphosis and distribution of the Swedish asilids in relation to the whole family of asilids. *Zool. Bidr. Uppsala* 8:1–317.

Merritt, R. W., and K. W. Cummins. 1996. *An Introduction to the Aquatic Insects of North America,* 3d ed. Dubuque, Iowa: Kendall-Hunt.

Mesler, M. R., J. A. Ackerman, and K. L. Lu. 1980. The effectiveness of fungus-gnats as pollinators. *Am. J. Bot.* 67:564–567.

Metcalf, C. L. 1929. The mouthparts of insects. *Trans. Ill. Acad. Sci.* 21:109–125.

Meyer, J. 1987. *Plant Galls and Gall Inducers.* Berlin: Gebrüder Borntraeger.

Michelsen, V. 1988. A world revision of *Strobilomyia* gen. n.: The anthomyiid seed pests of conifers (Diptera: Anthomyiidae). *Syst. Entomol.* 13:271–314.

Miki, S. 1937. Plant fossils from the *Stegodon* Beds and the *Elephas* Beds near Akashi. *Jpn. J. Bot.* 8:303–341.

Miller, M., and C. C. Labandeira. 2002. Invertebrates' slow crawl across the salinity divide. *Geol. Soc. Am. Today* 12:4–10.

Mitter, J. L. 1918. Note on the method of feeding of *Corizoneura* (*Pangonia*) *longirostris* Hardwick, with a description of the mouth parts. *Ind. J. Med. Res.* 5:523–528.

Möhn, E. 1960. Eine neue Gallmücke aus der niederrheinschen Braunkohle, *Sequoiomyia kraeuseli* n. g., n. sp. (Diptera, Itonididae). *Senck. Leth.* 41:513–522.

Morge, G., and N. Nanu. 1981. *Earomyia impossibile* Morge und *Earomyia grusia* Morge (Diptera: Lonchaeidae), Schädlinge der Tannenzapfen und-samen (*Abies alba* Mill.) in Rumänien. *Beitr. Entomol.* 31:17–25.

Mostovski, M. B. 1998. A revision of the nemestrinid flies (Diptera, Nemestrinidae) described by Rohdendorf, and a description of new taxa of the Nemestrinidae from the Upper Jurassic of Kazakhstan. *Paleontol. J.* 32:369–375.

Mostovski, M. B., and X. Martínez-Delclòs. 2000. New Nemestrinoidea (Diptera: Brachycera) from the Upper Jurassic–Lower Cretaceous of Eurasia, taxonomy and palaeobiology. *Entomol. Prob.* 31:137–148.

Moussel, B. 1980. Gouttelette réceptrice du pollen et pollinisation chez l'*Ephedra distachya* L. Observations sur le vivant et en microscopies photonique et électronique. *Rev. Cytol. Biol Végét.-Bot.* 3:65–89.

Müller, H. 1883. *The Fertilisation of Flowers.* New York: Arno Press.

Nagatomi, A., and K. Soroida. 1985. The structure of the mouthparts of the orthorrhaphous Brachycera (Diptera) with a special reference to blood-sucking. *Beitr. Entomol.* 35:263–368.

Nagatomi, A., Yang, C.-K., and Yang, D. 1999. The Chinese species and the world genera of Vermileonidae (Diptera). *Tropics* 1:1–154.

Nartshuk, E. P. 1985. Adaptations of Cyclorrhapha larvae (Diptera) for the inhabitation of living plants. In O. A. Skarlato, ed., *Systematics of Diptera (Insecta): Ecological and Morphological Principles*, pp. 97–101. New Delhi: Amerind Publishing.

Neff, S. E., and J. B. Wallace. 1969. Observations on the immature stages of *Cordilura (Achaetella) deceptiva* and *C. (A.) varipes*. *Ann. Entomol. Soc. Am.* 62:775–785.

Nessimian, J. L., A. M. Sanseverino, and A. L. H. de Oliveira. 1999. Relações tróficas de larvas de Chironomidae (Diptera) e sua importância na rede alimentar em um brejo no litoral do Estado do Rio de Janeiro. *Rev. Brasil. Entomol.* 43:47–53.

Nicholson, S. W. 1994. Pollen feeding in the eucalypt nectar fly, *Drosophila flavohirta*. *Physiol. Entomol.* 19:58–60.

Nilsson, L. A. 1992. Orchid pollination biology. *Trends Ecol. Evol.* 7:255–259.

Nilsson, L. E. 1988. The evolution of flowers with deep corolla tubes. *Nature* 334:147–149.

Nishida, H., and N. Hayashi. 1996. Cretaceous coleopteran larva fed on a female fructification of extinct gymnosperm. *J. Plant. Res.* 109:327–330.

Nitzulescu, V. 1927. L'appareil buccal de *Corizoneura schwetzi* et de *C. inornata*. *Rev. Zool. Afr.* 15: 200–213.

Okada, T. 1963. Caenogenetic differentiation of mouth hooks in drosophilid larvae. *Evolution* 17:84–98.

Oldroyd, H. 1964. *The Natural History of Flies*. New York: Norton.

Ollerton, J. 1996a. Reconciling ecological processes with phylogenetic patterns: The apparent paradox of plant-pollinator systems. *J. Ecol.* 84:767–769.

Ollerton, J. 1996b. Interactions between gall midges (Diptera: Cecidomyiidae) and inflorescences of *Piper novae-hollandiae* (Piperaceae) in Australia. *Entomologist* 115:181–184.

Osche, G. 1983. Optische Signale in der Coevolution von Pflanze und Tier. *Ber. Deut. Bot. Ges.* 96:1–27.

Otter, G. W. 1938. On the morphology of the larvae of three species of Cecidomyiidae (Diptera) from knapweed (*Centaurea*) flowers. *Trans. R. Entomol. Soc. Lond.* 87:39–68.

Owens, J. N., G. L. Catalano, S. J. Morris, and J. Aitken-Christie. 1995. The reproductive biology of kauri (*Agathis australis*). 1. Pollination and prefertilization development. *Int. J. Plant Sci.* 156:257–269.

Owens, J. N., T. Takaso, and C. J. Runions. 1998. Pollination in conifers. *Trends Plant Sci.* 13:479–485.

Paramonov, S. J. 1953. A review of Australian Nemestrinidae (Diptera). *Australian J. Zool.* 1:242–290.

Pechuman, L. L. 1943. Notes on the biology of *Chyliza notata* (Diptera: Psilidae). *Bull. Brooklyn Entomol. Soc.* 38:97.

Pellmyr, O. 1992. The phylogeny of a mutualism: Evolution and coadaptation between *Trollius* and its seed-parasitic pollinators. *Biol. J. Linn. Soc.* 47:337–365.

Pellmyr, O., W. Tang, I. Groth, G. Bertström, and L. B. Thien. 1991. Cycad cone and angiosperm floral volatiles: Inferences for the evolution of insect pollination. *Biochem. Syst. Ecol.* 19:623–627.

Peterson, A. 1916. The head-capsule and mouth-parts of Diptera. *Ill. Biol. Mongr.* 3:1–112.

Pettit, R. H. 1900. A leaf-miner, *Cheironomus* sp., in water lilies. *Rept. Mich. Acad. Sci.* 1:109–111.

Pfretzschner, H. U. 1998. Ein weiteres Exemplar von *Lethe? corbieri* Nel 1993 (Lepidoptera, Satyridae) aus dem Unter-Oligozän von Céreste (Süd-Frankreich). *Paläont. Z.* 72:59–64.

Pijl, L. van der. 1953. On the flower biology of some plants from Java, with general remarks on fly-traps (species of *Annona, Artocarpus, Typhonium, Gnetum, Arisaema* and *Abroma*). *Ann. Bogor.* 1:77–99.

Pike, E. M. 1995. *Amber Taphonomy and the Grassy Lake, Alberta, Amber Fauna*. Dissertation. Calgary, Canada: University of Calgary.

Pocock, S. A. J., G. Vasanthy, and B. S. Venkatachala. 1990. Pollen of Circumpolles—An enigma or morphotrends showing evolutionary adaptation? *Rev. Palaeobot. Palynol.* 65:179–193.

Ponomarenko, A. G. 1996. Evolution of continental aquatic ecosystems. *Paleontol. J.* 30:705–709.

Pont, A. C. 1993. Observations on anthophilous Muscidae and other Diptera (Insecta) in Abisko National Park, Sweden. *J. Nat. Hist.* 27:631–643.

Poort, R. J., H. Visscher, and D. L. Dilcher. 1996. Zoidogamy in fossil gymnosperms: The centenary of a concept, with special reference to prepollen of late Paleozoic conifers. *Proc. Natl. Acad. Sci. USA* 93:11713–11717.

Porsch, O. 1958. Alte Insektentypen als Blumenausbeuter. *Öster. Bot. Zeit.* 104:115–163.

Powell, J. A., C. Mitter, and B. Farrell. 1999. Evolution of larval food preferences in Lepidoptera. In N. P. Kristensen, ed., *Lepidoptera, Moths and Butterflies, vol. 1: Evolution, Systematics, and Biogeography,* pp. 403–422. *Handbuch der Zoologie.* Berlin: Walter de Gruyter.

Prévost, Y. H. 1990. Spruce cone axis midge, *Dasineura rachiphaga* Tripp (Diptera: Cecidomyiidae), in cones of black spruce, *Picea mariana* (Mill.) B.S.P. *Can. Entomol.* 122:441–447.

Proctor, M., P. Yeo, and A. Lack. 1996. *The Natural History of Pollination.* Portland, Ore.: Timber Press.

Pschorn-Walcher, H., and E. Altenhofer. 1989. The parasitoid community of leaf-mining sawflies (Fenusini and Heterarthrini): A comparative analysis. *Zool. Anz.* 222:37–56.

Rasnitsyn, A. P., and V. A. Krassilov. 1996. Pollen in the gut contents of fossil insects as evidence of co-evolution. *Paleontol. J.* 30:716–722.

Rasnitsyn, A. P., and A. J. Ross. 2000. A preliminary list of arthropod families present in the Burmese amber collection at the Natural History Museum, London. *Bull. Nat. Hist. Mus. Lond. (Geol.)* 56:21–24.

Rayner, R. J., and S. B. Waters. 1991. Floral sex and the fossil insect. *Naturwissenschaften* 78:280–282.

Regal, P. J. 1977. Ecology and evolution of flowering plant dominance. *Science* 196:622–629.

Ren, D. 1998. Flower-associated Brachycera flies as fossil evidence for Jurassic angiosperm origins. *Science* 280:85–88.

Ridley, H. N. 1890. On the method of fertilization in *Bulbophyllum macranthum,* and allied orchids. *Ann. Bot.* 4:327–336.

Roberts, M. J. 1969. Structure of the mouthparts of the larvae of the flies *Rhagio* and *Sargus* in relation to feeding habits. *J. Zool.* 159:381–398.

Roberts, M. J. 1971. The structure of the mouthparts of some calypterate dipteran larvae in relation to their feeding habits. *Acta Zool.* 12:171–188.

Rohdendorf, B. B. 1968. New Mesozoic nemestrinids (Diptera, Nemestrinidae). In B. B. Rohdendorf, ed., *Jurassic Insects of Karatau,* pp. 180–189. Moscow: Izdatelstvo Nauka.

Root, R. B. 1967. The niche exploitation pattern of the blue-gray gnatcatcher. *Ecol. Mongr.* 37:317–350.

Root, R. B. 1973. Organization of a plant-arthropod association in simple and diverse habitats: The fauna of collards (*Brassica oleracea*). *Ecol. Mongr.* 67:95–124.

Roskam, J. C. 1992. Evolution of the gall-inducing guild. In J. D. Shorthouse and O. Rohfritsch, eds., *Biology of Insect-Induced Galls,* pp. 34–49. New York: Oxford University Press.

Rotheray, G. E. 1988. Morphology and feeding behaviour of the leaf-mining larva of *Cheilosia semifasciata* (Diptera: Syrphidae). *J. Nat. Hist.* 22:865–873.

Rothwell, G. W. 1977. Evidence for a pollination-drop mechanism in Paleozoic pteridosperms. *Science* 198:1251–1252.

Rozefelds, A. C. 1988. Lepidoptera mines in *Pachypteris* leaves (Corystospermaceae: Pteridospermophyta) from the Upper Jurassic/Lower Cretaceous Battle Camp Formation, north Queensland. *Proc. R. Soc. Queensland* 99:77–81.

Sargent, O. H. 1934. Pollination in *Pterostylis. Vict. Nat.* 51:82–84.

Sasakawa, M. 1997. New leaf miner and stem borer of Sciaridae (Diptera). *Japan. J. Entomol.* 65:171–176.

Saunders, L. G. 1959. Methods for studying *Forcipomyia* midges, with special reference to cacao-pollinating species (Diptera, Ceratopogonidae). *Can. J. Zool.* 37:33–51.

Scheffer, S. J., and B. M. Wiegmann. 2000. Molecular phylogenetics of the holly leafminers (Diptera: Agromyzidae: *Phytomyza*): Species limits, speciation, and dietary specialization. *Mol. Phylo. Evol.* 17: 244–255.

Scheirs, J., I. Vandevyvere, and L. De Bruyn. 1997. Influence of monocotyl leaf anatomy on the feeding pattern of a grass-mining agromyzid (Diptera). *Ann. Entomol. Soc. Am.* 90:646–654.

Schick, B. 1982. Zur Morphologie, Entwicklung, Feinstruktur und Funktion des Translators von *Periploca* L. (Asclepiadaceae). *Trop. Subtrop. Pflanz.* 40:507–553.

Schiemenz, H. 1957. Vergleichende funktionellanatomische Untersuchungen der Kopfmuskulatur von *Theobaldia* und *Eristalis* (Dipt. Culicid und Syrphid.). *Deut. Entomol. Z.* 4:268–331.

Schlee, D. 1977. Florale und extraflorale Nektarien sowie Insektenkot als Nahrungsquelle für Chironomidae-Imagines (und andere Diptera). *Stutt. Beit. Naturk. A* 300:1–16.

Schremmer, F. 1951. Die Mundteile der Brachycerenlarven und der Kopfbau de Larve von *Stratiomys chamaeleon* L. *Öster. Zool. Z.* 3:326–397.

Schuhmacher, H., and H. Hoffmann. 1982. Zur Funktion der Mundwerkzeuge von Schwebfliegen bei Nahrungsaufnahme (Diptera: Syrphidae). *Entomol. Gen.* 7:327–342.

Séguy, E. 1950. *La Biologie des Diptères.* Paris: Paul Lechevalier.

Séguy, E. 1957. Note sur l'anthophilie des diptères hématophages. *Rev. Fran. Entomol.* 24:380–395.

Simberloff, D., and T. Dayan. 1991. The guild concept and the structure of ecological communities. *Annu. Rev. Ecol. Syst.* 22:115–143.

Sinclair, B. J. 1992. A phylogenetic interpretation of the Brachycera (Diptera) based on the larval mandible and associated mouthpart structures. *Syst. Entomol.* 17:233–252.

Sinichenkova, N. D. 2001. The main ecological events in aquatic insects history. *Abstracts of the 2nd International Congress on Palaeoentomology,* pp. 67–68. Kraków: Polish Academy of Sciences.

Sinichenkova, N. D., and V. V. Zherikhin. 1996. Mesozoic lacustrine biota: Extinction and persistence of communities. *Paleont. J.* 30:710–715.

Sittig, W. 1927. Blattminierende Insektenlarven. *Natur Mus.* 57:348–350.

Skevington, J. H., and P. T. Dang, eds. 2002. Exploring the diversity of flies (Diptera). *Biodiversity* 3(4):3–27.

Slansky, F., Jr., and J. G. Rodriguez. 1987. *Nutritional Ecology of Insects, Mites, Spiders, and Related Invertebrates.* New York: John Wiley.

Smart, J. 1943. *Simulium* feeding on ivy flowers. *Entomologist* 76:20–21.

Smith, K. G. V. 1967. The biology and taxonomy of the genus *Stylogaster* Macquart, 1835 (Diptera: Conopidae, Stylogasterinae) in the Ethiopian and Malagasy regions. *Trans. R. Entomol. Soc. Lond.* 119: 47–69.

Smith, P. E., N. M. Evensen, D. York, M.-M. Chang, F. Jin, J.-L. Li, S. Cumba, and D. Russell. 1995. Dates and rates in ancient lakes: ^{40}Ar-^{39}Ar evidence for an Early Cretaceous age from the Jehol Group, northeast China. *Can. J. Earth Sci.* 32:1426–1431.

Solomon, J. D. 1995. *Guide to Insect Borers in North American Broadleaf Trees and Shrubs.* U.S. Department of Agriculture Forest Service Agricultural Handbook AH-706, pp. 1–735. Washington, D.C.: U.S. Government Printing Office.

Sonderstrom, R. T., and C. E. Calderón. 1971. Insect pollination in tropical rain forest grasses. *Biotropica* 3:1–16.

Spencer, K. A. 1990. *Host Specialization in the World Agromyzidae (Diptera).* Dordrecht: Kluwer.

Srivastava, A. K., S. R. Abbas, R. C. Mehrotra, and R. Srivastava. 2000. Cecidomyiid leaf galls in Palaeocene leaves from north-eastern India. *Acta Palaeobot.* 40:43–47.

Ssymank, A., and F. Gilbert. 1993. Anemophilous pollen in the diet of syrphid flies with special reference to the leaf feeding strategy occurring in Xylotini (Diptera, Syrphidae). *Deut. Entomol. Z.* 40:245–258.

Steffan, W. A. 1981. Sciaridae. In J. F. McAlpine, B. V. Peterson, G. E. Shewell, H. J. Teskey, J. R. Vockeroth, and D. M. Wood, eds., *Manual of Nearctic Diptera,* vol. 1. *Res. Branch Agric. Can. Mongr.* 27:247–255.

Stelleman, P. 1981. Anthecological relations between reputedly anemophilous flowers and syrphid flies. V. Some special aspects of the visiting of *Plantago media* and *P. lanceolata* by insects. *Beitr. Biol. Pflanzen* 55:157–167.

Straus, A. 1977. Gallen, minen und andere Fraβspuren im Pliokän von Willershausen am Harz. *Verhand. Bot. Ver. Prov. Brandenburg* 113:43–80.

Strong, D. R., J. H. Lawton, and T. R. E. Southwood. 1984. *Insects on Plants: Community Patterns and Mechanisms.* Cambridge, Mass.: Harvard University Press.

Stuckenberg, B. R. 1996. A revised generic classification of the wormlion flies of Southern Africa previously placed in *Lampromyia* Macquart, with reinstatement of *Leptynoma* Westwood 1876, and descriptions of a new subgenus and two new species (Diptera, Vermileonidae). *Ann. Natal Mus.* 37:239–266.

Stuckenberg, B. R. 1999. Antennal evolution in the Brachycera (Diptera), with a reassessment of terminology relating to the flagellum. *Studia Dipt.* 6:33–48.

Stuckenberg, B. R. 2000. *Namamyia,* a new genus of Vermileonidae (Diptera) from Namaqualand, with a conspectus of the southern African genera and an account of the adaptations for anthophily. *Ann. Natal Mus.* 41:181–202.

Stuckenberg, B. R. 2001. Pruning the tree: A critical review of classifications of the Homeodactyla (Diptera, Brachycera), with new perspectives and an alternative classification. *Studia Dipt.* 8:3–41.

Sun, G., Q. Ji, D. L. Dilcher, S. Zheng, K. C. Nixon, and X. Wang. 2002. Archaefructaceae, a new basal angiosperm family. *Science* 296:899–904.

Süss, H. 1979. Durch *Protophytobia cupressorum* gen. nov. sp. nov. (Agromyzidae, Diptera) verursachte Markflecken in einem Holz von *Juniperoxylon* aus dem Tertiär von Süd-Limburg (Niederlande) und der Nachweis von Markflecken in einer rezenten *Callitris*-Art. *Feddes Repertorium* 90:165–172.

Süss, H. 1980. Fossile Kambium-Minierer der familie Agromyzidae (Diptera) in tertiären Laub- und Nadelholzresten. *Z. Geol. Wiss.* 8:1217–1225.

Süss, H., and W. R. Müller-Stoll. 1975. Durch *Palaeophytobia platani* n.g. n.sp. (Agromyzidae, Diptera) verursachte Markflecken im Holz fossiler Platanen aus dem ungarischen Miozän. *Wissenschaft. Z. Humboldt-Universität Berlin (Mathemat.-Naturwiss. Reihe)* 24:515–519.

Süss, H., and W. R. Müller-Stoll. 1980. Das fossile Holz *Pruninium gummosum* Platen emend. Süss u. Müller-Stoll aus dem Yellowstone Nationalpark und sein Parasit *Palaeophytobia prunorum* sp. nov. nebst Bemerkungen über Markflecken. In W. Vent, ed., *Jubiläumsschrift 100 Jahre Arboretum Berlin,* pp. 343–364. Berlin: Humboldt Universität.

Swezey, O. H. 1915. A leaf mining cranefly in Hawaii. *Proc. Hawaiian Entomol. Soc.* 3:87–89.

Swisher, C. C. III, Y.-Q Wang, X.-L. Wang, X. Xu, and Y. Wang. 1999. Cretaceous age for the feathered dinosaurs of Liaoning, China. *Nature* 400:58–61.

Szadziewski, R. 1988. Biting midges (Diptera, Ceratopogonidae) from Baltic amber. *Pol. Pis. Entomol.* 58: 3–283.

Szucsich, N. U., and H. W. Krenn. 2000. Morphology and function of the proboscis in Bombyliidae (Diptera, Brachycera) and implications for proboscis evolution in Brachycera. *Zoomorphology* 120:79–90.

Terry, I. 2001. Thrips and weevils as dual, specialist pollinators of the Australian cycad *Macrozamia communis* (Zamiaceae). *Int. J. Plant Sci.* 162:1293–1305.

Teskey, H. J. 1976. Diptera larvae associated with trees in North America. *Mem. Entomol. Soc. Can.* 100: 1–53.

Teskey, H. J. 1981. Morphology and terminology—Larvae. In J. F. McAlpine, B. V. Peterson, G. E. Shewell, H. J. Teskey, J. R. Vockeroth, and D. M. Wood, eds., *Manual of Nearctic Diptera,* vol. 1. *Res. Branch Agric. Can. Mongr.* 27:65–88.

Tetley, H. 1918. The structure of the mouth-parts of *Pangonia longirostris* in relation to the probable feeding-habits of the species. *Bull. Entomol. Res.* 8:253–267.

Thien, L. B. 1969. Mosquitoes can pollinate orchids. *Morris Arbor. Bull.* 20:19–23.

Thien, L. B. 1982. Fly pollination in *Drimys* (Winteraceae), a primitive angiosperm. *Mongr. Biol.* 42:529–533.

Thien, L. B., D. A. White, and L. Y. Yatsu. 1983. The reproductive biology of a relict—*Illicium floridanum* Ellis. *Am. J. Bot.* 70:719–727.

Thien, L. B., H. Azuma, and S. Kawano. 2000. New perspectives on the pollination biology of basal angiosperms. *Int. J. Plant Sci.* 161:S225–S235.

Thomas, I. 1938. On the bionomics and structure of some dipterous larvae infesting cereals and grasses. III. *Geomyza* (*Balioptera*) *tripunctata* Fall. *Ann. Appl. Biol.* 25:181–196.

Thorpe, W. H. 1934. Observations on the structure, biology and systematic position of *Pantophthalmus tabaninus* Thunb. (Diptera Pantophthalmidae). *Trans. R. Entomol. Soc. Lond.* 82:5–22.

Tomlinson, P. B., J. E. Braggins, and J. A. Rattenbury. 1991. Pollination drop in relation to cone morphology in Podocarpaceae: A novel reproductive mechanism. *Am. J. Bot.* 78:1289–1303.

Totland, Ø. 1993. Pollination in alpine Norway: Flowering phenology, insect visitors, and visitation rates in two plant communities. *Can. J. Bot.* 71:1072–1079.

Tsacas, L. 1963. Recherches sur la structure et le fonctionnement de la tête et des pièces buccales larvaires des Rhagionidae (Diptères). *Mém. Mus. Natl. Hist. Nat. Zool.* 27:147–236.

Turgeon, J. J., A. Roques, and P. de Groot. 1994. Insect fauna of coniferous seed cones: Diversity, host plant interactions, and management. *Annu. Rev. Entomol.* 39:179–212.

Valley, K., J. A. Novak, and B. A. Foote. 1969. Biology and immature stages of *Eumetopiella rufipes. Ann. Entomol. Soc. Am.* 62:227–234.

Velde, G. van der, and R. Hiddink. 1987. Chironomidae mining in *Nuphar lutea* (L.) Sm. (Nympheaceae). *Entomol. Scand. Suppl.* 29:255–264.

Villalta, J. F. de. 1957. Dos zoocecidías fósiles del Mioceno de Cerdaña (prov. De Lérida). *Curs. Conf. Inst. "Lucas Mallada"* 4:63–64.

Vogel, S. 1954. *Blütenbiologische Typen als Elemente der Sippengliederung.* Jena: Gustav Fischer.

Vogel, S. 1973. Fungus gnat flowers and fungus mimesis. In N. B. M. Brantjes and H. F. Linskens, eds., *Pollination and Dispersal,* pp. 13–18. Nijmegen, Netherlands: University of Nijmegen.

Vogel, S. 1978a. Pilzmückenblumen als Pilzmimeten. Erster Teil. *Flora* 167:329–366.

Vogel, S. 1978b. Evolutionary shifts from reward to deception in pollen flowers. In A. J. Richards, ed., *The Pollination of Flowers by Insects,* pp. 89–96. London: Academic Press.

Vogel, S. 1997. Remarkable nectaries: Structure, ecology, organophyletic perspectives. I. Substitutive nectaries. *Flora* 192:305–333.

Vogel, S. 1998. Remarkable nectaries: Structure, ecology, organophyletic perspectives. II. Nectarioles. *Flora* 193:1–29.

Wacht, S., K. Lunau, and K. Hansen. 2000. Chemosensory control of pollen ingestion in the hoverfly *Eristalis tenax* by labellar taste hairs. *J. Comp. Physiol. A* 186:193–203.

Walker, M. V. 1938. Evidence of Triassic insects in the Petrified Forest National Monument, Arizona. *Proc. U.S. Natl. Mus.* 85:137–141.

Walshe, B. M. 1951. The feeding habits of certain chironomid larvae (Subfamily Tendipedinae). *Proc. Zool. Soc. Lond.* 121:63–79.

Walter, D., and H. Proctor. 1999. *Mites: Ecology, Evolution and Behaviour.* Sydney: University of New South Wales Press.

Waser, N. M., L. Chittka, M. V. Price, N. M. Williams, and J. Ollerton. 1996. Generalization in pollination systems, and why it matters. *Ecology* 77:1043–1060.

Watanabe, M., and K. Kamimura. 1975. [Nectar sucking behavior of *Tabanus iyoensis* (Diptera: Tabanidae).] *Jpn. J. Sanit. Zool.* 26:41–47. [In Japanese]

Watson, J. 1988. The Cheirolepidiaceae. In C. B. Beck, ed., *Origin and Evolution of Gymnosperms,* pp. 382–447. New York: Columbia University Press.

Webber, A. C. 1981. Alguns aspectos da biologia floral de *Annona sericea* Dun. (Annonaceae). *Acta Amazonica* 11:61–65.

Wetschnig, W., and B. Depisch. 1999. Pollination biology of *Welwitschia mirabilis* Hook. F. (Welwitschiaceae, Gnetopsida). *Phyton* 39:167–183.

White, D. A., and L. B. Thien. 1985. The pollination of *Illicium parviflorum* (Illicaceae). *J. Elisha Mitchell Sci. Soc.* 101:15–18.

Wichard, W., W. Arens, and G. Eisenbeis. 2002. *Biological Atlas of Aquatic Insects.* Stenstrup, Denmark: Apollo.

Wiegmann, B. M., S.-C. Tsaur, D. W. Webb, D. K. Yeates, and B. K. Cassel. 2000. Monophyly and relationships of the Tabanomorpha (Diptera: Brachycera) based on 28S ribosomal gene sequences. *Ann. Entomol. Soc. Am.* 93:1031–1038.

Williams, P. H., and K. J. Gaston. 1994. Measuring more of biodiversity: Can higher-taxon richness predict wholesale species richness? *Biol. Conserv.* 67:211–217.

Wilson, B. H., and M. Lieux. 1972. Pollen grains in the guts of field collected tabanids in Louisiana. *Ann. Entomol. Soc. Am.* 65:1264–1266.

Wing, S. L., and L. D. Boucher. 1998. Ecological aspects of the Cretaceous flowering plant radiation. *Annu. Rev. Earth Planet. Sci.* 26:379–421.

Yeates, D. K., and B. M. Wiegmann. 1999. Congruence and controversy: Toward a higher-level phylogeny of Diptera. *Annu. Rev. Entomol.* 44:397–428.

Yeo, P. F. 1972. Miscellaneous notes on pollination and pollinators. *J. Nat. Hist.* 6:667–686.

Young, A. M. 1984. Mechanism of pollination by Phoridae (Diptera) in some *Herrania* species (Sterculiaceae) in Costa Rica. *Proc. Entomol. Soc. Wash.* 86:503–518.

Young, A. M., B. J. Erickson, E. H. Erickson, Jr., and M. A. Strand. 1986. Pollination biology of *Theobroma* and *Herrania* (Sterculiaceae). I. Floral biology. *Insect Sci. Appl.* 8:151–164.

Zaitzev, V. F. 1982. Microstructure of the labella of the fly proboscis. 1. Structure of the apparatus closing the pseudotracheae. *Entomol. Rev.* 61(3):60–69.

Zaitzev, V. F. 1984. Anthophilie und Rüssellabellenstruktur der Dipteren. In Z. Kaszab, ed., *Verhandlungen des Zehnten Internationalen Symposiums über Entomofaunistik Mitteleuropas,* pp. 169–172. Budapest.

Zaitzev, V. F. 1992. Evolutionary tends of morphofunctional adaptations exemplified by the superfamily Bombylioidea. *Trud. Zool. Inst. St. Petersburg* 185:1–240.

Zherikhin, V. V., M. B. Mostovski, P. Vrsansky, V. A. Blagoderov, and E. D. Lukashevich. 1999. The unique Lower Cretaceous locality Baissa and other contemporaneous fossil insect sites in North and West Transbaikalia. *Proceedings of the First International Palaeoentomology Conference,* pp. 185–191. Bratislava: AMBA Projects.

Zhou, Z.-Y., and B. Zhang. 1988. A sideritic *Protocupressinoxylon* with insect borings and frass from the Middle Jurassic, Henan, China. *Rev. Palaeobot. Palynol.* 59:133–143.

CHAPTER TEN

Biogeographic Patterns in the Evolution of Diptera

Peter Cranston

Biogeography is the study of the distribution patterns of plants and animals, and the explanations of such patterns. Biogeographic studies of Diptera have been particularly influential, notably those of Hennig (1960), Brundin (1966), Munroe (1974), and Matile (1990). Studies of *Rhagoletis* fruit flies and *Drosophila* vinegar flies have been highly influential in elucidating the role of geographic distribution in speciation (Bush 1975 et seq.).

Biogeography has unique concepts and a terminology: thus, every taxon, whether species, genus or whatever rank, occupies a particular geographic area, termed its "distribution," "range," or "area of endemism." This area of distribution might be discontinuous (disjunct). Within a range, a taxon may be scarce or abundant, permanent, seasonal or ephemeral. Distributions are not fixed, but may change seasonally, or over shorter or longer periods of geologic time, by expansion, contraction, or shift. Long-distance dispersal may allow establishment in a new area, and human activity causes extinction, leading to global impoverishment or redistribution of synanthropic pests or global homogeneity.

Several macroregions of distinctive endemism are recognized in six biogeographic regions; namely, the Afrotropical, Australasian, Nearctic, Neotropical, Oriental, and Palaearctic regions (Fig. 10.1). The Nearctic and Palaearctic regions are often united as the Holarctic region, a view supported by extensive shared generic and species-level Diptera distributions and relatively low levels of endemism in each area.

Faunistic surveys, combined with recognition of areas of endemism, explanation of their genesis, and interpretation of the interrelationship of faunas and the areas they occupy, are integral to biogeography, the discipline that unites biology with geography, geology, and ecology.

Here I consider data sources for Diptera distributions, followed by an outline of the different approaches to biogeographic analysis together with a review of case studies involving Diptera associated with selected common patterns. A section briefly considering the value of palaeontologic evidence in biogeography precedes a final summary and suggestions for future biogeographic studies for dipterists.

Distributional Data

A fundamental requirement for any biogeographic analysis is distributional data, with detail depending upon scale, which relates to the questions being considered. Here local scale, likely to be affected by current ecological factors, is distinguished from broader-scale patterns, in which history plays a major role.

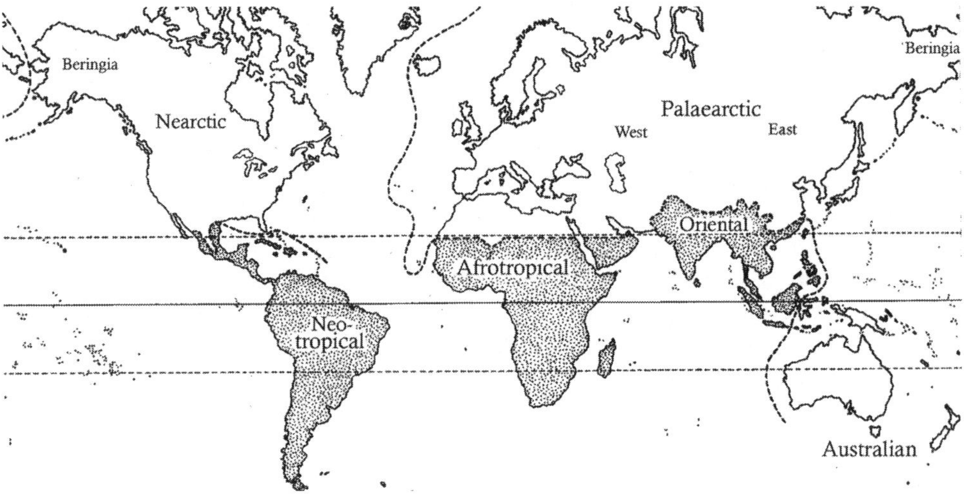

FIGURE 10.1. Geographic regions of the globe. After Cranston and Naumann (1991).

DISTRIBUTIONAL DATA FOR LOCAL-SCALE ECOLOGICAL BIOGEOGRAPHY

A biogeographic comparison of the dipteran fauna associated with use of specific larval substrates as a pabulum would require that faunas be known in detail; for example, on a fruit-by-fruit (host species) basis, such as the study undertaken for tropical Australian *Drosophila* by van Klinken and Walter (2001). Likewise, inter- and intraforest patch comparisons require detailed knowledge of the fauna, verified by regular and widespread survey to elucidate seasonal and interannual variability. Sampling and analysis must be statistically valid, to allow testing of correlation of faunal assemblages with the environmental factors that are postulated to control distribution in space. The faunal classification schemes proposed for lakes using dipteran biota are ecological biogeographic schemes, as are observations on local responses of taxa to anthropogenic impacts, such as pollutants.

DISTRIBUTIONAL DATA FOR BROAD-SCALE BIOGEOGRAPHY

Moving to broader geographic scale, regional inventories provide foundations of biogeographic knowledge, as Hennig (1960) pointed out for New Zealand. Unlike local-scale ecological biogeography, in which the boundaries include pertinent geographic features (e.g., a single drainage, suite of lakes, forest type), at the broader scale, the region is predominantly chosen on a nonbiological basis. Regions and countries are political constructs that may fail to coincide with significant biogeographic entities. Austria is unlikely to have a very different fauna to that generally found in central Europe, and Maine probably has no species not found in similar contiguous habitat. Exceptions to the biological arbitrariness of such political division are islands that are single political units, such as Hawaii, Iceland, and some islands of the Caribbean or the Pacific. Even Australia, which represents uniquely a single political unit and a continent, perhaps ought to include parts of New Guinea in its biogeographic region. Some countries cover more than one biogeographic region; for example,

China and Japan combine some of the Palaearctic and Oriental regions, Mexico contains Nearctic and Neotropical areas, and Saudi Arabia links the southern Palaearctic and Afrotropical regions. Despite these reservations, inventory data remain fundamental to the documentation of diversity at wider geographic scales. Data acquisition for these broader-scale studies may derive from local ecological biodiversity studies, but differs usually in the statistical rigor of sampling. Furthermore, such studies often seek congruence of distributions only with current ecological environmental data. Compilers of regional inventories seek to include all taxa ever recorded from an area—an exercise that of necessity includes interpretation of historical data. However, such inventories are dynamic, often reflecting our incomplete understanding even of well-studied families. Thus, the Diptera fauna of Europe has been studied in variable detail for over 250 years, often on a country-by-country basis, with inconsistent nomenclatural equivalence between different inventories. Comparisons of variations in species richness through space (regionally) and across taxonomic groups (usually within and between subfamilies) often are made from such inventories. Indeed, this "phenetic" pattern (see below) seems to be the most frequently proposed in dipteran biodiversity studies.

SAMPLING BIASES AND ERRORS

Between-area comparisons (beta-diversity) may be compromised seriously by variation in sampling intensity and systematic study between regions. For many families of Diptera, the Palaearctic region is thoroughly studied, followed by the Nearctic, Afrotropical, and Australasian, whereas the Neotropical and Oriental regional faunas are poorly studied. Although recent decades have seen greater activity concerning the less-well–known regional faunas, notably in association with the production of regional catalogs, the observation remains generally true.

Actual erroneous distributional data have been promulgated ever since some early entomologists were vague or in error concerning labeling the origin of specimens they recorded, collected, and had described. The literature is replete with reconsiderations of localities that often emanated from long oceanic voyages, including South America, southern Africa, and Van Diemens Land (e.g., Drew et al. 1994). An additional complication is that compilation and publication of faunal lists often are considered to be outside mainstream science. The biogeographer thus has difficulties, including those that arise from infrequent updating of ink-on-paper publications, despite the continuing acquisition of novel information.

EXISTING DISTRIBUTIONAL INFORMATION

Regional Catalogs. Regional faunal catalogs are primary sources of distributional data for use in biogeographic studies. Updating of checklists of Diptera at the national level continue to be the objective of biogeographically inclined dipterists (e.g., Petersen and Meier 2001), with scores of papers published each year, adding and emending regional faunas. Increasingly, such information is becoming available electronically, and the massive task of computerizing previous regional catalogs has been undertaken. Online searchable databases provide access to nomenclatural data, upon which biogeographic distributions are compiled. The ongoing production of such easily accessible Diptera databases appears to fulfill the criteria for advancing biosystematics (including biogeographical) studies as advocated by Godfray (2002).

World-Wide Overviews. Completing this review of publications on dipteran biogeography, excellent reviews have been derived from world-wide phylogeny and distributional information. Brundin's (1966) publication was an early compendious treatment of the phylogeny and distributional patterns exhibited by the predominantly austral Chironomidae subfamilies Podonominae and Aphroteniinae and the austral clade Heptagyiae in the globally distributed tribe Diamesinae.

The phylogeny and zoogeography of the Ditomyiinae, treated as a subfamily within Mycetophilidae, was investigated by Munroe (1974). Another clade of the Mycetophiloidea, the Keroplatidae, was the subject of monographic and historical biogeographic treatment by Matile (1990). Also among Lower Diptera, the Anisopodoidea have been subjected to phylogenetic and biogeographic analysis by Amorim and Tozoni (1994). Few major lineages of higher Diptera have been subjected to global phylogenetic and biogeographic reviews. Papavero's (1977) treatment of the Oestridae is suspect in its conclusions, in the light of a very different phylogeny proposed by Pape (2001), although the implications of the new view of relationships does not address alternative biogeographic and coevolutionary hypotheses. Summary and exemplar findings from some of these studies and others are provided below.

In the following sections of this chapter, I examine how explanations of distributional patterns are deduced, with coverage of the principal methods, followed by some examples from particular dipteran studies.

Methods in Biogeographic Analysis

Examination of just a few papers on dipteran biogeography shows that many approaches to biogeography have been essayed. Four nonexclusive categories can be recognized, namely: (1) ecological (physical) biogeography, (2) phenetic biogeography, (3) dispersal biogeography, and (4) vicariance biogeography. The term "vicariance biogeography" has been argued by Humphries (2000) as a metaphor for historical biogeography; it is a method for classification of areas of endemism at all scales, in terms of their historical relationships. Obviously, current ecologically based explanations must take cognizance of historical biotic and environmental changes.

Ideally, biogeography requires a phylogeny of a group under study, although it is possible (and even prevalent) for speculations to lack robust hypotheses of taxonomic relationships. Hennig himself (1960) argued that the minimal requirement was assertion of the local monophyly of the clade relative to its sister group in a different area. It is important to note that distributional data ought not be incorporated into genesis of a phylogenetic hypothesis, to avoid compromise if there is to be subsequent biogeographic analysis.

ECOLOGICAL BIOGEOGRAPHY

Ecological biogeographers seek explanations of the suitability of disparate geographic regions to particular organisms through an understanding of the physical and climatic requirements of the taxa in question (Cranston and Naumann 1991). Obviously, different species of Diptera thrive in different environments: some eurytopic species have widespread ecological tolerances, whereas stenotopic ones are more restricted. It may be possible to determine the optimal conditions for certain populations and, by inference, for the species; for example, experimental manipulation of environmental parameters can allow inferences of physiologic responses of individuals and populations. However, even if we could do this for more few species of flies, extrapolation from laboratory to field conditions is troublesome. Among other untested effects, species interactions, such as competition, may restrict particular taxa to a narrow band of their potential physicochemical, edaphic, and other abiotic environmental tolerances. Nonetheless, even without quantitative laboratory-based studies, larval Diptera evidently live in preferred ranges governed by biotic and abiotic environmental variables. For example, most *Chironomus* live as larvae in nutrient-rich, oxygen-deprived aquatic sediments; all *Fergusoniina* live (naturally) in galls induced in Australian Myrtaceae; all *Cryptochaetum* are internal parasitoids of Coccoidea; all streblids are ectoparasitic on bats; and all oestrids develop as subcutaneous parasites of mammals. If a dipteran lives in a closely restricted mutualistic, phoretic, or ectoparasitic host association, then its distribution will reflect that of the host, although it need not fully occupy the host range.

Ecological biogeographers examine the broad requirements of study organisms in terms of physical and climatic factors and discuss how different regions of the globe fulfill these criteria. However, these models explain only a limited number of biogeographic observations and do not address the frequent phenomenon of areas with comparable climate, soil, and topography supporting quite disparate plant and animal communities. Furthermore, the relationship between ecological factors and the distribution patterns of higher taxa (e.g., species groups, genera, families) is far from clear. If species diversification in monophyletic lineages (clades) is predominantly allopatric, with speciation in isolation (Bush 1975) and little or no ecological divergence, then models of ecology, particularly those incorporating climatic influences, might predict the distributions of these multitaxon clades. However, if cladogenesis produces sister taxa that diverge in their ecologies, or speciation is sympatric without accompanying ecological differentiation, environmental parameters would not correlate with distribution.

CLIMATIC MODELING OF DIPTERA DISTRIBUTIONS

The many environmental factors that affect dipteran development can allow predictive modeling of abundance and distribution. The abundance of any poikilothermic species is determined largely by interactions between proximate ecological factors, including climate, habitat, food availability, and the presence of predators and competitors. Distribution results not only from such ecological factors, but also includes a historical component. Thus, ecology determines whether a species can continue to live in an area; history determines whether it does, or ever got the chance to live there. This difference is related to timing; given enough time, an ecological factor will be considered a historical factor. Thus, in studies of where pest flies occur and what the limits of their spread might be, history accounts for the original or native distribution. However, ecological understanding may allow improved prediction of potential or future distributions under changed environmental conditions (e.g., the greenhouse effect) or as a result of accidental or intentional human relocation.

Many computer-based models pertain to the population biology of economic insects, especially those affecting major crop systems in Western countries, and have been used to predict the potential relative abundance and distribution of several Diptera around the world, using biological data and the recorded geographic distribution. Commonly, the existing geographic distribution and seasonal incidence of a pest species are known, but biological data pertaining to climatic effects on development are scant. Fortunately, the limiting effects of climate on a species usually can be estimated reliably from observations on the geographic distribution. Climatic tolerances of the species are inferred from the climate of sites where the species is known to occur. Of course, other information on the climatic tolerances of the species should be used if possible, because procedures reliant on the assumption that climate limits the present distribution might be an oversimplification.

Such modeling has been carried out for screw-worm flies (Stuart et al. 1995), biting flies of *Haematobia* species (Sutherst and Maywald 1985), tephritid fruit flies (Worner 1988; Yonow and Sutherst 1998; Sutherst et al. 2000), and *Drosophila* (Jenkins and Hoffman 2001). Output from these models obviously has great utility in applied entomology, such as epidemiology, quarantine, and management of dipteran pests. Biogeographic inferences have been made less widely (at least explicitly), despite the obvious potential for understanding the current controlling factors governing distribution. Notably, Jenkins and Hoffman (2001) could address the question of why *Drosophila serrata* distribution in southeastern Australia was restricted at a southerly limit. They observed that the border shifted seasonally, and concluded that this was not resource limitation or interspecies competition, but was influenced by cold stress parameters. These and similar conclusions have important implications for models of distributional change associated with climate change ("global warming"). Thus far, much modeling of dipteran distributions under altered climate scenarios relates to vectors of disease (e.g., de Wet et al. 2001; Hoff and Foley 2001; Wittman et al. 2001), notwithstanding criticisms of the applicability of such techniques (Davis et al. 1998; Reiter 2001).

We have detailed information on ecological performances of only a few species of flies, although such data are essential. Nonetheless, models of distribution are required in the absence of detailed ecological performance data. Given these practical constraints, a class of modeling has been developed that accepts distribution point data as surrogates for "performance (process) characteristics" of organisms. These points are defined bioclimatically, and models can estimate potential distributions using some flexible modeling procedures. Analyses depend upon the assumption that current distributions of species are restricted (constrained) by bioclimatic factors. Using correlative range-modeling programs developed in Australia (BIOCLIM, Nix 1986; DOMAIN, Carpenter et al. 1993), potential constraints on species distribution are estimated in a stepwise process. First, for each site at which a species is known to occur, the climate is estimated using a set of bioclimatic indicators based on an irregular network of weather stations across the region under consideration. Annual precipitation, seasonality of precipitation, precipitation of driest quarter, minimum temperature of coldest period, maximum temperature of warmest period, and elevation are influential and likely to have broad significance in determining the distributions of poikilothermic organisms. A bioclimatic profile is developed from pooled climate per site estimates, providing a profile of the range of climatic conditions at all sites for the species. Next, the bioclimatic profiles so produced are matched with climate estimates at other sites across a regional grid to identify all other locations with similar climates. Specialized software can then be used to measure the similarity of sites, with comparisons being made via a digital elevation model with fine resolution. All locations within the grid with similar climates to the species profile form a predicted bioclimatic domain. This is represented spatially on a map as a "predicted potential distribution" for the taxon under consideration, with isobars (colors) representing different degrees of confidence in the prediction of presence.

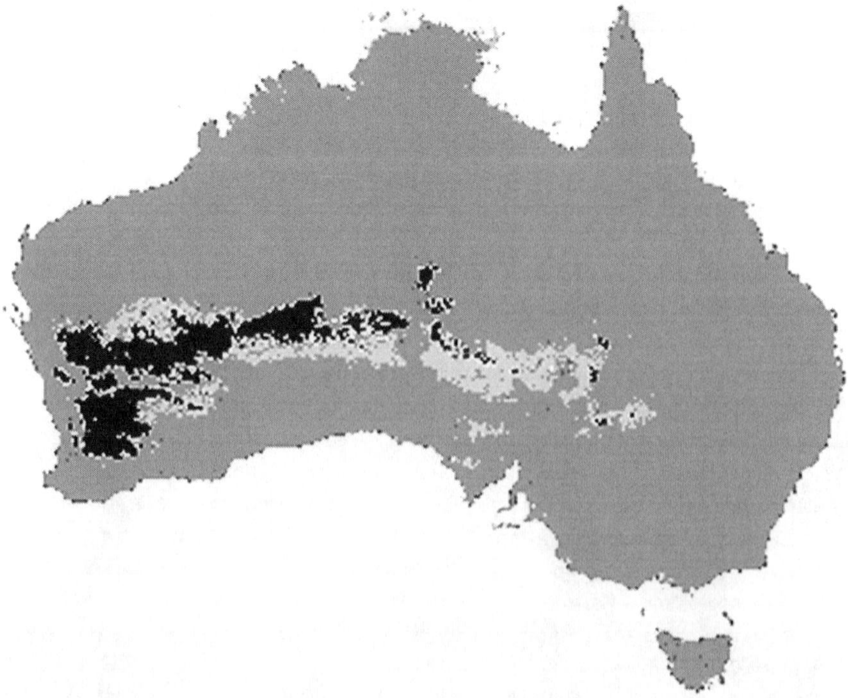

FIGURE 10.2. Modeled distribution for *Austrochlus* species based on presence data. Black, predicted presence within 98% confidence limits; pale gray, within 95% confidence limits.

The estimated potential distribution of the genus *Archaeochlus* (Chironomidae) based on known data points in southwestern Australia is shown in Fig. 10.2. Based on climatic parameters (predominantly seasonal rainfall levels), dark distribution shows a high probability of occurrence; light gray show a lesser likelihood. The model, based on two partially sympatric species well surveyed in southwestern Western Australia, predicts the occurrence of an ecologically related taxon in central Australia, subsequently discovered (Cranston et al. 2002). The effectiveness of bioclimatic modeling in predicting distributions of sister taxa, as shown here and in other studies, implies that much speciation has been by vicariance, with little or no ecological divergence.

The technique has also been used in back-tracking, to reconstruct past distributions using models of past climate and/or reconstruction of past climates, using postglacial fossil remains to provide information on past distributions. Although environmental determinants clearly play an important role in constraining the distribution of organisms, historical factors are vital in determining more fundamental patterns, including those that may relate more directly to species formation.

PHENETIC BIOGEOGRAPHY

Process explanations in dipteran faunistic studies, where they are given, often utilize taxa shared between areas. In such interpretations, widespread taxa are understood to convey biogeographic information, because they allow assessment of estimates of overall similarity between regions. Endemics, particularly narrow endemics, are not considered relevant. Typically, widespread species are considered as providing evidence that all occupied areas have some common history, with the number of taxa in common being related proportionately to the recency of faunal connection between areas. Phylogenetic relationships between taxa in the areas under study are either ignored or (more typically) unavailable. Although allowing description and comparison of biotas, these studies reveal little history. A major difficulty concerns the emphasis on widespread species, which can be taxa that have failed to respond by speciation to earth-history events, and the ignoring of the restricted distribution of often allopatric endemics, whose evolutionary history can be tracked with a phylogeny.

DISPERSAL BIOGEOGRAPHY

The frequently observed disjunct distributions of related organisms are interpreted in dispersal biogeography as having arisen in the following manner: groups originate in one place, diffuse (range expand) until some kind of barrier is reached, disperse ("jump") across these pre-existing barriers, then differentiate subsequently in isolation. Darwin (1859) in *The Origin of Species* and subsequently Wallace (1876) promoted this view, which was important in the formulation of Darwin's and Wallace's ideas on species formation.

The view that organisms have identifiable centers of origin from which they spread to other parts of the globe is central to the dispersalist approach. Proposed criteria suggested for recognizing a center of origin of a group of organisms have included: the area (1) most ecologically suitable, or (2) harboring the greatest number of extant species of the group, (3) circumscribing the greatest morphological diversity for the group, (4) containing the greatest number of advanced forms with primitive forms "pushed" to the periphery, or (5) containing the greatest number of primitive forms as advanced forms "progressively" disperse from the origin (the reverse of (4), known as "Hennig's progression rule"). Palaeontologic evidence may be sought, in the belief that the site of the oldest fossils indicates the origin of the group. However, reliance upon the vagaries of the fossil record has its problems (see below). Following identification of the center(s) of origin, a route is traced for the colonization of the remainder of the modern range, involving assessment of relative vagility; that is, the tendency or ability to disperse or diffuse from a center of origin. Subsequently, barriers that prevent the organisms from becoming universal are identified or postulated.

Assessment of dispersivity tends to be somewhat subjective and arbitrary, but entomologists make some common observations. First, brachypterous and apterous adult insects are presumed to have limited powers of dispersal relative to winged forms. However, evidence of the ability of other life history stages (e.g., eggs) to survive dispersal can counter a sedentary adult. The occurrence of particular insects in the aerial planktonic drift supports dispersal powers, as does the regular arrival (not necessarily establishment) of nonnative taxa on the opposite shores of bodies of water, such as the British Channel or the Tasman Sea. The discovery of particular insects on newly formed islands, or faunistically denuded areas, such as that following volcanic activity or marine retreat after inundation of terrestrial systems, is taken to indicate their ready dispersivity. For nonvolant insects for which distributions transcend physical barriers, such as oceans, dispersivity may be inferred to have arisen through contiguous land connections either extant, ancient, or postulated, in combination with various ad hoc scenarios, such as the propensity to transoceanic rafting, elevation in cyclonic winds, and recent transport by humans, either intentionally or involuntarily.

Given powers of dispersal, restrictions are postulated to curtail radiation of all taxa from their point of origin. Commonly three kinds of filters or barriers have been recognized (cf. Cranston and Naumann 1991): (1) Corridors or "bottlenecks" that are variably narrow constrictions of suitable habitats that may be impediments to dispersal between larger areas at each end. Ecologically similar areas connected by corridors may have very similar biotas. (2) Filter bridges that allow limited transgressions compared with corridors but more than the sweepstake routes described below. Chains of oceanic or ecological islands are often cited as examples of filter bridges. (3) Sweepstakes routes are major barriers, such as great expanses of ocean for terrestrial and freshwater biota, or tropical regions for cold stenothermic organisms. Generally it is held to be a matter of chance which organisms will survive a sweepstakes route and successfully colonize the new area. Dispersal via sweepstakes routes is commonly invoked to explain the depauperate, unbalanced mixture of taxa found on remote islands.

Dispersal has always been central to explanations of island biotas. Macarthur and Wilson's (1967) theory of island biogeography is a mathematical elaboration of dispersal biogeography. Among other things, this theory postulates a relationship between species and area, such that larger islands support a more diverse biota than do smaller ones. An island biota is visualized as being in or approaching dynamic equilibrium (Macarthur and Wilson 1967). At equilibrium, immigration and extinction rates are equal, so that the total number of species on an island is constant. However, the actual composition of the biota changes with time, through species interactions, often in relation to geologic and/or climatic changes. Island biotas, whether derived by dispersal or vicariance, have wide application in biogeography—a remote coral atoll, an isolated lake, or a fruiting tree or fungal fruiting body—all can be envisaged as "islands" isolated by surrounding unsuitable habitats. Examples of dipteran distributions on islands are discussed below.

BIOGEOGRAPHY OF DISPERSAL, FOUNDING, AND SPECIATION

Dispersalist explanations of disjunct biotic relationships originated when the present land masses were believed to be immobile. With fixed land masses, observed biotic disjunctions must have arisen by migration—and mechanisms outlined above were postulated. All observed distributions were explained by one or a combination of the above scenarios of faunal relocation, with differential extinction explaining absences.

Although dispersal scenarios have been made for groups with understood phylogenies, few attempts have been made to relate phylogeny with distribution—each taxon could be treated in isolation, and, generally, congruence between distribution and phylogenetic relationship was not sought. Repeated patterns might imply similar dispersal histories for the organisms investigated, but repeated ad hoc explanations could account for most distribution patterns.

Dispersal across a barrier followed by establishment implies dispersal into some kind of ecological "vacuum," often termed an "empty niche." Ecological vacuums certainly occur, and the evidence from newly formed or denuded areas, such as postvolcanic Mount St. Helens, Washington (Anderson 1992), or Krakatoa in the Pacific Ocean (Thornton and New 1988), shows rapid colonization. Furthermore, on a rather longer geologic time scale, dispersal must have been involved in the acquisition of characteristic postglacial floras and faunas by New Zealand, Patagonia, and vast areas of the cool temperate zones of the northern hemisphere, including the British Isles, by post-Pleistocene colonization from nonglaciated refugial areas. Range changes have occurred in the past, and such mechanisms influence current distributions. For example, the tephritid fly *Urophora cardui* in Europe is highly mobile and has repeatedly recolonized some suboptimal European regions since the Pleistocene after retreats during "little ice ages," as shown by patterns of allozyme variation (Eber and Brandl 1997), in contrast to the previous suspicion that its patchy distribution represented ongoing reimmigration from Pleistocene refugia. Such patterns have also been recognized from subfossil larval head capsules of dipterans, notably of Chironomidae and less often from Chaoboridae (Walker 1994). Such paleoentomological studies have been based predominantly in the northern hemisphere, where Pleistocene glaciations indubitably caused major range shifts (but seemingly little or no morphological speciation; Coope 1979). In the southern hemisphere, subfossil lacustrine Chironomidae also show dramatic temperature-related shifts in the past 10 kyr for Australia (Dimitriadis and Cranston 2001) and southern Chile (Massafero and Brooks 2002).

VICARIANCE BIOGEOGRAPHY

Croizat (1958) argued that disjunct taxic distributions, interpreted as due to mobile biota between static continents, can be viewed equally in terms of mobile land masses and static biotas. These ideas do not originate with Croizat, for there were many early converts among austral biologists who embraced continental mobility as a causative agent in biogeography. These far-sighted individuals included the Australian dipterist Ian Mackerras, who recognized that the regularity of "precisely parallel data" and the long-recognized "reality of certain faunal provinces" required a biogeographic hypothesis to explain the origins of the Australian Diptera. He speculated on widespread primitive taxa being "distributed before the continents spread widely" (Mackerras 1950: 2), building in turn on the early speculations by Harrison (1928) on the derivation of the Australian fauna by the "Wegener hypothesis" (plate tectonism).

Croizat (1958) promulgated the view that present-day biotic distributions represent ancient patterns that have been disrupted (vicariated) in the past by such factors as altered geology and climate. Compendious mappings documented distributions of many organisms and connected distributions of related taxa by lines (tracks). Repeated lines (generalized tracks) connecting occurrences across the globe bore only indirect relationships to modern geography. In the southern hemisphere, generalized tracks cross oceans, linking distant land masses. In the northern hemisphere, tracks traverse mountain ranges. Croizat (1964) argued that the congruence (or regularity) of so many distributions to form generalized tracks could not be explained by dispersal, but provided evidence for the appearance of impassable barriers that divided many taxa to produce congruent patterns of speciation, whose geographic relationships were identical. In the terminology of historical biogeography, widespread ancestral species are divided into vicariant populations (incipient species) by a vicariance event. An event may be geologic or climatologic occurrences, such as sea level alteration, ocean formation, orogeny, aridity, or glaciation. In contrast to a dispersalist view, in which a barrier predates the disjunction, in vicariance biogeography, the barrier causes the disjunction. Although Croizat gave no explicit means of assessing taxonomic relationships, his views on the interrelationship of life, space, and time were incorporated within the cladistic method (Croizat et al. 1974), thereby becoming exposed to a wider audience.

The melding of cladistic phylogenetics and vicariance biogeography formed a framework espoused by many subsequent historical biogeographers. Hennig's formulation of an explicit method for phylogenetic reconstruction provided a framework within which biogeographic patterns could be examined. Thus, a phylogenetic background to biogeography has provided the framework for dipterist biogeographers, including Hennig himself (1960), Brundin (1966), Munroe (1974), and Zwick (1977). For those who adopted a Hennigian view of the relationship between phylogeny and distribution, the search for centers of origin of clades was a central element. Thus, for Brundin (1966), Hennig's "progression rule" stemmed from a belief that taxic plesiomorphy indicated a center of origin, with increasing apomorphy expected toward the periphery of the range. Because concepts of plesiomorphy and apomorphy are referable only to character states and not taxa, such searches for centers of origin have languished in recent decades, although as modeled by Humphries (2000), a revisit of the theoretical basis is overdue.

The vicariant biogeographic method can be summarized as follows: taxon names in a cladogram are replaced by their respective areas of endemism to give an area cladogram. This shows the historical relationships between the areas, as evidenced by the biota studied. Identical geographic areas with repetitious presence of taxa ("redundant areas") are then removed to give a reduced area cladogram. An example from the postulated relationships among 16 species of the monophyletic *Parochlus araucanus* group of podonomine chironomid midges is depicted in the cladogram shown in Fig. 10.3. Substitution of the areas of endemism for the terminal taxon names gives an area cladogram showing distribution over southern land masses, except for *P. kiefferi* in North America. One clade (*P. novaezelandiae, P. longicornis,* and *P. glacialis*) contains redundancy, as all are New Zealand species. Removal provides the reduced area cladogram shown in Fig. 10.4, apparently depicting several New Zealand–South America vicariance events and one Australia–South America vicariance event occurring after the initial event involving Australia–South America–New Zealand. Brundin argued that these sequences were compatible with the known earth history involving late Cretaceous Gondwanan fragmentation.

Detailed re-examination by Nelson and Ladiges (1996) of Brundin's data and its subsequent elaborations shows a change in interpretation. Brundin's (1966) initially proposed dual pattern (South Africa (New Zealand, South America)) and (South Africa (southeastern Australia, South America)) came to include a dual southern South America, in which Australia + South America "1" are more closely related to one another than to South America "2"; the three areas then combined to form the sister to New Zealand (e.g., Brundin 1972, 1988).

Nelson and Ladiges (1996) suggest that reduced area cladograms such as Fig. 10.4 over-interpret events because a geographically restricted lineage splitting has taken place deeper in the tree. This is seen as analogous to the concept of paralogy in molecular biology, in which a gene duplicates prior to speciation—a clade duplicates prior to a vicariance event that affects each group (subtree) equally. They argue that the only information relevant to cladistic biogeography is "paralogy-free," with each branch containing only unique (non-overlapping, nonredundant) geographic data. Thus the paralogy-free tree for *P. araucanus* group is (New Zealand (South America, Australia)), with the anomalous Laurasian *P. kiefferi* difficult to interpret.

Three possible explanations exist for the position of *P. kiefferi:* (1) faulty phylogenetic relationships (however, even if true, the problem recurs wherever *P. kiefferi* lies within the otherwise austral clade); (2) dispersal from a cold stenothermic southern hemisphere location to its present North American localities; (3) the clade predates early Cretaceous Gondwana,

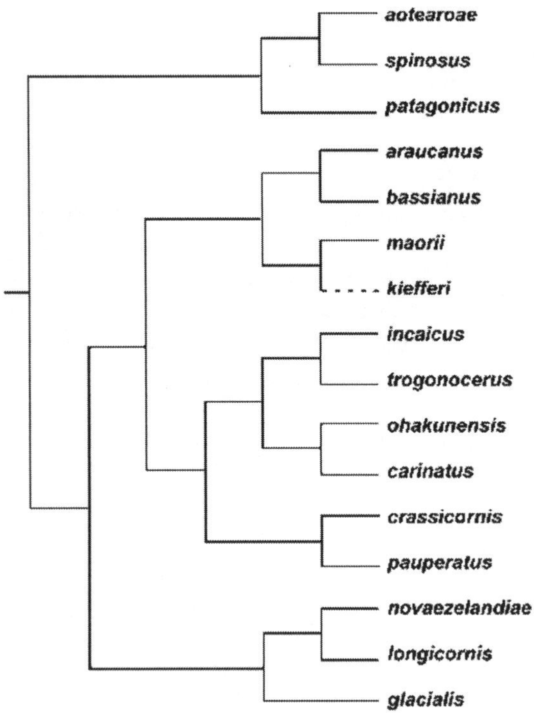

FIGURE 10.3. Postulated phylogeny (derived from Brundin 1966) for *Parochlus araucanus* group.

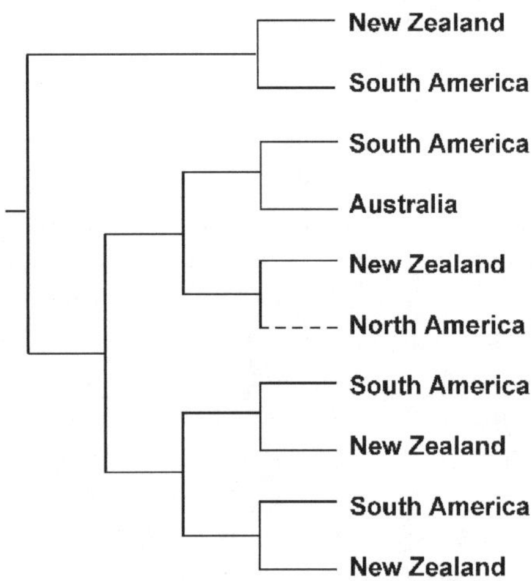

FIGURE 10.4. Reduced area cladogram for *Parochlus araucanus* group.

having originated in Pangaea, in which North and South America were contiguous, with either failure to radiate or differential extinction in the North, leading to asymmetric diversity. The possibility of a composite New Zealand (e.g., Craw 1988; Heads 1999) suggests these patterns might profitably be re-examined.

Comparison of a reduced area cladogram or paralogy-free tree with theories of earth history can allow inference of deterministic processes. Such explanations have greater generality if identical patterns recur multiple times among different taxa. Implicit in a vicariance explanation is continental mobility rather than fixity (Wegener 1915). Continental drift implies that land masses move and disjunct distributions are explicable by ancient contiguity or at least proximity, with subsequent range disruption. Thus Croizat's generalized tracks identify links between faunas that were formerly united, and instead of mobile faunas dispersing on a fixed geography, immobile faunas were dispersed on moving plates of land. Thus, Brundin argued that austral chironomid distributions explained early observations concerning the similarity of the plant faunas of all southern continents (e.g., Hooker 1853). The dispersalist faunal interpretations of Darwin and Wallace, and many successors who believed in a stable geography, were challenged. However, just as the distribution of a single organism or pair of species can be interpreted by past dispersal, so can a single cladistic biogeographic analysis always be interpreted. The likelihood of a common explanation increases as more geographically congruent patterns are revealed, such that as more similar patterns are discovered, the stronger is the evidence that disjunct distributions are deterministic rather than stochastic.

Broad-Scale Dipteran Biogeographic Patterns

Fossil evidence implies that major diversification, including ancestors of extant families, was well in place by the Late Triassic, with earliest definitive records from the early mid-Triassic, and the dating of the putative bittacid-like Mecopteran sister group of Diptera to the Permian. The global aerial land at this time comprised a single landmass—termed "Pangaea," itself made up of previously isolated components that preceded the current continents. We have little understanding of contemporary distribution of the Insecta, although by extrapolation from contemporaneous plants and fish, very widespread distributions across ecologically appropriate areas might be expected. Given that many recent dipteran families are globally distributed now, it is tempting to argue for past Pangaeic distributions, with successive vicariance-induced diversification induced by continental rearrangements, orogeny, fluctuating ocean levels, and associated climatic changes. A notable postulated early event was the early sundering of Pangaea into a northern Laurasian component and a southern Gondwana landmass. However, although appropriately dated and well-identified Jurassic/Early Cretaceous fossils of extant higher taxa undoubtedly exist, such as a mid-Jurassic prosimuliine Simuliidae (Crosskey 1990), an early Cretaceous aphrotenine chironomid (Brundin 1976), and many others are being discovered, detailed dating of diversification from fossils is unlikely. Use of phylogenies of extant dipteran clades to discern Pangaeic groups is controversial: the time period involved must have seen much extinction; roots and early branching events lie very deep and may not be amenable to identification using molecular techniques. Present-day "Pangaeic" distributions may reflect only the effects of subsequent stochastic intrahemispheric dispersal. Nonetheless, the existence of higher taxon sister clades, each restricted to one of the major Jurassic land masses of Laurasia and Gondwana, may reflect deep historical association with the sundering of Pangaea. Examples are found (inter alia) among the Anisopodoidea (Olbiogastridae, Anisopodidae, Mycetobiinae, and Valeseguyinae; Amorim and Tozoni 1994), the Chironomidae subfamily Diamesinae (subtribes Heptagyiae and Diamesae; Brundin 1966), and the Apioceridae and megasceline Mydidae (Yeates and Irwin 1996).

Although most dipterists have worked in the northern hemisphere, biogeographic studies of dipteran distributions have tended to emphasize the southern, Gondwanan, continents. Thus, Nagatomi (1991) combined phylogenetic and amber fossil data to claim that most Lower Brachycera originated in Gondwana. In contrast to some very striking patterns in the south, which can apparently readily be interpreted in a vicariance paradigm, northern hemisphere patterns tend to be complex and difficult to interpret. Furthermore, for most dipterans, northern taxa tend to be distributed widely across the Nearctic and Palaearctic, sometimes lacking even distinction at species level. In discussing distributions of Mycetophiloidea, Matile (1988) elaborated a northern trans-Atlantic track (Fig. 10.5) for Keroplatidae (Matile 1990), including the genus *Hesperodes,* known also from early Oligocene Baltic amber, and demonstrated by all Bolitophilidae, Diadocidiidae, and many Ditomyiidae and Mycetophilidae, and many other Baltic amber fossils. This is the most commonly observed track followed by northern hemisphere Diptera.

The relative recency of these patterns (perhaps Eocene/Oligocene) and the lack of subsequent extinction tend to lead to poorly differentiated radiations of species, including cryptic taxa. Thus, morphological species-level phylogenies tend to have poor support, and

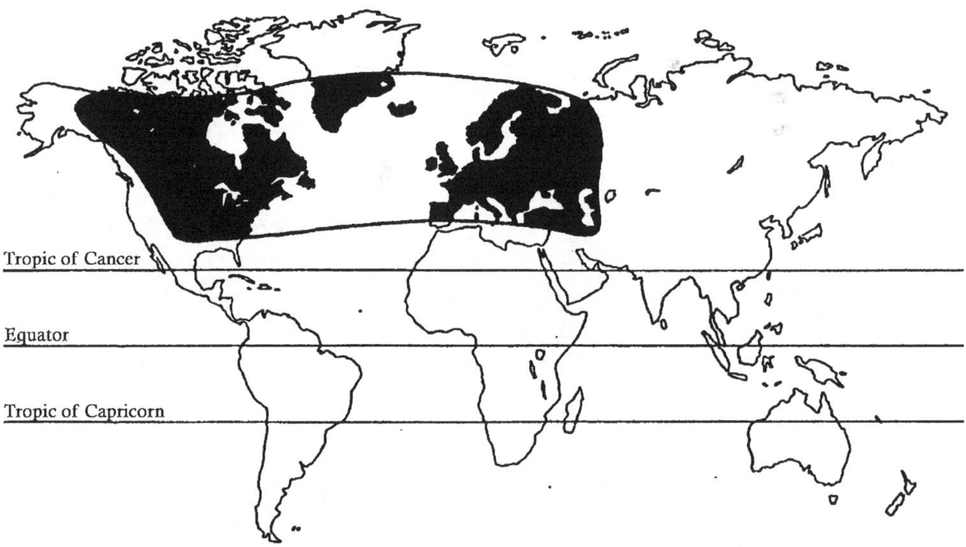

FIGURE 10.5. The northern trans-Atlantic track. Adapted from Matile (1990).

biogeographic patterns may be obscured by ambiguities in phylogenetic reconstruction. Furthermore, in the middle and higher latitudes, there may be no concerted patterns to detect, as the biota presumably largely derives from stochastic recolonization processes following the ravages of the Pleistocene glaciations.

Nonetheless, entomologists have disentangled some interesting broader-scale patterns; for example, identifying roles of Beringia and the Mediterranean in generating geographic vicariant patterns. First identified among currently boreal insects, including many carabid beetles, congruent Beringian patterns link eastern Asia with nonglaciated, northwestern Nearctic. For the tipulid *Nephrotoma dorsalis* group, Tangelder (1988) proposed that all early interactions between Palaearctic and Nearctic involved Beringia, with two main vicariances dated as Oligocene and late Pliocene—with the Pleistocene influencing only local diversity and distribution, perhaps with two exceptions, for which there was renewed trans-Beringian movement. Tertiary events in Beringia appear to be associated with development of species of the simuliid genus *Gymnopias* (Wood 1978) and anthomyiid cone seed pests in the genus *Strobilomyia* (Michelsen 1988).

A source area for the therevid tribe Cyclotelini in the Late Cretaceous, termed "Westamerica," is identified by Gaimari and Irwin (2000). Possibly three separate clades were proposed to have migrated from Asia through Beringia (or were present already, undiversified, throughout Westamerica), in a land with boreotropical to paratropical climate, with ensuing diversification in the Americas. Subsequent major barriers associated with Cyclotelini speciation in the Americas include the existence of the Mid-Continental sea, the formation of north-south mountain chain (the Rockies) preventing substantial eastward dispersal, and the Eocene formation of the proto-Antillean Archipelago, allowing dispersal and radiation in the Caribbean and Central and South America (Gaimari and Irwin 2000). This Beringian pattern (or some close approximation, whatever its biogeographic origin) is well understood by botanists, and describes the distribution of certain host-specific aphids (Moran 1989): host-specific phytophagous dipterans might profitably be studied in this context.

The influence of the western Mediterranean on dipteran diversification is addressed for many Tipulidae (Tangelder 1988; Oosterbroek and Arntzen 1992); de Jong (1998) incorporated data from these and other insects in a metaanalysis. Thirteen areas of endemism can be recognized: the Alps, Apennines, Atlas, Corsica, Iberia, Palaearctic, Pyrenees, Sardinia, Sicily, and High Atlas in the western Mediterranean, and Anatolia, Balkan, and Caucasus-Elburz added from the eastern Mediterranean. These areas could be related in a hierarchic geological framework associated with the Jurassic-onward collision and/or shearing of Africa with Eurasia; loss of the shallow Tethys; infilling of the Mediterranean; and tectonism associated with the Atlas, Alpine, and Apennine orogonies. As others have observed, the fauna of Atlantic islands of volcanic origin (e.g., the Azores, Canaries, Madeira) must have been derived by dispersal from continental Mediterranean areas. Reliable faunal relationships between Mediterranean areas of endemism were not recovered, not least because so many species are widespread. More recently, Léger and Depaquit (2002) assessed the patterns seen in the psychodid genus *Phlebotomus* against de Jong's (1998) geologic history, but without an explicit phylogeny. Subsequently, the history of the *Phlebotomus* subgenus *Larrousius* was assessed with an explicit phylogeny and a molecular clock estimate of dating (Esseghir et al. 2000). The calculated tempo of diversification implied that postulated pre-Pliocene tectonic events espoused by Léger and Depaquit (2002) did not play a major role in diversification, the authors proposing instead climate-driven speciation associated with more recent aridification (Esseghir et al. 2000).

In contrast to the sometimes confused, perhaps often disrupted, patterns of Laurasian taxa, in the southern hemisphere, the results of Gondwanan fragmentation are evident in the modern biota.

Gondwanan Tracks

Gondwana comprised Africa, Madagascar + India + Sri Lanka, New Zealand, Australia, and South America. Smaller continental fragments (slivers) termed "terranes" are numerous, and of importance in the evolution of the austral biota, and include, for example, New Caledonia, Norfolk Island, and Lord Howe Island, as well as a disputed number of Pacific islands and parts of Southeast Asia. Break-up of Gondwana followed three stages: initial rifting commenced about 180 Mya in the Late Jurassic with the formation of a seaway between Africa + South America (west Gondwana) and Antarctica + Australia + India + Madagascar and New Zealand (east Gondwana); the oldest sea floor is dated to 156 Mya in the Mozambique, Somali, and Weddell Sea basins. South America separated from Africa around 130 Mya (Early Cretaceous), as Africa + India separated from Antarctica. The final major breaks took place in the Late Cretaceous (100–90 Mya), as Australia and New Zealand detached from the core Antarctic.

Gondwanan distribution patterns in which phylogenies and documented distributions are partly or fully concordant with an orderly sequence of fragmentation of Gondwana are widespread among Australian insects and provide the fundamental explanation for patterns in Ephemeroptera, Plecoptera, many Hemiptera, and Holometabola (Cranston and Naumann 1991). Dipteran taxa are numerous and include among the Lower Diptera (="Nematocera") Psychodidae, Tanyderidae, Tipulidae (sensu latu), Dixidae, Thaumaleidae, Chironomidae, Blephariceridae, Canthyloscelidae, Scatopsidae, Anisopodoidea, and many elements of the Mycetophiloidea, including Ditomyiidae and Keroplatidae and some Mycetophilidae sensu stricto (Hennig 1960; Brundin 1966; Colless 1970; Munroe 1974; Hutson 1977; Zwick 1977; Matile 1990; Amorim and Tozoni 1994; Amorim and Pires 1996; Amorim 2000). Among the Brachycera, the pattern is evident, at least partially, among the orthorrhaphan Brachycera; Stratiomyiidae, Tabanidae, Coenomyiidae, Exeretonevridae, Heterostomidae, Pelecorhynchidae, Xylomyidae, Rachiceridae, Xylophagidae, Athericidae, Vermileonidae, Empididae, Rhagionidae, Nemestrinidae, Asilidae, and Bombyliidae (Mackerras 1950, 1962; Hennig 1960; Nagatomi 1982, 1991; Matile 1990; Colless and McAlpine 1991), and perhaps only Sciadoceridae and Anthomyiidae among the cyclorrhaphans.

The separately encircled area of Southeast Asia in Fig. 10.6 is undoubtedly polyphyletic, including multiple terranes derived from Gondwana embedded in a Laurasian-derived matrix (see the Australo-Oriental and Pacific track discussions below). Matile (1990) cites the *Paramacrocera* group (Keroplatidae: Macrocerinae) as following this track, with genera in eastern Africa and Madagascar, the Afrotropics, Oriental, and Australasia, having a sister group relationship to *Paramacrocera* distributed in the Neotropics and Australia via an amphinotal track (see below). In the keroplatid tribe Orfelini, *Isoneuromyia* occupies all Gondwanan masses, but includes some Laurasian (Palaearctic) species, proposed to derive from central Asian relatives. The implied minimum origin date for groups showing this

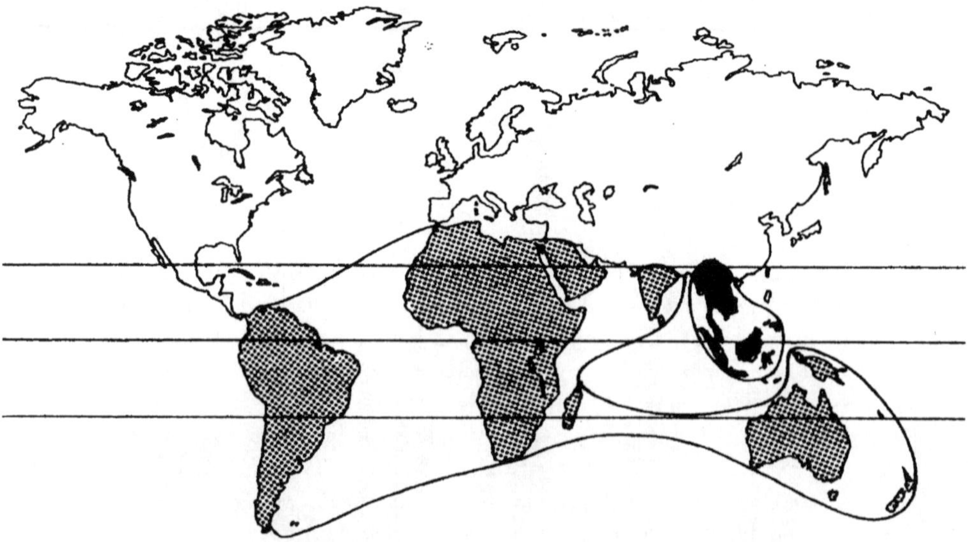

FIGURE 10.6. Generalized Gondwanan track. Adapted from Matile (1990).

distribution is Late Jurassic. Somewhat similar patterns are seen in many groups of Diptera, but they vary and/or differ in:

- The presence and extent of any Holarctic representation;
- The presence of only a subset of Gondwanan land masses; and
- The restriction on Gondwanan landmasses to certain broad ecological zones (tropical, temperate).

GONDWANAN SUBTRACKS

A temperate transantarctic track (Fig. 10.7) links essentially temperate areas of the southern hemisphere. Since the now-miniscule temperate zone of South Africa is involved, and the area was the first to fragment from Gondwana, this track dates minimally to the Early Cretaceous (about 140 Mya). Recognition of this pattern (e.g., by Hooker 1845 for gymnosperms, including Araucariaceae and Podocarpaceae, and Proteaceae) has a distinguished history.

Brundin (1966 et seq.) recognized this track for certain elements of the Chironomidae subfamilies Podonominae, Aphroteniinae, and Diamesinae, observations that have been confirmed and extended (Cranston and Edwards 1992; Cranston et al. 2002). Predictive possibilities from partially known taxa exist: on the basis of the previously known austral distribution of the keroplatid subgenus *Chiasmoneura* (*Prochiasmoneura*), Matile predicted its distribution ought to include Australia, as was confirmed later (Matile 1990). As a generalization, but requiring further study, organisms showing this transantarctic track tend in South Africa to be refugial and localized in the Afrotropical forest close to the temperate south coast, rather than associated with Afromontane habitats or the very recently evolved feinbos (southern restionaceous heathlands).

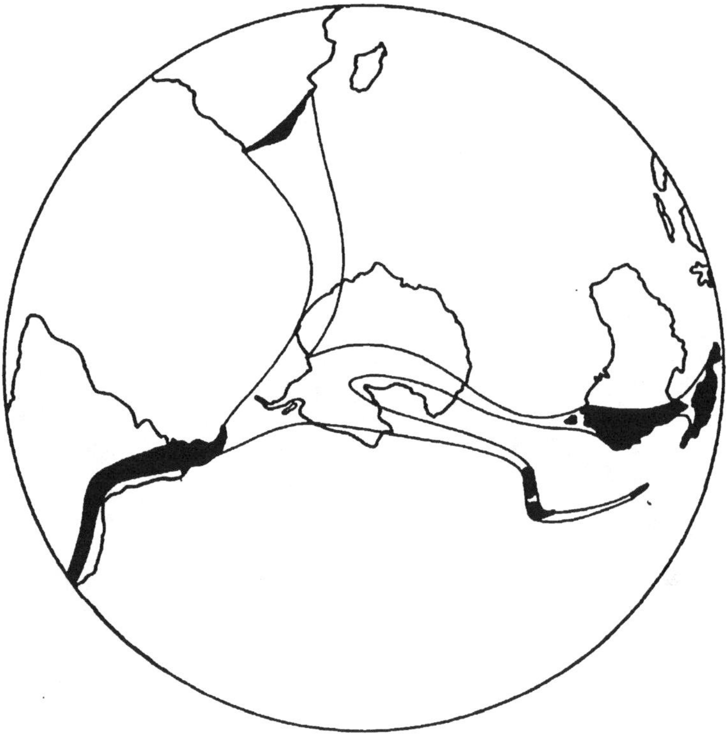

FIGURE 10.7. Temperate transantarctic track. Adapted from Matile (1990).

TEMPERATE AMPHINOTIC TRACK

The austral Gondwanan track, which resembles the transantarctic track but excludes South Africa (Fig. 10.8), is termed the "amphinotic track" (Matile 1990). The ecological correlate of the track remains cool temperate, perhaps even more so than the transantarctic track. Early dates for vicariance involved in this track derive from the timing of the separation of New Zealand from Gondwanan Antarctica (Marie Byrd Land) in the mid-Late Cretaceous (100–65 Mya), with geographic isolation from Australia by the opening of the Tasman Sea (about 80 Mya). Subsequently, southern South America separated from the Antarctic, dated at 38 Mya, which may have been the final severing of the possibility of further major vicariance on this track, although it may have been severed much earlier by fissure of Antarctica into west and east. The track was recognized by early biogeographers, notably by such botanists as Hooker, but also by entomologists, including Erichson (1842) and Tillyard (1926). Among Diptera showing this track are certain Prosimuliini (austral "*Cnephia*," Davies and Györkös 1988; Crosskey 1990; Roig Junent and Coscarón 2001); many Mycetophiloidea (Munroe 1974; Matile 1990), including Ditomyiidae (*Nervijuncta* and *Australosymmerus*, Amorim and Pires (1996), Scatopsidae (*Diamphidicus*, Amorim 1989; Canthyloscelidae [*Canthyloscelus*], Hennig 1960; Amorim 2000), and many Chironomidae (Brundin 1966).

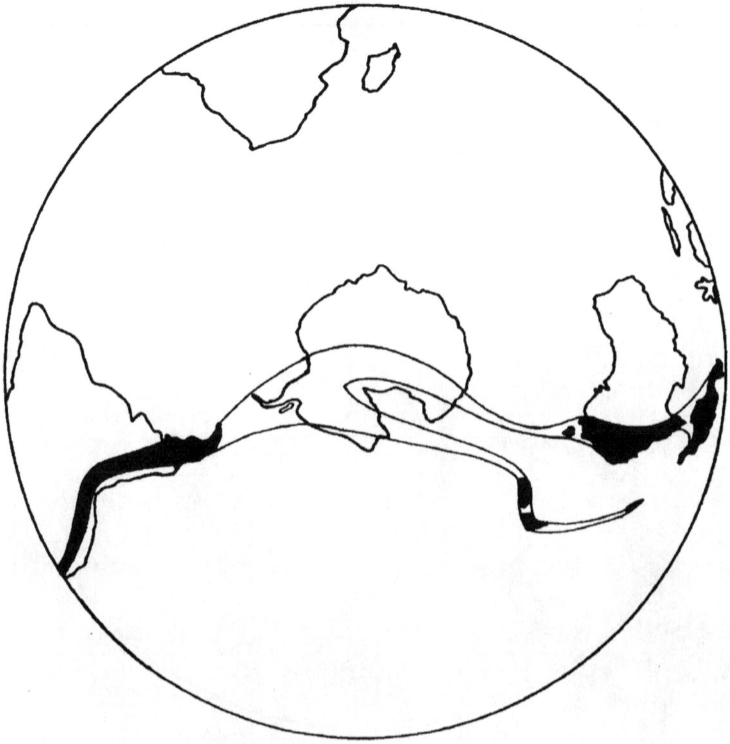

FIGURE 10.8. Temperate amphinotic track. Adapted from Matile (1990).

Brundin (1966), in assessing distribution patterns in australly distributed subclades of Chironomidae, proposed that:

- The sister group of a New Zealand clade always lies in South America, or South America + Tasmania + Australia;
- There are no direct connections between a New Zealand clade and a Tasman + Australian clade; and
- A Tasman + Australian clade is always more apomorph than a South American clade.

Evidence for the family Ditomyiidae supported the first two points (Colless 1970; Munroe 1974; Matile 1990) but not the third, as seen by exemplars of trans-Tasman tracks (see below).

TROPICAL GONDWANAN TRACK

Among Diptera, especially higher Diptera and more late-branching nodes in Lower Diptera phylogenies, the most commonly observed Gondwanan subtrack perhaps is tropical rather than the cool temperate of the Antarctic and amphinotic tracks. Distributions in South America include Brazil and most other South and Central American countries bearing tropical forests; in Africa, much of the central continent, excluding the temperate south and the Afromontane (and Saharan and North Africa); certain terranes of Southeast Asia; and if extending to Australia, including usually only New Guinea and the northern and northeastern parts of Australia. This is essentially the generalized Gondwanan track (see Fig. 10.6; a historical entity) lacking the temperate-to-cold distributed taxa of the Antarctic and amphonotic tracks (see Figs. 10.7, 10.8). The track, which is dated minimally to the early Cretaceous, was recognized by Harrison (1928), a perceptive early adopter of the Wegener paradigm. Inclusion of India (and Sri Lanka) in this track may be justified on geological grounds, yet few presumptive Gondwanan biotic distributions include India. The explanations are several:

- After departure from its connection with Madagascar at 92–89 Mya (Torsvik et al. 2000), India was submerged and lost its terrestrial biota;
- India (plus Madagascar) left Gondwanan (by the opening of the Mozambique channel, dated at 140 Mya)—too early to have carried a recognizably Gondwanan biota;
- Gondwanan taxa are present in India, unrecognized, in elevated areas, such as the Ghats and central Sri Lanka;
- India was Gondwanan but the elements are extinct; or
- India never was Gondwanan: tectonic reconstructions are incorrect.

All Matile's (1990) examples of the tropical Gondwanan track include sporadic "recent extensions" into the Holarctic; these include Lygistorrhinidae, *Proceroplatus* + *Xenoplatyura* (Orfelini), and *Heteropterna*. As Matile observed, some taxa, such as Lygistorrhinidae, that show this track among extant members have fossil material from the northern hemisphere, and in the Anisopodidae, this track is seen in the pairs of sister genera *Eogaster* (Afro-Oriental)/*Olbiogaster* (Neotropical), and *Mesochria* (also Afro-Oriental)/*Neomesochria* (Neotropical/Australian) (Amorim and Pires 1996).

The track (in part if not in toto) is common among verified monophyletic genera of Chironomidae, especially in the tribe Chironomini (e.g., *Conochironomus, Skusella, Fissimentum*), but also including *Djalmabatista* (Tanypodinae). The inferred physiologies of larvae of these clades tend to suggest warm eurythermy—an ecological correlate on the historical template.

SOUTHERN TRANS-ATLANTIC TRACK

A southern Atlantic track linking South America with Africa (Fig. 10.9) dates from no later than the mid-Cretaceous. The opening of the southern Atlantic commenced in the Late Cretaceous (about 130 Mya), but the date of completion of the separation of the two continents depends upon the model used—the Turonian/Cretaceous on a constant dimensions globe (Howarth 1981); or earlier, at the start of the Turonian in an expanding earth model (Owen 1983; Matile 1990). Whatever the case may be, the opening of the southern Atlantic was protracted and postdated the opening of the Indian Ocean, which started earlier by rift faulting in the Oxfordian Late Jurassic (about 155 Mya). Separation of Africa from Madagascar + Sri Lanka + India and Antarctica continued throughout the Kimmeridgian Late Jurassic, and Africa was separated completely from the Antarctic by oceanic crust by the Hauterivian Early Cretaceous (about 120 Mya). Whether this southern Atlantic track is an attenuated tropical Gondwanan track, with either extinction, primitively restricted distribution, or unrecognized sister-group relations in Indo-Australia, can be determined only with large-scale phylogenetic analyses, of which there are few. Certainly the track applies to

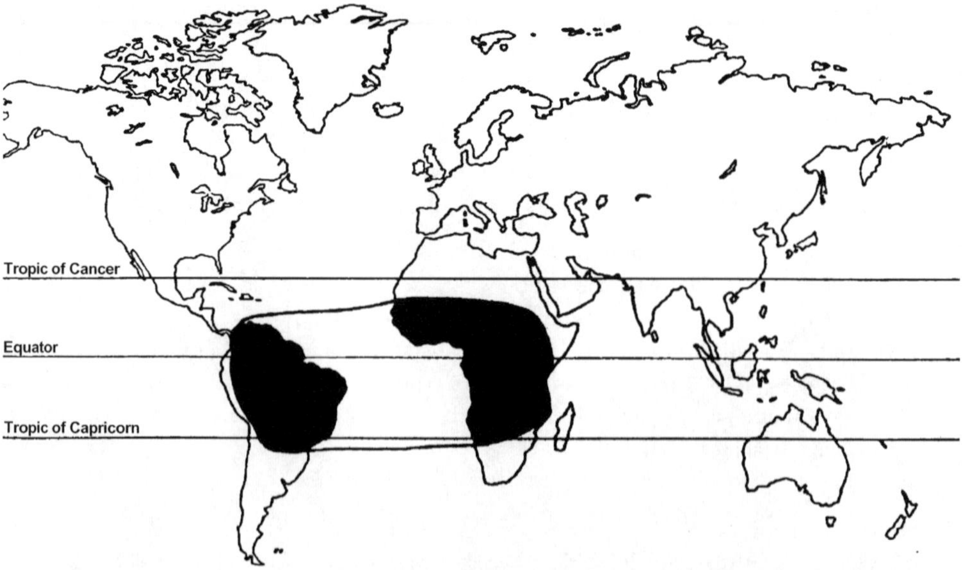

FIGURE 10.9. Southern Atlantic track. Adapted from Matile (1990).

certain angiosperms (e.g., cacti), but Matile (1990) detected the track only in a few groups of Keroplatini—the *Hikanoptilon* group, *Ctenoceradion*, and certain sister species in *Keroplatus*. *Lyprauta* among the Ofeliini has Brazilian-African sister taxa, but also includes Madagascar, and thus provides an earlier date for the lineage. In the phlebotomine sandflies, the major division between *Lutzomyia* + *Brumptomyia* (South American) and *Phlebotomus, Sergentomyia*, and relatives is associated with the opening of the southern Atlantic, suggesting to Léger and Depaquit (2002) a dating of 120 Mya for the vicariance.

Precise relationships of internal areas of South America seem to vary with taxa and nuances of analyses. Although frequently a major vicariance between northern (Amazonian/tropical) and southern (Antarctic/Patagonian) areas is identified, the Guyanan Shield does not always appear as early in the evolution of the biota as expected from its geology, and the "recent" Amazonian area may be deeper than expected. Unifying vicariance-inducing features include marine ingression to form intercontinental seas (final ingression Early Miocene, 20 Mya); the raising of the Puna and then the Andes, causing climate-induced drying of Patagonia; and fluctuations in temperature of the Antarctic (southernmost) biota. Thus, taxa associated with Gondwana via the transantarctic or amphinotic track tend to lie in Antarctic South America, associated with both slopes of the Andes and some relictual areas farther north, where they abut desert to the north and east, separating from the southern Atlantic track derived biota (Roig Junent and Coscarón 2001).

TRANS-TASMAN TRACK

The trans-Tasman track links eastern Australia (from the tropics to the temperate Tasmania) with New Zealand (both islands) and New Caledonia (Fig. 10.10). The New Caledonian

FIGURE 10.10. Trans-Tasman track. Adapted from Matile (1990).

biota is incompletely known for Diptera, but the biota resembles that of northern Australia, including New Guinea, more than that of New Zealand or temperate Australia. Examples of New Zealand/Australian disjunctions with New Caledonian representation include Keroplatidae (e.g., *Pseudoplatyura, Neoplatyura monticola* group), Mycetophilidae (*Eudicrana*) (Matile 1991), Blephariceridae (Apistomyini in part, Zwick 1977), and certain Chironomidae. The predominatly Australian phlebotomine genus *Australophlebotomus* (Psychodidae) has a "derived" species pair in New Caledonia, argued by Léger and Depaquit (2002), perhaps resulting from vicariance in the Palaeocene (about 60 Mya).

Although relevant robust Diptera phylogenies are few, strictly Australian/New Zealand sister group relationships undoubtedly exist, inter alia, among Tipulidae (sensu lato); Simuliidae (*Austrosimulium,* Crosskey 1990); Keroplatidae (including *Arachnocampa* and *Paramacrocera* sensu stricto, Matile 1990); Mycetophilidae (e.g., *Morganiella/Paramorganiella,* Matile 1990); Chironomidae (e.g., Boothroyd and Cranston 1995; contra Brundin 1966); Tabanidae (*Paranopsis/Ectenopsis*); Empididae (*Thinempis,* Bickel 1996; *Clinocera,* Sinclair 2000); *Sematopoda, Monodromia, Pseudoscelobates;* Dolichopodidae (*Scorpiurus*); Helosciomyzidae (total clade); and Lauxaniidae (*Poecilohetaerus;* also Lord Howe and Norfolk Island) (after Hennig 1960; MacFarlane and Andrew 2001, unless stated otherwise).

Geological events associated with the distribution patterns involve New Zealand's departure from its Gondwanan connection to Antarctica, severing the biotic link to Australia + Antarctica + South America. The precise timing of the loss of contiguous, above-sea, connections between Australia and New Zealand (via Tasmania and/or the Campbell Plateau, or the Lord Howe Rise) is unclear because of variations in marine regressions in the Late Cretaceous (100–65 Mya), but 80 Mya is agreed for complete continental separation and new sea floor creation. This vicariance event surely was associated with the deeper (older) speciation patterns seen in several New Zealand taxa. Younger biotic links to Australia may indicate a hybrid origin of modern New Zealand, with a later (or longer-lasting) connection through the Lord Howe Rise (to central-eastern Australia), and/or the Norfolk Island Rise, via New Caledonia and the northern Australia + New Guinea + Melanesian arc. Alternatively (or complementarily), more recent sister group relationships between Australia and New Zealand + New Caledonia may result from dispersal and/or a combination of extinction and differential ecological responses, as the terranes have moved northward since the Jurassic. Eventually, molecular evidence for dating could provide a means of distinguishing between these processes.

AUSTRALO-ORIENTAL TRACK

An Australo-Oriental track (Fig. 10.11) is postulated by Matile (1990) to connect the Australian and Indian subcontinent via Wallacea (essentially Southeast Asia and traversing the Wallace and Weber lines). This track resembles that described by Mackerras (1950) as Lemurian (of postulated Pliocene age) or Indo-Malayan (Pleistocene), citing *Aedes* (*Ochlerotatus*); some Tipulidae, *Lomatia* (Bombyliidae; Lemurian), and Anophelinae (Culicidae); *Silvius* (Tabanidae); *Chrysomyia* (Calliphoridae); and perhaps many Cyclorrhapha (Indo-Malayan). For Matile, an Australo-Oriental track is exemplified by the mycetophilid genus *Stenphragma,* with extension as far eastward as New Caledonia. The histories of taxa showing this distribution are highly dependent on their phylogenetic relationships, especially the relationships of Southeast Asian taxa to relatives in the undoubtedly Gondwanan areas of India and Australia—that is, whether Antarctica played a role in a (Cretaceous) pattern, or Southeast Asian elements were introduced by allochthonous terranes derived from farther south and of Gondwanan origin. Such distribution patterns evidently challenge the reality and limits of traditional "Oriental" and "Australian" regions according to taxa studied.

Many geologists argue that Southeast Asia should be considered part of Gondwana, and although once disputed, certainly tropical components of Southeast Asia fauna and flora connect with other Gondwanan areas. Whether such disjunctions arose through vicariance of primitively widely distributed taxa or by dispersal depends very much on the ages of the clades and areas. Thus, Metcalfe (1998) describes the sequential arrival of Gondwanan-originating terranes becoming embedded in Southeast Asia in waves dating between the Devonian and Late Jurassic, accompanying the opening and closure of three successive Tethys Oceans. Given such dates, Austro-Oriental taxa representing shallow nodes in phylogenies are too young to have arisen by vicariance associated with such ancient events. However younger terranes, including those derived from northwestern Australia, evidently are incorporated in a "jigsaw puzzlelike" matrix of inner and outer Melanesian ridges and arcs of differing but younger ages (e.g., Hall and Holloway 1998; Metcalfe 1998).

For some researchers, the faunal associations of the Australo-Oriental track challenge the existence of a Jurassic Tethyean oceanic divide between Gondwanan Australia and Laurasian Asia (McCarthy 2003). Additionally, one may question whether this track is any more than an attenuated tropical Gondwanan track, with extended relationships to Africa and South America lacking through primitive absence, subsequent extinction, or systematists' failure to recognize sister groups located there.

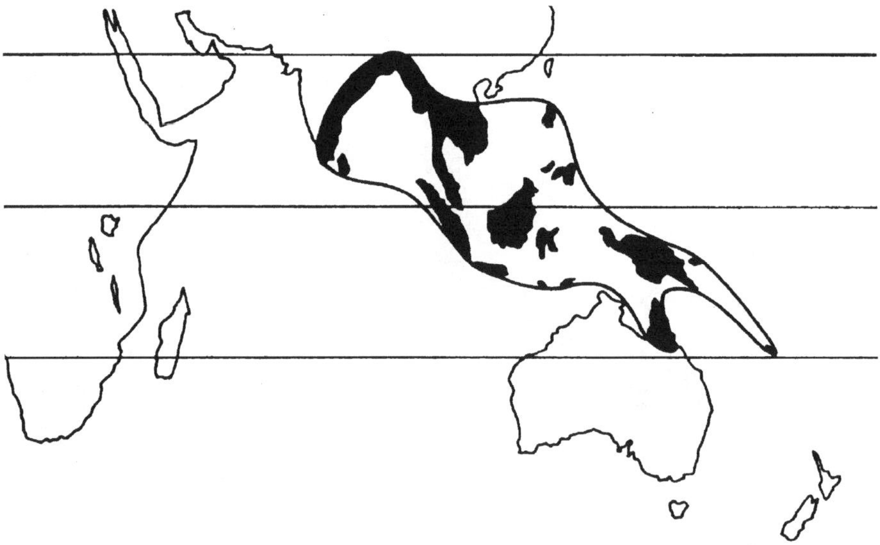

FIGURE 10.11. Australo-Oriental track. Adapted from Matile (1990).

TRANS-PACIFIC TRACKS

Ever since their postulation by Croizat (1958), the interpretation of trans-Pacific tracks is one of the most contentious issues in historical biogeography (McCarthy 2003). In conventional tectonic reconstructions with a fixed-dimensions globe, a Panthalassan Ocean occupied the noncontinental half of the Pangaeic globe (e.g., Howarth 1981; Owen 1981). In such a reconstruction, the Jurassic Panthalassa was about twice the width of the present-day Pacific, and trans-Pacific biotic tracks, such as those described below, must derive either from very long-distance dispersal or vicariance of prior ranges (including either Antarctica —amphinotic or temperate antarctic tracks—or Beringia). There was subsequently extensive extinction in low latitudes, producing the present disjunctions.

Under Matile's (1990) modification (Fig. 10.12), a northern Pacific track links western North America (south of Beringia) with the southern Oriental region, including parts of Southeast Asia; a central track links tropical South America with the Sunda arcs, and Macronesian and Melanesian rifts, arcs, and microterranes. The keroplatid tribe Robsonomyiini follows the northern track with *Robsonomyia* in the western United States, *Srilankana* in Sri Lanka, and *Micrepimera* on Christmas Island. A central Pacific track, or at least some elements of it, finds support, including Michaux and White's (1999) explanation of patterns in *Bactrocera* (Tephritidae), which is discussed below. The keroplatids *Platyroptilon* and *Setostylus*, the *Euceroplatus* group, and also the deeper-branching radiation of *Heteropterus* are cited as evidence for this track (Matile 1990). The chironomine genus *Nandeva* occurs in Brazil, Patagonia, and central-northern Australia (Cranston 1999), and *Fissimentum* shows a similar distribution (Cranston and Nolte 1996); otherwise, this tropical track is rare in

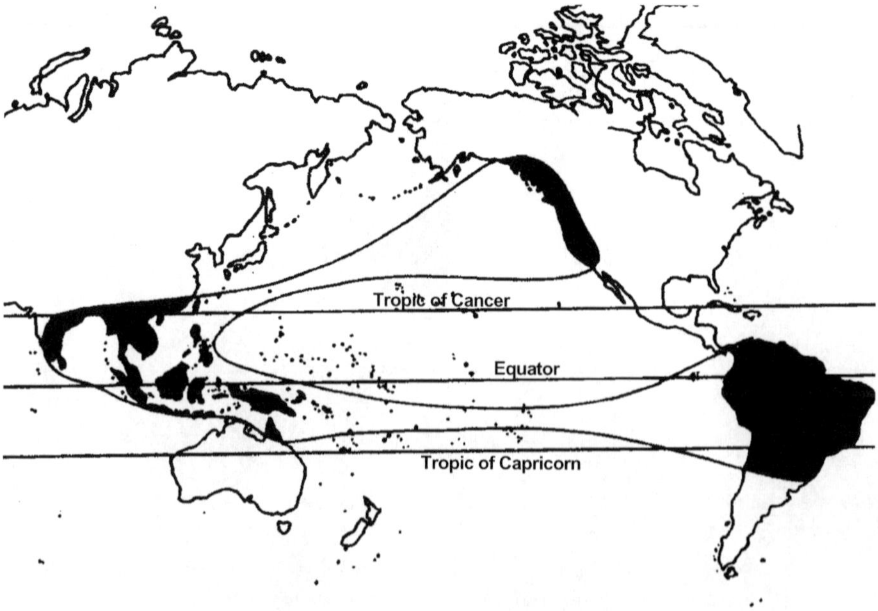

FIGURE 10.12. North and central trans-Pacific tracks. Adapted from Matile (1990).

other chironomid subfamilies. A more southerly "Pacific" track with sister species relationships between Tasmania and Valdivian Chile + Patagonia involves (inter alia) *Aphroteniella* and *Paraphrotenia* (Brundin 1966; Cranston and Edwards 1992), *Botryocladius* (Cranston and Edwards 1999), *Pirara* (Boothroyd and Cranston 1995; Cranston 1999), and *Austrobrillia* (Cranston 2000). The relictual genus *Valeseguya* (Anisopodidae) from a single recent species in Australia (Colless 1990) and a common species in Dominican amber (Grimaldi 1991) apparently fits a Pacific track.

Although dipteran evidence for trans-Pacific track(s) is modest, certainly there are many such tracks among other insects. For example, the Trichopteran sister families Hydrobiosidae + Rhyacophilidae exemplify this track, with Rhyacophilidae typically Laurasian, whereas Hydrobiosidae is present in the Neotropics and along the west coast of the Pacific, from New Zealand to Japan (Mey 1998). The track is known to botanists (Heads 1999) and vertebrate biologists (McCarthy 2003), and we may expect elucidation of the pattern in many more taxa, including for Diptera.

Perhaps Matile's restriction of the Pacific track to the central/north Pacific, with a division into central and northern tracks, is somewhat restrictive, as there may have been in the past extensive continuity between the eastern and western margins of the flanking land masses, thereby producing numerous trans-Pacific vicariance events all the way south to Tasmania and Antarctic Patagonia. As noted above, the conventional earth history explanation for such observed disjunctions relies on concerted long-distance dispersal or unparsimoniously reconstructed major range contractions since the Jurassic. A contrarian perspective is introduced with the concept of the Triassic/Jurassic globe being of smaller dimensions, either lacking the broad Tethys but with a wide Parathalassia (Carey 1975, 1976; Owen 1981), or lacking Parathalassia and with a Jurassic opening of the Pacific (Shields 1998; McCarthy 2003; and see below).

Dipteran Speciation and the Pacific

The study of island biotas inevitably leads to recognition of the significance of founding colonization events and subsequent radiation, none more so than the studies of Hawaiian Drosophilidae (DeSalle 1995; Kaneshiro et al. 1995). Molecular studies identify monophyletic clades derived from single colonization events, although morphology appeared to indicate two or more events. The lack of widespread species (all are single-island endemics) makes analysis rather straightforward, and thus, substituting island for taxon provides species distributions that are generally congruent with the geology, such that the colonists of older islands and older volcanoes (those of Oahu and Molokai) gave rise to descendants that have radiated more recently on the younger islands and younger volcanoes of Maui and Hawaii (Wagner and Funk 1995). A similar scenario of an older colonization with recent radiation associated with island age is seen in the Hawaiian Pipunculidae (de Meyer 1996). However, there are enough examples of back-founder events (recolonization from younger islands to older) to indicate a greater complexity of the evolution of island endemics. A somewhat comparable situation exists with the *Simulium* subgenus *Inseliellum* (Simuliidae) studied by Craig et al. (2002), which is distributed across western Pacific archipelagos and atolls of various ages. The subgenus occurs disparately on Guam (northern Mariana Islands); Chuuk (Caroline Islands); Rarotonga (Cook Islands); and the Marquesas, Society, and Austral Islands (French Polynesia), a range of some 9,000 km. With a well-corroborated phylogeny, Craig et al. (2001) argued that patterns of deep ("basal") branches associated with older islands and the complex younger branching pattern indicate colonization of younger islands, with an ecological restriction imposed by the availability of adequate elevation to allow persistent running water. With erosion and valley development—in particular, on Tahiti—species radiated into specialized habitats, such as cascades, with extinction presumed to have followed elimination of lotic habitats as senescence proceeds with continued erosion.

Somewhat contrasting with the island age, dispersal-to-appropriate habit perspective of Craig et al. (2001), phylogenies of several subgenera of *Bactrocera* (dacine Tephritidae) have been used to deduce southwestern Pacific island biogeography, interpreted in a more exclusively vicariant mode (Michaux and White 1999). Treating widespread species, which formed a modest proportion of the included taxa as informative of monophyly of all occupied areas, a scenario of east Gondwanan fragmentation was proposed by Michaux and White. This involved separation from the eastern Australian/New Guinean craton at the end of the Mesozoic (78–63 Mya) of a complex region of continental crust comprising the Melanesian rift and arc plus the easternmost islands of the southwestern Pacific; namely, the Cooks, Marquesas, Pitcairn, Australs, Society, and Tuamoto, termed "Pacifica," following Nur and Ben-Avraham (1989). Relationships between *Bactrocera* species indicate a sequence of events commencing with separation of the Melanesian rift from the Melanesian arc and Pacifica. Next, the central Melanesian arc (Vanuatu) separated from the southern arc, which comprises (((Fiji, Samoa) Tonga) Niue) plus Pacifica. Finally, Pacifica separated from the southern Melanesian arc, at about 7 Mya. Michaux and White (1999) suggest the hypothesis is tested, because if Pacifican *Bactrocera* is sister to both Melanesian rift and arc (rather than only to southern arc islands), then the area cladogram would look very different, with the dating of Pacifica being substantially older. Melanesian rift is identified as comprising slivers of continental origin (therefore incorrectly termed an "inner Melanesian arc"), including Lord Howe Rise, Norfolk Ridge, New Caledonia, and the Challenger Plateau connected to at least part of New Zealand, plus, at the northern end, some elements of New Guinean central mountains and the Solomons. The occurrence of an endemic species of *Bactrocera* in Micronesia is not readily explained. The morphological phylogeny on which the rationale is based is weakly supported, requiring successive approximations weighting for resolution, and yet it provides a geologically coherent framework for understanding the evolution of this clade of tephritids.

On the other side of the Pacific Ocean, the dipteran fauna of the Galapagos apparently presents a different story from that of Hawaii and Melanesia. The Ceratopogonid fauna of 11 species apparently derived from ten independent founding events (Borkent 1991), a pattern confirmed by Mathis (1995) for the Ephydridae and Bickel and Sinclair (1997) for the Dolichopodidae. Widespread species predominantly are shared with Meso- or South America, and the few endemic species have sister group relationships (where known) either with the Cocos or, more usually, the nearest South American mainland, as is postulated for much of the fauna (Peck 1996). In contrast to the inferred completely dispersal-derived fauna (Peck 1996), a contrarian argument is proposed by Grehan (2001) that associates the Galapagos with tectonism and infers that at least some of the biota, including subgenera of *Paracanace* (Canacidae; Mathis and Wirth 1978), are associated with an early Caribbean plate and others, including *Cymatopus* (Dolichopodidae; Meuffels and Grootaert 1984; Bickel and Sinclair 1997), show a Galapagos–trans-Pacific distribution, possibly Pacifican in the sense of Nur and Ben-Avraham (1989).

Arrival per se at an isolated island does not necessarily result in endemic species radiation, as can be seen, for example, by the dipteran biota of Henderson Island, an isolated elevated coral atoll in the eastern South Pacific. Among 37 species, only three are believed to be endemic—a *Bactrocera* (Tephritidae), a *Gaurax* (Chloropidae), and an undescribed *Dasyhelea* (Ceratopogonidae). Other taxa show pantropical/cosmopolitan distributions, or are either widespread Pacifican or Polynesian (Benton 1995).

Do the Diptera Recognize a Tethys?

Dipteran biogeographic studies have contributed data to a discussion on the past composition of the area dividing Eurasia from Africa and perhaps farther eastward. It is postulated in conventional biogeographic reconstructions on a globe of constant dimensions that in Jurassic/Cretaceous times, non-African Arabia formed the western end of a postulated Tethyan sea/ocean that separated a northern Laurasia from a southern Gondwana. The ocean has been envisaged as narrow, probably epicontinental in the west, broadening out to a huge, deep oceanic expanse at the eastern end, where it separated the Australian plate from Asia and the developing terranes of Southeast Asia. In studying the historical biogeography of Arabia, Cranston and Judd (1989) used a predominantly dispersalist explanation within a phenetic biogeographic paradigm. Geologic evidence suggests that the Arabian peninsula predominantly lay beneath the Tethys, excepting high ground following the Rift Valley along the north-west trending Red Sea shore. Sea-level change associated with the retreat of the Tethyian Sea in the Late Palaeogene exposed the present-day, arid, predominantly lowland landscape. The chironomid fauna of this once-inundated land contains many widely distributed species, which therefore are understood to have colonized the newly exposed habitats. Chironomid colonizers were identified predominantly as having diffused (range expanded) from proximate, previously noninundated Palaearctic areas. Two predominant groups of colonists were evident: those with current-day Mediterranean distributions and those with wider Palaeotropical distributions. Afrotropical species are restricted to the historically noninundated montane areas along the Red Sea. Endemism is low, and in only one case is there postulated vicariant speciation across the Red Sea opening—all other Afrotropical species in Arabia are morphologically undifferentiated from populations on the opposite (Ethiopian) side of the Red Sea. The submerged nature of proto-Arabia in a western Tethys (Adams et al. 1983), or at least a shallow, epicontinental sea prior to the formation of the Mediterranean, is central to this scenario. However, the existence of a broad, deep Tethys separating eastern Gondwana from Laurasia is challenged by many taxon distributions, some of which are discussed above under the sections on the Austro-Oriental and Pacific tracks.

Thus, Matile (1990) presents several distributions, such as that for *Platyroptilon*, for which, if reconstructed on a conventional Jurassic globe, large disjunctions remain (Fig. 10.13A). A postulated broad Tethys separates the Southeast Asian "kirkspriggsi" group from its Australian sister "collessi" group, and this clade is disjunct from the "neotropical" group by Antarctica, Chile-Patagonia, and southern Africa. In contrast, the smaller Jurassic globe implicit in an inconstant dimensions earth—whether with slow (Carey 1975, 1976; Owen 1981) or rapid (Shields 1979, 1983, 1998) expansion—lacks a broad Tethys, and biotic reconstruction on the reduced globe shows impressive contiguity of the clade in the Jurassic (Fig. 10.13B). Furthermore, *Platyroptilon* distribution on an expanding earth model requires no postulated extinction in Antarctica + Patagonia to unite New and Old World clades via a trans-Antarctic track—postulating that ancient contiguity was vicariated by continental separation caused by tectonism, or expansion, or both, with modern ranges representing no more or less than the Jurassic/Cretaceous ranges.

Not only is there no broad Tethys in such a reconstruction, neither is there a Panthalassia, and the Pacific does not commence opening until the Jurassic (Shields 1998; McCarthy 2003). In past reconstructions, biotas in the Americas abut or are contiguous with those of Asia, Southeast Asia, Australia, New Zealand, and the Campbell Plateau, from Beringia to the sub-Antarctic. The opening of the Pacific, which can be likened to the opening of a zipper, creates the vicariance events that produce the patterns identified above as trans-Pacific tracks. Tectonism, such as rifting and the formation of terranes and island arcs in Micronesia and Melanesia, causes additional vicariance and redistribution of subsamples from the original Jurassic biota throughout the enlarging Pacific, Coral, and China Seas. The rapid expanding earth model of Shields (1998) implies that there could be as many trans-Pacific tracks as contiguous ancestral populations existed in the Jurassic/Cretaceous, reflecting ambient ecologies at the opening of the Pacific. The role of ecological stasis in the face of such a mobilist view of earth history needs study.

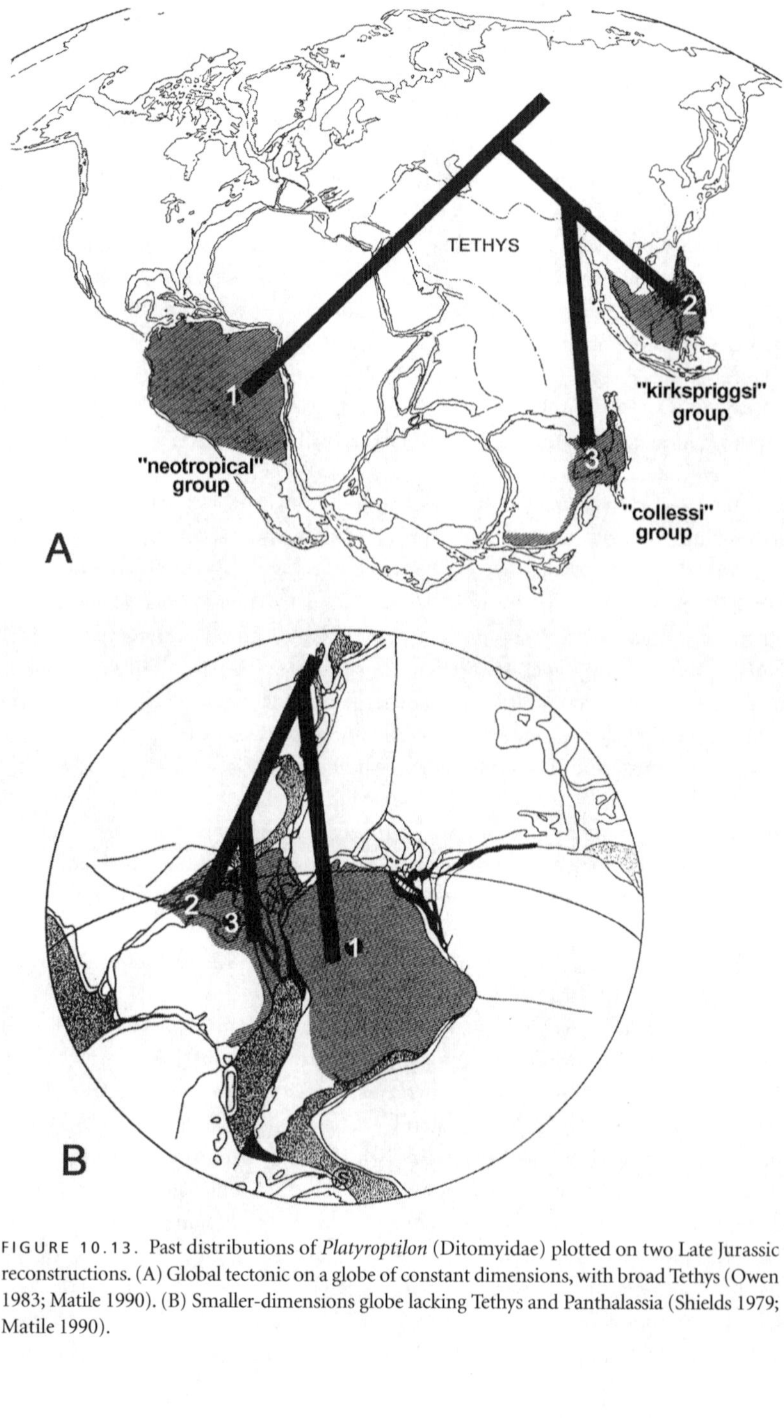

FIGURE 10.13. Past distributions of *Platyroptilon* (Ditomyidae) plotted on two Late Jurassic reconstructions. (A) Global tectonic on a globe of constant dimensions, with broad Tethys (Owen 1983; Matile 1990). (B) Smaller-dimensions globe lacking Tethys and Panthalassia (Shields 1979; Matile 1990).

Significance of Paleontology to Dipteran Biogeography

ROLE OF PALAEONTOLOGY

Just as palaeontology was once considered to be central to phylogenetics, it has been argued that it is central to biogeography in the determination of historic distribution patterns and processes. Thus, many biologists state that the biogeography (and evolutionary history) of their study group cannot be reconstructed because of a paucity of fossils. As seen above, biogeographic reconstruction does not rely on, or even require, a knowledge of fossils. Although palaeontology provides the only direct evidence of past biotic distributions, the imperfections and incompleteness of the fossil record are well known. With the development of phylogenetic reconstruction through cladistics and the development of vicariance methods of biogeographic analysis, even such eminent paleontologists as Patterson (1981) have admitted that fossils were too often uninformative, providing too few characters for analysis, or too vague in their relationships to other fossil and extant taxa. As Amorim and Silva (2002) demonstrated, lack of precision or resolution in positioning fossils in the phylogeny of recent groups generates underestimated ages of groups and an inability to detect biogeographic patterns actually present in the data. Furthermore, assumptions that the site of the oldest fossil of a group or the most "primitive" member marked its center of origin were challenged as untenable. Notwithstanding, fossils can (1) provide new morphological and ontogenetic data for systematics; (2) provide additional taxa to add to the known range of a higher taxon; and, of significance for our discussion below, (3) establish the minimum age for a taxon and (4) allow analysis of biogeographic patterns.

As far as Diptera are concerned, some of Patterson's (1981) reservations are removed, thanks to the excellent preservation of fossils in amber. Fossiliferous amber dates from the Early Cretaceous (Neocomian) of Lebanon; perhaps contemporaneous amber of Burma; through the Cretaceous of New Jersey, Siberia, and Spain; the Late Cretaceous of Canada; the Eocene/Oligocene of the Baltic; and the Miocene of Saxony, Sicily, and Mexico. Unfortunately, amber is rarely present in the southern hemisphere.

FOSSILS AND VICARIANCE-BASED DATING

Evidence supporting vicariance-dated minimum ages for the origins of clades can derive from the fossil record; notably, from insect inclusions in the amber mentioned above. With new discoveries in the past decade and an associated revival of interest in palaeoentomology, many dates for the origin of groups are being taken back much deeper into time. Such dates tend to substantiate vicariance-based dates, especially those established from tectonic vicariance, such as Gondwanan fragmentation. Thus, two vicariance-based biogeographic dating proposals within the Chironomidae, which imply that *Aphrotenia* (Aphroteniinae) and the *Archaeochlus/Austrochlus* clade (Podonominae) date from at least the Jurassic can be substantiated with fossil evidence.

Substantiation for the dating of *Archaeochlus* by South African/Australian vicariance comes from the presence of a quite late-branching podonomine, *Libanoclites* (Brundin 1976), from Early Cretaceous Lebanese amber. Despite reservations concerning some aspects of the phylogenetic relations of the Podonominae, *Libanoclites* is evidently closely related to the modern Laurasian *Boreochlus* and *Paraboreochlus*. Whatever the present geographic relationships of Lebanon in the Late Jurassic–Early Cretaceous, as stated by Brundin (1976), *Libanoclites* confirms that the Podonominae had already radiated in morphology and distribution by the Early Cretaceous.

Another Cretaceous dating is available for the chironomid subfamily Diamesinae, with Kalugina (1976) describing *Cretodiamesa taimyrica* from Late Cretaceous amber from Siberia, allocating it to the new tribe Cretodiamesini. As yet, the phylogenetic placement of this tribe is unknown relative to the modern Diamesinae, whose other tribes show a vicariant Laurasian-Gondwanan pattern.

Fossils have another important role in biogeography: they may refute vicariance-based biogeographic scenarios. For example, fossil evidence of a presently exclusively Gondwanan-distributed taxa from post-Cretaceous (i.e., post break-up) non-Gondwanan facies would require re-examination of an exclusively austral vicariance hypothesis. Thus, the presence in Dominican amber (Late Oligocene/Early Miocene) of *Valeseguya* (Anisopodidae), extant only from Australia, evidently requires a different hypothesis of origination, as does the finding of otherwise Old World tropical *Mesochria* (Anisopodidae) in the same fossil facies (Grimaldi 1991). The tipulid genus *Brachypremna*, known from Australia and the Neotropics regions, has its sister group in Baltic amber (Collucci 2000). Siberian Late Cretaceous amber revealed *Electrotenia brundini* (Kalugina 1980), clearly possessing all adult synapomorphies of the chironomid subfamily Aphroteniinae, which is now confined to the amphinotic Gondwanan track.

The fossil infers the existence of morphological radiation in Pangaea, supporting a role for continental fragmentation plus elucidation of an earlier Pangaean aphroteniine clade prior to Pangaean break-up. The northern Laurasian part appears to have become extinct and the southern section was vicariated by the Jurassic rifting of Gondwana.

Such a model of Pangaeic distribution prior to the Jurassic—as evidenced by fossils, vicariant taxa, or both—with asymmetric modern distributions is not unusual (Brundin 1976; Matile 1990; Grimaldi 1991; Yeates and Irwin 1996; Amorim 1998, 2000). A scenario of differential extinction is suggested, due to climate change in the second half of the Tertiary, with colder temperatures in areas of lower latitude, evidently bringing massive extinction to tropical groups of Laurasian affinities (Amorim and Silva 2002). This model suggests that relictual distributions might be expected in Mediterranean and southern Nearctic areas. Alternatively, a hypothesis of differential extinction in the north (Laurasia) due to the end-of-Cretaceous, Chicxulub (Mexico) bolide impact (e.g., McKinnon 1992) is worth pursuing.

Summary and Suggestions for Future Directions

Diptera and the dipterists who study their phylogeny and biogeography demonstrate much about the processes that have led to diversification through time. Many of these observations and interpretations have greater generality: most patterns observed here, from recent ecological to deep time vicariant, influence much of the earth's terrestrial biota, irrespective of taxonomic group or vagility. However, before suggesting that all patterns discussed here are equiprobable, some cautionary factors must be mentioned.

Vicariant biogeographic relationships postulated between areas are only as good as the phylogenetic hypotheses from which they are derived. Although systematists may have confidence in a phylogeny derived for their own particular study organisms, vicariance biogeography demands comparisons of area relationships derived from many unrelated groups. These may be incongruent with one another, and some incongruence may derive from the different approaches to deriving the phylogenies compared. Furthermore, few cladograms have been tested for robustness and apparently incongruent area relationships may be no more robust than fully congruent ones.

Further difficulties in interpretation arise when well-substantiated area relationships remain contradictory. Genuine incongruence can arise when studies lack comparable areas of endemism due to the effects of redundancy of data, extinction, existence of unique taxa (or areas), and/or the occurrence of widespread taxa that overlap endemic areas defined by other taxa. Furthermore, area relationships may be trivial in providing no more than biotic documentation of the geophysicists' views (e.g., their views of the break-up of the main plates of Gondwana). Detailed relationships (e.g., those concerning the biota of terranes of Southeast Asia) are scarce and when available, their complexities make comparisons very difficult. Vicariance paradigms assume that the earth's history is reflected in the patterns of evolutionary relationships and geographic distribution, yet congruent dispersal could explain similar patterns, although the mechanism that might produce such concerted movements through time is not evident. Vicariance biogeography ought not ignore dispersal, but should reject its primacy. Dispersal that produces incongruence in area cladograms is the equivalent of homoplasious character states in a phylogenetic analysis of characters. Nonetheless, there is evidence of the historical removal of barriers, with range expansion providing the template of ancestral distributions that are disrupted by vicariance events.

The hazards of an inadequate biogeographic database have been alluded to, but if absolute knowledge of the distributions of all Diptera is required before speculating on the mechanisms producing their evolutionary relationships and the reasons for their distributions, we would make no progress. Problems arise from human activity: (1) species are being lost from their native ranges because of anthropogenic impacts, meaning that we may well be interpreting recent "paleontological" patterns rather than current ones; (2) increasingly, we cannot recognize natural distributions, because of the substantial number of synanthropic species that appear to be becoming world-wide as part of the universal faunal homogenization that goes hand-in-hand with enhanced extinction rates. A further difficulty lies in the variation in taxonomic knowledge between zoogeographic regions: a single species may reside under different names in different places. Few taxonomists can take a broad perspective and review faunas on a world-wide basis, although this approach is fundamental to biogeographic studies.

Accepting the limitations of our present knowledge, at least we can frame biogeographic hypotheses in a scientific manner, making them amenable to testing against further data. This implies being explicit about the relevance of new data, allowing new data to be sought in such a way as to provide an explicit test of a well-formulated hypothesis. Willassen and Cranston (1986) used phylogeny to postulate that the eastern African montane *Diamesa* were derived by dispersal from their northern Caucasian sister taxa, and were not part of the Gondwanan radiation of the subfamily Diamesinae. The scenario was suggested to be testable in two ways: (1) by congruence of the observed pattern with other organisms, but more particularly, (2) by the predicted phylogenetic relationship between other Gondwanan *Diamesa* species. In the latter test, a Gondwanan, but non-African, *Diamesa* with a closer relationship to Afrotropical taxa than to a northern (Laurasian) group, was seen as a potential falsifier of their biogeographic hypothesis. A similar test of the proposed relative recency of the separation of "Pacifica" from the southern Micronesian arc, in which sister group relationships exist between Pacifican taxa and Melanesian rift, rather than southern Melanesian arc, is proposed by Michaux and White (1999).

In an analogous manner, distributions postulated as reflecting the temporal sequence of Gondwanan fragmentation may be amenable to testing. First, fossil evidence may demonstrate the previous presence of the taxon in an area unassociated with Gondwana. Second, the discovery of crucial taxa lying outside the regions of Gondwanan origin might serve to refute a particular hypothesis. However, three problems arise in defining testable hypotheses concerning Gondwanan distributions. Some of these are exemplified by *Parochlus kiefferi*, as discussed above and by Cranston (1994): (1) dispersal may have occurred from a Gondwanan area at any time in the past, thereby disrupting a purely vicariance-derived pattern; (2) the "non-Gondwanan" area actually may be of Gondwanan origin—the fragmentation history of small blocks (terranes) on the margins of Gondwana is complex and still somewhat unclear; (3) the clade may predate an early Cretaceous Gondwana, perhaps having originated in earlier geologic time in megacontinental Pangaea; (4) the distinction between Laurasia and Gondwana may be an artifact of past continental reconstruction on a constant dimensions globe (i.e., there never was a broad Tethys).

Besides erecting testable hypotheses concerning biogeographic scenarios, certain other matters could be profitably pursued; namely, the ecological basis, if there is one, between different Gondwanan tracks. The generally accepted Gondwanan fragmentation, which leaves vicariant daughter taxa in the cooler parts of the southern hemisphere, the transantarctic and amphinotic tracks, is represented by organisms that seem typically cool temperate in their physiology. In contrast, equally notable tropical Gondwanan pattern (trans-Indian Ocean, south Atlantic, and Pacific tracks) involves warm eurythermic species. Studies of a some purportedly cool temperate "Gondwanan" chironomids show that they are more widely distributed than recognized by Brundin (1966), and although their ecophysiological responses actually reflect more mesothermy than stenothermy (McKie et al. in press), their history and ecology are strongly intertwined. Just how much ecological stasis of the biota, including dipterans, through geologic time has constrained past and current distributions evidently is of great importance, but is not readily amenable to incorporation into evolutionary biogeographic studies.

ACKNOWLEDGMENTS

This contribution is dedicated to the memory of Loïc Matile, entomologist, biogeographer, anglophile, bon-vivant, and scientific sceptic. I am happy that Dalton Amorim de Souza and Dennis McCarthy provoked in me a deeper re-investigation of some of the more iconoclastic literature. Any taxonomic bias toward the "Lower Diptera" (=Nematocera) results from my greater familiarity with the grade, but may also reflect a genuine bias among biogeographically orientated dipterists.

REFERENCES

Adams, C. G., A. W. Gentry, and P. J. Whybrow. 1983. Dating the terminal Tethyan event. *Micropaleontol. Bull.* 30:273–298.

Amorim, D. S. 1989. A new species of the genus *Diamphidicus* Cook (Diptera: Bibionomorpha: Scatopsidae) from Chile. *Rev. Bras. Entomo.* 33:477–482.

Amorim, D. S. 1998. Amber fossil Scatopsidae (Diptera: Psychodomorpha): I. Considerations on described taxa, *Procolobostema roseni,* new species, from Dominican amber, and the position of *Procolobostema* in the family. *Am. Mus. Novit.* 3227:1–17.

Amorim, D. S. 2000. A new phylogeny and phylogenetic classification for the Canthyloscelidae (Diptera: Psychodomorpha). *Can. J. Zool.* 78:1067–1077.

Amorim, D. S., and M. R. S. Pires. 1996. Neotropical biogeography and a method for maximum biodiversity estimation. In C. E. M. Bicudo and N. A Menezes, eds., *Biodiversity in Brazil: A First Approach,* pp. 183–219. São Paulo: CNPq.

Amorim, D. S., and V. C. Silva. 2002. How far advanced was Diptera evolution in the Pangaea? *Ann. Soc. Entomol. France* 38:177–200.

Amorim, D. S., and S. H. S. Tozoni. 1994. Phylogenetic and biogeographic analysis of the Anisopodoidea (Diptera, Bibionomorpha), with an area cladogram for intercontinental relationships. *Rev. Bras. Entomol.* 38:517–543.

Anderson, N. 1992. Influence of disturbance on insect communities in Pacific Northwest streams. *Hydrobiologia* 248:79–92.

Benton, T. G. 1995. Biodiversity and biogeography of Henderson Island's insects. *Biol. J. Linn. Soc.* 56: 245–259.

Bickel, D. J. 1996. *Thinempis,* a new genus from Australia and New Zealand (Diptera: Empididae), with notes on the tribal classification of the Empidinae. *Syst. Entomol* 21:115–128.

Bickel, D. J., and B. J. Sinclair. 1997. The Dolichopodidae (Diptera) of the Galapagos Islands, with notes on the New World fauna. *Entomol. Scand.* 28:241–270.

Boothroyd, I. K. G., and P. S. Cranston. 1995. Two Orthocladiinae (Chironomidae) genera common to New Zealand and Australia: *Pirara* n. gen. and *Eukiefferiella* Thienemann. In P. S. Cranston, ed., *Chironomids: From Genes to Ecosystem,* pp. 389–408. Melbourne: CSIRO.

Borkent, A. 1991. The Ceratopogonidae (Diptera) of the Galapagos Islands, Ecuador, with a discussion of their phylogenetic relationships and zoogeographic origins. *Entomol. Scand.* 22:97–122.

Brundin, L. 1966. Transantarctic relationships and their significance, as evidenced by chironomid midges with a monograph of the subfamilies Podonominae and Aphroteniinae and the austral Heptagyiae. *Kung. Sven. Vetenskaps. Hand.* 11:1–472.

Brundin, L. 1972. Phylogeny and biogeography. *Syst. Zool.* 21:69–79.

Brundin, L. 1976. A Neocomian chironomid and Podonominae-Aphroteniinae (Diptera) in the light of phylogenetics and biogeography. *Zool. Script.* 5:139–160.

Brundin, L. 1988. Phylogenetic biogeography. In A. A. Myers and P. S. Giller, eds., *Vicariance Biogeography: An Integrated Approach to the Study of Animal and Plant Distributions,* pp. 343–369. London: Chapman and Hall.

Bush, G. L. 1975. Modes of animal speciation. *Annu. Rev. Ecol. Syst.* 6:339–364.

Carey, S. W. 1975. The expanding Earth—An essay review. *Earth Sci. Rev.* 11:105–143.

Carey, S. W. 1976. The expanding Earth. *Dev. Geotect.* 10:1–488.

Carpenter G., A. N. Gillison, and J. Winter. 1993. DOMAIN: A flexible modelling procedure for mapping potential distributions of plants and animals. *Biodivers. Conserv.* 2:667–680.

Colless, D. H. 1970. The Mycetophilidae (Diptera) of Australia, part 1. Introduction, key to subfamilies, and review of the Ditomyiinae. *J. Aust. Entomol. Soc.* 9:83–99.

Colless, D. H. 1990. *Valeseguya rieki,* a new genus and species of dipteran from Australia (Nematocera: Anisopodidae). *Ann. Soc. Entomol. France* 26:351–353.

Colless, D. H., and D. K. McAlpine. 1991. Diptera (flies). In CSIRO, *The Insects of Australia,* 2nd ed., pp. 717–786. Carlton: Melbourne University Press.

Collucci, E. 2000. *Análise cladística do gênero Brachypremna Osten Sacken 1886 (Diptera: Tipulidae).* Ph.D. thesis, Universidade de São Paulo, Ribeirão Preto.

Coope, G. R. 1979. Late Coenozoic fossil Coleoptera: Evolution, biogeography and ecology. *Annu. Rev. Ecol. Syst.* 10:247–267.

Craig, D. A., D. C. Currie, and D. Joy. 2001. A geographical history of the central-western Pacific black fly subgenus *Inseliellum* (Diptera: Simuliidae: *Simulium*) based on a reconstructed phylogeny of the species, hot-spot archipelagoes and hydrological considerations. *J. Biogeog.* 28:1101–1127.

Cranston, P. S. 1994. Biogeography. In P. Armitage, P. S. Cranston, and L. C. V. Pinder, eds., *Chironomidae: Biology and Ecology of Non-Biting Midges,* pp. 62–84. London: Chapman and Hall.

Cranston, P. S. 1999. Two unusual Chironomini (Diptera: Chironomidae) from Australian rainforest streams: One new genus and a neotropical genus new for the region. *Aust. J. Entomol.* 38:291–299.

Cranston, P. S. 2000. *Austrobrillia* Freeman: Immature stages, and a new species from the Neotropics. *Spixiana* 23:101–111.

Cranston, P. S., and D. H. D. Edwards. 1992. A systematic reappraisal of the Australian Aphroteniinae (Chironomidae) with dating from vicariance biogeography. *Syst. Entomol.* 17:41–54.

Cranston, P. S., and D. H. D. Edwards. 1999. *Botryocladius* gen. nov.: A new transantarctic genus of orthocladiine midge (Diptera:Chironomidae). *Syst. Entomol.* 24:305–333.

Cranston, P. S., and D. D. Judd. 1989. Diptera: Fam. Chironomidae of the Arabian Peninsula. *Fauna Saudi Arabia* 10:236–89.

Cranston P. S., and I. Naumann. 1991. Biogeography. In CSIRO, *The Insects of Australia* 2nd ed., pp. 180–197. Carlton: Melbourne University Press.

Cranston, P. S., and U. Nolte. 1996. *Fissimentum,* a new genus of drought-tolerant chironomini (Diptera: Chironomidae) from the Americas and Australia. *Entomol. News* 107:1–15.

Cranston, P. S., D. H. D. Edwards, and L. G. Cook. 2002. New status, distribution records and phylogeny for Australian mandibulate Chironomidae (Diptera). *Aust. J. Entomol.* 41:357–366.

Craw, R. C. 1988. Panbiogeography: Method and synthesis in biogeography. In A. A. Myers and P. S. Giller, eds., *Analytical Biogeography. An Integrated Approach to the Study of Animal and Plant Distributions,* pp. 405–435, London, New York: Chapman and Hall.

Croizat, L. 1958. *Panbiogeography, xxii.* Caracas: privately published by the author.

Croizat, L. 1964. *Space, Time, Form: The Biological Synthesi.* Caracas: privately published by the author.

Croizat, L., G. Nelson, and D. Rosen. 1974. Centres of origin and related concepts. *Syst. Zool.* 23:265–287.

Crosskey, R. W. 1990. *The Natural History of Blackflies.* London: British Museum of Natural History.

Darwin, C. 1859. *On the Origin of Species by Means of Natural Selection.* London: John Murray.

Davies, D. M., and H. Györkös. 1988. Two new Australian species of Simuliidae (Diptera). *J. Aust. Entomol. Soc.* 27:105–115.

Davis, A. J., L. S Jenkinson, J. H. Lawton, B. Shorrocks, and S. Wood. 1998. Making mistakes when predicting shifts in species range in response to global warming. *Nature* 391:783–786.

de Jong, H. 1998. In search of historical biogeographic patterns in the western Mediterranean terrestrial fauna. *Biol. J. Linn. Soc.* 65:99–164.

de Meyer, M. 1996. Cladistic and biogeographic analyses of Hawaiian Pipunculidae (Diptera) revisited. *Cladistic* 12:291–303.

DeSalle, R. 1995. Molecular approaches to biogeographic analysis of Hawaiian Drosophilidae. In W. L. Wagner and V. A. Funk, eds., *Hawaiian Biogeography: Evolution on a Hot Spot Archipelago,* pp. 72–89. Washington, D.C.: Smithsonian Institution Press.

de Wet, N., W. Ye, S. Hales, R. Warrick, A. Woodward, and P. Weinstein. 2001. Use of a computer model to identify potential hotspots for dengue fever in New Zealand. *New Zeal. Med. J.* 114:420–422.

Dimitriadis, S., and P. S. Cranston. 2001. An Australian Holocene climate reconstruction using Chironomidae from a tropical volcanic maar lake. *Palaeogeogr. Palaeoclim.* 176:109–131.

Drew, R. A. I., D. L. Hancock, and I. M. White. 1994. The identity of *Dacus limbipennis* Macquart (Diptera: Tephritidae). *J. Aust. Ento. Soc.* 33:376.

Eber, S., and R. Brandl. 1997. Genetic differentiation of the tephritid fly *Urophora cardui* in Europe as evidence for its biogeographical history. *Molec. Ecol.* 6:651–660.

Erichson, W. F. 1842. Beitrag zur Fauna von Vandiemensland mit besonderer Rücksicht auf die geographische Verbreitung der Insekten. *Wiegmann Arc.* 8:83–287.

Esseghir, S., P. D. Ready, and R. Ben-Ismail. 2000. Speciation of *Phlebotomus* sandflies of the subgenus *Larroussius* coincided with the Late Miocene–Pliocene aridification of the Mediterranean subregion. *Biol. J. Linn. Soc.* 70:189–219.

Gaimari, S. D., and M. E. Irwin. 2000. Phylogeny, classification, and biogeography of the cycloteline Therevinae (Insecta: Diptera: Therevidae). *Zool. J. Linn. Soc.* 129:129–240.

Godfray, H. C. J. 2002. How might more systematics be funded? *Antenna* 21:11–17.

Grehan, J. 2001. Biogeography and evolution of the Galapagos: Integration of biological and geological evidence. *Biol. J. Linn. Soc.* 74:267–287.

Grimaldi, D. A. 1991. Mycetobiine woodgnats (Diptera: Anisopodidae) from the Oligo-Miocene amber of the Dominican Republic and Old World affinities. *Am. Mus. Novit* 3014:1–24.

Hall, R., and J. D. Holloway. 1998. *Biogeography and Geological Evolution of SE Asia.* Leiden: Backhuys.

Harrison, L. 1928. The composition and origins of the Australian fauna, with special reference to the Wegener hypothesis. *Report of the 18th Meeting of the Australasian Association for the Advancement of Science* 28:332–396.

Heads, M. 1999. Vicariance biogeography and terrane tectonics in the South Pacific: Analyses of the genus *Abrotanella* (Compositae). *Biol. J. Linn. Soc.* 67:391–432.

Hennig, W. 1960. Die Dipteren-Fauna von Neuseeland als systematisches und tiergeographisches Problem. *Beitr. Entomol.* 10:221–329.

Hoff, M. J., and J. A. Foley. 2001. Global-scale relationships between climate and the dengue fever vector, *Aedes aegypti. Clim. Change* 48:441–463.

Hooker, J. D. 1845. On the Huon Pine, and on *Microcachrys*, a new genus of Coniferae from Tasmania, with remarks on the geographical distribution of that order in the southern hemisphere. *J. Botany* 4: 137–157.

Hooker, J. D. 1853. *The Botany of the Antarctic Voyage of H.M. Discovery Ships* Erebus *and* Terror *in the Years 1839–1843. II. Flora Novae-Zelandiae.* London: Lovell Reeve.

Howarth, M. K. 1981. Palaeogeography of the Mesozoic. In L. C Cocks, ed., *The Evolving Earth,* pp. 197–220. London: British Museum of Natural History.

Humphries, C. J. 2000. Vicariance biogeography. In S. A. Levin, ed., *Encyclopedia of Biodiversity,* vol. 5, pp. 767–779. San Diego: Academic Press.

Hutson, A. M. 1977. A revision of the families Synneuridae and Canthyloscelidae (Diptera). *Bull. Brit. Mus. Nat. Hist. Entomol.* 35:65–100.

Jenkins, N. L., and A. A. Hoffmann. 2001. Distribution of *Drosophila serrata* Malloch (Diptera: Drosophilidae) in Australia with particular reference to the southern border. *Aust. J. Entomol.* 40:41–48.

Kalugina, N. S. 1976. Non-biting midges of the subfamily Diamesinae (Diptera, Chironomidae) from the Upper Cretaceous of the Taimyr. *Paleontol. J.* 10(1):78–83. [English translation of Russian]

Kalugina, N. S. 1980. Cretaceous Aphroteniinae from North Siberia (Diptera: Chironomidae). *Electrotenia brundini* gen. nov., sp. nov. *Acta U. Carol. Biol.* 1978:89–93.

Kaneshiro K. Y., R. G. Gillespie, and H. L. Carson. 1995. Chromosomes and male genitalia of Hawaiian *Drosophila.* In W. L. Wagner and V. A. Funk, eds., *Hawaiian Biogeography: Evolution on a Hot Spot Archipelago,* pp. 57–71. Washington, D.C.: Smithsonian Institution Press.

Léger, N., and J. Depaquit. 2002. Systématique et biogéographie des Phlébotomes (Diptera: Psychodidae). *Ann. Soc. Entomol. France* 38:163–175.

Macarthur, R. H., and E. O. Wilson. 1967. *The Theory of Island Biogeography.* Princeton: Princeton University Press.

MacFarlane, R. P., and I. G. Andrew. 2001. New Zealand Diptera identification, diversity and biogeography: A summary. *Rec. Canterbury Mus.* 15:33–72.

Mackerras, I. M. 1950. The zoogeography of the Diptera. *Aust. J. Sci.* 5:1–6.

Mackerras, I. M. 1962. Speciation in Australian Tabanidae. In G. W. Leeper, ed., *The Evolution of Living Organisms,* pp. 328–358. Carlton: Melbourne University Press.

Massafero, J., and S. J. Brooks. 2002. Response of chironomids to Late Quaternary environmental change in the Taito Peninsula, southern Chile. *J. Quatern. Sci.* 17:101–111.

Mathis, W. N. 1995. Shore flies of the Galapagos Islands (Diptera: Ephydridae). *Ann. Entomol. Soc. Am.* 88:627–640.

Mathis, W. N., and W. W. Wirth. 1978. A new genus near *Canaceoides* Cresson, three new species and notes on their classification (Diptera: Canacidae). *P. Entomol. Soc. Wash.* 80:524–537.

Matile, L. 1988. *Rocetelion,* a new holarctic genus of the Keroplatidae (Diptera: Mycetophiloidea): Description, phylogenetic and biogeographical notes. *Suomen hyönt. Akak* 54:107–113.

Matile, L. 1990. Recherches sur la systématique et l'évolution des Keroplatidae (Diptera, Mycetophiloidea). *Mém. Mus. Natl. Hist. Nat. A Zool.* 148:1–682.

Matile, L. 1991. Diptera Mycetophiloidea from New Caledonia South Pacific Ocean 4. Mycetophilidae; Mycomyinae, Sciophilinae and Gnoristinae. *Mém. Mus. Natl. Hist. Nat. A Zool.* 149:233–250.

McCarthy, D. 2003. The trans-Pacific zipper effect: Disjunct sister taxa and matching geological outlines that link the Pacific margins. *J. Biogeogr.* 30:1545–1561.

McKie B. G. L., P. S. Cranston, and R. G. Pearson. In press. Gondwanan mesotherms and widespread eurytherms: Effects of temperature on the development and survival of several Australian Chironomidae (Diptera). *Austral Ecol.* in press.

McKinnon, W. B. 1992. Killer acid at the K/T boundary. *Nature* 357:15–16.

Metcalfe, I. 1998. Palaeozoic and Mesozoic geological evolution of the SE Asian region: Multidisciplinary constraints and implications biogeography. In R. Hall and J. D. Holloway, eds., *Biogeography and Geological Evolution of SE Asia,* pp. 25–41. Leiden: Backhuys.

Meuffels, H. J. G., and P. Grootaert. 1984. Dolichopodidae (Diptera) from Papua New Guinea I: The genus *Cymatopus* Kertész with a discussion on *Abatetia* Miller and *Cemocarus* gen. nov. *Indo-Malayan Zool.* 1:141–158.

Mey, W. 1998. The distribution of *Apsilochorema* Ulmer, 1907: Biogeographic evidence for the Mesozoic accretion of a Gondwana microcontinent to Laurasia. In R. Hall and J. D. Holloway, eds., *Biogeography and Geological Evolution of SE Asia,* pp. 91–98. Leiden: Backhuys.

Michaux, B., and I. M. White. 1999. Systematics and biogeography of the south-west Pacific *Bactrocera* (Diptera: Tephritidae: Dacini). *Palaeogeogr. Palaeoclim.* 153:337–351.

Michelsen, V. 1988. A world revision of *Strobilomyia* new-genus the anthomyiid seed pests of conifers (Diptera: Anthomyliidae). *Syst. Entomol* 13:271–314.

Moran, N. A. 1989. A 48-million-year-old aphid-host plant association and complex life cycle: Biogeographic evidence. *Science* 245:173–175.

Munroe, D. D. 1974. The systematics, phylogeny, and zoogeography of *Symmerus* Walker and *Australosymmerus* Freeman (Diptera: Mycetophilidae: Ditomyiinae). *Mem. Entomol. Soc. Can.* 92:1–183.

Nagatomi, A. 1982. Geographical distribution of the Lower Brachycera. *Pac. Insects* 24:139–150.

Nagatomi, A. 1991. History of some families of Diptera, chiefly those of the Lower Brachycera (Insecta: Diptera). *Bull. Biogeogr. Soc. Jpn.* 46:1–22.

Nelson, G., and P. Ladiges. 1996. Paralogy in cladistic biogeography and analysis of paralogy-free subtrees. *Am. Mus. Novit.* 3167:1–58.

Nix, H. 1986. A biogeographic analysis of Australian elapid snakes. In R. Longmore, ed., *Atlas of Elapid Snakes of Australia,* pp. 4–15. Canberra: Australian Government Publishing Service.

Nur, A., and Z. Ben-Avraham. 1989. Oceanic plateaus and the Pacific Ocean margins. In Z. Ben-Avraham, ed., *The Evolution of the Pacific Ocean Margins,* pp. 7–19. Oxford Monograph of Geology and Geophysics 8. Oxford: Clarendon.

Oosterbroek, P., and J. W. Arntzen. 1992. Area-cladograms of circum-Mediterranean taxa in relation to Mediterranean palaeobiogeography. *J. Biogeog.* 19:3–20.

Owen, H. G. 1981. Constant dimensions or an expanding earth? In L. C. Cocks, ed., *The Evolving Earth,* pp. 179–192. London: British Museum of Natural History.

Owen, H. G. 1983. *Atlas of Continental Displacement: 200 Million Years to the Present.* Cambridge: Cambridge University Press.

Papavero, N. 1977. *The World Oestridae (Diptera), Mammals and Continental Drift.* The Hague: W. Junk.

Pape, T. 2001. Phylogeny of the Oestridae. *Syst. Entomol.* 26:133–171.

Patterson, C. 1981. The development of the North American fish fauna—A problem of historical biogeography. In P. L. Forey, ed., *The Evolving Biosphere,* pp. 265–281. London: British Museum of Natural History.

Peck, S. B. 1996. Origin and development of an insect fauna on a remote archipelago: The Galapagos Islands, Ecuador. In A. Keast and S. E. Miller, eds., *The Origin and Evolution of Pacific Island Biotas, New Guinea to Eastern Polynesia: Patterns and Processes,* pp. 91–122. Amsterdam: SPB Academic Publishing.

Petersen, F. T., and R. Meier. 2001. A preliminary list of the Diptera of Denmark. *Steenstrupia* 26:119–276.

Reiter, P. 2001. Climate change and mosquito-borne disease. *Environ. Health Persp. Suppl. 1.* 109:141–146.

Roig Junent, S., and C. Coscarón. 2001. Biogeographical history of the Neotropical and Neantarctic Simuliidae (Diptera). *Rev. Mus. Argen. Cien. Nat.* 3:119–134.

Shields, O. 1979. Evidence for initial opening of the Pacific ocean in the Jurassic. *Palaeogeogr. Palaeoclim* 26:181–220.

Shields, O. 1983. Trans-Pacific biotic links that suggest Earth expansion. In S. W. Carey, ed., *Expanding Earth Symposium, Sydney, 1981,* pp. 199–205. Hobart: University of Tasmania.

Shields, O. 1998. Upper Triassic Pacific vicariance as a test of geological theories. *J. Biogeogr.* 25:203–211.

Sinclair, B. J. 2000. Revision of the genus *Clinocera* Meigen from Australia and New Zealand (Diptera: Empididae: Clinocerinae). *Invert. Tax* 14:347–361.

Stuart, M. A., D. G. Mayer, M. G. Atzeni, G. M. McKeon, and J. P. Spradbery. 1995. Determination of a relationship between soil moisture and screwworm fly (Diptera: Calliphoridae) activity. *J. Aust. Entomol. Soc.* 34:1–5.

Sutherst, R. W., and Maywald, G. F. 1985. A computerised system for matching climates in ecology. *Agr. Ecosyst. Environment* 13:281–299.

Sutherst, R. W., B. S. Collyer, and T. Yonow. 2000. The vulnerability of Australian horticulture to the Queensland fruit fly, *Bactrocera (Dacus) tryoni,* under climate change. *Aust. J. Agr. Res.* 51:467–480.

Tangelder, I. R. M. 1988. The biogeography of the Holarctic *Nephrotoma-dorsalis* species-group Diptera: Tipulidae. *Beauforti* 38:1–35.

Thornton, I. W. B., and T. R. New. 1988. Krakatau invertebrates: The 1980s fauna in the context of a century of recolonisation. *Philos. Trans. R. Soc. Lond.* 322:493–522.

Tillyard, R. J. 1926. *The Insects of Australia and New Zealand.* Sydney: Angus and Robertson.

Torsvik, T. H., R. D. Tucker, L. D. Ashwal, L. M. Carter, B. Jamtveit, K. T. Vidyadharan, and P. Venkataramana. 2000. Late Cretaceous India-Madagascar fit and timing of break-up related magmatism. *Terra Nov* 12:220–224.

van Klinken, R. D., and G. H. Walter. 2001. Larval hosts of Australian Drosophilidae (Diptera): A field survey in subtropical and tropical Australia. *Aust. J. Entomol.* 40:163–179.

Wagner W. L., and V. A. Funk. 1995. Biogeographic patterns in the Hawaiian Islands. In W. L. Wagner and V. A. Funk, eds., *Hawaiian Biogeography: Evolution on a Hot Spot Archipelago,* pp. 379–419. Washington, D.C.: Smithsonian Institution Press.

Walker, I. 1994. Chironomids as indicators of past environmental change. In P. Armitage, P. S. Cranston, and L. C. V. Pinder, eds., *Chironomidae: Biology and Ecology of Non-Biting Midges,* pp. 405–422. London: Chapman and Hall.

Wallace, A. R. 1876. *The Geographical Distribution of Animals,* vol. 1. London: Macmillan.

Wegener, A. 1915. *Die Entstehumg der Kontinente und Ozeane.* Vieweg: Braunschweig.

Willassen, E., and P. S. Cranston. 1986. African montane midges (Diptera, Chironomidae, *Diamesa*); biosystematics and biogeography. *Zool. J. Linn. Soc.* 87:91–123.

Wittmann, E. J., P. S. Mellor, and M. Baylis. 2001. Using climate data to map the potential distribution of *Culicoides imicola* (Diptera: Ceratopogonidae) in Europe. *Rev. Sci. Tech. Office Int. Epizoot.* 20:731–740.

Wood, D. M. 1978. Taxonomy of the Nearctic species of *Twinnia* and *Gymnopais* (Diptera: Simuliidae) and a discussion of the ancestry of the Simuliidae. *Can. Entomol* 110:1297–1338.

Worner, S. P. 1988. Ecoclimatic assessment of potential establishment of exotic pests. *J. Econom. Entomol.* 81:973–983.

Yeates, D. K., and M. E. Irwin. 1996. Apioceridae (Insecta: Diptera): Cladistics and biogeography. *Zool. J. Linn. Soc.* 116:247–301.

Yonow, T., and R. W. Sutherst. 1998. The geographical distribution of the Queensland fruit fly, *Bactrocera (Dacus) tryoni,* in relation to climate. *Aust. J. Agr. Res.* 49:935–953.

Zwick, P. 1977. The Australian Blephariceridae (Diptera). *Aust. J. Zool. Suppl.* 46:1–121.

CHAPTER ELEVEN

Sexual Selection and the Evolution of Mating Systems in Flies

Gerald S. Wilkinson and Philip M. Johns

Flies are a model group for the study of mating systems because of their extensive evolutionary diversification and ecological variation (see Yeates and Wiegmann, Chapter 2). For example, adults of some species feed on nectar, pollen, or other exudates; others suck blood from vertebrate hosts or are exclusively predaceous; and a few are even kleptoparasitic (Sivinski 1999). Larvae of many species are detritovores in semiaquatic or aquatic environments, but parasitic larval forms have also evolved repeatedly, as have plant-feeding forms. Based on classical mating system theory (Emlen and Oring 1977), one would expect that the ecological diversity exhibited by adult flies would influence the form of the mating system. As will become clear in this chapter, to a large extent, this prediction is supported. In addition, many species of flies in two families, Tephritidae and Drosophilidae, have received extensive study due to their economic importance and use as model organisms, respectively. The scope, detail, and diversity of studies on these groups provide fertile ground to identify patterns and establish causal mechanisms. Finally, due in part to the development of *Drosophila melanogaster* as a model system for the study of genetics, Mendelian markers have been available for many years and have been used to quantify parentage and sperm precedence. The combination of genetically confirmed parentage with detailed anatomical, physiological, and molecular studies on the factors influencing fertilization success have allowed for unprecedented insight into the mechanisms of sperm competition in *Drosophila* and a few other species of flies.

In this review, we adopt the perspective that an animal's mating system encompasses all activities that influence zygote formation of males, females, and their gametes. Thus, after briefly reviewing recent theoretical work on sexual selection, we have organized this chapter according to the events that transpire before, during, or after mating that influence fertilization success. We divide these events into four categories: obtaining mates, precopulatory activity, copulation, and postcopulatory activity. We further assume that if variation in morphological, behavioral, physiological, or biochemical factors influences fertilization success, then sexual selection will occur and, if genetic variation is present, result in evolutionary change. Throughout the chapter, we use this evolutionary logic to interpret variation within and among species of flies.

Our goal in this chapter is to review what is known about the mating systems of flies and identify where additional study would be particularly fruitful to advance mating system theory. We also consider the degree to which ecological factors have influenced pre- and post-copulatory activity by identifying traits that have likely undergone rapid evolutionary change.

Ideally, this chapter would contain a number of comparative analyses in which, for example, the evolution of mating systems and related traits were inferred from parsimonious reconstruction of character change on a phylogenetic tree of all Diptera. We could then test the role of ecology and other factors by examining the significance of correlated change among phylogenetically independent contrasts (Felsenstein 1985). Unfortunately, relationships among many groups of flies remain controversial (Yeates and Wiegmann 1999), so such analyses across all Diptera would be premature. However, to facilitate discussion of possible evolutionary patterns, we use Lower and Higher Diptera to indicate the groups of families conventionally referred to as the Nematocera and Brachycera, respectively (Yeates and Wiegmann 1999). Moreover, we do discuss and present trait evolution for a few groups of flies for which phylogenetic information is available.

Sexual Selection and Sexual Conflict

One of the earliest experimental studies demonstrating the process of sexual selection was conducted using flies. By combining three males and three females each with recognizable dominant mutations into replicate vials, Bateman (1948) showed that the variance in reproductive success among male *D. melanogaster* was greater than among females. He inferred from these results that traits that enhance male mating ability will be favored by sexual selection. In contrast, the reproductive success of females is limited more by egg production than by the number of mates. Thus in flies, as well as in other animals, male and female reproductive interests rarely coincide, because relative investment in gamete size and number differ (Trivers 1972). When males produce many small gametes, they are expected to expend energy searching and competing for mates. In contrast, females that produce a few large gametes are expected to base mating decisions on their ability to acquire resources and produce offspring. These ideas have been used successfully to explain how variation in the distribution of resources influences the distribution of females, which, in turn, influences how males distribute themselves to obtain matings, both among animals in general (Emlen and Oring 1977) and insects in particular (Parker 1978).

Sexual selection can lead to the evolution of sexual dimorphism if a trait influences a male's ability to acquire and defend resources or to attract females. For example, males often have larger bodies than females when there is resource defense (e.g., Borgia 1980; Dodson 1986; Day et al. 1987; Ding and Blanckenhorn 2002). Sexually dimorphic traits also can be the result of sexual selection acting via female choice. Although females may prefer larger males (Gilburn et al. 1992) or males with head modifications (Fig. 11.1) that facilitate fighting or assessment (Wilkinson and Dodson 1997), females of some species choose males on the basis of traits not associated with fighting. In flies, such traits include food, pheromones, courtship behaviors, or morphological features. For example, the midlegs of the culicid mosquito, *Sabethes cyaneus,* form large, iridescent paddles that males wave while hanging to attract females (Hancock et al. 1990).

Considerable controversy revolves around what females gain from choosing males. In some cases, benefits are direct because female fecundity is enhanced. For example, in several dance fly species (family Empididae), females choose males that offer them prey (Cumming 1994). In many other cases, however, females receive only sperm from their mates, so benefits can only be indirect in the sense that the offspring, rather than the female, benefit.

FIGURE 11.1. Examples of flies with head projections. (A) *Teleopsis breviscopium* from Brunei; (B) *Phytalmia cervicornis* from Papua New Guinea; (C) *Zygothrica exuberans* from Ecuador; (D) *Drosophila heteroneura* from Hawaii. Reproduced with permission.

One possibility is that offspring of choosy females simply inherit genes for attractive male traits and for preferences. Such a scenario leads to rapid coevolution of trait and preference in a runaway process until natural selection acts against the ornament (Fisher 1958; Lande 1981; Kirkpatrick 1982). Alternatively, offspring might inherit genetic factors that increase their survival (Pomiankowski 1988) or otherwise enhance the propagation of their genes, such as by distorting the sex ratio (Lande and Wilkinson 1999). Distinguishing between direct and indirect benefits (Tregenza and Wedell 2000; Bussiere 2002; Wedell et al. 2002), as well as between runaway and good genes models of sexual selection (e.g., Jones et al. 1998; Wilkinson et al. 1998; Gilburn and Day 1999), is an especially active area of research for which many species of flies undoubtedly will continue to provide valuable information.

Sexual conflict is exemplified not only by differences in courtship and mating behavior between males and females, but also can occur after the termination of copulation (Alexander et al. 1997; Partridge and Hurst 1998; Birkhead and Pizzari 2002). For example, studies on *D. melanogaster* have shown that males may attempt to influence female fertilization and oviposition decisions by transferring small peptides, referred to as accessory proteins, during mating. These proteins typically decrease female receptivity and increase egg-laying rate (Simmons 2001), but also can reduce female lifespan (Chapman et al. 1995) and alter the relative fertilizing ability of sperm (Wolfner 1997; Price et al. 1999; Chapman 2001). These interactions are not confined to *Drosophila* (Chapman et al. 1998), and have been reported for many species of Diptera (for a review, see Simmons 2001).

Sexual conflict is important because it can lead to a dynamic evolutionary process in which each player is continually under selection to compete with an opponent. Such an arms race need not lead to a static equilibrium, as predicted by early game theory models of mate searching (Parker 1978). Instead, male and female traits should exhibit correlated patterns of evolution, as have been observed across species for sperm length and size of female sperm storage organs in *Drosophila* (Pitnick et al. 1999) and diopsid stalk-eyed flies (Presgraves et al. 1999), and between testes size and female accessory gland size in yellow dung flies, *Scatophaga stercoraria* (Hosken et al. 2001). Similarly, the rapid molecular evolution of accessory protein genes among *Drosophila* species is best explained by positive selection mediated by sexual conflict (Begun et al. 2000). Evolutionary change in male accessory proteins can also occur within species, as has been demonstrated by experiments using *D. melanogaster* in which chromosomes have been allowed to undergo evolution only in males (Rice 1996).

It is important to realize, though, that sexual conflict is not inevitable. Sexual conflict is not expected when the reproductive interests of males and females coincide. Among vertebrates, sexual conflict is not expected in genetically monogamous species with biparental care because investment by both the male and the female is necessary for successful reproduction. Such a mating system is unknown for flies. However, sexual conflict would also not be expected in species with very short adult life spans if males and females only mate a single time. Some chironomid midges, for example, may approach this situation (Armitage 1995).

Obtaining a Mate

As a consequence of low expenditure on gametes, males are expected to maximize encounter rates with females by searching, displaying, or defending resources (Parker 1978). Among flies, males may wait for females to visit them in an aerial swarm or while they display from a substrate, which may or may not be near a feeding or oviposition site. In some species, males search for eclosing females and inseminate them immediately after they emerge from their puparia. Cross-species comparisons suggest that the encounter site convention adopted by a species is strongly influenced by the distribution of defensible resources required by females. How males encounter females influences, in turn, the degree to which males fight and exhibit courtship behavior. Below we describe these patterns and discuss how sexual selection has influenced the morphology and behavior of males and females that adopt each encounter site convention.

AERIAL SWARMING

Aerial swarms are the most common mating site convention in the Lower Diptera (Downes 1969; Sullivan 1980; Sivinski and Petersson 1997), but are also found in species throughout the Higher Diptera (Table 11.1). This phylogenetic distribution is consistent with aerial swarming being the ancestral mating system of the Diptera. Mating swarms typically contain 10 to 1,000 individuals (Sullivan 1980; Shelly and Whittier 1997) that hover within a few cubic meters over a conspicuous landmark, such as a ridge, hill, tree top, fence post, church steeple, or even a plume of smoke (Chandler 2001). One species of chironomid midge, *Abiskomyia virgo,* even swarms on the surface of the ocean (Armitage 1995). Aerial swarms usually consist predominantly of males, although there are a few noteworthy cases in which swarm composition is strongly female-biased; for example, the chironomid *Palpomyia brachialis*

TABLE 11.1. Distribution of Mate Encounter Strategies among Families of Flies

Family	Species and Reference	Swarm	Mate Searching	Resource Defense	Substrate Lek
Lower Diptera					
Bibionidae	*Plecia nearctica* (Thornhill 1976)	X			
Cecidomyiidae	*Anarete prichardi* (Chiang 1968)	X			
Ceratopogonidae	*Serromyia femorata, Culicoides brevitarsus* (Sullivan 1980)	X			
Chironomidae	*Abiskomyia virgo, Smittia extrema* (Armitage 1995), *Stictochironomus crassiforceps* (Shelly and Whittier 1997), *Chironomus plumosus* (Neems et al. 1992), *C. strenzkei, C. riparius* (Sullivan 1980), *Palpomyia brachialis* (Sivinski and Petersson 1997), *Allochironomus crassiforceps, Glyptotendipes paripes, Chricotopus* sp. (Sullivan 1980)	X			
	Clunio marinus, Diamesa japonica, Oliveridia hugginsi, Pontomyia sp. (Armitage 1995)		X		
Culicidae	*Aedes communis, Ae. excrucians, Ae. caspius, Ae. dorsalis, Ae. cataphylla, Ae. hexodontus, Ae. taeniorhynchus, Anopheles franciscanus, An. punctipennis, An. maculipennis, Culex pipiens, Mansonia fuscopennata, Psorophora ferax* (Sullivan 1980)	X			
Simulidae	*Austrosimulium pestilens* (Shelly and Whittier 1997)	X			
Limoniidae	*Erioptera gemina* (Shelly and Whittier 1997)	X			
Pyschodidae	*Lutzomyia longipalpis* (Jarvis and Rutledge 1992; Jones and Hamilton 1998)				X
Tipulidae	*Pseudolimnophila inornata* (Sullivan 1980), *Erioptera taenionota* (Shelly and Whittier 1997)	X			
Higher Diptera					
Bombyliidae	*Comptosia tutela* (Yeates and Dodson 1990)				X
	Lordotus pulichrissimus (Shelly and Whittier 1997)	X			
Mydidae	*Mydas ventralis, M. xanthopterus* (Shelly and Whittier 1997)				X
Empididae	*Aplomera, Empis, Hilara, Rhamphomyia* (Downes 1969; Cumming 1994)	X			
Stratiomyidae	*Hermetia comstocki* (Alcock 1990), *H. illucens* (Tomberlin and Sheppard 2001)				X
	Platyna hastata, Beris morrisii (Oldroyd 1964)	X			
Streblidae	*Ascodipteran jonesi* (Oldroyd 1964)		X		
Rhagionidae	*Symphoromyia sackenia* (Shelly and Whittier 1997)	X			
Tabanidae	*Chrysops atlanticus, Tabanus bishoppi* (Shelly and Whittier 1997)	X			
Phoridae	*Megaselia aurea* (Sullivan 1980; Sivinski 1988)	X			
	Puliciphora boringuensis (Miller 1984), *Phalacrotophora* sp. (Disney 1994)		X		
	Dohrniphora maddisoni, Borophaga incrassata (Disney 1994)				X
Platypezidae	*Calotarsa pallipes, C. calceata, C. insignis* (Sivinski and Petersson 1997; Chandler 2001)	X			
Syrphidae	*Syrphus ribesii* (Shelly and Whittier 1997)	X			
	Somula decora (Maier and Waldbauer 1979)			X	

Coelopidae	*Coelopa frigida* (Day and Gilburn 1997), C. ursina (Crean and Gilburn 1998), *C. nebularum* (Weall and Gilburn 2000), *C. vanduzeei, C. pilipes, Gluma musgravi, G. nitida* (Crean et al. 2000)		X		
Dryomyzidae	*Dryomyza anilis* (Otronen 1984)			X	
Drosophilidae	*Drosophila adunca, D. cnecopleura, D. comatifemora, D. conformis, D. crucigera, D. cyrtoloma, D. differens, D. engyochraceae, D. formella, D. grimshawi, D. heteroneura, D. imparisitae, D. lasiopoda, D. mimica, D. mixtura, D. mycetophaga, D. nigribasis, D. obscuripes, D. percnosoma, D. planitibia, D. polypori, D. pullipes, D. silvestris* (Shelly and Whittier 1997)				X
	D. melanderi (Spieth and Heed 1975)			X	
Chloropidae	*Lipara lucens* (Mook and Bruggemann 1968)		X		
Otitidae	*Physiphora demandata* (Shelly and Whittier 1997)				X
Neriidae	*Odontoloxozus longicornis* (Mangan 1979)			X	
Piophilidae	*Protopiophila litigata* (Bonduriansky and Brooks 1999b), *P. latipes, Stearibia nigriceps, Liopiophila varipes, Prochyliza xanthostoma, Parapiophila* sp. (Bonduriansky and Brooks 1999a)			X	
Sepsidae	*Sepsis cynipsea* (Parker 1972)			X	
	Microsepsis armillata (Eberhard 2000)				X
Tephritidae	*Anastrepha fraterculus, A. ludens, A. obliqua, A. pseudoparella, A. serpentina, A. sororcula, A. suspensa, Batrocerca cucurbitae, B. dorsalis, B. tryoni, Procecidochares* sp. (Shelly and Whittier 1997), *Ceratitis capitata* (Field et al. 2002)				X
	Aciurina trixa, Valentibulla dodsoni (Dodson 1987), *Phytalmia mouldsi, P. biarmata, P. cervicornis, P. alcicornis* (Dodson 1997, 2000), *Toxotrypana curvicauda* (Landolt and Hendrichs 1983), *Anastrepha bistrigata* (Morgante et al. 1993), *Rhagoletis boycei, R. juglandis* (Papaj 1994), *Achias australis* (McAlpine 1979)			X	
Lonchaeidae	*Lonchaea* sp. (McAlpine and Monroe 1968)	X			
Gasterophilidae	*Gasterophilus intestinalis* (Shelly and Whittier 1997)				X
Cuterebridae	*Cuterebra austeni, C. grisea, C. latifrons, C. lepivora, C. polita, C. tenebrosa* (Shelly and Whittier 1997)				X
Fannidae	*Fannia canicularis* (Shelly and Whittier 1997)	X			
Glossinidae	*Glossina morsitans* (Wall and Langley 1993; Leak 1999)		X		
Muscidae	*Ophyria leucostoma* (Shelly and Whittier 1997)	X			
Oestridae	*Cephenemyia apicata, C. jellisoni, Hypoderma lineatum* (Shelly and Whittier 1997)				X
Scathophagidae	*Scatophaga stercoraria* (Parker 1970a)			X	

(Sivinski and Petersson 1997), the phorid *Megaselia aurea* (Sivinski 1988), and several species of empidid dance flies of genus *Rhamphomyia* (Cumming 1994; Funk and Tallamy 2000).

Aerial swarming is thought to be the most efficient strategy for obtaining a mate when the distribution of resources is homogeneous or unpredictable (Sivinski and Petersson 1997) or when population density is low (Kon 1987). In either case, males cannot reliably encounter females by defending or waiting near resources. Instead, they aggregate near conspicuous features at locations where females pass. Swarm sites, such as hilltops, sunbeams, or branch tips, have been hypothesized to be landmarks used for navigation (Sullivan 1980). Females appear to have limited opportunity to select males within a swarm and most species only mate once. Male courtship behavior has not been reported for any swarming species. Rather than display, males quickly converge on any approaching target that resembles a female. In some Lower Diptera, males detect and respond to the frequency of wing vibration produced by females (Wishart and Riordan 1959; Armitage 1995). Copulation occurs in the air in some species, whereas others copulate on the ground after males intercept females (Armitage 1995; Chandler 2001). Those that copulate in the air transfer sperm rapidly in a few seconds, whereas those that copulate on the ground may take several minutes or, in the case of the lovebug, *Plecia nearctica,* up to three days (Thornhill 1976; Hieber and Cohen 1983).

Aerial swarms are well known to attract predators, such as dragonflies, nighthawks, and bats (Sullivan 1980). Consequently, the risks associated with revisiting a swarm likely outweigh the benefits to females of remating. Moreover, a short life span, low density, and weak flying ability all favor mating in the first swarm encountered. If predators do not recruit to swarms and they do not last long, then the per capita risk of predation should decline with swarm size as a consequence of predator dilution. Consequently, females should prefer to mate in larger swarms, which has been observed in *Chironomus plumosus* (Neems et al. 1992) and in *Empis borealis* (Svensson and Petersson 1992).

In many swarm-forming species, the morphology of the compound eye exhibits sexual dimorphism. The eyes of males are typically larger than those of females and converge or join at the top of the head (Wenk 1987). Furthermore, the size of the ommatidia on the dorsal surface of male eyes is often larger and the rhabdomere is longer, which permits increased resolution and light sensitivity (Kirschfeld and Wenk 1976). These modifications allow males to detect moving objects flying overhead at greater distances than can females (Zeil 1983). Body size typically does not differ between males that do or do not acquire mates, although in the bibionid lovebug, *P. nearctica,* larger males hover lower in the swarm and are first to intercept and mate with newly eclosed females that enter the swarm (Thornhill 1980). Some investigators have also argued that successful males are more maneuverable and therefore smaller males are more successful than larger ones (McLachlan and Allen 1987). Males of species of *Calotarsa,* family Platypezidae, have highly enlarged hind tarsi with silver markings that flash in the sunlight, presumably to attract females (Sivinski 1997; Chandler 2001).

Female-biased aggregations occur in species that exhibit sex role reversal and reversed sexual dimorphism. For example, in the empidid *Rhamphomyia longicauda,* females have larger wings, inflatable sacs on the abdomen, and pinnate scales along the lateral margins of their hind legs. These traits give a flying female a saucer appearance. In this species, males capture and carry prey to a female swarm at dusk, approach a female, and exchange the prey item while copulating in the air. In this and related empidids, males provide the only protein source for adult females (Cumming 1994). Experiments with model females show that males preferentially approach females with larger abdomens (Funk and Tallamy 2000). Because abdomen size predicts egg size in a species without eversible sacs, but not in *R. longicauda*, Funk and Tallamy (2000) argue that inflated abdomens of female *R. longicauda* represent a deceptive signal to males. Alternatively, this modified morphology could decrease the energetic cost of hovering flight. Sex role reversal has also been reported for *Empis borealis*. In this species, the distal margins of female wings are expanded and males prefer to mate with the largest females (Svensson et al. 1989), which is consistent with the flight performance hypothesis.

MATE SEARCHING

Another common method of locating mates is for males to search for females at potential encounter sites, resulting in scramble competition polygyny. This mate searching strategy is typical of species for which males mate with recently eclosed females, which is often associated with (1) harsh environments not conducive to aerial swarming, (2) early receptivity of females, and (3) either single mating or first male sperm precedence (Thornhill and Alcock 1983). These situations do not favor high overlap of female home ranges (Wickman and Rutowski 1999). For example, male Himalayan wingless glacier midges search for females near emergence sites and then mate on the snow (Kohshima 1984). Similarly, males in the chironomid midge, *Diamesa japonica*, mate with newly emerged females on the surface of fast-flowing streams (Kon 1987). In the marine genus *Clunio*, males skate over intertidal pools searching for wingless, vermiform females. Copulation occurs either on shore at low tide or on the water surface (Armitage 1995). In two species of culicid mosquitos that either develop in crab holes, *Deinocerites cancer*, or rock pools, *Opifex fuscus*, males search the water surface for pupae containing females, grab them by a pupal horn, and hold them until eclosion in order to mate with them upon emergence (Provost and Haeger 1967). In the phorid fly, *Puliciphora borinquensis*, males search for and mate with wingless females (Miller 1984). In the parasitic streblid fly, *Ascodipteron jonesi*, adult females embed in the wings of bats, lose all appendages, undergo tremendous abdominal swelling and then drop mature larvae to the floor of a cave (Oldroyd 1964). Presumably, males, which retain wings and legs, search out females for mating.

Males search for females when females must visit an unpredictable or mobile food source. For example, in seaweed flies (e.g., *Coelopa frigida*), males search for females on wracks of seaweed that have recently washed ashore. Males are larger than females due to the ability of larger individuals to withstand female rejection attempts and resist takeovers by smaller males (Day and Gilburn 1997). Similarly, in many species of flies in which females feed on blood, males will follow potential vertebrate hosts and intercept females coming for a blood meal. Such mate finding behavior has been reported for some ceratopogonid midges (Wirth 1952), *Aedes* mosquitoes (Downes 1958), several species of blackflies (Wenk 1987), several species of hippoboscid flies (Bequaert 1953), and tsetse flies, *Glossina* (Wall and Langley 1993; Leak 1999). In tsetse flies, females can mate multiple times (Dame and Ford 1968). Similarly, males of some parasitoid phorid flies, *Megaselia* spp. and *Phalacrotophora* spp., approach and mate with females just before they oviposit into beetle pupae or prepupae (Disney 1994).

Species in which males wait, but do not fight, for females typically exhibit reverse sexual dimorphism (i.e., females are larger than males). As female fecundity is usually a function of adult body size, larger females are able to produce more offspring than smaller females. In some species, males emit signals to attract females. For example, in the gall-forming chloropid fly, *Lipara lucens,* males search for dispersed females on reeds and emit vibrations when they land. Receptive females respond by vibrating (Mook and Bruggemann 1968).

RESOURCE DEFENSE

Resources that are predictable and sufficiently uncommon that they attract multiple females, such as oviposition or feeding sites, provide another location for mating encounters to occur. Typically, females will mate multiple times and exhibit last male sperm precedence (Simmons 2001). Mating occurs at oviposition sites in many resource defense systems (see Table 11.1), such as the dung fly, *Scatophaga stercoraria* (Parker 1970a); several species of antler flies, *Phytalmia* spp. (Dodson 1997); the moose antler fly, *Protophila litigata* (Bonduriansky and Brooks 1999b); the neriid cactus fly, *Odontoloxozus longicornis* (Mangan 1979); many fungi feeding *Drosophila,* such as *D. melanderi* (Spieth and Heed 1975); and some syrphid flies, such as *Somula decora* (Maier and Waldbauer 1979). Mating occurs at feeding sites in several species of blackflies, where males wait on flowers for females to visit (Wenk 1987); this also holds for many species of *Drosophila* (Markow 1996). In the diopsids *Cyrtodiopsis whitei* (Lorch et al. 1993) and *C. dalmanni* (Wilkinson and Reillo 1994), mating occurs primarily at nocturnal roosting sites, which are located on rootlets underneath stream embankments.

Resource defense mating systems are typically associated with little or no precopulatory courtship activity (Prokopy 1980; Burk 1981). Where male signaling occurs, it often consists solely of repetitive actions in a single sensory modality. Examples of this type of behavior include waving patterned wings in tephritid flies (Burk 1981) and producing wing vibrations in *Drosophila* (Ewing 1977). Resource defense mating systems are also often associated with male fighting. Among *Rhagoletis* fruit flies, males defend food or oviposition sites by wing-waving, charging, foreleg kicking, and boxing (Prokopy and Papaj 2000). One consequence of such male-male competition is that sexual selection favors larger male body size. Larger males typically monopolize territorial sites (Borgia 1980; Wilkinson and Dodson 1997). Moreover, in flies that exhibit resource defense mating systems, males often have larger body sizes than females (Dodson 1986; Dodson 1987; Alcock 1990).

Males of some species that engage in resource defense also possess conspicuous sexually dimorphic structures. Such traits include eye stalks, enlarged mouthparts or bristles on the head, and antlerlike projections of the head (Grimaldi and Fenster 1989; Sivinski and Dodson 1992; Sivinski 1997; Wilkinson and Dodson 1997; Dodson 2000; Han 2000). In most cases, flies use these projections in direct male combat (McAlpine 1975; McAlpine 1979; Dodson 1987; Burla 1990; Dodson 2000; Han 2000). For example, in *Phytalmia mouldsi* (Dodson 1997), males guard scattered oviposition sites and compete with one another by engaging in pushing matches using head projections. However, in some cases, such as diopsid stalk-eyed flies and some *Phytalmia* species, the dimorphic trait is only used for size assessment during confrontations (Wilkinson and Dodson 1997; Panhuis and Wilkinson 1999).

SUBSTRATE LEKS

Lek mating refers to a mating system in which males aggregate at traditional sites and display to females (Bradbury and Gibson 1980; Höglund and Alatalo 1995). Typically, few males at a lek successfully copulate with any females. Because females come to males and males avoid or fight with each other, females are presumed to choose their mating partner. Vertebrate leks are characteristically disassociated from food or other limiting resources. Among flies, aggregations of displaying males can occur on and away from food or oviposition sites at different times. Similar switching between resource defense and lek mating has been described for fallow deer (Clutton-Brock et al. 1988) and has been attributed to differences in the amount and distribution of food over time. Substrate lek mating has been described for one family of Lower Diptera and ten families of Higher Diptera (see Table 11.1), with the greatest number of examples occuring in the Tephritidae and Drosophilidae (Spieth 1974; Shelly and Whittier 1997; Aluja and Norrbom 2000; Sivinski et al. 2000).

Leks have been hypothesized to occur when resources are relatively abundant and homogeneously distributed (Bradbury and Gibson 1980; Höglund and Alatalo 1995). Consequently, resource defense by males is unlikely to be a profitable method of encountering females. Thus, as with aerial swarms, males are thought to form aggregations in areas where females are likely to pass. For example, bombyliid flies of the genus *Comptosia* lek on a hilltop location (Dodson and Yeates 1990; Yeates and Dodson 1990). In some species, males gather on a substrate, such as above or below the surface of a leaf. Because these locations are relatively cryptic, males typically produce signals to attract females. Moreover, aggregation size is typically smaller for substrate leks than for aerial swarms (Shelly and Whittier 1997). Consequently, females are likely to have more freedom to choose mates in substrate leks than in aerial swarms.

Among tephritid flies, two general mating patterns have been described that are associated with differences in feeding and oviposition behavior. Monophagous species, which are typically temperate in distribution, usually mate on the host plant where oviposition occurs. In contrast, polyphagous species, most of which are tropical, usually mate in aggregations that form on the surface of leaves on or near a host plant (Prokopy 1980; Burk 1981). Although some exceptions to this pattern have been described—for example, lekking occurs in some monophagous species (Headrick and Goeden 1994)—they likely represent cases in which the host is sufficiently abundant that male defense is unlikely to be profitable (Sivinski et al. 2000). Resource dispersion has also been implicated in the occurrence of substrate leks among Hawaiian *Drosophila* (Droney 1996).

In a few species, lek mating alternates with resource defense at different times of the day or year. For example, Carib flies (*Anastrepha suspensa*) court on fruit in the morning and signal from the underside of leaves in the late afternoon by emitting pheromones and producing wing vibrations. In this species, most matings (85%) occur on leaves (Burk 1983). In the medfly, *Ceratitis capitata,* males signal from leaves between mid-morning and mid-afternoon, and court on fruit earlier and later. Over twice as many successful copulations occurred under leaves than on fruit (Prokopy and Hendrichs 1979). However, calling leads to higher predation (Hendrichs and Hendrichs 1998), which provides an explanation for the occurrence of alternative mating strategies. In the monophagous fruit fly, *Rhagoletis pomenella,* male mating strategy changes seasonally, with males mating on apple tree foliage early in the season and on fruit later in the season. Mating attempts on the two locations were equally successful (Smith and Prokopy 1980).

The factors favoring lek mating have received considerable study in the medfly, *Ceratitis capitata* (Yuval and Hendrichs 2000). As is typical of other lekking species, only a small fraction of males in an aggregation mate successfully (Arita and Kaneshiro 1985; Whittier et al.

1994). In experimental chambers, longer wing length predicted male mating success (Hunt et al. 2002). However, other studies have failed to find any effect of body size on male mating success (Whittier et al. 1994; Whittier and Kaneshiro 1995). Aggregations of males in trees may form in locations that increase encounter rates with females (Field et al. 2002), a process known as the hotspot effect (Bradbury and Gibson 1980). The hotspot does not, however, provide a compelling explanation for why two to eight males gather together to display. Some evidence indicates that females actively choose larger aggregations (Shelly 2001b), perhaps to reduce potential predation (Field et al. 2002). Little evidence supports an alternative possibility that males gather around a preferred mate, referred to as a hotshot (Beehler and Foster 1988).

Lek mating species exhibit the most elaborate and complex courtship behaviors that have been described for flies. For example, some male tephritids produce visual (Sivinski et al. 2000), chemical (Heath et al. 2000), and acoustic signals (Webb et al. 1984; Aluja 1993; Aluja et al. 2000) to attract females. In the lekking tephritid *Batrocerca dorsalis,* males provided with either synthetic methyl eugenol or with flowers containing methyl eugenol-like compounds have higher visitation rates and mating success in field and lab experiments (Shelly 2001a). Both males and females recruit to this substance, and males almost certainly use metabolites of it in their pheromones. Males aggregate on the surface of leaves in several species of phorids and either fan their wings (*Dohrniphora maddisoni*), drum on the leaf with their hind coxae (*Borophaga incrassata*), or perform complex wing and leg movements (Disney 1994). Four species of Hawaiian *Drosophila* in the planitibia subgroup, *D. differens, D. planitibia, D. silvestris,* and *D. heteroneura,* exhibit lek mating. Males perch on tree fern fronds and perform complicated courtship movements, including waving patterned wings, purring, and emitting wing vibrations (Spieth 1978; Hoy et al. 1988; Hoikkala et al. 1994; Hoikkala and Welbergen 1995) to attract females.

Mating success in leks can be influenced by diet, presumably because diet influences male courtship activity and, therefore, attractiveness to females. For example, a high protein diet increased courtship display and mating success of *Drosophila grimshawi* in experimental leks (Droney 1996). However, in medflies, diet did not affect male display activity, even though males fed protein mated more successfully than males fed sugar (Shelly et al. 2002). Females are, though, more likely to remate after mating with a male fed sugar versus a male fed protein (Yuval et al. 2002), suggesting that males may transfer some proteinaceous material to females during mating in this species.

Copulation and Postcopulatory Activity

Over the past 30 years, considerable evidence has revealed that fertilization success is not determined solely by copulation success. In flies, sperm is first transferred to female storage organs and then subsequently allowed to move to the micropyles, which are holes in the egg shell, or chorion, through which sperm must pass to reach the vitelline membrane and effect fertilization. The number of sperm transferred by a male can vary, depending on the duration of a copulation and the presence of a spermatophore. If females remate, then subsequent males can attempt to displace or interfere with previously deposited sperm. Females can also eject spermatophores or utilize stored sperm selectively. Sperm size, sperm number, and female storage organs have undergone tremendous diversification among species of flies. Although the functional significance of much of this behavioral and morphological diversity remains to be elucidated, a number of experimental studies using flies, especially involving species of *Drosophila* and the dung fly, *Scatophaga stercoraria,* have begun to shed light on the evolutionary processes responsible for the variation that has been described.

FEMALE REMATING

Whether females mate once or multiple times depends, in part, on what males provide to them during a copulation. For example, in some species of *Drosophila,* males transfer nutrients during mating and females mate repeatedly. In others, males transfer harmful substances and females mate infrequently (Markow 2002). Multiple mating by females is expected when gamete incompatibility or genomic conflict is possible (Haig and Bergstrom 1995; Zeh and Zeh 1996; Tregenza and Wedell 2000). Support for this hypothesis comes from studies showing conspecific sperm precedence in *Drosophila* (Markow 1997; Price 1997; Price et al. 2000) and from studies of mating rate and meiotic drive frequency among populations of diopsid stalk-eyed flies (Wilkinson et al. 2003). Nevertheless, for many species of flies, the most likely benefit to females for remating is increased offspring production (Ridley 1988; Arnqvist and Nilsson 2000; Baker et al. 2001). This conclusion seems surprising, given the conventional assumption that males pass more sperm per copulation than needed. However, this assumption is clearly false for some species of flies (Pitnick 1996).

The frequency of female remating can influence the intensity of both precopulatory and postcopulatory sexual selection. In species for which females remate frequently, the operational sex ratio should be less male-biased, which should reduce precopulatory competition for mates but increase postcopulatory competition for fertilizations. Comparisons among taxa suggest that female remating does influence pre- vs. postcopulatory competition among males. Species of *Drosophila,* for example, can be split into two groups: species for which females are likely to mate more than once per day, and species for which females are not. Frequently remating species are less likely to show secondary sexual dimorphism or special courtship tactics, such as nuptial feeding, than are species that mate infrequently. However, frequently remating species are much more likely to have exaggerated male ejaculate traits, such as giant sperm and ejaculate donations (Markow 2002). The frequency of remating also influences whether females can store or sort sperm. In the dung fly, *Scatophaga stercoraria,* a week between copulations affords females the opportunity to sort sperm among three spermathecae, allowing females to control which males' sperm fertilizes eggs (Ward and Hauschteck-Jungen 1993).

Remating by females can also be influenced by the behavior of males. For example, males can transfer material that either hinders access to sperm storage organs by subsequent mates (Lorch et al. 1993) or directly alters female behavior. In some ceratopogonid flies, males sacrifice part of their abdomen to block female gential openings (Downes 1978). In some species of *Drosophila,* mating induces a postinsemination reaction, which causes females to become reluctant to mate (Markow and Ankney 1988). An alternative strategy to reduce remating is for the male to remain in contact or near the female even after copulation has terminated. This behavior is typically described as mate guarding (Alcock 1994) and has been observed in a nereid fly (Mangan 1979) and the dung fly, *Scatophaga stercoraria* (Parker

1970b). Finally, in some species, copulations persist for periods exceeding the time necessary for sperm transfer. For example, prolonged copulations by *Plecia nearctica* have been interpreted as mate guarding (Thornhill 1976).

COPULATION DURATION

Among species of flies, copulation duration varies enormously and independently from how males encounter females. For example, the chironomid, *Glyptotendipes paripes,* copulates for 5 seconds (Nielsen 1959), whereas the bibionid, *Plecia nearctica,* copulates for up to 56 hours (Hieber and Cohen 1983). Among species of *Drosophila,* copulations can last from 30 seconds in *D. robusta* to over 2 hours in *D. acanthoptera* (Markow 1996). Reconstruction of copulation duration on phylogenies for three groups of flies (Fig. 11.2) reveals that short copulations have been derived recently in each group. Moreover, copulation duration has repeatedly increased and decreased among closely related species. These patterns suggest that copulation duration can evolve rapidly and must, therefore, be exposed to strong selection pressure.

Some of the variation in copulation duration can be attributed to differences in the amount of sperm and seminal material transferred from the male to the female (see the next section below). For example, in most diopsid stalk-eyed flies, males transfer sperm in a spermatophore and the size of the spermatophore correlates with copulation duration across species (Kotrba 1996). Similarly, copulation duration correlates with nuptial prey size in *Empis borealis* (Svensson et al. 1990), presumably to enable successful males to transfer more sperm. In some species, copulation duration depends on whether females have mated previously. In these cases, longer copulations may reflect an increase in seminal material that will either dilute or displace previously stored sperm (Snook and Karr 1998).

In some species, copulation duration varies due to male behaviors that have been referred to as copulatory courtship (Eberhard 1991). For example, males may tap or rub the abdomen of the female while in copula (Kotrba 1996; Bonduriansky and Brooks 1998). Eberhard (1996) has championed the view that females base their selection of sperm on these behaviors. These behaviors may, however, also provide information to males about the number of eggs carried by a female or her mating history. The best evidence that such behavior influences fertilization success comes from studies on *Dryomyza anilis.* In this species, the male taps the female's genitalia with his claspers while squeezing her abdomen with his hind legs (Otronen 1990). The number of tapping sequences made by a male during copulation correlates strongly with fertilization success. Some of this effect can be attributed to body size, as larger males tap more and have higher fertilization success (Otronen 1994; Otronen 1997).

FIGURE 11.2. Squared change parsimony reconstruction (Maddison and Maddison 1997) of copulation duration for fruit flies in (A) the genus *Anastrepha,* family Tephritidae, and (B) genus *Drosophila,* family Drosophilidae. We illustrate the variation in copulation duration for *Anastrepha* (Aluja et al. 2000) and *Drosophila* (Markow 1996) by taking natural logarithms of minutes. The *Anastrepha* phylogeny is based on a neighbor-joining analysis of 175 phylogenetically informative sites from16S rRNA mitochondrial DNA (McPheron et al. 2000). The *Drosophila* phylogeny is derived from a combination of morphological and molecular characters, including 2.7 Kb of nuclear sequence and 1.5 Kb of mitochondrial sequence (Pitnick et al. 1999).

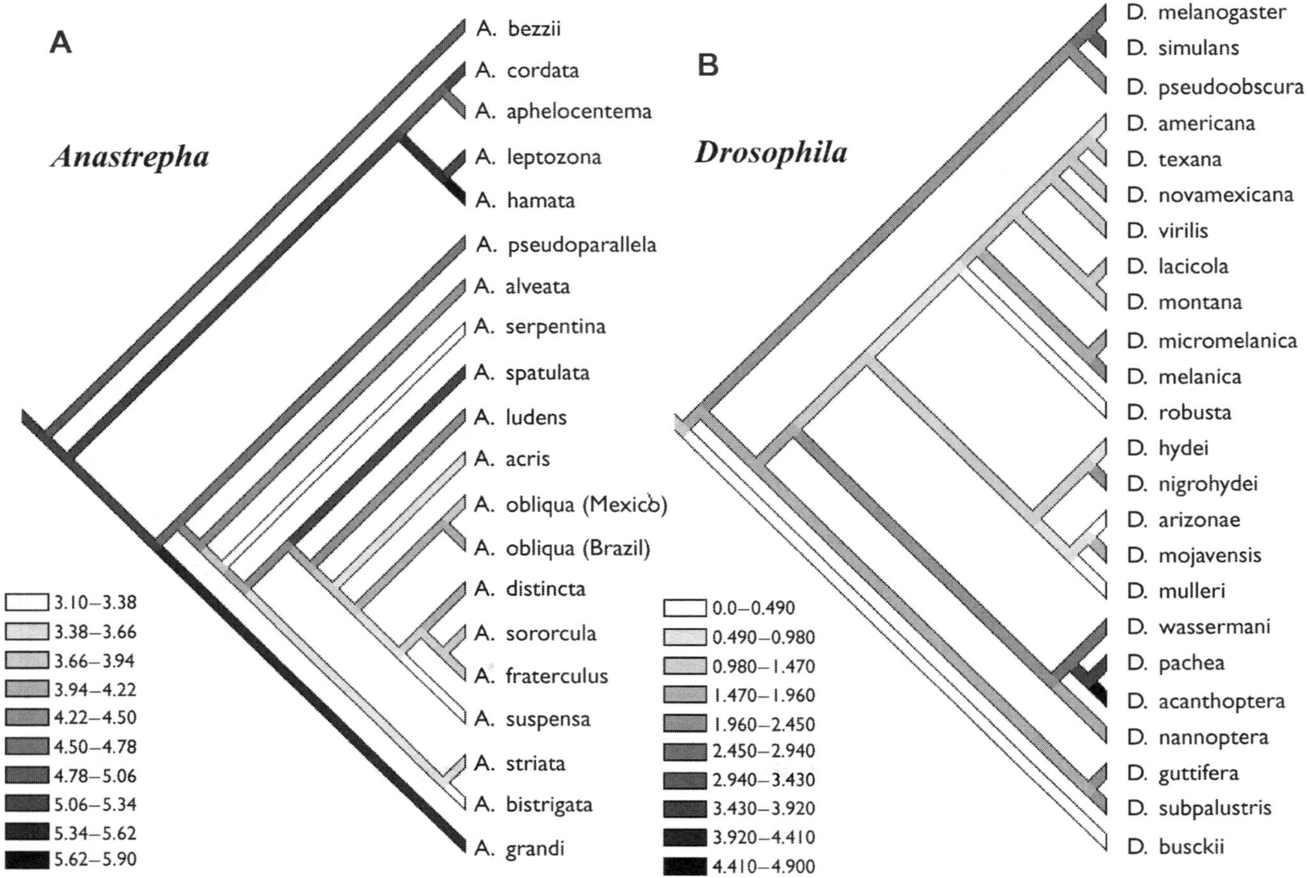

A
Anastrepha
A. bezzii
A. cordata
A. aphelocentema
A. leptozona
A. hamata
A. pseudoparallela
A. alveata
A. serpentina
A. spatulata
A. ludens
A. acris
A. obliqua (Mexico)
A. obliqua (Brazil)
A. distincta
A. sororcula
A. fraterculus
A. suspensa
A. striata
A. bistrigata
A. grandi
3.10–3.38
3.38–3.66
3.66–3.94
3.94–4.22
4.22–4.50
4.50–4.78
4.78–5.06
5.06–5.34
5.34–5.62
5.62–5.90
B
Drosophila
D. melanogaster
D. simulans
D. pseudoobscura
D. americana
D. texana
D. novamexicana
D. virilis
D. lacicola
D. montana
D. micromelanica
D. melanica
D. robusta
D. hydei
D. nigrohydei
D. arizonae
D. mojavensis
D. mulleri
D. wassermani
D. pachea
D. acanthoptera
D. nannoptera
D. guttifera
D. subpalustris
D. busckii
0.0–0.490
0.490–0.980
0.980–1.470
1.470–1.960
1.960–2.450
2.450–2.940
2.940–3.430
3.430–3.920
3.920–4.410
4.410–4.900

In addition, males with more asymmetrical and shorter small claspers have higher fertilization success (Otronen 1998).

Considerable theory has been developed to predict the optimal copulation duration in the dung fly, *Scatophaga stercoraria,* when there is sperm competition (reviewed in Simmons 2001). Females of this species store sperm in three spherical, scleritized spermathecae. Sperm counts from previously unmated females have revealed that sperm number increases asymptotically over time as the spermathecae fill. Maximum storage capacity appears to be about 950 sperm, which can be transferred by one male in a single copulation. Males gather on freshly deposited dung and wait to intercept and mate with females who arrive to oviposit. Since most arriving females have already mated and have full spermathecae, males must displace previously stored sperm. Because sperm mixes as displacement occurs, males that continue to transfer sperm will displace their own sperm. The marginal value theorem can predict the optimal copulation time that maximizes male fitness (Parker and Stuart 1976). Males should stop copulating when the marginal gain from mating with the current female drops below that expected from finding and copulating with another female. By this logic, Parker and Stuart (1976) predicted that males should copulate for 42 minutes, which is reasonably close to the observed average copulation duration of 35 minutes.

Including additional factors, such as phenotypic differences in males and females, refines such estimates. For example, larger males displace sperm faster than smaller males and, therefore, have shorter copulation times (Simmons and Parker 1992). In addition, larger males are more likely to take over females from smaller males, which both reduces large males' search times and also leads to shorter copulation durations (Parker and Simmons 1994). Assuming that females mediate displacement by allowing sperm to move from the bursa copulatrix to the spermathecae allows an even better fit between observed and predicted copulation durations (Parker and Simmons 2000). Finally, copula duration increases with female size, primarily because larger females carry more eggs, which increases the fertilization benefit to the male (Parker et al. 1999).

Much of the theoretical work on copulation duration assumes that males control mating. This assumption provides a reasonable fit to the data for yellow dung flies but may not be appropriate for other species. For example, in the dung fly, *Sepsis cynipsea,* females are reluctant to mate and attempt to dislodge mounted males by vigorous shaking (Blanckenhorn et al. 2000). Experiments have revealed that male genitalia injure females internally during copulation and these injuries increase with mating frequency (Blanckenhorn et al. 2002). Mated females show higher mortality than unmated females. Thus, copula duration seems to be another example of sexual conflict. Which sex controls mating may depend on relative body size, densities of each sex, predation rates, and other ecological factors.

SPERM TRANSFER AND SPERMATOPHORES

Many flies have closed copulatory systems, in which males pump free sperm, often from bifid or trifid penes, directly to spermathecal duct openings. However, in many species of Lower Diptera and a few species of Higher Diptera, sperm is transferred in spermatophores, packages of accessory gland secretions that enclose the sperm mass (Pollack 1972; Kotrba 1996). The use of spermatophores is ambiguous in some taxa, including species of Culicidae, Phoridae, and Spheroceridae (Kotrba 1996), in which clear gelatinous material surrounds the sperm mass after sperm transfer to the female. Whether flies employ spermatophores is not clear even in some *Drosophila* species, including *D. melanogaster,* which have varying reports of a "sperm sac" (DeVries 1964; Gromko and Gilbert 1984; Alonso-Pimentel and Tolbert 1994).

Why males in some fly taxa package sperm in spermatophores whereas others pump free sperm into their mates seems to depend on several selection pressures. Spermatophores might provide a competitive advantage to males that mate repeatedly (Gerber 1970). Mating aggregations allow the opportunity to mate several times in rapid succession and provide selective pressure for small, rapidly deployed spermatophores. Males that package and deposit their sperm quickly are free to copulate with other females. Among flies, some evidence bears out this hypothesis. Several taxa that form mating aggregations also transfer spermatophores, including chironomid midges (Nielsen 1959); black flies (Wenk 1987); tsetse flies, *Glossina austeni* (Pollack 1970); and mosquitos, *Anopheles gambiae* (Giglioli and Mason 1966).

Spermatophores do not always permit rapid sperm deployment, however. To allow males to mate quickly, spermatophores need only be small, simple sacks packed with sperm. In contrast, diopsid stalk-eyed fly spermatophores range from small sacs in the genus *Sphyracephala* to larger and elongate (*Cyrtodiopsis*), club-shaped (*Diasemopsis*), or tear-shaped (*Teleopsis*) structures that fill the bursa (Kotrba 1996). Nor do all flies that transfer spermatophores mate quickly. Flies that transfer large spermatophores form them inside the females' vaginal cavity during copulation. Such behavior takes time. For example, male dung flies, *Sepsis cynipsea,* and stalk-eyed flies in the genus *Teleopsis,* may remain in copula for more than 40 minutes while transferring spermatophores (Kotrba 1996; Martin and Hosken 2002). Moreover, male lovebugs, *Plecia nearctica,* both form mating swarms and transfer spermatophores (Pollack 1972), but remain in copula for days (Thornhill 1976). Thus, spermatophores do not function solely for rapid sperm transfer in Diptera. Large spermatophores may evolve in response to sperm competition where males do not have the opportunity to remate quickly (Pollack 1972; Kotrba 1996). In some taxa, like the dungfly, *Sepsis cynipsea,* males mate longer and transfer more sperm to females likely not to be virgins (Martin and Hosken 2002). Thus, males of this species can adjust their spermatophore size to account for the risk of sperm competition.

Spermatophores potentially have other effects than rapid deployment. In diopsids and *Anopheles gambiae,* spermatophores can act as mating plugs, preventing other males from inserting sperm into the same spermathecal duct (Giglioli and Mason 1966; Lorch et al. 1993). However, packaging sperm has drawbacks. Males may displace previously deposited spermatophores (Kotrba 1996) and females can reject spermatophores and consequently fertilize eggs selectively with sperm from some males over others. The secondary reduction and loss of spematophores in the stalk-eyed fly genus *Diopsis* could, for example, be a response to displacement of whole spermatophores (Kotrba 1996).

SPERM LENGTH

In addition to mode of transfer, sperm morphology varies enormously among fly species and can influence fertilization success. Sperm length varies 60-fold among Diopsidae (Presgraves et al. 1999) and 200-fold within *Drosophila* (Pitnick et al. 1995). For example, *D. bifurca* have sperm almost 58 mm long, or twenty times their body length. Not surprisingly, in *Drosophila,* longer testes and longer sperm evolve together (cf. Pitnick 1996: Fig. 2). In experimental lines alleviated from sexual selection through random, monogamous mating, males had fewer sperm and shorter testes (Pitnick et al. 2001). Thus, long sperm seem to have evolved for postcopulatory competition. This notion is at odds with the usual evolutionary theories of anisogamy, in which ejaculates of many tiny sperm evolved to compete with one another to fertilize eggs (Parker 1982). Sperm size should come at the expense of sperm number, and across *Drosophila,* species that produce long sperm tend not to produce as many (Pitnick 1996; Snook and Markow 2001). Females in species of *Drosophila* with long sperm tend to remate frequently (Markow 2002), yet display lesser degrees of second-male sperm precedence than species with small sperm (Simmons 2001). The reason for this pattern is not obvious, although a numbers game may be involved. In species with longer sperm, ejaculates may not contain enough sperm for high second-male paternity patterns. However, these are conditions that should favor ejaculates of many small sperm (Parker 1982). Clearly something else is involved (see the next section below).

Sperm length also can vary among males within a species. For example, male *D. melanogaster* and dung flies, *Scatophaga stercoraria,* show heritable variation in sperm length, independent of body size (Ward 2000b; Miller and Pitnick 2002). Longer sperm in dung flies give males a postcopulatory advantage because long sperm are more likely to enter spermathecal ducts, although this effect is influenced by other factors, such as male size and sperm number (Ward and Hauschteck-Jungen 1993; Otronen et al. 1997). Sperm length can vary within individuals, as well. Many insects have heteromorphic sperm (Swallow and Wilkinson 2002). In Diptera, both the long and short sperm in sperm-heteromorphic species are nucleated, although only the long sperm morph have been found to fertilize eggs (Snook and Karr 1998). As a general rule, sperm in heteromorphic species are shorter than sperm in monomorphic species (Swallow and Wilkinson 2002). For example, Markow (2002) makes a case that "giant" sperm in *Drosophila* are over 6 mm long, and the longest sperm for heteromorphic species of *Drosophila* are 1.8 mm in *D. azteca* (Bircher and Hauschteck-Jungen 1997). The role short sperm play is not entirely clear, although they may function to eliminate the sperm of rival males from female spermathecae (see below). Conditions that would favor the evolution of sperm that displace rival males' sperm and improve the odds of fertilization are generally consistent with dipteran mating systems (Swallow and Wilkinson 2002).

FEMALE REPRODUCTIVE ANATOMY

Evolutionarily, female flies are not static vessels for sperm storage. Instead, recent evidence clearly indicates that female anatomy can evolve rapidly to create a dynamic battleground on which sperm must compete to fertilize eggs. Flies may store sperm in seminal receptacles, spermathecae, or both, before sperm have the opportunity to fertilize eggs. Females in some species may also use other organs for sperm storage. For example, female *Drosophila nigricruria* store sperm in paired parovaria, which are about twice as large as those of other species, as well as in the spermathecae and seminal receptacles (Pitnick et al. 1999). The relative size of these female reproductive structures also varies hugely among taxa. For example, just as with sperm length, there is 200-fold variation in seminal receptacle length in *Drosophila* (Pitnick et al. 1999). Spermathecae number can vary among taxa, too. Whereas *Drosophila* tend to have two spermathecae (Pitnick et al. 1999), most Diopsidae have three (Presgraves et al. 1999), and dung flies (*Scatophaga stercorariai*) usually have three, but sometimes four (Ward 2000a). Therevidae and their relatives have one, two, or three spermathecae and may have an associated spermathecal sac. The function of the sac is unknown, but it does contain sperm after mating (Winterton et al. 1999).

Having multiple sperm storage organs can allow females to manipulate which sperm they use to fertilize eggs. For example, larger yellow dung fly males, *Scatophaga stercoraria,* have a postcopulatory advantage (Ward and Hauschteck-Jungen 1993), siring a greater proportion of progeny than smaller males if they are second to mate with a female. Examining which male's sperm winds up in which spermatheca suggests that females may allow sperm to distribute evenly among spermathecae during the first mating, but eject the sperm of the first male if the second male is larger (Otronen et al. 1997). Complicating this effect is sperm number and sperm length: larger sperm tend to be less numerous but are more likely to arrive in spermathecae. Females may preferentially use sperm with different phosphoglucomutase alleles to best match the environmental conditions in which larvae grow (Ward 1998).

The presence of monomorphic and heteromorphic sperm is also correlated with female reproductive tract morphology. For example, whereas some species of diopsid stalk-eyed flies exhibit sperm dimorphism and three well-developed spermathecae, one genus, *Diasemopsis,* has large (0.9–2.9 mm) monomorphic sperm, only two spermathecae, and in at least one species, these are degenerate. Across Diopsidae, spermatheca size correlates with the length of the short rather than the long sperm morph (Presgraves et al. 1999). In sperm-heteromorphic *Drosophila* species, short sperm tend to arrive in spermathecae more quickly than do long sperm (Snook and Markow 2001), and receptive females tend to have a lower proportion of short sperm in their ventral receptacle than nonreceptive females (Snook 1998). These findings bolster the claim that short sperm function in postcopulatory competition, perhaps by displacing rival males' functional sperm while stored in the spermathecae (Swallow and Wilkinson 2002).

Reproductive tract morphology, especially seminal receptacle length, correlates strongly with sperm length across taxa. In *Drosophila,* species with longer sperm also have longer seminal receptacles (Fig. 11.3; cf. Pitnick et al. 1999: Fig. 2). In diopsid stalk-eyed flies, too, ventral (seminal) receptacle length correlates with sperm length (Presgraves et al. 1999). Interestingly, the *Drosophila* reported to have heteromorphic sperm do not appear to mate as frequently as *Drosophila* with giant sperm (Markow 2002). Although short sperm almost certainly function in postcopulatory competition (Presgraves et al. 1999), in *Drosophila,* the conditions that favor intense postcopulatory competition—namely, rapid remating by females—tend to lead to giant rather than heteromorphic sperm (Markow 2002). Recently Miller and Pitnick (2002) selected *D. melanogaster* lines for either female seminal receptacle length or for male sperm length. They then allowed males of short- or long-sperm strains to compete for fertilizations within females of short- or long-receptacle strains. Males with long sperm sired a greater proportion of offspring when competing for fertilizations within long-seminal receptacle females. That is, males with long sperm have an advantage if female receptacles are long. Furthermore, in lines selected for longer seminal receptacles in females, males evolved longer sperm (Miller and Pitnick 2002). Thus, longer sperm receptacles afford a competitive advantage to longer sperm. Miller and Pitnick (2002: 1230) conclude that giant sperm are the "cellular equivalent of the peacock's tail" because female reproductive tracts bias paternity for males with longer tails.

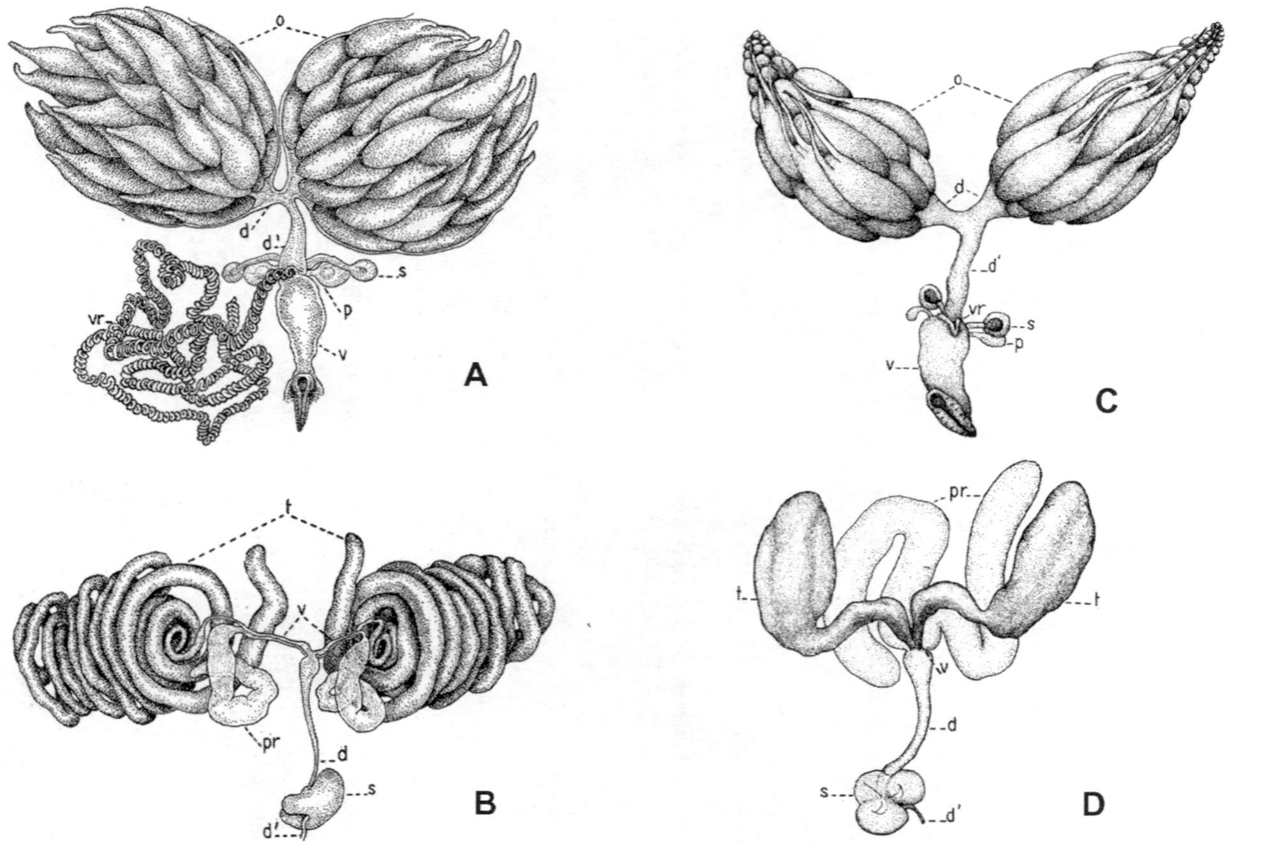

FIGURE 11.3. Reproductive tracts of *Drosophila hydei* (A) females and (B) males, and of *D. victoria* (C) females and (D) males. *D. hydei* males produce sperm that are 23.3 mm in length. Female anatomy: d, lateral oviduct; d′, common oviduct; o, ovary; p, parovarium (female accessory gland); s, spermatheca; v, vagina; vr, ventral (seminal) receptacle. Male anatomy: d, testicular duct; d′, posterior ejaculatory duct; pr, paragonium (male accessory gland); s, sperm pump; t, testis; v, vas deferens. Adapted with permission from Patterson (1943).

Conclusions

At least two conclusions can be drawn from this review. First, the link between the distribution of resources, especially food and oviposition sites, and type of mate encounter convention is strong, whereas that between ecological factors and postmating activities is less clear and almost certainly more complex. Much of this complexity arises because we now know that flies do not fit the stereotypic image of females as passive sperm recipients and males as prolific producers of tiny sperm. Interestingly, the tremendous morphological variation in male and female reproductive traits of *Drosophila* has been known for many years (e.g., Patterson 1943). Nevertheless, the functional significance of this variation was not appreciated until formal comparative studies revealed that male and female reproductive characters exhibit correlated evolutionary change (Pitnick et al. 1999; Presgraves et al. 1999). This result suggests that old taxonomic treatments of other dipteran groups may provide a treasure trove of information on reproductive trait morphology. Rather than rely on these traits to infer systematic relationships, reconstructing an independent hypothesis for the evolutionary relationships among species may help reveal the functional significance of structure variation and identify taxa worthy of observational and experimental study.

A second conclusion that can be drawn is that some behaviors related to mating have evolved very rapidly among species in some closely related groups of flies. For example, dramatic differences in precopulatory courtship, male fighting, and copulation duration have been reported for species of piophilid carrion flies (Bonduriansky and Brooks 1999a), *Anastrepha* fruit flies (Aluja et al. 2000), and Hawaiian *Drosophila* (Spieth 1978; Hoikkala et al. 1994; Boake et al. 2000; Boake 2002). These observations suggest that the form of the mating system may influence the rate of speciation. In particular, those species that rely on courtship behavior to choose mates may be more likely to become reproductively isolated as a consequence of small differences in courtship behaviors that might arise from genetic drift or selection. Such a scenario has long been thought to be important for the evolution and radiation of Hawaiian *Drosophila* (Ringo 1977; Boake 2000; Boake 2002). A corollary of this possibility is that those species that do not exhibit precopulatory courtship may be less likely to evolve premating isolation, but more likely to evolve postmating isolation mechanisms as a consequence of epistatic incompatibilies accumulating due to local coadaptation within isolated populations. Comparative studies of species diversity across taxa with different mating systems would be useful in assessing this possibility.

ACKNOWLEDGMENTS

This review was written with support from National Science Foundation grant DEB-0077878 to GSW. We thank Mark Moffett for allowing us to reproduce Fig. 11.1A,B and David Grimaldi for Fig. 11.1C,D.

REFERENCES

Alcock, J. 1990. A large male competitive advantage in a lekking fly, *Hermetia comstocki* Williston (Diptera: Stratiomyidae). *Psyche* 97:267–279.

Alcock, J. 1994. Postinsemination associations between males and females in insects: The mate-guarding hypothesis. *Annu. Rev. Entomol.* 39:1–21

Alexander, R. D., D. C. Marshall, and J. R. Cooley. 1997. Evolutionary perspectives on insect mating. In J. C. Choe and B. J. Crespi, eds., *The Evolution of Mating Systems in Insects and Arachnids,* pp. 4–31. Cambridge: Cambridge University Press.

Alonso-Pimentel, H., and L. Tolbert. 1994. Ultrastructure examination of the insemination reaction in Drosophila. *Cell Tissue Res.* 275:467–479.

Aluja, M. 1993. Unusual calling behavior of *Anastrepha robusta* (Diptera: Tephritidae) flies in nature. *Fla. Entomol.* 76:391–395.

Aluja, M., and A. L. Norrbom. 2000. *Fruit Flies (Tephritidae): Phylogeny and Evolution of Behavior.* Boca Raton, Fla.: CRC Press.

Aluja, M., J. Pinero, I. Jacome, F. Diaz-Fleischer, and J. Sivinski. 2000. Behavior of flies in the genus *Anastrepha* (Trypetinae: Toxotrypanini). In M. Aluja and A. L. Norrbom, eds., *Fruit Flies (Tephritidae): Phylogeny and Evolution of Behavior,* pp. 375–408. Boca Raton, Fla.: CRC Press.

Arita, L. H., and K. Y. Kaneshiro. 1985. The dynamics of the lek system and mating success in males of the Mediterranean fruit fly, *Ceratitis capitata. P. Hawaii Entomol. Soc.* 25:39–48.

Armitage, P. D. 1995. Behavior and ecology of adults. In P. D. Armitage, P. S. Cranston, and L. C. V. Pinder, eds., *The Chironomidae: Biology and Ecology of Non-Biting Midges,* pp. 194–224. New York: Chapman and Hall.

Arnqvist, G., and T. Nilsson. 2000. The evolution of polyandry: Multiple mating and female fitness in insects. *Animal Behav.* 60:145–164.

Baker, R. H., R. I. S Ashwell, T. A. Richards, K. Fowler, T. Chapman, and A. Pomiankowski. 2001. Effects of multiple mating and male eye span on female reproductive output in the stalk-eyed fly, *Cyrtodiopsis dalmanni. Behav. Ecol.* 12:732–739.

Bateman, A. J. 1948. Intra-sexual selection in *Drosophila. Heredity* 2:349–368.

Beehler, B. M., and M. S. Foster. 1988. Hotshots, hotspots, and female preference in the organization of lek mating systems. *Am. Nat.* 131:203–219.

Begun, D. J., P. Whitley, B. L. Todd, H. M. Waldrip-Dail, and A. G. Clark. 2000. Molecular population genetics of male accessory gland proteins in *Drosophila. Genetics* 156:1879–1888.

Bequaert, J. C. 1953. The Hippoboscidae or louse-flies (Diptera) of mammals and birds. Part I. Structure, physiology and natural history. *Entomol. Am.* 32:1–209.

Bircher, U., and E. Hauschteck-Jungen. 1997. The length of the sperm nucleus in *Drosophila obscura* group species is dependent on the total length of the sperm. *Invertebr. Reprod. Dev.* 32:225–229.

Birkhead, T. R., and T. Pizzari. 2002. Postcopulatory sexual selection. *Natl. Rev. Genet.* 3:262–273.

Blanckenhorn, W. U., C. Muhlhauser, C. Morf, T. Reusch, and M. Reuter. 2000. Female choice, female reluctance to mate and sexual selection on body size in the dung fly *Sepsis cynipsea. Ethology* 106:577–593.

Blanckenhorn, W. U., D. J. Hosken, O. Y. Martin, C. Reim, Y. Teuschl, and P. I. Ward. 2002. The costs of copulating in the dung fly *Sepsis cynipsea. Behav. Ecol.* 13:353–358.

Boake, C. R. B. 2000. Flying apart: Mating behavior and speciation. *Bioscience* 50:501–508.

Boake, C. R. B. 2002. Sexual signaling and speciation, a microevolutionary perspective. *Genetica* 116: 205–214.

Boake, C. R. B., D. K. Andreadis, and A. Witzel. 2000. Behavioural isolation between two closely related Hawaiian *Drosophila* species: The role of courtship. *Animal Behav.* 60:495–501.

Bonduriansky, R., and R. J. Brooks. 1998. Copulation and oviposition behaviour of *Protopiophila litigata* (Diptera: Piophilidae). *Can. Entomol.* 130:399–405.

Bonduriansky, R., and R. J. Brooks. 1999a. Reproductive allocation and reproductive ecology of seven species of Diptera. *Ecol. Entomol.* 24:389–395.

Bonduriansky, R., and R. J. Brooks. 1999b. Why do male antler flies (*Protopiophila litigata*) fight? The role of male combat in the structure of mating aggregations on moose antlers. *Ethol. Ecol. Evol.* 11:287–301.

Borgia, G. 1980. Sexual competition in *Scatophaga stercoraria:* Size- and density-related changes in male ability to capture females. *Behaviour* 75:185–206.

Bradbury, J. W., and R. Gibson. 1980. Leks and mate choice. In P. P. G. Bateson, ed., *Mate Choice,* pp. 109–138. Cambridge: Cambridge University Press.

Burk, T. 1981. Signalling and sex in acalyptrate flies. *Fla. Entomol.* 64:30–43.

Burk, T. 1983. Behavioral ecology of mating in the Caribbean fruit fly, *Anastrepha suspensa* (Loew) (Diptera: Tephritidae): Territorial fights and signaling stimulation. *Fla. Entomol.* 66:330–344.

Burla, H. 1990. Lek behavior in hypercephalic *Zygothrica dispar* Wiedemann (Diptera, Drosophilidae). *Z. Zool. Syst. Evol.* 28:69–77.

Bussiere, L. F. 2002. A model of the interaction between "good genes" and direct benefits in courtship-feeding animals: When do males of high genetic quality invest less? *Philos. Trans. R. Soc. B* 357:309–317.

Chandler, P. J. 2001. *The flat-footed flies (Diptera: Opetiidae and Platypezidae) of Europe.* Leiden: Koninklijke Brill.

Chapman, T. 2001. Seminal fluid-mediated fitness traits in Drosophila. *Heredity* 87:511–521.

Chapman, T., L. F. Liddle, J. M. Kalb, M. F. Wolfner, and L. Partridge. 1995. Cost of mating in *Drosophila melanogaster* females is mediated by male accessory gland products. *Nature* 373:241–244.

Chapman, T., T. Miyatake, H. K. Smith, and L. Partridge. 1998. Interactions of mating, egg production and death rates in females of the Mediterranean fruit fly, *Ceratitis capitata. Proc. R. Soc. Lond. B* 265:1879–1894.

Chiang, H. C. 1968. Ecology of insect swarms. V. Movement of individual midges, *Anarete pritchardi,* within a swarm. *Ann. Entomol. Soc. Am.* 71:584–587.

Clutton-Brock, T. H., D. Green, M. Hiraiwa-Hasegawa, and S. D. Albon. 1988. Passing the buck: Resource defence, lek breeding and mate choice in fallow deer. *Behav. Ecol. Sociobiol.* 23:281–296.

Crean, C. S., and A. S. Gilburn. 1998. Sexual selection as a side-effect of sexual conflict in the seaweed fly, *Coelopa ursina* (Diptera: Coelopidae). *Animal Behav.* 56:1405–1410.

Crean, C. S., D. W. Dunn, T. H. Day, and A. S. Gilburn. 2000. Female mate choice for large males in several species of seaweed fly (Diptera: Coelopidae). *Animal Behav.* 59:121–126.

Cumming, J. M. 1994. Sexual selection and the evolution of dance fly mating systems (Diptera: Empididae; Empidinae). *Can. Entomol.* 126:907–920.

Dame, D. A., and H. R. Ford. 1968. Multiple mating of *Glossina morsitans* (Westwood) and its potential effect on the sterile male technique. *Br. Entomol. Res.* 58:213–219.

Day, T. H., and A. S. Gilburn. 1997. Sexual selection in seaweed flies. *Adv. Study Behav.* 26:1–57.

Day, T. H., S. Miles, M. D. Pilkington, and R. K. Butlin. 1987. Differential mating success in populations of seaweed flies (*Coelopa frigida*). *Heredity* 58:203–212.

DeVries, J. 1964. Insemination and sperm storage in *Drosophila melanogaster. Evolution* 18:271–282.

Ding, A., and W. U. Blanckenhorn. 2002. The effect of sexual size dimorphism on mating behaviour in two dung flies with contrasting dimorphism. *Evol. Ecol. Res.* 4:259–273.

Disney, R. H. L. 1994. *Scuttle Flies: The Phoridae.* New York: Chapman and Hall.

Dodson, G. 1986. Lek mating system and large male aggressive advantage in a gall-forming tephritid fly (Diptera: Tephritidae). *Ethology* 72:99–108.

Dodson, G. 1987. The significance of sexual dimorphism in the mating system of two species of tephritid flies (*Aciurina trixa* and *Valentibulla dodsoni*) (Diptera: Tephritidae). *Can. J. Zool.* 65:194–198.

Dodson, G. 2000. Behavior of the Phytalmiinae and the evolution of antlers in Tephritid flies. In M. Aluja and A. L. Norrbom, eds., *Fruit Flies (Tephritidae): Phylogeny and Evolution of Behavior,* pp. 175–184. Boca Raton, Fla.: CRC Press.

Dodson, G. N. 1997. The resource defense mating system of antlered flies, *Phytalmia* spp. (Diptera: Tephritidae). *Ann. Entomol. Soc. Am.* 90:496–504.

Dodson, G. N., and D. K. Yeates. 1990. The mating system of a bee fly (Diptera: Bombyliidae): 2. Factors affecting individual male success. *Insect Behav.* 3:619–636.

Downes, J. A. 1958. Assembly and mating in the biting nematocera. *Proceedings of the 10th International Congress of Entomology* 2:425–434.

Downes, J. A. 1969. The swarming and mating flight of Diptera. *Annu. Rev. Entomol.* 14:271–298.

Downes, J. A. 1978. Feeding and mating in the insectivorous Ceratoponidae (Diptera). *Mem. Entomol. Soc. Can.* 104:1–62.

Droney, D. C. 1996. Environmental influences on male courtship and implications for female choice in lekking Hawaiian *Drosophila. Animal Behav.* 51:821–830.

Eberhard, W. G. 1991. Copulatory courtship and cryptic female choice in insects. *Biol. Rev.* 66:1–31.

Eberhard, W. G. 1996. *Female Control: Sexual Selection by Cryptic Female Choice.* Princeton: Princeton University Press.

Eberhard, W. G. 2000. Behavior and reproductive status of *Microsepsis armillata* (Diptera: Sepsidae) flies away from oviposition sites. *Ann. Entomol. Soc. Am.* 93:966–971.

Emlen, S. T., and L. W. Oring. 1977. Ecology, sexual selection and the evolution of mating systems. *Science* 197:215–222.

Ewing, A. W. 1977. Communication in Diptera. In T. A. Sebeok, ed., *How Animals Communicate*, pp. 403–417. Bloomington: Indiana University Press.

Felsenstein, J. 1985. Phylogenies and the comparative method. *Am. Nat.* 126:1–25.

Field, S. A., R. Kaspi, and B. Yuval. 2002. Why do calling medflies (Diptera:Tephritidae) cluster? Assessing the empirical evidence for models of medfly lek evolution. *Fla. Entomol.* 85:63–72.

Fisher, R. A. 1958. *The Genetical Theory of Natural Selection.* New York: Dover.

Funk, D. H., and D. W. Tallamy. 2000. Courtship role reversal and deceptive signals in the long-tailed dance fly, *Rhamphomyia longicauda. Animal Behav.* 59:411–421.

Gerber, G. H. 1970. Evolution of the methods of spermatophore formation in pterygotan insects. *Can. Entomol.* 102:358–362.

Giglioli, M., and G. Mason. 1966. The mating plug of anopheline mosquitoes. *Proc. R. Entomol. Soc. Lond.* 41:123–129.

Gilburn, A. S., and T. H. Day. 1999. Evolution of female choice in seaweed flies: Fisherian and good-genes mechanisms operate in different populations. *Proc. R. Soc. Lond. B* 255:159–165.

Gilburn, A. S., S. P. Foster, and T. H. Day. 1992. Female mating preference for large size in *Coelopa frigida. Heredity* 69:209–216.

Grimaldi, D., and G. Fenster. 1989. Evolution of extreme sexual dimorphisms: Structural and behavioral convergence among broad-headed Drosophilidae (Diptera). *Am. Mus. Novit.* 2939:1–25.

Gromko, M. H., and D. Gilbert. 1984. Sperm transfer and use in the multiple mating system of *Drosophila.* In R. L. Smith, ed., *Sperm Competition and the Evolution of Animal Mating Systems*, pp. 372–427. New York: Academic Press.

Haig, D., and C. T. Bergstrom. 1995. Multiple mating, sperm competition and meiotic drive. *J. Evol. Biol.* 8:265–282.

Han, H.-Y. 2000. Phylogeny and behavior of flies in the tribe Trypetini (Trypetinae). In M. Aluja and A. L. Norrbom, eds., *Fruit Flies (Tephritidae): Phylogeny and Evolution of Behavior*, pp. 253–297. Boca Raton, Fla.: CRC Press.

Hancock, R. G., A. F. Woodbridge, and W. L. Yee. 1990. Courtship behavior of the mosquito *Sabethes cyaneus* (Diptera: Culicidae). *J. Insect Behav.* 3:401–416.

Headrick, D. H., and R. D. Goeden. 1994. Reproductive behavior of California fruit flies and the classification and evolution of Tephritidae (Diptera) mating systems. *Stud. Dipterol.* 1:194–252.

Heath, R. R., P. Landolt, D. C. Robacker, B. D. Dueben, and N. D Epsky. 2000. Sexual pheromones of tephritid flies: Clues to unravel phylogeny and behavior. In M. Aluja and A. L. Norrbom, eds., *Fruit Flies (Tephritidae): Phylogeny and Evolution of Behavior*, pp. 793–809. Boca Raton, Fla.: CRC Press.

Hendrichs, M. A., and J. Hendrichs. 1998. Perfumed to be killed: Interception of Mediterranean fruit fly (Diptera: Tephritidae) sexual signalling by predatory foraging wasps (Hymenoptera: Vespidae). *Ann. Entomol. Soc. Am.* 91:228–234.

Hieber, C. S., and J. A. Cohen. 1983. Sexual selection in the lovebug, *Plecia nearctica:* The role of male choice. *Evolution* 37:987–992.

Höglund, J., and R. V. Alatalo. 1995. *Leks.* Princeton: Princeton University Press.

Hoikkala, A., and P. Welbergen. 1995. Signals and responses of females and males in successful and unsuccessful courtships of three Hawaiian lek-mating *Drosophila* species. *Animal Behav.* 50:177–190.

Hoikkala, A., K. Y. Kaneshiro, and R. R. Hoy. 1994. Courtship songs of the picture-winged *Drosophila planitibia* subgroup species. *Animal Behav.* 47:1363–1374.

Hosken, D. J., T. W. J. Garner, and P. I. Ward. 2001. Sexual conflict selects for male and female reproductive characters. *Curr. Biol.* 11:489–493.

Hoy, R. R., A. Hoikkala, and K. Kaneshiro. 1988. Hawaiian courtship songs—Evolutionary innovation in communication signals of *Drosophila. Science* 240:217–219.

Hunt, M. K., E. A. Roux, R. J. Wood, and A. S. Gilburn. 2002. The effect of supra-fronto-orbital (SFO) removal on male mating success in the Mediterranean fruit fly (Diptera: Tephritidae). *Fla. Entomol.* 85:83–88.

Jarvis, E. K., and L. C. Rutledge. 1992. Laboratory observations on mating and leklike aggregations in *Lutzomyia longipalpis* (Diptera, Psychodidae). *J. Med. Entomol.* 29:171–177.

Jones, T. M., and J. G. C. Hamilton. 1998. A role for pheromones in mate choice in a lekking sandfly. *Animal Behav.* 56:891–898.

Jones, T. M., R. J. Quinnell, and A. Balmford. 1998. Fisherian flies: Benefits of female choice in a lekking sandfly. *Proc. R. Soc. Lond. B* 265:1651–1657.

Kirkpatrick, M. 1982. Sexual selection and the evolution of female choice. *Evolution* 36:1–12.

Kirschfeld, K., and P. Wenk. 1976. The dorsal compound eye of Simuliid flies: An eye specialized for the detection of small, rapidly moving objects. *Z. Naturforsch.* 31:764–765.

Kohshima, S. 1984. A novel cold-tolerant insect found in a Himalayan glacier. *Nature* 310:225–227.

Kon, M. 1987. The mating system of chironomid midges (Diptera: Chironomidae): A review. *Mem Fac. Sci. Kyoto Uni.* 12:129–134.

Kotrba, M. 1996. Sperm transfer by spermatophore in Diptera: New results from the Diopsidae. *Zool. J. Linn. Soc.* 117:305–323.

Lande, R. 1981. Models of speciation by sexual selection on polygenic traits. *Proc. Natl. Acad. Sci. USA* 78: 3721–3725.

Lande, R., and G. S. Wilkinson. 1999. Models of sex-ratio meiotic drive and sexual selection in stalk-eyed flies. *Genet. Res.* 74:245–253.

Landolt, P. J., and J. Hendrichs. 1983. Reproductive behavior of the papaya fruit fly, *Toxotrypana curvicauda* Gerstaecker (Diptera: Tephritidae). *Ann. Entomol. Soc. Am.* 76:413–417.

Leak, S. G. A. 1999. *Tsetse Biology and Ecology.* New York: CABI Publishing.

Lorch, P., G. S. Wilkinson, and P. R. Reill. 1993. Copulation duration and sperm precedence in the Malaysian stalk-eyed fly, *Cyrtodiopsis whitei* (Diptera: Diopsidae). *Behav. Ecol. Sociobiol.* 32:303–311.

Maddison, W. P., and D. R.. Maddison. 1997. *MacClade: Analysis of phylogeny and character evolution.* Sunderland, Mass.: Sinauer.

Maier, C. T., and G. P. Waldbauer. 1979. Dual mate-seeking strategies in male syrphid flies (Diptera: Syrphidae). *Ann. Entomol. Soc. Am.* 72:54–61.

Mangan, R. L. 1979. Reproductive behavior of the cactus fly, *Odontoloxozus longicornis,* male territoriality and female guarding as adaptive strategies. *Behav. Ecol. Sociobiol.* 4:265–278.

Markow, T. A. 1996. Evolution of *Drosophila* mating systems. *Evol. Biol.* 29:73–106.

Markow, T. A. 1997. Assortative fertilization in *Drosophila. Proc. Natl. Acad. Sci. USA* 94:7756–7760.

Markow, T. A. 2002. Perspective: Female remating, operational sex ratio, and the arena of sexual selection in *Drosophila* species. *Evolution* 56:1725–1734.

Markow, T. A., and P. F. Ankney. 1988. Insemination reaction in *Drosophila:* Found in species whose males contribute materials to oocytes before fertilization. *Evolution* 48:1097–1101.

Martin, O. Y., and D. J. Hosken. 2002. Strategic ejaculation in the common dung fly *Sepsis cynipsea. Animal Behav.* 63:541–546.

McAlpine, D. K. 1975. Combat between males of *Pogonortalis doclea* (Diptera, Platystomatidae) and its relation to structural modification. *Aust. Entomol. Mag.* 2:104–107.

McAlpine, D. K. 1979. Agonistic behavior in *Achias australis* (Diptera, Platystomatidae) and the significance of eyestalks. In M. S. Blum, and N. A. Blum, eds., *Sexual Selection and Reproductive Competition in Insects,* pp. 221–230. New York: Academic Press.

McAlpine, D. K., and D. Monroe. 1968. Swarming of lonchaeid flies and other insects, with descriptions of four new species of Lonchaeidae (Diptera). *Can. Entomol.* 100:1154–1178.

McLachlan, A. J., and D. F. Allen. 1987. Mating success in Diptera: Advantages of small size. *Oikos* 48:11–49.

McPheron, B. A., H.-Y. Han, G. Silva, and A. L. Norrbom. 2000. Phylogeny of the genera *Anastrepha* and *Toxotrypana* (Trypetinae: Toxotrypanini) based upon 16S rRNA mitochondrial DNA sequences. In M. Aluja and A. L. Norrbom, eds., *Fruit Flies (Tephritidae): Phylogeny and Evolution of Behavior,* pp. 343–361. Boca Raton, Fla.: CRC Press.

Miller, G. T., and S. Pitnick. 2002. Sperm-female coevolution in *Drosophila. Science* 298:1230–1233.

Miller, P. L. 1984. Alternative reproductive routines in a small fly, *Puliciphora borinquensis* (Diptera: Phoridae). *Ecol. Entomol.* 9:293–302.

Mook, J. H., and C. G. Bruggemann. 1968. Acoustical communication by *Lipara lucens* (Diptera, Chloropidae). *Entomol. Exp. Appl.* 11:397–402.

Morgante, J. S., D. Selivon, V. N. Solferini, and S. R. Matioli. 1993. Evolutionary patterns in specialist and generalist species of *Anastrepha.* In M. Aluja and P. Liedo, eds., *Fruit Flies: Biology and Management,* pp. 15–20. New York: Springer-Verlag.

Neems, R., J. Lazarus, and A. J. McLachlan. 1992. Swarming behavior in male chironomid midges: A cost-benefit analysis. *Behav. Ecol.* 3:285–290.

Nielsen, E. 1959. Copulation of *Glyptotendipes (Phytotendipes) paripes* Edwards. *Nature* 184:1252–1253.

Oldroyd, H. 1964. *The Natural History of Flies.* New York: W. W. Norton.

Otronen, M. 1984. Male contests for territories and females in the fly *Dryomyza anilis. Animal Behav.* 32:891–898.

Otronen, M. 1990. Mating behavior and sperm competition in the fly, *Dryomyza anilis. Behav. Ecol. Sociobiol.* 26:349–356.

Otronen, M. 1994. Fertilisation success in the fly *Dryomyza anilis* (Dryomyzidae): Effects of male size and the mating situation. *Behav. Ecol. Sociobiol.* 35:33–38.

Otronen, M. 1997. Variation in sperm precedence during mating in male flies, *Dryomyza anilis. Animal Behav.* 53:1233–1240.

Otronen, M. 1998. Male asymmetry and postcopulatory sexual selection in the fly *Dryomyza anilis. Behav. Ecol. Sociobiol.* 42:185–192.

Otronen, M., P. Reguera, and P. I. Ward. 1997. Sperm storage in the yellow dung fly *Scatophaga stercoraria:* Identifying the sperm of competing males in separate female spermathecae. *Ethology* 103:844–854.

Panhuis, T. M., and G. S. Wilkinson. 1999. Exaggerated male eye span influences contest outcome in stalk-eyed flies. *Behav. Ecol. Sociobiol.* 46:221–227.

Papaj, D. R. 1994. Oviposition site guarding by male walnut flies and its possible consequences for mating success. *Behav. Ecol. Sociobiol.* 34:187–195.

Parker, G. A. 1970a. The reproductive behaviour and the nature of sexual selection in *Scatophaga stercoraria* L. (Diptera: Scatophagidae). II. The fertilization rate and the spatial and temporal relationships of each sex around the site of mating and oviposition. *J. Animal Ecol.* 39:205–228.

Parker, G. A. 1970b. The reproductive behaviour and the nature of sexual selection in *Scatophaga stercoraria* L. (Diptera: Scatophagidae). VII. The origin and evolution of the passive phase. *Evolution* 24: 774–788.

Parker, G. A. 1972. Reproductive behaviour of *Sepsis cynipsea* (L.) (Diptera: Sepsidae). I. A preliminary analysis of the reproductive strategy and its associated behaviour patterns. *Behaviour* 41:172–206.

Parker, G. A. 1978. Evolution of competitive mate searching. *Annu. Rev. Entomol.* 23:173–196.

Parker, G. A. 1982. Why are there so many tiny sperm? Sperm competition and the maintenance of two sexes. *J. Theor. Biol.* 96:281–294.

Parker, G. A., and L. W. Simmons. 1994. Evolution of phenotypic optima and copula duration in dungflies. *Nature* 370:53–56.

Parker, G. A., and L. W. Simmons. 2000. Optimal copula duration in the yellow dung flies: Ejaculatory duct dimensions and size-dependent sperm displacement. *Evolution* 54:924–935.

Parker, G. A., and R. A. Stuart. 1976. Animal behavior as a strategy optimizer: Evolution of resource assessment strategies and optimal emigration thresholds. *Am. Nat.* 110:1055–1076.

Parker, G. A., L. W. Simmons, P. Stockley, D. M. McChristy, and E. L. Charnov. 1999. Optimal copula duration in yellow dung flies: Effects of female size and egg content. *Animal Behav.* 57:795–805.

Partridge, L., and L. D. Hurst. 1998. Sex and conflict. *Science* 281:2003–2008.

Patterson, J. T. 1943. Studies in the genetics of *Drosophila.* III. The Drosophilidae of the southwest. *U. Texas Publ.* 4313:7–203.

Pitnick, S. 1996. Investment in testes and the cost of making long sperm in *Drosophila. Am. Nat.* 148:57–80.

Pitnick, S., T. A. Markow, and G. S. Spicer. 1995. Delayed male maturity is a cost of producing large sperm in *Drosophila. Proc. Natl. Acad. Sci. USA* 92:10614–10618.

Pitnick, S., T. Markow, and G. S. Spicer. 1999. Evolution of multiple kinds of female sperm-storage organs in *Drosophila. Evolution* 53:1804–1822.

Pitnick, S., G. T. Miller, J. Reagan, and B. Holland. 2001. Males' evolutionary responses to experimental removal of sexual selection. *Proc. R. Soc. Lond. B* 268:1071–1080.

Pollack, J. N. 1970. Sperm transfer by spermatophores in *Glossina austeni* Newstead. *Nature* 225:1063–1064.

Pollack, J. N. 1972. The evolution of sperm transfer mechanisms in the Diptera. *J. Entomol. A* 47:29–35.

Pomiankowski, A. 1988. The evolution of female mate preferences for male genetic quality. *Oxford Surv. Evol. Biol.* 5:136–184.

Presgraves, D. C., R. H. Baker, and G. S. Wilkinson. 1999. Coevolution of sperm and female reproductive tract morphology in stalk-eyed flies. *Proc. R. Soc. Lond. B* 266:1041–1047.

Price, C. S. C. 1997. Conspecific sperm precedence in *Drosophila. Nature* 388:663–666.

Price, C. S. C., K. A. Dyer, and J. A. Coyne. 1999. Sperm competition between *Drosophila* males involves both displacement and incapacitation. *Nature* 400:449–452.

Price, C. S. C., C. H. Kim, J. Posluszny, and J. A. Coyne. 2000. Mechanisms of conspecific sperm precedence in *Drosophila. Evolution* 54:2028–2037.

Prokopy, R. 1980. Mating behavior of frugivorous Tephritidae in nature. In J. Koyama, ed., *XVI International Congress of Entomology,* pp. 37–46, Kyoto, August 1980.

Prokopy, R., and J. Hendrichs. 1979. Mating behavior of *Ceratitis capitata* on a field caged host tree. *Ann. Entomol. Soc. Am.* 72:642–648.

Prokopy, R. J., and D. R. Papaj. 2000. Behavior of flies of the genera *Rhagoletis, Zonosemata,* and *Carpomya* (Trypetinae: Carpomyina). In M. Aluja and A. L. Norrbom, eds., *Fruit Flies (Tephritidae): Phylogeny and Evolution of Behavior,* pp. 219–252. Boca Raton, Fla.: CRC Press.

Provost, M. W., and J. S. Haeger. 1967. Mating and pupae attendance in *Deinocerites cancer* and comparisons with *Opifex fuscus* (Diptera: Culicidae). *Ann. Entomol. Soc. Am.* 60:565–574.

Rice, W. R. 1996. Sexually antagonistic male adaptation triggered by experimental arrest of female evolution. *Nature* 381:232–235.

Ridley, M. 1988. Mating frequency and fecundity in insects. *Biol. Rev.* 63:509–549.

Ringo, J. M. 1977. Why 300 species of Hawaiian *Drosophila?* The sexual selection hypothesis. *Evolution* 31:695–696.

Shelly, T. E. 2001a. Feeding on methyl eugenol and *Fagraea berteriana* flowers increases long-range female attraction by males of the oriental fruit fly (Diptera: Tephritidae). *Fla. Entomol.* 84:634–641.

Shelly, T. E. 2001b. Lek size and female visitation in two species of tephritid fruit flies. *Animal Behav.* 62: 33–40.

Shelly, T. E., and T. S. Whittier. 1997. Lek behavior in insects. In J. C. Choe and B. J. Crespi, eds., *The Evolution of Mating Systems in Insects and Arachnids,* pp. 273–293. Cambridge: Cambridge University Press.

Shelly, T. E., S. S. Kennelly, and D. O. McInnis. 2002. Effect of adult diet on signaling activity, mate attraction, and mating success in male Mediterranean fruit flies (Diptera: Tephritidae). *Fla. Entomol.* 85: 150–156.

Simmons, L. W. 2001. *Sperm Competition and Its Evolutionary Consequences in the Insects.* Princeton: Princeton University Press.

Simmons, L. W., and G. A. Parker. 1992. Individual variation in sperm competition success of yellow dung flies, *Scatophaga stercoraria. Evolution* 46:366–375.

Sivinski, J. 1988. Unusual female-aggregated mating systems in phorid flies. *J. Insect Behav.* 1:123–128.

Sivinski, J. 1997. Ornaments in the Diptera. *Fla. Entomol.* 80:144–164.

Sivinski, J. 1999. Kleptoparasitism and phoresy in the Diptera. *Fla. Entomol.* 82:179–197.

Sivinski, J. M., and G. Dodson. 1992. Sexual Dimorphism in *Anastrepha suspensa* (Loew) and other tephritid fruit flies (Diptera, Tephritidae)—Possible roles of developmental rate, fecundity, and dispersal. *J. Insect Behav.* 5:491–506.

Sivinski, J. M., and E. Petersson. 1997. Mate choice and species isolation in swarming insects. In J. C. Choe and B. J. Crespi, eds., *The Evolution of Mating Systems in Insects and Arachnids,* pp. 294–309. Cambridge: Cambridge University Press.

Sivinski, J., M. Aluja, G. Dodson, A. Freidberg, D. Headrick, K. Y. Kaneshiro, and P. Landolt. 2000. Topics in the evolution of sexual behavior in the Tephritidae. In M. Aluja and A. L. Norrbom, eds., *Fruit Flies (Tephritidae): Phylogeny and Evolution of Behavior,* pp. 751–792. Boca Raton, Fla.: CRC Press.

Smith, D. C., and R. J. Prokopy. 1980. Mating behavior of *Rhagoletis pomenella.* VI. Site of early season encounters. *Can. Entomol.* 112:585–590.

Snook, R. R. 1998. The risk of sperm competition and the evolution of sperm heteromorphism. *Animal Behav.* 56:1497–1507.

Snook, R. R., and T. A. Karr. 1998. Only long sperm are fertilization-competent in six sperm-heteromorphic *Drosophila* species. *Curr. Biol.* 8:291–294.

Snook, R. R., and T. A. Markow. 2001. Mating system evolution in sperm-heteromorphic *Drosophila. J. Insect Physiol.* 47:957–964.

Spieth, H. T. 1974. Courtship behavior in Drosophila. *Annu. Rev. Entomol.* 19:385–405.

Spieth, H. T. 1978. Courtship patterns and evolution of the *Drosophila adiastola* and *planitibia* species subgroups. *Evolution* 32:435–451.

Spieth, H. T., and W. B. Heed. 1975. The *Drosophila pinicola* species group (Diptera: Drosophilidae). *Pan-Pac. Entomol.* 51:287–295.

Sullivan, R. T. 1980. Insect swarming and mating. *Fla. Entomol.* 64:44–65.

Svensson, B. G., and E. Petersson. 1992. Why insects swarm: Testing the models for lek mating systems on swarming *Empis borealis* females. *Behav. Ecol. Sociobiol.* 31:253–261.

Svensson, B. G., E. Petersson, and E. Forsgren. 1989. Why do males of the dance fly *Empis borealis* refuse to mate? The importance of female age and size. *J. Insect Behav.* 2:387–395.

Svensson, B. G., E. Petersson, and M. Frisk. 1990. Nuptial gift size prolongs copulation duration in the dance fly *Empis borealis. Ecol. Entomol.* 15:225–229.

Swallow, J. G., and G. S. Wilkinson. 2002. The long and short of sperm polymorphisms in insects. *Biol. Rev.* 77:153–182.

Thornhill, R. 1976. Reproductive behavior of the lovebug, *Plecia nearctica* (Diptera: Bibionidae). *Ann. Entomol. Soc. Am.* 69:843–847.

Thornhill, R. 1980. Sexual selection within mating swarms of the lovebug *Plecia nearctica* (Diptera: Bibionidae). *Animal Behav.* 28:405–412.

Thornhill, R., and J. Alcock. 1983. *The Evolution of Insect Mating Systems.* Cambridge, Mass.: Harvard University Press.

Tomberlin, J. K., and D. C. Sheppard. 2001. Lekking behavior of the black soldier fly (Diptera: Stratiomyidae). *Fla. Entomol.* 84:729–730.

Tregenza, T., and N. Wedell. 2000. Genetic compatibility, mate choice and patterns of parentage: Invited review. *Molec. Ecol.* 9:1013–1027.

Trivers, R. L. 1972. Parental investment and sexual selection. In B. Campbell, ed., *Sexual Selection and the Descent of Man,* pp. 136–179. Chicago: Aldine.

Wall, R., and P. A. Langley. 1993. The mating behaviour of tsetse flies (*Glossina*): A review. *Physiol. Entomol.* 18:211–218.

Ward, P. I. 1998. A possible explanation for cryptic female choice in the yellow dung fly, *Scatophaga stercoraria* (L.). *Ethology* 104:97–110.

Ward, P. I. 2000a. Cryptic female choice in the yellow dung fly *Scathophaga stercoraria* (L.). *Evolution* 54: 1680–1686.

Ward, P. I. 2000b. Sperm length is heritable and sex-linked in the yellow dung fly (*Scathophaga stercoraria*). *J. Zool.* 251:349–353.

Ward, P. I., and E. Hauschteck-Jungen. 1993. Variation in sperm length in the yellow dung fly *Scathophaga stercoraria* (L). *J. Insect Physiol.* 39:545–547.

Weall, C. V., and A. S. Gilburn. 2000. Factors influencing the choice of female mate rejection strategies in the seaweed fly *Coelopa nebularum* (Diptera: Coelopidae). *J. Insect Behav.* 13:539–552.

Webb, J. C., J. Sivinski, and C. Litzkow. 1984. Acoustical behavior and sexual success in the Caribbean fruit fly, *Anastrepha suspensa* (Loew) (Diptera: Tephritidae). *Env. Entomol.* 13:650–656.

Wedell, N., M. J. G. Gage, and G. A. Parker. 2002. Sperm competition, male prudence and sperm-limited females. *Trends Ecol. Evol.* 17:313–320.

Wenk, P. 1987. Swarming and mating behavior of black flies. In K. C. Kim and R. W. Merritt, eds., *Black Flies.* University Park, Penn.: Pennsylvania State University.

Whittier, T. S., and K. Y. Kaneshiro. 1995. Intersexual selection in the Mediterranean fruit fly: Does female choice enhance fitness? *Evolution* 49:990–996.

Whittier, T. S., F. Y. Nam, T. E. Shelly, and K. Y. Kaneshiro. 1994. Male courtship success and female discrimination in the Mediterranean fruit fly (Diptera: Tephritidae). *J. Insect Behav.* 7:159–170.

Wickman, P. O., and R. L. Rutowski. 1999. The evolution of mating dispersion in insects. *Oikos* 84:463–472.

Wilkinson, G. S., and G. Dodson. 1997. Function and evolution of antlers and eye stalks in flies. In J. C. Choe and B. J. Crespi, eds., *The Evolution of Mating Systems in Insects and Arachnids,* pp. 310–328. Cambridge: Cambridge University Press.

Wilkinson, G. S., and P. R. Reillo. 1994. Female preference response to artificial selection on an exaggerated male trait in a stalk-eyed fly. *Proc. R. Soc. Lond. B* 255:1–6.

Wilkinson, G. S., D. C. Presgraves, and L. Crymes. 1998. Male eye span in stalk-eyed flies indicates genetic quality by meiotic drive suppression. *Nature* 391:276–278.

Wilkinson, G. S., J. G. Swallow, S. J. Christensen, and K. Madden. 2003. Phylogeography of sex ratio and multiple mating in stalk-eyed flies from Southeast Asia. *Genetica* 117:37–46.

Winterton, S. L., D. J. Merritt, A. O'Toole, M. E. Irwin, and D. K. Yeates. 1999. Morphology and histology of the spermathecal sac, a novel structure in the reproductive system of Therevidae (Diptera). *Int. J. Insect Morphol.* 28:273–279.

Wirth, W. W. 1952. The Heleidae of California. *U. Calif. Publ. Entomol.* 9:95–266.

Wishart, G., and D. F. Riordan. 1959. Flight responses to various sounds by adult males of *Aedes agypti* (L.). *Can. Entomol.* 91:181–191.

Wolfner, M. F. 1997. Tokens of love: Functions and regulation of *Drosophila* male accessory gland products. *Insect Biochem. Mol.* 27:179–192.

Yeates, D. K., and G. N. Dodson. 1990. The mating system of a bee fly (Diptera: Bombyliidae): 1. Non-resource based hilltop territoriality and a resource-based alternative. *Insect Behav.* 3:603–617.

Yeates, D. K., and B. M. Wiegmann. 1999. Congruence and controversy: Toward a higher-level phylogeny of Diptera. *Annu. Rev. Entomol.* 44:397–428.

Yuval, B., and J. Hendrichs. 2000. Behavior of flies in the genus *Ceratitis* (Dacinae: Ceratitidini). In M. Aluja and A. L. Norrbom, eds., *Fruit Flies (Tephritidae): Phylogeny and Evolution of Behavior,* pp. 429–457. Boca Raton, Fla.: CRC Press.

Yuval, B., R. Kaspi, S. A. Field, S. Blay, and P. Taylor. 2002. Effects of post-teneral nutrition on reproductive success of male Mediterranean fruit flies (Diptera: Tephritidae). *Fla. Entomol.* 85:165–170.

Zeh, J. A., and D. W. Zeh. 1996. The evolution of polyandry I: Intragenomic conflict and genetic incompatibility. *Proc. R. Soc. Lond. B* 263:1711–1717.

Zeil, J. 1983. Sexual dimorphism in the visual system of flies: The free flight behaviour of male Bibionidae (Diptera). *J. Comp. Physiol.* 150:395–412.

CHAPTER TWELVE

Ecological Genetics of Host Use in Diptera

Kenneth E. Filchak, Bill J. Etges,

Nora J. Besansky, and James L. Feder

The order Diptera represents one of the most species rich and taxonomically diverse groups on Earth (Yeates and Wiegmann, Chapter 2). Among insects, flies tend to be relatively small; however, their morphologies are quite varied, ranging from mosquitoes and crane flies to bee mimics, stalk-eyed flies, and fruit flies (Merritt et al. 2003). Dipterans are widespread and can be found in almost every terrestrial and fresh water habitat. They have colonized even the seemingly most inhospitable of environments. Indeed, there are species of Antarctic flies (e.g., *Belgica antarctica*), as well as those that survive in pools of crude petroleum (e.g., *Helaeomyia petrolei*) and at the margins of boiling springs (e.g., *Ephydra bruesi, Ephedra thermophila*) (Thorpe 1930; Oldroyd 1964; Brock et al. 1969). The life styles and diets of flies are as broad and varied as their distributions. A plethora of flies feed on various parts of plants (e.g., fruit, leafs, stems, roots), whereas many others prefer animals, fungi, or their byproducts (Oldroyd 1964). Diptera can be frugivorous, gall forming, leaf mining, or root boring, as well as predaceous, parasitic, bloodsucking, biting, and coprophagous (Oldroyd 1964; Muirhead-Thomson 1982; New 1991; Merritt et al. 2003). Moreover, many flies are host specific, restricting their attention to a relatively limited subset of potential resources.

In this chapter, we explore the issue of dipteran host use and specificity. Host use and specificity are terms generally associated with the ecology of phytophagous insects, referring to the range of plant species eaten by particular taxa (Evans 1984; Strong et al. 1984). However, the same issues that pertain to plant use are also relevant to flies that use animal hosts (Muirhead-Thomson 1982; New 1991). Phytophagous insects can simply be viewed as parasites feeding on plants, rather than animals. We take this broad view of host specificity here, highlighting several phytophagous, mycophagous, and parasitic flies that have provided insight into the question of host use.

Why is understanding the genetics of dipteran host use important? One reason is that several flies are vectors of human diseases, such as malaria, yellow fever, filariasis, dengue, and sleeping sickness (Kettle 1994). More than a million deaths per year result from malaria alone (Breman et al. 2001). A second reason is that many Diptera are economically important pests of cultivated crops (Merritt et al. 2003). A single outbreak of the Hessian fly (*Mayetiola destructor*) in North Dakota caused $28 million in damage (Hudson et al. 1991). Thus, insight into fly-host relationships can have profound implications for disease management and biocontrol. Finally, there is the question of basic scientific interest. Elucidating the physiological and molecular mechanisms responsible for host recognition and specificity touches on all aspects of biology. Why and how do organisms see, smell, hear, touch, and respond in positive ways to certain stimuli? Moreover, understanding how recognition and performance traits change and the role they play in host shifts is central to understanding the genesis and maintenance of biodiversity (Bush 1975).

The chapter is organized into four sections. We begin with a discussion of anthropophilic mosquitoes for two reasons. First, some of the most fruitful and meaningful investigations have and will be done on these flies. Second, mosquitoes represent a general category, animal ectoparasites, which are relatively poorly known because of their taxonomic and ecological diversity (Muirhead-Thomson 1982). For example, *Lutzomyia* sand flies (Psychodidae) are made up of approximately 400 species, with generally distinctive blood feeding host preferences. However, even the alpha taxonomy for this group is poorly resolved, and knowledge of their ecological genetics is rudimentary (L. E. Munstermann pers. comm.; Young and Duncan 1994). In the next section, we discuss *Drosophila*. Given the status of these flies as a model genetic system, almost any *Drosophila* species possessing a resolvable and interesting ecology should be experimentally revealing. We follow this with a discussion of two tephritids, the apple maggot fly, *Rhagoletis pomonella*, and the goldenrod gall fly, *Eurosta solidaginis*, that have contributed significantly to our understanding of the relationship between host shifts and speciation. We conclude with a synopsis of the Hessian fly, *M. destructor*, a paradigm for antagonistic plant-insect interactions, and mention a few other species, such as the pitcher plant mosquito, *Wyeomyia smithii*, that have provided insights or are promising subjects for the study of host use.

Host Specificity in Mosquitoes

Adult females of most mosquito species are temporary ectoparasites, preying on vertebrates for the blood meal required to produce eggs. However, not all vertebrates are equal game, and some degree of host specialization is typical. Blood feeding patterns indicate that most mosquitoes prefer mammals and birds. Within these classes, the subset of targeted host species, as well as the degree of host specificity, vary widely and are idiosyncratic for each mosquito species. *Culex salinarius*, with a relatively unrestricted host range, feeds willingly on birds, horses, and dogs (Lehane 1991). *Cx. nigripalpus* and *Cx. tarsalis* feed on birds in the winter and mammals in the spring, a feature that contributes to arbovirus transmission in humans (Lehane 1991). Mosquitoes that specialize on a single host species are extremely few in number. The most prominent examples are *Anopheles gambiae* and *An. funestus*, whose anthropophilic ("human-loving") and anthropophagic ("human-biting") tendencies make them the most efficient vectors of human malaria in the world.

The factors underlying patterns of blood feeding are numerous and interact in complex ways. In addition to any innate, genetically based preferences for particular hosts, host seeking is influenced by mosquito flight patterns, mosquito physiological state, host defensive behavior, host availability, and other chance environmental events (Clements 1999; McCall and Kelly 2002). Sympatric species that seek blood meals at different heights from the ground or at different times of the day or night may attack nonoverlapping sets of hosts. Those that feed at night may readily bite humans outdoors but fail to enter small apertures that would gain them access to humans sleeping indoors. In the absence of a preferred host, opportunistic species will attack alternate hosts, sometimes at the price of reduced fecundity

(Clements 1999). Shifts in the relative proportion of one host to another can also affect feeding behavior. For anopheline vectors with more catholic tastes than *An. gambiae* and *An. funestus,* decreasing the ratio of humans to nearby domestic animals can divert enough bites toward these dead-end hosts of malaria parasites to provide effective protection against malaria (zooprophylaxis; Burkot 1988; Hadis et al. 1997; Killeen et al. 2001). Finally, there is evidence that learning and memory can influence host choice, leading to host preference based on previously successful feeding (McCall and Kelly 2002).

Host preference has been defined as the innate tendency to respond to particular host cues (Clements 1999). Understanding the unique contribution of innate tendencies to the overall pattern of blood feeding requires that other confounding factors be removed or minimized to the extent possible. Such confounding factors arise in field-based studies not only from uncontrollable environmental events, but also from experimental perturbations inherent to the method of mosquito collection and systematic biases introduced by human collectors. Moreover, host seeking is a complex sequence of behaviors. Methods that target different phases of this sequence may not produce directly comparable results. For example, two traditional methods for assessing host preference in the field are blood meal analysis and comparison of landing rates on alternative hosts. Identification of the blood meal source provides data on those mosquitoes that have successfully completed the multistep process from attraction at a distance through landing and biting a host. Landing catches on alternative hosts focus on a much earlier stage of host seeking, usually before biting. Recognition of the importance of olfaction in host selection has led to the development of field-based olfactometers that can measure relative responses to odors from two hosts. Odor-baited entry traps (OBETs) capture mosquitoes during upwind flight in response to host odors, in the absence of confounding visual cues or human collectors (Costantini et al. 1993). The recently developed OBET captures mosquitoes when host preferences are first expressed. The OBET improves on existing techniques by eliminating perturbations due to human collectors, removing confounding visual cues, and minimizing accidental capture of mosquitoes not actively seeking blood meals (Costantini et al. 1993). Odors from hosts—concealed beneath opaque tents—are blown out of the tent and into the trap via tubing, creating a current against which entering mosquitoes must fly. Standardization of this or other traps, such as the Mbita trap (Mathenge et al. 2002), can make results easier to interpret across different mosquito populations and geographic locations, and perhaps can resolve contradictions that may be more apparent than real.

A comprehensive review of behavioral studies of mosquito host preference can be found in Clements (1999). Here, we focus on the Afrotropical *An. gambiae* sibling species complex (see also Takken and Knols 1999). Of special interest in this medically important and recently diverged group of species is the range in response to human odor by different species, an observation that hints at the evolutionary lability of host preference.

There are six formally named species in the *An. gambiae* complex: *An. gambiae, An. arabiensis, An. merus, An. melas, An. Bwambae,* and *An. quadriannulatus* (White 1974). Based on blood meal analysis, these species show an impressive diversity of blood feeding patterns (White 1974; Gillies and Coetzee 1987). The nominal species *An. gambiae* is one of the most highly anthropophilic mosquitoes known, whereas *An. quadriannulatus* feeds almost exclusively on domestic animals, such as cattle. *An. arabiensis* is an opportunistic species that feeds preferentially on humans in many parts of Africa, but can be diverted to domestic animals as animal density increases, particularly in East Africa. The remaining species, although considered predominantly zoophilic (tending to bite domestic animals), readily bite humans.

Evidence from field studies suggests that species-specific differences in blood feeding behavior are rooted in species-specific host preferences and mediated by differential responses to host odors (Table 12.1). When given a choice of human odor and either CO_2 or calf odor, *An. gambiae* and *An. arabiensis* were the only two species that generally preferred human odor, consistent with numerous blood meal analyses of members of the *An. gambiae* complex. Results from laboratory experiments using colonized mosquitoes have largely agreed with field studies, with some notable exceptions (Table 12.2). Whereas the SUA strain of *An. gambiae* strongly preferred human odor over CO_2 or cow odor, as expected, two laboratory strains of *An. quadriannulatus* did not discriminate between calf and human (Pates et al. 2001a). Although the SKUQUA strain established in 1995 had been maintained in the laboratory on blood from a human arm for 3.5 years, the SANGQUA strain had been established more recently (1999) and had been fed on cattle blood through membrane feeders for the 4 months preceding the experiments. Without further investigation, it is difficult to reconcile differences between field and laboratory studies. However, such disagreements can be anticipated, given the unknown yet probably severe genetic and behavioral consequences of adaptation to laboratory culture, the pronounced differences in trap entry behavior between species (Pates et al. 2001a), and numerous other departures from natural environmental conditions.

Complicating efforts to understand how blood feeding patterns differ among closely related species is that host preference, even within the same species, is not rigid. In *An. gambiae* and *An. arabiensis,* unusually low rates of anthropophily have been reported from widely separated parts of Africa, from the dry sahel of Senegal on the continent to the islands of São Tomé and Madagascar off the west and east coasts, respectively (Lemasson et al. 1997; Diatta et al. 1998; Duchemin et al. 2001; Sousa et al. 2001). Although these locations are not physically close, they share a relative isolation from the rest of the mainland due to physical or climatic barriers. The *An. gambiae* and/or *An. arabiensis* populations from parts of Senegal, São Tomé, and Madagascar are not only unexpectedly zoophilic, but are also unusually exophilic and exophagic (resting and biting outdoors). Because parts of these islands were subject to insecticide spray campaigns that targeted indoor-resting mosquitoes by treating interior walls of dwellings, it is conceivable that either the endophilic component was eliminated from a population that was originally polymorphic in resting habits, or the spraying caused a behavioral shift that remains, despite more than 50 years since cessation of treatments (Duchemin et al. 2001). Another possible explanation can be found in the relatively low ratio of humans to domestic animals (cattle in Senegal and Madagascar, dogs in São Tomé) in these locations. It has been known for some time that *An. gambiae* will bite bovids more frequently if they outnumber human hosts, as reflected in increased numbers of indoor resting bovid-fed females (White and Rosen 1973; Garrett-Jones et al. 1980). This phenomenon may explain why dogs are the predominant host for *An. gambiae* on São Tomé (Sousa et al. 2001). However, host availability alone cannot explain the markedly zoophilic behavior of both *An. gambiae* and *An. arabiensis* on Madagascar. As the experimental design in the Madagascar study involved a choice test using OBETs, the results suggest that the preference for cattle odor was intrinsic (Duchemin et al. 2001). Thus, variation in the extent of host specialization is apparent not only among species in the *An. gambiae* complex, but

TABLE 12.1. Host Preference of Natural Populations of *Anopheles gambiae* sensu lato as Measured in the Field

Reference	Region	Site	Method	Bait(s)	Species: Preference
Gillies 1967	East Africa, Tanzania	Indoors	Release into screened room with alternative bait	Human, calf	*An. gambiae:* human *An. merus:* calf
Bryan et al. 1987	West Africa, The Gambia	Outdoors	Landing catch	Human, calf	*An. gambiae:* human *An. melas:* calf
Costantini et al. 1996	West Africa, Burkina Faso	Outdoors	Dual choice OBET	Human, CO_2	*An. gambiae:* human *An. arabiensis:* human
Diatta et al. 1998	West Africa, Senegal	Outdoors	Baited net	Human, calf	*An. gambiae:* calf *An. arabiensis:* calf
Dekker and Takken 1998	South Africa	Outdoors	Baited net	Human, CO_2, calf	*An. arabiensis:* human *An. quadriannulatus:* calf, CO_2
Duchemin et al. 2001	Madagascar	Outdoors	Dual choice OBET	Human, calf	*An. gambiae:* calf *An. arabiensis:* calf

TABLE 12.2. Host Preference of Laboratory Strains of *Anopheles gambiae* sensu lato as Measured in the Laboratory or under Nonnative Conditions

Reference	Method	Bait(s)	Species: Preference
Dekker et al. 2001	Dual port olfactometer	Human odor vs. no odor	*An. gambiae* SUA: human odor *An. quadriannulatus* SKUQUA: no odor
		Cow odor vs. no odor	*An. gambiae* SUA: no odor *An. quadriannulatus* SKUQUA: no preferente
Pates et al. 2001b	Dual port olfactometer	(1) Human odor vs. no odor (2) Human odor vs. cow odor (3) Human + cow vs. no odor (4) Human + cow vs. cow odor	*An. gambiae* SUA: human odor, human + cow odor
Pates et al. 2001a	Release into screened room with alternative bait, outdoors	Human, calf	*An. gambiae* SUA: human *An. quadriannulatus* SKUQUA: no preference *An. quadriannulatus* SANGQUA: no preference

within species as well, suggesting that its genetic underpinning is evolutionarily labile and perhaps relatively simple.

Simple or not, our understanding of the genetic basis of host preference in *An. gambiae* is in its infancy. Until recently, the only evidence bearing on a relationship between genotype and host-seeking behavior came from two sources: the distribution of chromosomal inversions in natural populations and genetic selection experiments. In the case of chromosomal inversions, the relationship between genotype and host-seeking behavior is indirect. *An. gambiae* and *An. arabiensis* are highly polymorphic for chromosome-2 inversions, recognized as rearrangements of the banding pattern of polytene chromosomes. There is compelling evidence suggesting that carriers of different combinations of chromosome-2 inversions are differentially adapted to variation in environments, particularly with respect to aridity (Toure et al. 1998; Powell et al. 1999; Coluzzi et al. 2002). Along north-south geographic transects of ecoclimatic zones, there are stable clines in frequencies of inversions, from absence in humid forest through fixation in arid sahel. The same inversions associated with these geographic clines are nonrandomly distributed across rainy and dry seasons, as well as microspatially. For example, inversion frequencies are higher in samples collected inside houses, where there is a nocturnal saturation deficit relative to outdoors. Within *An. gambiae* and *An. arabiensis* populations, mosquitoes carrying inverted chromosomal arrangements are more likely to be found resting inside houses at night than are mosquitoes with the standard arrangements. Therefore, where local populations of *An. gambiae* and *An. arabiensis* are highly polymorphic for inversions, the probability of inversion carriers encountering and biting humans sleeping indoors is higher, as blood feeding occurs at night (Coluzzi et al. 1977; Powell et al. 1999).

Only one selection experiment for host preference has been reported in *An. gambiae* (Gillies 1964). This experiment provides the first direct evidence of a genetic basis for host preference. The experiment was conducted in Tanzania, at a locale characterized by low domestic animal density and highly anthropophilic *An. gambiae* populations. A collection of wild gravid females was allowed to oviposit en masse and the female offspring were released into an experimental hut designed with a central release room flanked by two communicating rooms. On one side was a "man-room" occupied by a human volunteer, and on the other a "calf-room." Females recaptured in the man-room with fresh blood meals (assumed to have fed on the human) founded the parental man-strain, whereas females recaptured in the calf-room with fresh blood meals founded the parental calf-strain. For subsequent generations, females from both lines were released simultaneously, having been marked with distinguishing fluorescent dusts. Two replicate experimental series were conducted, running for six and four generations. Significant differences in host preference were detected by the F_2 generation in the first series, and by the F_4 in the second. The efficiency with which differential behavior was selected, all the more surprising in that only females were targeted, indicates that *An. gambiae* is polymorphic for host preference, and suggests that only a few genes of large effect (or several genes protected from recombination by an inversion) may be involved.

Genetic and environmental factors interact to define the behavioral sequence of events involved in mosquito host choice. An understanding of the process has been elusive, not only because the problem is inherently difficult to study, but also because of lack of attention and necessary tools. Yet the potential benefits of such research, beyond theoretical interest, range from novel repellants to more effective means of vector control. A major obstacle to behavioral research on *An. gambiae* and other members of cryptic species complexes was overcome when rapid molecular methods became available to distinguish the isomorphic and often sympatric species. An equally serious obstacle to progress in genetic research on most mosquitoes, including *An. gambiae,* has until recently been the paucity of morphological or molecular markers on a highly resolved genetic map (see Collins et al. 1986). The completed *An. gambiae* genome sequence (Holt et al. 2002) and associated resources, together with new and anticipated sequencing projects for *Aedes aegypti* and *Culex pipiens,* are changing this outlook rapidly (e.g., Biessmann et al. 2002; Hill et al. 2002; Merrill et al. 2002). Already, a female-specific member of a family of odorant receptors from *An. gambiae* has been found to respond to a component of human sweat when expressed in a *Drosophila* olfactory receptor neuron (Hallem et al. 2004). Using a comparative approach, with *An. gambiae* as a model, it should now be possible to advance our understanding of the genetic basis of host preference apace with more sophisticated laboratory and field experimentation.

Drosophila

The ecology of *Drosophila* breeding sites has long been of central interest to evolutionary biologists because of the prominent roles several members of this genus have played in the history of genetics and evolution (Patterson and Stone 1952; Carson 1971; Powell 1997). Those species that were easily captured, usually by baiting with fermenting fruit, and cultured in the laboratory became models for understanding genetics and the forces of evolution. However, it soon became clear that *D. melanogaster* and flies like it were largely cosmopolitan, human commensals and were not as useful for evolutionary studies as truly wild species might be (Sturtevant 1921). Without knowledge of their ecology, or more specifically, the conditions that had shaped their evolutionary history, interpretation of population genetic studies would always be uncertain. Ironically, many wild *Drosophila* species could not be cultured in the laboratory, and those that adapted easily to laboratory conditions, such as *D. pseudoobscura, D. robusta,* and *D. subobscura,* were difficult to find breeding in nature. Some tropical species, like those in the *D. willistoni* group, were found to use a variety of fruits as oviposition sites (Heed 1957), and many of these neotropical species remain the focus of systematic ecological analysis (Sevenster and Alphen 1996; Krijger et al. 2001). The many frustrating searches for *D. pseudoobscura* and *D. robusta* breeding sites are legendary among *Drosophila* workers, even though sap fluxes were identified as oviposition sites over 50 years ago (Carson 1951; Carson and Stalker 1951). Despite the enormous roles these species have had in population genetics research in the twentieth century (Lewontin et al. 1981; Levitan 1992), manipulative studies of these larval resources have been impractical, thus preventing an understanding of the ecological genetics of host use, and in particular, the consequences of host use on genetic polymorphism.

An exemplary case-study emphasizing the frustration that the ecology of *D. pseudoobscura* presents researchers involves the sex-ratio chromosome found in this and other species (Wu and Beckenbach 1983; Jaenike 2002). Sex-ratio (SR) is a naturally occurring X-linked meiotic drive system that in males, eliminates most Y-bearing sperm. Once SR reaches fixation, local population extinction is assured. In *D. pseudoobscura,* SR reaches frequencies of 15–25% in nature, implying antagonistic forces are at work that prevent fixation. Summarizing the extensive theoretical and experimental work of the fitness effects of SR throughout the life cycle, Beckenbach (1996: 787) concluded, "most selection opposing SR appears to be operating at the larval stages in nature." Thus, the maintenance of this classical meiotic drive system will not be resolved until the ecology of *D. pseudoobscura* can be manipulated to show how natural breeding sites reduce frequencies of larvae or pupae carrying SR.

There is an impressive amount of ecological information about drosophilid breeding sites for species groups using fermenting or rotting vegetation, tree bark, sap fluxes, mushrooms, flowers, fruits, cacti, dripping wine barrels, and highly specialized resources, such as other insects, spider egg masses, and aquatic frogs' eggs (Ashburner 1981). Fallen and decaying herbaceous plants have also been shown to be frequently used breeding sites by many temperate, woodland *Drosophila* in Japan (Kimura et al. 1977) and may be an important, but overlooked, resource for these species in North America (H. L. Carson, pers. comm.). Characterization of the ecology of the huge, endemic Hawaiian *Drosophila* fauna has revealed significant host shifts among species groups (Heed 1968; Carson 1971; Heed 1971; Montgomery 1975). These host shifts have recently been put into a phylogenetic perspective by Kambysellis et al. (1995). However, the genetic basis of host use has only been indirectly investigated in a few Hawaiian picture-wing species. Ohta (1980) revealed the pattern of inheritance in crosses between ecological generalists and specialists among populations of *D. grimshawii* and *D. pullipes*. Populations using a variety of hosts, characterized as ecological generalists, were collected from the islands of the Maui nui complex and those described as specialists, reared only from a single host, the rotting bark of *Wikestroemia* trees, from Oahu, Kauai, and Hawaii. Comparisons of male sterility between parental, F_1, F_1 backcross, and F_2 generations showed a rather simple mode of inheritance that Ohta (1980) suggested had resulted from adaptive differences in degree of host specialization. Nevertheless, *Drosophila* using fermenting cacti and mushrooms have been most heavily studied, and genetic analysis of host use preference and performance in these species is discussed in detail below.

A few notable studies in other *Drosophila* systems bear mentioning, including natural substrates in studies of genetic polymorphisms, adaptation to novel hosts, or direct examination of the genetic architecture of host use. Brncic (1962, 1968) and Carson (2001) discovered the flower-breeding species *D. flavipilosa* in the valleys of Chile and documented the effects of temperature on karyotype frequencies of flies reared in their natural flower substrates. Motivated by the observation of parallel evolution of three different species of *Drosophila* that use different parts of land crabs as breeding sites (Carson 1974), including Caribbean *D. carcinophila*, whose larvae feed on the nitrogenous waste products of *Gecarcinus ruricola* green glands, Wallace (1978) constructed a laboratory "crab" containing a microbe-innoculated nitrogenous waste product (human urine) as an energy source. He introduced several *Drosophila* laboratory stock species into the chamber, and after 13 months, only a small population of *D. virilis* remained that could no longer survive on standard laboratory food. An apparent irreversible genetic response to selection had occurred, but unfortunately, no further genetic analysis of this stock was performed.

In another unique system, the genetic basis of oviposition preference for *Morinda citrifolia* fruits by *D. sechellia* has been investigated. Other *Drosophila* species avoid this resource as a breeding site due to the fruits' high concentrations of octanoic acid (R'Kha et al. 1991). Backcrosses between *D. simulans* and *D. sechellia* demonstrated that *D. sechellia*'s oviposition preference for *Morinda* was recessive to *D. simulans*'s avoidance of this medium, with a large effect of the second chromosome (Higa and Fuyama 1993). Further backcross data narrowed this effect to the left arm of the second chromosome and the right arm of the third chromosome (C. D. Jones, unpubl. data), although earlier deletion mapping studies suggested an effect of the distal end of the right arm of chromosome 2 (Sugaya et al. 1995). Results of crosses with marked strains of *D. simulans* suggested that the genetic basis of adult resistance to octanoic acid in *D. sechellia* involves about five genes distributed across the X, second, and third chromosomes (Jones 1998), and even fewer genes when larval resistance was analyzed (Jones 2001). Thus, the genes responsible for *Morinda* preference and performance may eventually be isolated and identified.

The degree to which local populations of *Drosophila* can specialize on different hosts or resources will depend on available genetic variance for host use, host acceptability, host persistence, population structure, and the strength of selection imposed by novel or multiple hosts (Futuyma and Peterson 1985; Jaenike 1990). Evidence suggests that genetic polymorphism can easily be maintained by habitat selection (Taylor and Powell 1977; Hedrick 1990), and that resource use in natural populations is, in general, heritable (Hoffman and O'Donnell 1990). Field experiments have documented the significance of adult experience and learning on host preference (Hoffman 1985; Jaenike 1985), but studies describing the kinds of genetic changes that accompany host specialization in *Drosophila* are still uncommon.

CACTOPHILIC *DROSOPHILA*

The arid lands of the New World are home to hundreds of species of succulent cacti, many of which are hosts to a variety of *Drosophila* species, principally the members of the large *D. repleta* group (Patterson and Stone 1952; Heed 1982, 1989). The best-known relationships have been studied in the Sonoran Desert, the adjoining arid lands, and in two South American regions, the northern arid zones of Venezuela and Colombia, and to the south in Argentina and Brazil. Several *D. repleta* group species have been introduced elsewhere owing to human transportation of their host plants, principally in Australia, the Mediterranean region, and Hawaii.

Four species of *Drosophila* are endemic to the Sonoran and Mojave Deserts of northwestern Mexico, southern California, and Arizona (Heed and Mangan 1986). *D. nigrospiracula* and *D. mettleri* are typically restricted to saguaro cactus, *Carnegiea giganta,* cardon cactus, *Pachycereus pringlei,* and occasionally hecho, or *P. pectin-aboriginum.* Although adults of both fly species congregate around the dripping rots of these hosts, female *D. mettleri* oviposit in the cactus juice–soaked soil at the base of host plants (Heed 1977; Fogleman et al. 1981; Fogleman and Williams 1987), and *D. nigrospiracula* uses the fermenting stem tissues. Both of these flies are restricted to these columnar cacti (although *D. mettleri* has been reared out of soaked soils beneath organ pipe cacti, *Stenocereus thurberi*) because other host species contain secondary compounds that are toxic to them, including alkaloids, triterpene glycosides, and sterol diols (Fogleman and Heed 1989; Fogleman and Abril 1990; Frank and Fogleman 1992).

In a classic study of ecological chemistry, Heed and Kircher (1965) demonstrated that *D. pachea,* a member of the cactophilic *D. nannoptera* group, uses only senita cactus, *Lophocereus schottii,* endemic to the Sonoran Desert, because the larvae require a unique intermediary sterol to complete development. As this host cactus contains high concentrations of alkaloids that exclude other *Drosophila* species, *D. pachea* has effectively no other congeneric competitors (Kircher et al. 1967). The other endemic species, *D. mojavensis,* uses a variety of host cacti, but tends to specialize on single species in different parts of its species range. Throughout the lower two-thirds of Baja California, agria cactus, *Stenocereus gummosus,* is sympatric with organ pipe, *S. thurberi,* yet *D. mojavensis* almost exclusively uses agria (Downing 1985; Newby and Etges 1998). A large endemic columnar species in north-central Baja, *Myrtillocactus cochal,* is occasionally used. Agria is also found on the islands in the Gulf of California and restricted to a small patch in coastal Sonora (Heed 1978). Organ pipe cactus is sympatric in these areas, too, and becomes the major host plant throughout mainland Sonora, Sinaloa, and southern Arizona. In southern Sonoran and Sinaloa, sina cactus, *S. alamosensis,* is sometimes shared with a sibling species, *D. arizonae* (Fellows and Heed 1972; Ruiz and Heed 1988). In the Mojave Desert in southern California and northern Arizona, *D. mojavensis* is restricted to California barrel cactus, *Ferocactus cylindraceous.* An isolated population of *D. mojavensis* has also been found on Santa Catalina Island near Los Angeles, California, using *Opuntia demissa.* Thus, the causal factors determining host plant use in *D. mojavensis* have been the most intensively studied given the variety of hosts used throughout its range. The other endemic Sonoran Desert species use either single hosts or genetically similar and ecologically "equivalent" ones (i.e., saguaro and cardon; Heed and Mangan 1986; Fogleman and Abril 1990; Hartmann et al. 2002).

The major host plant transition among cactophilic *Drosophila* has been the shift from chemically benign *Opuntia* species to the more chemically complex columnar cacti (Etges et al., in prep.). This transition is characterized by the members of the *D. mojavensis* cluster, *D. mojavensis, D. arizonae,* and *D. navojoa* (Ruiz and Heed 1988; Ruiz et al. 1990), with the latter, more ancestral species restricted to *Opuntia* and the former two, derived sibling species using a variety of *Opuntia* and columnar cacti to carry out their life cycles. So far, studies of the genetics of host use in *D. mojavensis* have focused on the switch from agria to organ pipe and *Opuntia* cacti. Based on the genetic affinities and biogeography of its closest relatives, Ruiz et al. (1990) proposed that *D. mojavensis* and *D. arizonae* split from their direct ancestors in southern Mexico, allowing *D. mojavensis* to become reproductively isolated in Baja California. These ancestral populations possess considerable inversion polymorphism, including a rare ancestral gene arrangement, and use agria cactus (Etges et al. 1999). Once *D. mojavensis* invaded the mainland, it diverged onto organ pipe, sina, California barrel cacti, and *Opuntia* on Santa Catalina Island.

Field studies have demonstrated that mainland *D. mojavensis* prefer agria, even in areas where agria is not present, such as southern Arizona (Fellows and Heed 1972). Laboratory preference tests employing synthetic "cocktails" of volatiles based on gas chromatographic profiles of naturally occurring agria and organ pipe rots also demonstrated preference for agria, the ancestral host (Downing 1985). Sufficient within-population genetic variation for acceptance of agria by ovipositing females from a Santa Catalina Island population restricted to *Opuntia* was revealed in a half sib–full sib experimental design (Lofdahl 1986). In subsequent choice tests, a comparison of five populations from throughout the species range indicated preference for agria volatiles, particularly a population from southern California that uses barrel cactus in nature. Only one organ pipe cactus–using population from Sonora showed significant preference for organ pipe volatiles, suggesting genetic divergence for host preference (Newby and Etges 1998). In all, the data suggest evolutionary retention of a "sensory bias" for an ancestral host, but genetic variation in preference for other hosts awaits further study, particularly in relation to host performance.

The host shift to organ pipe cactus once *D. mojavensis* invaded mainland Mexico and Arizona has had significant influence on the evolution of life history characters, physiological traits associated with host use, and gene and inversion frequency differences. The relevant ecological differences between hosts are the larger stem diameters of organ pipe, slower rates of tissue fermentation, and approximately 40 times lower measured stem densities compared to agria (Heed 1981, 1982; Heed and Mangan 1986; Etges 1989). Common garden experiments with fermenting tissues of agria and organ pipe cacti have revealed that Baja populations express shorter egg-to-adult development times, higher viabilities, and shorter thorax sizes than mainland populations when reared on agria vs. organ pipe (Etges and Heed 1987). Results from nested ANOVAs revealed the presence of region by cactus interaction terms, suggesting that agria-using Baja populations and organ pipe–using mainland populations are host races (sensu Jaenike 1981; Etges 1990). Genetic variances for these traits, significant genotype by cactus interactions, and positive across-host genetic correlations indicate ongoing life history evolution in both Baja and mainland populations and evidence for ecological generalism within populations (Etges 1993).

Also tied to these life history shifts is decreased developmental homeostasis with increasing larval densities in mainland populations; population crosses revealed that higher genetic homeostasis in Baja populations is dominant over lower homeostasis in a mainland population for development time, viability, and thorax size. Such genetic differences in developmental homeostasis are concordant with the increased tissue fermentation rates and smaller stem diameters of agria cactus, and thus decreased predictability for larval survival in natural populations (Etges 1989).

Evolution of adult fitness components is also patterned around the ecology of agria and organ pipe rots. Mainland *D. mojavensis* are genetically larger as adults, have higher metabolic rates (standardized O_2 consumption min^{-1} g^{-1} live mass), have higher ovariole numbers (Heed, unpubl. data) and lifetime fecundities, but younger ages at first reproduction than Baja populations (Etges and Klassen 1989). As larger, more active adults can disperse farther, mainland *D. mojavensis* have evolved a life history "syndrome" associated with lower (40-fold) organ pipe rot densities, larger distances between rots, and thus a decreased predictability of finding new breeding sites. Modes of cactus reproduction are largely responsible: organ pipe cacti have self-incompatible, bird-pollinated flowers (Fleming et al. 1996) and are propagated by seed dispersal. Agria reproduces mainly by vegetative growth, producing huge clonal thickets consisting of many arms, so that local stem densities are high (Mangan 1982). The contrast between decreased predictability for larval survival in agria-using populations and decreased predictability for adult survival in organ pipe–using populations has produced a classic life history tradeoff (Stearns 1977) evident in the patterns of genetic differences in life history characters between Baja and mainland populations of *D. mojavensis* (Fig. 12.1).

Shifts in the frequencies of several allozymes are also concordant with the shift from agria to organ pipe cactus (Zouros 1973; Starmer et al. 1977), but the most notable differences are in the frequencies of both second and third chromosome gene arrangements from Baja to the mainland. Most mainland populations are homokaryotypic, compared to the highly polymorphic, ancestral populations along the Baja peninsula. Host plant density is positively

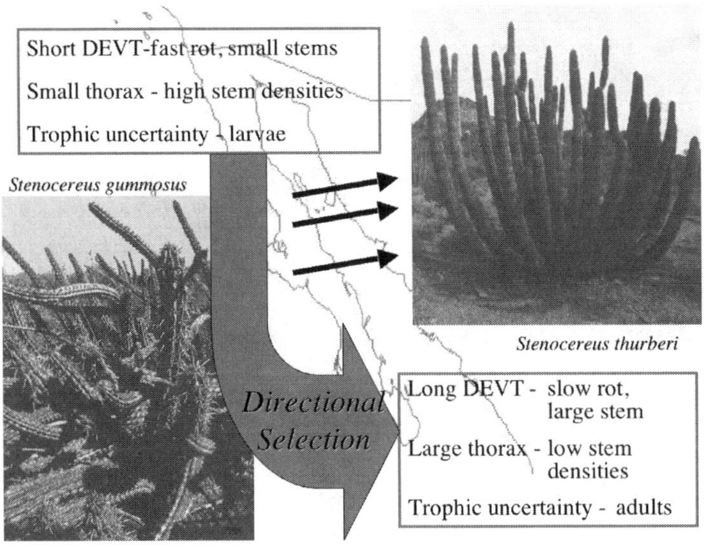

FIGURE 12.1. Host plant shift from agria cactus, *Stenocereus gummosus,* in Baja California to organ pipe cactus, *S. thurberi,* in mainland Sonora, Sinaloa, and Arizona is shown as *D. mojavensis* colonized the mainland. Directional natural selection on both larval and adult stages is hypothesized to have caused life history evolution due to contrasting regimes of trophic unpredictability in agria- vs. organ pipe–using populations. Abbreviation: DEVT, developmental time.

correlated, and statistically significant, with the degree of inversion polymorphism across populations (Heed and Etges, unpubl. data). Although no experiments have been performed to determine the influences of karyotypic diversity on patterns of host use, variation among karyotypes conforms strongly to the use of different host cacti and geographic structure of plant communities or phytogeographic provinces of the Sonoran Desert. These chromosomal polymorphisms, like those in other *Drosophila* species, have been shaped by local environmental conditions (Etges et al. 1999).

The South American *D. repleta* group includes the monophyletic *D. buzzatii* complex containing at least 11 species (Ruiz and Wasserman 1993; Tidon-Sklorz and Sene 1995; Ruiz et al. 2000; Tidon-Sklorz and Sene 2001) and a number of other more basal members in the group (Durando et al. 2000). All members of the *D. buzzatii* complex are known to be cactophilic. Host cactus use, but not its genetic basis, has been thoroughly studied in the northerly distributed *D. martensis* cluster (*D. starmeri, D. uniseta, D. venezolana,* and *D. martensis;* Cerda 1989). Studies have shown that both interspecific competition and ant predation are important determinants of cactus use (Benado and Montero 1988; Benado 1989; Escalante and Benado 1990). In the south, *D. buzzatii* is perhaps the most intensively studied species in this group, as it has been transported around the world with the introduction of its host plant, *Opuntia.* Within its ancestral range, it is also almost exclusively restricted to *Opuntia,* whereas its closest relative, *D. koepferae,* has a much more restricted geographic range and primarily uses *Trichocereus terschekii.* In areas of sympatry (Argentina), *D. buzzatii* expresses higher fitness on two species of endemic *Opuntias*, whereas *D. koepferae* performs better (egg-to-adult viability, development time) on *T. terschekii* (Fanara et al. 1999), concordant with patterns of oviposition preference (Fanara and Hasson 2001).

Studies of the rich inversion polymorphism in *D. buzzatii* have revealed ample evidence of selection operating at various stages of the life cycle (Ruiz et al. 1986; Hasson et al. 1991), but conflicting results concerning inversion-related changes due to use of different substrates in ancestral and derived populations. Ruiz and Fontdevila (1985) reported gene arrangement shifts in laboratory populations from Argentina flies exposed to *Opuntia ficus-indica* fruits vs. the fermenting pads. However, Fanara et al. (1996) found no chromosome frequency differences among individuals using *O. ficus-indica* or *O. quimilo* in another Argentinian population. Despite the number of different *Opuntia* species introduced into Australia, habitat selection by *D. buzzatii* involves genotypic preferences for specific cactophilic yeast species inhabiting fermenting *Opuntia* tissues rather than different cactus species (Barker et al. 1994). Thus, the more chemically benign species in the genus *Opuntia*, the ancestral host plants for so many members of the *D. repleta* group, do not seem to pose an adaptive challenge for the *Drosophila* species using them, unlike the many species of more chemically specialized columnar cacti (Ruiz and Heed 1988).

MYCOPHAGOUS *DROSOPHILA*

A number of temperate North American and European drosophilids inhabiting fungal substrates have played important roles in studies of habitat selection and the genetics of host specialization. Most are members of the *immigrans-Hirtodrosophila* radiation of the subgenus *Drosophila* (Throckmorton 1975). Two of three nearctic *Mycodrosophila* species (*M. claytoni* and *M. stalkeri*) specialize on Polyporaceae bracket fungi, whereas *M. dimidiata* uses a number of Basidiomycetes, including coral fungi (Clavariaceae), jelly fungi (Heterobasidiomycetes), and species of Ascomycetes (Lacy 1984). Other species groups that originated in the tropics and spread into temperate forests include the *D. quinaria* group (*D. quinaria, D. guttifera, D. recens,* and *D. falleni*) and the *D. testacea* group (*D. testacea, D. neotestacea,* and *D. putrida*) (Grimaldi et al. 1992). Many of these species appear to be broad generalists, feeding and ovipositing on an array of mushrooms, including *Amanita muscaria, A. flavorubescens, A. rubescens,* and *Russula subfoetans* (Jaenike 1978a). However, *D. quinaria* has only been reared out of skunk cabbage, *Symplocarpus foetidus* (Araceae) (Jaenike 1978b). In addition, several related, temperate, fungus breeding *Drosophila* in northern Japan appear to specialize on particular groups of mushrooms and decaying plants (Kimura et al. 1977). A single species from the large *D. tripunctata* radiation, *D. tripunctata,* is commonly encountered around mushrooms. Populations of this species are characterized by significant genetic variation for preference for mushrooms or fruits (Jaenike and Grimaldi 1983).

Among mushroom breeding species, considerable overlap in resource use has been documented (Wertheim et al. 2000), but little genetic evidence exists for host specialization on particular mushroom species (Jaenike 1978a; Jaenike and Selander 1979). However, substantial genetic variation does exist for accepting novel hosts in some cases (Courtney and Chen 1988). *D. putrida, D. recens,* and *D. tripunctata* are quite tolerant of mushrooms containing alpha-amanitin, a potent eukaryotic toxin due to its affinity for binding RNA polymerase II (Jaenike et al. 1983). Amanitin tolerance is a pleisiomorphy (ancestral state), at least within the *D. quinaria* group (Spicer and Jaenike 1996). Lack of host preference in mark-recapture trials and little extant genetic variation at several polymorphic enzyme loci among geographically isolated populations or populations using different mushroom species suggest that the unpredictability of mushroom hosts over space and time has prevented specialization for different host species (Jaenike 1978a; Lacy 1983).

Substantial genetic variability exists within and among populations of *D. tripunctata* for host preference (Jaenike and Grimaldi 1983). Jaenike (1986) performed mark-recapture experiments with two isofemale lines that strongly differed in long-distance orientation to mushrooms vs. tomatoes or settling behavior, and oviposition site preference, as measured by the number of eggs laid on either substrate in the laboratory. He released marked females from parental lines that preferred tomato or mushrooms for oviposition, F_2 flies from crossing these lines, and tested all recaptured adults in the lab for oviposition preference. He found no association between settling behavior and oviposition preference in F_2, indicating that these behaviors are influenced by different gene loci. Crosses of isofemale lines that preferred mushrooms with those that preferred tomatoes for oviposition sites revealed that host preference was influenced by a number of autosomal genes exerting dominance and interaction effects (Jaenike 1987). Furthermore, isofemale lines derived from a variety of populations across the species range expressed significant differences for oviposition site preference and larval development time on mushrooms vs. tomatoes. Little geographic differentiation was apparent for oviposition site preference and larval performance, and within populations, correlations between preference and performance were near zero (Jaenike 1989). Thus, host preference in *D. tripunctata* is a composite trait with a somewhat complex genetic basis.

True Fruit Flies: Tephritidae

Tephritids offer great promise for the genetic study of host use (see Aluja and Norrbom 2000). Efforts to taxonomically describe and record host use for tephritids indicate that many of these true fruit flies are specialists, restricting their host use to relatively few plant taxa (Foote et al. 1993). In addition, many tephritids are serious economic pests of commercially grown vegetables and fruits (Foote et al. 1993; Merritt et al. 2003). Thus, understanding the biology of tephritid host use is important from both basic scientific and applied biocontrol perspectives.

Despite their economic and scientific significance, relatively little is known about the genetics of tephritid host use. Two tephritids that have received significant study are the apple maggot fly, *Rhagoletis pomonella,* and the goldenrod gall fly, *Eurosta solidaginis.* These flies have been at the center of a long-standing debate concerning sympatric speciation—the idea that certain host-specific insects speciate in the absence of geographic isolation in the process of shifting and adapting to new host plants (Bush 1966, 1969, 1975, 1992). Indeed, as early as 1867, Benjamin Walsh cited the shift of *R. pomonella* from its native host hawthorn (*Crataegus* spp.) to introduced domesticated apple (*Malus pumila*) as an example of sympatric speciation in action. The term "host race" has subsequently been coined to describe this putative incipient stage in the sympatric divergence process. Host races are considered to be partially reproductively isolated populations that owe their isolation primarily to host-related adaptation (Diehl and Bush 1984).

RHAGOLETIS POMONELLA

The apple maggot fly is a member of an endemic North American sibling species complex. Four species in the group, *R. pomonella* [Walsh]; the blueberry maggot, *R. mendax* [Curran]; the snowberry fly, *R. zephyria* [Snow]; and the *Cornus amomum* fly, *R. cornivora* [Bush]; as well as the undescribed flowering dogwood fly that infests *Cornus florida,* are well characterized (Berlocher 2000). A number of other host specialist populations in the complex (e.g., sparkleberry fly, plum fly, mayhaw fly) are of uncertain taxonomic status. As emphasized by Bush (1966, 1969), the *R. pomonella* complex has several attributes consistent with sympatric divergence. The geographic distributions of these species are broadly sympatric across the eastern United States and Canada, and each taxon infests a unique, nonoverlapping set of host plants.

Enthusiasm for sympatric speciation has waxed and waned over the years. Conventional wisdom for most of the twentieth century held that geographic isolation was a prerequisite for animal speciation. Under this "Mayrian" view, complete allopatry was necessary for populations to overcome the homogenizing effects of gene flow and to genetically diverge (Mayr 1963). However, renewed interest in sympatric speciation was generated by the discovery of genetic differentiation (allozyme frequency differences) between the apple and hawthorn host races of *R. pomonella* (Feder et al. 1988; McPheron et al. 1988). This discovery, coupled with the findings of additional host races/cryptic species in other phytophagous insects (Wood 1980; Wood and Keese 1990; Via 1991; Carroll and Boyd 1992; Abrahamson et al. 1994), as well as evidence from fish (Schliewen et al. 1994; Schluter 1996), helped triggered a reassessment of the general role of ecological specialization (adaptation) in speciation (Morell 1999).

Studies investigating the genetics of host use in *R. pomonella* have focused on traits involved in host identification and the overwintering pupal diapause. There is little evidence for host fruit–related survivorship (larval performance) differences between the races (Prokopy et al. 1988). Host identification is important because *Rhagoletis* flies mate exclusively on or near the fruit of their host plants (Prokopy et al. 1971, 1972; Feder and Bush 1989). Hence, host choice translates directly into mate choice and prezygotic reproductive isolation. Mark-release-recapture experiments have demonstrated that apple and haw flies tend to return to their natal host species to mate and oviposit (Feder et al. 1994, 1998). This "host fidelity" reduces gene flow to about 4–6% per generation between the races. Similar studies with *R. mendax* have shown that host fidelity has the potential to completely isolate fly taxa (Feder and Bush 1989).

Rhagoletis use specific visual, olfactory, and tactile cues for recognizing their host plants (Prokopy 1968, 1972, 1977; Prokopy and Bush, 1973; Prokopy et al. 1973; Moericke et al. 1975; Prokopy et al. 1982; Owens and Prokopy, 1986). Field and laboratory manipulations have indicated that fruit odor is a particularly important signal for discriminating among hosts (Prokopy et al. 1973; Aluja and Prokopy 1992). Indeed, apple and haw flies have been found to respond positively in wind tunnel assays to blends of volatile compounds isolated from their respective fruits (C. Linn, pers. comm.). Quantitative trait locus (QTL) mapping is currently under way to assess the genetics for this behavioral difference. In addition to odor, fruit shape, size, and texture are also used in host discrimination. To this end, crosses between apple flies that prefer larger-sized fruit and *Cornus florida* flies that like smaller-sized fruit have demonstrated that fruit-size preference has a largely additive genetic basis (Smith 1986).

The timing of pupal diapause is a second key trait differentiating the *R. pomonella* host races. *Rhagoletis* flies are univoltine and have limited adult longevity (Boller and Prokopy 1976). Flies must therefore eclose in synchrony with the phenologies of their respective host plants to ensure maximal fruit availability for mating and oviposition. Apple varieties favored by *R. pomonella* generally fruit 3–4 weeks earlier in the field season than do hawthorn trees. As a result, apple flies eclose earlier than haw flies (Feder et al. 1993; Feder 1995; Feder and Filchak 1999). This developmental difference causes apple and haw fly larvae and pupae to experience different environmental conditions affecting diapause regulation, resulting in a postmating barrier to gene flow (Feder et al. 1997; Filchak et al. 2000). The significance of diapause to the host race story is highlighted by the fact that all three regions of the genome currently known to differentiate apple and haw flies affect the timing of adult eclosion. Moreover, genetic crosses have revealed that those three chromosomal regions are subsumed by inversion polymorphism (Feder et al. 2003). The "diapause hypothesis" has been tested by manipulating environmental rearing conditions simulating those experienced by flies infesting apple vs. haw fruit (Feder and Filchak 1999; Filchak et al. 1999, 2001). These experiments have documented that warmer, longer prewinter periods, typically faced by apple flies in nature, selectively favor alleles more common to the apple than to the hawthorn race. In contrast, cooler rearing conditions favor haw race alleles, producing a gene × environment interaction (Feder et al. 1997; Filchak et al. 2000). The effect of this gene × environment interaction is a fitness trade-off maintaining genetic differentiation between the host races (Filchak et al. 2000).

EUROSTA

Eurosta goldenrod gall flies have also contributed greatly to our understanding of the genetics of host specialization and speciation. The genus *Eurosta* consists of seven North American species of host specialists that form galls on various members of the Asteraceae (Ming 1989). Taxonomic relationships of *Eurosta* flies are uncertain at present; phylogenies based on morphological and molecular data are not in complete agreement (Ming 1989; Abrahamson and Weis 1997). The eastern subspecies *E. solidaginis solidaginis* is of particular interest, however, because it displays several interesting parallels with *R. pomonella.*

E. s. solidaginis infests the goldenrod *Solidago altissima* over much of its North American range. However, in the Midwest and Eastern United States, as well as in Canada, the fly also attacks *S. gigantea* (Abrahamson et al. 1994). Genetic differentiation among populations is consistent with host use, implying that altissima and gigantea flies represent host races (Waring et al. 1990; Itami et al. 1998). Patterns of genetic differentiation suggest that *S. altissima* is the ancestral host (Brown et al. 1996; Itami et al. 1998). One of the allozyme loci showing frequency differences between gigantea and altissima flies also differs between the haw and apple *R. pomonella* races (*Hbdh* in *Eurosta, Had* in *Rhagoletis*). There has been no study to determine whether *Hbdh* or linked loci affect the timing of diapause in *Eurosta,* as it does in *Rhagoletis.*

Variation in host preference, performance, and developmental timing all appear to contribute to reproductively isolating the goldenrod host races. *E. s. solidaginis* use their host plants as rendezvous sites for mating and oviposition. Like the *R. pomonella* races, altissima and gigantea flies exhibit a high degree of host fidelity, resulting in positive assortative mating (Craig et al. 1993, 2001). The exact cues used for host discrimination in *Eurosta* are uncertain. However, goldenrod bud shape, size, and tactile chemical cues all appear to influence host choice (Abrahamson et al. 1989; Horner and Abrahamson 1992; Abrahamson et al. 1994). Genetic crosses have indicated that host choice is genetically controlled (Craig et al. 2001). Preference for *S. gigantea* has been shown to be dominant, and possibly controlled by only a few genes. In addition, male and female flies have been found to display different patterns of host preference. Therefore, certain genes affecting host preference are sex influenced (Craig et al. 2001). These same experiments failed to detect any effect of nongenetic factors on host choice (e.g., larval conditioning, adult experience). However, not all genotypes of a given host species are equally preferred (Horner and Abrahamson 1992; Craig et al. 1999; Cronin and Abrahamson 1999; Horner and Abrahamson 1999; Craig et al. 2000).

Unlike the *R. pomonella* host races, altissima and gigantea flies display performance differences on the alternate host plant. As expected, survivorship is greater for both races on their natal host (Craig et al. 1997). Crosses between altissima and gigantea flies indicated that performance is genetically controlled, and host plant dependent (Craig et al. 1997). Data from these crosses imply that loci responsible for survivorship are autosomal and act in a generally additive manner. Like preference, performance appears to be influenced by plant genotype (Craig et al. 1999, 2000; Cronin and Abrahamson 1999; Horner and Abrahamson 1999). There are certain plant genotypes in which F_1 survivorship is higher than others. It is tempting to speculate that relatively benign host plant genotypes aided in the initial stages of the host shift.

Temporal adaptations to seasonal differences in *S. altissima* and *S. gigantea* abundance also appear to isolate the *E. s. solidaginis* races (Abrahamson et al. 1994). *S. gigantea* is generally available earlier in the summer than *S. altissima*, and postdiapause emergence patterns of the races differ in accordance with host phenology (Craig et al. 1993; Itami et al. 1998; Horner et al. 1999). Genetic control of developmental differences has been addressed by rearing both host races under similar conditions (Craig et al. 1993; Itami et al. 1998; Horner et al. 1999). Results from these experiments imply that the developmental differences have a genetic component. However, further work (e.g., crosses or correlations with genetic variation) is required to substantiate the developmental data.

BLEPHARONEURA

Blepharoneura flies are a genus of perhaps 200 or more highly host-specific Neotropical Tephritids species, all of which infest plants of the Cucurbitaceae (Condon and Norrbom 2000). *Blepharoneura* flies specialize not only on different species of host plants, but also on different parts of the same host species. For example, these flies have been found to specialize on male and female flowers, seeds, and stems of the same plant (Condon and Norrbom 1994; Condon and Steck 1997). Few genetic studies have been done to determine whether *Blepharoneura* flies radiate primarily by shifting hosts, by specializing on different structures within hosts, or by a combination of both processes. However, Condon and Steck (1997) found that two highland and two lowland Costa Rican populations of flies utilizing the same flowering vine, *Gurania costaricensis*, represent four distinct species based on allozyme data. In both locations, species boundaries coincided with specialization on male vs. female flowers. Moreover, the data suggest that lowland populations are possible sister species, consistent with sympatric speciation on the different flower sexes. In contrast, the highland populations were not sister taxa. The fascinating pattern of host-related diversification for *Blepharoneura* flies presents a great opportunity for the genetic study of host use and specialization.

Hessian Fly, *Mayetiola destructor* Say (Cecidomyiidae)

The Hessian fly is a model for antagonistic plant-insect interactions. Hessian flies belong to a taxonomically diverse group of gall forming midges (Harris et al. 2003). *M. destructor* parasitizes a variety of wild grasses, but has gained notoriety as the world's most destructive insect pest of wheat (*Triticum aestivum* L.) (Gallun 1977). Losses from Hessian flies can reach $100 million or more in some years (Ratcliffe and Hatchett 1997).

The genetic study of *M. destructor* host use has concentrated on plant resistance and fly virulence (Gallun 1977). Genetic resistance to *M. destructor* attack has been found in common and durum wheat, wild wheat (*Triticum tauschii* Cus), and rye (*Secal cereale* L.), and is a principal source of Hessian fly control efforts. To date, 27 separate resistance genes (*H1–H27*) have been identified (Ratcliffe and Hatchett 1997). Resistance is generally dominant and determined by a single locus. However, there are also examples in which resistance is partially dominant, recessive, and multigenic. The effect of resistance genes on *M. destructor* varies, but larval death is the typical result.

Genetic variation exists within Hessian fly populations to counter the effects of resistance genes (Gallun 1977). Loci responsible for fly survival are referred to as "virulence" (*vH* or *avr*) genes. Fly virulence genes have been found to overcome all but nine of the resistance plant biotypes. Virulence tends to have a monogenetic basis involving recessive alleles that are either sex-linked or autosomal (Harris et al. 2003). Interestingly, fly oviposition preference appears to be unrelated to the presence or absence of resistant genes in the host plant (Harris et al. 2001). The pattern of virulence and resistance is consistent with a gene-for-gene relationship. However, whether this relationship arose in a strictly co-evolutionary manner is difficult to establish.

Pitcher-Plant Mosquito, *Wyeomyia smithii*

Genetic analysis of host use in populations of *W. smithii* has focused on the evolutionary consequences of using the specialized aquatic habitat provided by North American pitcher plants of the genus *Sarracenia*. Of the more than 80 tropical and subtropical *Wyeomyia* species, most are bromeliad breeders (Dyar 1927; Lane 1953). An exception is *W. smithii*, which is adapted to the cooler and more seasonal environments of temperate North America, where it ordinarily inhabits the purple pitcher plant *S. purpurea* from the Gulf of Mexico to northern Canada (Bradshaw 1983). Of the eight species of *Sarracenia*, only *S. psittacina* and *S. purpurea* have evergreen pitchers that last through winter. The other six species form water holding, pitcher-shaped leaves in the summer, but these leaves usually die back in winter, when the plants may develop flat, blade-shaped phyllodia (Bell 1952; McDaniel 1971). In disturbed habitats, *S. flava*, *S. leucophylla*, *S. rubra*, and *S. alata* can hybridize with *S. purpurea* and may support low densities of *W. smithii* in mild, but not cold winters (Bradshaw

1983). Thus, *S. purpurea* is the principal host of *W. smithii* and, in southern states, may serve as a source for colonizing the sink habitats presented by the other species of *Sarracenia* and their hybrids with *S. purpurea*.

Following the last glacial retreat, *W. smithii* dispersed from the southeastern coastal plain to northern Canada and, in the process, adapted to cooler, more seasonal environments. As in many overwintering arthropods, *W. smithii* enters a diapause stage, induced and maintained by short critical daylengths, and overwinters as dormant third instar larvae inside the frozen pitchers (Smith 1904; Bradshaw and Lounibos 1972). Variation in egg-to-adult development time and tendency to diapause varies considerably within populations (Bradshaw and Lounibos 1972; Evans and Brust 1972) and increases with latitude and altitude of origin (R^2 is repeatedly >90) (Bradshaw 1976; Hard et al. 1993a; Lair et al. 1997; Bradshaw and Holzapfel 2001a).

In a classic series of studies, Istock (Istock et al. 1975, 1976a,b; Istock 1978, 1981) demonstrated that genetic variation in egg-to-adult development time for a New York population of mosquitoes was likely maintained by fluctuating natural selection. In addition, development time was genetically correlated with the tendency to diapause under warm temperature conditions. Because only diapaused third instar larvae can survive winter, populations must track late-season photoperiods to avoid extinction: all individuals not in diapause at the onset of the first hard freeze will not survive. In years with earlier than average hard freezes, most slow-developing diapausing individuals survive winter, whereas all nondiapausers perish. However, in years when the onset of winter is late, fast developers may be able to complete another generation before the first hard freeze and thus increase their relative fitnesses. Therefore, normal year-to-year fluctuation in the timing of the first hard, late-season freeze maintains genetic variation in the population, accounting for substantial heritability for development time ($h^2 = 0.33$) in the New York population (Istock et al. 1976b). Within this same population, the genetic correlation between development time and tendency to diapause constrains response to short-term selection (Scheiner and Istock 1991). In other populations, realized heritability for development time has been found to be lower (from 0.0 to 0.10) and increasing with latitude (Bradshaw et al. 1997). Moreover, in most populations, development time is positively genetically correlated with critical photoperiod (Hard et al. 1993b). Likewise, in most populations, there is a positive genetic relationship between critical photoperiod, depth of diapause, and the number of long days required to avert diapause (Campbell and Bradshaw 1992). Hence, within populations of *W. smithii*, there is a genetic continuum between fast developing, diapause averse and slow developing, diapause prone individuals, maintaining evolutionary flexibility (Hard et al. 1993b).

The genetic architecture of development varies across the range of *W. smithii*. Allozyme frequencies show a deep split among northern populations near the maximum extent of the Laurentide Ice Sheet, as well as divergence between Gulf and Atlantic Coast populations. This pattern of genetic variation suggests that multiple northward expansions of mosquitoes occurred during the recent glacial history of North America (Armbruster et al. 1998). Critical photoperiod, a factor determining the tendency to diapause, has been shown to be positively correlated with latitude ($r^2 = 0.98$, $n = 6$) among populations from 30–49° N (Wegis et al. 1997). Joint scaling tests of crosses between populations show an overwhelming effect of epistasis on genetic divergence of critical photoperiod among populations (Hard et al. 1992), suggesting that multiple founder events may have been responsible for the release of additive from epistatic genetic variance (Hard et al. 1993a; Lair et al. 1997; Armbruster et al. 1998; Bradshaw and Holzapfel 2000, 2001b). Positive genetic correlations between development time and critical photoperiod are not found in all populations, implying the possible evolution of different life history trajectories in northern populations associated with founder events and/or reduced gene flow among semi-isolated demes (Hard et al. 1993b). Patterns of allozyme variation and line-cross experiments among geographically isolated populations showed that divergence and differentiation are largely due to strong natural selection and genes with dominance and epistatic effects unique to particular crosses. Thus, *W. smithii* populations has been cited as a complex gene interaction system making up an "adaptive landscape" consistent with a Wrightian view of evolutionary change (Istock and Weisburg 1989; Armbruster et al. 1997). However, the role of drift in contributing to this pattern, as opposed to strong natural selection generating alternative adaptive solutions in different isolated demes, still needs to be clarified.

Conclusion

Our understanding of the genetics of host use is still in its infancy. Although our review of the topic is not meant to be exhaustive, it is clear that even the most basic issues, such as the genetic architecture of host discrimination traits, still need to be addressed in most systems. Moreover, in many cases, the alpha-ecology of host use has yet to be fully resolved, even for flies with substantial societal impact.

One interesting trend to emerge from many studies of dipteran host use is the apparent association of chromosomal inversions with ecological adaptation and speciation. Whether this relationship is real or illusionary still needs to be confirmed. It is possible that inversions facilitate and protect the evolution of co-adapted blocks of genes involved in utilizing different hosts. However, it is also possible that the emphasis on inversions reflects a bias in detection and genetic scoring. Given that several genetic surveys of fly populations concentrated on a gross, chromosomal-level analysis of variation, it is not surprising that inversions were beacons for differentiation. The same problem also potentially affects allelic-based surveys of variation in natural populations. Neutral marker loci inside inversions will tend to be in disequilibrium with a large number of loci, some of which may affect host specialization. In contrast, linkage equilibrium will be the rule for loci outside inversions. Thus, random surveys of genes looking for genetic correlations with host usage will disproportionately identify makers residing in inversions or other chromosomal rearrangements restricting recombination. Elucidation of a specific role for inversions in host utilization will have to await more detailed QTL-based mapping studies to confirm a disproportionate concentration of preference and performance loci within rearrangements.

Regardless, the question remains: why do we not have a better understanding of the genetics of dipteran host use? Clearly this phenomenon is important on many levels. The answer is not a lack of potential study systems, as numerous model systems already exist, and many more wait to be investigated. Although not all of these flies are ideal candidates for laboratory husbandry, many can be reared with trial and error. Moreover, medical and economic concerns associated with many flies should help provide resources to solve these problems.

Perhaps the biggest confounding factor is a shortage of readily available genetic tools (e.g., cDNA and BAC libraries, linkage maps, large-scale sequencing) with which to examine a particular fly model. *Drosophila* and mosquito genomics can serve as an important foundation from which to build. However, *D. melanogaster* and *An. gambiae* alone will not provide sufficient genomics to resolve questions of host use in the Diptera. These two species do not adequately represent the taxonomic diversity of flies and mosquitoes. We are entering the beginning of the genomics era for the Diptera, which will herald great advances for understanding many basic and applied questions. To realize this potential, however, increased resources must be placed in developing the genetics of many species across the Diptera. Further attention must be devoted to gaining a more thorough understanding of the ecology and demography of natural fly populations to utilize these genomic resources to their fullest. Great opportunity and reward await students of dipteran host use in the future.

ACKNOWLEDGMENTS

We thank Uwe Stolz, Sarah Smyczek, Sebastian Velez, W. Abrahamson, Leonard Munstermann, Jeff Stuart, Roger Butlin, and Jack Duman. E. Bradshaw and H. L. Carson made substantial improvements to an earlier version of this manuscript. We also thank J. Jaenike, C. Costantini, T. Craig, and M. Condon for comments, and C. Jones for his unpublished data. WJE was funded by grants from the National Science Foundation. Research support to (NJB) laboratory from the National Institutes of Health (AI44003, AI48842) and the State of Indiana (CRTF 042700-0207) is gratefully acknowledged. JLF was supported by the National Science Foundation, U.S. Department of Agriculture, and the State of Indiana 21st Century grants.

REFERENCES

Abrahamson, W. G., and A. E. Weis. 1997. *The Evolutionary Ecology of a Tritrophic-Level Interaction: Goldenrod, the Stem Gallmaker and Its Natural Enemies.* Monographs in Population Biology 29. Princeton: Princeton University Press.

Abrahamson, W. G., K. D. McCrea, and S. S. Anderson. 1989. Host preference and recognition by the goldenrod ball gallmaker *Eurosta solidaginis* (Diptera: Tephritidae). *Am. Midl. Nat.* 121:322–330.

Abrahamson, W. G., W. M. Brown, S. K. Roth, D. V. Sumerford, J. D. Horner, M. D. Hess, S. T. How, T. P. Craig, R. A. Packer, and J. K. Itami. 1994. Gallmaker speciation: An assessment of the roles of host-plant characters, phenology, gallmaker competition and natural enemies. In P. Price, W. Mattson, and Y. Baranchilov, eds., *Gall-Forming Insects,* pp. 226–240. General Technical Report NC-174. St. Paul, Minn.: U.S. Department of Agriculture Forest Service, North Central Experimental Station.

Aluja, M., and A. L. Norrbom, eds. 2000. *Fruit Flies (Tephritidae): Phylogeny and Evolution of Behavior.* Boca Raton, Fla.: CRC Press.

Aluja, M., and R. J. Prokopy. 1992. Host search behavior by *Rhagoletis pomonella* flies: Inter-tree movement patterns in response to wind-borne fruit volatiles under field conditions. *Physiol. Entomol.* 17:1–8.

Armbruster, P., W. E. Bradshaw, and C. M. Holzapfel. 1997. Evolution of the genetic architecture underlying fitness in the pitcher-plant mosquito, *Wyeomyia smithii. Evolution* 51:451–458.

Armbruster, P., W. E. Bradshaw, and C. M. Holzapfel. 1998. Effects of postglacial range expansion on allozyme and quantitative genetic variation of the pitcher-plant mosquito, *Wyeomyia smithii. Evolution* 52:1697–1704.

Ashburner, M. 1981. Entomophagous and other bizarre Drosophilidae. In M. Ashburner, H. L. Carson, and J. N. Thompson, Jr., eds., *The Genetics and Biology of* Drosophila, pp. 395–429. New York: Academic Press.

Barker, J. S. F., W. T. Starmer, and J. C. Fogleman. 1994. Genotype-specific habitat selection for oviposition sites in the cactophilic species of *Drosophila buzzatii. Heredity* 72:384–395.

Beckenbach, A. T. 1996. Selection and the "sex-ratio" polymorphism in natural populations of *Drosophila pseudoobscura. Evolution* 50:787–794.

Bell, C. R. 1952. Natural hybrids in the genus *Sarracenia*. *J. Elisha Mitchell Sci. Soc.* 68:55–79.

Benado, M. 1989. Competitive release in the cactophilic fly, *Drosophila venezolana*. *Ecotropicos* 2:45–48.

Benado, M., and C. Montero. 1988. Competition between cactophilic species *Drosophila starmeri* and *D. uniseta*. *Rev. Chil. Hist. Nat.* 61:187–190.

Berlocher, S. 2000. Radiation and divergence in the *Rhagoletis pomonella* species group: Inferences from allozymes. *Evolution* 54:543–557.

Biessmann, H., M. F. Walter, S. Dimitratos, and D. Woods. 2002. Isolation of cDNA clones encoding putative odourant binding proteins from the antennae of the malaria-transmitting mosquito, *Anopheles gambiae*. *Insect Mol. Biol.* 11:123–132.

Boller, E. F., and R. J. Prokopy. 1976. Bionomics and management of *Rhagoletis*. *Annu. Rev. Entomol.* 21: 223–246.

Bradshaw, W. E. 1976. Geography of photoperiodic response in a diapausing mosquito. *Nature* 262:384–386.

Bradshaw, W. E. 1983. Interaction between the mosquito, *Wyeomyia smithii*, the midge, *Metriocnemus knabi*, and their carnivorous host *Sarracenia purpurea*. In J. H. Frank and L. P. Lounibos, eds., Phytotelmata: *Terrestrial Plants as Hosts of Aquatic Insect Communities*, pp. 161–189. Medford, N.J.: Plexus.

Bradshaw, W. E., and C. M. Holzapfel. 2000. The evolution of genetic architectures and the divergence of natural populations. In J. B. Wolf, E. D. Brodie III, and M. J. Wade, eds., *Epistasis and the Evolutionary Process*, pp. 245–263. New York: Oxford University Press.

Bradshaw, W. E., and C. M. Holzapfel. 2001a. Genetic shift in photoperiodic response correlated with global warming. *Proc. Natl. Acad. Sci. USA* 98:14509–14511.

Bradshaw, W. E., and C. M. Holzapfel. 2001b. Phenotypic evolution and the genetic architecture underlying photoperiodic time measurement. *J. Insect Physiol.* 47:809–820.

Bradshaw, W. E., and T. P. Lounibos. 1972. Photoperiodic control of development in the pitcher-plant mosquito, *Wyeomyia smithii*. *Can. J. Zool.* 50:713–719.

Bradshaw, W. E., C. M. Holzapfel, C. A. Kleckner, and J. J. Hard. 1997. Heritability of development time and protandry in the pitcher-plant mosquito, *Wyeomyia smithii*. *Ecology* 78:969–976.

Breman J. G., A. Egan, and G. T. Keusch. 2001. The intolerable burden of malaria: A new look at the numbers. *Am. J. Trop. Med. Hyg.* 64(suppl.):1.

Brncic, D. 1962. Chromosomal structure of populations of *Drosophila flavopilosa* studied in larvae collected in their natural breeding sites. *Chromosoma* 13:183–195.

Brncic, D. 1968. The effects of temperature on chromosomal polymorphism of *Drosophila flavopilosa* larvae. *Genetics* 59:427–432.

Brock, M. L., R. G. Wiegert, and T. D. Brock. 1969. Feeding by *Paracoenia* and *Ephydra* (Diptera: Ephydridae) on micro-organisms of hot springs. *Ecology* 50:192–200.

Brown, J. M., W. G. Abrahamson, and P. A. Way. 1996. Mitochondrial DNA phylogeography of host races of the goldenrod gall maker *Eurosta solidaginis* (Diptera: Tephritidae). *Evolution* 50:777–786.

Bryan, J. H., V. Petrarca, M. A. Di Deco, and M. Coluzzi. 1987. Adult behavior of members of the *Anopheles gambiae* complex in The Gambia with special reference to *An. melas* and its chromosomal variants. *Parassitologia* 29:221–249.

Burkot, T. R. 1988. Non-random host selection by anopheline mosquitoes. *Parasitol. Today* 4:156–162.

Bush, G. L. 1966. The taxonomy, cytology and evolution of the genus *Rhagoletis* in North America. *Mus. Comp. Zool. Cambr.* 134:431–562.

Bush, G. L. 1969. Sympatric host race formation and speciation in frugivorous flies of the genus *Rhagoletis* (Diptera: Tephritidae). *Evolution* 23:237–251.

Bush, G. L. 1975. Sympatric speciation in phytophagous parasitic insects. In P. W. Price, ed., *Evolutionary Strategies of Parasitic Insects and Mites*, pp. 187–206. New York: Plenum Press.

Bush, G. L. 1992. Host race formation and sympatric speciation in *Rhagoletis* fruit flies (Diptera: Tephritidae). *Psyche* 99:335–358.

Campbell, M. D., and W. E. Bradshaw. 1992. Genetic coordination of diapause in the pitcher plant mosquito *Wyeomyia smithii* (Diptera; Culicidae). *Ann. Entomol. Soc. Am.* 85:445–451.

Carroll, S. P., and C. Boyd. 1992. Host race radiation in the soapberry bug: Natural history with the history. *Evolution* 46:1052–1069.

Carson, H. L. 1951. Breeding sites of *Drosophila pseudoobscura* and *D. persimilis* in the transition zone of the Sierra Nevada. *Evolution* 5:91–96.

Carson, H. L. 1971. The ecology of *Drosophila* breeding sites. *Harold L. Lyon Arboretum Lecture* 2:1–27.

Carson, H. L. 1974. Three flies and three islands: Parallel evolution in *Drosophila*. *Proc. Natl. Acad. Sci. USA* 71:3517–3521.

Carson, H. L. 2001. Danko Brncic and the flower flies. *Rev. Chil. Hist. Nat.* 74:33–40.

Carson, H. L., and H. D. Stalker. 1951. Natural breeding sites for some wild species of *Drosophila* in the eastern United States. *Ecology* 32:317–330.

Cerda, H. G. 1989. Ecologia y genetica evolutiva del cluster *martensis* (Grupo *repleta*) de *Drosophila* asociadas a cactaceas de desiertos en el norte Sudamericano (Venezula y Colombia). Ph.D. thesis, Universidad Autonoma de Barcelona, Barcelona.

Clements, A. N. 1999. *The Biology of Mosquitoes*, vol. 2: *Sensory Reception and Behavior.* New York: CABI Publishing.

Collins, F. H., R. K. Sakai, K. D. Vernick, S. Paskewitz, D. C. Seeley, L. H. Miller, W. E. Collins, C. C. Campbell, and R. W. Gwadz. 1986. Genetic selection of a *Plasmodium*-refractory strain of the malaria vector *Anopheles gambiae. Science* 234:607–610.

Coluzzi, M., A. Sabatini, V. Petrarca, and M. A. Di Deco. 1977. Behavioural divergences between mosquitoes with different inversion karyotypes in polymorphic populations of the *Anopheles gambiae* complex. *Nature* 266:832–833.

Coluzzi, M., A. Sabatini, A. Della Torre, M. A. Di Deco, and V. Petrarca. 2002. A polytene chromosome analysis of the *Anopheles gambiae* species complex. *Science* 298:1414–1418.

Condon, M. A., and A. L. Norrbom. 1994. Three sympatric species of *Blepharoneura* (Diptera: Tephritidae) on a single species of host (*Gurania spinulosa,* Cucurbitaceae): New species and new taxonomic methods. *Syst. Entomol.* 19:297–304.

Condon, M. A., and A. L. Norrbom. 2000. Behavior of flies in the genus *Blepharoneura* (Blepharoneurinae). In M. Aluja and A. L. Norrbom, eds., *Fruit Flies (Tephritidae): Phylogeny and Evolution of Behavior,* pp. 158–174. Boca Raton, Fla.: CRC Press.

Condon, M. A., and G. J. Steck. 1997. Evolution of host use in fruit flies of the genus *Blepharoneura* (Diptera: Tephritidae): Cryptic species on sexually dimorphic host plants. *Biol. J. Linn. Soc.* 60:443–466.

Costantini, C., G. Gibson, J. Brady, L. Merzagora, and M. Coluzzi. 1993. A new odour-baited trap to collect host-seeking mosquitoes. *Parasitologia* 35:5–9.

Costantini, C., G. Gibson, N. Sagnon, A. Della Torre, J. Brady, and M. Coluzzi. 1996. Mosquito responses to carbon dioxide in a west African Sudan savanna village. *Med. Vet. Entomol.* 10:220–227.

Costantini, C., N. F. Sagnon, A. della Torre, M. Diallo, J. Brady, G. Gibson, and M. Coluzzi. 1998. Odor-mediated host preferences of West African mosquitoes, with particular reference to malaria vectors. *Am. J. Trop. Med. Hyg.* 58:56–63.

Courtney, S. P., and G. K. Chen. 1988. Genetic and environmental variation in oviposition behavior in the mycophagous *Drosophila suboccidentalis* spcr. *Functional Ecol.* 2:521–528.

Craig, T. P., J. K. Itami, W. G. Abrahamson, and J. D. Horner. 1993. Behavioral evidence for host-race formation in *Eurosta solidaginis. Evolution* 47:1696–1710.

Craig, T. P., J. D. Horner, and J. K. Itami. 1997. Hybridization studies on the host races of *Eurosta solidaginis:* Implication for sympatric speciation. *Evolution* 51:1552–1560.

Craig, T. P., J. K. Itami, W. G. Abrahamson, and J. D. Horner. 1999. Oviposition preference and offspring performance of *Eurosta solidaginis* on genotypes of *Solidago altissima. Oikos* 86:119–126.

Craig, T. P., J. K. Itami, C. Schantz, W. G. Abrahamson, and J. D. Horner. 2000. The influence of host plant variation and intraspecific competition on oviposition preference in the host races of *Eurosta solidaginis. Ecol. Entomol.* 25:7–18.

Craig, T. P., J. D. Horner, and J. K. Itami. 2001. Genetics, experience, and host-plant preference in *Eurosta solidaginis:* Implications for host shifts and speciation. *Evolution* 55:773–782.

Cronin, J. T., and W. G. Abrahamson. 1999. Host-plant genotype and other herbivores influence goldenrod stem galler preference and performance. *Oecologia* 121:392–404.

Dekker, T., and W. Takken. 1998. Differential responses of mosquito sibling species *Anopheles arabiensis* and *An. quadriannulatus* to carbon dioxide, a man or a calf. *Med. Vet. Entomol.* 12:136–140.

Dekker, T., W. Takken, and M. A. Braks. 2001. Innate preference for host-odor blends modulates degree of anthropophagy of *Anopheles gambiae* sensu lato (Diptera: Culicidae). *J. Med. Entomol.* 38:868–871.

Diatta, M., A. Spiegel, L. Lochouarn, and D. Fontenille. 1998. Similar feeding preferences of *Anopheles gambiae* and *A. arabiensis* in Senegal. *Trans. R. Soc. Trop. Med. Hyg.* 92:270–272.

Diehl, S. R., and G. L. Bush. 1984. An evolutionary and applied perspective of insect biotypes. *Annu. Rev. Entomol.* 29:471–504.

Downing, R. J. 1985. The chemical basis for host plant selection in *Drosophila mojavensis.* Masters thesis, University of Denver, Denver, Colo.

Duchemin, J. B., J. M. Tsy, P. Rabarison, J. Roux, M. Coluzzi, and C. Costantini. 2001. Zoophily of *Anopheles arabiensis* and *An. gambiae* in Madagascar demonstrated by odour-baited entry traps. *Med. Vet. Entomol.* 15:50–57.

Durando, C. M., R. H. Baker, W. J. Etges, W. B. Heed, M. Wasserman, and R. DeSalle. 2000. Phylogenetic analysis of the *repleta* species group of the genus *Drosophila* using multiple sources of characters. *Mol. Phylog. Evol.* 16:296–307.

Dyar, H. G. 1927. Water bearing plants of Panama which harbor mosquitoes, with a synopsis of *Wyeomyia* (Diptera: Culicidae). *Proc. Entomol. Soc. Wash.* 30:110–112.

Escalante, A., and M. Benado. 1990. Predation on the cactophilic fly, *Drosophila starmeri,* in the columnar cactus, *Pilosocereus lanuginosus. Biotropica* 22:48–50.

Etges, W. J. 1989. Evolution of developmental homeostasis in *Drosophila mojavensis. Evol. Ecol.* 3:189–201.

Etges, W. J. 1990. Direction of life history evolution in *Drosophila mojavensis.* In J. S. F. Barker, W. T. Starmer, and R. J. MacIntyre, eds., *Ecological and Evolutionary Genetics of* Drosophila, pp. 37–56. New York: Plenum Press.

Etges, W. J. 1993. Genetics of host-cactus response and life-history evolution among ancestral and derived populations of cactophilic *Drosophila mojavensis. Evolution* 47:750–767.

Etges, W. J., and W. B. Heed. 1987. Sensitivity to larval density in populations of *Drosophila mojavensis:* Influences of host plant variation on components of fitness. *Oecologia* 71:375–381.

Etges, W. J., and C. S. Klassen. 1989. Influences of atmospheric ethanol on adult *Drosophila mojavensis:* Altered metabolic rates and increases in fitness among populations. *Physiol. Zool.* 62:170–193.

Etges, W. J., W. R. Johnson, G. A. Duncan, G. Huckins, and W. B. Heed. 1999. Ecological genetics of cactophilic *Drosophila.* In R. Robichaux, ed., *Ecology of Sonoran Desert Plants and Plant Communities,* pp. 164–214. Tucson: University of Arizona Press.

Evans, H. E. 1984. *Insect Biology.* Reading, Mass.: Addison-Wesley.

Evans, K. W., and R. A. Brust. 1972. Induction and termination of diapause in *Wyeomyia smithii* (Diptera: Culicidae) and larval survival studies at subzero temperatures. *Can. Entomol.* 104:1937–1950.

Fanara, J. J., and E. Hasson. 2001. Oviposition acceptance and fecundity schedule in the cactophilic sibling species *Drosophila buzzatii* and *D. koepferae* on their natural hosts. *Evolution* 55:2615–2619.

Fanara, J. J., E. Hasson, C. Rodriguez, M. Santos, and A. Fontdevila. 1996. The evolutionary history of *D. buzzatii.* XXXIII. Are *Opuntia* hosts a selective factor for the inversion polymorphism? *Heredity* 77:500–508.

Fanara, J. J., A. Fontdevila, and E. Hasson. 1999. Oviposition preference and life history traits in cactophilic *Drosophila koepferae* and *D. buzzatti* in association with their natural hosts. *Evol. Ecol.* 13:173–190.

Feder, J. L. 1995. The effects of parasitoids on sympatric host races of the apple maggot fly, *Rhagoletis pomonella* (Diptera: Tephritidae). *Ecology* 76:801–813.

Feder, J. L., and G. L. Bush. 1989. Gene frequency clines for host races of *Rhagoletis pomonella* (Diptera: Tephritidae) in the midwestern United States. *Heredity* 63:245–266.

Feder, J. L., and K. E. Filchak. 1999. It's about time: The evidence for host plant–mediated selection in the apple maggot fly, *Rhagoletis pomonella,* and its implications for fitness trade-offs in phytophagous insects. *Entomol. Exp. Appl.* 91:211–225.

Feder, J. L., C. A. Chilcote, and G. L. Bush. 1988. Genetic differentiation between sympatric host races of *Rhagoletis pomonella. Nature* 336:61–64.

Feder, J. L., T. A. Hunt, and G. L. Bush. 1993. The effects of climate, host plant phenology and host fidelity on the genetics of apple and hawthorn infesting races of *Rhagoletis pomonella. Entomol. Exp. Appl.* 69:117–135.

Feder, J. L., S. Opp, B. Wlazlo, K. Reynolds, W. Go, and S. Spisak. 1994. Host fidelity is an effective premating barrier between sympatric races of the apple maggot fly. *Proc. Natl. Acad. Sci. USA* 91:7990–7994.

Feder, J. L., J. B. Roethele, B. Wlazlo, and S. H. Berlocher. 1997. Selective maintenance of allozyme differences between sympatric host races of the apple maggot fly. *Proc. Natl. Acad. Sci. USA* 94:11417–11421.

Feder, J. L., S. H. Berlocher, and S. B. Opp. 1998. Sympatric host race formation and speciation in *Rhagoletis* (Diptera: Tephritidae): A tale of two species for Charles D. In S. Mopper and S. Strauss, eds., *Genetic Structure and Local Adaptation in Natural Insect Populations,* pp. 408–441. New York: Chapman and Hall.

Feder, J. L., J. B. Roethele, K. Filchak, J. Niedbalski, and J. Romero-Seversen. 2003. Evidence for inversion polymorphism related to sympatric host race formation in the apple maggot fly, *Rhagoletis pomonella. Genetics* 163:939–953.

Fellows, D. P., and W. B. Heed. 1972. Factors affecting host plant selection in desert-adapted cactophilic *Drosophila. Ecology* 53:850–858.

Filchak, K. E., J. L. Feder, J. B. Roethele, and U. Stolz. 1999. A field test for host-plant dependent selection on larvae of the apple maggot fly, *Rhagoletis pomonella. Evolution* 53:187–200.

Filchak, K. E., J. B. Roethele, and J. L. Feder. 2000. Natural selection and sympatric divergence in the apple maggot *Rhagoletis pomonella. Nature* 407:739–742.

Filchak, K. E., J. B. Roethele, and J. L. Feder. 2001. Effects of photoperiod and light intensity on the genetics of diapause in the apple maggot (Diptera; Tephritidae). *Ann. Entomol. Soc. Am.* 94:902–908.

Fleming, T. H., M. D. Tuttle, and M. A. Horner. 1996. Pollination biology and the relative importance of nocturnal and diurnal pollinators in three species of Sonoran Desert columnar cacti. *Southwestern Nat.* 41:257–269.

Fogleman, J., and J. Williams. 1987. Oviposition site preference of cactophilic *Drosophila* in the *eremophila* complex. Drosophila *Info. Serv.* 66:51–52.

Fogleman, J. C., and J. R. Abril. 1990. Ecological and evolutionary importance of host plant chemistry. In J. S. F. Barker, W. T. Starmer, and R. J. MacIntyre, eds., *Ecological and Evolutionary Genetics of* Drosophila, pp. 121–141. New York: Plenum.

Fogleman, J. C., and W. B. Heed. 1989. Columnar cacti and desert *Drosophila:* The chemistry of host plant specificity. In J. O. Schmidt, ed., *Special Biotic Relationships of the Southwest,* pp. 1–24. Albuquerque, N.M.: University of New Mexico Press.

Fogleman, J. C., K. R. Hackbarth, and W. B. Heed. 1981. Behavioral differentiation between 2 species of cactophilic *Drosophila.* 3. Oviposition site preference. *Am. Nat.* 118:541–548.

Foote, R. H., F. L. Blanc, and A. L. Norrbom. 1993. *Handbook of the Fruit Flies (Diptera: Tephritidae) of America North of Mexico.* New York: Cornell University Press.

Frank, M. R., and J. C. Fogleman. 1992. Involvement of the cytochrome P450 in host-plant utilization by Sonoran Desert *Drosophila. Proc. Natl. Acad. Sci. USA* 89:11998–12002.

Futuyma, D. J., and S. C. Peterson. 1985. Genetic variation in the use of resources by insects. *Annu. Rev. Entomol.* 30:217–238.

Gallun, R. L. 1977. Genetic basis of Hessian fly epidemics. *Ann. New York Acad. Sci.* 287:223–229.

Garrett-Jones, C., P. F. L. Boreham, and C. P. Pant. 1980. Feeding habits of anophelines (Diptera: Culicidae) in 1971–78, with reference to the human blood index: A review. *Bull. Entomol. Res.* 70:165–185.

Gillies, M. T. 1964. Selection for host preference in *Anopheles gambiae. Nature* 203:852–854.

Gillies, M. T. 1967. Experiments on host selection in the *Anopheles gambiae* complex. *Ann. Trop. Med. Parasitol.* 61:68–75.

Gillies, M. T., and M. Coetzee. 1987. *A Supplement to the Anophelinae of Africa South of the Sahara.* Johannesburg: South African Institute for Medical Research.

Grimaldi, D., A. C. James, and J. Jaenike. 1992. Systematics and modes of reproductive isolation in the holarctic *Drosophila testacea* species group (Diptera: Drosophilidae). *Ann. Entomol. Soc. Am.* 85:671–685.

Hadis, M., M. Lulu, Y. Makonnen, and T. Asfaw. 1997. Host choice by indoor-resting *Anopheles arabiensis* in Ethiopia. *Trans. R. Soc. Trop. Med. Hyg.* 91:376–378.

Hallem, E. A., A. N. Fox, L. J. Zwiebel, and J. R. Carlson. 2004. Olfaction: Mosquito receptor for human-sweat odorant. *Nature* 427:212–213.

Hard, J. J., W. E. Bradshaw, and C. M. Holzapfel. 1993a. The genetic basis of photoperiodism and its evolutionary divergence among populations of the pitcher-plant mosquito, *Wyeomyia smithi. Am. Nat.* 142: 457–473.

Hard, J. J., W. E. Bradshaw, and C. M. Holzapfel. 1993b. Genetic coordination of demography and phenology in the pitcher-plant mosquito, *Wyeomyia smithii. J. Evol. Biol.* 6:707–723.

Hard, J. J., W. E. Bradshaw, and C. M. Holzapfel. 1992. Epistasis and the genetic divergence of photoperiodism between populations of the pitcher-plant mosquito, *Wyeomia smithii. Genetics* 131:389–396.

Harris, M. O., M. Sandanayaka, and W. Griffin. 2001. Oviposition preferences of the Hessian fly and their consequences for the survival and reproduction potential of offspring. *Ecol. Entomol.* 26:473–486.

Harris, M. O., J. J. Stuart, M. Mohan, S. Nair, R. J. Lamb, and O. Rohfritsch. 2003. Grasses and gall midges: Plant defense and insect adaptation. *Annu. Rev. Entomol.* 48:549–577.

Hartmann, S., J. D. Nason, and D. Bhattacharya. 2002. Phylogenetic origins of *Lophocereus* (Cactaceae) and the senita cactus–senita moth pollination mutualism. *Am. J. Bot.* 89:1085–1092.

Hasson, E., J. C. Vilardi, H. Naviera, J. J. Fanara, C. Rodriguez, O. A. Reig, and A. Fontdevila. 1991. The evolutionary history of *Drosophila buzzatii.* XVI. Fitness component analysis in an original natural population from Argentina. *J. Evol. Biol.* 4:209–225.

Hedrick, P. W. 1990. Theoretical analysis of habitat selection and the maintenance of genetic variation. In J. S. F. Barker, W. T. Starmer, and R. J. MacIntyre, eds., *Ecological and Evolutionary Genetics of* Drosophila, pp. 209–227. New York: Plenum Press.

Heed, W. B. 1957. Ecological and distributional notes on the Drosophilidae (Diptera) of El Salvador. *Univ. Texas Publ.* 5721:62–78.

Heed, W. B. 1968. Ecology of the Hawaiian Drosophilidae. *Univ. Texas Publ.* 6818:387–419.

Heed, W. B. 1971. Host plant specificity and speciation in Hawaiian *Drosophila. Taxon* 20:115–121.

Heed, W. B. 1977. A new cactus-feeding but soil-breeding species of *Drosophila* (Diptera: Drosophilidae). *Proc. Entomol. Soc. Wash.* 79:649–654.

Heed, W. B. 1978. Ecology and genetics of Sonoran Desert *Drosophila.* In P. F. Brussard, ed., *Ecological Genetics: The Interface,* pp. 109–126. New York: Springer-Verlag.

Heed, W. B. 1981. Central and marginal populations revisited. Drosophila *Info. Serv.* 56:60–61.

Heed, W. B. 1982. The origin of *Drosophila* in the Sonoran Desert. In J. S. F. Barker and W. T. Starmer, eds., *Ecological Genetics and Evolution: The Cactus-Yeast-*Drosophila *Model System,* pp. 65–80. Sydney: Academic Press.

Heed, W. B. 1989. Origin of *Drosophila* of the Sonoran Desert revisited: In search of a founder event and the description of a new species in the *eremophila* complex. In L. V. Giddings, K. Y. Kaneshiro, and W. W. Anderson, eds., *Genetics, Speciation and the Founder Principle,* pp. 253–278. Oxford: Oxford University Press.

Heed, W. B., and H. W. Kircher. 1965. Unique sterol in the ecology and nutrition of *Drosophila pachea. Science* 149:758–761.

Heed, W. B., and R. L. Mangan. 1986. Community ecology of the Sonoran Desert *Drosophila.* In M. Ashburner, H. L. Carson, and J. J. N. Thompson, eds., *The Genetics and Biology of* Drosophila, pp. 311–345. New York: Academic Press.

Higa, I., and Y. Fuyama. 1993. Genetics of food preference in *Drosophila sechellia.* I. Responses to food attractants. *Genetica* 88:129–136.

Hill, C. A., A. N. Fox, R. J. Pitts, L. B. Kent, P. L. Tan, M. A. Chrystal, A. Cravchik, F. H. Collins, H. M. Robertson, and L. J. Zwiebel. 2002. G protein–coupled receptors in *Anopheles gambiae. Science* 298:176–178.

Hoffman, A. A. 1985. Effects of experience on oviposition and attraction in *Drosophila:* Comparing apples and oranges. *Am. Nat.* 126:41–51.

Hoffman, A. A., and S. O'Donnell. 1990. Heritable variation in resource use in *Drosophila* in the field. In J. S. F. Barker, W. T. Starmer, and R. J. MacIntyre, eds., *Ecological and Evolutionary Genetics of* Drosophila, pp. 177–193. New York: Plenum Press.

Holt, R. A., G. M. Subramanian, A. Halpern, G. G. Sutton, R. Charlab, D. R. Nusskern, P. Wincker, A. G. Clark, J. M. Ribeiro, R. Wides, S. L. Salzberg, B. Loftus, M. Yandell, W. H. Majoros, D. B. Rusch, Z. Lai, C. L. Kraft, J. F. Abril, V. Anthouard, P. Arensburger, P. W. Atkinson, H. Baden, V. de Berardinis, D. Baldwin, V. Benes, J. Biedler, C. Blass, R. Bolanos, D. Boscus, M. Barnstead, S. Cai, A. Center, K. Chatuverdi, G. K. Christophides, M. A. Chrystal, M. Clamp, A. Cravchik, V. Curwen, A. Dana, A. Delcher, I. Dew, C. A. Evans, M. Flanigan, A. Grundschober-Freimoser, L. Friedli, Z. Gu, P. Guan, R. Guigo, M. E. Hillenmeyer, S. L. Hladun, J. R. Hogan, Y. S. Hong, J. Hoover, O. Jaillon, Z. Ke, C. Kodira, E. Kokoza, A. Koutsos, I. Letunic, A. Levitsky, Y. Liang, J. J. Lin, N. F. Lobo, J. R. Lopez, J. A. Malek, T. C. McIntosh, S. Meister, J. Miller, C. Mobarry, E. Mongin, S. D. Murphy, D. A. O'Brochta, C. Pfannkoch, R. Qi, M. A. Regier, K. Remington, H. Shao, M. V. Sharakhova, C. D. Sitter, J. Shetty, T. J. Smith, R. Strong, J. Sun, D. Thomasova, L. Q. Ton, P. Topalis, Z. Tu, M. F. Unger, B. Walenz, A. Wang, J. Wang, M. Wang, X. Wang, K. J. Woodford, J. R. Wortman, M. Wu, A. Yao, E. M. Zdobnov, H. Zhang, Q. Zhao, S. Zhao, S. C. Zhu, I. Zhimulev, M. Coluzzi, A. della Torre, C. W. Roth, C. Louis, F. Kalush, R. J. Mural, E. W. Myers, M. D.

Adams, H. O. Smith, S. Broder, M. J. Gardner, C. M. Fraser, E. Birney, P. Bork, P. T. Brey, J. C. Venter, J. Weissenbach, F. C. Kafatos, F. H. Collins, and S. L. Hoffman. 2002. The genome sequence of the malaria mosquito *Anopheles gambiae*. *Science* 298:129–149.

Horner, J. D., and W. D. Abrahamson. 1992. Influence of plant genotype and environment on oviposition preference and offspring survival in a gallmaking herbivore. *Oecologia* 90:323–332.

Horner, J. D., and W. D. Abrahamson. 1999. Influence of plant genotype and early-season water deficits on oviposition preference and offspring performance in *Eurosta solidaginis* (Diptera; Tephritidae). *Am. Midl. Nat.* 142:162–172.

Horner, J. D., T. P. Craig, and J. K. Itami. 1999. The influence of oviposition phenology on survival in host races of *Eurosta solidaginis*. *Entomol. Exp. Appl.* 93:121–129.

Hudson, R. D., G. D. Buntin, and D. J. Isenhour. 1991. Small grain insects. In G. K. Douce and R. M. Mcpherson, eds., *Summary of Losses from Insect Damage and Costs of Control in Georgia, 1989*, p. 29. Special Publication 70. Athens, Ga.: Georgia Agricultural Experimental Station.

Istock, C. A. 1978. Fitness variation in a natural population. In H. Dingle, ed., *Evolution of Insect Migration and Diapause*, pp. 171–190. New York: Springer-Verlag.

Istock, C. A. 1981. Natural selection and life history variation: Theory plus lessons from a mosquito. In R. F. Denno and H. Dingle, eds., *Insect Life History Patterns: Habitat and Geographic Variation*, pp. 113–127. New York: Springer-Verlag.

Istock, C. A., and W. G. Weisburg. 1989. Strong habitat selection and the development of population structure in a mosquito. *Evol. Ecol.* 1:348–362.

Istock, C. A., S. A. Wasserman, and H. Zimmer. 1975. Ecology and evolution of the pitcher-plant mosquito. Part 1. Population dynamics and laboratory responses to food and population density. *Evolution* 29:296–312.

Istock, C. A., K. J. Vavra, and H. Zimmer. 1976a. Ecology and evolution of the pitcher-plant mosquito. 3. Resource tracking by a natural population. *Evolution* 30:548–557.

Istock, C. A., J. Zisfein, and K. J. Vavra. 1976b. Ecology and evolution of the pitcher-plant mosquito. 2. The substructure of fitness. *Evolution* 30:535–547.

Itami, J. K., T. P. Craig, and J. D. Horner. 1998. Factors affecting gene flow between the host races of *Eurosta solidaginis*. In S. Mopper and S. Straus, eds., *Genetic Structure and Local Adaptations in Insect Populations: Effects of Ecology, Life History, and Behavior*, pp. 375–377. London: Chapman and Hall.

Jaenike, J. 1978a. Host selection by mycophagous *Drosophila*. *Ecology* 59:1286–1288.

Jaenike, J. 1978b. Resource predictability and niche breadth in the *Drosophila quinaria* species group. *Evolution* 32:676–678.

Jaenike, J. 1981. Criteria for ascertaining the existence of host races. *Am. Nat.* 117:830–834.

Jaenike, J. 1985. Genetic and environmental determinants of food preference in *Drosophila tripunctata*. *Evolution* 39:362–369.

Jaenike, J. 1986. Genetic complexity of host-selection behavior in *Drosophila*. *Proc. Natl. Acad. Sci. USA* 83:2148–2151.

Jaenike, J. 1987. Genetics of oviposition-site preference in *Drosophila tripunctata*. *Heredity* 59:363–370.

Jaenike, J. 1989. Genetic population structure of *Drosophila tripunctata* patterns of variation and covariation of traits affecting resource use. *Evolution* 43:1467–1482.

Jaenike, J. 1990. Host specialization in phytophagous insects. *Annu. Rev. Ecol. Syst.* 21:243–273.

Jaenike, J. 2002. Sex chromosome meiotic drive. *Annu. Rev. Ecol. Syst.* 32:25–49.

Jaenike, J., and D. Grimaldi. 1983. Genetic variation for host preference within and among populations of *Drosophila tripunctata*. *Evolution* 37:1023–1033.

Jaenike, J., and R. K. Selander. 1979. Ecological generalism in *Drosophila falleni* genetic evidence. *Evolution* 33:741–748.

Jaenike, J., D. A. Grimaldi, A. E. Sluder, and A. L. Greenleaf. 1983. Alpha amanitin tolerance in mycophagous *Drosophila*. *Science* 221:165–167.

Jones, C. 1998. The genetic basis of *Drosophila sechellia*'s resistance to a host plant toxin. *Genetics* 149:1899–1908.

Jones, C. D. 2001. The genetic basis of larval resistance to a host plant toxin in *Drosophila sechellia*. *Genet. Res. Camb.* 78:225–233.

Kambysellis, M. P., K.-F. Ho, E. M. Craddock, F. Piano, M. Paris, and J. Cohen. 1995. Pattern of ecological shifts in the diversification of Hawaiian *Drosophila* inferred from a molecular phylogeny. *Curr. Biol.* 5:1129–1139.

Kettle, D. S. 1994. *Medical and Veterinary Entomology,* 2nd ed. Cambridge: Cambridge University Press.

Killeen, G. F., F. E. McKenzie, B. D. Foy, C. Bogh, and J. C. Beier. 2001. The availability of potential hosts as a determinant of feeding behaviours and malaria transmission by African mosquito populations. *Trans. R. Soc. Trop. Med. Hyg.* 95:469–476.

Kimura, M. T., M. J. Toda, K. Beppu, and H. Watabe. 1977. Breeding sites of drosophilid flies in and near Sapporo, northern Japan, with supplementary notes on adult feeding habits. *Kontyû* 45:571–582.

Kircher, H. W., W. B. Heed, J. S. Russell, and J. Grove. 1967. Senita cactus alkaloids: Their significance to Sonoran Desert *Drosophila* ecology. *J. Insect Phsiol.* 13:1869–1874.

Krijger, C. L., Y. C. Peters, and J. G. Sevenster. 2001. Competitive ability of neotropical *Drosophila* species predicted from larval development times. *Oikos* 92:325–332.

Lacy, R. C. 1983. Structure of genetic variation within and between populations of mycophagous *Drosophila. Genetics* 104:81–94.

Lacy, R. C. 1984. Predictability toxicity and trophic niche breadth in fungus feeding Drosophilidae (Diptera). *Ecol. Entomol.* 9:43–54.

Lair, K. P., W. E. Bradshaw, and C. M. Holzapfel. 1997. Evolutionary divergence of the genetic architecture underlying photoperiodism in the pitcher-plant mosquito, *Wyeomia smithii. Genetics* 147:1873–1883.

Lane, J. 1953. *Neotropical Culicidae.* São Paulo: University of São Paulo.

Lehane, M. J. 1991. *Biology of Blood-Sucking Insects.* London: HarperCollins Academic.

Lemasson, J. J., D. Fontenille, L. Lochouarn, I. Dia, F. Simard, K. Ba, A. Diop, M. Diatta, and J. F. Molez. 1997. Comparison of behavior and vector efficiency of *Anopheles gambiae* and *An. arabiensis* (Diptera: Culicidae) in Barkedji, a Sahelian area of Senegal. *J. Med. Entomol.* 34:396–403.

Levitan, M. 1992. Chromosomal variation in *Drosophila robusta* Sturtevant. In C. B. Krimbas and J. R. Powell, eds., Drosophila *Inversion Polymorphism,* pp. 221–338. Boca Raton, Fla.: CRC Press.

Lewontin, R. C., J. A. Moore, W. B. Provine, and B. Wallace. 1981. *Dobzhansky's Genetics of Natural Populations,* I–XLIII. New York: Columbia University Press.

Lofdahl, K. L. 1986. A genetical analysis of habitat selection in the cactophilic species *Drosophila mojavensis.* In M. D. Huettel, ed., *Evolutionary Genetics of Invertebrate Behavior: Progress and Prospects,* pp. 153–162. New York: Plenum Press.

Mangan, R. L. 1982. Adaptations to competition in cactus breeding *Drosophila.* In J. S. F. Barker and W. T. Starmer, eds., *Ecological Genetics and Evolution: The Cactus-Yeast-*Drosophila *Model System,* pp. 257–272. Sydney: Academic Press.

Mathenge, E. M., G. F. Killeen, D. O. Oulo, L. W. Irungu, P. N. Ndegwa, and B. G. Knols. 2002. Development of an exposure-free bednet trap for sampling Afrotropical malaria vectors. *Med. Vet. Entomol.* 16:67–74.

Mayr, E. 1963. *Animal Species and Evolution.* Cambridge, Mass.: Harvard University Press.

Mboera, L. E. G., B. G. J. Knols, W. Takken, and A. della Torre. 1997. The response of *Anopheles gambiae* s. l. and *A. funestus* (Diptera: Culicidae) to tents baited with human odour or carbon dioxide in Tanzania. *Bull. Entomol. Res.* 87:173–178.

McCall, P. J., and D. W. Kelly. 2002. Learning and memory in disease vectors. *Trends Parasitol.* 18:429–433.

McDaniel, S. T. 1971. The genus *Sarracenia* (Sarraceniaceae). *Bull. Tall Timbers Res. Station* 9:1–36.

McPheron, B. A., D. C. Smith, and S. H. Berlocher. 1988. Genetic differences between *Rhagoletis pomonella* host races. *Nature* 336:64–66.

Merrill, C. E., J. Riesgo-Escovar, R. J. Pitts, F. C. Kafatos, J. R. Carlson, and L. J. Zwiebel. 2002. Visual arrestins in olfactory pathways of *Drosophila* and the malaria vector mosquito *Anopheles gambiae. Proc. Natl. Acad. Sci. USA* 99:1633–1638.

Merritt, R. W., G. W. Courtney, and J. B. Keiper. 2003. Diptera (flies, mosquitoes, midges, gnats). In V. H. Resch and R. T. Cardé, eds., *Encyclopedia of Insects,* pp. 324–340. San Diego: Academic Press.

Ming, Y. 1989. A revision of the genus *Eurosta* Loew with scanning electron microscopic study of taxonomic characters (Diptera: Tephritidae). Masters thesis, Washington State University, Pullman, Wash.

Moericke, V., R. J. Prokopy, S. Berlocher, and G. L. Bush. 1975. Visual stimuli eliciting attraction of *Rhagoletis pomonella* (Diptera: Tephritidae) flies to trees. *Entomol. Exp. Appl.* 18:497–507.

Montgomery, S. L. 1975. Comparative breeding site ecology and the adaptive radiation of picture-winged *Drosophila* (Diptera: Drosophilidae) in Hawaii. *Proc. Haw. Entomol. Soc.* 22:65–103.

Morell, V. 1999. Ecology returns to speciation studies. *Science* 284:2106–2108.

Muirhead-Thomson, E. C. 1982. *Behavior Patterns of Blood-Sucking Flies.* New York: Pergamon Press.

New, T. R. 1991. *Insects as Predators.* Kensington, Australia: New South Wales University Press.

Newby, B. D., and W. J. Etges. 1998. Host preference among populations of *Drosophila mojavensis* that use different host cacti. *J. Insect Behav.* 11:691–712.

Ohta, A. T. 1980. Coadapted gene complexes in incipient species of Hawaiian *Drosophila. Am. Nat.* 115: 121–132.

Oldroyd, H. 1964. *The Natural History of Flies.* London: Weidenfeld and Nicolson.

Owens, E. D., and R. J. Prokopy. 1986. Relationship between reflectance spectra of host plant surfaces and visual detection of host fruit by *Rhagoletis pomonella* flies. *Physiol. Entomol.* 11:297–307.

Pates, H. V., W. Takken, C. F. Curtis, P. W. Huisman, O. Akinpelu, and G. S. Gill. 2001a. Unexpected anthropophagic behaviour in *Anopheles quadriannulatus. Med. Vet. Entomol.* 15:293–298.

Pates, H. V., W. Takken, K. Stuke, and C. F. Curtis. 2001b. Differential behaviour of *Anopheles gambiae* sensu stricto (Diptera: Culicidae) to human and cow odours in the laboratory. *Bull. Entomol. Res.* 91:289–296.

Patterson, J. T., and W. S. Stone. 1952. *Evolution in the Genus* Drosophila. New York: Macmillan.

Powell, J. R. 1997. *Progress and Prospects in Evolutionary Biology: The* Drosophila *Model.* New York: Oxford University Press.

Powell, J. R., V. Petrarca, A. della Torre, A. Caccone, and M. Coluzzi. 1999. Population structure, speciation, and introgression in the *Anopheles gambiae* complex. *Parassitologia* 41:101–113.

Prokopy, R. J. 1968. The influence of photoperiod, temperature and food on the initiation of diapause in the apple maggot. *Can. Entomol.* 100:318–329.

Prokopy, R. J. 1972. Evidence for a marking pheromone deterring repeated oviposition in apple maggot flies. *Env. Entomol.* 1:326–332.

Prokopy, R. J. 1977. Attraction of *Rhagoletis* flies (Diptera: Tephritidae) to red spheres of different sizes. *Can. Entomol.* 109:593–596.

Prokopy R. J., and G. L. Bush. 1973. Ovipositional responses to different sizes of artificial fruit by flies of *Rhagoletis pomonella* species group. *Ann. Entomol. Soc. Am.* 66:927–929.

Prokopy, R. J., E. W. Bennett, and G. L. Bush. 1971. Mating behavior in *Rhagoletis pomonella* (Diptera: Tephritidae). I. Site of assembly. *Can. Entomol.* 103:1405–1409.

Prokopy, R. J., E. W. Bennett, and G. L. Bush. 1972. Mating behavior in *Rhagoletis pomonella* (Diptera: Tephritidae). II. Temporal organization. *Can. Entomol.* 104:97–104.

Prokopy, R. J., V. Moericke, and G. L. Bush. 1973. Attraction of apple maggot flies to odor of apples. *Env. Entomol.* 2:743–749.

Prokopy, R. J., A. L. Averill, S. S. Cooley, and C. A. Roitberg. 1982. Associative learning in egglaying site selection by apple maggot flies. *Science* 218:76–77.

Prokopy, R. J., S. R. Diehl, and S. S. Cooley. 1988. Behavioral evidence for host races in *Rhagoletis pomonella* flies. *Oecologia* 76:138–147.

Ratcliffe, R. H., and J. H. Hatchett. 1997. Biology and genetics of the Hessian fly and resistance in wheat. In K. Bondari, ed., *New Developments in Entomology,* pp. 47–56. Trivandurm, India: Research Signpost, Scientific Information Guild.

R'Kha, S., P. Capy, and J. R. David. 1991. Host-plant specialization in the *Drosophila melanogaster* species complex: A physiological, behavioral, and genetical analysis. *Proc. Natl. Acad. Sci. USA* 88:1835–1839.

Ruiz, A., and A. Fontdevila. 1985. The evolutionary history of *Drosophila buzzatii* VI. Adaptive chromosomal changes in experimental populations with natural substrates. *Genetica* 66:63–71.

Ruiz, A., and W. B. Heed. 1988. Host-plant specificity in the cactophilic *Drosophila mulleri* species complex. *J. Animal Ecol.* 57:237–249.

Ruiz, A., and M. Wasserman. 1993. Evolutionary cytogenetics of the *Drosophila buzzatii* species complex. *Heredity* 70:582–596.

Ruiz, A., A. Fontdevila, M. Santos, M. Seoane, and E. Torroja. 1986. The evolutionary history of *D. buzzatii.* VIII. Evidence for endocyclic selection acting on the inversion polymorphism in a natural population. *Evolution* 40:740–755.

Ruiz, A., W. B. Heed, and M. Wasserman. 1990. Evolution of the *mojavensis* cluster of cactophilic *Drosophila* with descriptions of two new species. *J. Hered.* 81:30–42.

Ruiz, A., A. M. Cansian, G. C. S. Kuhn, M. A. R. Alves, and F. M. Sene. 2000. The *Drosophila serido* speciation puzzle: Putting new pieces together. *Genetica* 108:217–227.

Scheiner, S. M., and C. A. Istock. 1991. Correlational selection on life history traits in the pitcher-plant mosquito. *Genetica* 84:123–128.

Schliewen, U. K., D. Tautz, and S. Pabbo. 1994. Sympatric speciation suggested by monophyly of crater lake cichlids. *Nature* 368:629–632.

Schluter, D. 1996. Ecological causes of adaptive radiation. *Am. Nat.* 148:S40–S64.

Sevenster, J. G., and J. J. M. V. Alphen. 1996. Aggregation and coexistence. II. A Neotropical *Drosophila* community. *J. Animal Ecol.* 65:308–324.

Smith, D. C. 1986. Genetics and reproductive isolation of Rhagoletis flies. Chapter V: Genetic differences in size preference for artificial fruit by flies of the *Rhagoletis pomonella* species group. Ph.D. thesis, University of Illinois, Urbana-Champaign, Ill.

Smith, J. B. 1904. *Report of the New Jersey State Agricultural Experiment Station upon the Mosquitoes Occurring within the State, Their Habits, Life Histories, etc.* Trenton: MacCrellish and Quigley.

Sousa, C. A., J. Pinto, A. P. Almeida, C. Ferreira, V. E. do Rosario, and J. D. Charlwood. 2001. Dogs as a favored host choice of *Anopheles gambiae* sensu stricto (Diptera: Culicidae) of São Tomé, West Africa. *J. Med. Entomol.* 38:122–125.

Spicer, G. S., and J. Jaenike. 1996. Phylogenetic analysis of breeding site use and alpha-amanitin tolerance within the *Drosophila quinaria* species group. *Evolution* 50:2328–2337.

Starmer, W. T., W. B. Heed, and E. S. Rockwood-Sluss. 1977. Extension of longevity in *Drosophila mojavensis* by environmental ethanol: Differences between subraces. *Proc. Natl. Acad. Sci. USA* 74:387–391.

Stearns, S. C. 1977. The evolution of life history traits: A critique of the theory and a review of the data. *Annu. Rev. Ecol. Syst.* 8:145–171.

Strong, D. R., J. H. Lawton, and R. Southwood. 1984. *Insects on Plants.* Cambridge, Mass.: Harvard University Press.

Sturtevant, A. H. 1921. *The North American Species of Drosophila.* Washington, D.C.: Carnegie Institute.

Sugaya, S., I. Higa, and Y. Fuyama. 1995. Food preference of *Drosophila sechellia* VI. *Jpn. J. Genet.* 70:739.

Takken, W., and B. G. Knols. 1999. Odor-mediated behavior of Afrotropical malaria mosquitoes. *Annu. Rev. Entomol.* 44:131–157.

Taylor, C. E., and J. R. Powell. 1977. Microgeographic variation of chromosomal and enzyme polymorphisms of *Drosophila persimilis. Genetics* 85:681–695.

Thorpe, W. H. 1930. The biology of the petroleum fly (*Psilopa petrolii* Coq.). *Trans. Entomol. Soc. Lond.* 78:331–344.

Throckmorton, L. 1975. The phylogeny, ecology, and geography of *Drosophila.* In R. C. King, ed., *Invertebrates of Genetics Interest,* pp. 421–469. New York: Plenum Press.

Tidon-Sklorz, R., and F. M. Sene. 1995. *Drosophila seriema* n. sp.: New member of the *Drosophila serido* (Diptera: Drosophilidae) superspecies taxon. *Ann. Entomol. Soc. Am.* 88:139–142.

Tidon-Sklorz, R., and F. M. Sene. 2001. Two new species of the *Drosophila serido* sibling set (Diptera, Drosophilidae). *Iheringia* (Ser. Zool.) 90:141–146.

Toure, Y. T., V. Petrarca, S. F. Traore, A. Coulibaly, H. M. Maiga, O. Sankare, M. Sow, M. A. DiDeco, and M. Coluzzi. 1998. The distribution and inversion polymorphism of chromosomally recognized taxa of the *Anopheles gambiae* complex in Mali, West Africa. *Parassitologia* 40:477–511.

Via, S. 1991. The genetic structure of host plant adaptation in a spatial patchwork: Demographic variability among reciprocally transplanted pea aphid clones. *Evolution* 45:827–852.

Wallace, B. 1978. The adaptation of *Drosophila virilis* to life on an artificial crab. *Am. Nat.* 112:971–973.

Walsh, B. D. 1867. The apple-worm and the apple maggot. *J. Hort.* 2:338–343.

Waring, G. L., W. G. Abrahamson, and D. L. Howard. 1990. Genetic differentiation among host-associated populations of the gall-maker *Eurosta solidaginis* (Diptera: Tephritidae). *Evolution* 44:1648–1655.

Wegis, M. C., W. E. Bradshaw, T. E. Davison, and C. M. Holzapfel. 1997. Rhythmic components of photoperiodic time measurement in the pitcher-plant mosquito, *Wyeomyia smithii. Oecologia* 110:32–39.

Wertheim, B., J. G. Sevenster, I. E. M. Eijs, and J. J. M. V. Alphen. 2000. Species diversity in a mycophagous insect community: The case of spatial aggregation versus resource partitioning. *J. Animal Ecol.* 69: 335–351.

White, G. B. 1974. *Anopheles gambiae* complex and disease transmission in Africa. *Trans. R. Soc. Trop. Med. Hyg.* 68:278–301.

White, G. B., and P. Rosen. 1973. Comparative studies on sibling species of the *Anopheles gambiae* complex (Dipt., Culicidae). II. Ecology of species A and B in savanna around Kaduna, Nigeria, during transition from wet to dry season. *Bull. Entomol. Res.* 62:613–625.

Wood, T. K. 1980. Divergence in the *Enchenopa binotata* complex (Homoptera: Membracidae) effected by host plant adaptation. *Evolution* 34:147–160.

Wood, T. K., and M. C. Keese. 1990. Host-plant-induced assortative mating in *Enchenopa* treehoppers. *Evolution* 44:619–628.

Wu, C.-I., and A. T. Beckenbach. 1983. Evidence for extensive genetic differentiation between the sex-ratio and the standard arrangement of *Drosophila pseudoobscura* and *D. persimilis* and identification of hybrid sterility factors. *Genetics* 105:71–86.

Young, D. G., and M. A. Duncan. 1994. Guide to the identification and geographic distribution of *Lutzomyia* sand flies in Mexico, the West Indies, Central and South America (Diptera: Psychodidae). Gainsville, Fla.: Associated Publishers.

Zouros, E. 1973. Genic differentiation associated with the early stages of speciation in the *mulleri* subgroup of *Drosophila. Evolution* 27:601–621.

CHAPTER THIRTEEN

Invasive Diptera: Using Molecular Markers to Investigate Cryptic Species and the Global Spread of Introduced Flies

Sonja J. Scheffer

Given the great diversity in both the larval and adult habits of Diptera (Oldroyd 1964; McAlpine et al. 1981, 1987; Ferrar 1987), it comes as no surprise that many flies are, to a greater or lesser extent, associated with humans. In addition to feeding directly on humans, various species of flies are associated with our refuse, our animals, and/or our crops. Many synanthropic flies undoubtedly have a long history of association with humans. As humans have invaded existing habitats and created new ones, the fly fauna has adapted accordingly, with some species, such as the housefly and some mosquitoes, becoming nearly obligately associated with human societies. Flies are also important components of natural ecosystems and can have significant impact as pollinators, predators, parasites, necrophages, and disease vectors.

The recent explosion in global movement of people and goods has caused a tremendous increase in the number of organisms introduced to new locations as a result of human activity (Enserink 1999; McMichael and Bouma 2000; Huber et al. 2002). This doubtless includes many Diptera, although exact figures are not known. For fly species of economic or medical importance, determining whether a species is introduced may be relatively straightforward because such species are likely to be noticed shortly after they arrive (e.g., Peyton et al. 1999). For other dipteran species, it may be more difficult to determine whether a species is native or introduced, depending on our knowledge of the indiginous dipteran fauna of a given region.

Understanding the patterns and processes involved in the colonization of new areas by invasive Diptera is critical to managing introduced populations. Although in most cases, traditional taxonomy remains important for species identifications, the use of molecular markers, such as allozymes, microsatellites, or sequence data, provides scientists with new tools for investigating population history, genetic structure, and patterns of movement of flies. With these tools, we can answer the questions of whether a population is introduced and, if so, from where. Knowing the place of origin of an invasive species enables managers to look for natural enemies (i.e., biocontrol agents) in appropriate locations. Knowing the means of spread allows managers to strengthen quarantine pathways to prevent further invasions.

In this chapter, I review information on invasive Diptera and the molecular methods used to investigate such species. Investigations into the spread of invasive mosquitos, drosophilids, med flies, and agromyzid leafminers are discussed in detail.

Invasion Pathways of Diptera

The majority of human-aided movement of Diptera occurs during the egg, larval, or pupal stage, as a result of the transport of larval substrates. The most successful invasive mosquitoes are those that can reproduce in water-holding containers, allowing global movement of substantial populations via commercial ships (Lounibos 2002). The yellow fever mosquito, *Aedes aegypti* (Culicidae), is believed to have traveled from Africa to the New World aboard slave ships. More recently, several mosquito species have been transported from Asia to other regions of the world in shiploads of used tires, which hold water used by the mosquitoes for breeding (Moore and Mitchell 1997; Fonseca et al. 2001). After establishment in a new area, human-aided movement can be important to the subsequent spread of an invasive species; after an initial population was established in Texas, new populations of the invasive Asian tiger mosquito *Aedes albopictus* (Culicidae) in the United States were for some time associated with the interstate highway system, implicating human activities in the regional as well as the global spread of this species (Moore and Mitchell 1997). In some instances, movement of adult, rather than larval or pupal, flies has been implicated, as in cases of "airport malaria," in which residents near an international airport in a malaria-free region contract the disease despite not having traveled to a region harboring malaria (Lounibos 2002).

Parasitic flies are often spread as their animal or plant hosts are moved around the world. Flies attacking livestock may reach new locations by traveling with host animals; the sheep blowfly, *Lucilia cuprina* (Calliphoridae) is believed to have traveled from its native range in Africa to Western Australia in the 1800s in shiploads of infested sheep (Norris 1990). Plant-feeding flies are commonly transported as their larval substrates are shipped around the world as part of global agricultural trade (Minkenberg 1988; Bartlett 1993; Cheek et al. 1993). Many governments have elaborate inspection systems in place to screen imported plant products to prevent the importation of potential plant pests (Bartlett 1993; Cheek et al. 1993; Headrick and Goeden 1996). Plant-feeding flies may also be transported to new regions by vacationers returning home with infested exotic fruits, vegetables, seeds, and even plants in their luggage (Carey 1991).

In addition to direct human-mediated dispersal of flies, habitat modification by humans may encourage the evolution of anthrophilic traits leading to further range expansions. As humans began to congregate in larger and larger groupings, the opportunities for a variety of flies expanded. The establishment of large and predictable concentrations of human-associated fly resources (e.g., livestock herds, refuse piles, crop monocultures) may have selected for fly genotypes drawn to human environments. Domestication and the evolution of anthrophilic traits in houseflies and some drosophilids have allowed certain species to attain cosmopolitan status. The important anthrophilic malaria vector *Anopheles gambiae* sensu stricto (Culicidae) probably evolved from a more generalized African ancestor within the past several thousand years and spread with agriculture to other Old World regions (Tabachnick 1991; Coluzzi et al. 2002). Molecular data from *An. gambiae* are consistent with the hypothesis of a population expansion that was contemporaneous with the agricultural expansion in sub-Saharan Africa (Donnelly et al. 2001). Interestingly, molecular data on the malaria parasite, *Plasmodium falciparum*, also indicate a population bottleneck and subsequent expansion within the past several thousand years (Rich et al. 1998). Even today, new habitats are being created for species of flies, leading to invasions and range expansions (Walsh et al. 1993). Several tree- or rock hole–inhabiting mosquitoes have recently adapted to human-modified habitats, particularly to used automobile tires and discarded containers (Lounibos 2002). This has resulted in range expansions at the periphery of the original ranges, as well as long-distance dispersal associated with global shipment of goods (Urbanelli et al. 2000; Lounibos 2002). Recently, the potential impact of global warming on the distributions of invasive species has become a topic of interest and some controversy (Dukes and Mooney 1999; Rogers and Randolph 2000).

Molecular Variation and Geographic Origins

The use of molecular data is often essential to understanding the history, structure, and spread of invasive species. Even if morphological characters are sufficient for determining the identity of an invasive species, in most cases, morphological characters do not exhibit enough variation to be used to determine geographic origins and pathways of spread. Typically, introduced populations exhibit considerably less molecular variation than do endemic populations because of the bottlenecks experienced during the introduction process (Nei et al. 1975; Villablanca et al. 1998; Scheffer and Grissell 2003). For cases in which there is little pre-existing information regarding invasion status, the pattern of reduced variation in introduced populations can be used to determine whether a population is introduced or endemic to a particular area (Tabachnick 1991; Scheffer and Grissell 2003). Virtually any of the commonly used molecular markers (e.g., allozymes, randomly amplified polymorphic DNAs [RAPD], microsatellites, sequence data) can be used in this way. Because of smaller effective population sizes for mitochondrial vs. nuclear markers, the reduction in variation in mitochondrial markers is expected to be greater than that exhibited by nuclear loci (Villablanca et al. 1998). However, if founding populations are large or if multiple introductions from different source populations have occurred, a reduction in genetic variation in the introduced populations may not be apparent (Black et al. 1988; Urbanelli et al. 2000).

Phylogeography, the study of the geographic distribution of closely related lineages (Avise 2000), provides a powerful approach to understanding historical processes that have occurred within a species. A phylogeographic approach relies on the phylogenetic analysis of DNA sequence data (generally from the mitochondrion) from geographically distributed samples. The relationships uncovered by such studies can suggest the presence of previously unsuspected historical processes, such as dispersal, host race formation, and speciation events (Avise 2000). Various insect groups have long been known to contain cryptic species (Busvine 1980; Diehl and Bush 1984), and the relatively new techniques of phylogeographic analysis are uncovering additional evidence of cryptic lineages (Collins and Paskewitz 1996; Avise 2000; Scheffer 2000; Molbo et al. 2003). Among flies, some of the groups that have been shown to contain species complexes include mosquitoes (Collins and Paskewitz 1996; Krzywinski and Besansky 2003), sand flies (Psychodidae; Soto et al. 2000), true fruit flies (Tephritidae; Berlocher 2000), and leafmining flies (Agromyzidae; Scheffer 2000; Scheffer and Wiegmann 2000; Scheffer and Lewis, unpub. data); it is likely that when studied with molecular markers used at the species level, many other fly groups will exhibit similar patterns. It should be noted, however, that within some species groups, speciation may have occurred so recently that neutral molecular markers have not yet become differentiated; for example, *Rhagoletis pomonella* (Tephritidae) group (Smith and Bush 2000). In these cases, differences in ecology or behavior may be the only easily identifiable traits defining cryptic species.

Identifying cryptic species or highly diverged lineages when dealing with invasive or pest species can be critical to successful management and control (Rosen 1978; Schauff and LaSalle 1998; Lounibos and Conn 2000). Biological features that have been shown to differ among closely related cryptic lineages of flies include host preference, mating preference, diapause, phenology, voltinism, salinity tolerance, and ability to transmit malaria (Bush 1969; White 1974; Rosen 1978; Lounibos and Conn 2000; Reitz and Trumble 2002; Scheffer and Hawthorne, unpubl. data). Management efforts that fail to recognize cryptic variation and associated differences in biological features may prove ineffective, sometimes after considerable effort and/or expense (Rosen 1978; Busvine 1980; Schauff and LaSalle 1998). With an accurate understanding of the systematics of species complexes and the origins of invasive populations, control efforts can be based on specific biological characteristics and more precisely tailored to suit the problem at hand (Busvine 1980; Roderick and Villablanca 1996; Schauff and LaSalle 1998).

Once a molecular marker is available that exhibits phylogeographic variation across the native range of a species, that marker can then be used to determine the geographic origin(s) of invasive populations. In the case of a new population resulting from a single introduction, the molecular signature may be very strong and easy to interpret. In the case of multiple introductions from multiple source populations, it can be considerably more difficult to determine the history of introductions. Additionally, once introduced populations are established and become potential source populations themselves, it can be extremely difficult to sort out primary vs. secondary introductions (Villablanca et al. 1998; Davies et al. 1999; Scheffer and Grissell 2003). Although in many cases, it would be highly desirable to be able to determine the precise sequence of invasion events, phylogeography generally cannot provide this information because the rate of mitochondrial sequence evolution is too slow to track the generally short intervals between multiple invasion events (Davies et al. 1999). Alternative approaches use other types of markers, such as microsatellites, that tend to maintain more variation through bottlenecks. Analysis of microsatellites for determining the source(s) of invading populations can be accomplished using multilocus assignment tests (Rannala and Mountain 1997; Cornuet et al. 1999; Davies et al. 1999), but this approach has only recently been applied to a dipteran invasion (Bonizzoni et al. 2001).

Some Invasive Flies

MOSQUITOES (CULICIDAE)

Among all of the invasive Diptera, mosquitoes are the group that has had the greatest impact on human and animal populations. Mosquito-borne malaria, yellow fever, dengue, filariases, and arboviruses kill and incapacitate millions of people every year (Tabachnick 1991). When a vector mosquito and the disease it transmits are introduced into a new area, the effects can be devastating (Bryan 1999; Lounibos 2002). The spread of the mosquito *Ae. aegpti* and yellow fever to the New World, probably via the slave trade (Bryan 1999; Lounibos 2002), led to several centuries of epidemics with extensive mortality (Lounibos 2002). The introduction of the African malaria vector, *An. gambiae* sensu lato, into Brazil led to an epidemic having a mortality rate reaching 25% (Lounibos 2002). In this case, the malaria parasite was already present, but the endemic mosquito vectors were inefficient transmitters of the disease; the introduction of *An. gambiae* sensu lato allowed the disease to flourish and reach epidemic levels (Lounibos 2002). Invasive mosquitoes can also have tremendous effects on native wildlife populations. The introduction of *Culex quinquefasciatus* into the Hawaiian Islands in the nineteenth century appears to have allowed avian malaria to decimate local bird populations, leading to the extinction of several species and restricting native forest birds to high-elevation "mosquito-free zones" (van Riper et al. 1986; Fonseca et al. 2000).

The genetic structure of disease-vectoring mosquitoes is the subject of an extensive literature primarily focused on taxonomy and the question of gene flow between populations. In some cases, the resolution of taxonomic issues has proved quite difficult, with many species belonging to species complexes (Collins and Paskewitz 1996; Krzywinski and Besansky 2003). For example, *An. gambiae* sensu lato is the primary vector of malaria in Africa. It has been shown to be comprised of seven morphologically indistinguishable cryptic species that differ in ecological, behavioral, and genetic traits (Coluzzi et al. 1979), including vectoring capability (White 1974). One of these seven species, *An. gambiae* sensu stricto, is further comprised of five "forms" based on chromosomal inversions and two "types" based on ribosomal sequence data. Although there is some evidence of reproductive isolation between chromosomal forms and between molecular types, there does not seem to be an entirely straightforward correspondence of forms and types, making taxonomic delimitations difficult (Gentile et al. 2002; Krzywinski and Besansky 2003; Lehmann et al. 2003). Fluctuating selection and periodic gene flow in different ecological regions across Africa may be contributing to the complex patterns observed in this group (Lehmann et al. 2003). The question of gene flow between mosquito populations is itself of considerable interest to better understand the potential movement of insecticide resistance genes from one population or species to another (Guillemaud et al. 1996; Besansky et al. 1997; Donnelly et al. 2001, 2002).

The geographic origins of introduced mosquito populations have been investigated using a variety of molecular markers. Allozyme analysis of *Ae. albopictus* suggested that introduced mosquitoes in the United States and Brazil originated in Japan (Kambhampati et al. 1991), but further investigation using mitochondrial sequence data revealed that U.S. and Brazilian populations harbor different mitochondrial haplotypes, indicating independent colonizations (Birungi and Munstermann 2002). This finding is consistent with a diapause difference observed between U.S. and Brazilian populations (Hawley et al. 1987; Birungi and Munstermann 2002). Another Asian container-breeding species recently found in several U.S. locations, *Ae. japonicus,* also appears to have been introduced multiple times, according to an analysis of RAPD and mitochondrial sequence data. In both cases of these invasive container-breeding species, the nuclear markers (allozymes or RAPDs) do not exhibit the reduced variation expected following severe bottlenecks. This finding suggests that founding populations may have been large (Kambhampati et al. 1991; Fonesca et al. 2001), consistent with the presumed pathway of introduction of these species via the used tire trade (Fonseca et al. 2001). However, mitochondrial results from invasive populations of both species exhibit reduced variation, suggestive of bottlenecks. Additional investigation in both native and introduced ranges will be necessary to fully understand the invasion history in these species.

DROSOPHILIDS (DROSOPHILIDAE)

Some drosophilids also appear to be adept at colonizing new regions, with several erstwhile geographically restricted species becoming essentially cosmopolitan because of human-aided movement (Lachaise et al. 1988; Irvin et al. 1998). *Drosophila melanogaster,* the most famous drosophilid, has been the primary workhorse in the field of genetics for a century, but it is only within the past several decades that we have begun to understand the natural history of this species in the wild (David and Capy 1988). Currently cosmopolitan, *D. melanogaster* is Afrotropical in origin, with spread to new regions having occurred in two stages: the ancient spread of flies from Africa to Eurasia following the last glaciation, and recent spread via human transport to the Americas, Australia, and oceanic islands within the past several centuries (David and Capy 1988; Singh and Long 1992). Although it was once thought that human-aided movement would cause this highly anthrophilic species to be essentially panmictic, it now appears that African populations differ extensively in behavioral, physiological, and genetic traits, as do colonized populations. In fact, multiple lines of evidence suggest that within the Americas, tropical populations originated from tropical Africa, whereas temperate North American populations came from cold-tolerant European populations (David and Capy 1988).

Among the most thoroughly investigated dipteran invasions is *D. subobscura,* a well-studied Palearctic species that was discovered in Chile in 1978 and in the Pacific Northwest of North America in 1982 (Ayala et al. 1989; Balanya et al. 1994). Subsequent to its introduction at these locations, this fly spread extensively in North and South America, providing scientists with an opportunity to investigate replicated invasion events contemporaneously. Molecular data collected from a variety of marker systems show that the colonizing New World populations harbor considerably less genetic variation than do European populations (Balanya et al. 1994; Pascual et al. 2001). Additionally, the South and North American populations exhibit genetic similarities that indicate that they are not independent introductions and perhaps are derived from the same source population (Ayala et al. 1989; Balanya et al. 1994). Although European populations contain more variation than introduced populations, the variation is not sufficient to allow identification of a particular source population for the New World colonizations, even using microsatellite data (Pascual et al. 2001).

MEDITERRANEAN FRUIT FLY (TEPHRITIDAE)

The Mediterranean fruit fly, *Ceratitis capitata* (Tephritidae), attacks more than 180 fruits and causes considerable economic damage where it is established (Sheppard et al. 1992; Headrick and Goeden 1996). Originally a native of sub-Saharan Africa, the medfly has expanded its range during the past 200 or so years to include the Mediterranean region, western Australia, Hawaii, and much of the tropical and subtropical Americas (Sheppard et al. 1992; Gasperi et al. 2002). Considerable effort has been made in an attempt to understand the history of medfly colonization patterns, current population structure, and the origins of new populations. As with other invasive species, the ancestral source populations exhibit more molecular variation than do introduced populations (Gasparich et al. 1997; Villablanca et al. 1998; Gasperi et al. 2002). Within the United States, the question of whether the occasional medflies captured in California are the result of repeated independent introductions or represent a small but established resident population has important consequences in terms of both eradication efforts and agricultural exports (Carey 1991; Siebert and Cooper 1995; Headrick and Goeden 1996). He and Haymer (1999) found that polymerase chain reaction–restriction fragment length polymorphism (PCR-RFLP) variation suggests that the medflies found in different locations in California do not belong to a single homogeneous population. At least some of the flies, particularly those found in southern California, appear to have a close relationship to introduced flies in Guatemala (Gasparich et al. 1997; Haymer et al. 1997; He and Haymer 1999; Bonizzoni et al. 2001). Flies captured within the Los Angeles basin from 1992 to 1997 exhibit substantial homogeneity and share a private microsatellite allele, making independent origins for these flies unlikely; it may be that a continuous population persisted during the years 1992–1997 (Bonizzoni et al. 2001). Most recently, Bohonak et al. (2001) used a simulation approach to test several colonization scenarios for established Latin American medfly populations that may themselves be a source for California infestations. They find that, given certain starting conditions, they can reject some hypothesized colonization scenerios (e.g., that Latin American populations are the result of single invaders from multiple sources), although they stress that other additional approaches are also critical to fully understanding the invasion biology of medflies.

LEAFMINING FLIES (AGROMYZIDAE)

Agromyzid leafminers are small flies that spend their larval stage feeding in and tunneling through leaves. As each larva ingests mesophyll, it leaves behind a conspicuous tunnel or "leaf mine." Leafminers usually do not remove enough tissue to have a large impact on their hosts, but in agricultural settings, they may become so abundant that they cause significant plant damage (Spencer 1973; Shepard et al. 1998).

Among the leafminers that are the most damaging to agricultural crops are species in the genus *Liriomyza* (Spencer 1973). Although most *Liriomyza* leafminers are highly host-specific, there are three polyphagous morphospecies that are major vegetable and cut-flower pests that have spread around the world within the past several decades. Although each has its own history, several common patterns have emerged regarding their distribution and spread. Mitochondrial data suggest that all three morphospecies contain highly diverged lineages, indicative of the presence of cryptic species. Additionally, within each morphospecies, only certain lineages appear to be invasive, even though all are pests within their native ranges.

Liriomyza huidobrensis/L. langei. *Liriomyza huidobrensis* is a polyphagous pest of dozens of vegetable and flower crops in at least 14 plant families (Spencer 1990a). Prior to late the 1980s, this leafminer was considered a relatively minor secondary pest that rarely caused economic damage. During the 1980s, *L. huidobrensis* evolved resistance to a number of insecticides, acquired new hosts, and spread to new locations around the globe (Weintraub and Horowitz 1995; Steck 1996; He et al. 2002). Invasive populations of *L. huidobrensis* are particularly prone to outbreaks, and entire crops may be destroyed during population explosions (Shepard et al. 1998). Currently, *L. huidobrensis* is present in many of the world's major vegetable growing regions and is considered a major pest, constituting a significant threat to a variety of crops (Weintraub and Horowitz 1995; Scheffer et al. 2001).

Until recently, it was believed that the endemic distribution of *L. huidobrensis* was in western North America, Central America, and South America (Spencer 1973; Parrella 1982). Phylogenetic analysis of 941 bp of mitochondrial cytochrome oxidase (CO) I and II sequence data indicated that what was once considered one species is actually comprised of two allopatric sibling species; the North American populations form a morphologically cryptic species that is highly distinct from the South and Central American populations (Scheffer 2000). Confirmation of the divergence detected by mitochondrial data was obtained from sequence data from two nuclear genes (EF-1a and a B-tubulin intron) (Scheffer and Lewis 2001). The name "*L. langei*" was revived for the cryptic species present in California and Hawaii and that of "*L. huidobrensis*" was restricted to the South and Central American form (Scheffer and Lewis 2001). Both species appear to be polyphagous, although it is possible that their most preferred hosts may differ.

Even though both *L. huidobrensis* and *L. langei* are important crop pests in their native ranges, it is only *L. huidobrensis* that has become invasive. The phylogenetic analysis of sequence data identified *L. huidobrensis* from Central or South America as the source of invasive *L. huidobrensis* populations sampled to date from around the world, including populations in Italy, Israel, South Africa, Sri Lanka, the Philippines, Malaysia, China, and Canada (Scheffer 2000; Scheffer et al. 2001; He et al. 2002). All invasive *L. huidobrensis* sampled from around the world belong to an "invasive haplotype" group, in which flies carry identical or nearly identical haplotypes. Increased sampling of individuals within several invasive populations confirmed that these populations are comprised entirely of flies belonging to the invasive haplotype group (He et al. 2002; Scheffer unpubl. data).

Determining the geographic origins of invasive *L. huidobrensis* more precisely requires increased sampling of populations from within both South and Central America and the collection of additional sequence data. Analysis of 2,152 bp spanning most of mitochondrial cytochrome oxidase I and II indicates that there is considerable phylogeographic structuring within this species across South America (Fig. 13.1). Samples from Argentina are distinct from all other samples. Some of the flies from Ecuador cluster with those from Colombia, whereas others from Ecuador form a clade with all Peruvian and all Guatemalan specimens. This clade also contains all flies sampled from invasive populations.

Despite the presence of phylogeographic structure across South America, that all Guatemalan and all Peruvian specimens belong to the invasive haplotype group indicates some movement of flies between Central and South America. From the phylogenetic reconstruction, it appears that the invasive haplotype carried by some Peruvian and Ecuadorian samples fits well within the phylogeographic structure seen within South America, suggesting that these samples may represent an endemic population. The *L. huidobrensis* present in Guatemala may be the result of an introduction from South to Central America. The diversity of the sequence data collected to date supports this hypothesis; all sampled South American populations exhibit some variation in mitochondrial haplotypes, whereas the equivalently sampled Guatemalan population exhibits no variation (Fig. 13.2; Scheffer unpubl. data). The first reported records of *L. huidobrensis* in Central America are from Costa Rica in 1982 (Spencer 1983), a somewhat late date for the discovery of an endemic pest. Additionally, *L. huidobrensis* has yet to be reported from Mexico, although vegetable growing regions of the country suffer from other leafminers commonly associated with *L. huidobrensis* in other areas.

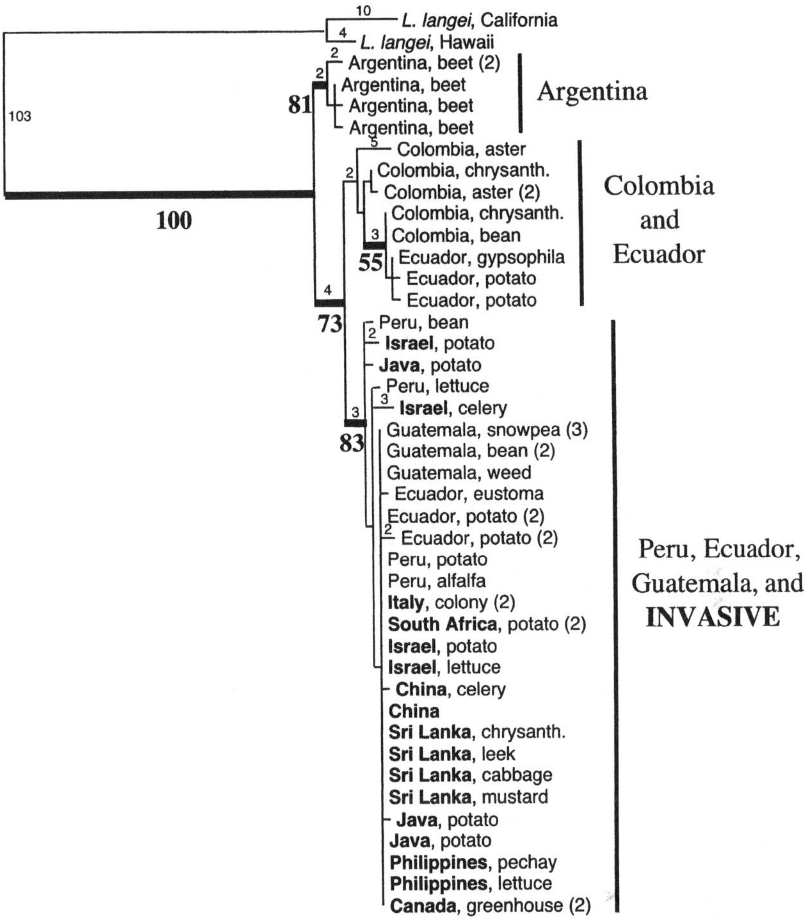

FIGURE 13.1. Phylogeographic structure within *L. huidobrensis* from analysis of 2,152 bp of mitochondrial COI and COII sequence data. One of many equally parsimonious trees showing estimated branch lengths above branches and bootstrap values (based on 100 pseudoreplicates) below. Branches in bold are those recovered in the strict consensus. Numbers in parentheses indicate the number of specimens from a given host and location that share a haplotype. Specimens from invasive populations are indicated in bold.

Taken together, these lines of evidence suggest that the presence of *L. huidobrensis* in Central America is the result of one or several introductions from South America. The hypothesis that *L. huidobrensis* is not endemic to Central America can be tested by sampling this species from additional locations across Central America. Finding only the invasive haplotype in flies from other regions will indicate that current populations have resulted from introductions. Finding variation in haplotypes, particularly if the variation is extensive or phylogeographically structured, will suggest that at least some populations of *L. huidobrensis* are endemic to Central America. Given the limited sampling that has been conducted to date, either scenario remains viable.

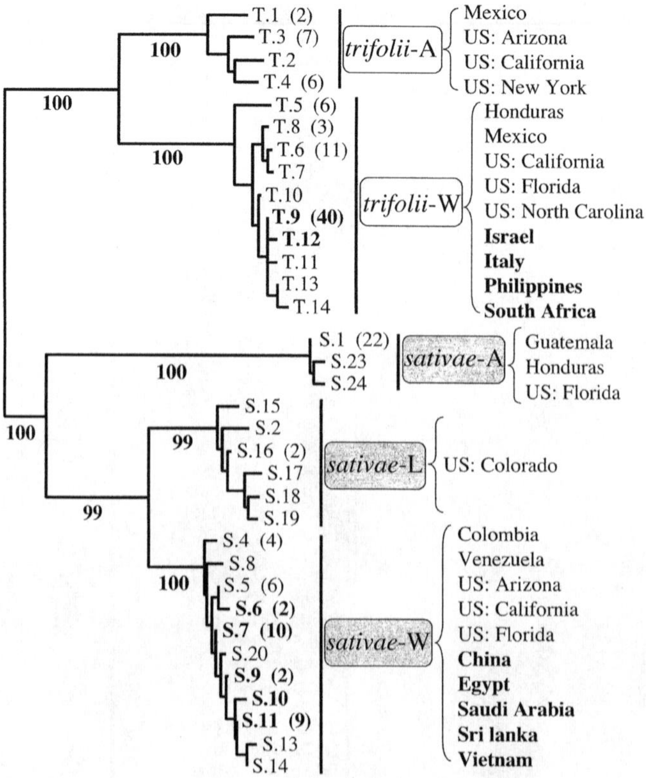

FIGURE 13.2. Phylogeographic structure within the *L. trifolii–L. sativae* species complex from analysis of 500 bp of mitochondrial COI data. Results of neighbor-joining analysis with bootstrap values (based on 100 pseudoreplicates) below branches. Numbers in parentheses indicate the number of specimens that share a particular haplotype. Haplotypes present in invasive populations are indicated in bold. Outgroup taxa were removed following analysis to conserve space in the figure.

Liriomyza trifolii *and* Liriomyza sativae. *L. trifolii* and *L. sativae* are also polyphagous leafmining pests of vegetables and flowers that are currently found in various locations throughout the world (Spencer 1973, 1990a; Parrella 1982, 1987). Both species are endemic to the Americas and appear to have had early distributions including both North and South America (Spencer 1973). Invasive populations have appeared in agricultural regions of Europe, Africa, the Middle East, and Asia within the past several decades (Spencer 1973; Parrella 1987; Minkenberg 1988). Within both species, phylogenetic analysis of 500 bp of mitochondrial cytochrome oxidase I sequence data uncovered highly diverged lineages, suggestive of cryptic species (Fig. 13.2; Scheffer and Lewis, unpubl. data). It should be noted that conclusions drawn from the neighbor-joining tree presented in the figure are consistent with those from a parsimony analysis of 1,392 bp of COI from exemplars from the major lineages (Scheffer and Lewis, unpubl. data).

Within *L. trifolii*, analysis of 82 specimens resulted in 14 mitochondrial haplotypes with frequencies ranging from one to 40. Phylogenetic analysis resulted in a tree containing two highly distinct lineages (as well as a very shallow host race: see Morgan et al. 2000; Reitz and Trumble 2002; also Fig. 13.2; Scheffer and Lewis, unpubl. data). The two lineages *trifolii*-A and *trifolii*-W differ by 4.7–5.7% uncorrected pairwise divergence. The clade *trifolii*-A consists of samples from California, Arizona, Mexico, and from a recent outbreak in New York. The *trifolii*-W clade consists of flies from California, Florida, Puerto Rico, and includes all *L. trifolii* sampled to date from outside of the Americas, including *L. trifolii* from the Philippines, South Africa, Italy, and Israel (Scheffer and Lewis, unpubl. data).

Within *L. sativae*, analysis of 65 specimens resulted in 20 mitochondrial haplotypes with frequencies ranging from one to 18. Phylogenetic analysis resulted in three distinct lineages (Fig. 13.2). Uncorrected pairwise distances between the lineages ranged from 2.5% to 8.8% (*sativae*-A to *sativae*-L: 7.6–8.4%; *sativae*-A to *sativae*-W: 7.6–8.8%; *sativae*-L to *sativae*-W: 2.5–3.8%). From samples collected to date, one clade (*sativae*-A) has only been found in Florida and Guatemala. The second clade (*sativae*–L) is from a single population from Colorado. The third clade (*sativae*-W) is found in California, Arizona, Florida, and Colombia, and includes all *L. sativae* so far sampled outside of the Americas, including specimens from Israel, Sri Lanka, Vietnam, the Philippines, and China. It is possible that increased sampling within North America will reduce the level of apparent divergence between clades *sativae*-L and *sativae*-W, but the greater divergence between these clades and *sativae*-A may reflect the presence of cryptic species (see Scheffer and Lewis, unpubl. data). However, as with *L. trifolii*, the mitochondrial divergences within *L. sativae* have not yet been corroborated by data from additional loci.

Similarities in Leafminer Invasions. Phylogeographic analysis of all three *Liriomyza* morphospecies (*L. huidobrensis*/*L. langei*, *L. trifolii*, and *L. sativae*) has uncovered striking similarities. All three initially appeared to have widespread endemic distributions spanning North and South America, and all three were found to have deep phylogenetic structures, suggesting the presence of cryptic species. In all three cases, the samples from geographically diverse invasive populations trace to only one of the pest-containing lineages. In fact, in all three cases, the invasive samples trace to only a portion of one of the cryptic lineages. It seems clear that within these polyphagous leafminers, haplotypes differ in their probability of being present in invasive populations.

Several mechanisms might cause some haplotypes to be more commonly represented than others in invasive populations. Leafminers are typically spread to new areas as their host plants are shipped around the world (Minkenberg 1988; Spencer 1990b). If leafminer haplotypes are geographically structured and exports come primarily from one geographic region, only flies carrying one or a few related haplotypes will be exported. If flies with different haplotypes also differ in abundance, there may be a difference in propagule pressure (Williamson 1999) such that flies carrying particular haplotypes will be more likely to reach new areas. This might be especially important if the evolution of insecticide resistance causes some populations to become highly abundant. Finally, differences among haplotypes in the probabilities of establishment and spread once a new region is reached could also lead to the prevalence of certain haplotypes in colonizing populations.

For *Liriomyza* leafminers, available evidence suggests that several processes might account for the differential spread of haplotypes. In the case of *L. huidobrensis*, preliminary implications of the molecular data collected so far are consistent with what is known concerning its history. For decades, this species was present in South America as a minor pest of little importance (Spencer 1973; Cisneros and Mujica 1998). Following heavy insecticide use for the control of other crop pests, large populations of insecticide resistant *L. huidobrensis* began to appear in Peru (Cisneros and Mujica 1998). Outbreak populations exhibited extraordinary leafminer densities, and invasive populations of *L. huidobrensis* began to appear outside of the Americas in the 1980s (Weintraub and Horowitz 1995). This scenario suggests that the evolution of insecticide resistance in South America may have led to large numbers of leafminers, increasing the propagule pressure such that invasions into new regions became more likely. Because *L. huidobrensis* populations in new areas commonly reach outbreak densities, these new populations are probably also serving as source populations for the rapid spread of this species that has been observed in the past decade. With the data collected so far, secondary spread from previously introduced populations cannot be distinguished from primary spread coming directly from endemic populations. Further investigation with highly variable markers, such as microsatellites, may provide additional information regarding the pattern of spread.

The mitochondrial sequence data presented here for *L. sativae* and *L. trifolii* represent only a first step in exploring molecular variation and the history of global expansion in these leafminers. Doubtless, there has been considerable movement of these flies within the Americas, as well as from the Americas to other regions. Additional sampling of both pest and nonpest populations throughout the Americas is critical to fully assessing endemic phylogeographic structure. However, broad patterns regarding global spread discerned from the current dataset are consistent with the known history of these species. Specifically, *L. trifolii*'s recent past is similar to that of *L. huidobrensis* and involves both repeated evolution of insecticide resistance and rapid spread to new locations (Parrella et al. 1984; Minkenberg 1988). *L. sativae* has been a pest for a longer period (Zehnder and Trumble 1984a; Palumbo et al. 1994) and is not quite as prone to evolve insecticide resistance (Zehnder and Trumble 1984b; Trumble 1990; Palumbo et al. 1994). This difference between the morphospecies is possibly reflected by differences in the mitochondrial variation exhibited by the invasive populations: from invasive *L. trifolii* populations, only two different haplotypes were recovered from 41 specimens sequenced, whereas within invasive *L. sativae*, there were eight haplotypes from 24 specimens (Fig. 13.2; Scheffer and Lewis, unpubl. data).

Conclusions

Understanding the patterns and processes involved in the colonization of new areas by invasive species is critical to managing introduced populations and preventing the spread and establishment of additional populations. Molecular markers are useful in several respects; molecular markers can be used to differentiate endemic from introduced populations, can uncover biologically significant cryptic variation, and can determine the geographic origins of introduced populations. Whether this last goal can be accomplished depends on the degree of molecular differentiation within the native range or source populations of the species, and it also can depend on the nature of the introduction(s). Data from a population resulting from a single introduction from one location will be much easier to interpret than data from a population created by multiple sets of founders from different locations. Studies of invasive Diptera illustrate the utility of molecular markers in investigating the patterns and processes involved in the colonization of new areas. As new tools for the analysis of invasive populations arise, we can expect invasive Diptera to continue to be at the forefront of our understanding of both applied and basic questions in invasion biology.

ACKNOWLEDGMENTS

I thank Matt Lewis for his assistance collecting sequence data on invasive leafminers. Kevin Omland, Tom Henry, Dug Miller, Al Norrbom, Chris Thompson, and Norm Woodley provided helpful comments on the manuscript. I also thank Brian Wiegmann and David Yeates for the opportunity to contribute to this volume.

REFERENCES

Avise, J. C. 2000. *Phylogeography, the History and Formation of Species.* Cambridge, Mass.: Harvard University Press.

Ayala, F. J., L. Serra, and A. Prevosti. 1989. A grand experiment in evolution: The *Drosophila subobscura* colonization of the Americas. *Genome* 31:246–255.

Balanya, J., C. Segarra, A. Prevosti, and L. Serra. 1994. Colonization of America by *Drosophila subobscura:* The founder event and a rapid expansion. *J. Hered.* 85:427–432.

Bartlett, P. W. 1993. Plant quarantine experience of *Liriomyza* spp. in England and Wales. In "Liriomyza" *Conference on Leafmining Flies in Cultivated Plants,* pp. 23–30. Montpellier, France: CIRAD.

Berlocher, S. H. 2000. Radiation and divergence in the *Rhagoletis pomonella* species group: Inferences from allozymes. *Evolution* 54:543–557.

Besansky, N. J., T. Lehmann, G. T. Fahey, D. Fontenille, L. E. O. Braack, W. A. Hawley, and F. H. Collins. 1997. Patterns of mitochondrial variation within and between African malaria vectors, *Anopheles gambiae* and *An. arabiensis,* suggest extensive gene flow. *Genetics* 147:1817–1828.

Birungi, J., and L. E. Munstermann. 2002. Genetic structure of *Aedes albopictus* (Diptera: Culicidae) populations based on mitochondrial ND5 sequences: Evidence for an independent invasion into Brazil and United States. *Ann. Entomol. Soc. Am.* 95:125–132.

Black, W. C. IV, J. A. Ferrari, K. S. Rai, and D. Sprenger. 1988. Breeding structure of a colonising species: *Aedes albopictus* (Skuse) in the United States. *Heredity* 60:173–181.

Bohonak, A. J., N. Davies, F. X. Villablanca, and G. K. Roderick. 2001. Invasion genetics of New World medflies: Testing alternative colonization scenarios. *Biol. Invasions* 3:103–111.

Bonizzoni, M., L. Zheng, C. R. Guglielmino, D. S. Haymer, G. Gasperi, L. M. Gomulski, and A. R. Malacrida. 2001. Microsatellite analysis of medfly bioinfestations in California. *Mol. Ecol.* 10:2515–2524.

Bryan, R. T. 1999. Alien species and emerging infectious diseases: Past lessons and future implications. In O. T. Dandlund, P. J. Schei, and A. Viken, eds., *Invasive Species and Biodiversity Management,* pp. 163–175. Dordrecht: Kluwer Academic.

Bush, G. L. 1969. Sympatric host race formation and speciation in frugivorous flies of the genus *Rhagoletis* (Diptera. Tephritidae). *Evolution* 23:237–251.

Busvine, J. R. 1980. Cryptic species of insect disease vectors and their importance. *Endeavour* 4:108–112.

Carey, J. R. 1991. Establishment of the Mediterranean fruit fly in California. *Science* 253:1369–1373.

Cheek, S., O. C. MacDonald, and P. W. Bartlett. 1993. Statutory action against *Liriomyza huidobrensis* (Blanchard) in the United Kingdom. In "Liriomyza" *Conference on Leafmining Flies in Cultivated Plants,* pp. 79–86. Montpellier, France: CIRAD.

Cisneros, F., and N. Mujica. 1998. The Leafminer fly in potato: Plant reaction and natural enemies as natural mortality factors. In *CIP Program Report 1997–1998,* pp. 129–140. Lima: CIP (International Potato Center).

Collins, F. H., and S. M. Paskewitz. 1996. A review of the use of ribosomal DNA (rDNA) to differentiate among cryptic *Anopheles* species. *Insect Mol. Biol.* 5:1–9.

Coluzzi, M., A. Sabatini, V. Petrarca, and M. A. Di Deco. 1979. Chromosomal differentiation and adaptation to human environments in the *Anopheles gambiae* complex. *Trans. R. Soc. Trop. Med. Hyg.* 73: 483–497.

Coluzzi, M., A. Sabatini, A. della Torre, M. A. Di Deco, and V. Petrarca. 2002. A polytene chromosome analysis of the *Anopheles gambiae* species complex. *Science* 298:1415–1418.

Cornuet, J.-M., S. Piry, G. Luikart, A. Estoup, and M. Solignac. 1999. New methods employing multilocus genotypes to select or exclude populations as origins of individuals. *Genetics* 153:1989–2000.

David, J. R., and P. Capy. 1988. Genetic variation of *Drosophila melanogaster* natural populations. *Trends Genet.* 4:106–111.

Davies, N., F. X. Villablanca, and G. K. Roderick. 1999. Bioinvasions of the medfly *Ceratitis capitata:* Source estimation using DNA sequences at multiple intron loci. *Genetics* 153:351–360.

Diehl, S. R., and G. L. Bush. 1984. An evolutionary and applied perspective of insect biotypes. *Annu. Rev. Entomol.* 29:471–504.

Donnelly, M. J., M. C. Licht, and T. Lehmann. 2001. Evidence for recent population expansion in the evolutionary history of the malaria vectors *Anopheles arabiensis* and *Anopheles gambiae. Mol. Biol. Evol.* 18:1353–1364.

Donnelly, M. J., F. Simard, and T. Lehmann. 2002. Evolutionary studies of malaria vectors. *Trends Parasitol.* 18:75–80.

Dukes, J. S., and H. A. Mooney. 1999. Does global change increase the success of biological invaders? *TREE* 14:135–139.

Enserink, M. 1999. Biological invaders sweep in. *Science* 285:1834–1836.

Ferrar, P. 1987. A guide to the breeding habits and immature stages of Diptera Cyclorrhapha. *Entomonograph* 8(1–2):907.

Fonseca, D. M., D. A. LaPointe, and R. C. Fleischer. 2000. Bottlenecks and multiple introductions: Population genetics of the vector of avian malaria in Hawaii. *Mol. Ecol.* 9:1803–1814.

Fonseca, D. M., S. Campbell, W. J. Crans, M. Mogi, I. Miyagi, T. Toma, M. Bullians, T. G. Andreadis, R. L. Berry, B. Pagac, M. R. Sardelis, and R. C. Wilkerson. 2001. *Aedes* (*Finlaya*) *japonicus* (Diptera: Culicidae), a newly recognized mosquito in the United States: Analyses of genetic variation in the United States and putative source populations. *J. Med. Entomol.* 38:135–146.

Gasparich, G. E., J. G. Silva, H.-Y. Han, B. A. McPheron, G. J. Steck, and W. S. Sheppard. 1997. Population genetic structure of Mediterranean fruit fly (Diptera: Tephritidae) and implications for worldwide colonization patterns. *Ann. Entomol. Soc. Am.* 90:790–797.

Gasperi, G., M. Bonizzoni, L. M. Gomulski, V. Murelli, C. Torti, A. R. Malacrida, and C. R. Guglielmino. 2002. Genetic differentiation, gene flow and the origin of infestations of the medfly, *Ceratitis capitata. Genetica* 116:125–135.

Gentile, G., A. della Torre, B. Maegga, J. R. Powell, and A. Caccone. 2002. Genetic differentiation in the African malaria vector, *Anopheles gambiae* s.s., and the problem of taxonomic status. *Genetics* 161: 1561–1578.

Guillemaud, T., S. Rooker, N. Pasteur, and M. Raymond. 1996. Testing the unique amplification event and the worldwide migration hypothesis of insecticide resistance genes with sequence data. *Heredity* 77:535–543.

Hawley, W. A., P. Reiter, R. S. Copeland, C. B. Pumpuni, and G. B. Craig, Jr. 1987. *Aedes albopictus* in North America: Probable introduction in tires from northern Asia. *Science* 236:1114–1116.

Haymer, D. S., M. He, and D. O. McInnis. 1997. Genetic marker analysis of spatial and temporal relationships among existing populations and new infestations of the Mediterranean fruit fly, *Ceratitis capitata. Heredity* 79:302–309.

He, L., Y. Zhang, N. Xiao, J. Wei, and R. Kuang. 2002. *Liriomyza huidobrensis* in Yunnan, China: Current distribution an genetic structure of a recently established population. *Entomol. Exp. Appl.* 102:213–219.

He, M., and D. S. Haymer. 1999. Genetic relationships of populations and the origins of new infestations of the Mediterranean fruit fly. *Mol. Ecol.* 8:1247–1257.

Headrick, D. H., and R. D. Goeden. 1996. Issues concerning the eradication or establishment and biological control of the Mediterranean fruit fly, *Ceratitis capitata* (Wiedemann) (Diptera: Tephritidae), in California. *Biol. Cont.* 6:412–421.

Huber, D. M., M. E. Hugh-Jones, M. K. Rust, S. R. Sheffield, D. Simberloff, and C. R. Taylor. 2002. Invasive pest species: Impacts on agricultural production, natural resources, and the environment. Council for

Agricultural Science and Technology Issue Paper No. 20, pp. 1–18. Ames, Iowa: Council for Agricultural Science and Technology.

Irvin, S. D., K. A. Wetterstrand, C. M. Hunter, and C. F. Aquadro. 1998. Genetic variation and differentiation at microsatellite loci in *Drosophila simulans:* Evidence for founder effects in New World populations. *Genetics* 150:777–790.

Kambhampati, S., W. C. Black, IV, and K. S. Rai. 1991. Geographic origin of the US and Brazilian *Aedes albopictus* inferred from allozyme analysis. *Heredity* 67:85–94.

Krzywinski, J., and N. J. Besansky. 2003. Molecular systematics of *Anopheles:* From subgenera to subpopulations. *Annu. Rev. Entomol.* 48:111–139.

Lachaise, D., M.-L. Cariou, J. R. David, F. Lemeunier, L. Tsacas, and M. Ashburner. 1988. Historical biogeography of the *Drosophila melanogaster* species subgroup. *Evol. Biol.* 22:159–255.

Lehmann, T., M. Light, N. Elissa, B. T. A. Maega, J. M. Chimumbwa, F. T. Watsenga, C. S. Wondji, F. Simard, and W. A. Hawley. 2003. Population structure of *Anopheles gambiae* in Africa. *J. Hered.* 94:133–147.

Lounibos, L. P. 2002. Invasions by insect vectors of human disease. *Annu. Rev. Entomol.* 47:233–266.

Lounibos, L. P., and J. E. Conn. 2000. Malaria vector heterogeneity in South America. *Am. Entomol.* 46: 238–249.

McAlpine, J. F., B. V. Peterson, G. E. Shewell, H. J. Teskey, J. R. Vockeroth, and D. M. Wood, eds. 1981. *Manual of Nearctic Diptera,* vol. 1. *Res. Branch Agric. Can. Mongr.* 27:1–674.

McAlpine, J. F., B. V. Peterson, G. E. Shewell, H. J. Teskey, J. R. Vockeroth, and D. M. Wood, eds. 1987. *Manual of Nearctic Diptera,* vol. 2. *Res. Branch Agric. Can. Mongr.* 28:1–674.

McMichael, A. J., and M. J. Bouma. 2000. Global changes, invasive species, and human health. In H. A. Mooney and R. J. Hobbs, eds., *Invasive Species in a Changing World,* pp. 191–210. Washington, D.C.: Island Press.

Minkenberg, O. P. J. M. 1988. Dispersal of *Liriomyza trifolii. Bull. OEPP/EPPO Bull.* 18:173–182.

Molbo, D., C. A. Machado, J. G. Sevenster, L. Keller, and E. A. Herre. 2003. Cryptic species of fig-pollinating wasps: Implications for the evolution of the fig-wasp mutualism, sex allocation, and precision of adaptation. *Proc. Natl. Acad. Sci. USA* 100:5867–5872.

Moore, C. G., and C. J. Mitchell. 1997. *Aedes albopictus* in the United States: Ten-year presence and public health implications. *Emerg. Infect. Dis.* 3:329–334.

Morgan, D. J. W., S. R. Reitz, P. W. Atkinson, and J. T. Trumble. 2000. The resolution of California populations of *Liriomyza huidobrensis* and *Liriomyza trifolii* (Diptera: Agromyzidae) using PCR. *Heredity* 85:53–61.

Nei, M., T. Maruyama, and R. Chakraborty. 1975. The bottleneck effect and genetic variability in populations. *Evolution* 29:1–10.

Norris, K. R. 1990. Evidence for the multiple exotic origin of Australian populations of the sheep blowfly, *Lucilia cuprina* (Wiedemann) (Diptera: Calliphoridae). *Aust. J. Zool.* 38:635–648.

Oldroyd, H. 1964. *The Natural History of Flies.* New York: W. W. Norton.

Palumbo, J. C., C. H. Mukkis, Jr., and F. J. Reyes. 1994. Composition, seasonal abundance, and parasitism of *Liriomyza* (Diptera: Agromyzidae) species on lettuce in Arizona. *J. Econ. Entomol.* 87:1070–1077.

Parrella, M. P. 1982. A review of the history and taxonomy of economically important serpentine leafminers (*Liriomyza* spp.) in California (Diptera: Agromyzidae). *Pan-Pac. Entomol.* 58:302–308.

Parrella, M. P. 1987. Biology of *Liriomyza. Annu. Rev. Entomol.* 32:201–224.

Parrella, M. P., C. B. Keil, and J. G. Morse. 1984. Insecticide resistance in *Liriomyza trifolii. Calif. Agr.* 38:22–23.

Pascual, M., C. F. Aquadro, V. Soto, and L. Serra. 2001. Microsatellite variation in colonizing and Palearctic populations of *Drosophila subobscura. Mol. Biol. Evol.* 18:731–740.

Peyton, E. L., S. R. Campbell, T. M. Candeletti, M. Romanowski, and W. J. Crans. 1999. *Aedes (Finlaya) japonicus japonicus* (Theobald), a new introduction into the United States. *J. Am. Mosq. Control Assoc.* 15:238–241.

Rannala, B., and J. L. Mountain. 1997. Detecting immigration by using multilocus genotypes. *Proc. Natl. Acad. Sci. USA* 94:9197–9201.

Reitz, S. R., and J. T. Trumble. 2002. Interspecific and intraspecific differences in two *Liriomyza* leafminer species in California. *Entomol. Exp. Appl.* 102:101–113.

Rich, S. M., M. C. Licht, R. R. Hudson, and F. J. Ayala. 1998. Malaria's Eve: Evidence of a recent population bottleneck throughout the world populations of *Plasmodium falciparum. Proc. Natl. Acad. Sci. USA* 95: 4425–4430.

Roderick, G. K., and F. X. Villablanca. 1996. Genetic and statistical analysis of colonization. In B. A. McPheron and G. J. Steck, eds., *Fruit Fly Pests: A World Assessment of Their Biology and Management*, pp. 281–290. Delray Beach, Fla.: St. Lucie Press.

Rogers, D. J., and S. E. Randolph. 2000. The global spread of malaria in a future, warmer world. *Science* 289:1763–1766.

Rosen, D. 1978. The importance of cryptic species and specific identifications as related to biological control. In J. A. Romberger, ed., *Biosystematics in Agriculture*, pp. 23–35. Montclair, N.J.: Allanheld, Osmun & Co.

Schauff, M. E., and J. LaSalle. 1998. The relevance of systematics to biological control: Protecting the investment in research. In *Pest Management—Future Challenges*, vol. 1, pp. 425–436. Proceedings of the 6th Australian Applied Entomological Conference, Brisbane, September 29–October 2, 1998.

Scheffer, S. J. 2000. Molecular evidence of cryptic species within the *Liriomyza huidobrensis* (Diptera: Agromyzidae). *J. Econ. Entomol.* 93:1146–1151.

Scheffer, S. J., and E. E. Grissell. 2003. Tracing the geographic origin of *Megastigmus transvaalensis* (Hymenoptera: Torymidae): An African wasp feeding on a South American plant in North America. *Mol. Ecol.* 12:405–414.

Scheffer, S. J., and M. L. Lewis. 2001. Two nuclear genes confirm mitochondrial evidence of cryptic species within *Liriomyza huidobrensis* (Diptera: Agromyzidae). *Ann. Entomol. Soc. Am.* 94:648–653.

Scheffer, S. J., and B. M. Wiegmann. 2000. Molecular phylogenetics of the holly leafminers (Diptera: Agromyzidae: *Phytomyza*): Species limits, speciation, and dietary specialization. *Mol. Phylogenet. Evol.* 17:244–255.

Scheffer, S. J., A. Wijesekara, D. Visser, and R. H. Hallett. 2001. Polymerase chain reaction-restriction fragment-length polymorphism method to distinguish *Liriomyza huidobrensis* from *L. langei* (Diptera: Agromyzidae) applied to three recent leafminer invasions. *J. Econ. Entomol.* 94:1177–1182.

Shepard, B. M., Samsudin, and A. R. Braun. 1998. Seasonal incidence of *Liriomyza huidobrensis* (Diptera: Agromyzidae) and its parasitoids on vegetables in Indonesia. *Int. J. Pest Manag.* 44:43–47.

Sheppard, W. S., G. J. Steck, and B. A. McPheron. 1992. Geographic populations of the medfly may be differentiated by mitochondrial DNA variation. *Experientia* 48:1010–1013.

Siebert, J. B., and T. Cooper. 1995. Embargo on California produce would cause revenue, job loss. *Calif. Agr.* 49:7–12.

Singh, R. S., and A. D. Long. 1992. Geographic variation in *Drosophila:* From molecules to morphology and back. *TREE* 7:340–345.

Smith, J. J., and G. L. Bush. 2000. Phylogeny of the subtribe Carpomyina (Trypetinae), emphasizing relationships of the genus *Rhagoletis.* In M. Aluja and A. L. Norrbom, eds., *Fruit Flies (Tephritidae): Phylogeny and Evolution of Behavior*, pp. 187–217. Boca Raton, Fla.: CRC Press.

Soto, S. I. U., T. Lehmann, E. D. Rowton, I. D. Velez, and C. Porter. 2000. Speciation and populations structure in the morphospecies *Lutzomyia longipalpis* (Lutz & Neiva) as derived from the mitochondrial *ND4* gene. *Mol. Phylog. Evol.* 18:84–93.

Spencer, K. A. 1973. *Agromyzidae (Diptera) of Economic Importance.* The Hague: W. Junk.

Spencer, K. A. 1983. Leaf mining Agromyzidae (Diptera) in Costa Rica. *Revta Biol. Trop.* 31:41–67.

Spencer, K. A. 1990a. *Host Specialization in the World Agromyzidae (Diptera).* Dordrecht: Kluver Academic.

Spencer, K. A. 1990b. Keeping track of *L. huidobrensis. Grower* 113:15.

Steck, G. J. 1996. *Pea Leaf Miner,* Liriomyza huidobrensis *(Diptera: Agromyzidae)*, pp. 1–3. Gainesville, Fla.: Division of Plant Industry, Florida Department of Agriculture and Consumer Services.

Tabachnick, W. J. 1991. Evolutionary genetics and arthropod-borne disease. The yellow fever mosquito. *Am. Entomol.* 37:14–24.

Trumble, J. T. 1990. Vegetable insect control with minimum use of insecticide. *HortScience* 25:159–164.

Urbanelli, S., R. Bellini, M. Carrieri, P. Sallicandro, and G. Celli. 2000. Population structure of *Aedes albopictus* (Skuse): The mosquito which is colonizing Mediterranean countries. *Heredity* 84:331–337.

van Riper, C. S., G. van Riper, M. L. Goff, and M. Laird. 1986. The epizootiology and ecological significance of malaria in Hawaiian land birds. *Ecol. Monogr.* 56:327–344.

Villablanca, F. X., G. K. Roderick, and S. R. Palumbi. 1998. Invasion genetics of the Mediterranean fruit fly: Variation in multiple nuclear introns. *Mol. Ecol.* 7:547–560.

Walsh, J. F., D. H. Molyneux, and M. H. Birley. 1993. Deforestation: Effects on vector-borne disease. *Parasitology* 106:S55–S75.

Weintraub, P. G., and A. R. Horowitz. 1995. The newest leafminer pest in Israel, *Liriomyza huidobrensis*. *Phytoparasitica* 23:177–184.
White, G. B. 1974. *Anopheles gambiae* complex and disease transmission in Africa. *Trans. R. Soc. Trop. Med. Hyg.* 68:278–301.
Williamson, M. 1999. Invasions. *Ecography* 22:5–12.
Zehnder, G. W., and J. T. Trumble. 1984a. Intercrop movement of leafminers. *Calif. Agr.* 38:7–8.
Zehnder, G. W., and J. T. Trumble. 1984b. Host selection of *Liriomyza* species (Diptera: Agromyzidae) and associated parasites in adjacent plantings of tomato and celery. *Envir. Entomol.* 13:492–496.

CHAPTER FOURTEEN

Guild Analyses of Dipteran Assemblages: A Rationale and Investigation of Seasonality and Stratification in Selected Rainforest Faunas

Roger L. Kitching, Daniel J. Bickel, and Sarah Boulter

It is commonplace in studies of insect biodiversity based on mass sampling to analyze the samples not only on the basis of their taxonomic composition, but also using the same data rearranged into trophic guilds (Southwood et al. 1982; Stork 1987; Hammond 1990; Hammond et al. 1996; Krüger and McGavin 2001). In this regard, a guild is defined as a set of species from a larger insect assemblage, the members of which have similar feeding habits. The level of detail involved in such designations varies from study to study, but basic trophic information (e.g., herbivore, predator, saprophage) may be modified by size (e.g., large, medium, small), by feeding habit (e.g., sucking, chewing) or by basic environmental factors (e.g., aquatic, terrestrial). As more qualifications are added to the guild designation, more categories emerge. This process may add to the "realism" of the analysis but may come at the cost of statistical power, as more and more categories become poorly represented or even absent within a particular data matrix.

Most guild analyses of insect assemblages have been carried out on samples of Coleoptera following the sorting of large samples to family and, where appropriate, subfamily. The Coleoptera have the advantage, in this respect, of being trophically diverse as an order but trophically more or less uniform within families or subfamilies (although there are a few taxa that are notable exceptions to this rule). There are more or less generally accepted designations of feeding guilds for each family or subfamily (Hammond 1990). This has led to a disproportionate representation of the Coleoptera in biodiversity studies of arthropods. Of course there is nothing wrong with this situation, and the Coleoptera will—and should—remain favored targets for such analysis. The Coleoptera are, however, but one of several "megadiverse" insect orders, and insights on the structure and functioning of species assemblages are also likely to be reached through studies of one or more of the others.

Table 14.1 is an analysis of the Australian fauna in this connection. For each very broad (in this case) guild designation, we have partitioned the Australian fauna as described in the second edition of *Insects of Australia* (CSIRO 1991) by numbers of species. We present data on the four very large endopterygote orders and the largest of the exopterygote taxa, the Hemiptera. This analysis is useful in several contexts. First, it shows that two of the orders, the Hemiptera and Lepidoptera, are remarkably uniform trophically, with a large percentage of their members being phytophagous (82% in the case of the Hemiptera; 91% of Lepidoptera). This uniformity does not mean that they should be excluded from ecological

TABLE 14.1. Macroguild Structure of the Five Largest Insect Orders[a]

Statistic	Hemiptera	Coleoptera	Lepidoptera	Diptera	Hymenoptera
Total species categorized	5509	28234	20840	6218	14781
Phytophages[b]	81.8	47.4	90.9	11.8	11.2
Predators & parasites[c]	15.1	22	0.4	27.7	67.4
Decomposers[d]	0	24.3	8.7	54.1	0
Fungivores[e]	3.2	6.3	<0.001	6.5	0
Ants	—	—	—	—	20.3

[a] Based on the percentages of species listed in Colless and McAlpine (1991).
[b] Includes folivores, xylophages, frugivores, pollenophages, and gall formers but not ants.
[c] Includes endo-, ecto-, and kleptoparasites but not ants.
[d] Includes necrophages and scatophages but not ants.
[e] Does not include ants.

analyses, particularly when these explicitly address plant/animal interactions (see, e.g., Kitching et al. 2000b). However, their trophic uniformity will reduce their utility when a surrogate for overall food-web structure is required. The Hymenoptera can be sidelined for comparable reasons. Sixty-seven percent of Hymenoptera are either predators or parasites (sensu lato) of one kind or another. Indeed this percentage is an underestimate, as a substantial proportion of Australia's 3,000 ant species may also be classed as predatory. Following earlier workers, we retain the "ants" as a feeding guild in themselves, as a reflection both of their omnivory and diversity. This leaves the Coleoptera and Diptera as appropriate targets for general guild analysis at the level of the assemblage. The Coleoptera are distributed much more evenly over four general feeding guilds—even the most dominant of these represents less than 50% of the entire fauna. It is worth remarking that, the remarks of Farrell (1998) notwithstanding, only 48% of beetle species are phytophagous: a very substantial 24% are decomposers of one sort or another. For the Diptera, again, no one guild is overwhelmingly dominant. The order is very rich in decomposers (54%) and predators (28%), but other guilds are also represented.

This rather coarse analysis clearly supports the use of either the Coleoptera or the Diptera in community guild analyses. Indeed the complementary nature of the relative abundance of species in guilds in each of the orders suggests that both orders may be combined for a powerful analysis. From the perspective of this chapter, we believe that the protocols and designations for the Coleoptera are reasonably well established. Here we attempt to provide the same rationale for the Diptera.

The Diptera contains insects that, as larvae and/or adults, may be herbivorous, fungivorous, saprophagous, xylophagous, myrmecophilous, termitophilous, parasitoidal, or predatory. This basic feeding classification may be further subdivided by taking into account the propensity for gall forming, together with the distinction between aquatic and nonaquatic forms (see below). The order is readily sampled in large numbers (Kitching et al. 1993, 2001), and its analysis at the guild level can throw light on a number of questions relating biodiversity to ecosystem functioning, levels of disturbance, and/or management regimes. Comparisons of the collections resulting from standard sampling protocols will show how the different ecological histories of contrasted locations affect the trophic structure of dipteran assemblages. Different disturbance regimes, for example, may be reflected in different proportions of predators and parasites, or different levels of saprophagous taxa. Such an approach adds a level of analysis related to the functionality of the order being sampled and may be of more interest and utility to ecosystem managers, for example, than one based on taxa alone.

We have been engaged in the taxonomic and guild analysis of dipteran assemblages from a set of rainforest plots arranged on a latitudinal gradient from southeastern Queensland to Southeast Asia (Kitching 2000). To date, these results have been analyzed taxonomically in a biogeographic context (Kitching et al. 2004). This chapter sets the scene for a further analysis at the guild level.

In this chapter, we review knowledge of feeding behaviors (and consequent guild designations) for the families of Diptera encountered in rainforest samples. We explore what an appropriate guild classification might be at both the larval and adult stages. We address the issue of what proportion of any sample might be confidently designated into feeding guilds. We conclude that there *is* adequate information to carry out analyses of dipteran assemblages of at least comparable accuracy to those that are commonplace in studies using Coleoptera.

To demonstrate the utility of this approach, we apply the guild designations and analyze differences both on temporal and spatial scales for subsets of our rainforest data. We compare samples from subtropical rainforest in southeastern Queensland for both summer and winter assemblages of Diptera using the results from three different sampling techniques. In a second analysis, we compare Malaise trap catches (only) from ground level with those obtained in the canopy from several Australasian rainforest sites. Future applications of the approach are discussed.

The Database

We consider here the data derived from extensive samples of Diptera collected using three sampling methods—canopy knockdown, Malaise traps, and yellow pan (water) traps—from seven rainforest locations in Australasia. In each case, these were derived from "snapshot" surveys of the arthropod biodiversity of 1-ha sites carried out during the wet season at each location. Sites ranged from Lamington National Park, southeastern Queensland (28° S) to Baitabag, Papua New Guinea (5° S) (Table 14.2). More details of the trapping protocol are provided in Kitching et al. (2000a), and full descriptions of the study sites are available in Kitching et al. (2004). The latter paper also discusses differences among these sites both in terms of ecology and the associated fauna.

SAMPLING METHODS AND SORTING

Three different sampling methods were used in our surveys, selected from the seven or eight methods used in our wider surveys as those that generated the largest numbers of Diptera. These were yellow pan (water) traps, canopy knockdown, and Malaise traps (see Kitching et al. 2000a).

Malaise Traps. Malaise traps are fabric intercept traps constructed on a tentlike principle. They were first proposed by Malaise (1937) and have become widely accepted as a means of mass sampling of flying insects at ground level. Juillet (1963) has suggested they are unbiased in their trapping of Diptera (in contrast to some other orders), but Roberts (1970) has pointed out that catches, even of Diptera, are influenced by the trap's shape and color. We used three

TABLE 14.2. Seven Study Sites from Which Diptera Were Sorted

Survey Location	Date	Position	Altitude[a] (m)	Forest Type	Number of Stems >5 cm	Number of Tree Species	Annual Mean Rainfall (mm)	Annual Temperatures Maximum (°C)	Annual Temperatures Minimum (°C)	Reference
Lamington National Park, southeastern Queensland	January 1995, July 1996	28°13′ S 153°07′ E	600	Complex notophyll vine forest	1,278	75	1,623	31.2	2.8	Laidlaw 1999; Laidlaw et al. 2000
Conondales National Park, southeastern Queensland	January 1998	26°43′50″ S 152°36′00″ E	550	Complex notophyll vine forest	1,313	51	1,345	34.2	3.7	Laidlaw 1999
Eungella National Park, central Queensland	January 1997	22°01′ S 148°37′ E	720	Simple/complex notophyll vine forest	1,983	49	1,699	34.8	6.4	Laidlaw 1999
Robson Creek, far north Queensland	January 1996	17° S 145° E	686	Complex notophyll vine forest	1,163	195	1,395	29.9	7.6	Laidlaw 1999
Cape Tribulation, far north Queensland	March 2000	16°06′ S 145°27′ E	81	Complex mesophyll vine forest[b]	1,538	137	2,500	28	22	Kitching et al. 1993; McIntyre et al. 1994
Baitabag, Papua New Guinea	June/July 1999	5°08′31″ S 145°46′37″ E	100	Complex mesophyll vine forest[b]	1,042	152	1,972	30	23.1	Laidlaw et al. in press
Oomsis, Papua New Guinea	July 2000	6°41′ S 146°48′ E	65	Medium crowned lowland hill forest[c]	1,020	121	1,979	32.3	21.6	Laidlaw et al. in press

[a] Above mean sea level.
[b] Considerable evidence of shifting cultivation.
[c] Paijmans (1970).

commercially available traps set up on the forest floor at three different random locations within the hectare on each sampling occasion. These traps have a collector at one end. Trapped insects were killed in situ, using a small block of dichlorvos-impregnated plastic. Traps were emptied daily for 4 days during each survey. Analyses have been carried out on the catches summed across days for each of the three traps used.

Canopy Knockdown. The use of short-lived insecticides dispensed into the forest canopy to knock down arboreal insects has become the technique of choice for mass-sampling of canopy arthropods (Martin 1966; Roberts 1973; Erwin 1990). We carried out three such canopy knockdowns within the study plot on each survey occasion. In each instance, a rope and pulley was secured over as high a branch as possible at a preselected, randomized location. This rope was used to haul a Stihl™ backpack mister (Stihl-Waiblingen, Germany) into the canopy to disperse a mixture of pyrethrum insecticides into the surrounding foliage for 5 minutes in each case. Insects and other arthropods knocked out of the canopy in this fashion were collected in circular collecting funnels each .5 m^2 in area. Twenty such funnels were suspended within a 10 × 10-m area around the sampling point. The arthropods collected over the 3–4-hour period following the insecticide application made up each sample.

Yellow Pan Traps. Yellow pan traps are water-based devices that collect small to very small airborne arthropods making up the so-called aerial "plankton." They are simply shallow pans of water containing a droplet of detergent to ensure that incident insects break the water tension and are retained within the water body. First used by Moericke (1951), those painted yellow have been shown to be more effective than traps of other colors (Harper and Story 1962). We used plastic food containers each 1675 × 120 × 40 mm in size painted yellow using an aerosol spray paint. Six of these traps were placed randomly on the ground within our study plot during the summer survey and ten during the winter survey. Each was emptied and reset for 4 days during each survey. Again, analyses were made on the summed catches of each trap.

Following collection, each of the samples was sorted to order using binocular microscopes in an adjacent field laboratory. Diptera were transported to the Australian Museum, Sydney, for subsequent sorting to family. The sorted samples are retained in the Australian Museum.

DIPTERA

The three sampling methods applied at seven locations generated 28,647 adult Diptera that were sorted to family under the direction of one of us (DJB). At four sites, Malaise traps were also placed in the forest canopy, generating an additional 10,812 specimens. Fifty-six families were encountered in this survey (Table 14.3). We note immediately that three families alone, the Phoridae, Cecidomyidae and Chironomidae, made up more than 50% of the total number of individual flies sampled (excluding the canopy Malaise samples). A further nine families are needed before more than 90% of the total sample are accounted for—and a further 12 to bring the score to 99%. Looking at these figures in a different way, of the total of 56 families encountered, 32 taken together made up less than 1% of the total sample.

Why is this important? Simply that we really do not need confident guild designations on *all* of the families encountered to be able to carry out a robust statistical analysis based on the distribution of numbers of individuals within each guild. Hence, lack of information on

TABLE 14.3. Diptera Encountered in a Sampling Program in Seven Australasian Rainforest Sites[a]

Cut-off Level	Family	Abundance	Cumulative Percentage	Larval Guild: Designation	Larval Guild: Aquatic/Terrestrial[b]	Larval Guild: Reference	Adult Guild: Designation	Adult Guild: Reference
	Phoridae	7,791	0.2720	Saprophage/ parasitoid	Terrestrial	Disney 1994	Nectarivores[c]	Disney 1994
50%	Cecidomyidae	4,349	0.4238	Gall former	—	Colless and McAlpine 1991	No trophic impact	
	Chironomidae	3,660	0.5515	Saprophage[d]	Aquatic	Berg 1995	No trophic impact	Armitage et al. 1995
	Sciaridae	2,124	0.6257	Saprophage	Terrestrial	Colless and McAlpine 1991	No trophic impact	
	Psychodidae	1,611	0.6819	Saprophage	Aquatic	Oldroyd 1964	Detritivores	
	Drosophilidae	1,464	0.7330	Fungivore/ frugivore	—	Colless and McAlpine 1991	Fungivore/frugivore	van Klinken and Walter 2001
	Mycetophilidae	1,246	0.7765	Fungivore	—	Colless and McAlpine 1991	No trophic impact	
	Ceratopogonidae	1,102	0.8150	Saprophage	Aquatic	Oldroyd 1964	Haematophage	Oldroyd 1964
	Dolichopodidae	992	0.8496	Predator	—	Oldroyd 1964	Predator	Colless and McAlpine 1991
90%	Sphaeroceridae	825	0.8784	Saprophage	Terrestrial	Colless and McAlpine 1991	Saprophage	Colyer and Hammond 1951
	Tipulidae	809	0.9066	Saprophage	Aquatic/ terrestrial	Oldroyd 1964	Nectarivores	Oldroyd 1964
	Chloropidae	779	0.9338	Various[e]	—	Ferrar 1987	No trophic impact	
	Empididae	381	0.9471	Predator	—	Smith 1969	Predator	Colless and McAlpine 1991
95%	Muscidae	314	0.9581	Saprophage	Terrestrial	Skidmore 1985	Saprophage	Colless and McAlpine 1991
	Pyrgotidae	152	0.9634	Parasites	—	Colless and McAlpine 1991	Nectarivores	McAlpine 1978
	Calliphoridae	124	0.9681	Saprophage	Terrestrial	Colless and McAlpine 1991	Saprophage	Colless and McAlpine 1991

continued

TABLE 14.3. *Continued*

Cut-off Level	Family	Abundance	Cumulative Percentage	Larval Guild			Adult Guild	
				Designation	Aquatic/Terrestrial[b]	Reference	Designation	Reference
	Micropezidae	127	0.9725	Saprophage	Terrestrial	Oldroyd 1964	Predator	Colyer and Hammond 1951
	Sepsidae	110	0.9763	Saprophage	Terrestrial	Colless and McAlpine 1991	Saprophage	Colless and McAlpine 1991
	Lauxanidae	96	0.9797	Saprophage	Terrestrial	Ferrar 1987	Saprophage/ fungivore	Kim 1994
	Milichiidae	92	0.9829	Saprophage	Terrestrial	Colless and McAlpine 1991	Predator/ kleptoparasite	Colyer and Hammond 1951
	Culicidae	79	0.9857	Saprophage	Aquatic	Oldroyd 1964	Haematophage	Oldroyd 1964
	Platystomatidae	64	0.9879	Saprophage/ fungivore	Terrestrial	Colless and McAlpine 1991	Saprophage	Colless and McAlpine 1991
99%	Stratiomyidae	39	0.9892	Saprophage	Aquatic	Bozkosnv 1982	Unknown	
	Tabanidae	36	0.9905	Saprophage	Aquatic/ terrestrial	Oldroyd 1964	Haematophage	Colless and McAlpine 1991
	Asilidae	31	0.9916	Predator	—	Colless and McAlpine 1991	Predators	Oldroyd 1964
	Tachinidae	30	0.9926	Parasite	—	Colless and McAlpine 1991	Nectarivore	Colyer and Hammond 1951
	Anisopodidae	25	0.9935	Saprophage	Terrestrial	Colyer and Hammond 1951	No trophic impact	
	Ephydridae	23	0.9943	Phytophage	—	Colless and McAlpine 1991	Predators	Colless and McAlpine 1991
	Clusiidae	20	0.9950	Saprophage	Terrestrial	McAlpine 1960	Unknown	
	Teratomyzidae	15	0.9955	Phytophage	—	Colless and McAlpine 1991	Phytophages	Colless and McAlpine 1991
	Scatopsidae	14	0.9960	Saprophage	Terrestrial	Oldroyd 1964	Unknown	
	Syrphidae	13	0.9965	Various[e]	—	Colless and McAlpine 1991	Pollenophage	various
	Periscelidae	12	0.9969	Saprophage	Aquatic/ terrestrial	Colless and McAlpine 1991	Saprophage	Colyer and Hammond 1951
	Platypezidae	12	0.9973	Fungivore	—	Colless and McAlpine 1991	Unknown	

	Pseudopomyzidae	9	0.9976	Unknown	—	Colless and McAlpine 1991	Unknown	
	Rhagionidae	9	0.998	Xylophage	—	Colless and McAlpine 1991	Predator	Colyer and Hammond 1951
	Simulidae	8	0.998	Saprophage	Aquatic	Colless and McAlpine 1991	Haematophage	Oldroyd 1964
99.90%	Conopidae	6	0.998	Parasite	—	Colless and McAlpine 1991	Nectarivores	Colyer and Hammond 1951
	Neriidae	6	0.999	Saprophage	Terrestrial	Colless and McAlpine 1991	Unknown	
	Lonchaeidae	5	0.999	Various[e]	—	Colless and McAlpine 1991	Frugivores?	
	Tephritidae	5	0.999	Frugivore	—	Colless and McAlpine 1991	Frugivores	
	Pipunculidae	4	0.999	Parasite	—	Colless and McAlpine 1991	Unknown	
	Bibionidae	3	0.999	Saprophage	Terrestrial	Colless and McAlpine 1991	Nectarivore	Colyer and Hammond 1951
	Heleomyzidae	3	0.999	Various[e]	—	Colless and McAlpine 1991	Saprophage	Colyer and Hammond 1951
	Therevidae	3	0.999	Predator	Terrestrial	Colless and McAlpine 1991	Unknown	Colless and McAlpine 1991
	Agromyzidae	2	0.999	Phytophage	—	Colless and McAlpine 1991	No trophic impact	
	Anthomyiidae	2	0.999	Phytophage	—	Colless and McAlpine 1991	Nectarivore	Colyer and Hammond 1951
	Cypselosomatidae	2	0.999	Saprophage	Terrestrial	Colless and McAlpine 1991	Unknown	
	Sacrophagidae	2	0.999	Saprophage	Terrestrial	Colless and McAlpine 1991	Saprophage	
	Trichoceridae	2	0.999	Saprophage	Terrestrial	Colless and McAlpine 1991	No trophic impact	
	Asteiidae	1	1.000	Fungivore	—	Colless and McAlpine 1991	Fungivorous?	

continued

TABLE 14.3. *Continued*

Cut-off Level	Family	Abundance	Cumulative Percentage	Larval Guild			Adult Guild	
				Designation	Aquatic/Terrestrial[b]	Reference	Designation	Reference
	Bombylidae	1	1.000	Parasite	—	Colless and McAlpine 1991	Nectarivore	Colyer and Hammond 1951
	Cryptochetidae	1	1.000	Parasite	—	Colless and McAlpine 1991	Unknown	
	Dixidae	1	1.000	Saprophage	Aquatic	Colless and McAlpine 1991	No trophic impact	
	Fergusoniidae	1	1.000	Gall former	—	Colless and McAlpine 1991	Unknown	
	Xylomyiidae	1	1.000	Unknown	—	Colless and McAlpine 1991	Unknown	

[a] Samples taken using a combination of Malaise traps, yellow pan traps, and pyrethrum knockdown sampling. Where an authoroity for a designation is not given, it has been based on the knowledge of the authors.
[b] Designated for saprophages only.
[c] Includes honeydew feeders.
[d] Includes facultative necrophagy and feeding on morbid prey.
[e] Cannot be placed into larval guilds unambiguously.

the biology of, for example, the Xenasteiidae or the Asteiidae (to select two "rare" families at random) will be of little hindrance to our analysis.

Guild Designation

We have attempted in Table 14.3 to make designations of the trophic guild for both larvae and adults of the families that make up our sample. We have done this in a number of ways.

LARVAE

There a number of excellent general accounts of the Diptera, or large sections of them, which collate information on the larval habits of families. These include Oldroyd (1964), McAlpine et al. (1981, 1987), Ferrar (1987), McAlpine and Wood (1989), and Colless and McAlpine (1991). Other works describe the fauna associated with particular resources (Teskey 1976; Lawrence and Milner 1996). Where consensus exists as to the feeding habits of taxa, this will be apparent from these works. In other instances, single families have been monographed in ways that summarize large amounts of available data on the biology of their subjects. Notable examples are the works of Rozkosny (1982) on Stratiomyidae, Skidmore (1985) on Muscidae, Disney (1994) on Phoridae, and Armitage et al. (1995) on Chironomidae—the list could be extended. Some decisions are inherent in the guild designations assigned to larvae in Table 14.3.

First, where "unusual" feeding relationships exist within an otherwise ecologically uniform family, we have designated the family according to the habits of the majority. So, for example, a very few Muscidae have aquatic or predatory larvae, although the majority are terrestrial decomposers. It makes sense in this case to designate the family as terrestrial decomposers unless there is some special reason to suppose that the exceptional life styles are somehow likely to have been favored by a particular sampling method or location. In general, trying to take account of the minority life styles would be no more sensible than classifying Lepidoptera as partly predatory on the basis of the minute proportion of that order known to be aphytophagous. We are aware, in this context, that not all Cecidomyidae are gall formers, especially in forest systems. We retain them in this guild rather than assigning some of them as terrestrial saprophages because the majority of published life histories recount their gall-forming activities. As more information comes to hand, this designation may need to be revised and the numbers partitioned accordingly.

Second, in a few cases, families seem to be more or less equally split between two related life styles. Rainforest Drosophilidae, for example, seem to be either frugivorous or fungivorous (van Klinken and Walter 2001), although some species also exploit flowers or detritus. In this situation, especially given the background information available from studies in Australasian rainforests specifically, adult samples, most prudently, can be assigned 50:50 to a frugivorous or fungivorous habit. A second important instance is the Phoridae, for which the vast majority of known feeding behaviors of the larvae indicate them to be either saprophages or parasitoids (R. H. L. Disney 1994, pers. comm.). A small number of species, however, are known to be termitophilous or mymecophilous, and at least one is ectoparasitic (Colless and McAlpine 1991). So in this case, we have arbitrarily designated the Phoridae encountered to the saprophagous and parasitoidal guilds on a 50:50 basis.

The designation of larval Diptera as saprophages raises two additional problems. There has long been difficulty in differentiating between larvae that are truly saprophagous—feeding on the substance of dead and dying plant and animal materials—and those that gain sustenance from the microorganisms associated with such detritus (see the extended discussion in Kitching 2000). We have taken the conventional approach in this chapter of designating as saprophages those larvae that feed on dead and dying organic matter and associated microorganisms. In addition, Diptera may have aquatic or terrestrial larvae, and it may well be generally useful to separate the saprophagous groups into an aquatic and a terrestrial guild. So, for example, Culicidae and Chironomidae will be designated as aquatic, Sphaeroceridae and Muscidae as terrestrial. In making designations in this fashion, we have been obliged to use the predominant habit in some instances (e.g., the Ceratopogonidae, designated here as aquatic; the Phoridae, as terrestrial). In other cases, the guild must be split more or less equally between aquatic and terrestrial (e.g., Tipulidae, Tabanidae, Periscelidae). We explore the advantages and disadvantages of this additional characterization in the analysis that follows.

Finally, in at least one important case (the Chloropidae), larval habits are simply too disparate to make any single designation sensible. A defensible guild classification of this family would, presumably, require sorting to a finer level of taxonomic resolution. As is the case with several dipteran families, this effort could not be sustained by the current taxonomic knowledge of the taxon, even if resources for this finer sorting were available. In this instance the sensible course of action is to include this family in an "other" category in guild analysis. We note that authors have made similar decisions in guild analyses of other orders retaining, for example, the "ants" as a guild in themselves to take account of the heterogeneous and often general trophic habits of this family.

ADULTS

The designation of adult flies to guilds is more problematical. Some authors do deal with the adult habits of some families (see, e.g., the review by Hövemeyer 2000). The predatory habits of adult dolichopodids and asilids, for example, are well accepted. Similarly, the attraction to and use of standing or fallen fruit of tephritids and drosophilids is well documented. Blood sucking flies, such as tabanids, culicids, and simulids, are likewise well known.

As is the case with larvae, a number of pragmatic decisions need to be made if a majority of adult flies in a sample are to be classified into guilds. First, many flies visit, and have an impact on, several different resources during their adult lives. Calliphorids, sarcophagids, phorids, and muscids will seek both protein and sugar (and water) during their adult lives (see, e.g., Roberts and Kitching 1974). In these instances, we designate the flies to a guild that reflects the interaction with the limiting resource—protein in this case, rather than with the widely available sugar source. Such protein feeders should be considered as saprophages in this context. In other instances, of course, adult flies seek nectar and honeydew as their primary goal, and assignment to a "nectar"-(loosely defined) based guild is most appropriate.

Many small flies, however, feed little as adults, and even when they are attracted to flowers or fruits, it is unclear just what the interaction with the insect might be. Adults of such families as the Chironomidae and Cecidomyiidae fall into this category (Hövemeyer 2000: Table 11.1). In any case, the minute biomass of many of these flies and their weak mouthparts suggest that their ecological impact will likely be small. An argument can thus be made for the maintenance of a, albeit less euphonious, "no trophic impact" guild, comparable with the "tourist" guild of earlier writers (Southwood et al. 1982; Stork 1987), indicating the presence of these insects within the assemblage but their essential lack of impact as adults on their surroundings. We prefer the designation "no trophic impact," as this underscores the observation that these insects remain part of local faunal assemblages in other respects and may well have impacts as, say, pollinators or as prey for other organisms.

SAMPLE SIZE, ACCURACY, AND POWER

Fig. 14.1A–C presents summaries of the larval guild structure for our the sample of 28,647 flies from the three sampling methods applied at seven locations (see above) based on the entire dataset, 50% of the dataset, and 90% of the dataset. The free-living phytophagous (including pollenophagous) and xylophagous guilds are represented by so few individuals (0.0014% and 0.0010% of the sample, respectively) that these do not show on the pie charts. It is clear from the charts that the guild structure based on 90% of the sample (i.e., on results from just nine of the 56 families) is an excellent representation of the structure obtained from the complete dataset. Indeed, the structure based on 90% of the dataset is highly significantly correlated with that based on all the data ($r = 0.995$). Moving to a 95% cut (chart not shown) introduces four more families but is virtually indistinguishable from the 90% version (increasing the r value to 0.998). The chloropids (which cannot readily be designated as a single larval guild) introduce a small "other" category (representing 2.7% of the overall sample), but this is of no statistical consequence.

A comparable analysis for the adult guilds is presented in Fig. 14.1D–F. The folivore guild (0.0008% of the sample) and the "unknown" category (0.0039%) do not show up on the pie charts. Results are very similar to those observed for the larval guilds. The guild distribution based on 90% of the data correlates highly with that based on the total dataset ($r = 0.998$). We note here that the inclusion of major families either in the "no trophic impact," or the "flower-feeding" categories has involved interpretation and some assumptions on our part. We return to this point in the discussion.

These guild structures, of course, are based on the total sample of flies obtained using three sampling methods across seven sites. The proposition that guild analyses is a useful approach hinges on the fact that interpretable differences will appear in guild structures from different sites or times.

Applications of Guild Analyses of Dipteran Assemblages from Australasian Rainforests

The second half of this chapter presents examples of guild analyses, based on the rationale discussed above, to selected subsets of our samples from Australian rainforests. The survey methodology used has been discussed at length by Kitching et al. (2000a) and analyses at the ordinal level presented by Kitching et al. (2001). Restricted taxonomic analyses of results on the ants (Majer et al. 2001), the Tipulidae (Kitching and Theischinger 1996), the Coeloptera (Hammond et al. 1996), and the Diptera (Kitching et al. 2004) have been published for all or part of the wider dataset.

We have selected two subsets of data to illustrate the guild analyses that we advocate in this chapter. In each case, we present results on the family-level composition of our samples

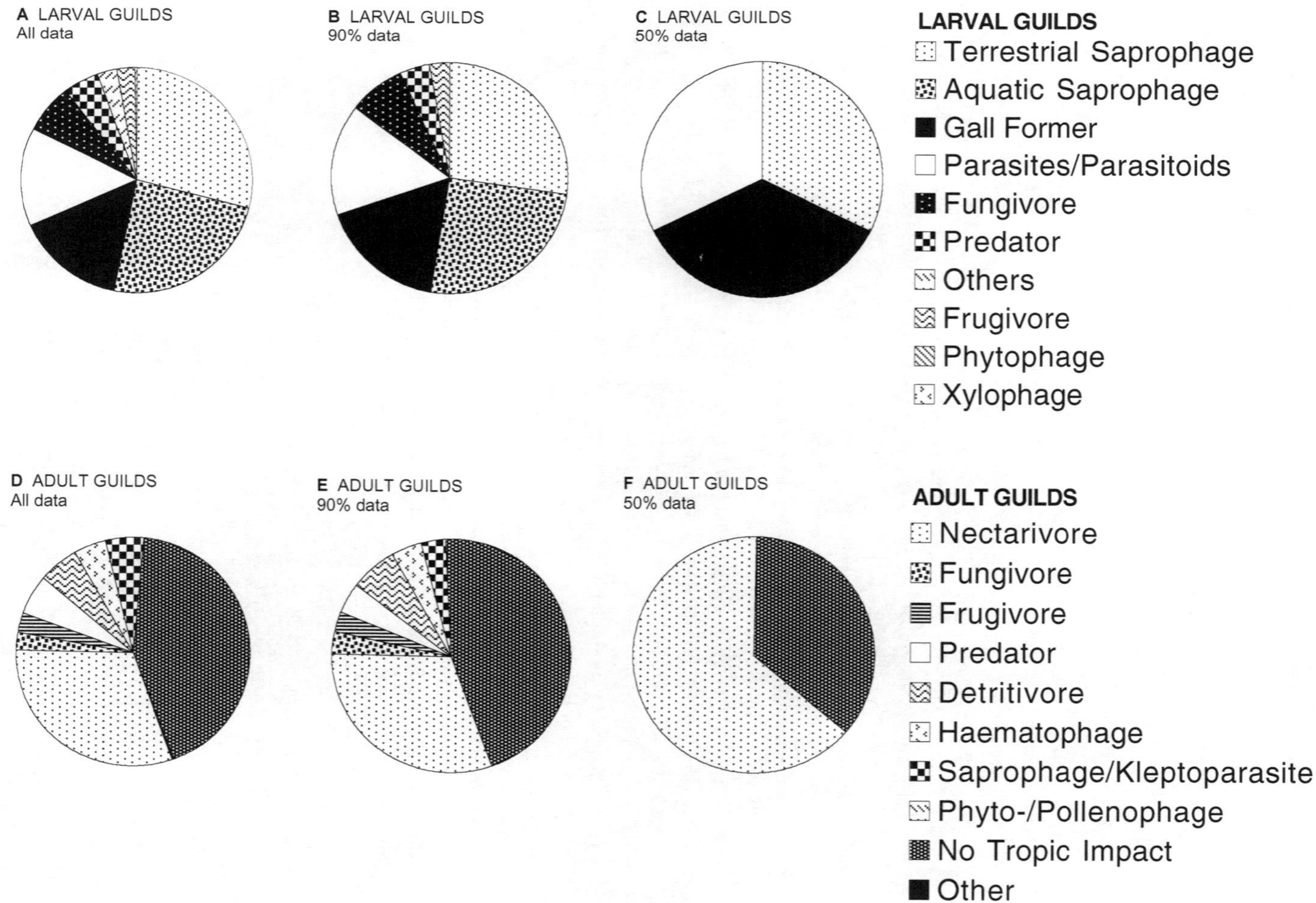
A LARVAL GUILDS
All data
B LARVAL GUILDS
90% data
C LARVAL GUILDS
50% data
LARVAL GUILDS
Terrestrial Saprophage
Aquatic Saprophage
Gall Former
Parasites/Parasitoids
Fungivore
Predator
Others
Frugivore
Phytophage
Xylophage
D ADULT GUILDS
All data
E ADULT GUILDS
90% data
F ADULT GUILDS
50% data
ADULT GUILDS
Nectarivore
Fungivore
Frugivore
Predator
Detritivore
Haematophage
Saprophage/Kleptoparasite
Phyto-/Pollenophage
No Tropic Impact
Other

and then the guild structure derived from these family profiles. For these analyses, we have used the larval guilds. In general, this is because the larval guilds are the most functionally important part of the life cycle of the flies, as well as the longest lived.

First, we examine dipteran assemblages based on yellow pan (water) traps, canopy pyrethrum knockdown samples, and Malaise traps collected during the wet (summer) season and the dry (winter) season in the subtropical broadleaved rainforest of Lamington National Park, southeastern Queensland. We hypothesis that there will be, in general, a lower diversity of Diptera in both taxonomic and guild-based terms in the cooler, drier winter samples. Furthermore, we suggest that this difference would be most apparent in reductions in the numbers of predators (which may depend on increased summer productivity to provide an abundance of prey) and in the phytophages (including gall formers and frugivores), whose activities may track leaf flushes that occur principally in the wetter, warmer months. Drier winter months may also be less suitable for frugivores. Seasonal impacts are less likely to be evident among the decomposers, for whom a rich resource base may be available throughout the year (in rainforest in particular).

Second, we have made collections using Malaise traps set out simultaneously at ground level and suspended in the canopy at four forests sites in Australasian rainforests (two in Queensland and two in Papua New Guinea) during successive wet seasons. We compare the ground and canopy dipteran guilds from these four sites.

We hypothesize that the decomposer and fungivore guilds will be more abundant at ground level, phytophages at mid- or upper canopy, and that predators should be unaffected by stratum. In this instance, however, we are aware that observed differences in numbers of adults may be based on postemergence relocation rather than breeding sites. This may complicate the interpretation of guilds when these are based on larval feeding habits.

SEASONALITY IN TAXONOMIC AND GUILD STRUCTURES OF AUSTRALIAN RAINFOREST DIPTERAN ASSEMBLAGES

To compare dipteran assemblages collected during the wet season and dry season, the results of collection using the three sampling methods (Malaise traps, canopy knockdown, and yellow pans) at Lamington National Park are presented here. A detailed description of the study site is available in Kitching et al. (2004) and a summary is provided in Table 14.2. The wet season sample described in this chapter was collected in January 1995 and the dry season one in July 1996.

Analysis. We have compared summer and winter samples on a method-by-method basis. The relative importance of particular families within the overall sample from a particular season has been estimated by calculating the Berger-Parker Index for each family—the ratio of the numbers of individuals in the family in question to the overall sample total on that occasion (Berger and Parker 1970).

FIGURE 14.1. Distribution of Diptera into larval (A–C) and adult (D–F) guilds based on percentage of samples after ranking into family order. (A) larval, 100%; (B) larval, 90%; (C) larval, 50%; (D) adult, 100%; (E) adult, 90%; and (F) adult, 50% of sample.

Simple comparisons of the family structure of the wet and dry season samples were made by rank correlation analysis. In addition, we carried out a series of one-way analyses of variance on both the taxonomic and guild composition indicated in the data. In this way, we analyzed the total numbers of flies, the total families in the sample, the Berger-Parker dominance measure, and the abundances of each of the most abundant families caught by each sampling method. For the yellow pan traps, these were the Cecidomyidae, Chloropidae, Phoridae, and Sciaridae; for the canopy knockdown samples, the Cecidomyidae, Ceratopogonidae, Chironomidae, Chloropidae, Phoridae, and Sciaridae; and for the Malaise traps, the Cecidomyiidae, Ceratopogonidae, Chironomidae, Chloropidae, Mycetophilidae, Phoridae, Psychodidiae, and Sciaridae. For the guild analysis, all flies were included in the designations. Again, one-way analyses of variance were carried out on both the abundances and the (arcsine square-root transformed) proportions represented by each guild. In all analyses, Bonferroni corrections (0.05/number of nonindependent analyses) were applied to the conventional $P = 0.05$ level of acceptable significant differences.

Results. Fig. 14.2A–C presents the results of the seasonal comparisons of both taxonomic and guild structure for the assemblages sampled by each of the three trapping methods. Table 14.4 presents the results of the analysis of variance. Only significant or near-significant results are included in the table.

The season-to-season similarities based on rank correlation varied substantially across methods. A high value of 0.80 showed that wet and dry season samples based on canopy knockdown were essentially similar. In contrast, the value of 0.37 for the yellow pan samples indicates little correspondence between the target assemblages for that trapping method across seasons. The value for the Malaise trap profiles was intermediate (0.67).

Yellow-pan trap catches (Fig. 14.2A) were dominated in the wet season by Cecidomyiidae (Berger-Parker Index, 0.15), the Chironomidae (0.13), the Phoridae (0.14), the Sciaridae (0.24), and the Sphaeroceridae (0.15). These five families made up 81% of the entire wet season catch. In the dry season, a comparable 81% of the sample was made up of just three families: the Chloropidae (Berger-Parker Index, 0.20), the Phoridae (0.19), and the Sciaridae (0.42).

Canopy knockdown catches (Fig. 14.2B) were dominated in the wet season by Cecidomyiidae (Berger-Parker Index, 0.26), the Chironomidae (0.37), and the Sciaridae (0.16), with these three families making up 79% of the entire wet season sample. In the dry season, only two families showed Berger-Parker dominance values greater than 0.10: the Ceratopogonidae (0.14) and the Sciaridae (0.44). These two families, then, made up 58% of the entire dry season sample.

Malaise trap catches (Fig. 14.2C) were dominated in both seasons by Cecidomyiidae (Berger-Parker Indexes: wet season, 0.38; dry season, 0.19), the Mycetophilidae (0.18; 0.25), the Phoridae (0.20; 0.25), and the Sciaridae (0.10; 0.25). These dominance values mean that these four families made up more than 86% of the wet season sample and 94% of the dry season sample.

FIGURE 14.2. Distribution of dipteran families from subtropical rainforest in Lamington National Park in wet (stippled) and dry (gray) seasons together with the distribution of the same samples into guilds, as sampled by (A) yellow pan traps, (B) canopy knockdown sampling, and (C) Malaise traps at ground level. Families represented by a single individual across both seasons are not shown here.

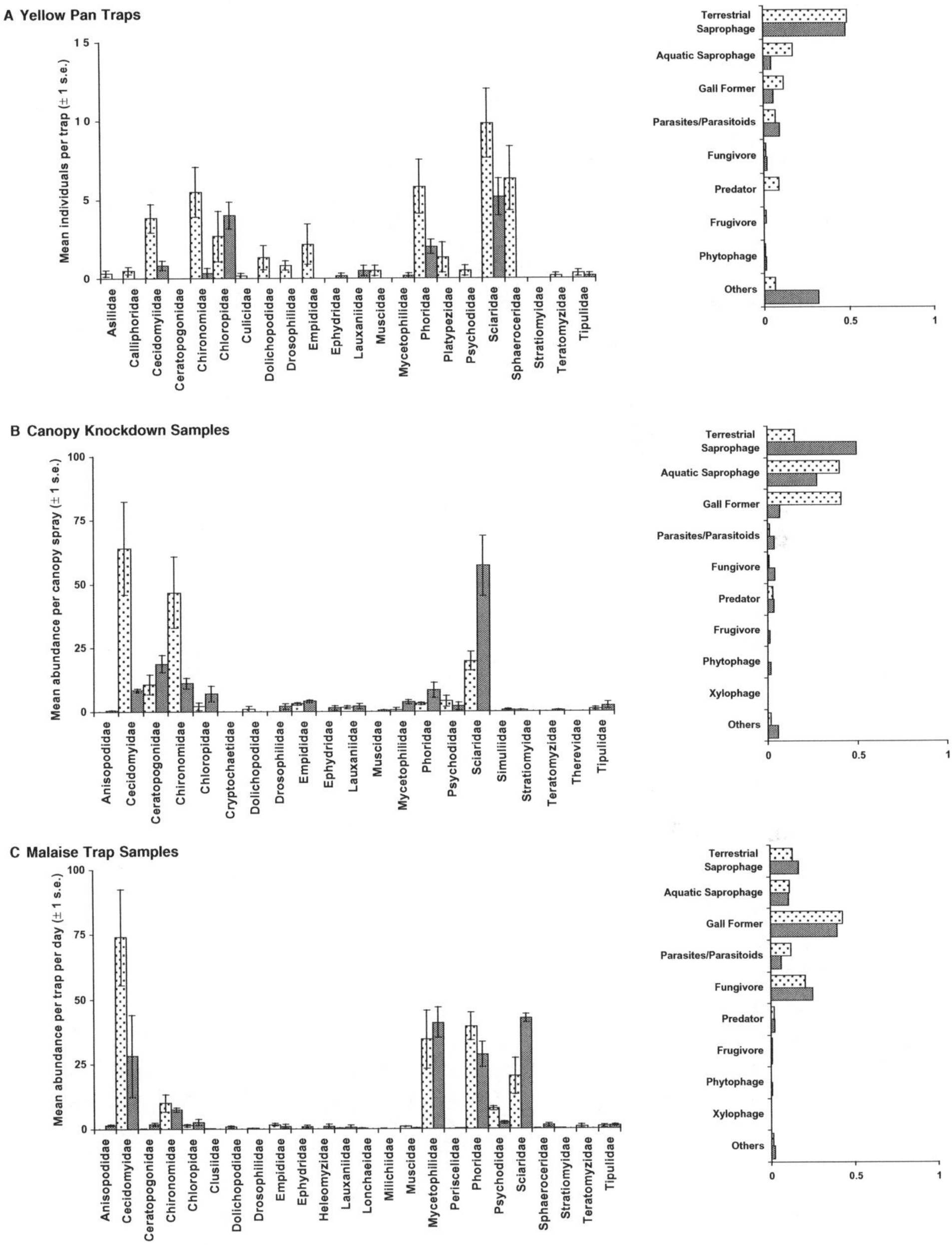
A Yellow Pan Traps
Mean individuals per trap (± 1 s.e.)
15
10
5
0
Asilidae
Calliphoridae
Cecidomyiidae
Ceratopogonidae
Chironomidae
Chloropidae
Culicidae
Dolichopodidae
Drosophilidae
Empididae
Ephydridae
Lauxaniidae
Muscidae
Mycetophilidae
Phoridae
Platypezidae
Psychodidae
Sciaridae
Sphaeroceridae
Stratiomyidae
Teratomyzidae
Tipulidae
Terrestrial Saprophage
Aquatic Saprophage
Gall Former
Parasites/Parasitoids
Fungivore
Predator
Frugivore
Phytophage
Others
0
0.5
1
B Canopy Knockdown Samples
Mean abundance per canopy spray (± 1 s.e.)
100
75
50
25
0
Anisopodidae
Cecidomyiidae
Ceratopogonidae
Chironomidae
Chloropidae
Cryptochaetidae
Dolichopodidae
Drosophilidae
Empididae
Ephydridae
Lauxaniidae
Muscidae
Mycetophilidae
Phoridae
Psychodidae
Sciaridae
Simuliidae
Stratiomyidae
Teratomyzidae
Therevidae
Tipulidae
Terrestrial Saprophage
Aquatic Saprophage
Gall Former
Parasites/Parasitoids
Fungivore
Predator
Frugivore
Phytophage
Xylophage
Others
0
0.5
1
C Malaise Trap Samples
Mean abundance per trap per day (± 1 s.e.)
100
75
50
25
0
Anisopodidae
Cecidomyiidae
Ceratopogonidae
Chironomidae
Chloropidae
Clusiidae
Dolichopodidae
Drosophilidae
Empididae
Ephydridae
Heleomyzidae
Lauxaniidae
Lonchaeidae
Milichiidae
Muscidae
Mycetophilidae
Periscelidae
Phoridae
Psychodidae
Sciaridae
Sphaeroceridae
Stratiomyidae
Teratomyzidae
Tipulidae
Terrestrial Saprophage
Aquatic Saprophage
Gall Former
Parasites/Parasitoids
Fungivore
Predator
Frugivore
Phytophage
Xylophage
Others
0
0.5
1

TABLE 14.4. Results of One-Way Analyses of Variance Comparing Wet and Dry Season Assemblages of Diptera from Lamington National Park[a]

Abundance Measures			Guild Abundances and Proportions		
Quantity	F	p[b]	Quantity	F	p
Yellow pan traps					
Total flies	10.69	0.008	Predator (abundance and proportion)	None in dry season	
Number of families	24.15	0.0006	Frugivore (abundance and proportion)	None in dry season	
Dominance	7.35	0.02	Others (proportion)	16.78	0.0022[c]
Chloropidae	9.76	0.011			
Phoridae	4.65	0.05 (ns)[c]			
Canopy knockdown					
Number of families	48.	0.0023[d]	Gall former (proportion)	62.92	0.0014[c]
Sciaridae	9.73	0.035 (ns)[d]	Frugivore	None in wet season	
			Phytophage	None in wet season	
Malaise traps					
Ceratopogonidae	8.82	0.04 (ns)[e]	Frugivore (abundance and proportion)	None in dry season	
Psychodidae	34.03	0.004			
Sciaridae	9.67	0.034 (ns)[e]			

[a] Significant or near significant results only are shown.
[b] ns, Not significant.
[c] Bonferroni correction for these values requires $p < 0.004$.
[d] Bonferroni correction for these values requires $p < 0.008$.
[e] Bonferroni correction for these values requires $p < 0.007$.

The analysis of variance for the taxonomic measures (Table 14.4) show few significant differences except for the yellow pan traps (as is to be expected from the correlation analyses). For the yellow pan traps, the total number of flies caught was significantly higher in the wet season, as was the number of families and the individual numbers of Chloropidae. The number of phorids caught in the wet season was substantially higher, although not formally significant. For the canopy knockdown samples, only the number of families and the number of Sciaridae approached significance, in both cases with numbers favoring the wet season. The number of Psychodidae caught in the Malaise traps was significantly higher in the wet season. Numbers of Ceratopogonidae and Sciaridae were also higher in the wet season, although neither showed formal significance in this difference.

The guild profiles from the yellow pan traps (Fig. 14.2A) show relatively few differences between the two seasons. The key difference was that "others" (principally the Chloropidae) formed a significantly higher proportion of the catches in the wet season. The dry season catches contained neither frugivores nor predators. For the canopy knockdown samples (Fig. 14.2B), the principal difference in guild structure was that gall formers were a significantly greater proportion of the catch in the wet season than in the dry. Both frugivores and phytophages were absent from the wet season catches. Guild profiles from the Malaise traps (Fig. 14.2C) were remarkably similar across the two seasons, the only difference of note being the absence of frugivores from the dry season samples.

DISCUSSION

In interpreting wet/dry season differences in subtropical rainforest sites, it must be recalled that this wet/dry difference also coincides with the summer/winter contrast. Winter (and, hence, dry season) temperatures are substantially lower than those in the wet summer. Daytime temperatures still fall well within the range suitable for insect activity, and nighttime temperatures do not reach lethal levels. Nevertheless, insect activity in general is reduced, contributing to the overall reduced numbers of Diptera trapped in the dry season compared with the wet. Accordingly we might predict that a "standard" pattern of seasonality would see a decline in numbers or even an absence in the cooler, drier season. This would reflect lower biological activity levels in both the flies themselves—as ectotherms—and in the general availability of resources due to reduced levels of productivity in both the producer and decomposer food chains.

Such a standard pattern is observed in our data for the Cecidomyiidae, Chironomidae, Psychodidae, Dolichopodidae, and Platypezidae across all trapping methods. The last two families in this list are actually absent from our dry season samples. The Muscidae and Drosophilidae show comparable patterns, although a very few individuals reverse the pattern in one of the three trapping methods. This coincides with our expectations vis-à-vis predators and phytophages (as represented by gall formers and frugivores), but suggests that the decomposers are not responding in the uniform manner anticipated. Terrestrial and aquatic decomposers clearly need to be considered separately. Resource availability for larval aquatic decomposers may well be reduced in the much drier winter season.

Other families show the reverse pattern, with an increase in numbers in the drier, cooler season. Such patterns are more difficult to explain. The principal families involved are the Chloropidae, Ceratopogonidae, Lauxaniidae, Anisopodidae, Heliomyzidae, and Ephydridae. Indeed, the last three families occurred only in the dry season samples. In guild terms, these are a mixed bag, dominated by decomposers, herbivores, and fungivores. With the possible exception of a small proportion of chloropids, this set includes no predators or fully aquatic decomposers. We postulate, therefore, that the resource base for these groups is maintained throughout the dry season. The possible decline in numbers or activities of predators may also have a reciprocal effect on numbers of families that are prey items. We suggest that at least one family, the Anisopodidae, has exploited this situation to become a dry season, cool weather specialist. This preference for winter activity has been noted by Colless and McAlpine (1991) and, for temperate species, by Freeman (1950). The Heliomyzidae (fungivores) and Ephydridae (aquatic herbivores) also fall into this category. The Heliomyzidae are known to be a more temperate group (Colless and McAlpine 1991), perhaps appearing in the cooler months in the northern part of their range. The winter predominance of the Anisopodidae and Heliomyzidae may be due to the phenomenon first described by Mackerras (1950), who observed that species of Diptera that are part of the cool temperate fauna will emerge in winter or very early spring toward the northern limits of their ranges but appear later in the season in higher and more southerly localities.

Other numerically abundant families appear from our samples to thrive in both seasons. These include three of the four most abundant families—the Phoridae, Mycetophilidae, and Sciaridae, together with the less abundant Tipulidae. This implies that the resources on which they depend remain readily available throughout the year. Their roles as principally terrestrial decomposers and fungivores perhaps predispose them to show little season-to-season change. The remaining families in our samples are not considered further, as their overall numbers are so low as to make any pattern interpretation meaningless.

CANOPY/GROUND CONTRASTS IN TAXONOMIC AND GUILD STRUCTURES OF MALAISE-TRAPPED ASSEMBLAGES OF DIPTERA FROM AUSTRALASIAN RAINFORESTS

The comparison of ground vs. canopy taxonomic and guild structures is derived from Malaise trap catches from four of the seven survey sites. The four 1-ha plots were located at the Conondales National Park, southeastern Queensland (Conondales), Cape Tribulation, north Queensland (Cape Tribulation), Oomsis Experimental Forest, Morobe Province, Papua New Guinea (Oomsis), and the Kau Wildlife Area, Baitabag, Madang Province, Papua New Guinea (Baitabag). Again, full descriptions of these sites are in Kitching et al. (2004) with a summary of information in Table 14.2.

Malaise traps of the design already described were used. Three traps were dispersed within the study hectares, based on random coordinates. In addition, a further three Malaise traps were erected within aluminium frames and positioned in the canopy using ropes over high branches (of emergent trees wherever possible). Location of the canopy traps was random within the hectare, but this had to be modified by the availability of suitable branches from which traps could be suspended. Trapping at each site was carried out over 4 or 5 days. All analyses are based on catches per day of sampling.

Analyses. Analyses of variance were carried out as in the seasonal study, but two-way comparisons were made in which the response variable was tested against both site and height of the sample. A series of taxonomically based analyses were carried out, followed by guild comparisons both based on abundances of each guild and the same data represented as (transformed) proportions.

Results. Fig. 14.3 presents the taxonomic results from each of the four sites and contrasts canopy (stippled bars) with ground (gray bars) samples. Fig. 14.4 represents the same data converted into guild categories and represented as pie charts. The significant outcomes of the two-way analyses of variance are included in Table 14.5.

Neither the total number of flies caught nor the number of families differed significantly between the canopy and the ground samples, although these statistics did differ significantly from site to site (which is not surprising, given the huge latitudinal gradient involved). Berger-Parker dominance values were significantly higher in the canopy samples, as well as differing significantly from site to site. The dominant families present an interesting set of contrasts. Phoridae were only numerically dominant in tropical ground samples (from Baitabag, Oomsis, and Cape Tribulation). Chironomidae or Psychodidae dominated the canopy samples in Cape Tribulation (all traps) and Baitabag (two of three traps), respectively. Other samples in both the canopy and on the ground were dominated by the ubiquitous Cecidomyiidae (one of three canopy traps and two of three ground traps at Baitabag; all three canopy traps and one of three ground traps at Oomsis; two of three canopy traps and all three ground traps at the Conondales). One of the canopy traps at the Conondales was dominated by Chironomidae.

On a family-by-family basis, the analysis of variance showed significant effects of both site and height for the Chironomidae, Phoridae, and Sphaeroceridae. Height but not site had a significant impact on numbers of Dolichopodidae, whereas there was a site but no height effect on the Ceratopogonidae and the Tipulidae. Summarizing the impact of height from these results: Chironomidae were generally at higher abundances in the canopy; Sphaeroceridae, Phoridae, and Dolichopodidae at ground level.

Guild analyses showed remarkable cross-site and cross-height differences (Table 14.5). In general, these differences were apparent when analyses were done on either simple abundances or on the transformed proportional data. Terrestrial saprophages were significantly more abundant at ground level, whereas aquatic saprophages dominated the canopy samples, particularly at Cape Tribulation (where Chironomidae were by far the dominant family). Parasitic groups (particularly represented by the Phoridae) were predominantly a ground-level phenomenon, especially in the three tropical sites. Fungivores (primarily Mycetophilidae) were never dominant but did represent a higher proportion of the ground fauna than they did of the canopy. Predatory groups (principally Dolichopodidae and Empididae) were significantly more abundant and dominant at the ground level.

DISCUSSION

The dramatic, statistically supported differences between the canopy and ground faunas are at first sight rather surprising. They certainly are contrary to the view that the Diptera are not masters of their own location and are mixed through air movements over which they have no control.

In seeking explanations for the patterns, we need to remember first that we are using adult specimens as indicators of larval presence. Even so, it is natural to explain the presence of large numbers of a particular guild or fly family by suggesting an abundance of key resources in either the canopy or the ground zone. Finally, we need to remember that the statistical patterns seen are just that; they become apparent only for those groups that have large numbers of individuals spread across many samples: at best, using samples aggregated at the

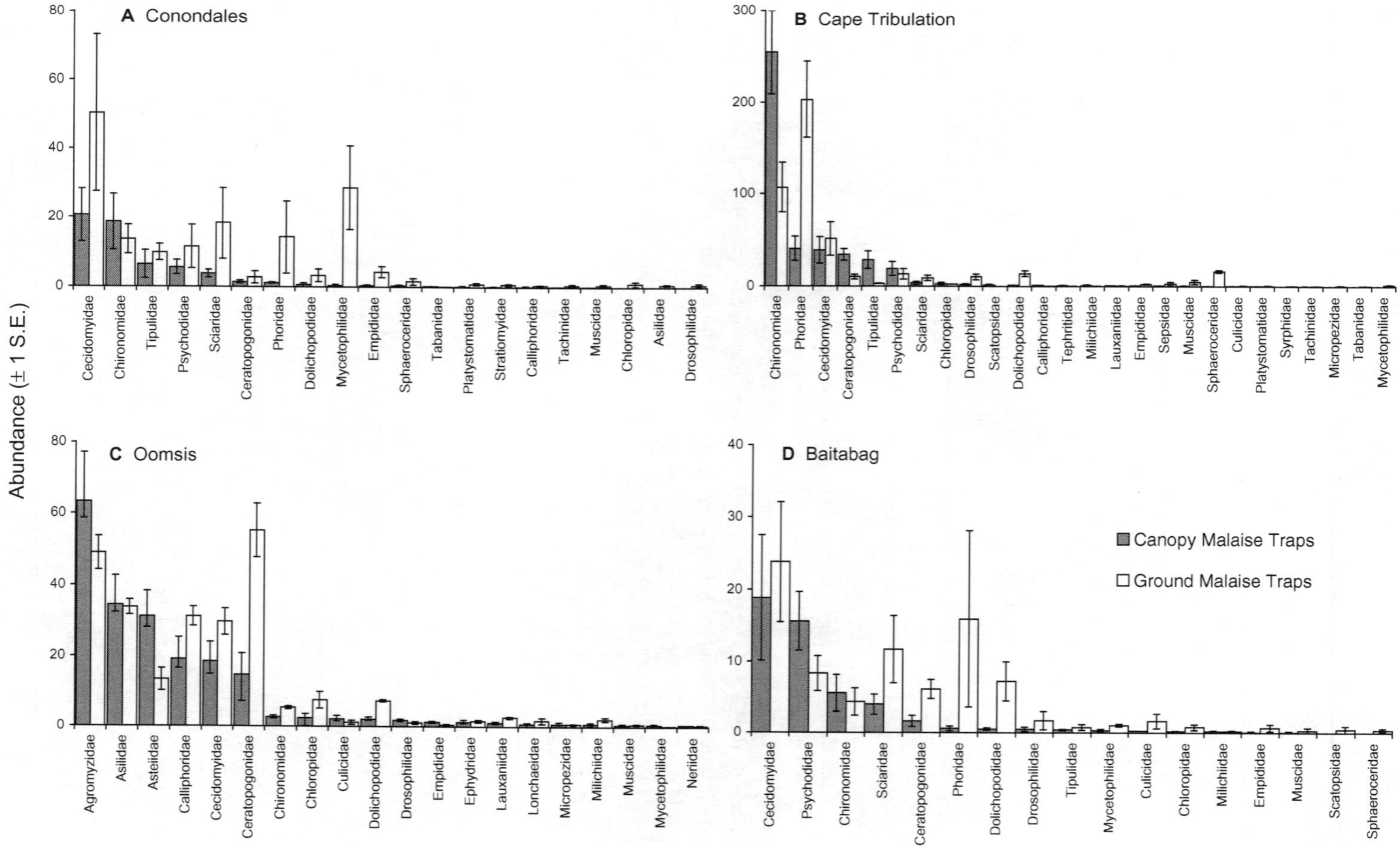

Abundance (± 1 S.E.)
A Conondales
80
60
40
20
0
Cecidomyiidae
Chironomidae
Tipulidae
Psychodidae
Sciaridae
Ceratopogonidae
Phoridae
Dolichopodidae
Mycetophilidae
Empididae
Sphaeroceridae
Tabanidae
Platystomatidae
Stratiomyidae
Calliphoridae
Tachinidae
Muscidae
Chloropidae
Asilidae
Drosophilidae
B Cape Tribulation
300
200
100
0
Chironomidae
Phoridae
Cecidomyiidae
Ceratopogonidae
Tipulidae
Psychodidae
Sciaridae
Chloropidae
Drosophilidae
Scatopsidae
Dolichopodidae
Calliphoridae
Tephritidae
Milichiidae
Lauxaniidae
Empididae
Sepsidae
Muscidae
Sphaeroceridae
Culicidae
Platystomatidae
Syrphidae
Tachinidae
Micropezidae
Tabanidae
Mycetophilidae
C Oomsis
80
60
40
20
0
Agromyzidae
Asilidae
Asteiidae
Calliphoridae
Cecidomyiidae
Ceratopogonidae
Chironomidae
Chloropidae
Culicidae
Dolichopodidae
Drosophilidae
Empididae
Ephydridae
Lauxaniidae
Lonchaeidae
Micropezidae
Milichiidae
Muscidae
Mycetophilidae
Neriidae
D Baitabag
40
30
20
10
0
Canopy Malaise Traps
Ground Malaise Traps
Cecidomyiidae
Psychodidae
Chironomidae
Sciaridae
Ceratopogonidae
Phoridae
Dolichopodidae
Drosophilidae
Tipulidae
Mycetophilidae
Culicidae
Chloropidae
Milichiidae
Empididae
Muscidae
Scatopsidae
Sphaeroceridae

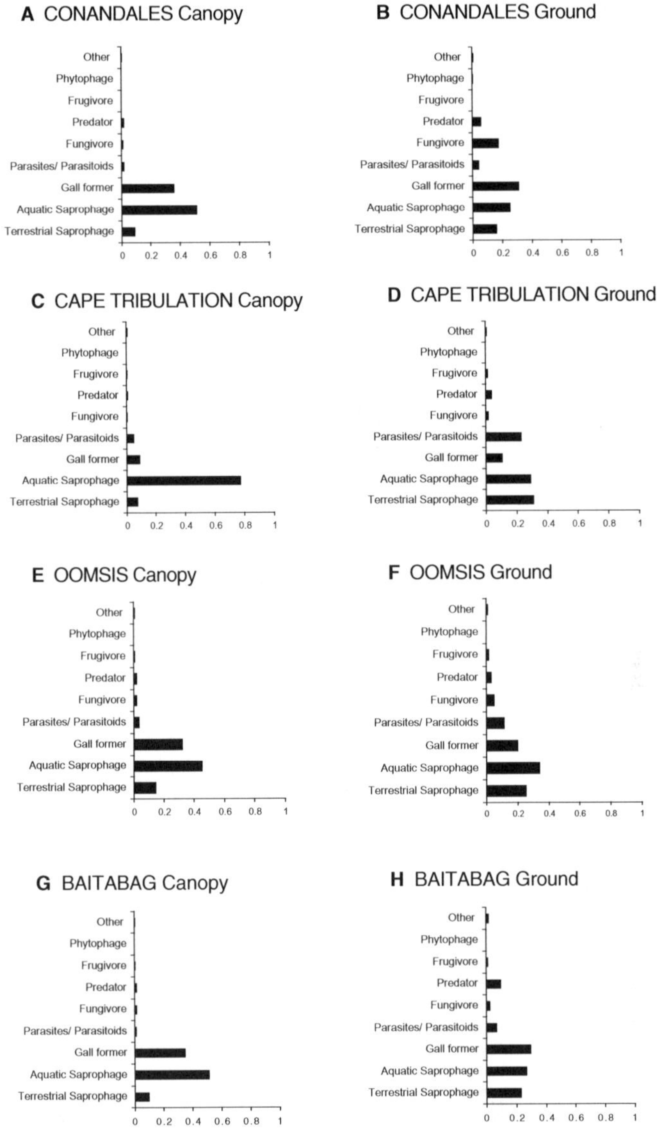

FIGURE 14.4. Distribution of Diptera into larval guilds from canopy (A, C, E, G) and ground zone (B, D, F, H). Malaise samples from (A, B) the Conondales Ranges, southeastern Queensland; (C, D) Cape Tribulation, north Queensland; (E, F) Oomsis, Morobe Province, Papua New Guinea; and (G, H) Baitabag, Madang Province, Papua New Guinea.

FIGURE 14.3. Distribution of dipteran families collected by canopy (gray) and ground zone (white) Malaise traps at four Australasian rainforest sites. (A) Conandales Ranges, southeastern Queensland: (B) Cape Tribulation, north Queensland; (C) Baitabag, Madang Province, Papua New Guinea; and (D) Oomsis, Morobe Province, Papua New Guinea.

TABLE 14.5. Results of Two-Way Analyses of Variance on the Structure of Assemblages of Diptera from Four Old-World Tropical Rainforests[a]

	Site Effect		Canopy vs. Ground		Interaction	
Response Variable	*F*	*p*[b]	*F*	*p*[b]	*F*	*p*[b]
Analyses of numbers						
Total number of flies caught	21.76	<0.0001	2.07	ns	0.26	ns
Total families in sample	8.48	0.0013	4.13	ns	1.79	ns
Berger-Parker dominance measure	12.41	0.0002	18.19	0.00059	0.52	ns
Dominant families[c]						
Ceratopogonidae	30.56	<0.0001	2.80	ns	5.42	0.0092
Chironomidae	37.53	<0.0001	9.91	0.0062	6.61	0.0041
Dolichopodidae	4.28	ns	29.65	<0.0001	4.14	ns
Phoridae	20.45	0.00001	23.68	0.00017	8.86	0.0011
Sphaeroceridae	65.95	<0.0001	103.52	<0.0001	62.92	<0.0001
Tipulidae	6.70	0.0039	3.50	ns	6.46	0.0045
Analyses of guild structure[d]						
Terrestrial saprophages (abundance)	13.18	<0.0001	24.36	0.00014	5.24	ns
Terrestrial saprophages (proportion)	3.11	ns	54.98	<0.0001	3.53	ns
Aquatic saprophages (abundance)	44.53	<0.0001	12.17	0.003	11.82	0.00024
Aquatic saprophages (proportion)	3.87	ns	53.96	<0.0001	4.36	ns
Parasites/parasitoids (abundance)	20.35	<0.0001	23.55	0.00017	8.82	0.0011
Parasites/parasitoids	15.46	<0.0001	52.24	<0.0001	3.38	ns
Fungivores (proportion)	22.07	<0.0001	76.28	<0.0001	19.21	<0.0001
Predators (abundance)	2.41	ns	33.39	<0.0001	2.27	ns
Predators (proportion)	8.11	0.0016	99.98	<0.0001	7.39	0.0025
Frugivores (proportion)	8.30	0.0014	8.22	ns	0.45	ns

[a] Sampled at canopy and ground level using Malise traps. Only taxa or guilds showing significant differences are shown.
[b] ns, Not significant.
[c] Critical *p* value after Bonferroni correction = 0.005.
[d] Critical *p* value after Bonferroni correction = 0.0028.

family level can be ecological indicators of function. They are not and cannot be indicators of species diversity. Bearing these caveats in mind, consider each of the significant outcomes indicated in Table 14.5. Note we discuss here only the canopy ground contrasts, not the site-to-site patterns.

Also note that although we had hypothetical expectations that related to the vertical distribution of phytophagous families, the numbers obtained were simply too small to allow satisfactory statistical analysis.

DOMINANCE

Significantly higher dominance levels in the canopy suggest that the canopy samples show a greater unevenness than do the ground assemblages. Overall, the canopy samples contain 42 families compared with 43 in the ground zone samples. As the dominance results suggest, the overall evenness (a measure of the distribution of abundance of all families in the sample) of the ground samples is somewhat higher (Shannon Evenness Index, 0.58) for the ground samples than for the canopy samples (0.51). This difference is so small, however, that it is more likely the consequence of the dominance value (which is based on the single most numerous family in each sample) than the cause of it. Accordingly, we must seek explanation for the significantly higher dominance values in the canopy samples in the biology of the dominant families. The dominant canopy families were the Chironomidae, Psychodidae, and Cecidomyiidae. All are small and have very high rates of potential population growth (through high fecundity or even paedogenesis). That they may reach very high numbers, and hence dominate samples, is not surprising. Why they should dominate more in the canopy than at ground level (where Phoridae and Cecidomyiidae are likely to be the dominant families) can only remain a matter for speculation (see below).

TERRESTRIAL SAPROPHAGES; SPHAEROCERIDAE

The terrestrial saprophages are a heterogeneous group, including detritus, dung, and carrion feeders. Although saprophagous resources do occur in the canopy (see, e.g., Rodgers 1999; Kitching 2000), there is no doubt that richer picking will be available on the ground. Finding this guild (and selected families within it) in significantly greater numbers and proportion on the ground supports our initial hypothesis concerning the vertical distribution of saprophages.

The sphaerocerids are generally supposed to be dung-associated flies and, as such, might find more resources on the forest floor for gravitational reasons alone. Their dominance in ground zone samples accordingly is not surprising.

AQUATIC SAPROPHAGES; CHIRONOMIDAE

The aquatic saprophages, in general, are more homogeneous than their terrestrial analogs. Essentially they are comminuters of allochthonous detritus in streams, temporary or permanent pools, and phytotelmata. As indicated earlier, our expectation is that they will be more abundant at ground level. Accordingly, their significantly higher numbers in the canopy requires further debate. Numerically they are dominated by the Chironomidae, Psychodidae, and Ceratopogonidae, which together make up a substantial proportion of any sample of rainforest Diptera that we have examined. Chironomid midges breed in fresh water or waterlogged soil and are well-known as adults for forming huge male swarms. It is likely that their dominance in canopy samples, particularly at Cape Tribulation, is a behavioral rather than an ecological phenomenon. Although some midges breed in water-filled tree holes (Kitching 2001) and others in perched litter associated with epiphytes, neither of these two habitats is likely to generate the large numbers required to dominate the canopy samples. It seems more likely that the samples represent ground-level emergences producing male swarms that either move to or are lifted to the canopy subsequently. The other very abundant families in this guild show significant site-to-site differences but are found across both strata in the forest.

PREDATORS; DOLICHOPODIDAE

The predatory guild comprises the three families Asilidae, Dolichopodidae, and Empidae. None of these reach high numbers, but the Dolichopodidae certainly dominate the guild, with at least three times the abundance levels of the empids and 30 times that of the asilids. This pattern is maintained at both canopy and ground levels, but numbers at the ground level for each family are at least four times as high as in the canopy samples. This is somewhat contrary to our expectations. The Dolichipodidae are relatively robust predatory flies whose adults actively seek prey. Their significantly higher numbers at the ground level probably reflect the greater numerical availability of prey at that level. The physical environment at lower levels may also be more conducive to their predatory tactics. Similar arguments can be adduced for the other two predatory families, but the Dolichopodidae emerge as significant separately because of their numerical dominance.

PARASITIC GROUPS; PHORIDAE

The Phoridae emerge from our studies as quintessentially ground-dwelling flies occurring in all sites at between two and a hundred times the frequency at the ground level than in the canopy. This reflects on the guild analysis, as we currently assign them 50:50 to the parasitic guild (see above). Phorids are remarkably ecologically diverse (Disney 1994) and any generalizations about their abundance or distribution must be made with great circumspection. As terrestrial decomposers, they are more likely to find resources on the forest floor, and as parasites, frequently of ants or termites, they are also more likely to encounter host nests at ground level. Little more than this can be said until we obtain information at a finer level of resolution about the taxonomic and ecological composition of our particular samples. At the current level of specialist engagement with the family in Australasia, this is unlikely to occur, at least within the lifetime of the senior author!

FUNGIVORES

Fungivores (principally the Mycetophilidae), as we hypothesized, are predominantly ground zone animals. Given the saprophytic nature of fungal substrates, this distribution within the forest is not surprising. Epiphytic, saproxylic, and ground-based fungi will be more abundant at ground level. Adult fungivorous flies, both as emergents and when questing for oviposition sites, are accordingly likely to be commoner at the ground level than in the canopy.

Conclusions

NUMBERS VS. SPECIES

All the analyses of guilds discussed in this chapter (with the exception of Table 14.1) have been based on partitioning of numbers of *individuals* within samples, using prior sorting to the family level. This was pragmatic, as few of the families have been sorted to species or putative species at this stage. Nevertheless it is, in principle, possible to carry out such a sorting task, and a guild analysis then could be carried out on the basis of the numbers of *species* within each guild rather than the number of individuals. In fact, the two analyses would address different classes of individuals. An analysis of numbers of individuals (as presented here) is appropriate when ecological questions are being posed: what is the relative importance of detritivores, predators, or others within the ecosystem being sampled and how might this differ from elsewhere? In contrast, a species-level analysis will be more appropriate for biogeographic analysis, in which questions are being posed about the evolutionary history of particular taxa and, in consequence, particular guilds: what might this say about the structural characteristics of the environmental space represented by particular regions, and how have processes of adaptive radiation responded to or, indeed, participated in formation of this space?

In Australia, in practice, this will be a frustrating exercise, as most of the dominant families of Diptera (in rainforests at least) lack a resident (or even extant) specialist taxonomist (Kitching et al. 2004).

DIPTERA AS INDICATORS OF ECOLOGICAL CONTRASTS

We have demonstrated clear season-to-season and layer-to-layer differences from a family-level analysis of the Diptera. In addition, the ready classification of the Diptera at the family level into feeding guilds provides a powerful tool for erecting hypotheses concerning the actual differences between samples from different seasons and strata. The interpretation of such guild structures is predicated on the idea that the adult flies are direct indicators of the larval assemblages, which, for the most part, are the ones actually having direct ecological impacts within the forest. We suggest that the generally short lives and low vagility of adults—at least within the closed forest—make this a reasonable supposition.

The overall levels of abundance, the wide diversity of taxa, the ease of sampling, and the variety of ecological roles played by the Diptera in rainforest systems rank them alongside the Coleoptera as potentially useful indicators for ecological analyses. Indeed, they are considerably more suitable for such analysis than are, say, the frequently used Lepidoptera.

Of course the conversion of taxonomic data into guild data must be done both carefully and adaptively. For some families, we clearly have only just begun to understand their ecological roles. As fresh data or reviews come to hand, the basis for guild assignment will need to be reviewed. No doubt some workers will disagree with our assignments set out in Table 14.3. Indeed, we have already modified this several times as specialist dipterists have told us more about their taxa. This process needs to be ongoing. However, the lack of perfect knowledge should not deter development of new analytical tools and, in this regard, we point to the development and argument that has accompanied the use of Coleoptera in guild analyses. What has emerged from the latter is a robust tool that has been used constructively in biodiversity analyses. Our goal in this chapter is to begin the process that will allow the equally important and perhaps even more ubiquitous Diptera to play their role in these analyses.

ACKNOWLEDGMENTS

We are grateful to the Field Studies Centre (Earthwatch) and the Australian Centre for International Agricultural Research (ACIAR) for supporting the field work involved in these studies. They provided the financial support for the teams of students, volunteers, and staff who assisted in our surveys. The collections were made under permits from the Queensland National Parks and Wildlife Service and the Department of Agriculture of Papua New Guinea. Earlier versions of this chapter and its antecedents have benefited greatly from discussions with Don Colless, David Yeates, Dick Drew, and Henry Disney.

REFERENCES

Armitage, P., P. S. Cranston, and L. C. V. Pinder. 1995. *The Chironomidae—The Biology and Ecology of Non-Biting Midges.* London: Chapman and Hall.

Berg, M. B. 1995. Larval food and feeding behavior. In P. D. Armitage, P. S. Cranston, and L. C. V. Pinder, eds., *The Chironomidae: The Biology and Ecology of Non-Biting Midges,* pp. 136–168. London: Chapman and Hall.

Berger, W. H., and F. L. Parker. 1970. Diversity of planktonic Foraminifera in deep sea sediments. *Science* 168:1345–1347.

Colless, D. H., and D. K. McAlpine. 1991. Diptera (Flies). In CSIRO, *The Insects of Australia,* 2nd ed., pp. 717–786. Carlton: Melbourne University Press.

Colyer, C. N., and C. O. Hammond. 1951. *Flies of the British Isles.* London: Warne.

CSIRO. 1991. *The Insects of Australia,* 2nd ed. Carlton: Melbourne University Press.

Disney, R. H. L. 1994. *The Scuttle Flies: The Phoridae.* London: Chapman and Hall.

Erwin, T. L. 1990. Canopy arthropod diversity: A chronology of sampling techniques and results. *Rev. Peru. Entomol.* 32:71–77.

Farrell, B. D. 1998. "Inordinate fondness" explained: Why are there so many beetles? *Science* 281:555–559.

Ferrar, P. 1987. *A Guide to the Breeding Habits and Immature Stages of Diptera Cyclorrhapha.* Leiden: E. J. Brill/Scandinavian Science Press.

Freeman, P. 1950. Family Anisopodidae (Rhyphidae). 9:70–72.

Hammond, P. M. 1990. Insect abundance and diversity in the Dumoga-Bone National Park, North Sulawesi, with special reference to the beetle fauna of lowland rainforest in the Toraut region. In W. J. Knight and J. D. Holloway, eds., *Insects and the Rain Forests of South-east Asia,* pp. 197–254. London: Royal Entomological Society.

Hammond, P. M., R. L. Kitching, and N. E. Stork. 1996. The biodiversity of arthropods from Australian rainforest canopies: Coleoptera from subtropical tree-crowns. *Ecotropica* 2:99–108.

Harper, A. M., and T. P. Story. 1962. Reliability of trapping in determining emergence period and sex ratio of the sugar beet maggot *Tetanops myopaeformis* (Röder) (Diptera: Tephritidae). *Can. Entomol* 94:268–271.

Hövemeyer, K. 2000. Ecology of diptera. In L. Papp and B. Darvas, eds., *Contributions to a Manual of Palaearctic Diptera,* vol. 1: *General and Applied Dipterology,* pp. 437–489. Budapest: Science Herald.

Juillet, J. A. 1963. A comparison of four types of trap used for capturing flying insects. *Can. J. Zool* 41: 219–223.

Kim, S. P. 1994. *Australian Lauxaniid Flies.* Monographs on Invertebrate Taxonomy 1. Australia: CSIRO.

Kitching, R. L. 2000. *Food Webs and Container Habitats.* Cambridge: Cambridge University Press.

Kitching, R. L. 2001. Food webs in phytotelmata: "Bottom-up" and "top-down" explanations for community structure. *Annu. Rev. Entomol.* 46:729–760.

Kitching, R. L., and G. Theischinger. 1996. The biodiversity of arthropods from Australian rainforest canopies: Tipulidae, with a description of the new species *Leptotarsus (Meromastix) alfie* Theischinger. *Entomolog* 115:140–153.

Kitching, R. L., J. M. Bergelsen, M. D. Lowman, S. McIntyre, and G. Carruthers. 1993. The biodiversity of arthropods from Australian rainforest canopies: General introduction, methods, sites and ordinal results. *Aust. J. Ecol* 18:181–191.

Kitching, R. L., G. Vickerman, K. Hurley, and M. Laidlaw. 2000a. *The Comparative Assessment of Arthropod and Tree Biodiversity in Old-World Rainforests: The CRC-TREM/Earthwatch Protocol.* Cairns, Australia: Cooperative Research Centre for Tropical Rainforest Ecology and Management.

Kitching, R. L., A. G. Orr, L. Thalib, H. Mitchell, M. S. Hopkins, and A. W. Graham. 2000b. Moth assemblages as indicators of environmental quality in remnants of upland Australian rain forest. *J. Appl. Ecol.* 37:284–297.

Kitching, R. L., D. Q. Li, and N. E. Stork. 2001. Assessing biodiversity "sampling packages": How similar are arthropod assemblages in different tropical rainforests? *Biodiv. Cons.* 10:793–813.

Kitching, R. L., D. Bickel, A. C. Creagh, K. Hurley, and C. Symonds. 2004. The biodiversity of Diptera in Old-World rainforest surveys: A comparative faunistic analysis. *J. Biogeogr.* 31:1185–1200.

Krüger, O., and G. C. McGavin. 2001. Predator-prey ratio and guild constancy in a tropical insect community. *J. Zool.* 253:265–273.

Laidlaw, M. J. 1999. *Variations in rainforest vegetation along a latitudinal gradient.* B.S. honors thesis, Griffith University, Brisbane.

Laidlaw, M. J., M. J. Olsen, R. L. Kitching, and M. Greenway. 2000. Tree floristics and structural characteristics of one hectare of subtropical rainforest in Lamington National Park, Queensland. *Proc. R. Soc. Queensland* 109:91–105.

Laidlaw, M. J., R. L. Kitching, K. Damas, and R. Kiapranis. In press. Tree floristic and structural attributes of two lowland rainforest plots in Papua New Guinea. *Biotropica.*

Lawrence, J. F., and R. J. Milner. 1996. Associations between arthropods and fungi. In *Fungi of Australia,* vol. 1B: *Introduction—Fungi and the Environment,* pp. 137–202. Canberra: Australian Biological Resources Study.

Mackerras, I. M. 1950. The zoogeography of Diptera. *Aust. J. Sci.* 12:157–161.

Majer, J. D., R. L. Kitching, B. E. Heterick, K. Hurley, and K. E. C. Brennan. 2001. North-South patterns within arboreal ant assemblages from rainforests in eastern Australia. *Biotropica* 33:643–661.

Malaise, R. 1937. A new insect trap. *Entomol. Tidjschr.* 58:148–160.

Martin, J. L. 1966. The insect ecology of red pine plantations in central Ontario. IV. The crown fauna. *Can. Entomol.* 98:10–27.

McAlpine, D. K. 1960. A review of the Australian species of Clusiidae (Diptera, Acalyptra). *Rec. Aust. Mus.* 25:263–283.

McAlpine, D. K. 1978. Notes on nomenclature and type-specimens of Australian Pyrgotidae (Diptera, Schizophora) *Aust. ent. Mag.* 5:27–33.

McAlpine, J. F., and D. M. Wood, eds. 1989. *Manual of Nearctic Diptera,* vol. 3. *Res. Branch Agric. Can. Mongr.* 32:1333–1581.

McAlpine, J. F., B. V. Peterson, G. E. Shewell, H. J. Teskey, J. R. Vockeroth, and D. M. Wood, eds. 1981. *Manual of Nearctic Diptera,* vol. 1. *Res. Branch Agric. Can. Mongr.* 27:1–674.

McAlpine, J. F., B. V. Peterson, G. E. Shewell, H. J. Teskey, J. R. Vockeroth, and D. M. Wood, eds. 1987. *Manual of Nearctic Diptera,* vol. 2. *Res. Branch Agric. Can. Mongr.* 28:1–674.

McIntyre, S., R. L. Kitching, and L. W. Jessup. 1994. Structure and composition of tropical rainforest in plots at Cape Tribulation, Queensland. *Proc. R. Soc. Queensland* 104:25–41.

Moericke, V. 1951. Über den Farbensinnder Pfirschblattlaus *Myzodes persicae* Sulz. *Z. Tierpsychol.* 7:265–274.

Oldroyd, H. 1964. *The Natural History of Flies.* London: Weidenfeld and Nicolson.

Paijmans, K. 1970. An analysis of four tropical rain forest sites in New Guinea. *J. Ecol.* 8:77–101.

Roberts, H. R. 1973. Arboreal Orthoptera in the rain forests of Costa Rica collected with insecticide: A report on grasshoppers (Acrididae) including new species. *Proc. Acad. Nat. Sci. Philadelphia* 125:46–66.

Roberts, J. A., and R. L. Kitching. 1974. Ingestion of sugar, protein and water by adult *Lucilia cuprina* (Wied.) (Diptera, Calliphoridae). *Bull. Entomol. Res.* 64:81–88.

Roberts, M. J. 1970. The structure of the mouthparts of syphid larvae (Diptera) in relation to feeding habits. *Acta Zool.* 51:43–65.

Rodgers, D. J. 1999. Structural dynamics of rainforest collembolan assemblages in leaf litter suspended in epiphytes and on the forest floor. Ph.D. thesis, Griffith University, Brisbane.

Rozkosny, R. 1982. *Biosystematic Study of the European Stratiomyidae.* Dordrecht: Kluwer Academic.

Skidmore, P. 1985. *The Biology of the Muscidae of the World.* Dordrecht: W. Junk.

Smith, K. G. V. 1969. The Empididae of southern Africa (Diptera). *Ann. Natal. Mus.* 19:1–347.

Southwood, T. R. E., V. C. Moran, and C. E. J. Kennedy. 1982. The richness, abundance and biomass of the arthropod communities on trees. *J. Animal Ecol.* 51:635–649.

Stork, N. E. 1987. Guild structure of arthropods from Bornean rain forest trees. *Ecol. Entomol.* 12:69–80.

Teskey, H. J. 1976. Diptera larvae associated with trees in North America. *Mem. Ent. Soc. Can.* 100:1–53.

van Klinken, R. D., and G. H. Walter. 2001. Subtropical drosophilids in Australia can be characterized by adult distribution across vegetation type and by height above forest floor. *J. Trop. Ecol.* 17:705–718.

CONTRIBUTORS

Michael Ashburner
Department of Genetics
University of Cambridge
Cambridge CB2 3EH
England

Nora J. Besansky
Department of Biological Sciences
University of Notre Dame
Notre Dame, Ind. 46556
USA

Daniel J. Bickel
Entomology Section
Australian Museum
6 College Street
Sydney, NSW 2010
Australia

Sarah Boulter
Department of Environmental Sciences
Griffith University
Nathan, Qld. 4111
Australia

Peter Cranston
Department of Entomology
University of California, Davis
Davis, Calif. 95616
USA

Neil Davies
Richard B. Gump South Pacific Research Station
University of California, Berkeley
BP 244-98728
Moorea
Polynésie Française

Rob DeSalle
Division of Invertebrates
American Museum of Natural History
79th Street at Central Park West
New York, N.Y. 10024
USA

Bill J. Etges
Department of Biological Sciences
University of Arkansas
Fayetteville, Ark. 72701
USA

James L. Feder
Department of Biological Sciences
University of Notre Dame
Notre Dame, Ind. 46556
USA

Kenneth E. Filchak
Department of Biological Sciences
Galvin Life Science Building
University of Notre Dame
Notre Dame, Ind. 46556
USA

Philip M. Johns
Department of Biology
University of Maryland
College Park, Md. 20742
USA

Margaret G. Kidwell
Department of Ecology and Evolutionary Biology
University of Arizona
Tucson, Ariz. 85721
USA

Roger L. Kitching
Department of Environmental Sciences
Griffith University
Nathan, Qld. 4111
Australia

Conrad C. Labandeira
Department of Paleobiology
Smithsonian Institution
Washington, D.C. 20013-7012
USA

Rudolf Meier
Department of Biological Sciences
National University of Singapore
14 Science Drive 4
Singapore 117543
and
Department of Entomology
Zoological Museum Copenhagen
Universitetsparken 15
2100 Copenhagen Oe
Denmark

David J. Merritt
Department of Zoology and Entomology
University of Queensland
St. Lucia, Qld. 4072
Australia

George K. Roderick
Environmental Science (ESPM)
201 Wellman Hall MC 3112
University of California, Berkeley
Berkeley, Calif. 94720-3112
USA

Sonja J. Scheffer
Systematic Entomology Laboratory
USDA-ARS-PSI
10300 Baltimore Avenue
Beltsville, Md. 20705
USA

Michael F. Whiting
Department of Integrative Biology
Brigham Young University
Provo, Utah 84602
USA

Brian M. Wiegmann
Department of Entomology
North Carolina State University
Raleigh, N.C. 27695-7613
USA

Gerald S. Wilkinson
Department of Biology
University of Maryland
College Park, Md. 20742
USA

David K. Yeates
CSIRO Entomology
P.O. Box 1700
Canberra, ACT 2601
Australia

INDEX

The suffix *f* on a page number indicates a figure; *t* indicates a table.